NUMERICAL SOLUTIONS OF PARTIAL DIFFERENTIAL EQUATIONS

NUMERICAL SOLUTIONS OF PARTIAL DIFFERENTIAL EQUATIONS

Proceedings of the 1981 Conference on the
Numerical Solutions of Partial Differential Equations
held at Queen's College, Melbourne University, Australia

Edited by

JOHN NOYE

Associate Dean
Faculty of Mathematical Sciences
The University of Adelaide
South Australia

N·H
P C

1982

NORTH-HOLLAND PUBLISHING COMPANY – AMSTERDAM • NEW YORK • OXFORD

ISBN: 0 444 86356 7

Publishers

NORTH-HOLLAND PUBLISHING COMPANY
AMSTERDAM · NEW YORK · OXFORD

Sole distributors for the U.S.A. and Canada

ELSEVIER NORTH-HOLLAND INC.
52 VANDERBILT AVENUE
NEW YORK, N.Y. 10017

Library of Congress Cataloging in Publication Data

Conference on the Numerical Solution of Differential
Equations (1981 : Melbourne University)
Numerical solutions of partial differential equations.

1. Differential equations, Partial--Numerical solutions--Congresses. I. Noye, John, 1930- . II. Title.
QA374.C654 1981 515.3'53 81-22411
ISBN 0-444-86356-7 AACR2

PRINTED IN THE NETHERLANDS

1981 Conference on

NUMERICAL SOLUTIONS OF PARTIAL DIFFERENTIAL EQUATIONS

Queen's College, University of Melbourne, Australia

23 - 27 August

ORGANISING COMMITTEE

Convenor:
Dr. John Noye,
Mathematics Department,
University of Adelaide,
Box 498, G.P.O.,
Adelaide, 5001,
Australia.

VICTORIAN BRANCH, DIVISION OF APPLIED MATHEMATICS, AUSTRALIAN MATHEMATICAL SOCIETY

Chairman:
Dr. Rhys Jones,
Aeronautical Research Laboratories,
Box 4331, G.P.O.,
Melbourne, 3001,
Australia.

Secretary:
Dr. Frank Barrington,
Mathematics Department,
University of Melbourne,
Parkville, 3052,
Australia.

Treasurer:
Dr. Alan Easton,
Mathematics Department,
Swinburne Institute of Technology,
John Street,
Hawthorn, 3122,
Australia.

PREFACE

This book is the outcome of the 1981 Conference on Numerical Solutions of Partial Differential Equations held at Queen's College, University of Melbourne, Australia, from August 23-27, and attended by 88 scientists. The aim of the conference was to present reviews of the most used methods of numerically solving partial differential equations and to give the opportunity to research workers to present their latest work on improving and applying these methods.

There were four invited contributions in the form of review presentations on the alternative techniques that can be used to solve problems involving partial differential equations. These papers were presented by John Noye of the University of Adelaide (Finite Difference Methods), Clive Fletcher of the University of Sydney (Galerkin Techniques), Josef Tomas of the Royal Melbourne Institute of Technology (Finite Element Methods) and Leigh Wardle of the C.S.I.R.O. Division of Applied Geomechanics (Boundary Integral Methods). There were also two invited contributions on methods of solving large sets of sparse linear algebraic equations. These were given by Len Colgan of the South Australian Institute of Technology (Iterative Methods) and Ken Mann of the Caulfield Institute of Technology (Direct Methods).

In addition, twenty papers were presented in which recent refinements of the various numerical techniques used to solve partial differential equations were described, or experiences in applying these methods to solve problems in engineering, physics, oceanography, hydrology or meteorology were reported.

A number of people spent time preparing and presenting material at the conference, but are not represented by a paper in the proceedings. These include Dr. W. Scott of Murdoch University, Mr. G. Tong of Swansea University (Wales), Dr. N. Stokes of the C.S.I.R.O. Division of Mathematics and Statistics and Dr. A. Raiche of the C.S.I.R.O. Division of Mineral Physics.

A special mention must be made of the work done by the Executive Committee of the Victorian Branch of the Division of Applied Mathematics of the Australian Mathematical Society, who organised the conference most efficiently. Those involved were Rhys Jones (Chairman), Frank Barrington (Secretary) and Alan Easton (Treasurer). Without their assistance, the conference could not have been held.

Finally, my personal thanks go to Drs. Arjen Sevenster (Mathematics Editor) and Mary Carpenter (Technical Editor) of the North-Holland Publishing Company in Amsterdam, for assistance in arranging the printing of this book. Without their help, the contents of this publication would not be available to scientists in other parts of the world.

John Noye
The University of Adelaide

TABLE OF CONTENTS

Part 2 : CONTRIBUTED PAPERS 397

PART 1:

INVITED PAPERS

NUMERICAL SOLUTIONS OF
PARTIAL DIFFERENTIAL EQUATIONS
J. Noye (editor)

FINITE DIFFERENCE METHODS FOR PARTIAL DIFFERENTIAL EQUATIONS

JOHN NOYE

University of Adelaide
South Australia

SUMMARY

Some of the more widely used finite difference techniques for solving partial differential equations are described. In particular the diffusion equation, the convection equation, the transport equation and the wave equation, in one, two and three dimensions have been considered. Laplace's, Poisson's and other equations are also considered as special cases of the abovementioned equations. Both explicit and implicit techniques are considered and compared on the basis of computational economy. Topics such as consistency of the finite difference equation with the corresponding partial differential equation, stability of the finite difference equation, and convergence of the solution of the finite difference equation to that of the corresponding partial differential equation, are considered in detail. The accuracy of the different methods, which depends on the discretisation error involved, is considered both theoretically and by means of illustrative numerical examples. The use of variable grid spacings is also considered, so that a finer grid can be used to give more detailed results in regions of interest.

CONTENTS

1. INTRODUCTION

1.1 General Considerations

This article describes some of the finite difference methods available for solving partial differential equations, with particular application to those equations which model heat and mass transfer. It also discusses the more common difficulties encountered by the mathematician, physicist or engineer when he first enters this field of numerical simulation.

Section 1.2 introduces the transport equation which is the example used to illustrate most of the various finite difference techniques in the following sections. For some initial and boundary conditions, analytic solutions can be found for this particular equation. These solutions are very useful for comparison, in order to get an idea of the accuracy of any numerical method.

However, since a numerical technique is employed only to solve problems to which an analytic solution is either very difficult or impossible to obtain, even with greatly simplified initial and boundary conditions, it is useful to know of methods which prove stability and convergence of such a technique. Unfortunately, such methods have not been developed for the more complicated (generally non-linear) problems and *ad hoc* approaches must be employed in which different methods are tested in numerical experiments.

In Section 2, the various finite difference approximations to derivatives at a point in the solution domain are described, and it is shown how to apply the forward time and centred space approximations to the derivatives in the one-dimensional diffusion equation (which is a typical example of parabolic type partial differential equations) to obtain approximate solutions of this equation at certain points in the region being considered.

In Section 3 the concepts of convergence, consistency and stability are considered in some detail, while in Section 4 some other finite difference methods, both explicit and implicit, of solving the one-dimensional diffusion equation are developed. Ways of incorporating derivative boundary conditions and of dealing with discontinuities between initial and boundary values are also discussed. Section 5 deals with methods of solving the transport equation in one space dimension. Section 6 deals with this equation in two and three space dimensions, and describes ways of dealing with boundary conditions on irregular boundaries. Methods of solving the wave equation are considered in Section 7. Sections 8 and 9 consider methods of improving the accuracy of solutions by varying the grid spacing, by transforming the coordinate system used, and by extrapolating the approximations found for different sized grid spacings to obtain the limiting value for a zero grid spacing. Section 10 summarises the various techniques considered.

It must be emphasised that the theoretical foundation of finite difference methods has not kept pace with their application. It is generally impossible to obtain a theoretical estimate of the accuracy of a numerical solution of a problem in which the non-linear terms in the equations are not small. Users of finite difference techniques must therefore apply every method available, such as successive grid refinement, comparison of the numerical solution with analytical solutions to simplified versions of the problem, and comparison with experimental results, in order to verify the quality of their results.

Readers who wish to use the methods described in this article should also study the original papers referenced and they should read more recent literature which discusses these methods. Publications such as Mathematics of Computation, Journal of Computational Physics, Computers and Fluids, Computer Methods in Applied Mechanics and Engineering, Numerical Methods in Engineering and the Journal of Fluid Mechanics, regularly include articles of relevance to finite difference techniques.

1.2 The Transport Equation

The behaviour of a fluid undergoing mass, vorticity or forced convective heat transfer, is described by a set of partial differential equations which are mathematical formulations of one or more of the conservation laws of physics. These laws include those of conservation of mass, momentum and energy.

For instance, if the fluid is incompressible and has constant thermal conductivity κ, the *heat equation*, a mathematical formulation of the law of conservation of thermal energy, is

$$\rho C_p \frac{DT}{Dt} = \kappa \nabla^2 T, \tag{1.2.1a}$$

where ρ is the density, C_p is the specific heat of the fluid at constant pressure, D/Dt is the operation of differentiation following the motion of the fluid, and T is its temperature. In Cartesian coordinates this can be written

$$\frac{\partial T}{\partial t} + u\frac{\partial T}{\partial x} + v\frac{\partial T}{\partial y} + w\frac{\partial T}{\partial z} = a\left(\frac{\partial^2 T}{\partial x^2} + \frac{\partial^2 T}{\partial y^2} + \frac{\partial^2 T}{\partial z^2}\right) \tag{1.2.1b}$$

where $a = \kappa/\rho C_p$ and u,v,w are the velocity components of the fluid at the point (x,y,z), at time t.

For fluids at rest, and for solids, $u=v=w=0$ and Equation (1.2.1b) reduces to the *diffusion (conduction) equation*

$$\frac{\partial T}{\partial t} = a\left(\frac{\partial^2 T}{\partial x^2} + \frac{\partial^2 T}{\partial y^2} + \frac{\partial^2 T}{\partial z^2}\right). \tag{1.2.1c}$$

Equation (1.2.1b) is a form of the transport equation: it involves transfer of heat due to the motion of the fluid (convection) and due to diffusion (conduction). The three terms $u\partial T/\partial x$, $v\partial T/\partial y$ and $w\partial T/\partial z$ are called convective terms, and the three terms $a\partial^2 T/\partial x^2$, $a\partial^2 T/\partial y^2$ and $a\partial^2 T/\partial z^2$ are usually called diffusive terms.

Similar cylindrical and spherical forms of Equation (1.2.1a) exist, but for the purpose of describing methods for their solution only the rectangular Cartesian forms will be considered.

In general, the transport equation (1.2.1a) may also describe the transfer of dissolved material or vorticity, as well as heat, to mention just three examples. In heat transfer problems T is the temperature, in material transfer problems it is the concentration and for vorticity transfer T represents the vorticity.

Throughout this article, Equation (1.2.1b) is used in its various forms such as:

the one-dimensional diffusion equation

$$\frac{\partial T}{\partial t} = a\frac{\partial^2 T}{\partial x^2}, \tag{1.2.2a}$$

the one-dimensional convection equation

$$\frac{\partial T}{\partial t} + u\frac{\partial T}{\partial x} = 0, \tag{1.2.2b}$$

the one-dimensional steady transport equation

$$u\frac{\partial T}{\partial x} = a\frac{\partial^2 T}{\partial x^2}, \tag{1.2.2c}$$

and so on. These forms will be used to illustrate the finite difference techniques described in later sections. For particular initial conditions and boundary values, analytic solutions (usually in terms of infinite series) can be found for these equations. The accuracy of the finite difference methods developed in this article can be estimated by comparison with these exact solutions.

In the examples which follow, the coordinates will be non-dimensionalised so that the boundaries in the x-direction will be located at $x=0$, $x=1$, the time will be non-dimensionalised so that a typical time period in the solution T will become 1, and the speed of the fluid flow will be non-dimensionalised accordingly.

1.3 Non-Dimensionalisation of the Transport Equation

The transport equation (1.2.1a) can have the space and time variables non-dimensionalised in the following way.

Let L,B,H be typical lengths, such as the distance between boundaries in the x,y, z-directions;for instance, the space region R over which (1.2.1a) is to be solved may be defined by $0 \leqslant x \leqslant L$, $0 \leqslant y \leqslant B$, $0 \leqslant z \leqslant H$. Replace the x,y,z variables by the dimensionless space variables

$$\left.\begin{aligned} x_* &= x/L, \\ y_* &= y/B, \\ z_* &= z/H, \end{aligned}\right\} \tag{1.3.1a}$$

so that the region R is now defined by $0 \leqslant x_* \leqslant 1$, $0 \leqslant y_* \leqslant 1$, $0 \leqslant z_* \leqslant 1$.

Let P be a characteristic interval of time associated with the given problem, such as the period of an oscillation in the dependent variable T. Replace the variable t by

$$t_* = t/P, \tag{1.3.1b}$$

so that in terms of the dimensionless time variable t_* the period of the oscillation is now 1.

Substitution of the new independent variables x_*, y_*, z_* and t_* into the transport equation (1.2.1b) gives

$$\frac{\partial T}{\partial t_*} + \frac{uP}{L}\frac{\partial T}{\partial x_*} + \frac{vP}{B}\frac{\partial T}{\partial y_*} + \frac{wP}{H}\frac{\partial T}{\partial z_*} = \frac{aP}{L^2}\frac{\partial^2 T}{\partial x_*^2} + \frac{aP}{B^2}\frac{\partial^2 T}{\partial y_*^2} + \frac{aP}{H^2}\frac{\partial^2 T}{\partial z_*^2} . \tag{1.3.2}$$

If T_0 is a typical value of the dependent variable T, this may also be non-dimensionalised by means of the substitution

$$T_* = T/T_0 . \tag{1.3.3}$$

Substitution of (1.3.3) into Equation (1.3.2) yields

$$\frac{\partial T_*}{\partial t_*} + u_*\frac{\partial T_*}{\partial x_*} + v_*\frac{\partial T_*}{\partial y_*} + w_*\frac{\partial T_*}{\partial z_*} = \alpha_x\frac{\partial^2 T_*}{\partial x_*^2} + \alpha_y\frac{\partial^2 T_*}{\partial y_*^2} + \alpha_z\frac{\partial^2 T_*}{\partial z_*^2} , \tag{1.3.4a}$$

where

$$\left.\begin{aligned} &u_* = uP/L,\ v_* = vP/B,\ w_* = wP/H, \\ &\alpha_x = \frac{aP}{L^2},\ \alpha_y = \frac{aP}{B^2},\ \alpha_z = \frac{aP}{H^2} . \end{aligned}\right\} \tag{1.3.4b}$$

Equation (1.3.4) is the non-dimensional form of the transport equation and appears in the same form as Equation (1.2.1b). It is to be solved on the space region R now defined by

$$0 \leqslant x_* \leqslant 1, \quad 0 \leqslant y_* \leqslant 1, \quad 0 \leqslant z_* \leqslant 1.$$

For convenience, in the following it will be assumed that all equations have been non-dimensionalised. The asterisks in Equation (1.3.4) will therefore be omitted. In general, we will consider the transport equation

$$\frac{\partial\bar{\tau}}{\partial t} + u\frac{\partial\bar{\tau}}{\partial x} + v\frac{\partial\bar{\tau}}{\partial y} + w\frac{\partial\bar{\tau}}{\partial z} = \alpha_x\frac{\partial^2\bar{\tau}}{\partial x^2} + \alpha_y\frac{\partial^2\bar{\tau}}{\partial y^2} + \alpha_z\frac{\partial^2\bar{\tau}}{\partial z^2}, \tag{1.3.5}$$

in which $\bar{\tau}$ is some suitably non-dimensionalised property such as temperature, concentration, or vorticity, t is a non-dimensional time variable with a characteristic value of 1, and x,y,z are non-dimensional space variables with maximum values of 1.

2. FINITE DIFFERENCE APPROXIMATIONS

2.1 The One-Dimensional Diffusion Equation

Many of the concepts associated with the numerical solution of partial differential equations by finite difference methods can be illustrated by considering the one-dimensional diffusion equation

$$\frac{\partial\bar{\tau}}{\partial t} = \alpha\,\frac{\partial^2\bar{\tau}}{\partial x^2}\,, \tag{2.1.1}$$

assumed to have been suitably non-dimensionalised so that $0 \leqslant \bar{\tau} \leqslant 1$ and $0 \leqslant x \leqslant 1$. For instance, the dependent variable $\bar{\tau}(x,t)$ may represent the non-dimensional temperature at time t at position x along a thin insulated rod of unit length with ends located at $x=0$ and $x=1$. If the rod is composed of homogeneous material then α, the non-dimensional coefficient of heat diffusion, is constant.

Certain initial and boundary conditions are required before Equation (2.1.1) can be solved, for example

$$\left.\begin{aligned} \bar{\tau}(x,0) &= f(x), \quad 0 \leqslant x \leqslant 1,\\ \bar{\tau}(0,t) &= g_0(t), \quad t \geqslant 0,\\ \bar{\tau}(1,t) &= g_1(t), \quad t \geqslant 0. \end{aligned}\right\} \tag{2.1.2}$$

Interpreted in terms of the problem defined above, the first of these conditions is a statement that the non-dimensional temperature everywhere along the rod is known to be $f(x)$ initially, the second is a statement that the non-dimensional temperature at the left end of the rod (that is at $x=0$) is prescribed for all time by $g_0(t)$, and the third is a statement that at the right end of the rod (that is at $x=1$) the non-dimensional temperature is $g_1(t)$.

Given the description of $f(x)$, $g_0(t)$ and $g_1(t)$, the solution $\bar{\tau}(x,t)$ is to be determined over the semi-infinite rectangle in x-t space defined by $0 < x < 1$, $t > 0$ (see Figure 2.1). This region, $\mathcal{D}$, on which the solution of the given differential equation (2.1.1) is to be found subject to the initial and boundary conditions (2.1.2), is called the solution domain.

For example, a thin insulated rod initially at a temperature T of 0°C suddenly has the temperature at each end raised to 100°C and held at that value for all time $t > 0$. Dividing the temperature in degrees C by 100°C gives a suitable non-dimensional temperature $\bar{\tau}(x,t)$, for which the initial and boundary conditions (2.1.2) become

$$\left.\begin{aligned} \bar{\tau}(x,0) &= 0, \quad 0 \leqslant x \leqslant 1,\\ \bar{\tau}(0,t) &= 1, \quad t \geqslant 0,\\ \bar{\tau}(1,t) &= 1, \quad t \geqslant 0. \end{aligned}\right\} \tag{2.1.3}$$

A problem arises when the initial temperatures at $x=0$ and $x=1$, namely $\bar{\tau}(0,0)$ and $\bar{\tau}(1,0)$, are to be prescribed. In practice the temperature at $x=0$ and $x=1$ cannot be altered instantaneously, but (2.1.3) requires

$$\bar{\tau}(0,0-) = 0 \ \text{ while } \ \bar{\tau}(0,0+) = 1,$$

$$\bar{\tau}(1,0-) = 0 \ \text{ while } \ \bar{\tau}(1,0+) = 1.$$

The simplest way of defining $\bar{\tau}(0,0)$ and $\bar{\tau}(1,0)$ so the resulting problem is tractable mathematically is to give them both the value of 1.

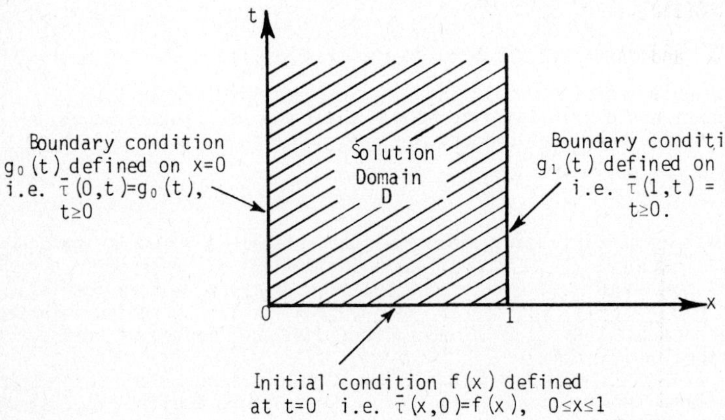

Figure 2.1: The solution domain in x-t space for the one-dimensional diffusion equation.

Equation (2.1.1) with initial and boundary conditions as now defined can be solved analytically by substituting $T = 1 - \bar{\tau}$ in (2.1.1) and (2.1.3), then applying the method of separation of variables (see Kreyszig, 1972, p. 430) giving the solution

$$\bar{\tau}(x,t) = 1 - \sum_{m=1}^{\infty} \frac{4}{(2m-1)\pi} \sin\{(2m-1)\pi x\}\exp\{-\alpha(2m-1)^2\pi^2 t\}, \tag{2.1.4}$$

which permits the value of the non-dimensional temperature $\bar{\tau}$ to be calculated at any point in the solution domain $0 < x < 1$, $t > 0$. For a given x and t, this requires the summation of the series (2.1.4) on a computer. This is done by adding more and more terms of the infinite series until successive partial sums remain constant with the required accuracy.

It should be noted that the method of separation of variables referred to above is restricted to the solution of linear partial differential equations with constant coefficients, and boundary values which are equivalent to reflection or periodicity conditions. It is difficult, and in some cases impossible, to find exact solutions to non-linear partial differential equations with variable coefficients and complicated boundary conditions.

An alternative method for solving the given partial differential equation is to use finite difference methods which are not restricted by such criteria as linearity, constant coefficients, and so forth.

2.2 The Finite Difference Grid

The solution domain in x-t space is covered by a rectangular grid (sometimes called a mesh or net), with grid spacings of Δx, Δt in the x,t directions respectively, the values of Δx and Δt being assumed uniform for the present. The grid consists of the set of lines parallel to the t-axis given by

$$x = x_j , \quad j=0(1)J, \tag{2.2.1}$$

where $x_j = j\Delta x$ and $\Delta x = 1/J$, and the set of lines parallel to the x-axis given by

$$t = t_n , \quad n=0,1,2,... \tag{2.2.2}$$

where $t_n = n\Delta t$, as shown in Figure 2.2.

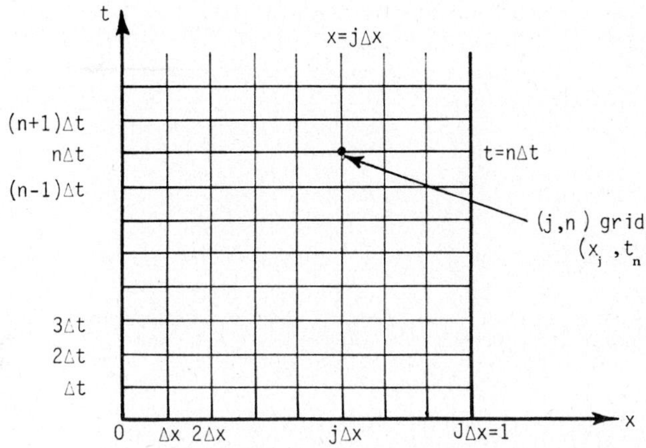

Figure 2.2: The finite difference grid in the solution region for the one-dimensional diffusion equation.

Finite difference methods will now be developed which determine approximately the values of $\bar{\tau}$ at interior points within the solution domain where the lines defined by (2.2.1) and (2.2.2) intersect. These points, with coordinates (x_j ,t_n), are called *grid-points* and are often denoted (j,n). The terms mesh-point or node are sometimes used instead of grid-point. Grid-points within the solution domain are termed interior; those on the boundary of the solution domain, that is those for which $j=0$ or J, or $n=0$, are called boundary grid-points. In the diffusion problem under consideration, the values of $\bar{\tau}$ on the boundaries of the solution region are defined by the initial and boundary conditions (2.1.3) (see Figure 2.1).

The following notation will be used for values of $\bar{\tau}$ and its derivatives at the (j,n) grid-point:

$$\left.\begin{aligned} \bar{\tau}_j^n &= \bar{\tau}(x_j ,t_n), \\ \left.\frac{\partial\bar{\tau}}{\partial t}\right|_j^n &= \frac{\partial\bar{\tau}(x_j ,t_n)}{\partial t} \end{aligned}\right\} , \tag{2.2.3}$$

and so on. Thus we denote the second spatial derivative at $x_{j-1} = (j-1)\Delta x$, $t_{n+1} = (n+1)\Delta t$, that is at the $(j-1,n+1)$ grid-point, by

$$\left.\frac{\partial^2\bar{\tau}}{\partial x^2}\right|_{j-1}^{n+1} .$$

Note that the superscript n in the left side of Equations (2.2.3) indicates the time level. In order to avoid confusion, if a superscript is to indicate a power, the number to be raised to that power will be enclosed in brackets. Thus, $(G)^n$ is the nth power of the number G. The brackets will be omitted when there is no possibility of misinterpretation.

2.3 Finite Difference Approximations to Derivatives

As a first step in developing a method of calculating the values of $\bar{\tau}$ at each interior grid-point, the space and time derivatives of $\bar{\tau}$ at the (j,n) grid-point must be expressed in terms of values of $\bar{\tau}$ at nearby grid-points. Taylor series expansions of $\bar{\tau}$ about the (j,n) grid-point will be used in this process, such as

$$\bar{\tau}(x_j+\Delta x,t_n) = \sum_{m=0}^{\infty} \frac{(\Delta x)^m}{m!} \frac{\partial^m \bar{\tau}(x_j,t_n)}{\partial x^m},$$

or, in the notation defined by (2.2.3),

$$\bar{\tau}_{j+1}^{n} = \sum_{m=0}^{\infty} \frac{(\Delta x)^m}{m!} \left.\frac{\partial^m \bar{\tau}}{\partial x^m}\right|_j^n$$

$$= \bar{\tau}_j^n + \Delta x \left.\frac{\partial \bar{\tau}}{\partial x}\right|_j^n + \frac{(\Delta x)^2}{2!} \left.\frac{\partial^2 \bar{\tau}}{\partial x^2}\right|_j^n + \frac{(\Delta x)^3}{3!} \left.\frac{\partial^3 \bar{\tau}}{\partial x^3}\right|_j^n + \dots . \qquad (2.3.1)$$

Similarly we will use

$$\bar{\tau}(x_j,t_n+\Delta t) = \sum_{m=0}^{\infty} \frac{(\Delta t)^m}{m!} \frac{\partial^m \bar{\tau}(x_j,t_n)}{\partial t^m},$$

which may be written

$$\bar{\tau}_j^{n+1} = \sum_{m=0}^{\infty} \frac{(\Delta t)^m}{m!} \left.\frac{\partial^m \bar{\tau}}{\partial t^m}\right|_j^n$$

$$= \bar{\tau}_j^n + \Delta t \left.\frac{\partial \bar{\tau}}{\partial t}\right|_j^n + \frac{(\Delta t)^2}{2!} \left.\frac{\partial^2 \bar{\tau}}{\partial t^2}\right|_j^n + \frac{(\Delta t)^3}{3!} \left.\frac{\partial^3 \bar{\tau}}{\partial t^3}\right|_j^n + \dots . \qquad (2.3.2)$$

These Taylor series expansions may be truncated after any number of terms, the resulting error being dominated by the next term in the expansion if $\Delta x \ll 1$ in (2.3.1), or if $\Delta t \ll 1$ in (2.3.2). Thus we may write

$$\bar{\tau}_{j+1}^n = \bar{\tau}_j^n + \Delta x \left.\frac{\partial \bar{\tau}}{\partial x}\right|_j^n + \frac{(\Delta x)^2}{2!} \left.\frac{\partial^2 \bar{\tau}}{\partial x^2}\right|_j^n + O\{(\Delta x)^3\}, \qquad (2.3.3)$$

and

$$\bar{\tau}_j^{n+1} = \bar{\tau}_j^n + \Delta t \left.\frac{\partial \bar{\tau}}{\partial t}\right|_j^n + \frac{(\Delta t)^2}{2!} \left.\frac{\partial^2 \bar{\tau}}{\partial t^2}\right|_j^n + O\{(\Delta t)^3\}. \qquad (2.3.4)$$

The term $O\{(\Delta t)^3\}$, for example, is interpreted as meaning there exists a positive constant κ, depending on $\bar{\tau}$, such that the difference between $\bar{\tau}$ at the (j,n+1) grid-point and the first three terms on the right side of (2.3.4), all evaluated at the (j,n) grid-point, is numerically less than $\kappa(\Delta t)^3$ for all sufficiently small Δt. When there is no possibility of confusion, the qualifying statement "as $\Delta t \to 0$" is omitted. It is clear that the error involved in this approximation reduces in magnitude as the grid spacing Δt decreases.

The value of the continuous derivative $\partial\bar{\tau}/\partial t$ at the (j,n) grid-point may be approximated in terms of values of $\bar{\tau}$ at nearby grid-points in a number of ways. For instance, suppose it is desired to express the derivative $\partial\bar{\tau}/\partial t|_j^n$ in terms of the values of $\bar{\tau}_j^n$ and $\bar{\tau}_j^{n+1}$, the latter being the value of $\bar{\tau}$ at the corresponding spatial position but one time level ahead. We therefore write

$$\left.\frac{\partial\bar{\tau}}{\partial t}\right|_j^n = -a_0\bar{\tau}_j^n + a_1\bar{\tau}_j^{n+1} + O\{(\Delta t)^m\} \ . \tag{2.3.5}$$

The term $O\{(\Delta t)^m\}$ will indicate the accuracy of the resulting approximation for $\partial\bar{\tau}/\partial t|_j^n$.

Since, by (2.3.4)

$$a_1\bar{\tau}_j^{n+1} = a_1\bar{\tau}_j^n + a_1\Delta t\left.\frac{\partial\bar{\tau}}{\partial t}\right|_j^n + a_1\frac{(\Delta t)^2}{2}\left.\frac{\partial^2\bar{\tau}}{\partial t^2}\right|_j^n + \dots ,$$

subtraction of $a_0\bar{\tau}_j^n$ gives

$$a_1\bar{\tau}_j^{n+1} - a_0\bar{\tau}_j^n = (a_1-a_0)\bar{\tau}_j^n + a_1\Delta t\left.\frac{\partial\bar{\tau}}{\partial t}\right|_j^n + a_1\frac{(\Delta t)^2}{2}\left.\frac{\partial^2\bar{\tau}}{\partial t^2}\right|_j^n + \dots \tag{2.3.6}$$

Clearly, we require $a_1\Delta t = 1$ and $a_1-a_0 = 0$ on the right side of this equation. Thus $a_1 = 1/\Delta t$ and $a_0 = a_1 = 1/\Delta t$. Substitution in Equation (2.3.6) yields

$$\left.\frac{\partial\bar{\tau}}{\partial t}\right|_j^n = \frac{\bar{\tau}_j^{n+1} - \bar{\tau}_j^n}{\Delta t} + O\{\Delta t\}. \tag{2.3.7a}$$

Thus we may write

$$\left.\frac{\partial\bar{\tau}}{\partial t}\right|_j^n \simeq \frac{\bar{\tau}_j^{n+1} - \bar{\tau}_j^n}{\Delta t} , \tag{2.3.7b}$$

which has an error of $O\{\Delta t\}$. Because the right hand side of this expression uses the value $\bar{\tau}_j^{n+1}$ which is forward of $\bar{\tau}_j^n$ in time, it is called the forward difference approximation to the time derivative evaluated at the (j,n) grid-point.

An alternative way of approximating the value of $\partial\bar{\tau}/\partial t$ at the (j,n) grid-point is to use the relation

$$\left.\frac{\partial\bar{\tau}}{\partial t}\right|_j^n = \frac{\bar{\tau}_j^n - \bar{\tau}_j^{n-1}}{\Delta t} + O\{\Delta t\}, \tag{2.3.8a}$$

which yields the approximation

$$\left.\frac{\partial\bar{\tau}}{\partial t}\right|_j^n \simeq \frac{\bar{\tau}_j^n - \bar{\tau}_j^{n-1}}{\Delta t} \ . \tag{2.3.8b}$$

Because the right side of (2.3.8b) uses the value $\bar{\tau}_j^{n-1}$ at a previous time level to $\bar{\tau}_j^n$ it is called the backward difference approximation to the time derivative. Like the forward difference approximation it has an error of $O\{\Delta t\}$.

Another way of approximating the value of $\partial\bar{\tau}/\partial t$ at the (j,n) grid-point is to use the relation

$$\left.\frac{\partial\bar{\tau}}{\partial t}\right|_j^n = \frac{\bar{\tau}_j^{n+1} - \bar{\tau}_j^{n-1}}{2\Delta t} + O\{(\Delta t)^2\}, \tag{2.3.9a}$$

which leads to the approximation

$$\left.\frac{\partial\bar{\tau}}{\partial t}\right|_j^n \simeq \frac{\bar{\tau}_j^{n+1} - \bar{\tau}_j^{n-1}}{2\Delta t}. \tag{2.3.9b}$$

The error in using the right side of (2.3.9b) for the value of $\partial\bar{\tau}/\partial t|_j^n$ is $O\{(\Delta t)^2\}$, which is smaller than $O\{\Delta t\}$ for $\Delta t << 1$, so this is a more accurate approximation than either the forward or backward difference approximations to $\partial\bar{\tau}/\partial t$. Because the time level $n\Delta t$ is centred between the time levels $(n+1)\Delta t$ and $(n-1)\Delta t$ at which the values $\bar{\tau}_j^{n+1}$ and $\bar{\tau}_j^{n-1}$ occur, the right hand side of (2.3.9b) is called the central difference approximation to the first-order time derivative $\partial\bar{\tau}/\partial t$ at the (j,n) grid-point.

In a similar way, finite difference approximations to the second-order time derivative $\partial^2\bar{\tau}/\partial t^2|_j^n$ can be derived. For example, suppose it is desired to express the derivative at the (j,n) grid-point in terms of the values of $\bar{\tau}$ at that grid-point and also at the grid-point one time level ahead and two time levels behind.

We therefore write

$$\left.\frac{\partial^2\bar{\tau}}{\partial t^2}\right|_j^n = a_{-2}\bar{\tau}_j^{n-2} + a_{-1}\bar{\tau}_j^{n-1} + a_0\bar{\tau}_j^n + a_1\bar{\tau}_j^{n+1} + O\{(\Delta t)^m\}, \tag{2.3.10}$$

where the coefficients a_{-2}, a_{-1}, a_0 and a_1 are to be determined, as well as the order of the error term.

Since

$$a_{-2}\bar{\tau}_j^{n-2} = a_{-2}\{\bar{\tau}_j^n - 2\Delta t\left.\frac{\partial\bar{\tau}}{\partial t}\right|_j^n + \frac{4(\Delta t)^2}{2!}\left.\frac{\partial^2\bar{\tau}}{\partial t^2}\right|_j^n - \frac{8(\Delta t)^3}{3!}\left.\frac{\partial^3\bar{\tau}}{\partial t^3}\right|_j^n + \frac{16(\Delta t)^4}{4!}\left.\frac{\partial^4\bar{\tau}}{\partial t^4}\right|_j^n + ..\},$$

$$a_{-1}\bar{\tau}_j^{n-1} = a_{-1}\{\bar{\tau}_j^n - \Delta t\left.\frac{\partial\bar{\tau}}{\partial t}\right|_j^n + \frac{(\Delta t)^2}{2!}\left.\frac{\partial^2\bar{\tau}}{\partial t^2}\right|_j^n - \frac{(\Delta t)^3}{3!}\left.\frac{\partial^3\bar{\tau}}{\partial t^3}\right|_j^n + \frac{(\Delta t)^4}{4!}\left.\frac{\partial^4\bar{\tau}}{\partial t^4}\right|_j^n + ..\},$$

$$a_0\bar{\tau}_j^n = a_0\bar{\tau}_j^n,$$

$$a_1\bar{\tau}_j^{n+1} = a_1\{\bar{\tau} + \Delta t\left.\frac{\partial\bar{\tau}}{\partial t}\right|_j^n + \frac{(\Delta t)^2}{2!}\left.\frac{\partial^2\bar{\tau}}{\partial t^2}\right|_j^n + \frac{(\Delta t)^3}{3!}\left.\frac{\partial^3\bar{\tau}}{\partial t^3}\right|_j^n + \frac{(\Delta t)^4}{4!}\left.\frac{\partial^4\bar{\tau}}{\partial t^4}\right|_j^n + ..\},$$

then their sum is

$$a_{-2}\bar{\tau}_j^{n-2} + a_{-1}\bar{\tau}_j^{n-1} + a_0\bar{\tau}_j^{n} + a_1\bar{\tau}_j^{n+1}$$

$$= (a_{-2}+a_{-1}+a_0+a_1)\bar{\tau}_j^n + \Delta t(-2a_{-2}-a_{-1}+a_1)\frac{\partial\bar{\tau}}{\partial t}\bigg|_j^n$$

$$+ \frac{(\Delta t)^2}{2}(4a_{-2}+a_{-1}+a_1)\frac{\partial^2\bar{\tau}}{\partial t^2}\bigg|_j^n + \frac{(\Delta t)^3}{6}(-8a_{-2}-a_{-1}+a_1)\frac{\partial^3\bar{\tau}}{\partial t^3}\bigg|_j^n$$

$$+ \frac{(\Delta t)^4}{24}(16a_{-2}+a_{-1}+a_1)\frac{\partial^4\bar{\tau}}{\partial t^4}\bigg|_j^n + \ldots \tag{2.3.11}$$

From (2.3.10) it is clear that the right side of (2.3.11) is to become

$$\frac{\partial^2\bar{\tau}}{\partial t^2}\bigg|_j^n + O\{(\Delta t)^m\} , \tag{2.3.12}$$

so we require

$$\left.\begin{aligned} &a_{-2} + a_{-1} + a_0 + a_1 = 0, \\ &-2a_{-2} - a_{-1} + a_1 = 0, \\ &4a_{-2} + a_{-1} + a_1 = 2(\Delta t)^{-2}. \end{aligned}\right\} \tag{2.3.13a}$$

Also, as there is still one degree of freedom left, the term in $(\Delta t)^3$ can be eliminated, giving

$$-8a_{-2} - a_{-1} + a_1 = 0. \tag{2.3.13b}$$

Solving the set of linear algebraic equations (2.3.13) gives

$$a_{-2} = 0,\ a_{-1} = (\Delta t)^{-2},\ a_0 = -2(\Delta t)^{-2},\ a_1 = (\Delta t)^{-2}, \tag{2.3.14}$$

so that (2.3.11) becomes

$$\frac{\partial^2\bar{\tau}}{\partial t^2}\bigg|_j^n = \frac{\bar{\tau}_j^{n-1} - 2\bar{\tau}_j^n + \tau_j^{n+1}}{(\Delta t)^2} + O\{(\Delta t)^2\}. \tag{2.3.15}$$

Note that a more accurate approximation to $\partial^2\bar{\tau}/\partial t^2$ is obtained by using values of $\bar{\tau}$ at only three consecutive time levels in the manner of (2.3.15), than by using values of $\bar{\tau}$ at four consecutive time levels.

In a similar manner finite difference approximations to the spatial derivatives of $\bar{\tau}$ at the (j,n) grid-point may be obtained. For the first-order spatial derivative the appropriate finite difference relations are:

$$\frac{\partial\bar{\tau}}{\partial x}\bigg|_j^n = \frac{\bar{\tau}_{j+1}^n - \bar{\tau}_j^n}{\Delta x} + O\{\Delta x\} , \tag{2.3.16}$$

which leads to a *forward difference approximation*;

$$\left.\frac{\partial\bar{\tau}}{\partial x}\right|_j^n = \frac{\bar{\tau}_j^n - \bar{\tau}_{j-1}^n}{\Delta x} + O\{\Delta x\} , \tag{2.3.17}$$

which leads to a *backward difference approximation;*

$$\left.\frac{\partial\bar{\tau}}{\partial x}\right|_j^n = \frac{\bar{\tau}_{j+1}^n - \bar{\tau}_{j-1}^n}{2\Delta x} + O\{(\Delta x)^2\}, \tag{2.3.18}$$

which leads to a *central difference approximation.*

The *central difference approximation* to the second-order spatial derivative is based on the relation

$$\left.\frac{\partial^2\bar{\tau}}{\partial x^2}\right|_j^n = \frac{\bar{\tau}_{j+1}^n - 2\bar{\tau}_j^n + \bar{\tau}_{j-1}^n}{(\Delta x)^2} + O\{(\Delta x)^2\}. \tag{2.3.19}$$

ASSESSING FINITE-DIFFERENCE APPROXIMATIONS TO DERIVATIVES OF WAVE-LIKE MOTIONS

Since wave-like motions are of interest in many problems, and since any motion may be considered as a sum of wave-like motions (that is, its functional form can be expressed as a Fourier series) it is of interest to determine the accuracy of finite difference approximations to derivatives of a simple harmonic wave.

Consider the progressive wave

$$\bar{\tau}(x,t) = \cos(\kappa x - \sigma t), \tag{2.3.20}$$

κ and σ being constants. This is a sinusoidal wave which moves in the positive x-direction with wave speed α/κ, and at a fixed point x_j it is periodic in time with period $P = 2\pi/\sigma$. It is also periodic in space with wave-length $\lambda = 2\pi/\kappa$.

At the point x_j at the n-th time level, the exact value of the time derivative of $\bar{\tau}$ is

$$\left.\frac{\partial\bar{\tau}}{\partial t}\right|_j^n = \sigma \sin(\kappa x_j - \sigma t_n). \tag{2.3.21}$$

However, the forward-time finite difference approximation to this derivative is

$$\frac{\bar{\tau}_j^{n+1} - \bar{\tau}_j^n}{\Delta t} = \frac{\cos(\kappa x_j - \sigma t_n - \sigma\Delta t) - \cos(\kappa x_j - \sigma t_n)}{\Delta t}$$

$$= \sigma\left\{\frac{\sin(\sigma\Delta t/2)}{(\sigma\Delta t/2)}\right\} \sin\left(\kappa x_j - \sigma t_n - \frac{\sigma\Delta t}{2}\right). \tag{2.3.22}$$

Clearly an error has been introduced in the amplitude and phase of the derivative if (2.3.22) is used as an approximation for (2.3.21). That is, the forward-time approximation for the time derivative of (2.3.20) is equal to the true value of the derivative given by Equation (2.3.21) multiplied by the factor $\{\sin(\sigma\Delta t/2)/(\sigma\Delta t/2)\}$ and altered in phase by $-\sigma\Delta t/2$, which is equivalent to a time lag of $\Delta t/2$. Clearly, as $\Delta t \to 0$ the factor multiplying the amplitude tends to 1 and the phase error tends to zero. Therefore the accuracy of the forward difference approximation in time increases with decreasing values of Δt.

The effect of using the central difference approximation for the second space derivative $\partial^2\bar{\tau}/\partial x^2|_j^n$ when $\bar{\tau}$ is given by the progressive wave form (2.3.20), is seen by comparing the exact value

$$\left.\frac{\partial^2\bar{\tau}}{\partial x^2}\right|_j^n = -\kappa^2\cos(\kappa x_j - \sigma t_n) \tag{2.3.23}$$

and the central difference approximation

$$\frac{\bar{\tau}_{j+1}^n - 2\bar{\tau}_j^n + \bar{\tau}_{j-1}^n}{(\Delta x)^2} = -\kappa^2\left\{\frac{\sin(\kappa\Delta x/2)}{(\kappa\Delta x/2)}\right\}^2 \cos(\kappa x_j - \sigma t_n). \tag{2.3.24}$$

This shows no phase error introduced, only a change to the amplitude of the derivative. Since the wavelength of (2.3.20) is $\lambda = 2\pi/\kappa$, then for small space steps so there are at least 20 in one wavelength, that is $\lambda \geqslant 20\Delta x$ so that $\kappa\Delta x/2 \leqslant \pi/20$, the amplitude of the derivative is multiplied by a factor between 0.992 and 1. However, when there are fewer than 2 grid spacings in one wavelength, so that $\Delta x \geqslant \lambda/2$ and $\kappa\Delta x/2 \geqslant \pi/2$, the amplitude of the derivative is less than 0.42 of its correct value.

2.4 An Approximate Solution to the Diffusion Equation

Approximate values τ_j^n to the true values $\bar{\tau}_j^n$ at the interior grid-points, that is at the (j,n) grid-points for $j=1(1)J-1$ and $n=1,2,3,\ldots$, may be found for the diffusion equation (2.1.1) by substituting finite difference approximations to the derivatives involved. At the (j,n) grid-point, the given equation is

$$\left.\frac{\partial\bar{\tau}}{\partial t}\right|_j^n = \alpha \left.\frac{\partial^2\bar{\tau}}{\partial x^2}\right|_j^n, \tag{2.4.1}$$

and any of the various approximations for $\partial\bar{\tau}/\partial t$ and $\partial^2\bar{\tau}/\partial x^2$ described in Section 2.3 can be used. One of the simplest finite difference equations which approximates the partial differential equation (2.4.1) is obtained by using the forward difference approximation to the time derivative with the central difference approximation to the space derivative.

Substitution from Equations (2.3.7a) and (2.3.19) into Equation (2.4.1) gives

$$\frac{\bar{\tau}_j^{n+1} - \bar{\tau}_j^n}{\Delta t} + O\{\Delta t\} = \alpha\left\{\frac{\bar{\tau}_{j+1}^n - 2\bar{\tau}_j^n + \bar{\tau}_{j-1}^n}{(\Delta x)^2} + O\{(\Delta x)^2\}\right\}. \tag{2.4.2}$$

Solving the finite difference equation

$$\frac{\tau_j^{n+1} - \tau_j^n}{\Delta t} = \alpha\left\{\frac{\tau_{j+1}^n - 2\tau_j^n + \tau_{j-1}^n}{(\Delta x)^2}\right\}, \tag{2.4.3}$$

with the given initial and boundary conditions therefore gives approximate values τ to the true solution $\bar{\tau}$ of the partial differential equation (2.4.1). Equation (2.4.3) is called the Forward-Time Centred-Space (FTCS) finite difference approximation to Equation (2.4.1). Clearly, the use of the finite difference approximation (2.4.3) involves a leading error of $O\{\Delta t,(\Delta x)^2\}$. The last statement is to be interpreted as meaning that there exist two positive constants κ_1 and κ_2, both depending on $\bar{\tau}$, such that the absolute value of the error in assuming (2.4.3) is the same as (2.4.1), is less than or equal to $\kappa_1\Delta t+\kappa_2(\Delta x)^2$, for all Δt and Δx which are sufficiently small.

Equation (2.4.3) may be rewritten in the form

$$\tau_j^{n+1} = s\tau_{j-1}^n + (1-2s)\tau_j^n + s\tau_{j+1}^n \tag{2.4.4}$$

where

$$s = \alpha\ \Delta t/(\Delta x)^2, \tag{2.4.5}$$

which is a formula giving the approximate value τ_j^{n+1} at the (n+1)th time level in terms of the approximate values τ_{j-1}^n, τ_j^n and τ_{j+1}^n at the nth time level.

Application of Equation (2.4.4) with $n=0$ gives the approximate values τ_j^1 at grid-points along the first time row $t=\Delta t$, in terms of the known boundary and initial values $\tau_j^0 = f(x_j), j=0(1)J$. Application of Equation (2.4.4) with $n=1$ then gives the approximate values τ_j^2 to the true solution $\bar{\tau}_{j_1}^2$ at grid-points along the second time level in terms of the computed values τ_j^1 at the grid-points along the first time level. An equation such as (2.4.4), which expresses one unknown value τ_j^{n+1} directly in terms of known values of τ at previous times is an explicit finite difference equation. Also, because it involves values along only two time levels, it is called a two-level formula.

Returning to the diffusion Equation (2.1.1) with $\alpha = 10^{-2}$ and boundary conditions given by (2.1.3), the explicit FTCS finite difference formula (2.4.4) can be used to obtain approximate solutions to the given problem at all interior grid-points. Experience shows that at least ten intervals are necessary to give reasonably accurate solutions to this problem, so $\Delta x=0.1$ will be chosen to illustrate the method.

Noting that Equation (2.4.4) takes the simple form

$$\tau_j^{n+1} = \tfrac{1}{2}(\tau_{j-1}^n + \tau_{j+1}^n) \tag{2.4.6}$$

when $s=\tfrac{1}{2}$, choose

$$\Delta t = \frac{(\Delta x)^2}{2\alpha} = 0.5.$$

In accordance with the initial conditions (2.1.3), the values of τ at grid-points along the time level $n=0$ are $\tau_0^0 = 1$, $\tau_1^0 = \tau_2^0 = \tau_3^0 = \tau_4^0 = \tau_5^0 = \tau_6^0 = \tau_7^0 = \tau_8^0 = \tau_9^0 = 0$, $\tau_{10}^0 = 1$. Equation (2.4.4) is then applied with $n=0$ and $j=1,2,\ldots,9$ to calculate the approximate solutions τ_j^1 at each of the internal grid-points along the time level $n=1$; that is,

$$\tau_j^1 = \tfrac{1}{2}(\tau_{j-1}^0 + \tau_{j+1}^0),\quad j=1(1)9,$$

is used. Values of τ_0^1 and τ_J^1, that is at grid-points on the boundaries $j=0$ and J, are specified by the boundary conditions of this problem and need not be calculated. The order in which the approximations τ_j^1 are calculated at the internal grid-points is immaterial when using this method, since Equation (2.4.4) gives τ_j^1 explicitly in terms of known values from previous time levels. The approximate values τ calculated at the first time level $t=0.5$ are therefore

$$\tau_0^1 = 1.0,\ \tau_1^1 = 0.5,\ \tau_2^1 = \tau_3^1 = \tau_4^1 = \tau_5^1 = \tau_6^1 = \tau_7^1 = \tau_8^1 = 0,$$

$$\tau_9^1 = 0.5,\ \tau_{10}^1 = 1.0.$$

Table 2.1 shows the results of these and calculations at further time levels. As the time increases, heat diffuses inwards and the internal temperatures are seen to be increasing steadily and it would be found, if the calculations were continued, that the temperatures gradually approach the equilibrium value of 1 at all internal grid-points. This is physically realistic. Note that it takes five time steps for the boundary temperature of 1 to have any effect on the temperature at the centre of the rod using this finite difference scheme with J=10.

Clearly, a large number of time steps will be needed before the values calculated at internal grid-points in the solution domain are close to 1. In order to reduce the number of arithmetic operations required, a larger time step is preferable. Choosing $\Delta t=1$, for which $s=1$, Equation (2.4.4) becomes

$$\tau_j^{n+1} = \tau_{j-1}^n - \tau_j^n + \tau_{j+1}^n . \tag{2.4.7}$$

Besides requiring only half the number of time levels to reach a particular time, this formula uses less computer time than application of (2.4.6). This is because multiplications and divisions take longer to carry out on a computer than do additions and subtractions. For instance, a multiplication or division on a CYBER 173 takes approximately 3 micro-seconds (one-millionth of a second) whereas

TABLE 2.1

Approximate solution of the diffusion problem using (2.4.4) with $s = \frac{1}{2}$

n	nΔt	Boundary Values										Boundary Values
6	3.0	1										1
5	2.5	1	0.6875	0.375	0.21875	0.0625	0.0625	0.0625	0.21875	0.375	0.6875	1
4	2.0	1	0.625	0.375	0.125	0.0625	0.0	0.0625	0.125	0.375	0.625	1
3	1.5	1	0.625	0.25	0.125	0.0	0.0	0.0	0.125	0.25	0.625	1
2	1.0	1	0.5	0.25	0.0	0.0	0.0	0.0	0.0	0.25	0.5	1
1	0.5	1	0.5	0.0	0.0	0.0	0.0	0.0	0.0	0.0	0.5	1
0 (Initial Values)	0.0	1	0	0	0	0	0	0	0	0	0	1
	jΔx	0.0	0.1	0.2	0.3	0.4	0.5	0.6	0.7	0.8	0.9	1.0
	j	0	1	2	3	4	5	6	7	8	9	10

an addition or subtraction takes approximately ½ micro-second. Therefore the arithmetic operations in (2.4.7) take about 1 micro-second to perform, while those in (2.4.6) take approximately 4 micro-seconds, four times as long.

Table 2.2 lists the results obtained by applying (2.4.7). Obviously the values shown are physically impossible; the effect of changing from $s=\frac{1}{2}$ to $s=1$ has been disastrous. Clearly the restrictions which apply to a particular finite difference method must be known to the user, otherwise he may accept unrealistic values as being accurate solutions to the given partial differential equation. In order to establish these restrictions three important concepts will be studied in

Section 3, those of convergence, consistency and stability.

TABLE 2.2

Approximate solution of the diffusion problem using (2.4.4) with $s = 1$

n	$n\Delta t$	Boundary Values										Boundary Values
6	6	1										1
5	5	1	7	-8	7	-3	2	-3	7	-8	7	1
4	4	1	-2	4	-2	1	0	1	-2	4	-2	1
3	3	1	2	-1	1	0	0	0	1	-1	2	1
2	2	1	0	1	0	0	0	0	0	1	0	1
1	1	1	1	0	0	0	0	0	0	0	1	1
0 (Initial Values)	0	1	0	0	0	0	0	0	0	0	0	1
	$j\Delta x$	0.0	0.1	0.2	0.3	0.4	0.5	0.6	0.7	0.8	0.9	1.0
	j	0	1	2	3	4	5	6	7	8	9	10

(axes: t vertical, x horizontal)

2.5 The Finite Difference Method - a Summary

The process of solving an initial value problem involving partial differential equations, using finite difference techniques, may be summarised as follows.

Firstly, a grid is defined which covers the solution domain. At the grid-points, the approximate solution to the problem will be found by algebraic processes.

As a second step, all dependent variables possible are initialised. These initial values may be exact, corresponding to a real initial situation for a transient problem, or they may be an approximation (perhaps a rough guess) for a steady (equilibrium) solution if the transient part is not of interest.

Once these two steps have been completed, the computational cycle begins. A finite difference equation, which is an analogue of the continuous partial differential equation is constructed. This is developed by replacing the continuous derivatives in the equation applied to an interior grid-point, by approximations in terms of values of the unknown dependent variable at nearby grid-points. The finite difference equation is used to calculate approximate values of the dependent variable at grid-points in the solution domain.

For instance, with the heat diffusion equation, the new temperature τ_j^{n+1} can be calculated approximately from the previously calculated temperatures τ_{j-1}^n, τ_j^n and τ_{j+1}^n using the FTCS formula (2.4.4). In this process, with time held constant at $t = n\Delta t$, j is varied from $1(1)J-1$ until the temperature at all points $j\Delta x$ in the grid are computed at the next time level $t = (n+1)\Delta t$. Once these values have been found, the time is again increased by Δt and the process repeated to get the temperatures at the new time level, $t = (n+2)\Delta t$, using the computed values at the time level $t = (n+1)\Delta t$.

This is continued until a predetermined time is reached, or until the dependent variables at each space position $j\Delta x$ remain unchanged when computed at successive time levels if an equilibrium solution is required.

3. CONVERGENCE, CONSISTENCY AND STABILITY

3.1 Convergence

A solution to a finite difference equation which approximates a given partial differential equation is said to be *convergent* if, at each grid-point in the solution domain, the finite difference solution approaches the solution of the partial differential equation as the grid spacings tend to zero. Formally, if we let $\bar{\tau}_j^n$ represent the exact solution of the partial differential equation and τ_j^n the exact solution of the approximating finite difference equation at the same point, then the finite difference solution is said to be convergent if

$$\tau_j^n \to \bar{\tau}_j^n \quad \text{as} \quad \Delta x \to 0,\ \Delta t \to 0.$$

The difference between the true solution of the partial differential equation and the exact solution of the approximating finite difference equation is called the *discretization error*, which will be denoted e_j^n; that is

$$e_j^n = \bar{\tau}_j^n - \tau_j^n . \tag{3.1.1}$$

The exact solution of the finite difference equation is obtained when no numerical errors of any sort, such as round-off errors, are introduced into the computations.

The magnitude of the error e_j^n at the (j,n) grid-point depends on the size of the grid-spacings Δx and Δt, and on the magnitude of the higher-order derivatives omitted in the finite difference approximations to the derivatives in the given differential equation. For instance, in the case of the FTCS method being used to solve Equation (2.1.1), rearrangement of Equation (2.4.2) gives

$$\bar{\tau}_j^{n+1} = s\bar{\tau}_{j-1}^n + (1-2s)\bar{\tau}_j^n + s\bar{\tau}_{j+1}^n + O\{(\Delta t)^2,\Delta t(\Delta x)^2\}. \tag{3.1.2}$$

By comparing Equation (2.4.4) with Equation (3.1.2) it is clear that the error term of $O\{(\Delta t)^2,\Delta t(\Delta x)^2\}$ is the contribution to the discretisation error e_j^{n+1} in the process of stepping from time level n to time level $(n+1)$ using the FTCS equation. However, it must be remembered that e_j^{n+1} also depends on the magnitude of the higher-order derivatives associated with the terms which contain $(\Delta t)^2,\Delta t(\Delta x)^2$ as well as the errors involved in previous time-steps.

The convergence of the FTCS method for solving the diffusion equation may be demonstrated by examining the behaviour of the discretization error e_j^n at the (j,n) grid-point and showing that $|e_j^n|$ is bounded by some number κ which tends to zero as $\Delta x \to 0$, $\Delta t \to 0$.

Considering the diffusion equation (2.1.1) with exact solution $\bar{\tau}_j^n$ at the (j,n) grid-point, and the explicit FTCS finite difference approximation (2.4.3) which has exact solution τ_j^n at the same point, Equation (3.1.1) gives

$$\tau_j^n = \bar{\tau}_j^n - e_j^n . \tag{3.1.3}$$

Substitution of (3.1.3) into Equation (2.4.4) yields

$$e_j^{n+1} = se_{j-1}^n + (1-2s)e_j^n + se_{j+1}^n + \bar{\tau}_j^{n+1} - \bar{\tau}_j^n - s\{\bar{\tau}_{j+1}^n + \bar{\tau}_{j-1}^n - 2\bar{\tau}_j^n\}. \tag{3.1.4}$$

Application of the truncated Taylor series expansion about the (j,n) grid-point, with a remainder term expressed in Lagrange's form (see Thomas, 1968, p. 639), gives

$$\bar{\tau}^n_{j+1} = \bar{\tau}^n_j + \Delta x \left.\frac{\partial\bar{\tau}}{\partial x}\right|^n_j + \frac{(\Delta x)^2}{2!}\left.\frac{\partial^2\bar{\tau}}{\partial x^2}\right|^n_{j+\theta_1},$$

$$\bar{\tau}^n_{j-1} = \bar{\tau}^n_j - \Delta x \left.\frac{\partial\bar{\tau}}{\partial x}\right|^n_j + \frac{(\Delta x)^2}{2!}\left.\frac{\partial^2\bar{\tau}}{\partial x^2}\right|^n_{j-\theta_2},$$

$$\bar{\tau}^{n+1}_j = \bar{\tau}^n_j + \Delta t \left.\frac{\partial\bar{\tau}}{\partial t}\right|^{n+\theta_3}_j ,$$

where $0 < \theta_1 < 1$, $0 < \theta_2 < 1$, $0 < \theta_3 < 1$.

Substitution of these expansions into Equation (3.1.4) gives

$$e^{n+1}_j = se^n_{j-1} + (1-2s)e^n_j + se^n_{j+1} + \Delta t \left.\frac{\partial\bar{\tau}}{\partial t}\right|^{n+\theta_3}_j - \frac{s(\Delta x)}{2}\left\{\left.\frac{\partial^2\bar{\tau}}{\partial x^2}\right|^n_{j+\theta_1} + \left.\frac{\partial^2\bar{\tau}}{\partial x^2}\right|^n_{j-\theta_2}\right\}. \qquad (3.1.5)$$

If $s \leqslant \frac{1}{2}$, so that $1-2s \geqslant 0$, and noting that $s = \alpha\Delta t/(\Delta x)^2 > 0$ since $\Delta t > 0$ and $\alpha > 0$ in any practical problem, Equation (3.1.5) may be written

$$|e^{n+1}_j| \leqslant s|e^n_{j-1}| + (1-2s)|e^n_j| + s|e^n_{j+1}| + \Delta t\left|\left.\frac{\partial\bar{\tau}}{\partial t}\right|^{n+\theta_3}_j - \frac{\alpha}{2}\left\{\left.\frac{\partial^2\bar{\tau}}{\partial x^2}\right|^n_{j+\theta_1} + \left.\frac{\partial^2\bar{\tau}}{\partial x^2}\right|^n_{j-\theta_2}\right\}\right| \qquad (3.1.6)$$

Let

$$M = \underset{\substack{j=0(1)J-1 \\ n=0(1)N-1}}{\text{Maximum}} \left|\left.\frac{\partial\bar{\tau}}{\partial t}\right|^{n+\theta_3}_j - \frac{\alpha}{2}\left\{\left.\frac{\partial^2\bar{\tau}}{\partial x^2}\right|^n_{j+\theta_1} + \left.\frac{\partial^2\bar{\tau}}{\partial x^2}\right|^n_{j-\theta_2}\right\}\right|,$$

where $t_N = N\Delta t$ is the last time for which τ is computed, and let

$$e^n_{max} = \underset{j=1(1)J-1}{\text{Maximum}} \ |e^n_j| .$$

That is, e^n_{max} is the maximum absolute value of the discretization error at interior grid-points at the nth time level. Then Equation (3.1.6) becomes

$$|e^{n+1}_j| \leqslant se^n_{max} + (1-2s)e^n_{max} + se^n_{max} + \Delta t\, M ,$$

or

$$|e^{n+1}_j| \leqslant e^n_{max} + \Delta t\, M.$$

Therefore, as this relation applies for all j,

$$e^{n+1}_{max} \leqslant e^n_{max} + \Delta t\, M. \qquad (3.1.7)$$

Repeated application of (3.1.7) with $n = 0(1)N-1$ yields

$$e^N_{max} \leqslant e^0_{max} + N\,\Delta t\,M\,. \tag{3.1.8}$$

But there is no initial discretization error because the initial values are known exactly. Therefore $e^0_{max} = 0$, giving

$$e^N_{max} \leqslant M\,t_N\,. \tag{3.1.9}$$

Now as $\Delta x \to 0$, $\Delta t = s(\Delta x)^2/\alpha \to 0$ for fixed s, and

$$M \to \underset{\substack{j=1(1)J-1\\ n=0(1)N-1}}{\text{Maximum}} \left| \left.\frac{\partial\bar{\tau}}{\partial t}\right|^n_j - \alpha \left.\frac{\partial^2\bar{\tau}}{\partial x^2}\right|^n_j \right|$$

$$\to 0,$$

because $\bar{\tau}^n_j$ is the exact solution to the diffusion equation (2.1.1). It therefore follows that, as $\Delta x \to 0$ for fixed values of s in the range $0 < s \leqslant \frac{1}{2}$,

$$|e^N_j| \to 0. \tag{3.1.10}$$

for fixed t_N. Therefore the solution to the finite difference approximation (2.4.4) converges to the exact solution of the given diffusion equation (2.1.1) as the grid becomes finer, provided $s \leqslant \frac{1}{2}$.

This method, unfortunately, is very difficult to apply when the given partial differential equation is even slightly more complicated than the diffusion equation (2.1.1). In such cases the only indication of convergence may be the accuracy of the finite difference solutions of simplified problems, similar to the one in question, for which exact solutions are known, or by the application of Lax's equivalence theorem if consistency and stability can be determined (see Sections 3.2 to 3.4).

3.2 Consistency

A finite difference equation is said to be *consistent* with a partial differential equation if in the limit as the grid spacings tend to zero, the difference equation becomes the same as the partial differential equation at each point in the solution domain.

Consider the FTCS finite difference approximation to the one-dimensional diffusion equation. The substitution of $\bar{\tau}^n_j$, the exact solution of the diffusion equation at the (j,n) grid-point, for the approximation τ^n_j in the FTCS formula gives

$$\bar{\tau}^{n+1}_j = s\bar{\tau}^n_{j-1} + (1-2s)\bar{\tau}^n_j + s\bar{\tau}^n_{j+1}\,. \tag{3.2.1}$$

We wish to determine how closely Equation (3.2.1) corresponds to the diffusion equation (2.4.1) at the (j,n) grid-point, that is, how close it is to

$$\left.\frac{\partial\bar{\tau}}{\partial t}\right|^n_j = \alpha \left.\frac{\partial^2\bar{\tau}}{\partial x^2}\right|^n_j\,.$$

Substitution of the Taylor series expansion about the (j,n) grid-point for each term of Equation (3.2.1) gives

$$\bar{\tau}_j^n + \Delta t \left.\frac{\partial\bar{\tau}}{\partial t}\right|_j^n + \frac{(\Delta t)^2}{2!}\left.\frac{\partial^2\bar{\tau}}{\partial t^2}\right|_j^n + \frac{(\Delta t)^3}{3!}\left.\frac{\partial^3\bar{\tau}}{\partial t^3}\right|_j^n + \ldots$$

$$= s\{\bar{\tau}_j^n + \Delta x \left.\frac{\partial\bar{\tau}}{\partial x}\right|_j^n + \frac{(\Delta x)^2}{2!}\left.\frac{\partial^2\bar{\tau}}{\partial x^2}\right|_j^n + \frac{(\Delta x)^3}{3!}\left.\frac{\partial^3\bar{\tau}}{\partial x^3}\right|_j^n + \frac{(\Delta x)^4}{4!}\left.\frac{\partial^4\bar{\tau}}{\partial x^4}\right|_j^n$$

$$+ \frac{(\Delta x)^5}{5!}\left.\frac{\partial^5\bar{\tau}}{\partial x^5}\right|_j^n + \frac{(\Delta x)^6}{6!}\left.\frac{\partial^6\bar{\tau}}{\partial x^6}\right|_j^n + \ldots\}$$

$$+ s\{\bar{\tau}_j^n - \Delta x \left.\frac{\partial\bar{\tau}}{\partial x}\right|_j^n + \frac{(\Delta x)^2}{2!}\left.\frac{\partial^2\bar{\tau}}{\partial x^2}\right|_j^n - \frac{(\Delta x)^3}{3!}\left.\frac{\partial^3\bar{\tau}}{\partial x^3}\right|_j^n + \frac{(\Delta x)^4}{4!}\left.\frac{\partial^4\bar{\tau}}{\partial x^4}\right|_j^n$$

$$- \frac{(\Delta x)^5}{5!}\left.\frac{\partial^5\bar{\tau}}{\partial x^5}\right|_j^n + \frac{(\Delta x)^6}{6!}\left.\frac{\partial^6\bar{\tau}}{\partial x^6}\right|_j^n - \ldots\}$$

$$+ (1-2s)\bar{\tau}_j^n \; . \tag{3.2.2}$$

Upon simplification, using $s = \alpha\Delta t/(\Delta x)^2$, Equation (3.2.2) becomes

$$\left.\frac{\partial\bar{\tau}}{\partial t}\right|_j^n = \alpha \left.\frac{\partial^2\bar{\tau}}{\partial x^2}\right|_j^n + E_j^n \; , \tag{3.2.3}$$

where

$$E_j^n = -\frac{\Delta t}{2}\left.\frac{\partial^2\bar{\tau}}{\partial t^2}\right|_j^n + \frac{\alpha(\Delta x)^2}{12}\left.\frac{\partial^4\bar{\tau}}{\partial x^4}\right|_j^n + O\{(\Delta t)^2,(\Delta x)^4\}. \tag{3.2.4}$$

The exact solution of the FTCS finite difference approximation to the diffusion equation (2.1.1) is therefore the solution of the partial differential equation (3.2.3), which differs from the diffusion equation by the inclusion of the extra terms contained in E_j^n.

This error E_j^n is called the *truncation error*: it is the discrepancy between the approximating finite difference equation with the exact values $\bar{\tau}_j^n$ substituted for τ_j^n, and the partial differential equation which the difference equation replaces at the (j,n) grid-point. Clearly, as the grid spacing gets smaller and smaller with the FTCS method, the truncation error gets smaller and smaller at a fixed point (x,t) in the solution domain. In the limit as $\Delta x \to 0$, $\Delta t \to 0$, the finite difference equation (3.2.1) is equivalent to the partial differential equation (2.4.1). This property is called *consistency* (or compatability).

For the FTCS finite difference approximation to the diffusion equation, the truncation error (3.2.4) is seen to be generally $O\{\Delta t,(\Delta x)^2\}$. However, since $\bar{\tau}$ satisfies the diffusion equation (2.1.1), it also satisfies the equation

$$\frac{\partial^2\bar{\tau}}{\partial t^2} = \alpha^2 \frac{\partial^4\bar{\tau}}{\partial x^4} \; . \tag{3.2.5}$$

Therefore the truncation error (3.2.4) may be written

$$E_j^n = -\frac{\alpha^2\Delta t}{2}\left(1 - \frac{1}{6s}\right)\left.\frac{\partial^4\bar{\tau}}{\partial x^4}\right|_j^n + O\{(\Delta t)^2,(\Delta x)^4\}. \tag{3.2.6}$$

If $s = 1/6$, the first term in the above expression vanishes and the truncation error is now $O\{(\Delta t)^2,(\Delta x)^4\}$, or, what is the same thing for fixed s, $O\{(\Delta x)^4\}$. In this case, the truncation error goes to zero faster than for any other value of

s when the grid spacing is made smaller and smaller. Therefore, the solution of the FTCS finite difference equation approaches the solution of the diffusion equation more rapidly when $s = 1/6$ than for other values of $s \leqslant \frac{1}{2}$, as the grid spacing is reduced.

Clearly, consistency is necessary if the finite difference solution is to converge to the solution of the partial differential equation being approximated. However, this is not a sufficient condition, for even though the finite difference equation might become equivalent to a certain partial differential equation as the grid spacing tends to zero, it does not follow that the *solution* of the finite difference equation approaches the *solution* of the partial differential equation. For instance, if $\alpha = 10^{-2}$, choosing $\Delta x = 0.1$ and $\Delta t = 1$ so $s = 1$, the FTCS explicit finite difference equation simplifies to (2.4.7) which can be solved as in Table 2.2. Continuing to reduce Δx and Δt, so that s remains unchanged at the value 1, the exact calculation of the finite difference solution at specific points in the solution domain rapidly diverge, as shown in Table 3.1, in contrast with solutions using $s = \frac{1}{2}$, $s = 1/6$.

TABLE 3.1
Finite difference approximations to $\bar{\tau}(0.4,8) = 0.4503963$

J	Δx	$\Delta t/s$	$s = 1$	$s = 1/2$	$s = 1/6$
5	0.2	4	$\tau_2^2 = 1.0000$	$\tau_2^4 = 0.5000000$	$\tau_2^{12} = 0.4684500$
10	0.1	1	$\tau_4^8 = 122.00$	$\tau_4^{16} = 0.4754791$	$\tau_4^{48} = 0.4549047$
15	$0.0\dot{6}$	$0.\dot{4}$	$\tau_6^{18} = 197\times10^6$	$\tau_6^{36} = 0.4661530$	$\tau_6^{108} = 0.4523997$
20	0.05	0.25	∞	$\tau_8^{64} = 0.4566436$	$\tau_8^{192} = 0.4515231$
25	0.04	0.16	∞	$\tau_{10}^{100} = 0.4524738$	$\tau_{10}^{300} = 0.4511174$
40	0.025	0.0625	∞	$\tau_{16}^{256} = 0.4519564$	$\tau_{16}^{768} = 0.4506780$
50	0.02	0.04	∞	$\tau_{20}^{400} = 0.4513947$	$\tau_{20}^{1200} = 0.4505766$
100	0.01	0.01	∞	$\tau_{40}^{1600} = 0.4506459$	$\tau_{40}^{4800} = 0.4504414$

Note that the functional dependence of the truncation error E_j^n on Δt and Δx not only indicates how closely the solution of the differential equation satisfies the approximating finite difference equation (that is, it shows *consistency* if it exists), it also indicates how rapidly the solution of the finite difference equation approaches the solution of the given differential equation as $\Delta x \to 0$, $\Delta t \to 0$ if the solution is convergent. For instance, with $s = 1/6$ the truncation error for the FTCS scheme is $O\{(\Delta t)^2,(\Delta x)^4\}$, whereas with $s = 1/2$ it is $O\{\Delta t,(\Delta x)^2\}$. Reversing the procedures involved in deriving (3.2.3) from Equation (3.2.1), we find that

$$\bar{\tau}_j^{n+1} = s\bar{\tau}_{j-1}^n + (1-2s)\bar{\tau}_j^n + s\bar{\tau}_{j+1}^n + \Delta t\, E_j^n , \tag{3.2.7}$$

which is equivalent to (3.1.2). Clearly, the contribution to the discretization error in going from one time level to the next is $\Delta t E_j^n$.

The contribution to the discretization error at each time step is therefore $O\{(\Delta t)^2,\Delta t(\Delta x)^2\}$ in general, but is $O\{(\Delta t)^3,\Delta t(\Delta x)^4\}$ when $s = 1/6$. Therefore there is a much greater rate of convergence of the FTCS finite difference equation to the solution of the diffusion equation if $s = 1/6$, than for other values of $0 < s \leqslant \frac{1}{2}$. This greater rate of convergence is seen in Table 3.1. However, note that with $s = 1$, even though $E_j^n = O\{\Delta t, (\Delta x)^2\}$ and the finite difference equation *is* consistent with the diffusion equation, the solution of the difference equation does *not* converge to the solution of the diffusion equation.

However, when assessing the economy of any particular numerical scheme, it is necessary to take into account the computational effort involved with each application of the formula, as well as the total number of applications. For example, even though the truncation and hence the discretization error is minimised by choosing $s = 1/6$, as evident in Table 3.1, the reduced number of applications of the FTCS formula to achieve a given accuracy may be more than offset by an increased computational effort at each application.

Consider the results obtained in Table 3.1 using the CYBER 173 computer, for which multiplications and divisions take about 3 micro-seconds while additions and subtractions take approximately ½ micro-second. With $s = 1/6$, the result τ_6^{108} is slightly more accurate than τ_{10}^{100} obtained with $s = \frac{1}{2}$. In the former case 14 applications of the FTCS formula (3.2.1) must be made at each of 108 time levels, a total of 1,512 applications. As each application of (3.2.1) requires approximately 7 micro-seconds if the calculations are carried out in the quickest way, which is to write it as

$$\tau_j^{n+1} = \frac{1}{6}(\tau_{j-1}^n + 4\tau_j^n + \tau_{j+1}^n),$$

then the computer time used in obtaining this result is about 10,584 micro-seconds. In the latter case, with $s = \frac{1}{2}$, 24 applications of (2.4.6), each taking 3½ micro-seconds must be made at each of 100 time-levels. Thus many more applications of the formula are required, namely 2,400, but the required computer time is actually less than the previous case, namely 8,400 micro-seconds.

The economy in choosing the FTCS scheme with $s = 1/6$ is really apparent in the evaluation of τ_{20}^{1200}, which required 58,800 applications of the formula in a time of 411,600 micro-seconds. This is less than the time taken to obtain the less accurate value τ_{40}^{1600} with $s = \frac{1}{2}$, which required 158,400 applications of the FTCS formula in a time of 554,400 micro-seconds.

3.3 Stability

A third important feature of a finite difference method of solving a partial differential equation is the stability of the associated finite difference equation which must be solved. The stability of such an equation is concerned with the growth, or decay, of errors produced in the finite difference solution by errors introduced in previously calculated values.

If it were possible to carry out all numerical operations to an infinite number of decimal places, the exact solution to the finite difference equation would be found. In practice, however, each calculation made by the computer is carried out to a finite number of significant figures which introduces a "round-off" error at every step. Hence the finite difference solution found to Equation (2.4.4) is not τ_j^{n+1}, but $*\tau_j^{n+1}$ (say). $*\tau_j^n$ is called the *numerical solution* of the finite difference equation in contrast with its exact solution τ_j^n.

A set of finite difference equations is said to be stable if the cumulative effect of all the rounding off errors is negligible. More specifically, consider the errors

$$\xi_j^n = *\tau_j^n - \tau_j^n \tag{3.3.1}$$

introduced at the (j,n) grid-points for $j = 1(1)J-1$ and for time levels up to

$N\Delta t$, all of which have absolute value $|\xi_j^n|$ less than some positive number δ (say). Then the finite difference equations are said to be *stable* if the maximum value of $|\xi_j^{n+1}|$ tends to zero as $\delta \to 0$. That is, the difference between the numerical solution $*\tau_j^{n+1}$ and the exact solution τ_j^{n+1} does not increase exponentially as the numbers of rows of calculations at successive time levels in the solution domain is increased. If the errors do not increase exponentially as n increases, but persists as linear combinations of the initial errors, they are usually numerically tolerable provided their sum remains much smaller than the exact finite difference solution τ_j^n.

It is usually not possible to determine the exact value of the numerical error $(*\tau_j^n - \tau_j^n)$ at the (j,n) grid-point for an arbitrary distribution of errors at other grid-points. However, it can be estimated using certain standard methods, some of which will be discussed in this section. Note that the numerical solutions are invariably more accurate than these estimates indicate, because stability analyses always assume the worst possible combination of individual errors. For instance, it may be assumed that all errors have a distribution of signs so their total effect is additive, which is not always the case.

Note that stability does not involve the difference between the exact solutions of the finite difference equation and the given partial differential equation. This difference is the discretization error, which is the province of convergence. Stability concerns only the calculation of the solution of the finite difference equation. The phenomenon shown in Table 2.2 is not due to instability, because the calculations were all performed exactly - no errors were introduced. Table 2.2 is an example of a non-convergent solution.

Each of the methods of stability analysis to be described give further insight into how a solution of a finite difference equation behaves when round-off errors effect the calculations. The two techniques considered here are:

(1) discrete perturbation stability analysis;
(2) von Neumann's method.

These methods will be illustrated by applications to the FTCS finite difference formula (2.4.4) which approximates the one-dimensional diffusion equation.

Note that in the calculation of τ_j^{n+1} using the FTCS formula we actually compute the numerical approximation

$$*\tau_j^{n+1} = \tau_j^{n+1} + \xi_j^{n+1}, \qquad (3.3.2a)$$

using numerical approximations from the previous time level $t = n\Delta t$, namely

$$*\tau_{j-1}^n = \tau_{j-1}^n + \xi_{j-1}^n, \qquad (3.3.2b)$$

$$*\tau_j^n = \tau_j^n + \xi_j^n, \qquad (3.3.2c)$$

$$*\tau_{j+1}^n = \tau_{j+1}^n + \xi_{j+1}^n. \qquad (3.3.2d)$$

It can be shown that, for linear finite difference equations, the corresponding error terms satisfy the same *homogeneous* finite difference equation as the values of τ. For instance, using Equation (2.4.4) means that we are actually calculating $*\tau_j^{n+1}$ using $*\tau_{j-1}^n$, $*\tau_j^n$ and $*\tau_{j+1}^n$, so that

$$*\tau_j^{n+1} = s(*\tau_{j-1}^n) + (1-2s)(*\tau_j^n) + s(*\tau_{j+1}^n). \qquad (3.3.3)$$

Since the exact numerical solutions τ satisfy the FTCS finite difference equation, subtraction of Equation (2.4.4) from Equation (3.3.3) followed by application of Equations (3.3.2) yields the homogeneous difference equation

$$\xi_j^{n+1} = s\xi_{j-1}^n + (1-2s)\xi_j^n + s\xi_{j+1}^n \quad . \tag{3.3.4}$$

The initial errors, ξ_j^0, $j=0(1)J$, and the boundary errors, ξ_0^n and ξ_J^n, $n=1,2,3,...$ for this equation, assuming given boundary and initial values of the form (2.1.2), will all be zero and, unless some error is introduced in calculating the finite difference solution τ_j^n at some interior grid-point, the resulting errors in the solution will remain zero.

(1) Discrete Perturbation Stability Analysis

In this method a discrete perturbation, or error, ε is introduced into the finite difference scheme in the determination of τ_j^n at an arbitrary (j,n) grid-point and its effect on the computation of subsequent values of the solution of the finite difference equation is examined. Stability is indicated if the perturbation eventually dies out.

For a given value of s, application of (3.3.4) with zero errors up to some time level $(n-1)$ and zero errors at grid-points everywhere along the nth time level except one, where an error ε is introduced, will show whether the introduced perturbation increases without bound and the scheme is unstable. For example, Table 3.2 shows that the introduction of an error $\xi_4^1 = \varepsilon$ in the value of τ_4^1 in the FTCS finite difference scheme considered in Section 2.4, with $s = \frac{1}{2}$, produces errors which decrease in magnitude at successive time levels.

TABLE 3.2

Error propagation due to introduction of error $\xi_4^1 = \varepsilon$ with $s = \frac{1}{2}$

n	nΔt	t										Errors in Boundary Values
7	3.5	0										
6	3.0	0	$\frac{1}{8}\varepsilon$	0	$\frac{7}{16}\varepsilon$	0	$\frac{7}{16}\varepsilon$	0	$\frac{5}{32}\varepsilon$	0	$\frac{1}{32}\varepsilon$	0
5	2.5	0	0	$\frac{1}{4}\varepsilon$	0	$\frac{3}{8}\varepsilon$	0	$\frac{1}{4}\varepsilon$	0	$\frac{1}{16}\varepsilon$	0	0
4	2.0	0	$\frac{1}{8}\varepsilon$	0	$\frac{3}{8}\varepsilon$	0	$\frac{3}{8}\varepsilon$	0	$\frac{1}{8}\varepsilon$	0	0	0
3	1.5	0	0	$\frac{1}{4}\varepsilon$	0	$\frac{1}{2}\varepsilon$	0	$\frac{1}{4}\varepsilon$	0	0	0	0
2	1.0	0	0	0	$\frac{1}{2}\varepsilon$	0	$\frac{1}{2}\varepsilon$	0	0	0	0	0
1	0.5	0	0	0	0	ε	0	0	0	0	0	0
0	0.0	0	0	0	0	0	0	0	0	0	0	0 → x
Errors in Initial Values	jΔx	0.0	0.1	0.2	0.3	0.4	0.5	0.6	0.7	0.8	0.9	1.0
	j	0	1	2	3	4	5	6	7	8	9	10

However, with $s = 1$, the introduction of the error $\xi_4^1 = \varepsilon$ into the FTCS finite difference equation produces a perturbation which rapidly increases in numerical value as in Table 3.3. Clearly, the finite difference equation (2.4.4) is unstable with $s = 1$.

TABLE 3.3
Error propagation due to introduction of error $\xi_4^1 = \varepsilon$ with $s = 1$

n	nΔt											Errors in Boundary Values
7	7	0										0
6	6	0	14ε	-30ε	45ε	-51ε	45ε	-30ε	15ε	-5ε	ε	0
5	5	0	-4ε	10ε	-16ε	19ε	-16ε	10ε	-4ε	ε	0	0
4	4	0	ε	-3ε	6ε	-7ε	6ε	-3ε	ε	0	0	0
3	3	0	0	ε	-2ε	3ε	-2ε	ε	0	0	0	0
2	2	0	0	0	ε	-ε	ε	0	0	0	0	0
1	1	0	0	0	0	ε	0	0	0	0	0	0
0	0	0	0	0	0	0	0	0	0	0	0	0
Errors in Initial Values	jΔx	0.0	0.1	0.2	0.3	0.4	0.5	0.6	0.7	0.8	0.9	1.0
	j	0	1	2	3	4	5	6	7	8	9	10

In general, if we wish to determine the values of s for which the finite difference equation (2.4.4) is stable, we must sweep through the grid as in Tables 3.2 and 3.3 using the formula for error propagation, namely Equation (3.3.4), with arbitrary s. The only permissible values of s are those for which the resulting errors produced at higher time levels have moduli less than or equal to $|\varepsilon|$. With the only error at the nth time level being $\xi_j^n = \varepsilon$, then the resulting errors at the (n+1)th time level are $\xi_{j-1}^{n+1} = s\varepsilon$, $\xi_j^{n+1} = (1-2s)\varepsilon$ and $\xi_{j+1}^{n+1} = s\varepsilon$. These errors are not greater in magnitude than ε if

$$|s| \leqslant 1 \quad \text{and} \quad |1-2s| \leqslant 1,$$

both of which are satisfied if

$$0 \leqslant s \leqslant 1. \tag{3.3.5}$$

Since $s = 0$ is impractical, as it implies $\Delta t = 0$, then the requirement for no increase in error from the nth to the (n+1)th time-level is

$$0 < s \leqslant 1. \tag{3.3.6}$$

The errors at the (n+2)th time level produced by the perturbation $\xi_j^n = \varepsilon$ are

$$\xi^{n+2}_{j-2} = s^2\varepsilon,\ \xi^{n+2}_{j-1} = (2s-4s^2)\varepsilon,\ \varepsilon^{n+2}_{j} = (1-4s+6s^2)\varepsilon,$$
$$\varepsilon^{n+2}_{j+1} = (2s-4s^2)\varepsilon,\ \xi^{n+2}_{j+2} = s^2\varepsilon,$$

where the superscripts on s indicate powers.

These errors are equal to or less than ε in magnitude if

$$|s^2| \leqslant 1,\ |2s-4s^2| \leqslant 1 \text{ and } |1-4s+6s^2| \leqslant 1;$$

that is, if

$$0 < s \leqslant 2/3. \qquad (3.3.7)$$

This process may be continued to higher time levels, but the algebra involved soon becomes prohibitively complicated. Consideration of the errors at time level $(n+1)$ lead to the requirement $0 < s \leqslant 1$ for errors not to increase in magnitude; consideration of the errors at time level $(n+2)$ leads to the requirement $0 \leqslant s \leqslant 2/3$ for errors not to increase in magnitude. The asymptotic limit of this restriction as higher time levels are considered is required.

The initial set of round-off errors at the first time level will propagate through the grid as described by Equation (3.3.4). To these errors will be added new errors produced by round-off in further calculations. It can be expected after many time steps that errors at any time level will be approximately the same in magnitude, either having the same sign or an alternating sign distribution along that time level.

If the errors at the nth time level all have the same sign, then

$$\xi^{n}_{j-1} = \varepsilon,\ \xi^{n}_{j} = \varepsilon,\ \xi^{n}_{j+1} = \varepsilon,$$

so that, by application of Equation (3.3.4),

$$\xi^{n+1}_{j} = \varepsilon,$$

which yields no restriction on s for stability.

If the distribution of errors at the nth time level has alternate positive and negative signs, such as

$$\xi^{n}_{j-1} = -\varepsilon,\ \xi^{n}_{j} = \varepsilon,\ \xi^{n}_{j+1} = -\varepsilon\ ,$$

application of (3.3.3) yields $\varepsilon^{n+1}_{j} = (1-4s)\varepsilon$. Stability therefore requires that

$$|1-4s| \leqslant 1$$

which yields the condition

$$0 < s \leqslant \tfrac{1}{2}. \qquad (3.3.8)$$

Examination of the error propagation at succeeding time levels does not change this restriction on s. The FTCS finite difference scheme is therefore both convergent and stable if (3.3.8) is satisfied.

For a given α and a fixed space interval Δx, condition (3.3.8) places a limitation on the time step Δt, namely

$$\Delta t \leqslant (\Delta x)^2/2\alpha. \tag{3.3.9}$$

The inequality (3.3.9) has important implications about the amount of computer time required to solve the FTCS finite difference approximation to the diffusion equation. Suppose the FTCS equation has been solved up to a certain time T with a grid size Δx_1, using the maximum possible time step of $\Delta t_1 = (\Delta x_1)^2/2\alpha$. If the computation is repeated with a halved grid spacing, say $\Delta x_2 = \Delta x_1/2$, in order to improve accuracy for example, then

$$\Delta t_2 = (\Delta x_2)^2/2\alpha = (\Delta x_1)^2/8\alpha = \Delta t_1/4.$$

To find solutions up to the same time T, the second case requires four times as many time levels in the solution domain. Furthermore, at each time level the calculations take twice as long to perform, since there are twice as many interior grid-points at one time level. The required computer time is therefore increased eight-fold. Obviously, finite difference methods of solving the diffusion equation which do not have a stability restriction of the type (3.3.8) are highly desirable.

(2) von Neumann Stability Analysis

This is the most commonly used method of stability analysis. However, it only provides information about the influence of derivative boundary conditions on the numerical stability of the finite difference scheme if it is applied separately to the algorithms used at the boundaries. The von Neumann (or Fourier series) method was developed in 1944 and it was rapidly revised and improved by von Neumann and his co-workers. The earliest complete description was given by O'Brien et al (1950).

In this method, the errors distributed along grid-points at one time level are expanded as a finite Fourier series. Then, the stability or instability of the finite difference equation is determined by considering whether separate Fourier components of the error distribution decay or amplify on progressing to the next time level. Thus the initial error vector ξ^0 is expressed as a finite complex Fourier series, so that at $x = j\Delta x$ the error is

$$\xi_j^0 = \sum_{m=1}^{J-1} a_m e^{i(m\pi j\Delta x)} ,$$

where $e^{i(m\pi j\Delta x)} = \exp\{i(m\pi j\Delta x)\}$ and $i = \sqrt{-1}$. That is,

$$\xi_j^0 = \sum_{m=1}^{J-1} a_m e^{i\beta_m j} , \quad j=0(1)J, \tag{3.3.10}$$

where $\beta_m = m\pi\Delta x$. Because the finite difference equation being studied is linear, then so is Equation (3.3.4) governing the error propagation, and it is sufficient to study the propagation of the error due to just the single term $e^{i\beta j}$ of the Fourier series representation (3.3.10). The coefficient a_m is a constant, so it has been omitted, and the subscript m has been dropped from β_m.

A solution of the error equation (3.3.4) is therefore sought in the variables separable form

$$\xi_j^n = (G)^n e^{i\beta j} , \tag{3.3.11}$$

where the time dependence of this Fourier component of the error at $x = j\Delta x$ is contained in the coefficient $(G)^n$, which is the nth power of the complex number G.

Substitution of the mth Fourier component of the errors at the nth time level into the error equation (3.3.4) gives the mth Fourier component of the errors at the (n+1)th time level from the equation

$$(G)^{n+1} e^{i\beta j} = s(G)^n e^{i\beta(j-1)} + (1-2s)(G)^n e^{i\beta j} + s(G)^n e^{i\beta(j+1)} .$$

This gives, on dividing throughout by $(G)^n e^{i\beta j}$,

$$G = se^{-i\beta} + (1-2s) + se^{i\beta}. \tag{3.3.12}$$

Clearly, G is the amplification factor for the mth Fourier mode of the error distribution as it propagates one step forward in time, since

$$\xi_j^{n+1}/\xi_j^n = G. \tag{3.3.13}$$

Note that G is a function of s and β. That is, $G = G(s,\beta)$ depends on the size of the grid spacings and the particular Fourier mode being considered. The errors will remain bounded if the absolute value of G, called the gain, is never greater than unity for all Fourier modes, that is, for all β. The stability requirement is therefore that

$$|G| \leqslant 1 \quad \text{for all} \quad \beta. \tag{3.3.14}$$

Note that, in general, the amplification factor G is a complex number and the amplitude of the mth Fourier component of the error is multiplied by $|G|$ and its phase is increased by arg G.

Equation (3.3.12) gives, on use of the relations $e^{i\beta} = \cos\beta + i\sin\beta$ and $\cos\beta = 1-2\sin^2(\beta/2)$, the amplification factor

$$G = 1 - 4s \sin^2(\beta/2). \tag{3.3.15}$$

Clearly, this amplification factor has a different value for each β, that is, for each Fourier component.

Our stability requirement for the FTCS finite difference equation is therefore, by (3.3.14),

$$-1 \leqslant 1 - 4s \sin^2(\beta/2) \leqslant 1, \quad \text{for all} \quad \beta . \tag{3.3.16}$$

The right-hand inequality is satisfied for all s and β, whereas the left-hand inequality requires

$$s \sin^2(\beta/2) \leqslant \tfrac{1}{2},$$

which is true for all β if

$$s \leqslant \tfrac{1}{2}. \tag{3.3.17}$$

This result is identical to that found by the discrete perturbation analysis.

The von Neumann method is the most commonly used way of determining stability criteria as it is generally the easiest to apply, the most straightforward and the most dependable. Unfortunately, it can only be used to establish necessary and sufficient conditions for stability of linear initial value problems with constant coefficients. Practical problems typically involve variable coefficients, non-linearities and complicated types of boundary conditions, in which case the method can only be applied locally. If the complete non-linear difference equation is linearized in a small part of the solution domain where its coefficients may be considered constant, then the conditions for the applicability of the method are satisfied locally even though they are not satisfied over the whole solution domain. And just as the von Neumann method can provide useful stability information when applied locally at interior grid-points, similar local application of the method at the boundaries can provide useful boundary stability information (see Trapp and Ramshaw, 1976).

Comparison of Stability Methods

There are many other methods of stability analysis in addition to the two described. Five others which deserve mention are the matrix method (see Smith, 1969) and those due to Hirt (1968), Friedrichs (see Hahn, 1958), Keller and Lax (see Richtmeyer and Morton, 1967) and Eddy (1949).

Both methods described in this article have particular merits. The discrete perturbation method of stability analysis is a rather hit or miss affair compared with the more methodical von Neumann method, and for this reason it is not widely used. However, in certain cases, it readily produces stability criteria obtained only by much effort using other methods. It focusses attention on the individual calculations performed, rather than on related abstractions, so that the magnitude of the errors can be estimated. It also provides insight about the best way of incorporating boundary conditions. The von Neumann analysis provides information not only on attenuation (or damping) produced by the finite difference approximation, but also on any phase effects produced. It therefore gives an insight into resulting dispersion errors (see Section 5).

In summary, both methods provide valuable information about the stability of a finite difference equation. However neither of them can be considered completely adequate, so when actually computing solutions to a finite difference equation they should be supplemented by numerical experimentation. Kusic and Lavi (1968) and Kusic (1969) present a non-iterative method of evaluating stability during the process of computing the solutions to a finite difference equation.

It must be noted that the application of the discrete perturbation technique can be extended to non-linear problems in order to get an estimate of stability, but the basis of the von Neumann stability analysis is that it applies only to linear systems with constant coefficients. It can however, give a rough estimate if applied to locally linearised forms of non-linear finite difference equations. John (1952) has shown that a mildly strengthened form of the von Neumann condition for boundaries is sufficient for stability of linear parabolic differential equations with variable coefficients. Lax (see Lax and Richtmeyer, 1956) has shown that a similar result holds for variable-coefficient hyperbolic differential equations.

3.4 Lax's Equivalence Theorem

There exist important connections between the stability of linear finite difference schemes and the convergence of their solutions to that of the linear partial differential equations to which they are consistent. An example of this connection is given by Lax's equivalence theorem.

This theorem, proved by Lax and described in Lax and Richtmeyer (1956), states: "Given a properly posed linear initial value problem and a finite difference approximation to it that satisfies the consistency condition, stability is the necessary and sufficient condition for convergence." This result is of great practical importance, for while it is relatively easy to show stability of a finite difference equation and its consistency with a partial differential equation, is usually very difficult to show convergence of its solution to that of the partial differential equation.

The conditions under which the theorem applies include:

(a) the initial-value problems must be well posed. This requires that the solution of the partial differential equation depends continuously on the given initial conditions.

(b) the problem must be linear. Because of this the errors propagate according to the homogeneous form of the given difference equation. It should be noted that attempts to find an equivalent theorem for non-linear problems have been unsuccessful.

Lax's equivalence theorem by-passes the need to prove convergence of the solution of the FTCS finite difference equation (2.4.4) to the solution of the partial differential equation (2.4.1). Because the diffusion equation and its associated initial and boundary conditions are linear and well-posed, and the FTCS finite difference equation is stable and consistent with the diffusion equation, it follows that the FTCS method is convergent.

4. SOLVING THE ONE-DIMENSIONAL DIFFUSION EQUATION

4.1 Explicit methods

The FTCS Method

By substituting the forward difference approximation for the time derivative and the central difference approximation for the space derivative in Equation (2.4.1), namely

$$\left.\frac{\partial\bar{\tau}}{\partial t}\right|_j^n = \alpha \left.\frac{\partial^2\bar{\tau}}{\partial x^2}\right|_j^n ,$$

the classical FTCS explicit finite difference equation (2.4.4) for solving the one-dimensional diffusion equation was obtained, namely

$$\tau_j^{n+1} = s\tau_{j-1}^n + (1-2s)\tau_j^n + s\tau_{j+1}^n ,$$

with s defined by Equation (2.4.5),

$$s = \alpha\Delta t/(\Delta x)^2$$

(see Section 2.4).

Then, in Section 3.1 it was shown that the solution τ_j^n of the finite difference equation (2.4.4) *converged* to the solution $\bar{\tau}_j^n$ of the diffusion equation (2.4.1) in the limit $\Delta x \to 0$, $\Delta t \to 0$, provided $s \leqslant \frac{1}{2}$. In Section 3.2 the finite difference equation (2.4.4), with $\bar{\tau}_j^n$ substituted for τ_j^n, was shown to be *consistent* with the one-dimensional diffusion equation; that is

$$\bar{\tau}_j^{n+1} = s\bar{\tau}_{j-1}^n + (1-2s)\bar{\tau}_j^n + s\bar{\tau}_{j+1}^n ,$$

was shown to be equivalent to (2.4.1) in the limit as $\Delta x \to 0$, $\Delta t \to 0$. Finally, in Section 3.3, it was shown that the finite difference equation (2.4.4) was *stable* provided $s \leqslant \frac{1}{2}$; that is, round-off errors in the calculation of τ_j^n at any time level produced bounded errors in values of τ computed at later times.

Equation (2.4.4) simplifies to $\tau_j^{n+1} = \frac{1}{2}(\tau_{j-1}^n + \tau_{j+1}^n)$ when $s = \frac{1}{2}$. That is, the value of τ at $x = j\Delta x$ at the $(n+1)$th time level is the arithmetic mean of the values of τ at $(j-1)\Delta x$ and $(j+1)\Delta x$ at the nth time level. The process of taking the arithmetic mean can be carried out either numerically or graphically, and in the latter form has been familiar for many years in works on heat transfer under the name of Schmidt's method (see Carslaw and Jaeger, 1962).

Since the one-dimensional diffusion equation is linear and is well-posed, Lax's equivalence theorem indicates that consistency and stability of the FTCS finite difference approximation is necessary and sufficient for the finite difference solution to converge to the solution of the diffusion equation. This theorem is also used to establish convergence of other finite difference methods of solving the diffusion equation developed in this Section, since stability and consistency are much easier to prove than convergence.

Once convergence has been proved, the solution to the given partial differential equation can be obtained to any desired degree of accuracy, provided the spacings Δx and Δt of the finite difference grid are made sufficiently small.

The FTCS method (2.4.4) is classified as explicit, because the value of τ_j^{n+1} at the $(n+1)$th time level may be calculated directly from known values of τ_j^n at previous time levels. It is a two-level method because values of τ at only two levels of time are involved in the approximating finite difference equation. Two other explicit methods will now be examined, those developed by Richardson

(1910) and DuFort and Frankel (1953).

Richardson's Method

This is a modification of the finite difference approximation (2.4.3) to Equation (2.4.1), in which the forward difference approximation for $\partial\bar{\tau}/\partial t$ is replaced by the central difference form. Equation (2.4.1) is written

$$\frac{\bar{\tau}_j^{n+1} - \bar{\tau}_j^{n-1}}{2\Delta t} + O\{(\Delta t)^2\} = \alpha\{\frac{\bar{\tau}_{j+1}^{n} - 2\bar{\tau}_j^{n} + \bar{\tau}_{j-1}^{n}}{(\Delta x)^2} + O\{(\Delta x)^2\}\}\ , \qquad (4.1.1)$$

which gives, on rearrangement,

$$\bar{\tau}_j^{n+1} = \bar{\tau}_j^{n-1} + 2s(\bar{\tau}_{j-1}^{n} - 2\bar{\tau}_j^{n} + \bar{\tau}_{j+1}^{n}) + O\{(\Delta t)^3,\ \Delta t(\Delta x)^2\}. \qquad (4.1.2)$$

This yields the explicit three-level finite difference equation

$$\tau_j^{n+1} = \tau_j^{n-1} + 2s(\tau_{j-1}^{n} - 2\tau_j^{n} + \tau_{j+1}^{n}), \qquad (4.1.3)$$

when the terms of $O\{(\Delta t)^3,\ \Delta t(\Delta x)^2\}$ are omitted from Equation (4.1.2).

Because τ_j^{n+1} is calculated using values at the *two* previous time levels, an alternative procedure, such as the FTCS method with $n = 0$, is required to start the computations and calculate the approximations τ_j^1, $j=1(1)J-1$, at time level $n = 1$ from the known initial values τ_j^0.

Comparison of Equations (4.1.2) and (4.1.3) indicates that there is a contribution of $O\{(\Delta t)^3,\Delta t(\Delta x)^2\}$ to the discretisation error in stepping from time levels (n-1) and n to the time level (n+1) with Equation (4.1.3). Also, Equation (4.1.1) indicates that the truncation error is generally $O\{(\Delta t)^2,(\Delta x)^2\}$. In order to determine the exact nature of the truncation error it is necessary to go through a formal consistency analysis of the method. For instance, without carrying out the formal consistency analysis of the FTCS method, the optimal case with minimal truncation and discretisation error for $s = 1/6$ would not have been evident.

To determine the *consistency* of the finite difference equation (4.1.3), the values τ_j^n, which satisfy this equation are replaced by values $\bar{\tau}_j^n$ which are solutions of the diffusion equation (2.4.1). Replacing these terms by their Taylor series expanded about the (j,n) grid-point yields

$$\bar{\tau}_j^n + \Delta t \left.\frac{\partial\bar{\tau}}{\partial t}\right|_j^n + \frac{(\Delta t)^2}{2!}\left.\frac{\partial^2\bar{\tau}}{\partial t^2}\right|_j^n + \frac{(\Delta t)^3}{3!}\left.\frac{\partial^3\bar{\tau}}{\partial t^3}\right|_j^n + \frac{(\Delta t)^4}{4!}\left.\frac{\partial^4\bar{\tau}}{\partial t^4}\right|_j^n + \cdots$$

$$= \bar{\tau}_j^n - \Delta t \left.\frac{\partial\bar{\tau}}{\partial t}\right|_j^n + \frac{(\Delta t)^2}{2!}\left.\frac{\partial^2\bar{\tau}}{\partial t^2}\right|_j^n - \frac{(\Delta t)^3}{3!}\left.\frac{\partial^3\bar{\tau}}{\partial t^3}\right|_j^n + \frac{(\Delta t)^4}{4!}\left.\frac{\partial^4\bar{\tau}}{\partial t^4}\right|_j^n - \cdots$$

$$+ 2s\left\{\begin{array}{l} \bar{\tau}_j^n - \Delta x \left.\frac{\partial\bar{\tau}}{\partial x}\right|_j^n + \frac{(\Delta x)^2}{2!}\left.\frac{\partial^2\bar{\tau}}{\partial x^2}\right|_j^n - \frac{(\Delta x)^3}{3!}\left.\frac{\partial^3\bar{\tau}}{\partial x^3}\right|_j^n + \frac{(\Delta x)^4}{4!}\left.\frac{\partial^4\bar{\tau}}{\partial x^4}\right|_j^n + \cdots \\ - 2\bar{\tau}_j^n \\ + \bar{\tau}_j^n + \Delta x \left.\frac{\partial\bar{\tau}}{\partial x}\right|_j^n + \frac{(\Delta x)^2}{2!}\left.\frac{\partial^2\bar{\tau}}{\partial x^2}\right|_j^n + \frac{(\Delta x)^3}{3!}\left.\frac{\partial^3\bar{\tau}}{\partial x^3}\right|_j^n + \frac{(\Delta x)^4}{4!}\left.\frac{\partial^4\bar{\tau}}{\partial x^4}\right|_j^n + \cdots \end{array}\right\}$$

This equation simplifies to

$$\left.\frac{\partial\bar{\tau}}{\partial t}\right|_j^n = \alpha \left.\frac{\partial^2\bar{\tau}}{\partial x^2}\right|_j^n + E_j^n , \tag{4.1.4a}$$

where

$$E_j^n = \frac{(\Delta t)^2}{6}\left.\frac{\partial^3\bar{\tau}}{\partial t^3}\right|_j^n + \frac{\alpha(\Delta x)^2}{12}\left.\frac{\partial^4\bar{\tau}}{\partial x^4}\right|_j^n + O\{(\Delta t)^4,(\Delta x)^4\}. \tag{4.1.4b}$$

Richardson's method is therefore consistent, because the truncation error $E_j^n \to 0$ as $\Delta x \to 0$, $\Delta t \to 0$, and it is of a higher order of accuracy than the FTCS method because it has truncation error $O\{(\Delta t)^2,(\Delta x)^2\}$ compared with $O\{\Delta t,(\Delta x)^2\}$.

In order to establish whether there is an optimum value of s for which E_j^n is a minimum, we can follow the same procedure that was used for the FTCS method in Section 3.2. In this case, the term $\partial^3\bar{\tau}/\partial t^3|_j^n$ may be replaced by $\alpha^3\ \partial^6\bar{\tau}/\partial x^6|_j^n$, which follows from (3.2.5) because $\bar{\tau}$ satisfies the diffusion equation (2.4.1). However, no matter what value of s is chosen, the term in $(\Delta x)^2$ cannot be eliminated.

To investigate the *stability* of Richardson's method the mth Fourier component of the error distribution at the nth time-level, namely $\xi_j^n = (G)^n e^{i\beta j}$ at the (j,n) grid-point, is substituted into the error equation corresponding to (4.1.3), namely

$$\xi_j^{n+1} = \xi_j^{n-1} + 2s(\xi_{j-1}^n - 2\xi_j^n + \xi_{j+1}^n), \tag{4.1.5}$$

giving

$$(G)^{n+1}e^{i\beta j} = (G)^{n-1}e^{i\beta j} + 2s\{(G)^n e^{i\beta(j-1)} - 2(G)^n e^{i\beta j} + (G)^n e^{i\beta(j+1)}\}. \tag{4.1.6}$$

Here G is the amplification factor by which the mth Fourier component of the error distribution at any time level is multiplied to get the corresponding Fourier component at the next time level. Dividing (4.1.6) by $(G)^n e^{i\beta j}$ yields

$$G - G^{-1} = 4s(\cos\beta - 1), \tag{4.1.7}$$

which gives the equation to be solved for G, namely

$$(G)^2 + 8s\gamma G - 1 = 0, \tag{4.1.8}$$

where $\gamma = \sin^2(\beta/2)$.

Since the discriminant of the quadratic equation (4.1.8) is positive for all s and β, then it has two *real* roots, G_1 and G_2, which satisfy the pair of equations

$$G_1 G_2 = -1 \tag{4.1.9a}$$

and

$$G_1 + G_2 = -8\gamma s. \tag{4.1.9b}$$

It follows from (4.1.9a) that either

$$|G_1| > 1 \quad \text{or} \quad |G_2| > 1,$$

in which case the errors at a given time level will contain some Fourier components which grow without bound as n increases; otherwise

$$G_1 = -G_2 = 1 \quad \text{(say)},$$

in which case (4.1.9b) indicates that $s = 0$, so $\Delta t = 0$, when the method has no practical value. Richardson's method is therefore unconditionally unstable and errors in the values computed using equation (4.1.3) become as large as the true values after only a few time steps. The method therefore cannot be used to solve the diffusion equation.

But, it must not be assumed that all methods which may involve steps in time or space for which the errors magnify, are of no practical use. Such steps may be part of a time-splitting method for an initial value problem, or a space-stepping procedure in a boundary value problem in which the errors are kept small relative to the true values.

DuFort-Frankel Method

Another explicit method for solving the diffusion equation is that developed by DuFort and Frankel (1953). This is a modification of Richardson's method in which the central grid-point value τ_j^n in the finite difference approximation for the diffusion term $\alpha\partial^2\bar{\tau}/\partial x^2$ in Equation (4.1.1) is replaced by its average at the $(n-1)$ and $(n+1)$ time levels. Since

$$\bar{\tau}_j^n = \tfrac{1}{2}(\bar{\tau}_j^{n+1} + \bar{\tau}_j^{n-1}) + O\{(\Delta t)^2\}, \tag{4.1.10}$$

it follows that Equation (4.1.1) can be rewritten

$$\frac{\bar{\tau}_j^{n+1} - \bar{\tau}_j^{n-1}}{2\Delta t} + O\{(\Delta t)^2\} = \alpha\{\frac{\bar{\tau}_{j+1}^n - (\bar{\tau}_j^{n+1} + \bar{\tau}_j^{n-1}) + \bar{\tau}_{j-1}^n}{(\Delta x)^2} + O\{(\Delta x)^2, (\frac{\Delta t}{\Delta x})^2\}\}. \tag{4.1.11}$$

Although a value of $\bar{\tau}$ at the $(n+1)$th time level appears on the right-hand side of this equation, it is at the jth space position, so Equation (4.1.11) can be rearranged to give

$$\bar{\tau}_j^{n+1} = \frac{2s}{1+2s}(\bar{\tau}_{j-1}^n + \bar{\tau}_{j+1}^n) + \frac{1-2s}{1+2s}\bar{\tau}_j^{n-1} + O\{(\Delta t)^3, \Delta t(\Delta x)^2, \alpha\Delta t/(\Delta x)^2\}. \tag{4.1.12}$$

Dropping the terms of $O\{(\Delta t)^3, \Delta t(\Delta x)^2, s\}$ gives the explicit finite difference equation

$$\tau_j^{n+1} = \frac{2s}{1+2s}(\tau_{j-1}^n + \tau_{j+1}^n) + \frac{1-2s}{1+2s}\tau_j^{n-1} . \tag{4.1.13}$$

Like Richardson's formula, the DuFort-Frankel formula involves values of the dependent variable at grid-points along three time levels and it requires special starting procedures. These two methods are three-level methods, in contrast with the FTCS explicit method which is a two-level method.

A stability analysis of the error equation corresponding to (4.1.13), namely

$$\xi_j^{n+1} = \frac{2s}{1+2s}(\xi_{j-1}^n + \xi_{j+1}^n) + \frac{1-2s}{1+2s}\xi_j^{n-1} , \tag{4.1.14}$$

by means of the von Neumann method gives the amplification factor from one time level to the next for the mth Fourier component of the error distribution to be

$$G = \frac{2s\cos\beta \pm (1-4s^2\sin^2\beta)^{1/2}}{1+2s} . \tag{4.1.15}$$

If s and β are such that $1 - 4s^2\sin^2\beta \geqslant 0$, then both terms in the numerator of (4.1.15) are real. Also, $1+2s > 0$ so that

$$|G| \leqslant \frac{|2s\cos\beta| + (1-4s^2\sin^2\beta)^{1/2}}{1+2s},$$

and as

$$|2s\cos\beta| \leqslant 2s, \; 0 \leqslant 1-4s^2\sin^2\beta \leqslant 1,$$

it follows that

$$|G| \leqslant \frac{2s+1}{1+2s} = 1. \tag{4.1.16}$$

Otherwise, $1-4s^2\sin^2\beta < 0$ so that

$$G = \frac{2s\cos\beta \pm i(4s^2\sin^2\beta-1)^{1/2}}{1+2s},$$

and

$$|G|^2 = \frac{4s^2\cos^2\beta+(4s^2\sin^2\beta-1)}{(1+2s)^2}$$

$$= \frac{4s^2-1}{4s^2+4s+1}$$

$$\leqslant 1. \tag{4.1.17}$$

Therefore it follows that this finite difference method is stable for all positive values of s.

Therefore Δt may be chosen as large as we like, provided the discretisation error does not become too large. The discretisation error depends on the truncation error, which can be established by a consistency analysis.

On checking the consistency of the finite difference equation (4.1.13) by substitution of $\bar{\tau}$, the solution of the diffusion equation, for the finite difference solution τ, followed by expansion of each term in Taylor series about the (j,n) grid-point, the following relation is obtained:

$$\left.\frac{\partial\bar{\tau}}{\partial t}\right|_j^n = \alpha\left.\frac{\partial^2\bar{\tau}}{\partial x^2}\right|_j^n - \alpha\left(\frac{\Delta t}{\Delta x}\right)^2\left.\frac{\partial^2\bar{\tau}}{\partial t^2}\right|_j^n - \frac{(\Delta t)^2}{6}\left.\frac{\partial^3\bar{\tau}}{\partial t^3}\right|_j^n + \frac{(\Delta x)^2}{12}\left.\frac{\partial^4\bar{\tau}}{\partial x^4}\right|_j^n + O\{(\Delta t)^3,(\Delta x)^3\}. \tag{4.1.18}$$

The DuFort-Frankel method is therefore consistent with the one-dimensional diffusion equation, with truncation error going to zero, only if $\Delta t/\Delta x \to 0$ as both Δx and $\Delta t \to 0$. Therefore, even though this method is unconditionally stable, $\Delta t \ll \Delta x$ is required for consistency. If Δt and $\Delta x \to 0$ in such a way that $\Delta t/\Delta x \to K$, a constant, then the finite difference equation (4.1.13) is consistent with the hyperbolic equation

$$\frac{\partial\bar{\tau}}{\partial t} = \alpha\left(\frac{\partial^2\bar{\tau}}{\partial x^2} - K^2\frac{\partial^2\bar{\tau}}{\partial t^2}\right), \tag{4.1.19}$$

and not with the parabolic one-dimensional diffusion equation (2.1.1).

In spite of the restriction $\Delta t \ll \Delta x$ required for consistency, the stabilising effect of the average value of τ_j^{n+1} and τ_j^{n-1} for τ_j^n in the finite difference approximation of the diffusion terms in various equations of fluid dynamics has been used successfully by many workers including Payne (1958), Fromm and Harlow (1963), Amsden and Harlow (1964), Hung and Macagno (1966), and Torrance (1968).

4.2 Implicit Methods

Each of the methods described previously is explicit. At the new time level, the finite difference equation contains only the one unknown value τ_j^{n+1}, which is calculated explicitly from values of τ known at previous time levels. These methods are easy to program and require few computations to determine the values of τ at each new time level. Unfortunately, each of the two usable methods already described is restricted to very small time steps Δt, the classical FTCS method because of stability requirements and the DuFort-Frankel method because of consistency requirements. For instance, if a numerical approximation at $t = 10$ is required and Δx must take the value 10^{-2}, then use of the FTCS method with $\alpha = 5$ implies that Δt must be chosen to satisfy

$$s = \frac{5\Delta t}{(10^{-2})^2} \leqslant \tfrac{1}{2}$$

so that

$$\Delta t \leqslant 10^{-5} .$$

To generate a numerical solution at $t = 10$ would require the computation of τ on at least one million rows of grid-points in the solution domain.

The Classical Implicit Method

By substituting the backward difference form (2.3.8a) for the time derivative and the central difference form (2.3.19) for the space derivative in the diffusion equation (2.1.1) evaluated at the $(j,n+1)$ grid-point, namely

$$\left.\frac{\partial\bar{\tau}}{\partial t}\right|_j^{n+1} = \alpha \left.\frac{\partial^2\bar{\tau}}{\partial x^2}\right|_j^{n+1} , \tag{4.2.1a}$$

gives

$$\frac{\bar{\tau}_j^{n+1} - \bar{\tau}_j^{n}}{\Delta t} + O\{\Delta t\} = \alpha\{\frac{\bar{\tau}_{j+1}^{n+1} - 2\bar{\tau}_j^{n+1} + \bar{\tau}_{j-1}^{n+1}}{(\Delta x)^2} + O\{(\Delta x)^2\}\}. \tag{4.2.1b}$$

The analogous finite difference approximation to (4.2.1a) is therefore

$$\frac{\tau_j^{n+1} - \tau_j^{n}}{\Delta t} = \alpha\{\frac{\tau_{j+1}^{n+1} - 2\tau_j^{n+1} + \tau_{j-1}^{n+1}}{(\Delta x)^2}\} , \tag{4.2.2a}$$

with an apparent truncation error of $O\{\Delta t,(\Delta x)^2\}$. Equation (4.2.2a) can be rearranged to yield the finite difference equation

$$-s\tau_{j-1}^{n+1} + (1+2s)\tau_j^{n+1} - s\tau_{j+1}^{n+1} = \tau_j^n . \tag{4.2.2b}$$

This equation is not explicit, since there are three unknown values of τ at the $(n+1)$th time level. With $n = 0$, substitution of $j=1(1)J-1$ in Equation (4.2.2b) gives $(J-1)$ simultaneous linear algebraic equations for the $(J-1)$ unknown values of τ at internal grid-points along the first time level, in terms of known initial and boundary values. These equations can be solved to give the values of τ_j^1, $j=1(1)J-1$. Similarly, putting $n = 1$ gives $(J-1)$

equations for the $(J-1)$ unknown values of τ at the second time level in terms of boundary values and the computed values of τ at the first time level, and so on. Methods such as this, in which the unknown values of τ at any time level are found by solving a set of algebraic equations are described as *implicit*. Equation (4.2.2b) defines what is known as the classical implicit method for solving the one-dimensional diffusion equation (see Laasonen, 1949, and O'Brien, 1950).

Assuming values are known at the nth time level we may substitute $j=1(1)J-1$ in Equation (4.2.2b) to obtain the following set of simultaneous linear algebraic equations with unknowns τ_j^{n+1}, $j=1(1)J-1$:

$$\left.\begin{array}{rl} (1+2s)\tau_1^{n+1} - s\tau_2^{n+1} & = d_1^n \\ -s\tau_1^{n+1} + (1+2s)\tau_2^{n+2} - s\tau_3^{n+1} & = d_2^n \\ -s\tau_2^{n+1} + (1+2s)\tau_3^{n+1} - s\tau_4^{n+1} & = d_3^n \\ \cdots & \vdots \\ -s\tau_{J-2}^{n+1} + (1+2s)\tau_{J-1}^{n+1} & = d_{J-1}^n \end{array}\right\} \quad (4.2.3)$$

where

$$d_1^n = \tau_1^n + sg_0^{n+1}, \; g_0^{n+1} = g_0(t_{n+1}),$$

$$d_j^n = \tau_j^n, \quad j=2(1)J-2,$$

$$d_{J-1}^n = \tau_{J-1}^n + sg_1^{n+1}, \; g_1^{n+1} = g_1(t_{n+1}).$$

This tri-diagonal system may be solved by the elimination process described by Thomas (1949) and outlined in the Appendix. The Thomas algorithm makes full use of the fact that most of the coefficients of the unknown values of τ at time level $(n+1)$ are zero's. It is much faster than using a Gauss elimination "package" which automatically goes through a process of making the coefficient of many terms zero, when they are already so.

The process of solving the set of Equations (4.2.3) must be repeated at every time step. However, compensation for the extra arithmetic required at each time step compared with that for the explicit methods occurs because the finite difference equation (4.2.2) is stable for all values of s, so that larger values of Δt may be used. In fact, application of the von Neumann stability analysis to the corresponding error equation

$$-s\xi_{j-1}^{n+1} + (1+2s)\xi_j^{n+1} - s\xi_{j+1}^{n+1} = \xi_j^n,$$

yields the amplification factor

$$G = \{1 + 2s(1 - \cos\beta)\}^{-1},$$

for the propagation of the mth Fourier mode of the error distribution from one time level to the next. Since $(1 - \cos\beta) \geqslant 0$ for all β,

$$1+2s(1-\cos\beta) \geqslant 1 \quad \text{for all} \quad s > 0.$$

Therefore $|G| \leqslant 1$ for all $s > 0$ and all β, so the classical implicit method is unconditionally stable.

The consistency of this implicit finite difference method clearly follows from the manner in which the finite difference equation (4.2.2b) was constructed from Equation (4.2.1b), the truncation error normally being of $O\{\Delta t,(\Delta x)^2\}$. However, the consistency should be formally established to determine whether any values of s occur for which the truncation error, and therefore the discretisation error, is of optimal order. This is done by substituting the solutions $\bar{\tau}$ of the given partial differential equation for the values of τ in the finite difference approximation (4.2.2), to obtain the relation

$$-s\bar{\tau}_{j-1}^{n+1} + (1+2s)\bar{\tau}_j^{n+1} - s\bar{\tau}_{j+1}^{n+1} = \bar{\tau}_j^n . \tag{4.2.4}$$

The difference between this equation and the one-dimensional diffusion equation is determined by expanding each term in a Taylor series about the (j,n) grid-point. The Taylor series expansion of $\bar{\tau}_{j+1}^{n+1}$ is

$$\bar{\tau}_{j+1}^{n+1} = \sum_{m=0}^{\infty} \frac{1}{m!} (\Delta t \frac{\partial}{\partial t} + \Delta x \frac{\partial}{\partial x})^m \bar{\tau}\Big|_j^n , \tag{4.2.5}$$

which becomes, on expansion,

$$\begin{aligned}\bar{\tau}_{j+1}^{n+1} = \bar{\tau}_j^n &+ \Delta t \frac{\partial\bar{\tau}}{\partial t}\Big|_j^n + \Delta x \frac{\partial\bar{\tau}}{\partial x}\Big|_j^n \\ &+ \frac{(\Delta t)^2}{2}\frac{\partial^2\bar{\tau}}{\partial t^2}\Big|_j^n + \Delta t\,\Delta x \frac{\partial^2\bar{\tau}}{\partial t\partial x}\Big|_j^n + \frac{(\Delta x)^2}{2}\frac{\partial^2\bar{\tau}}{\partial x^2}\Big|_j^n \\ &+ \frac{(\Delta t)^3}{6}\frac{\partial^3\bar{\tau}}{\partial t^3}\Big|_j^n + \frac{(\Delta t)^2\Delta x}{2}\frac{\partial^3\bar{\tau}}{\partial t^2\partial x}\Big|_j^n + \frac{\Delta t(\Delta x)^2}{2}\frac{\partial^3\bar{\tau}}{\partial t\partial x^2}\Big|_j^n \\ &+ \frac{(\Delta x)^3}{6}\frac{\partial^3\bar{\tau}}{\partial x^3}\Big|_j^n + \ldots\end{aligned} \tag{4.2.6a}$$

The Taylor series expansion of $\bar{\tau}_{j-1}^{n+1}$ may be found by substituting $-\Delta x$ for Δx in the above equation, giving

$$\begin{aligned}\bar{\tau}_{j-1}^{n+1} = \bar{\tau}_j^n &+ \Delta t \frac{\partial\bar{\tau}}{\partial t}\Big|_j^n - \Delta x \frac{\partial\bar{\tau}}{\partial x}\Big|_j^n \\ &+ \frac{(\Delta t)^2}{2}\frac{\partial^2\bar{\tau}}{\partial t^2}\Big|_j^n - \Delta t\,\Delta x \frac{\partial^2\bar{\tau}}{\partial t\partial x}\Big|_j^n + \frac{(\Delta x)^2}{2}\frac{\partial^2\bar{\tau}}{\partial x^2}\Big|_j^n \\ &+ \frac{(\Delta t)^3}{6}\frac{\partial^3\bar{\tau}}{\partial t^3}\Big|_j^n - \frac{(\Delta t)^2\Delta x}{2}\frac{\partial^3\bar{\tau}}{\partial t^2\partial x}\Big|_j^n + \frac{(\Delta t)^2\Delta x}{2}\frac{\partial^3\bar{\tau}}{\partial t\partial x^2}\Big|_j^n \\ &- \frac{(\Delta x)^3}{6}\frac{\partial^3\bar{\tau}}{\partial x^3}\Big|_j^n + \ldots .\end{aligned} \tag{4.2.6b}$$

Finally, by (2.3.2),

$$\bar{\tau}_j^{n+1} = \bar{\tau}_j^n + \Delta t \frac{\partial\bar{\tau}}{\partial t}\Big|_j^n + \frac{(\Delta t)^2}{2}\frac{\partial^2\bar{\tau}}{\partial t^2}\Big|_j^n + \frac{(\Delta t)^3}{6}\frac{\partial^3\bar{\tau}}{\partial t^3}\Big|_j^n + \ldots . \tag{4.2.6c}$$

Substitution of the expansions (4.2.6) into Equation (4.2.4) yields the equivalent relation

$$\left.\frac{\partial\bar{\tau}}{\partial t}\right|_j^n = \alpha \left.\frac{\partial^2\bar{\tau}}{\partial x^2}\right|_j^n + E_j^n \tag{4.2.7a}$$

where E_j^n is given by

$$E_j^n = -\frac{\Delta t}{2}\left.\frac{\partial^2\bar{\tau}}{\partial t^2}\right|_j^n + \alpha\Delta t \left.\frac{\partial^3\bar{\tau}}{\partial t\partial x^2}\right|_j^n + \frac{\alpha(\Delta x)^2}{12}\left.\frac{\partial^4\bar{\tau}}{\partial x^4}\right|_j^n + O\{(\Delta t)^2,\Delta t(\Delta x)^2\}. \tag{4.2.7b}$$

Since $\bar{\tau}$ is a solution of the one-dimensional diffusion equation (2.1.1) then

$$\frac{\partial^2\bar{\tau}}{\partial t^2} = \alpha\frac{\partial^3\bar{\tau}}{\partial t\partial x^2} = \alpha^2\frac{\partial^4\bar{\tau}}{\partial x^4}.$$

Hence, since $(\Delta x)^2 = s\Delta t/\alpha$, it follows that E_j^n may be expressed as

$$E_j^n = \frac{\Delta t}{2}\{1 + \frac{1}{6s}\}\left.\frac{\partial^2\bar{\tau}}{\partial t^2}\right|_j^n + O\{(\Delta t)^2\}, \quad \text{for fixed } s. \tag{4.2.8}$$

Clearly, as $\Delta t \to 0$ for a fixed value of s, $E_j^n \to 0$ and formula (4.2.4) approaches Equation (2.1.1), so the finite difference equation (4.2.2b) is consistent with the one-dimensional diffusion equation. Also, there is no optimal value of s for which the order of the truncation error may be reduced; for all $s > 0$, the truncation error E_j^n remains $O\{\Delta t\}$. Therefore, for a given s, taking a large value of Δt increases the error E_j^n in a manner approximately proportional to Δt. We must therefore compromise between taking large time steps, to save on the number of arithmetic operations needed, and increased inaccuracy produced by the larger truncation error and associated discretisation error which occurs. "Stability without accuracy has little to commend it" (Fox, 1962, p. 243).

An aspect of implicit finite difference methods seldom considered, is whether the resulting system of linear algebraic equations can be solved in practice. If the coefficient matrix of the system is diagonally dominant then any direct method of solving the corresponding linear algebraic system, such as that due to Thomas for tridiagonal systems or the Gauss elimination method for more general systems, is stable. That is, a round-off error in any computation does not magnify during the following calculations to make the results so inaccurate they cannot be taken as the true solutions of the original linear algebraic system of equations.

A system of linear algebraic equations is said to be diagonally dominant if along each row of the coefficient matrix the absolute value of the coefficient on the leading diagonal is greater than or equal to, the sum of the absolute values of the remaining coefficients along that row, and along at least one row there is strict inequality.

For the classical implicit method, in which the set of equations (4.2.3) must be solved, we have for the rows indicated

$$j=1,J-1 : \ |1+2s| > |-s| ,$$

$$j=2(1)J-2: \ |1+2s| > |-s| + |-s|,$$

since $s > 0$. Therefore the system (4.2.3) is solvable for all s.

The Crank-Nicolson Method

Another implicit method used to solve the one-dimensional diffusion equation is due to Crank and Nicolson (1947). This method uses centred finite difference approximations for both time and space derivatives at the point $(j\Delta x, n\Delta t+\frac{1}{2}\Delta t)$ which is halfway between the (j,n) and $(j,n+1)$ grid-points. The central difference approximation for the time derivative $\partial\bar{\tau}/\partial t$ at this point in the solution domain is the same as the forward difference approximation for this derivative at the (j,n) grid-point, but it is a more accurate approximation, the error being of $O\{(\Delta t)^2\}$. The spatial derivative $\partial^2\bar{\tau}/\partial x^2$ at $(j\Delta x, n\Delta t+\frac{1}{2}\Delta t)$ is approximated by the average of the central difference approximations of these spatial derivatives at the (j,n) and $(j,n+1)$ grid-points.

The diffusion equation applied at the point $(j\Delta x, n\Delta t+\frac{1}{2}\Delta t)$ is

$$\left.\frac{\partial\bar{\tau}}{\partial t}\right|_j^{n+\frac{1}{2}} = \alpha \left.\frac{\partial^2\bar{\tau}}{\partial x^2}\right|_j^{n+\frac{1}{2}} . \tag{4.2.9}$$

Replacing the right side of this equation by the average of the values at time-levels n and $(n+1)$ gives

$$\left.\frac{\partial\bar{\tau}}{\partial t}\right|_j^{n+\frac{1}{2}} = \frac{\alpha}{2}\left\{\left.\frac{\partial^2\bar{\tau}}{\partial x^2}\right|_j^{n} + \left.\frac{\partial^2\bar{\tau}}{\partial x^2}\right|_j^{n+1} + O\{(\Delta t)^2\}\right\} ,$$

which becomes, on applying central difference forms for each derivative,

$$\frac{\bar{\tau}_j^{n+1}-\bar{\tau}_j^{n}}{\Delta t} + O\{(\Delta t)^2\} = \frac{\alpha}{2}\left\{\frac{\bar{\tau}_{j+1}^{n+1}-2\bar{\tau}_j^{n+1}+\bar{\tau}_{j-1}^{n+1}}{(\Delta x)^2} + \frac{\bar{\tau}_{j+1}^{n}-2\bar{\tau}_j^{n}+\bar{\tau}_{j-1}^{n}}{(\Delta x)^2} + O\{(\Delta x)^2\}\right.$$
$$\left. + O\{(\Delta t)^2\}\right\}. \tag{4.2.10}$$

Neglecting the terms of $O\{(\Delta t)^2,(\Delta x)^2\}$, which is therefore the usual order of the truncation error, this becomes, on writing the values of τ at the $(n+1)$th time level on the left side and the known values of τ at the nth time level on the right side,

$$-\tfrac{1}{2}s\tau_{j-1}^{n+1} + (1+s)\tau_j^{n+1} - \tfrac{1}{2}s\tau_{j+1}^{n+1} = \tfrac{1}{2}s\tau_{j-1}^{n} + (1-s)\tau_j^{n} + \tfrac{1}{2}s\tau_{j+1}^{n} . \tag{4.2.11}$$

Clearly, the error involved in Equation (4.2.11) is $O\{(\Delta t)^3,\Delta t(\Delta x)^2\}$. As the values of τ at the $(n+1)$ time level are computed using this equation, the contribution to the discretisation error in stepping from one time level to the next is therefore $O\{(\Delta t)^3,\Delta t(\Delta x)^3\}$.

Assuming values of τ_j^n, $j=0(1)J$, are known, the right side of this equation is known for $j=1(1)J-1$. Application of the boundary conditions which define the values of τ_0^{n+1}, τ_J^{n+1}, then yields a set of simultaneous linear algebraic equations in the unknowns τ_j^{n+1}, $j=1(1)J-1$, similar to Equations (4.2.3), namely

$$\left.\begin{array}{lr}
(1+s)\tau_1^{n+1} - \tfrac{1}{2}s\tau_2^{n+1} & = d_1^n \\
-\tfrac{1}{2}s\tau_1^{n+1} + (1+s)\tau_2^{n+1} - \tfrac{1}{2}s\tau_3^{n+1} & = d_2^n \\
\quad - \tfrac{1}{2}s\tau_2^{n+1} + (1+s)\tau_3^{n+1} - \tfrac{1}{2}s\tau_4^{n+1} & = d_3^n \\
\qquad \cdot & \cdot \\
\qquad \cdot & \cdot \\
\qquad - \tfrac{1}{2}s\tau_{j-1}^{n+1} + (1+s)\tau_j^{n+1} - \tfrac{1}{2}s\tau_{j+1}^{n+1} & = d_j^n \\
\qquad\qquad \cdot & \cdot \\
\qquad\qquad - \tfrac{1}{2}s\tau_{J-2}^{n+1} + (1+s)\tau_{J-1}^{n+1} & = d_{J-1}^n
\end{array}\right\} \tag{4.2.12}$$

where $d_1^n = \frac{1}{2}s\tau_0^n + (1-s)\tau_1^n + \frac{1}{2}s\tau_2^n + \frac{1}{2}s\tau_0^{n+1}$,

$$d_j^n = \tfrac{1}{2}s\tau_{j-1}^n + (1-s)\tau_j^n + \tfrac{1}{2}s\tau_{j+1}^n, \quad j=2(1)J-2,$$

$$d_{J-1}^n = \tfrac{1}{2}s\tau_{J-2}^n + (1-s)\tau_{J-1}^n + \tfrac{1}{2}s\tau_J^n + \tfrac{1}{2}s\tau_J^{n+1},$$

are all known. Because it is tridiagonal this system can be solved efficiently using the Thomas algorithm. Commencing with $n=0$, then taking $n=1,2,3,\ldots$ in turn, values of $\tau_j^1, \tau_j^2, \tau_j^3, \tau_j^4, \ldots$, $j=1(1)J-1$, are found.

When the stability of the finite difference equation (4.2.11) is investigated, it is found that errors ξ_j^n in the computation of τ_j^n propagate according to the relation

$$-\tfrac{1}{2}s\xi_{j-1}^{n+1} + (1+s)\xi_j^{n+1} - \tfrac{1}{2}s\xi_{j+1}^{n+1} = \tfrac{1}{2}s\xi_{j-1}^n + (1-s)\xi_j^n + \tfrac{1}{2}s\xi_{j+1}^n, \tag{4.2.13}$$

for $j=1(1)J-1$. Application of the von Neumann method yields the amplification factor G for the mth Fourier component of the error distribution at any time level, namely

$$G = \frac{1-2s\sin^2(\beta/2)}{1+2s\sin^2(\beta/2)}. \tag{4.2.14}$$

Because

$$|G| \leqslant 1$$

for all positive s and all β, it is clear that this method applied to the diffusion problem with given boundary values is unconditionally stable.

The consistency of this finite difference scheme is investigated by substituting the exact solution $\bar{\tau}$ for the approximation τ in (4.2.11) and then expanding each term of the resulting equation in a Taylor series about the (j,n) grid-point, yielding

$$\left.\frac{\partial\bar{\tau}}{\partial t}\right|_j^n = \alpha \left.\frac{\partial^2\bar{\tau}}{\partial t^2}\right|_j^n + O\{(\Delta t)^2,(\Delta x)^2\}. \tag{4.2.15}$$

This gives Equation (4.1.1) in the limit as $\Delta x \to 0, \Delta t \to 0$, so the Crank-Nicolson equation is consistent with the original diffusion equation. It also has, as suggested previously, a truncation error which is $O\{(\Delta t)^2,(\Delta x)^2\}$.

As mentioned earlier, the tridiagonal set of equations (4.2.12) can readily be solved by the Thomas algorithm so long as the coefficient matrix of the linear algebraic system is diagonally dominant. This is true for all $s > 0$, since for the rows indicated, it is true that

$$j=1,J-1 \quad : \quad |1+s| > |-\tfrac{1}{2}s|,$$

$$j=2(1)J-2 \quad : \quad |1+s| > |-\tfrac{1}{2}s| + |-\tfrac{1}{2}s|.$$

Thus both the classical and Crank-Nicolson implicit methods are solvable for all $s > 0$.

Clearly, the solution of the set of Crank-Nicolson equations (4.2.12) requires more calculations per time-step than the classical implicit method to evaluate the d_j^n on the right side of the equations. Each d_j^n in equation (4.2.12) includes two values of τ more than in the d_j^n of the classical implicit equations (4.2.3). What then is the advantage of the Crank-Nicolson scheme over the classical implicit method? Because at each time step the contribution to the discretisation errors are $O\{(\Delta t)^3, \Delta t(\Delta x)^2\}$ then the Crank-Nicolson method is more accurate than the classical implicit method which contributes at each time step an error of $O\{(\Delta t)^2, \Delta t(\Delta x)^2\}$. For the same degree of accuracy, a larger time increment can be used for the Crank-Nicolson method and fewer time steps are necessary to compute values of τ up to a given time level.

It is of interest to note that the unconditionally stable Crank-Nicolson scheme may be considered as a combination of the conditionally stable FTCS explicit method and the unconditionally stable classical implicit method. Consider the FTCS method applied over the first half of the time interval from $n\Delta t$ to $(n+1)\Delta t$, producing values τ_j^* at the intermediate time $(n\Delta t + \frac{1}{2}\Delta t)$. The values of τ_j^* are therefore given in terms of τ_j^n by means of (2.4.4), namely

$$\tau_j^* = s^*\tau_{j-1}^n + (1-2s^*)\tau_j^n + s^*\tau_{j+1}^n, \tag{4.2.16a}$$

where $s^* = \alpha(\Delta t/2)/(\Delta x)^2 = s/2$. Application of the classical implicit equation over the second half of the time interval Δt gives the values of τ_j^{n+1} in terms of τ_j^* according to the relation

$$-s^*\tau_{j-1}^{n+1} + (1+2s^*)\tau_j^{n+1} - s^*\tau_{j+1}^{n+1} = \tau_j^* . \tag{4.2.16b}$$

The result of adding these two equations is to give the finite difference equation,

$$-s^*\tau_{j-1}^{n+1} + (1+2s^*)\tau_j^{n+1} - s^*\tau_{j+1}^{n+1} = s^*\tau_{j-1}^n + (1-2s^*)\tau_j^n + s^*\tau_{j+1}^n, \tag{4.2.17}$$

which directly connects values of τ at time level $(n+1)$ with those at time level n. Substitution of $s/2$ for s^* then yields the Crank-Nicolson finite difference equation for stepping the one-dimensional diffusion equation from one time level to the next.

Therefore, every application of the Crank-Nicolson finite difference equation is, in effect, an application of the FTCS explicit equation over the first half of each time interval followed by the classical implicit equation over the second half of the interval. This time-splitting of a stable finite difference equation into two equations, one of which may be unstable by itself, is used extensively to solve parabolic partial differential equations in two and three space dimensions. This is the basis of the Alternating Direction Implicit method (see Section 6.3).

The Crank-Nicolson method has been used to solve the problem presented in Section 2.1 and the results obtained are displayed in Table 4.1. Comparing these results with those in Table 3.1, obtained using the FTCS explicit method, it is clear that this implicit method gives more accurate answers. For the same grid-spacing Δx, a relatively large value of Δt and therefore of s, say $s = 4.0$, still yields a better answer than the FTCS method for $s = 1/2$, and for the optimal value $s = 1/6$.

However, this increased accuracy is acquired only with an increase in the number of arithmetic operations required to obtain the result. Since the time taken to perform a multiplication or division on a computer is generally much longer than that required to perform an addition or subtraction - for example, 2.9×10^{-6} sec compared with 0.55×10^{-6} sec on the CYBER 173 - only the number of multiplications and divisions required will be considered when comparing the computer times required for each of the above examples.

TABLE 4.1

Crank-Nicolson approximations to $\bar{\tau}(0.4,8) = 0.4503963$

J	Δx	Δt/s	s = 1/2	s = 1	s = 2	s = 4
10	0.1	1	$\tau_4^{16} = .4515527$	$\tau_4^{8} = .4517348$	$\tau_4^{4} = .4518655$	$\tau_4^{2} = .4568709$
20	0.05	0.25	$\tau_8^{64} = .4506646$	$\tau_8^{32} = .4506763$	$\tau_8^{16} = .4507236$	$\tau_8^{8} = .4508537$
40	0.025	0.0625	$\tau_{16}^{256} = .4504620$	$\tau_{16}^{128} = .4504628$	$\tau_{16}^{64} = .4504658$	$\tau_{16}^{32} = .4504777$
50	0.02	0.04	$\tau_{20}^{400} = .4504383$	$\tau_{20}^{200} = .4504386$	$\tau_{20}^{100} = .4504129$	$\tau_{20}^{50} = .4504447$
100	0.01	0.01	$\tau_{40}^{1600} = .4504068$	$\tau_{40}^{800} = .4504068$	$\tau_{40}^{400} = .4504069$	$\tau_{40}^{200} = .4504072$

If the formula used with the FTCS method is written

$$\tau_j^{n+1} = s(\tau_{j-1}^n + \tau_{j+1}^n - 2\tau_j^n) + \tau_j^n,$$

then only two multiplications are used for every application of the formula, unless $s = \frac{1}{2}$ when only one division is required. Therefore, computing all the values of τ at each time level requires $2(J-1)$, or approximately $2J$ multiplications if J is large. For example, the answer obtained in Table 4.1 when $J = 20$, $s = \frac{1}{2}$, required $64 \times 1 \times 20 = 1{,}280$ such operations. When using the Crank-Nicolson method with the Thomas algorithm, to compute values of τ at each new time level requires approximately $5J$ multiplications or divisions. In addition, at each time step the values d_j^n, for $j=1(1)J-1$, which appear on the right side of Equation (4.2.12), must be obtained and this requires approximately $2J$ multiplications, unless $s = 1$ when the number of multiplications is only J. Hence, a total of about $7J$ multiplications or divisions are required to advance one time step if $s \neq 1$, or approximately $6J$ multiplications if $s = 1$. For example, when $J = 20$ and $s = \frac{1}{2}$, 64 time steps had to be made to reach $t = 8$ and so about $64 \times 7 \times 20 = 8{,}960$ multiplications or divisions were required. Therefore, using the Crank-Nicolson method requires about 7 times the computational effort to reach $t = 8$ compared with the FTCS explicit method. However, the answer obtained is much more accurate.

A comparison of the computational efficiency of the explicit FTCS method and the implicit Crank-Nicolson method, is given in Table 4.2. With a fixed grid spacing in the x-direction, the value of Δt was varied, and the absolute error in the value of the approximation τ computed for $\bar{\tau}(0.4,8)$ was compared with the approximate number of arithmetic operations required to achieve the result. It took 160,000 such operations to obtain a result with an error of 25×10^{-5} with $s = \frac{1}{2}$ in the FTCS method; with $s = 1/6$ six times as many operations (960,000) reduced the error to one-sixth (4.5×10^{-5}). But using $s = 1$ with the Crank-Nicolson method took only 480,000 operations to give an error one-quarter of this, namely 1.05×10^{-5}.

With $s = 8$, the number of operations required with the implicit method decreased to 70,000 while the error increased only slightly. A direct comparison of the FTCS method with $s = \frac{1}{2}$ and the Crank-Nicolson method with $s = 8$ indicates the superiority of the latter; less than half the number of computations produced a result with less than one-twentieth of the error produced by the FTCS scheme.

TABLE 4.2
Comparison of computational efficiency of the FTCS and Crank-Nicolson methods for calculating $\bar{\tau}(0.4,8) = 0.4503963$ for the diffusion problem.

	FTCS Method		Crank-Nicolson Method	
J=100	s = 1/6	s = ½	s = 1	s = 8
Mult. and Divisions per time-level	200	100	600	700
Number of time-levels	4800	1600	800	100
Total number of Mult. and Divisions	960,000	160,000	480,000	70,000
ABSOLUTE ERROR	4.51×10^{-5}	24.96×10^{-5}	1.05×10^{-5}	1.21×10^{-5}

4.3 Iterative Techniques for Solving Implicit Finite Difference Equations

The stability restrictions which applied to the explicit methods described in Section 4.1 are largely overcome if implicit methods such as those described in Section 4.2 are used, thereby permitting the use of much larger time steps when using finite difference methods. For example, the Crank-Nicolson method for the one dimensional diffusion equation (2.4.1) is given by Equation (4.2.11), namely

$$-\tfrac{1}{2}s\tau_{j-1}^{n+1} + (1+s)\tau_j^{n+1} - \tfrac{1}{2}s\tau_{j+1}^{n+1} = \tfrac{1}{2}s\tau_{j-1}^{n} + (1-s)\tau_j^{n} + \tfrac{1}{2}s\tau_{j+1}^{n}, \quad j=1(1)J-1. \tag{4.3.1}$$

This can be written as the set of linear algebraic equations (4.2.12) in which the values of τ on the left side are at the (n+1)th time level, and the right side contains all the terms which are known, namely those from the previous time level and from the boundary at the (n+1) time level. Therefore, knowing the values of τ at the nth time level, those at the (n+1)th level can be computed. The best way of doing this is to use the very efficient Thomas algorithm for inverting tri-diagonal systems of linear algebraic equations. However, if the Crank-Nicolson method is applied to the two-dimensional diffusion equation (see Section 6.2), a penta-diagonal system of linear algebraic equations results. Such a system is much more difficult to solve than the tri-diagonal system of the one-dimensional case, since two of the non-zero diagonals in the coefficient matrix are separated from the three along the leading diagonal by a wide band of zero coefficients. Any simple elimination method is therefore much less computationally economical than the Thomas method. For such equations, iterative techniques may be employed. This method is illustrated by application to the Crank-Nicolson formula (4.3.1).

An iterative method for solving an equation or a set of equations is one in which a first approximation to the unknowns is used to determine a second approximation, which in turn is used to calculate a third estimate, and so on. Any iterative method is said to be convergent when the difference between the exact solution and the successive approximations tends to zero as the number of iterations increase.

Iterative methods may be used to advantage when solutions are to be found for a set of linear algebraic equations which contain a large number of unknowns, but in which the individual equations involve only a few of these unknowns. The main advantage is speed. This is because iterative methods automatically take advantage of the zeros which occur naturally in the matrix of coefficients for the system of equations, whereas many of the calculations of general elimination methods are concerned with introducing zero coefficients which may already exist.

In order to solve the set of equations (4.3.1) by iteration, it is rewritten in the form

$$\tau_j^{n+1} = \frac{1}{2(1+s)} \{s\tau_{j-1}^{n+1} + s\tau_{j+1}^{n+1} + 2D_j^n\} , \tag{4.3.2}$$

where

$$D_j^n = \tfrac{1}{2}s\tau_{j-1}^n + (1-s)\tau_j^n + \tfrac{1}{2}s\tau_{j+1}^n ,$$

for all interior grid-points, that is, for $j=1(1)J-1$. For $j=0,J$ the value of τ_j^{n+1} is either prescribed as the boundary conditions or must be found from some alternative scheme as described in Section 4.4 if derivative boundary conditions are prescribed. At any time level n, the value of D_j^n is known, while the remaining terms on the right hand side of Equation (4.3.2) can be taken, at a first estimate, to be their values at the time level n. Equation (4.3.2) is therefore used for all j for which τ is unknown at the $(n+1)$th time level to produce, successively, a set of values for τ of improving accuracy. The iterative form of this equation is

$$\tau_j^{(p+1)} = \frac{1}{2(1+s)} \{s\tau_{j-1}^{(p)} + s\tau_{j+1}^{(p)} + 2D_j^n\} , \tag{4.3.3}$$

for $p=0,1,2,\ldots,$ with $\tau_j^{(0)} = \tau_j^n$. For convenience, the superscript indicating the time level has been omitted. The bracketed superscript p denotes the iteration number. However, it is important to remember that the values of $\tau_j^{(p)}$, $p=0,1,2,\ldots,$ are successive estimates of the value of τ_j^{n+1}. A suitable solution is assumed to be found when successive estimates of τ_j^{n+1} differ by less than a specified small quantity for all j. It is then assumed that the process has converged sufficiently.

As p increases, the iterates $\tau_j^{(p)}$ obtained will only converge if certain conditions on the coefficients of Equation (4.3.2) are satisfied. As mentioned previously, the solutions of the implicit finite difference equations are convergent if the coefficient matrix is diagonally dominant. This is equivalent to requiring that the absolute value of the coefficient of τ on the left side of (4.3.2), and hence of (4.3.3), is greater than the sum of the absolute values of the coefficients of τ on the right side. Clearly, since

$$1 > \frac{|s|}{2|1+s|} + \frac{|s|}{2|1+s|} , \quad s > 0,$$

applies here, the iteration process defined by (4.3.3) is convergent. This particular process is an example of the *Jacobi method*. The order, in terms of j, in which the $(p+1)$th iterates for τ_j^{n+1} are calculated will not affect the values obtained or their rate of convergence, since they are calculated only in terms of the pth iterates.

However, this method is rather inefficient because it does not use the most up-to-date approximations for the values of τ in the right side of Equation (4.3.3). A better method is to use the most recent iterates as soon as they are available, that is to replace $\tau^{(p)}$ by $\tau^{(p+1)}$ as soon as it is calculated. When a computer is used, this replacement may be made systematically. For instance, if the next iterates are computed in order of increasing j then $\tau_{j-1}^{(p+1)}$ will have been found before $\tau_j^{(p+1)}$ is evaluated. Therefore it is more efficient to replace Equation

(4.3.3) by

$$\tau_j^{(p+1)} = \frac{1}{2(1+s)} \{s\tau_{j-1}^{(p+1)} + s\tau_{j+1}^{(p)} + 2D_j^n\} , \qquad (4.3.4)$$

with $j=1(1)J-1$, for each p. This procedure is called the *Gauss-Seidel* or *unextrapolated Liebmann* method and in this example it converges in about half the number of iterations for the Jacobi method. Unlike the Jacobi method the order, in terms of j, of calculating the $(p+1)$th iterates $\tau_j^{(p+1)}$ is fixed.

There is an even better iterative method available, based on a modification of the Gauss-Seidel method. Equation (4.3.4) is rewritten as

$$\tau_j^{(p+1)} = \tau_j^{(p)} + R_j^{(p)}, \qquad (4.3.5a)$$

where

$$R_j^{(p)} = \frac{1}{2(1+s)} \{s\tau_{j-1}^{(p+1)} + s\tau_{j+1}^{(p)} - 2(1+s)\tau_j^{(p)} + 2D_j^n\}. \qquad (4.3.5b)$$

The term $R_j^{(p)}$ may be considered as a correction to be added to $\tau_j^{(p)}$ to make it more exact, and as the process converges, $R_j^{(p)} \to 0$ as $p \to \infty$. The improvement to the Gauss-Seidel method is achieved by "over-relaxing"; that is, in Equation (4.3.5a) the correction term is multiplied by a factor ω, called the relaxation factor, where usually $\omega > 1$, producing

$$\tau_j^{(p+1)} = \tau_j^{(p)} + \omega R_j^{(p)} . \qquad (4.3.6)$$

The improved Gauss-Seidel method is generally called the *successive over-relaxation method* (after Young, 1954), or, alternatively, the *extrapolated Liebmann* method (Frankel, 1950). If the value of ω lies in the range $0 < \omega < 1$, then the process is called successive under-relaxation. This is not as widely used as successive over-relaxation.

With the successive over-relaxation method much computing time can be saved by determining ω_0, the optimum value of ω for convergence of the iterative scheme. A practical method for finding ω_0 has been given by Isenberg and de Vahl Davis (1975); they chose different values for ω, carried out five to ten test iterations, each time starting from the same initial set of values of τ, and selected the value of ω which gave the most rapid decrease in the difference between successive iterates.

It has been shown that using ω_0, the successive over-relaxation method is faster than the Gauss-Seidel method by a factor of the order of $2/\Delta x$. From this it is obvious that for small Δx it is important to be able to find ω_0.

Formulae for determining ω_0 can be found for certain finite difference methods. For example, solving the one-dimensional diffusion problem using successive over-relaxation on the Crank-Nicolson formula (4.2.11), which is analogous to the one-dimensional Poisson problem, it is found that

$$\omega_0 = 2(1 + \sqrt{1-\mu^2}),$$

where μ is the spectral radius of the corresponding Jacobi method, which in this case is given by

$$\mu = \frac{s}{1+s} \cos(\pi/J).$$

Here (J-1) is the total number of internal grid-points along a time level, the boundary values being known.

For equations whose solutions require a *large number* of iterations, the rate of convergence of the Jacobi, Gauss-Seidel and successive over-relaxation method with optimal ω are found theoretically to be proportional to

$$|\ln \mu|,\ 2|\ln \mu| \text{ and } |\ln(\omega_0-1)|,$$

respectively. For instance with $J = 10$ and $s = 2$ then $\mu = 0.63404$ and $\omega_0 = 1.1278$. Therefore $|\ln \mu| = 0.4556$, $2|\ln \mu| = 0.9112$, $|\ln(\omega_0-1)| = 2.0573$, showing that, in this case, the successive over-relaxation method converges more than twice as fast as the Gauss-Seidel method.

If possible, the successive over-relaxation method should be used with a value of ω as near as possible to the optimum value ω_0, because the rate of convergence can be very sensitive to small changes in ω. This is illustrated by a problem considered by Carre (1961) in which $\omega_0 = 1.9$. This value of the relaxation factor ω gave a rate of convergence forty times faster than that obtained using the Gauss-Seidel method ($\omega = 1$) and twice as fast as that given by reducing ω very slightly from 1.9 to 1.875.

Table 4.3 compares the number of iterations required to reach seven figure accuracy in the following example for the three iterative methods described, and the computation time on the University of Adelaide Cyber 173 computer. The diffusion equation (2.4.1) with $\alpha = 1$ and initial conditions $\bar{\tau}(x,0) = \sin(\pi x)$, $0 \leqslant x \leqslant 1$, and boundary conditions $\bar{\tau}(0,t) = \bar{\tau}(1,t) = 0$, $t \geqslant 0$, was solved using $\Delta x = 0.1$, $s = 2$, for one time step using the Crank-Nicolson equation. At each iteration the iterates were compared with the exact solution of the Crank-Nicolson equation obtained using the Thomas method.

Firstly, it is notable that the successive over-relaxation method is less sensitive to a small over-estimate of the optimal relaxation parameter ω_0, say $\omega = \omega_0 + 0.072 = 1.200$ when 1 extra iteration is required, than to an even smaller under-estimate, say $\omega = \omega_0 - 0.028 = 1.100$ when 2 extra iterations are required. Secondly, it is clear that Thomas's direct method is superior to all of the iterative methods described for solving this spatially one-dimensional problem. However, in problems involving more than one space dimension, direct methods become relatively more cumbersome and time consuming and it is often preferable to use iterative methods.

TABLE 4.3

Comparison of numbers of iterations and computer time for 7 figure accuracy, for various numerical methods of solving the implicit equation (4.3.1) to find τ_j^1.

Method	No. of iterations	Computer time (secs)
Jacobi iterative method	34	0.105
Gauss-Seidel (ω = 0)	19	0.085
S.O.R. (ω = 1.100)	13	0.052
S.O.R. (ω=ω_0=1.128)	11	0.045
S.O.R. (ω = 1.200)	12	0.053
Thomas Direct Method	-	0.012

Note that in these iterative methods, each approximation is found by a single calculation, and errors introduced do not prevent convergence of the process since the use of any incorrect values is equivalent to beginning with a new set of starting values. Hence the methods, if convergent, will converge to the correct result whatever the starting values.

4.4 Boundary Conditions

In the diffusion problem considered so far the boundary conditions consisted of known values of the dependent variable $\bar{\tau}$ at all time t. This is only one type of boundary condition that may apply.

Boundary conditions which may occur with a diffusion problem similar to the one-dimensional case described in Section 2.1 can be classified into three mathematical types.

Firstly, there is the Dirichlet type. In this problem the dependent variable $\bar{\tau}$ is given explicitly on all boundaries of the space region involved, even though the boundaries may not be regular. For the flow of heat along a thin insulated metal rod, this is the case where the temperature is given at each end, as described previously. For instance, at $x = 0$ the Dirichlet boundary condition for the diffusion problem may be written

$$\bar{\tau}(0,t) = c_0(t), \qquad (4.4.1)$$

where $c_0(t)$ is a known function of t.

Secondly, there is the Neumann type. In this case the normal derivative of $\bar{\tau}$ out from the boundary of the space region is given everywhere on the boundary. In general it may be written

$$\left.\frac{\partial\bar{\tau}}{\partial\eta}\right|_B = c_B(t) \qquad (4.4.2)$$

where η is the measure in the direction of the outward normal at B. For heat flow along the thin rod, this is the case where $\partial\bar{\tau}/\partial x$ is given at each end of the rod. At $x = 0$ the condition would be written

$$\left.\frac{\partial\bar{\tau}}{\partial(-x)}\right|_{(0,t)} = c_0(t). \qquad (4.4.3a)$$

The negative sign appears with the x coordinate, because at $x = 0$ the direction of the outward normal from the insulated rod is the negative x-direction. At $x = 1$, Equation (4.4.2) would be written

$$\left.\frac{\partial\bar{\tau}}{\partial x}\right|_{(1,t)} = c_1(t). \qquad (4.4.3b)$$

Thirdly, there is the mixed problem in which the value of a linear combination of $\bar{\tau}$ and $\partial\bar{\tau}/\partial\eta$ is known on the boundary B of the space region. This may be written

$$a_B(t)\bar{\tau}_B + b_B(t)\left.\frac{\partial\bar{\tau}}{\partial\eta}\right|_B = c_B(t), \qquad (4.4.4)$$

where the functions a_B, b_B and c_B may depend on the position B on the boundary and a_B, b_B have the same sign.

In particular, for the case of heat flow at the end $x = 0$ of the insulated rod of unit length, the boundary condition (4.4.4) becomes

$$a_0(t)\bar{\tau}(0,t) + b_0(t) \left.\frac{\partial\bar{\tau}}{\partial(-x)}\right|_{(0,t)} = c_0(t), \tag{4.4.5}$$

where a_0, b_0, c_0 are the coefficients applicable at $x = 0$. At the time level $t = n\Delta t$ this may be written

$$a_0^n\bar{\tau}_0^n - b_0^n \left.\frac{\partial\bar{\tau}}{\partial x}\right|_0^n = c_0^n \tag{4.4.6a}$$

where $a_0^n = a_0(n\Delta t)$, $b_0^n = b_0(n\Delta t)$, $c_0^n = c_0(n\Delta t)$.

At the end $x = 1$, the corresponding mixed boundary condition has the form

$$a_J^n\bar{\tau}_J^n + b_J^n \left.\frac{\partial\bar{\tau}}{\partial x}\right|_J^n = c_J^n, \tag{4.4.6b}$$

where a_J^n, b_J^n have the same sign.

Derivative Boundary Conditions

A typical derivative boundary condition at the end $x = 0$ for the diffusion problem (2.1.1), is

$$\left.\frac{\partial\bar{\tau}}{\partial x}\right|_{(0,t)} = c_0(t), \tag{4.4.7}$$

which is a particular example of the mixed boundary condition (4.4.5) with $a_0(t) \equiv 0$, $b_0(t) \equiv -1$.

Using the forward difference approximation in space at the $(0,n+1)$ grid-point Equation (4.4.7) becomes

$$\frac{\bar{\tau}_1^{n+1} - \bar{\tau}_0^{n+1}}{\Delta x} = c_0^{n+1} + O\{\Delta x\}, \tag{4.4.8}$$

where $c_0^{n+1} = c_0(n\Delta t+\Delta t)$.

The corresponding finite difference approximation to (4.4.8) is

$$\tau_0^{n+1} = \tau_1^{n+1} - \Delta x c_0^{n+1} . \tag{4.4.9}$$

Equation (4.4.9) can be used with the FTCS or any other finite difference approximation to (2.1.1) in order to obtain the boundary value τ_0^{n+1} once the interior value τ_1^{n+1} is known. Equation (4.4.8) shows that the truncation error of the finite difference approximation of the derivative boundary condition is $O\{\Delta x\}$, while the approximation for τ_0^{n+1} has a second order error in Δx which contributes to the discretisation error at each time step. In the case of the FTCS scheme, the truncation error is $O\{\Delta t,(\Delta x)^2\}$ and the contribution to the discretisation error at each time step is $O\{(\Delta t)^2,\Delta t(\Delta x)^2\}$, which is fourth order in Δx, for a given value of $s \neq 1/6$. Therefore, no matter how accurate is the finite difference method used in the interior of the solution domain, the overall accuracy is governed by the low order of accuracy of the boundary condition approximation.

Matching Truncation Errors

To overcome this anomaly, a method of higher order accuracy, such as one using the central difference approximation for the space derivative, could be used. Applying this method with Equation (4.4.7) at the (0,n) grid-point gives

$$\frac{\bar{\tau}_1^n - \bar{\tau}_{-1}^n}{2\Delta x} = c_0^n + O\{(\Delta x)^2\}\ , \qquad (4.4.10)$$

where $\bar{\tau}_{-1}^n$ is a fictitious value of $\bar{\tau}$ prescribed at the exterior point $(-\Delta x, n\Delta t)$ to the solution domain. Rearranging Equation (4.4.10) and dropping terms of $O\{(\Delta x)^3\}$ gives the corresponding finite difference approximation to the value $\bar{\tau}_{-1}^n$,

$$\tau_{-1}^n = \tau_1^n - 2\Delta x c_0^n. \qquad (4.4.11)$$

The computation of τ_{-1}^n by (4.4.11) has truncation error $O\{(\Delta x)^2\}$ which is consistent with the truncation error of the FTCS finite difference equation (2.4.4) used at interior grid points, provided $s \neq 1/6$.

It is assumed that the solution domain is extended to include the point $(-\Delta x, n\Delta t)$ so that fictitious values such as τ_{-1}^n can be incorporated with the FTCS method in the following way. Substitution of Equation (4.4.11) into Equation (2.4.4) with $j = 0$ gives

$$\tau_0^{n+1} = (1-2s)\tau_0^n + 2s\tau_1^n - 2s\Delta x\ c_0^n. \qquad (4.4.12)$$

The boundary value τ_0^{n+1} is therefore given in terms of the known values τ_0^n, τ_1^n from the previous time level, and the known derivative boundary value c_0^n.

The matching of truncation errors of the finite difference equation which represents the partial differential equation at gridpoints in the interior of the space region and the finite difference equation which represents the conditions applying at gridpoints on the boundary, is the usual criteria taken in order to obtain optimal finite difference solutions to a particular order of accuracy (see Smith, 1969, or Carnahan, Luther and Wilkes, 1969).

Results obtained in this way are better than those obtained using a lower order, and therefore larger, truncation error when approximating the boundary equations.

As an example, consider the thin insulated rod described in Section 2.1. The temperature, denoted by $\bar{\tau}(x,t)$, is initially 0°C throughout the rod. The right hand end of the rod is kept at 0°C for all time and at $t = 0$ a constant heat flux is suddenly imposed at the left hand end. The mathematical description of this problem is given by Equation (2.1.1) subject to the following boundary and initial conditions:

$$\left.\begin{aligned} \partial\bar{\tau}/\partial x|_{(0,t)} &= -1, \quad t > 0, \\ \bar{\tau}(1,t) &= 0, \quad t \geqslant 0, \\ \bar{\tau}(x,0) &= 0, \quad 0 \leqslant x \leqslant 1. \end{aligned}\right\} \qquad (4.4.13)$$

The negative sign on the right side of the first equation indicates that there is a steady flow of heat into the rod at $x = 0$.

The analytical solution to this non-homogeneous problem may be obtained using the method of separation of variables (see Myers, 1971, p. 103) and is

$$\bar{\tau}(x,t) = 1-x-\frac{8}{\pi^2}\sum_{m=1}^{\infty}(2m-1)^{-2}\cos\{(2m-1)\pi x/2\}\exp\{-(2m-1)^2\pi^2\alpha t/4\}. \qquad (4.4.14)$$

This problem has also been solved using the FTCS explicit method, firstly with the forward difference approximation to the spatial derivative in the boundary condition at $x = 0$ as given by Equation (4.4.9) and then with the central difference approximation which is given by Equation (4.4.11). The value of $\bar{\tau}(0.4,8)$ obtained using the analytic solution is 0.090711; the approximate value obtained using the FTCS finite difference method with J=20, s=0.2, and incorporating Equation (4.4. 9), is 0.099872 which has a relative error of 10.1%; the value obtained using the FTCS method with Equation (4.4.11) is 0.090468 which is in error by approximately 0.3%. Clearly, the use of the central difference approximation for the derivative boundary condition at x=0 gives a much more accurate result than using the forward difference approximation.

The relative accuracy achieved using the two approximations to the derivative boundary condition considered is more clearly shown in Figure 4.3, in which the distribution of percentage errors along the bar at time t=4.0 is shown for both schemes. In each case the percentage error introduced at x=0 increases from left to right along the bar. Also, for a fixed s it is found that the relative error decreases at a given position on the bar as time passes. This is due to the values of $\bar{\tau}(x,t)$ increasing as t increases, for a given x, while the errors remain about the same. For instance, using the central difference approximation to the boundary derivative at x=0, the errors at x=0.4 at the end of the fourth and sixteenth seconds are respectively 0.268% and 0.028%.

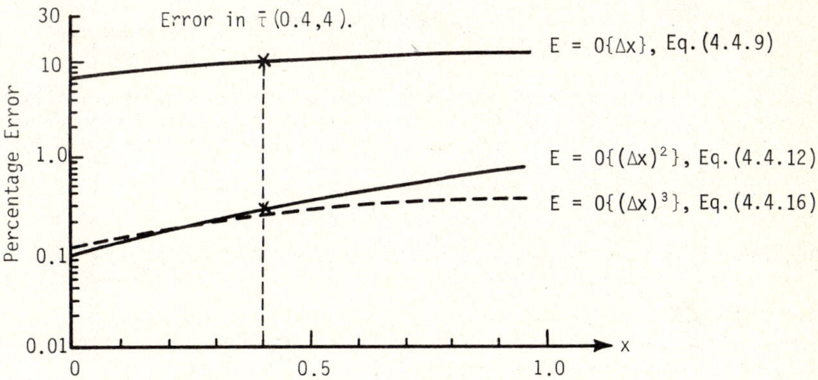

Figure 4.1 : Absolute error on $0 \leqslant x < 1$ using different finite difference approximations to a derivative boundary condition at $x = 0$.

Matching Discretisation Errors

There is another criterion which can be used for matching the accuracy of the finite difference approximations to the conditions given on the boundary and the partial differential equation which applies to the interior of the space region. Although truncation errors may be of the same order of accuracy, such as the finite difference equations (4.4.12) and (2.4.4), both of which have $E_j^n = O\{(\Delta x)^2\}$, the contribution to the discretisation error may be of quite different orders. Calculation of τ_0^{n+1} by means of (4.4.12) contributes an error of $O\{(\Delta x)^3\}$ to the discretisation error at each time step, whereas the calculation of τ_j^{n+1}, j=1(1)J-1, by means of (2.4.4) only contributes an error of $O\{(\Delta x)^4\}$. Since τ_0^{n+1} is used in the calculation of τ_j^{n+2}, j=0 and 1, which are in turn used in the calculation of τ_j^{n+3}, j=0,1 and 2, and so on, then the contributions to the discretisation error at each time step must overall be considered of $O\{(\Delta x)^3\}$.

Finite difference approximations to the boundary equations which produce boundary values with discretisation errors of the same order of magnitude as the discretisation errors of the finite difference equations used at interior grid-points, should give optimal results. Boundary conditions usually involve partial differential equations with derivatives of lower order than those in the partial differential equations which apply to the interior of the region. Therefore the truncation errors associated with the boundary equation approximations may need to be of a higher order than the truncation error associated with finite difference equations used at interior gridpoints if the discretisation errors are to be of the same order.

For instance, consider the relation

$$\left.\frac{\partial\bar{\tau}}{\partial x}\right|_j^n = \frac{-2\bar{\tau}_{j-1}^n - 3\bar{\tau}_j^n + 6\bar{\tau}_{j+1}^n - \bar{\tau}_{j+2}^n}{6\Delta x} + O\{(\Delta x)^3\}, \tag{4.4.15}$$

and the equation (4.4.7). From these we may obtain the finite difference equation for the fictitious value of τ at $(-\Delta x, n\Delta t)$,

$$\tau_{-1}^n = -3/2\tau_0^n + 3\tau_1^n - \tfrac{1}{2}\tau_2^n - 3\Delta x\ c_0^n ,$$

with an $O\{(\Delta x)^4\}$ contribution to the discretisation error at each time step.

Substitution of this value of τ_{-1}^n into Equation (2.4.4) with $j=0$, gives the finite difference equation,

$$\tau_0^{n+1} = (1 - 7s/2)\tau_0^n + 4s\tau_1^n - \tfrac{1}{2}s\tau_2^n - 3s\Delta x\ c_0^n, \tag{4.4.16}$$

with the same contribution to the discretisation error as τ_j^{n+1}, $j=1(1)J-1$, computed using Equation (2.4.4), if $s \neq 1/6$.

Results obtained using Equation (4.4.16) with (2.4.4) for $j=1(1)J-1$ to solve Equation (2.1.1) subject to conditions (4.4.13), are compared with those obtained using the less accurate Equations (4.4.9) and (4.4.12) in Figure 4.1.

Mixed Boundary Conditions

A general example of a mixed boundary condition at $x = 0$ for the heat conduction problem is Equation (4.4.6a), namely

$$a_0^n\bar{\tau}_0^n - b_0^n \left.\frac{\partial\bar{\tau}}{\partial x}\right|_0^n = c_0^n .$$

Approximating this derivative boundary condition using the forward difference equation for the spatial derivative at the $(0, n\Delta t)$ gridpoint gives

$$a_0^n\bar{\tau}_0^n - b_0^n\left(\frac{\bar{\tau}_1^n - \bar{\tau}_0^n}{\Delta x}\right) = c_0^n + O\{\Delta x\} . \tag{4.4.17}$$

Rearranging Equation (4.4.17) yields the finite difference approximation

$$\tau_0^n = (b_0^n\tau_1^n + c_0^n\ \Delta x)/(b_0^n + a_0^n\ \Delta x), \tag{4.4.18}$$

with an $O\{(\Delta x)^2\}$ contribution to the discretisation error at each time step.

Equation (4.4.18) can now be incorporated into the finite difference scheme being used so it covers the whole solution domain including grid-points on the boundary. However, even though the FTCS equation applied at interior grid-points only contributes an error of $O\{(\Delta t)^2, \Delta t(\Delta x)^2\}$ to the discretisation error at each time step, overall the results will be governed by the lower accuracy of the difference equations used at the boundary.

Matching Truncation Errors

To increase the order of accuracy when incorporating the boundary condition (4.4.6) at $x = 0$, the central difference approximation, which has a second order truncation error, may be used for the same derivative. Equation (4.4.7a) then yields the fictitious value of τ^n_{-1} for the point $(-\Delta x, n\Delta t)$,

$$\tau^n_{-1} = \tau^n_1 - (2a^n_0 \Delta x/b^n_0)\tau^n_0 + 2c^n_0 \Delta x/b^n_0, \tag{4.4.19}$$

provided $b^n_0 \neq 0$. This equation contributes an error of $O\{(\Delta x)^3\}$ at each time step. Using this expression for τ^n_{-1} the value of τ at the boundary $x = 0$ can be calculated using any of the finite difference schemes previously described, with $j = 0$. For example, substitution of Equation (4.4.19) into the FTCS equation (2.4.4) with $j = 0$, yields

$$\tau^{n+1}_0 = \{1-2s(1+a^n_0\Delta x/b^n_0)\}\tau^n_0 + 2s\tau^n_1 + 2s\Delta x\ c^n_0/b^n_0\ . \tag{4.4.20}$$

The boundary value τ^{n+1}_0 is therefore represented in terms of values of τ known at the previous time level and the known coefficients a^n_0, b^n_0, c^n_0 of Equation (4.4.6a) with a contribution of $O\{(\Delta x)^3\}$ to the discretisation error at each time step.

When $a_0(t) \equiv 0$ and $b_0(t) \equiv -1$, the mixed boundary condition (4.4.6a) reduces to the derivative boundary condition (4.4.7) and the resulting equations involving τ^{n+1}_0 reduce to the corresponding equations in the derivative boundary case; for example, Equation (4.4.20) reduces to Equation (4.4.12).

At $x = 1$, the relation corresponding to (4.4.20) is obtained from Equation (4.4.6b), namely

$$\tau^{n+1}_J = 2s\tau^n_{J-1} + \{1-2s(1+a^n_J\Delta x/b^n_J\}\tau^n_J + 2s\ x\Delta c^n_J/b^n_J, \tag{4.4.21}$$

$b^n_J \neq 0$. Again, the contribution to the discretisation error is $O\{(\Delta x)^3\}$.

Effect on Stability

Consider the diffusion equation being solved with the mixed boundary conditions (4.4.7) at $x = 0,1$ where the a,b and c's are constants, that is

$$a^n_0 \equiv a_0,\ b^n_0 \equiv b_0,\ c^n_0 \equiv c_0,$$

$$a^n_J \equiv a_J,\ b^n_J \equiv b_J,\ c^n_J \equiv c_J\ .$$

If Equation (4.4.20) is used to find τ^{n+1}_0, the FTCS equation (2.4.4) is used to find τ^{n+1}_j, $j=1(1)J-1$, and Equation (4.4.21) is used to find τ^n_J, then the propagation of the errors ξ^n_j is given by Equation (3.3.4) at interior grid-points,

$$\xi^{n+1}_0 = \{1-2s(1+a_0b_0^{-1}\Delta x)\}\xi^n_0 + 2s\xi^n_1 \tag{4.4.22}$$

at $x = 0$, and

$$\xi^{n+1}_J = 2s\xi^n_{J-1} + \{1-2s(1+a_Jb_J^{-1}\Delta x)\}\xi^n_J \tag{4.4.23}$$

at $x = 1$.

Application of the von Neumann stability analysis to the error equation (3.3.4), assumed to apply to all grid-points $j = 0(1)J$, gives the same stability criterion as that found in Section 3.3, namely

$$s \leqslant \tfrac{1}{2}\ .$$

Application of the analysis to the error equation (4.4.22), assuming it applies to all gridpoints $j = 0(1)J-1$, gives an amplification factor of

$$G = \{1-2s(1+a_0b_0^{-1}\Delta x)\} + 2se^{i\beta}$$

for the propagation of the m-th Fourier component of any error distribution from one time level to the next. Therefore

$$G = \{1-2(1+\gamma)s+2s\cos\beta\} - i(2s\sin\beta),$$

where $\gamma = a_0b_0^{-1}\Delta x > 0$, as a_0, b_0 have the same sign, so that

$$|G|^2-1 = 4s\{[(2+2\gamma+\gamma^2)s-(1+\gamma)] + [1-(2+2\gamma)s]\cos\beta\}. \qquad (4.4.24)$$

Defining the linear function

$$F(\chi) = [1-(2+2\gamma)s]\chi + [(2+2\gamma+\gamma^2)s-(1+\gamma)], \qquad (4.4.25)$$

then for stability we require $F(\chi) \leqslant 0$ for all χ in the closed interval $<-1,1>$.

If $1-(2+2\gamma)s \geqslant 0$, so that $s \leqslant 1/(2+2\gamma)$, then $F(\chi)$ will be non-positive on $-1 \leqslant \chi \leqslant 1$ so long as $F(1) \leqslant 0$, that is if

$$s \leqslant 1/\gamma.$$

Since $1/\gamma > 1/(2+2\gamma)$ for all $\gamma > 0$, it follows that $F(\chi) \leqslant 0$ on $<-1,1>$ if

$$s \leqslant 1/(2+2\gamma). \qquad (4.4.26)$$

If $1-(2+2\gamma)s \leqslant 0$, so that $s \geqslant 1/(2+2\gamma)$, then $F(\chi) \leqslant 0$ on $-1 \leqslant \chi \leqslant 1$ so long as $F(-1) \leqslant 0$, that is if

$$s \leqslant 1/(2+\gamma). \qquad (4.4.27)$$

Since for $\gamma > 0$ we have $1/(2+\gamma) > 1/2(1+\gamma)$, then combining (4.4.26) and (4.4.27) we find $F(\chi) \leqslant 0$ on $<-1,1>$ provided

$$0 < s \leqslant 1/(2+\gamma).$$

The finite difference equation (4.4.20) is therefore locally stable if

$$s \leqslant 1/(2+a_0b_0^{-1}\Delta x). \qquad (4.4.28a)$$

Similarly, application of the von Neumann method indicates that Equation (4.4.21) is locally stable if

$$s \leqslant 1/(2+a_Jb_J^{-1}\Delta x). \qquad (4.4.28b)$$

The stability of the FTCS finite difference method applied at interior grid points, with central difference approximations used for the mixed boundary conditions with constant coefficients, therefore requires

$$0 < s \leqslant \mathrm{Min}\{1/(2+a_0b_0^{-1}\Delta x),\ 1/(2+a_Jb_J^{-1}\Delta x)\}. \qquad (4.4.29)$$

4.5 Initial Conditions

Starting procedures

Given the initial conditions at $t = 0$ there is no difficulty in starting the FTCS method, since going from the time level $n = 0$ to the level $n = 1$ is explicitly determined by Equation (2.4.4). Because the FTCS method approximates the diffusion equation (2.1.1) with a truncation error of $O\{\Delta t,(\Delta x)^2\}$, the DuFort-Frankel method, which has a second-order truncation error in Δt is preferable (see Section 4.1). However, this explicit method is a three-level method; it calculates a value of τ at one time level using values from the previous two time levels. Therefore there is a problem with starting this method, because initial conditions are only given at the time level $t = 0$ and there are no values of $\bar{\tau}$ available for $t = -\Delta t$, the previous time level. To overcome this difficulty the FTCS method can be used to obtain values at the time level $n = 1$ from those at $n = 0$. Then, by using values at both these time levels, the DuFort-Frankel method can be used to continue the process. The FTCS method contributes an error of $O\{(\Delta t)^2,\Delta t(\Delta x)^2\}$ to the discretisation error, if $s \neq 1/6$, at the first time step. Thereafter, application of the DuFort-Frankel method contributes the smaller error of $O\{(\Delta t)^3,\Delta t(\Delta x)^2\}$ at each time step.

The classical implicit method and the Crank-Nicolson method are procedures by which values at any time level are computed in terms of known values at only one previous time level, so they require no special starting procedures. At each time step they contribute an error of $O\{(\Delta t)^3,\Delta t(\Delta x)^2\}$ to the discretisation error at any time level. They may also be used for the first time step to begin the DuFort-Frankel method, so it has a discretisation error of $O\{(\Delta t)^3\}$ in time at every time step.

However, not all implicit methods are "two-level" methods such as these. Consider the following finite difference formulation (Richtmeyer, 1957) for the diffusion equation, in which a weighted average of finite difference approximations used for the time derivative $\partial\bar{\tau}/\partial t$.

Since for $0 < \theta \leqslant 1$,

$$\bar{\tau}_j^n = (1+\theta)\bar{\tau}_j^n - \theta\bar{\tau}_j^{n-1} + O\{\Delta t\}, \tag{4.5.1}$$

then the diffusion equation at the $(j,n+1)$ grid-point, namely

$$\left.\frac{\partial\bar{\tau}}{\partial t}\right|_j^{n+1} = \alpha\left.\frac{\partial^2\bar{\tau}}{\partial x^2}\right|_j^{n+1},$$

may be written

$$(1+\theta)\left.\frac{\partial\bar{\tau}}{\partial t}\right|_j^{n+1} - \theta\left.\frac{\partial\bar{\tau}}{\partial t}\right|_j^{n} + O\{\Delta t\} = \alpha\left.\frac{\partial^2\bar{\tau}}{\partial x^2}\right|_j^{n+1}. \tag{4.5.2}$$

Using the backward difference forms for each of the time derivatives on the left side of this equation and the centred difference form for the second-order space derivative on the right side, one obtains

$$\begin{aligned}(1+\theta)\left(\frac{\bar{\tau}_j^{n+1} - \bar{\tau}_j^n}{\Delta t}\right) - \theta\left(\frac{\bar{\tau}_j^n - \bar{\tau}_j^{n-1}}{\Delta t}\right) + O\{\Delta t\}\\ = \alpha\left\{\frac{\bar{\tau}_{j+1}^{n+1} - 2\bar{\tau}_j^{n+1} + \bar{\tau}_j^{n+1}}{(\Delta x)^2}\right\} + O\{(\Delta x)^2\}.\end{aligned} \tag{4.5.3}$$

Dropping the terms of $O\{\Delta t,(\Delta x)^2\}$ gives a finite difference approximation to (4.5.3) which has a truncation error of that order. Rearrangement then gives the implicit "three-level" equation

$$-s\tau_{j-1}^{n+1} + (1+2s+\theta)\tau_j^{n+1} - s\tau_{j+1}^{n+1} = (2\theta+1)\tau_j^n - \theta\tau_j^{n-1}, \tag{4.5.4}$$

which, in general, contributes an error of $O\{(\Delta t)^2,\Delta t(\Delta x)^2\}$ to the discretisation error at each time step. The equation is also stable for all $s > 0$.

Richtmeyer has shown that this equation is extremely useful when dealing with rapidly varying or discontinuous initial data because it damps the resulting spurious short wave-length components very rapidly.

Choosing $\theta = 0$ yields the classical implicit equation (4.2.2b); choosing $\theta = \frac{1}{2}$ gives the formula

$$-s\tau_{j-1}^{n+1} + (3/2 + 2s)\tau_j^{n+1} - s\tau_{j+1}^{n+1} = 2\tau_j^n - 1/2\tau_j^{n-1} , \quad (4.5.5)$$

which is a special case with greater accuracy since it has truncation error $O\{(\Delta t)^2,(\Delta x)^2\}$ and contributes an error of only $O\{(\Delta t)^3,\Delta t(\Delta x)^2\}$ to the approximation for $\bar{\tau}_j^n$ at each time step.

Once again, because it is a three-level finite difference equation, there is a problem when starting the method. However, Equation (4.5.4) can be used after applying the FTCS method at the first time step and there is no loss of accuracy, unless $\theta = \frac{1}{2}$, when a more accurate method such as that due to Crank-Nicolson could be used at the first time step to retain an overall truncation error of $O\{(\Delta t)^2,(\Delta x)^2\}$.

Initial-value, boundary-value singularities

As noted in Section 2.1, singularities in the value of $\bar{\tau}$ may be encountered at the intersection of lines where initial and boundary values are specified in the solution domain. If the singularity is ignored in the application of a finite difference method, which was the case for the example involving heat diffusion considered in Section 2.4, then values of $\partial^4\bar{\tau}/\partial x^4$ and $\partial^2\bar{\tau}/\partial t^2$ are relatively very large near $x = 0$ and $x = 1$, so that, for a particular pair of values of Δx and Δt the truncation error of the finite difference approximation is much larger near the singularity than near the centre of the rod. As a result, the method is less accurate near the ends of the rod, and much smaller values of Δx may be required there than in the centre of the rod. The use of variable size grids to permit this is described in Section 8.

Application of special procedures for a short initial time

A better approach, however, is to employ an asymptotic expansion near the singularity, which usually results in the application of a simple analytic expression for the solution which is valid near the singularity.

Consider the example of the insulated rod at an initial temperature of zero with its ends suddenly heated to $\bar{\tau} = 1$, as described in Section 2.1. Due to symmetry, the region of interest is the interval $0 \leqslant x \leqslant \frac{1}{2}$. Initially only the region near $x = 0$ "feels" the sudden temperature disturbance so that, to start with, the problem is the same as that of a semi-infinite insulated rod initially at $0^{\circ}C$ and suddenly heated to $\bar{\tau} = 1$ at $x = 0$. Thus, for a short time the temperature may be considered to satisfy the diffusion equation (2.1.1) with initial condition

$$\bar{\tau}_L(x,0) = 0, \quad 0 < x < \infty, \quad (4.5.6a)$$

and boundary conditions

$$\bar{\tau}_L(0,t) = 1, \quad \bar{\tau}_L(\infty,t) = 0, \quad t \geqslant 0. \quad (4.5.6b)$$

With these conditions the diffusion equation yields the solution

$$\bar{\tau}_L(x,t) = \text{erfc}\{x/2\sqrt{\alpha t}\}, \quad (4.5.7)$$

where erfc(x) is the complementary error function (see Myers, 1971). This solution is valid for the original diffusion problem until the centre of the rod, at $x = 1/2$, begins to experience a significant increase in temperature. At this point in time, approximately $t = (100\alpha)^{-1}$, the finite difference method can be started using values for τ obtained by application of Equation (4.5.7) for $0 < x \leqslant \frac{1}{2}$, and of its mirror image in $x = \frac{1}{2}$, namely

$$\bar{\tau}_R(x,t) = \operatorname{erfc}\{(1-x)/2\sqrt{\alpha t}\}, \tag{4.5.8}$$

for $\frac{1}{2} < x < 1$.

Alternatively, we may use a very fine uniform grid to start with, and when $\partial^4\bar{\tau}/\partial x^4$ and $\partial^2\bar{\tau}/\partial t^2$ reduce to reasonable values, revert to a coarse grid. It may be noted that with the coarse grid, large time spacings may be used with explicit techniques such as the FTCS method.

5. THE ONE-DIMENSIONAL TRANSPORT EQUATION

5.1 The Convection Equation

The one-dimensional transport equation

$$\frac{\partial\bar{\tau}}{\partial t} + u\frac{\partial\bar{\tau}}{\partial x} = \alpha\frac{\partial^2\bar{\tau}}{\partial x^2}, \tag{5.1.1}$$

governs the change in the scalar property $\bar{\tau}$ caused by diffusion and convection acting together in a fluid moving with a speed u parallel to the x-axis. The term $u\,\partial\bar{\tau}/\partial x$ is the convection term and $\alpha\,\partial^2\bar{\tau}/\partial x^2$ is the diffusion term. Previously, in Sections 2,3 and 4, the pure diffusion problem was considered, namely Equation (2.1.1)

$$\frac{\partial\bar{\tau}}{\partial t} = \alpha\frac{\partial^2\bar{\tau}}{\partial x^2}.$$

This equation is the same as (5.1.1) with $u = 0$; that is, the diffusion is taking place in a stationary fluid or a solid.

In the following it will be assumed that the fluid moves with a *constant* speed $u > 0$, that is, in the positive x-direction.

Pure convection is described by the partial differential equation

$$\frac{\partial\bar{\tau}}{\partial t} + u\frac{\partial\bar{\tau}}{\partial x} = 0. \tag{5.1.2}$$

The general solution of Equation (5.1.2) is, for constant u,

$$\bar{\tau}(x,t) = f(x-ut) \tag{5.1.3}$$

where

$$\bar{\tau}(x,0) = f(x) \tag{5.1.4}$$

is the initial distribution of $\bar{\tau}$ for $-\infty < x < \infty$. That is, the initial distribution of $\bar{\tau}$ along the x-axis is translated without change as it travels along the axis at the constant speed u.

Evaluated at the (j,n) grid-point, the convection equation becomes

$$\left.\frac{\partial\bar{\tau}}{\partial t}\right|_j^n + u\left.\frac{\partial\bar{\tau}}{\partial x}\right|_j^n = 0. \tag{5.1.5}$$

Using the forward difference form (2.3.7a) for the time derivative and the central difference form (2.3.18) for the space derivative, Equation (5.1.5) becomes

$$\frac{\bar{\tau}_j^{n+1} - \bar{\tau}_j^n}{\Delta t} + O\{\Delta t\} + u\{\frac{\bar{\tau}_{j+1}^n - \bar{\tau}_{j-1}^n}{2\Delta x} + O\{(\Delta x)^2\}\} = 0. \tag{5.1.6}$$

Rearranging this equation gives

$$\bar{\tau}_j^{n+1} = \bar{\tau}_j^n - \tfrac{1}{2}c(\bar{\tau}_{j+1}^n - \bar{\tau}_{j-1}^n) + O\{(\Delta t)^2,\Delta t(\Delta x)^2\}, \tag{5.1.7}$$

where the constant

$$c = u\Delta t/\Delta x \tag{5.1.8}$$

is called the Courant number.

The explicit finite difference equation

$$\tau_j^{n+1} = \tau_j^n - \tfrac{1}{2}c(\tau_{j+1}^n - \tau_{j-1}^n), \tag{5.1.9}$$

which approximates (5.1.5) with truncation error $O\{\Delta t,(\Delta x)^2\}$, may then be used to compute approximate solutions to the convection Equation (5.1.2).

If Equation (5.1.9) is analysed for stability using the von Neumann method, it is found that the amplification factor for the mth Fourier component of the initial error distribution is

$$G = 1 - ic \sin \beta. \tag{5.1.10}$$

Hence the gain $|G|$ is given by

$$|G|^2 = 1 + (c \sin \beta)^2 \geqslant 1, \tag{5.1.11}$$

where the superscripts indicate powers. This result implies that Equation (5.1.9) is unstable for all values of Δx and Δt chosen. This method (which is analogous to the FTCS method for the pure diffusion problem) is therefore of no practical use.

An alternative method which uses *upwind differencing* instead of central differencing for the spatial derivative in (5.1.5) was first used by Courant, Isaacson and Rees (1952) in order to overcome the instability of Equation (5.1.9). Since then, the method has been used under various names with different rationales. For example, for some time meteorologists have known of the stabilising effect of "upwind" (Forsythe and Wasow, 1960) or "weather" (Frankel, 1956) differencing and have applied it to incompressible fluid flow problems. Mathematicians generally refer to the finite difference equations as having "positive coefficients" (Forsythe and Wasow, 1960).

In the upwind differencing method a one sided, rather than space centred, differencing is used for the space derivative in the convection term $u\, \partial\bar{\tau}/\partial x$, the direction of the difference being "upwind". That is, the backward-space finite difference form (2.3.17) is used for $\partial\bar{\tau}/\partial x$ at the (j,n) grid-point when $u > 0$. This has the following physical basis. When determining the value of τ_j^{n+1} from known values of τ at the nth time level, it must be remembered that the fluid flow carries information from the position x_{j-1} to x_j, so a backward difference form which involves values of τ at the $(j-1,n)$ and the (j,n) grid-points is chosen for the spatial derivative. To use a forward difference approximation would be unrealistic, since information cannot be carried from the position x_{j+1} to x_j in such a case. However, this choice would be appropriate if u was negative.

For $u > 0$, the forward time, backward space form of Equation (5.1.5) is

$$\frac{\bar{\tau}_j^{n+1} - \bar{\tau}_j^n}{\Delta t} + O\{\Delta t\} + u\left(\frac{\bar{\tau}_j^n - \bar{\tau}_{j-1}^n}{\Delta x} + O\{\Delta x\}\right) = 0. \tag{5.1.12}$$

This equation can be rewritten

$$\bar{\tau}_j^{n+1} = (1-c)\bar{\tau}_j^n + c\bar{\tau}_{j-1}^n + O\{(\Delta t)^2,\Delta t\Delta x\}, \tag{5.1.13}$$

which leads to the explicit finite difference equation for solving the convection equation,

$$\tau_j^{n+1} = (1-c)\tau_j^n + c\tau_{j-1}^n, \quad u > 0, \tag{5.1.14}$$

with an error of $O\{(\Delta t)^2, \Delta t\Delta x\}$ contributed to the discretisation error e_j^n at each of the n time steps.

Therefore, when u is a positive constant, the errors ξ_j^n propagate according to the relation

$$\xi_j^{n+1} = (1-c)\xi_j^n + c\xi_{j-1}^n . \tag{5.1.15}$$

Application of the von Neumann method of stability analysis to this error equation yields the amplification factor

$$G = 1 - c(1-\cos\beta) - ic\sin\beta, \tag{5.1.16}$$

so the gain $|G|$ for the propagation of the mth Fourier mode of the error distribution from one time level to the next is found by solving

$$|G|^2 = 1 - 4c(1-c)\sin^2(\beta/2). \tag{5.1.17}$$

When $c \leqslant 1$, $|G|^2$ has a maximum value of 1 when $\sin^2(\beta/2) = 0$, since the second term on the right hand side of Equation (5.1.17) is always positive. The method is then stable. When $c > 1$, $|G|^2$ has a maximum value of $1 + 4c(c-1) = (2c-1)^2 > 1$ when $\sin^2(\beta/2) = 1$, and the method is unstable. The finite difference Equation (5.1.14) is therefore stable so long as the Courant number is not greater than 1, that is if

$$c \leqslant 1. \tag{5.1.18}$$

The relation (5.1.18) is usually referred to as the Courant-Friedrichs-Lewy (CFL) condition. This condition applies generally to explicit finite difference approximations to hyperbolic partial differential equations. It requires $u\Delta t \leqslant \Delta x$, which means that the fluid should not travel more than one grid spacing in the x-direction in one time step.

If $c = 1$ when $u > 0$, Equation (5.1.14) becomes

$$\tau_j^{n+1} = \tau_{j-1}^n, \tag{5.1.19}$$

which is the exact solution to the convection equation. It is the exact solution, since by (5.1.3)

$$\bar{\tau}(x_{j-1}, t_n) = f((j-1)\Delta x - un\Delta t) \tag{5.1.20}$$

and

$$\begin{aligned} \bar{\tau}(x_j, t_{n+1}) &= f(j\Delta x - u(n+1)\Delta t) \\ &= f((j-1)\Delta x + \Delta x - un\Delta t - u\Delta t) \\ &= \bar{\tau}(x_{j-1}, t_n), \end{aligned} \tag{5.1.21}$$

as $c = 1$ implies that $\Delta x = u\Delta t$.

Further for $c < 1$, the method introduces an artificial damping, the values of τ_j^n being reduced in magnitude at successive time levels in the same way as the errors ξ_j^n are reduced, since Equations (5.1.14) and (5.1.15) are analogous.

The amplification factor G gives a quantitative estimate of the damping and phase shift produced in any infinitely long wave train propagating according to the partial differential equation (5.1.2). Consider the wave train given initially as

$$\bar{\tau}(x,0) = e^{i(\pi m x)} \quad \text{(real part intended)}, \tag{5.1.22}$$

where $e^{i(\pi m x)} = \exp(i\pi m x)$. This wave train has wave number πm and wavelength

$$\lambda = 2/m. \tag{5.1.23}$$

Therefore, at grid-points along the initial time level, the values of $\bar{\tau}$ are given by

$$\bar{\tau}_j^0 = e^{i\beta j}, \tag{5.1.24}$$

where, as in Section 3.3,

$$\beta = \pi m \Delta x. \tag{5.1.25}$$

An exact solution to the finite difference Equation (5.1.14) may be found in the variables separable form

$$\tau_j^n = (G)^n e^{i\beta j}, \tag{5.1.26}$$

which clearly satisfies the initial condition (5.1.24). Solving Equation (5.1.14) for G yields the same result as (5.1.16), namely

$$G = 1 - c(1-\cos\beta) - ic\sin\beta.$$

This happens because the form assumed for τ_j^n in (5.1.26) is the same as the error form $\xi_j^n = (G)^n e^{i\beta j}$ used in the von Neumann method, and the error equation (5.1.15) is the same as the homogeneous finite difference equation (5.1.14).

For the stable case, with $c \leqslant 1$, we may write the complex number G in polar form as

$$G = |G|\, e^{-i\psi} \tag{5.1.27}$$

where $|G|$ is given by (5.1.17) and

$$\psi = -\arg\{G\} = \arctan\left\{\frac{c\sin\beta}{1-c(1-\cos\beta)}\right\}. \tag{5.1.28}$$

Both $|G|$ and ψ are functions of β and c. Therefore, using the upwind differencing method, the exact finite difference solution to (5.1.2) with initial condition (5.1.22) is obtained by substituting (5.1.27) into (5.1.26). Using (5.1.25), the following finite difference solution is obtained,

$$\tau_j^n = |G|^n e^{i\pi m(j\Delta x - n\psi/\pi m)}, \tag{5.1.29}$$

where the superscripts on the right side represent powers.

However, by (5.1.3), the exact solution to the partial differential equation (5.1.2) with initial condition (5.1.22) is

$$\bar{\tau}(x,t) = e^{i\pi m(x-ut)},$$

so that the exact solution at the (j,n) grid-point is

$$\bar{\tau}(x_j, t_n) = e^{i\pi m(j\Delta x - n\, u\Delta t)} . \tag{5.1.30}$$

Clearly, (5.1.29) differs from the exact solution (5.1.30) by a factor of $|G|^n$ in amplitude. The wave speed has also changed from u to u_N, since $\psi/\pi m$ in (5.1.29) may be written as $u_N\Delta t$, similar to (5.1.30), where

$$u_N = \psi/(\pi m\Delta t). \tag{5.1.31}$$

Substitution of $\Delta t = c\Delta x/u$, from (5.1.8), into this result then yields the ratio of the wave speed of the numerical solution relative to the true wave speed: that is,

$$\frac{u_N}{u} = \frac{\psi}{\pi mc\Delta x} = \frac{\psi}{\beta c} . \tag{5.1.32}$$

Clearly, $|G|$ indicates the damping of an infinite wave-train produced in one time step by any two-level finite difference method used to solve (5.1.2), and $\psi = -\arg\{G\}$ indicates the error in the wave speed by application of Equation (5.1.32). In an ideal numerical method, the value of $|G|$ should be 1 for all β and c and the argument of G should be given by

$$\arg\{G\} = -\beta c, \tag{5.1.33}$$

when it follows that $u_N/u = 1$.

For $c = 1$ with upwind differencing, Equation (5.1.17) gives $|G| = 1$, so the method propagates a wave train with the correct amplitude. Also, Equation (5.1.28) gives $\arg\{G\} = -\psi(\beta,1) = -\arctan\{\tan\beta\} = -\beta$, so that (5.1.33) is satisfied. The method therefore propagates a wave train without change of wave speed.

Suppose the wavelength λ of a travelling wave is large compared with the grid spacing Δx. Let $\lambda = 20\Delta x$, when $m = 2/\lambda$ from (5.1.23), and it follows that $\beta = \pi m\Delta x = 2\pi\Delta x/\lambda = \pi/10$. If c is a little less than 1, say $c = 0.8$, then from (5.1.17) $|G| = 0.9921$ and from (5.1.28) $\psi = 0.0802\pi$, so that from (5.1.32) $u_N/u = 1.002$. In this case, the numerical method transmits the wave in slightly "damped" form, that is with its amplitude reduced to 0.9921 of its value at the previous time step. The method also propagates the wave train with a slightly increased speed, namely 1.002 of its true value.

If on the other hand, the wavelength is represented by only three grid-spacings, so that $\lambda = 3\Delta x$, then $\beta = 2\pi/3$. If $c = 0.8$, it is found that $|G| = 0.7211$ and $u_N = 1.105u$. Clearly there is considerable damping of the wave in this case since the amplitude decreases to 0.7211 of its value as the upwind difference method progresses through one time step. The wave speed is also increased, by the factor 1.105. These results indicate that with the upwind difference method the Courant number should be made as near to 1 as practicable and each wavelength should be represented by at least 20 grid-spacings.

The consistency of the difference equation (5.1.14) with the convection equation (5.1.2) follows from the manner in which it was derived from (5.1.12): the truncation error is $O\{\Delta t,\Delta x\}$ and tends to zero as $\Delta t \to 0$, $\Delta x \to 0$. Let us now formally check the consistency by deriving the actual truncation error E_j^n, in order to determine if there are ways of choosing the Courant number c so that E_j^n and therefore the contribution to the discretisation error at each time step, $\Delta t\, E_j^n$, can be minimised.

Substitution of the exact solution $\bar{\tau}$ for τ in (5.1.14) and expanding each term as a Taylor series evaluated at the (j,n) grid-point, yields the equivalent partial differential equation

$$\left.\frac{\partial\bar{\tau}}{\partial t}\right|_j^n + u\left.\frac{\partial\bar{\tau}}{\partial x}\right|_j^n = \tfrac{1}{2}u\Delta x\left.\frac{\partial^2\bar{\tau}}{\partial x^2}\right|_j^n - \tfrac{1}{2}\Delta t\left.\frac{\partial^2\bar{\tau}}{\partial t^2}\right|_j^n + O\{(\Delta t)^2,(\Delta x)^2\}. \tag{5.1.34}$$

Clearly, this is consistent with the pure convection equation (5.1.2) as $\Delta t \to 0$, $\Delta x \to 0$. However, in any practical situation $\tfrac{1}{2}u\Delta x$ is finite, so there is a diffusion type term, namely $\tfrac{1}{2}u\Delta x\ \partial^2\bar{\tau}/\partial x^2|_j^n$ in Equation (5.1.34). This is not the only such term as the following shows. Since $\bar{\tau}$ satisfies (5.1.2), then

$$\frac{\partial\bar{\tau}}{\partial t} = -u\frac{\partial\bar{\tau}}{\partial x} ,$$

so that

$$\frac{\partial^2\bar{\tau}}{\partial t^2} = u^2\frac{\partial^2\bar{\tau}}{\partial x^2} . \tag{5.1.35}$$

Substitution of (5.1.35) into Equation (5.1.34) yields

$$\left.\frac{\partial\bar{\tau}}{\partial t}\right|_j^n + u\left.\frac{\partial\bar{\tau}}{\partial x}\right|_j^n = \alpha'\left.\frac{\partial^2\bar{\tau}}{\partial x^2}\right|_j^n + O\{(\Delta t)^2,(\Delta x)^2\} , \tag{5.1.36}$$

where

$$\alpha' = \tfrac{1}{2}u\Delta x(1-c). \tag{5.1.37}$$

Therefore the upwind differencing method has introduced an artificial (or numerical) diffusion of $O\{\Delta t,\Delta x\}$ because of the non-physical coefficient α' of $\partial^2\bar{\tau}/\partial x^2$. For a given value of $c \neq 1$, large values of the speed u and of the space step Δx result in large values of α' and hence considerable artificial diffusion.

An example which illustrates this artificial diffusion and damping is shown in Figure 5.1. The function $\bar{\tau}(x,t)$ which is governed by the convection equation (5.1.2) is initially given by

$$\bar{\tau}(x,0) = \begin{cases} 20x & , \quad 0 \leqslant x \leqslant 0.05, \\ 20(0.1-x) & , \quad 0.05 \leqslant x \leqslant 0.1, \\ 0 & , \quad 0.1 \leqslant \quad \leqslant 1.0, \end{cases}$$

and satisfies boundary conditions

$$\bar{\tau}(0,t) = \bar{\tau}(1,t) = 0 \quad , \quad t \geqslant 0 .$$

On the interval $0 \leqslant x \leqslant 1$, the exact solution to this problem up to $t = 0.9u^{-1}$ is

$$\bar{\tau}(x,t) = \begin{cases} 0 & , \quad 0 \leqslant x \leqslant ut, \\ 20(x-ut) & , \quad ut \leqslant x \leqslant ut+0.05, \\ 20(0.1-x+ut), & \quad ut+0.05 \leqslant x \leqslant ut+0.1, \\ 0 & , \quad ut+0.1 \leqslant x \leqslant 1.0 \end{cases}$$

That is, the triangular pulse with a peak at $x = 0.05$ at time $t = 0$ in Figure 5.1 simply moves to the right with speed u. In the example considered, u was taken to be 0.1. Clearly, when $c = 1$, the correct solution was obtained using the upwind finite difference method. However, for other values of $c < 1$,

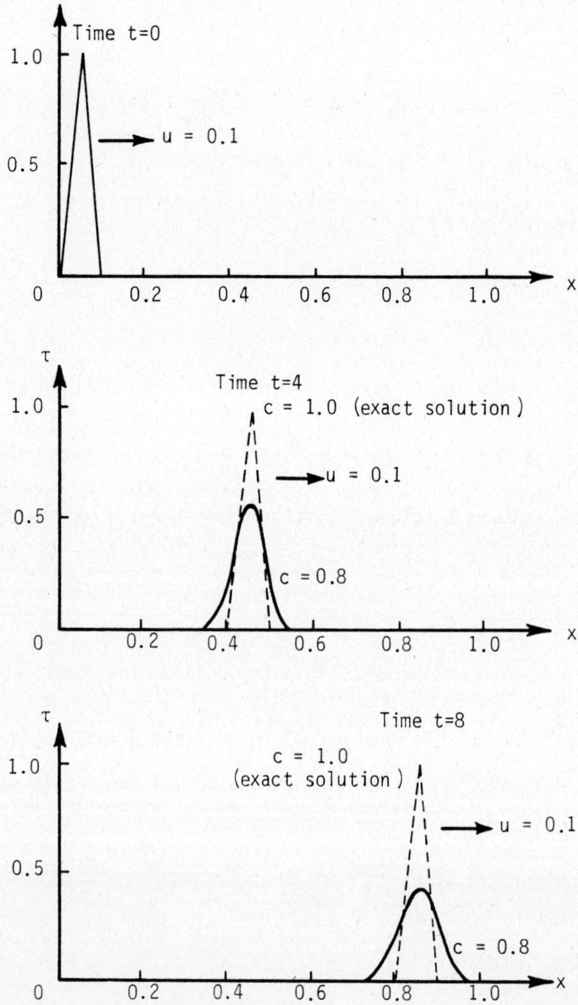

Figure 5.1: Solutions to the convection equation using upstream differencing with the Courant numbers $c = 1$ and 0.8.

this numerical method has the peak moving to the right with almost the correct speed, but it has undergone diffusion and damping. That is, the triangular pulse has spread out at its base and its height has decreased as it moved downstream; this effect increases as c gets smaller.

Another finite difference method of solving the convection equation is obtained if the central difference form (2.3.9a) is used for the time derivative in Equation (5.1.5) with the central difference form (2.3.18) for the space derivative, giving

$$\frac{\bar{\tau}_j^{n+1} - \bar{\tau}_j^{n-1}}{2\Delta t} + O\{(\Delta t)^2\} + u\left(\frac{\bar{\tau}_{j+1}^{n} - \bar{\tau}_{j-1}^{n}}{2\Delta x} + O\{(\Delta x)^2\}\right) = 0. \tag{5.1.38}$$

Rearrangement of this equation then yields

$$\bar{\tau}_j^{n+1} = \bar{\tau}_j^{n-1} + c(\bar{\tau}_{j-1}^{n} - \bar{\tau}_{j+1}^{n}) + O\{(\Delta t^3, \Delta t(\Delta x)^2\}, \tag{5.1.39}$$

which can be approximated by the three-level explicit finite difference equation

$$\tau_j^{n+1} = \tau_j^{n-1} + c(\tau_{j-1}^{n} - \tau_{j+1}^{n}), \tag{5.1.40}$$

with a contribution of $O\{(\Delta t)^3, \Delta t(\Delta x)^2\}$ to the discretisation error at each time step. This is much smaller than the contribution of $O\{(\Delta t)^2, \Delta t\Delta x\}$ to the discretisation error for each application of the upwind differencing method.

Because the formula (5.1.40) determines the value of τ_j^{n+1} from the corresponding value at τ_j^{n-1} by adding a change computed from values at the nth time level, this difference scheme is often called the *leapfrog method*.

The stability of (5.1.40) can be established using the von Neumann method. As the method proceeds from one time level to the next, the amplification factor G of the mth Fourier mode of the error distribution is found to satisfy

$$(G)^2 + (2ic \sin\beta)G - 1 = 0. \tag{5.1.41}$$

Solving the quadratic equation (5.1.41) gives

$$G_1, G_2 = -ic \sin\beta \pm \sqrt{1-(c \sin\beta)^2}\ . \tag{5.1.42}$$

If $c \leqslant 1$, then the second term on the right is real, and

$$|G_1|^2 = |G_2|^2 = (c \sin\beta)^2 + \{1-(c \sin\beta)^2\} = 1 \tag{5.1.43}$$

for all β. If $c > 1$, then for some β we have $c \sin\beta > 1$, when the quantity under the square root sign is negative and

$$G_1, G_2 = -i\{c \sin\beta \pm \sqrt{(c \sin\beta)^2 - 1}\}. \tag{5.1.44}$$

At least one of these values G_1, G_2 has a modulus greater than 1, so the corresponding Fourier modes of the error distribution magnify exponentially in modulus as values of τ at successive time levels are computed. This results in computational instability. The leap-frog method is therefore stable only if the Courant number $c \leqslant 1$.

As was seen with the upwind difference method for solving the convection equation, the amplification factor G of the von Neumann method gives more information than just a qualitative determination of stability of the finite difference equation. Because τ propagates according to the same difference equation as the errors, then seeking a solution to (5.1.40) with $c \leqslant 1$ and initial condition $\tau_j^0 = e^{i\beta j}$, in the variables separable form

$$\tau_j^n = (G)^n e^{i\beta j} \quad \text{(real part intended)},$$

it is found that G has the two possible values G_1, G_2 given by Equation (5.1.42). In its most general form, the exact solution to the leap-frog finite difference equation is therefore

$$\tau_j^n = M(G_1)^n e^{i\beta j} + E(G_2)^n e^{i\beta j}, \tag{5.1.45}$$

where M and E are arbitrary constants.

Firstly, substitution of $n = 0$ in (5.1.45) gives the initial condition, so that

$$(M+E)e^{i\beta j} = e^{i\beta j}$$

whence

$$M = 1 - E. \tag{5.1.46}$$

Also, for $c \leqslant 1$, $|G_1| = |G_2| = 1$, and $\arg\{G_1\} = -\psi$, $\arg\{G_2\} = \pi + \psi$, where

$$\psi = \arcsin\{c \sin\beta\}. \tag{5.1.47}$$

Therefore

$$\tau_j^n = (1-E)(e^{-i\psi})^n e^{i\beta j} + E(e^{i(\pi+\psi)})^n e^{i\beta j}, \tag{5.1.48}$$

and substitution of $\beta = \pi m\Delta x$, $e^{i\pi} = -1$, yields

$$\tau_j^n = (1-E)e^{i\pi m(j\Delta x - n\psi/\pi m)} + (-1)^n E e^{i\pi m(j\Delta x + n\psi/\pi m)} . \tag{5.1.49}$$

The real part of (5.1.49) is the exact solution of the leap-frog difference equation on the infinite interval $-\infty < x < \infty$ with initial condition $\tau_j^0 = e^{i\pi mj\Delta x}$, and consists of two waves rather than the one wave (5.1.30), namely

$$\bar{\tau}_j^n = e^{i\pi m(j\Delta x - nu\Delta t)},$$

which is the exact solution of the convection equation (5.1.2) with the given initial condition. The two waves occur in the solution of the leap-frog method because it involves three time levels, so the corresponding difference equation is second order.

Comparison of the exact finite difference solution (5.1.49) and the exact solution (5.1.30) of the convection equation shows that the amplitude of the first wave in the numerical solution is $(1-E)$, that of the second wave is E, while the exact solution to the convection equation has an amplitude of 1. Secondly, the speed of the first numerical wave compared with the true wave speed u is given by the ratio $\psi/(\beta c)$ (see Equation (5.1.32)), whereas that of the second wave is $-\psi/(\beta c)$; the first wave moves in the positive x-direction, the second wave moves in the negative x-direction. Thirdly, the second numerical wave undergoes a phase change every time step, because of the factor $(-1)^n$.

When $c = 1$, the speed of the first wave is the same as that of the true solution because

$$\psi/\beta = \beta^{-1}\arcsin(\sin\beta) = 1, \quad \text{for all } \beta.$$

For this reason, the first wave in the finite difference solution (5.1.49) is referred to as the *physical mode*. The second wave in the finite difference solution travels in the opposite direction to the true solution and changes phase at every time step. Because it has no counterpart in the exact solution and arises only from the finite difference method used, it is called the *computational mode*.

The same as any of the three-level schemes used to solve the diffusion equation, a special technique must be used to start this method. With $n = 0$, τ_j^1 cannot be computed using (5.1.40) since the values of τ_j^{-1} are not known.

For instance, any two-level method may be used for the first time step. Suppose it is the upwind differençing scheme (5.1.14) with $c \leqslant 1$, which gives for the initial condition $\tau_j^0 = e^{i\beta j}$, the numerical solution

$$\tau_j^1 = Ge^{i\beta j}, \tag{5.1.50}$$

namely (5.1.26) with $n = 1$, where the amplification factor G is given by (5.1.16), namely

$$G = 1 - c(1-\cos\beta) - ic \sin \beta.$$

After the first time step the leap-frog method (5.1.40) is used, so that (5.1.49) with $n = 1$ must have the same value as (5.1.50). Thus

$$(1-E)e^{i\pi mj\Delta x-i\psi} - Ee^{i\pi mj\Delta x+i\psi}$$
$$= \{1-c + c \cos\beta - ic \sin\beta\}e^{i\pi mj\Delta x}, \tag{5.1.51}$$

where ψ for the leap-frog scheme is given by (5.1.47). Division by $e^{i\pi mj\Delta x}$, and rearrangement, gives

$$(1-2E)\cos \psi - i\sin\psi = (1-c+c \cos \beta) - ic \sin \beta. \tag{5.1.52}$$

Since from (5.1.47),

$$\sin\psi = c \sin \beta,$$

it follows that the imaginary parts of (5.1.52) are equal. Also,

$$\cos\psi = \{1-(c \sin \beta)^2\}^{\frac{1}{2}}.$$

Equating the real parts of (5.1.52) gives

$$(1-2E)\{1-(c \sin \beta)^2\}^{\frac{1}{2}} = 1 - c + c \cos \beta, \tag{5.1.53}$$

whence

$$E = \tfrac{1}{2}\{1 - \frac{1-2c \sin^2(\beta/2)}{\sqrt{1-(c \sin \beta)^2}}\}. \tag{5.1.54}$$

Clearly, as $\Delta x \to 0$, $\beta \to 0$ and $E \to 0$. The amplitude of the physical mode then approaches the amplitude of the exact solution, while the amplitude of the computational mode tends to zero.

Now, for a given value of c, Equation (5.1.47) shows that the value of ψ will decrease with decreasing values of $\beta = \pi m\Delta x = 2\pi\Delta x/\lambda$. Therefore, when the wave-length λ is large compared to the grid-spacing Δx, β will be small and from (5.1.54) the value of E is small so the amplitude of the physical mode will be nearly that of the exact solution of the partial differential equation and the amplitude of the computational mode will be small. For example, if the wavelength equals 20 grid-spacings, then $\lambda = 20\Delta x$ and $\beta = \pi/10$. Choosing $c = 0.8$, then $E = 0.0042$, so the amplitude of the physical mode is about 0.9958 while that of the computational mode is only 0.0042. Also, the speed of the physical mode relative to the true wave speed is given by (5.1.32), namely,

$$u_N/u = \beta^{-1}c^{-1}\arcsin(c \sin \beta).$$

Here, the value of u_N/u is 0.9939. Therefore in this example, the amplitude and speed of the physical mode in the finite difference solution is very close to that of the true solution of this problem. On the other hand, if the grid-spacing is a large fraction of the wave length, say $\Delta x/\lambda = 0.5$ when $\beta = \pi$ and $E = 0.8$, then the amplitude of the spurious computational mode is 0.8 while the physical mode has an amplitude of only 0.2. Clearly the numerical solution bears little resemblance to the true solution of the convection problem in this case.

Implicit finite difference methods may also be used to solve the one-dimensional convection equation (5.1.2). Consider the following scheme, in which the partial differential equation is evaluated at the point $(j\Delta x, n\Delta t+\frac{1}{2}\Delta t)$ in the solution domain, giving

$$\left.\frac{\partial\bar{\tau}}{\partial t}\right|_j^{n+\frac{1}{2}} + u\left.\frac{\partial\bar{\tau}}{\partial x}\right|_j^{n+\frac{1}{2}} = 0. \tag{5.1.55}$$

This equation is now approximated in terms of values of $\bar{\tau}$ at nearby grid points. On writing the space derivative in terms of its average value at the nth and (n+1)th time levels, (5.1.55) becomes

$$\left.\frac{\partial\bar{\tau}}{\partial t}\right|_j^{n+\frac{1}{2}} + \frac{u}{2}\left\{\left.\frac{\partial\bar{\tau}}{\partial x}\right|_j^{n} + \left.\frac{\partial\bar{\tau}}{\partial x}\right|_j^{n+1} + O\{(\Delta t)^2\}\right\} = 0. \tag{5.1.56}$$

Writing the derivatives in this equation in their central-difference form gives

$$\frac{\bar{\tau}_j^{n+1} - \bar{\tau}_j^n}{\Delta t} + O\{(\Delta t)^2\} + \frac{u}{2}\left\{\frac{\bar{\tau}_{j+1}^n - \bar{\tau}_{j-1}^n}{2\Delta x} + \frac{\bar{\tau}_{j+1}^{n+1} - \bar{\tau}_{j-1}^{n+1}}{2\Delta x} + O\{(\Delta t)^2, (\Delta x)^2\}\right\} = 0. \tag{5.1.57}$$

Rearranging and then dropping the higher-order terms gives the two-level implicit finite difference equation

$$-\frac{c}{4}\tau_{j-1}^{n+1} + \tau_j^{n+1} + \frac{c}{4}\tau_{j+1}^{n+1} = \frac{c}{4}\tau_{j-1}^n + \tau_j^n - \frac{c}{4}\tau_{j+1}^n . \tag{5.1.58}$$

This can be used to solve the one-dimensional convection equation, with a truncation error of $O\{(\Delta t)^2, (\Delta x)^2\}$ and a contribution to the discretisation error of $O\{(\Delta t)^3, \Delta t(\Delta x)^2\}$ at each time step.

Knowing the values of τ_j^n at the nth time level and the boundary values τ_0^{n+1}, τ_J^{n+1} at the (n+1)th time level, the values τ_j^{n+1}, $j=1(1)J-1$, can be computed from the set of Equations (5.1.58), $j=1(1)J-1$, using the Thomas algorithm described in the Appendix.

Generally speaking, implicit methods are usually unconditionally stable. That is, round-off errors introduced at time level n are not magnified in modulus when values of the dependent variable are computed at time level $(n+1)$. We will now determine the stability criteria for this particular finite difference equation using the von Neumann method.

If the initial error distribution has an mth Fourier mode of the form $\xi_j^0 = e^{i\beta j}$, $\beta = \pi m \Delta x$, then at the nth time level it will have the form

$$\xi_j^n = (G)^n e^{i\beta j},$$

where G is found by substituting ξ_j^n in the error equation which is analogous to (5.1.58). Division by $(G)^n e^{i\beta j}$ then gives

$$G(-\frac{c}{4} e^{-i\beta} + 1 + \frac{c}{4} e^{i\beta}) = \frac{c}{4} e^{-i\beta} + 1 - \frac{c}{4} e^{i\beta}, \tag{5.1.59}$$

so that

$$G = \frac{1 - i\,(\frac{1}{2}c \sin \beta)}{1 + i\,(\frac{1}{2}c \sin \beta)}. \tag{5.1.60}$$

Clearly, $|G| = 1$ for all β and c, so that (5.1.58) is stable for all values of c.

Previously, we saw that the value of G also gave some quantitative information about the accuracy of the finite difference method, besides qualitative information about the stability of the finite difference equation. We saw that for two-level methods such as this, an infinite train of waves of length $\lambda = 2/m$, with initial form $\bar{\tau}_j^0 = e^{i\beta j}$, is propagated such that the amplitude of the wave is multiplied by $|G|$ with each step taken in time, and moves with a speed relative to the true speed given by $u_N/u = -\arg\{G\}/(\beta c)$. Since $|G| = 1$ for all β and c, the amplitude of such a wave train is not changed by this method of solution. However, since $\arg\{G\} = -2 \arctan\{\frac{1}{2}c \sin \beta\}$, then the relative wave speed is

$$u_N/u = 2 \arctan\{\tfrac{1}{2} c \sin \beta\}/(\beta c).$$

If the wavelength λ is much greater than Δx, then $\beta \ll 1$ and $u_N/u \simeq 2(\frac{1}{2}c\beta)/(\beta c) = 1$. Therefore the wave train found numerically has the same amplitude and very nearly the same speed as the true solution of the convection equation. Note that, in this method $u_N/u < 1$, so the speed of propagation of the numerical wave is always less than the true wave speed.

For this method, the contribution to the discretisation error at each time step is $O\{(\Delta t)^3, \Delta t(\Delta x)^2\}$, which is the same contribution as in the leap-frog method. Clearly, because the time derivative has been differenced about a point midway between two time levels to get an $O\{(\Delta t)^2\}$ truncation error in this case, this is a two-level method. Therefore the computational mode does not appear.

However, even though application of (5.1.58) does not amplify round-off errors so the finite difference equation is stable for all values of c, and the discretisation error is small so the method appears very accurate, there is a condition which must be placed on c in order to actually solve the system of linear algebraic equations using the Thomas algorithm. The linear algebraic system of equations can be solved without loss of accuracy because of round-off , if the coefficient matrix on the left side of (5.1.58) is diagonally dominant. This requires that

$$1 \geqslant |-\frac{c}{4}| + |\frac{c}{4}| = \frac{c}{2},$$

or

$$c \leqslant 2.$$

For values of $c > 2$, the loss of accuracy in the solution of the algebraic equations may make the results from an application of the Thomas algorithm quite worthless.

A further point which bears examination, is whether this numerical method introduces artificial diffusion. For example, the upwind difference method introduces some diffusion but the leap-frog method does not.

Carrying out a formal consistency analysis, by substituting $\bar{\tau}$ for τ in (5.1.58) and expanding all terms in their Taylor series, gives the corresponding partial differential equation

$$\frac{\partial\bar{\tau}}{\partial t} + u\frac{\partial\bar{\tau}}{\partial x} = -\frac{\Delta t}{2}\frac{\partial^2\bar{\tau}}{\partial t^2} - \tfrac{1}{2}u\Delta t\,\frac{\partial^2\bar{\tau}}{\partial t\partial x} + O\{(\Delta t)^2,(\Delta x)^2\}, \tag{5.1.61}$$

at the (j,n) grid-point. This equation is clearly consistent with the one-dimensional convection equation. Also, since $\bar{\tau}$ satisfies the convection equation then

$$\frac{\partial^2\bar{\tau}}{\partial t^2} = u^2\frac{\partial^2\bar{\tau}}{\partial x^2},$$

and

$$\frac{\partial^2\bar{\tau}}{\partial t\partial x} = -u\frac{\partial^2\bar{\tau}}{\partial x^2}.$$

Substitution of these equations in (5.1.61) gives

$$\frac{\partial\bar{\tau}}{\partial t} + u\frac{\partial\bar{\tau}}{\partial x} = O\{(\Delta t)^2,(\Delta x)^2\}. \tag{5.1.62}$$

Therefore this numerical method introduces no artificial diffusion and has a truncation error of $O\{(\Delta t)^2,(\Delta x)^2\}$.

5.2 The Transport Equation

Consider the complete one-dimensional transport equation (5.1.1) evaluated at the (j,n) grid-point, that is

$$\left.\frac{\partial\bar{\tau}}{\partial t}\right|_j^n + u\left.\frac{\partial\bar{\tau}}{\partial x}\right|_j^n = \alpha\left.\frac{\partial^2\bar{\tau}}{\partial x^2}\right|_j^n. \tag{5.2.1}$$

If the central difference form (2.3.18) is used for the spatial derivative in the convection term with the forward difference form (2.3.7a) for the time derivative and the central difference form (2.3.19) for the second order derivative in the diffusion term, this becomes

$$\frac{\bar{\tau}_j^{n+1} - \bar{\tau}_j^n}{\Delta t} + O\{\Delta t\} + u\{\frac{\bar{\tau}_{j+1}^n - \bar{\tau}_{j-1}^n}{2\Delta x} + O\{(\Delta x)^2\}\}$$
$$= \alpha\{\frac{\bar{\tau}_{j+1}^n - 2\bar{\tau}_j^n + \bar{\tau}_{j-1}^n}{(\Delta x)^2} + O\{(\Delta x)^2\}\}. \tag{5.2.2}$$

By dropping the terms of $O\{\Delta t,(\Delta x)^2\}$ from (5.2.2), an explicit finite difference equation is obtained which can be used to solve the one-dimensional transport equation. This equation is

$$\tau_j^{n+1} = (s+\tfrac{1}{2}c)\tau_{j-1}^n + (1-2s)\tau_j^n + (s-\tfrac{1}{2}c)\tau_{j+1}^n \,, \qquad (5.2.3)$$

which contributes an error of $O\{(\Delta t)^2, \Delta t(\Delta x)^2\}$ to the discretisation error at each time step.

Note that similar differencing produced a method which was always unstable in the absence of diffusion. We therefore first examine the stability of Equation (5.2.3). The amplification factor in the von Neumann stability analysis of this equation is found to be

$$G = 1 - 2s(1 - \cos\beta) - ic\sin\beta. \qquad (5.2.4)$$

The value of the gain $|G|$ is given by the expression

$$|G|^2 = (4s^2-c^2)\cos^2\beta - (8s^2-4s)\cos\beta + (4s^2+c^2-4s+1), \qquad (5.2.5)$$

where the superscripts are powers.

We wish to find the region of the s-c plane for which $|G|^2 - 1 \leqslant 0$ for all values of β in the interval $0 \leq \beta < 2\pi$. This is equivalent to finding the values of s and c for which the function $F(\chi) \leqslant 0$ on the interval $-1 \leqslant \chi \leqslant 1$, where

$$F(\chi) = (4s^2-c^2)\chi^2 - (8s^2-4s)\chi + (4s^2+c^2-4s), \qquad (5.2.6)$$

and $\chi = \cos\beta$. Firstly, if $4s^2-c^2 > 0$, that is if

$$0 < c < 2s, \qquad (5.2.7a)$$

then $F(\chi)$ is a quadratic function with a minimum value. Since $F(1) = 0$, then $F(\chi) \leqslant 0$ on $-1 \leqslant \chi \leqslant 1$ if $F(-1) \leqslant 0$; that is, if

$$16s(2s-1) \leqslant 0$$

or

$$0 \leqslant 2s \leqslant 1. \qquad (5.2.7b)$$

Combining the inequalities (5.2.7a,b) gives

$$0 < c < 2s \leqslant 1. \qquad (5.2.8)$$

If $c = 2s$, then $F(\chi)$ is a linear function of χ, and as $F(1) = 0$, then we also require $F(-1) \leqslant 0$ as above in order to have $F(\chi) \leqslant 0$ on $<-1,1>$. Thus (5.2.8) can be extended to

$$0 < c \leqslant 2s \leqslant 1. \qquad (5.2.9)$$

If $4s^2 - c^2 < 0$, that is if

$$0 < 2s < c, \qquad (5.2.10a)$$

then $F(\chi)$ is a quadratic function with a maximum value. Since $F(1) = 0$, then $F(\chi) \leqslant 0$ on $-1 \leqslant \chi \leqslant 1$ if the product of the roots of $F(\chi) = 0$ is not less than 1; that is, if

$$\frac{4s^2+c^2-4s}{4s^2-c^2} \geqslant 1.$$

Multiplying both sides of this inequality by the negative quantity $4s^2-c^2$, gives

$$2s \geqslant c^2. \tag{5.2.10b}$$

Combining (5.2.10a,b) gives

$$0 < c^2 \leqslant 2s < c. \tag{5.2.11}$$

Combining the results (5.2.9) and (5.2.11) it is seen that the finite difference equation (5.2.3) is stable if

$$c^2 \leqslant 2s \leqslant 1. \tag{5.2.12}$$

This is a larger region than that defined by (5.2.9) which is the stability criterion usually given for this method (see for example, Isenberg and de Vahl Davis, 1975).

By substituting $c = u\Delta t/\Delta x$ and $s = \alpha\Delta t/(\Delta x)^2$, the two inequalities in (5.2.12) lead to the following condition applying to the time-step Δt,

$$\Delta t \leqslant \min\{2\alpha/u^2, (\Delta x)^2/2\alpha\}. \tag{5.2.13a}$$

For instance, with $\alpha = 10^{-3}$, $u = 10^{-1}$, $\Delta x = 10^{-2}$, the requirement is that

$$\Delta t \leqslant \min\{2\times10^{-1},\ 5\times10^{-2}\},$$

so the time step should not exceed 5×10^{-2}.

Dividing (5.2.12) by $2c > 0$ and inverting gives

$$2c \leqslant r \leqslant 2c^{-1} \tag{5.2.13b}$$

where

$$r = c/s = u\Delta x/\alpha, \tag{5.2.14}$$

may be thought of a Reynolds number associated with the grid size.

It is clear from the way that the finite difference equation (5.2.3) is derived from (5.2.2) that the truncation error for this method is usually $O\{\Delta t,(\Delta x)^2\}$. A formal consistency analysis, commencing with (5.2.3) and replacing the approximation τ by $\bar{\tau}$, which is the exact solution of (5.1.1), yields the partial differential equation

$$\frac{\partial\bar{\tau}}{\partial t} + u\frac{\partial\bar{\tau}}{\partial x} = \alpha\frac{\partial^2\bar{\tau}}{\partial x^2} - \frac{\Delta t}{2}\frac{\partial^2\bar{\tau}}{\partial t^2} + O\{(\Delta t)^2,(\Delta x)^2\}, \tag{5.2.15}$$

evaluated at the (j,n) grid-point. Since $\bar{\tau}$ is the solution of the one-dimensional transport equation (5.1.1), it follows that

$$\frac{\partial}{\partial t}(\bar{\tau}) = \left(-u\frac{\partial}{\partial x} + \alpha\frac{\partial^2}{\partial x^2}\right)\bar{\tau},$$

whence

$$\frac{\partial^2\bar{\tau}}{\partial t^2} = \left(\frac{\partial}{\partial t}\right)^2\bar{\tau} = \left(u^2\frac{\partial^2}{\partial x^2} - 2\alpha u\frac{\partial^3}{\partial x^3} + \alpha^2\frac{\partial^4}{\partial x^4}\right)\bar{\tau}. \tag{5.2.16}$$

Substitution of (5.2.16) into (5.2.15) then gives

$$\frac{\partial\bar{\tau}}{\partial t} + u\frac{\partial\bar{\tau}}{\partial x} = (\alpha-\alpha')\frac{\partial^2\bar{\tau}}{\partial x^2} + O\{\Delta t,(\Delta x)^2\}, \qquad (5.2.17)$$

where $\alpha' = \frac{1}{2}u^2\Delta t$.

Even though (5.2.3) is consistent with the one-dimensional transport equation as $\Delta t \to 0$, $\Delta x \to 0$, in any practical situation $\Delta t \neq 0$ and it is the partial differential equation (5.2.17) which is actually being solved by this numerical method. Therefore, the solution of (5.2.3) is that of (5.1.1), to $O\{\Delta t,(\Delta x)^2\}$, in which the diffusion coefficient α has been reduced by the amount α'. The numerical method used has decreased the effect of the diffusion term $\alpha\,\partial^2\bar{\tau}/\partial x^2$ in (5.1.1) by an amount $\frac{1}{2}u^2\Delta t\,\partial^2\bar{\tau}/\partial x^2$, which can be large if both u and Δt are large.

In Section 4.1 it was seen that the major disadvantage of applying the Richardson method of using central time differencing with central space differencing to solve the one-dimensional diffusion equation, was that it gave a finite difference equation which was unstable for all s. However, the same procedure applied to the one-dimensional convection equation (5.1.2) in Section 5.1 gave a finite difference equation which was stable for $c \leqslant 1$. When this method is applied to the complete one-dimensional transport equation (5.2.1), the following finite difference form is obtained:

$$\frac{\bar{\tau}_j^{n+1} - \bar{\tau}_j^{n-1}}{2\Delta t} + O\{(\Delta t)^2\} + u\left(\frac{\bar{\tau}_{j+1}^{n} - \bar{\tau}_{j-1}^{n}}{2\Delta x} + O\{(\Delta x)^2\}\right)$$
$$= \alpha\left(\frac{\bar{\tau}_{j+1}^{n} - 2\bar{\tau}_j^{n} + \bar{\tau}_{j-1}^{n}}{(\Delta x)^2} + O\{(\Delta x)^2\}\right). \qquad (5.2.18)$$

Dropping terms of $O\{(\Delta t)^2,(\Delta x)^2\}$ yields the three-level explicit finite difference equation to solve (5.2.1),

$$\tau_j^{n+1} = (2s+c)\tau_{j-1}^{n} - 2s\tau_j^{n} + (2s-c)\tau_{j+1}^{n} + \tau_j^{n-1}, \qquad (5.2.19)$$

which contributes a term of $O\{(\Delta t)^2,\Delta t(\Delta x)^2\}$ to the discretisation error at each time step. Unfortunately, like the pure diffusion case, application of the von Neumann stability analysis shows that this method is also unstable for all values of $s \neq 0$. The method is stable for $s = 0$, that is for the pure convection equation, provided that $c \leqslant 1$. This case was investigated in Section 5.1.

However, application of the method of DuFort and Frankel (1953), produces a conditionally stable system. In this method the value $\bar{\tau}_j^n$ in the finite difference form used for the second-order space derivative $\partial^2\bar{\tau}/\partial x^2$ of the finite difference equation (5.2.18) is replaced by its average at time levels (n-1) and (n+1), giving

$$\frac{\bar{\tau}_j^{n+1} - \bar{\tau}_j^{n-1}}{2\Delta t} + O\{(\Delta t)^2\} + u\left(\frac{\bar{\tau}_{j+1}^{n} - \bar{\tau}_{j-1}^{n}}{2\Delta x} + O\{(\Delta x)^2\}\right)$$
$$= \alpha\left\{\frac{\bar{\tau}_{j+1}^{n} - (\bar{\tau}_j^{n+1} + \bar{\tau}_j^{n-1}) + \bar{\tau}_{j-1}^{n} + O\{(\Delta t)^2\}}{(\Delta x)^2} + O\{(\Delta x)^2\}\right\} \qquad (5.2.20)$$

Dropping terms of $O\{(\Delta t)^2,(\Delta x)^2,(\Delta t/\Delta x)^2\}$ gives the three-level explicit finite difference equation

$$\tau_j^{n+1} = \left(\frac{1-2s}{1+2s}\right)\tau_j^{n-1} + \left(\frac{c+2s}{1+2s}\right)\tau_{j-1}^{n} - \left(\frac{c-2s}{1+2s}\right)\tau_{j+1}^{n}, \tag{5.2.21}$$

which may be used to get approximate answers to the one-dimensional transport problem with a contribution to the discretisation error of $O\{(\Delta t)^3, \Delta t(\Delta x)^2, (\Delta t)^3/(\Delta x)^2\}$.

Application of the von Neumann stability analysis shows that the only stability restriction on Equation (5.2.21) is the requirement that $c \leqslant 1$. There is no restriction on the value of s.

Using a Taylor series expansion about the (j,n) grid-point with the exact value $\bar{\tau}$ replacing τ in Equation (5.2.21), produces the differential equation

$$\alpha\left(\frac{\Delta t}{\Delta x}\right)^2 \frac{\partial^2\bar{\tau}}{\partial t^2} + \frac{\partial\bar{\tau}}{\partial t} + u\frac{\partial\bar{\tau}}{\partial x} = \alpha\frac{\partial^2\bar{\tau}}{\partial x^2} + O\{(\Delta t)^2, (\Delta x)^2, (\Delta t)^4/(\Delta x)^2\} \tag{5.2.22}$$

at the (j,n) grid-point. The method is therefore consistent with the transport equation as $\Delta t \to 0$, $\Delta x \to 0$ only if Δt and Δx approach the limiting value of zero in such a way that $\Delta t/\Delta x \to 0$. This is an additional restriction on the size Δt can take, apart from the stability condition $\Delta t/\Delta x \leqslant u^{-1}$, which results from the condition $c \leqslant 1$.

DuFort and Frankel (1953) suggest that the stability of the method is due to the introduction of the first term in Equation (5.2.22), which makes the equation hyperbolic in nature.

It is clear that this scheme does have certain disadvantages, but it is a useful explicit conditionally stable method. For instance, Pearson (1965) has shown that the DuFort-Frankel method is more accurate than the previously described forward-time centred-space method (5.2.3) when applied to certain problems involving the transport equation.

Using the forward difference form for the time derivative and the central difference form for the space derivative in the diffusion term of Equation (5.2.1), together with the upwind difference form for the convection term gives, for $u > 0$,

$$\frac{\bar{\tau}_j^{n+1} - \bar{\tau}_j^{n}}{\Delta t} + O\{\Delta t\} + u\left(\frac{\bar{\tau}_j^{n} - \bar{\tau}_{j-1}^{n}}{\Delta x} + O\{\Delta x\}\right)$$

$$= \alpha\left\{\frac{\bar{\tau}_{j+1}^{n} - 2\bar{\tau}_j^{n} + \bar{\tau}_{j-1}^{n}}{(\Delta x)^2} + O\{(\Delta x)^2\}\right\}. \tag{5.2.23}$$

On dropping terms of $O\{\Delta t, \Delta x\}$ this yields the explicit finite difference equation

$$\tau_j^{n+1} = (s+c)\tau_{j-1}^{n} + (1-2s-c)\tau_j^{n} + s\tau_{j+1}^{n}, \tag{5.2.24}$$

which may be used to compute approximate solutions to the one-dimensional transport equation, with a truncation error of $O\{\Delta t, \Delta x\}$.

The amplification factor in the von Neumann stability analysis of Equation (5.2.24) is

$$G = 1 - (2s+c)(1-\cos\beta) - ic\sin\beta, \tag{5.2.25}$$

so that the gain $|G|$ is given by the solution of

$$|G|^2 = (4s^2+4sc)\cos^2\beta - (8s^2+8sc+2c^2-4s-2c)\cos\beta + (1+4s^2+4sc+2c^2-4s-2c), \quad (5.2.26)$$

where the indices represent powers. Consider the function $F(\chi) = |G|^2-1$, where $\chi = \cos\beta$. Then we wish to find the region of the s-c plane in which $F(\chi) \leqslant 0$ for all β, that is for all values of χ in <-1,1>. Now

$$F(\chi) = (4s^2+4sc)\chi^2 - (8s^2+8sc+2c^2-4s-2c)\chi + (4s^2+4sc+2c^2-4s-2c) \quad (5.2.27)$$

is a quadratic function, with a minimum value because the coefficient $(4s^2+4sc)$ of χ^2 is positive. But $F(1) = 0$ and

$$F(-1) = 16s^2 + 16sc + 4c^2 -8s - 4c$$
$$= 4(2s+c-1)(2s+c).$$

So long as

$$2s + c \leqslant 1, \quad (5.2.28)$$

then $F(-1) \leqslant 0$, and $F(\chi) \leqslant 0$ for $-1 \leqslant \chi \leqslant 1$. Therefore, the inequality (5.2.28) is the required stability criterion. This condition implies that the time step must satisfy the inequality

$$\Delta t \leqslant 2\{\frac{\alpha}{(\Delta x)^2} + \frac{u}{\Delta x}\}^{-1} . \quad (5.2.29)$$

The condition (5.2.28), on division by $c = u\Delta t/\Delta x$ and substitution of $s = \alpha\Delta t/(\Delta x)^2$, can be written

$$r \geqslant 2c/(1-c) \quad (5.2.29)$$

where $r = u\Delta x/\alpha$ is the Reynolds number based on the grid-spacing. This is less restrictive than the corresponding relation (5.2.13b) using the central space form (5.2.3) for the convective derivative.

In order to check the consistency of the finite difference equation (5.2.24), the approximation τ is replaced by the exact solution $\bar{\tau}$, and a Taylor series expansion of each term is taken about the (j,n) grid-point. Because $\bar{\tau}$ is the solution of (5.1.1), $\partial^2\bar{\tau}/\partial t^2$ can be replaced by (5.2.16), giving the corresponding partial differential equation

$$\frac{\partial\bar{\tau}}{\partial t} + u\frac{\partial\bar{\tau}}{\partial x} = (\alpha+\alpha')\frac{\partial^2\bar{\tau}}{\partial x^2} + O\{\Delta t \quad ,(\Delta x)^2\}, \quad (5.2.30)$$

at the (j,n) grid-point, where

$$\alpha' = \tfrac{1}{2}u\Delta x(1-c). \quad (5.2.31)$$

Equation (5.2.31) indicates that the method introduces numerical diffusion which will be minimized by taking c as close to 1 as possible. In general, accurate solutions are not possible unless $\alpha' \ll \alpha$ which, for a given value of $c < 1$, requires that $\tfrac{1}{2}u\Delta x \ll \alpha$ so the "grid" Reynold's number must satisfy $r \ll 2$.

Before leaving this section a word of caution must be given. By comparing Equations (5.2.28) and (5.1.18) it is seen that the presence of the diffusion term in the one-dimensional transport equation (5.1.1) when using the upwind finite difference approximation has a destabilising effect. This is in direct contrast to the findings when the central difference approximation to the convection term is used, and the presence of the diffusion term has a stabilising effect. However, it must be remembered that such effects are also closely related to the presence of artificial diffusion which tends to produce inaccurate results.

If α is chosen to be equal to $-\alpha'$, then Equation (5.2.30) becomes the convection equation with a truncation error of $O\{\Delta t \;, (\Delta x)^2\}$. This suggests that a possible method of modelling the convection equation using upwind differencing for the convective term, when the Courant number c is not equal to unity, is to model the transport equation (5.1.1) with $\alpha = -\alpha'$ at each grid point. Theoretically, this has the effect of cancelling the artificial diffusion term introduced by the method. This can be tested using the example considered previously, with the initial condition illustrated in the top diagram of Figure 5.1. Equation (5.2.30) is modelled using Equation (5.2.24) with

$$\alpha = -\alpha' = -\tfrac{1}{2}u\Delta x(1-c)$$

and $u = 0.1$, $c = 0.8$. The results obtained are compared with the exact solution of the convection equation in Figure 5.2. Clearly the use of this correction procedure results in a better solution to the problem in that the shape of the peak more closely resembles the exact solution; that is, less diffusion and damping have occurred.

However, some negative values of τ appear behind the peak as it moves from left to right, reflecting the effect of the terms of $O\{\Delta t \;, (\Delta x)^2\}$ in (5.2.30). The negative values of τ occur as part of an oscillation of decreasing amplitude which appears behind the peak in the numerical solution. A further disadvantage of using this correction procedure is that the maximum value of the computed peak now lags behind its true position. The phase error produced by the upwind difference method has produced a numerical wave speed less than that of the true speed.

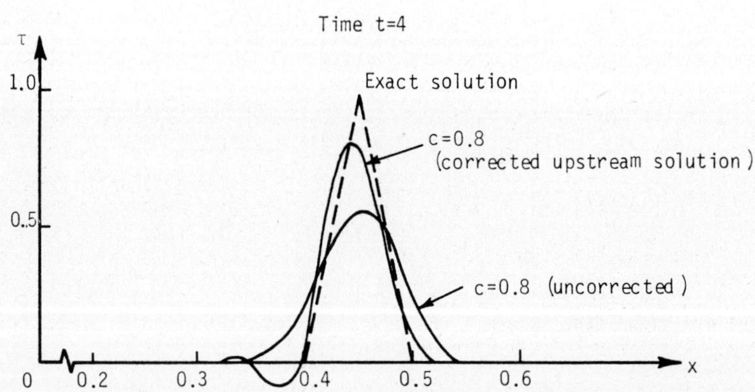

Figure 5.2: A comparison of the "corrected" upwind solution and the uncorrected solution, both for $c = 0.8$, and the exact solution, at time $t = 4$.

Implicit methods may also be used to solve the one-dimensional transport equation. Consider the scheme, similar to the Crank-Nicolson type for solving the one-dimensional diffusion equation, in which the given partial differential equation at the point $(j\Delta x, n\Delta t+\frac{1}{2}\Delta t)$ in the solution domain is approximated in terms of values of the dependent variable at nearby grid-points. Thus we have

$$\left.\frac{\partial\bar{\tau}}{\partial t}\right|_j^{n+\frac{1}{2}} + u\left.\frac{\partial\bar{\tau}}{\partial x}\right|_j^{n+\frac{1}{2}} = \alpha\left.\frac{\partial^2\bar{\tau}}{\partial x^2}\right|_j^{n+\frac{1}{2}}, \tag{5.2.32}$$

in which we write the two space derivatives in terms of the average of their values the time levels n and $(n+1)$. Thus (5.2.32) becomes

$$\left.\frac{\partial\bar{\tau}}{\partial t}\right|_j^{n+\frac{1}{2}} + \frac{u}{2}\left\{\left.\frac{\partial\bar{\tau}}{\partial x}\right|_j^{n} + \left.\frac{\partial\bar{\tau}}{\partial x}\right|_j^{n+1} + O\{(\Delta t)^2\}\right\}$$

$$= \frac{\alpha}{2}\left\{\left.\frac{\partial^2\bar{\tau}}{\partial x^2}\right|_j^{n} + \left.\frac{\partial^2\bar{\tau}}{\partial x^2}\right|_j^{n+1} + O\{(\Delta t)^2\}\right\}. \tag{5.2.33}$$

Writing all the derivatives in (5.2.33) in their centred-difference forms gives

$$\frac{\bar{\tau}_j^{n+1}-\bar{\tau}_j^n}{\Delta t} + O\{(\Delta t)^2\} + \frac{u}{2}\left\{\frac{\bar{\tau}_{j+1}^n-\bar{\tau}_{j-1}^n}{2\Delta x} + \frac{\bar{\tau}_{j+1}^{n+1}-\bar{\tau}_{j-1}^{n+1}}{2\Delta x} + O\{(\Delta t)^2,(\Delta x)^2\}\right\}$$

$$= \frac{\alpha}{2}\left\{\frac{\bar{\tau}_{j+1}^n-2\bar{\tau}_j^n+\bar{\tau}_{j-1}^n}{(\Delta x)^2} + \frac{\bar{\tau}_{j+1}^{n+1}-2\bar{\tau}_j^{n+1}+\bar{\tau}_{j-1}^{n+1}}{(\Delta x)^2} + O\{(\Delta t)^2,(\Delta x)^2\}\right\}. \tag{5.2.34}$$

On rearranging this equation with the terms at the time level $(n+1)$ on the left side and terms at the time level n on the right side, and dropping the terms of $O\{(\Delta t)^3,\Delta t(\Delta x)^2\}$, gives the implicit finite difference equation

$$\left(-\frac{s}{2}-\frac{c}{4}\right)\tau_{j-1}^{n+1} + (1+s)\tau_j^{n+1} + \left(-\frac{s}{2}+\frac{c}{4}\right)\tau_{j+1}^{n+1}$$

$$= \left(\frac{s}{2}+\frac{c}{4}\right)\tau_{j-1}^n + (1-s)\tau_j^n + \left(\frac{s}{2}-\frac{c}{4}\right)\tau_{j+1}^n. \tag{5.2.35}$$

This equation may be used to find approximate solutions to the one-dimensional transport equation, with a truncation error of $O\{(\Delta t)^2,(\Delta x)^2\}$ and a contribution of $O\{(\Delta t)^3,\Delta t(\Delta x)^2\}$ to the discretisation error at every time step.

Note that, if $u = 0$ so that $c = 0$, this becomes the Crank-Nicolson equation (4.2.11) for solving the one-dimensional diffusion equation (2.1.1). Also, when $\alpha = 0$ so that $s = 0$, this becomes the corresponding equation (5.1.58) used to solve the one-dimensional convection equation (5.1.2). Application of the Thomas method then gives the solutions τ_j^{n+1}, $j=1(1)J-1$, of the tri-diagonal set of algebraic equations (5.2.35), $j=1(1)J-1$, if all values of τ_j^n at the previous time level are known together with the boundary values τ_0^{n+1} τ_J^{n+1} (compare Equations (4.2.12) for the diffusion equation).

As seen previously, implicit finite difference equations are very stable. The stability of this method can be determined by applying the von Neumann analysis. The amplification factor is given by

$$G = \frac{(1-s+s\cos\beta) - i(\frac{1}{2}c\sin\beta)}{(1+s-s\cos\beta) + i(\frac{1}{2}c\sin\beta)}. \tag{5.2.36}$$

Therefore

$$|G|^2 = \frac{(1-s+s\cos\beta)^2 + (\tfrac{1}{2}c\sin\beta)^2}{(1+s-s\cos\beta)^2 + (\tfrac{1}{2}c\sin\beta)^2}, \tag{5.2.37}$$

where the superscripts are powers, and

$$|G|^2 - 1 = \frac{-4s(1-\cos\beta)}{(1+s-s\cos\beta)^2 + (\tfrac{1}{2}c\sin\beta)^2}. \tag{5.2.38}$$

Since the factor $(1-\cos\beta) \geqslant 0$ for all β, and the denominator of (5.2.38) is always positive, then

$$|G|^2 - 1 \leqslant 0$$

or

$$|G| \leqslant 1,$$

for all c, s, β. The equation (5.2.35) is therefore *unconditionally stable.*

The set of equations (5.2.35), $j=1(1)J-1$, may be solved using the Thomas algorithm or any other elimination technique provided the system is diagonally dominant. This requires

$$|1+s| \geqslant |-\tfrac{s}{2} - \tfrac{c}{4}| + |-\tfrac{s}{2} + \tfrac{c}{4}|,$$

or

$$1 + s \geqslant \frac{s}{2} + \frac{c}{4} + |\frac{c}{4} - \frac{s}{2}|. \tag{5.2.39}$$

If $s \leqslant \tfrac{1}{2}c$, this becomes $s \geqslant \tfrac{1}{2}c - 1$. If $s \geqslant \tfrac{1}{2}c$, (5.2.39) becomes $1 + s \geqslant s$ which is always true.

The system of equations (5.2.35) is therefore *solvable* so long as

$$s \geqslant \tfrac{1}{2}c - 1. \tag{5.2.40}$$

6. THE TRANSPORT EQUATION IN MULTI-DIMENSIONAL SPACE

6.1 Explicit Finite Difference Methods

Finite difference methods of solving the two- and three-dimensional transport equations are now considered. The transport equation in two-dimensional space is

$$\frac{\partial\bar{\tau}}{\partial t} + u\frac{\partial\bar{\tau}}{\partial x} + v\frac{\partial\bar{\tau}}{\partial y} = \alpha_x\frac{\partial^2\bar{\tau}}{\partial x^2} + \alpha_y\frac{\partial^2\bar{\tau}}{\partial y^2}, \qquad (6.1.1)$$

while in three dimensions it is given by

$$\frac{\partial\bar{\tau}}{\partial t} + u\frac{\partial\bar{\tau}}{\partial x} + v\frac{\partial\bar{\tau}}{\partial y} + w\frac{\partial\bar{\tau}}{\partial z} = \alpha_x\frac{\partial^2\bar{\tau}}{\partial x^2} + \alpha_y\frac{\partial^2\bar{\tau}}{\partial y^2} + \alpha_z\frac{\partial^2\bar{\tau}}{\partial z^2}. \qquad (6.1.2)$$

The Pure Diffusion Equation

The equation for the pure diffusion problem in two space dimensions is obtained from (6.1.1) by putting $u = v = 0$, giving

$$\frac{\partial\bar{\tau}}{\partial t} = \alpha_x\frac{\partial^2\bar{\tau}}{\partial x^2} + \alpha_y\frac{\partial^2\bar{\tau}}{\partial y^2}. \qquad (6.1.3)$$

The FTCS (forward-time centred-space) finite difference approximation for this equation is derived from Equation (6.1.3) evaluated at the (j,k,n) grid-point, that is at the point $(j\Delta x,k\Delta y,n\Delta t)$ in the solution domain in x-y-t space, with the time derivative being replaced by the forward-difference form and the space derivatives replaced by their centred-difference forms. The resulting equation is

$$\frac{\bar{\tau}^{n+1}_{j,k}-\bar{\tau}^{n}_{j,k}}{\Delta t} + O\{\Delta t\} = \alpha_x\left\{\frac{\bar{\tau}^{n}_{j+1,k}-2\bar{\tau}^{n}_{j,k}+\bar{\tau}^{n}_{j-1,k}}{(\Delta x)^2} + O\{(\Delta x)^2\}\right\}$$

$$+ \alpha_y\left\{\frac{\bar{\tau}^{n}_{j,k+1}-2\bar{\tau}^{n}_{j,k}+\bar{\tau}^{n}_{j,k-1}}{(\Delta y)^2} + O\{(\Delta y)^2\}\right\}, \qquad (6.1.4)$$

where $j=1(1)J-1$, $k=1(1)K-1$, and

$$\bar{\tau}^{n}_{j,k} = \bar{\tau}(j\Delta x,k\Delta y,n\Delta t).$$

Rearranging (6.1.4), and dropping the higher order terms, gives the finite difference approximation to equation (6.1.3),

$$\tau^{n+1}_{j,k} = s_y\tau^{n}_{j,k-1} + s_x\tau^{n}_{j-1,k} + (1-2s_x-2s_y)\tau^{n}_{j,k}$$

$$+ s_x\tau^{n}_{j+1,k} + s_y\tau^{n}_{j,k+1}, \quad j=1(1)J-1,\ k=1(1)K-1, \qquad (6.1.5)$$

where $s_x = \alpha_x\Delta t/(\Delta x)^2$, $s_y = \alpha_y\Delta t/(\Delta y)^2$. In general, this equation has a truncation error of $O\{\Delta t,(\Delta x)^2,(\Delta y)^2\}$ and contributes an error of $O\{(\Delta t)^2,\Delta t(\Delta x)^2,\Delta t(\Delta y)^2\}$ to the discretisation error e^n_j at each time step.

The stability of Equation (6.1.5) may be determined by application of the von Neumann method in the following way. Each Fourier component of the error distribution $\xi^n_{j,k}$ at the nth time level may be written

$$\xi^n_{j,k} = (G)^n e^{i\pi(m_x j\Delta x+m_y k\Delta y)}, \qquad (6.1.6)$$

where $(G)^n$ is the complex amplitude at time level n of the particular finite Fourier component whose wave numbers in the x and y directions are m_x and m_y, and where $i = \sqrt{-1}$. Defining x and y phase angles by $\beta_x = \pi m_x \Delta x$ and $\beta_y = \pi m_y \Delta y$, Equation (6.1.6) then becomes

$$\xi^n_{j,k} = (G)^n e^{i(j\beta_x + k\beta_y)} . \tag{6.1.7}$$

Substitution of (6.1.7) into the error equation corresponding to the finite difference equation (6.1.5) gives, on division by $(G)^n e^{i(j\beta_x + k\beta_y)}$,

$$G = 1 - 4s_x \sin^2(\beta_x/2) - 4s_y \sin^2(\beta_y/2). \tag{6.1.8}$$

Because G is real, the stability requirement $|G| \leqslant 1$ becomes

$$-1 \leqslant 1 - 4\{s_x \sin^2(\beta_x/2) + s_y \sin^2(\beta_y/2)\} \leqslant 1.$$

This is satisfied, and Equation (6.1.5) is stable, if

$$s_x + s_y \leqslant \tfrac{1}{2}. \tag{6.1.9}$$

If the dimensionless diffusion coefficients α_x, α_y and the grid spacings $\Delta x, \Delta y$ are such that $s_x = s_y = s$, then this relation becomes $s \leqslant \tfrac{1}{4}$, which is twice as restrictive on the time step Δt as in the one-dimensional case.

The three-dimensional diffusion equation is obtained from (6.1.2) by putting $u = v = w = 0$, giving

$$\frac{\partial \bar{\tau}}{\partial t} = \alpha_x \frac{\partial^2 \bar{\tau}}{\partial x^2} + \alpha_y \frac{\partial^2 \bar{\tau}}{\partial y^2} + \alpha_z \frac{\partial^2 \bar{\tau}}{\partial z^2} . \tag{6.1.10}$$

Using von Neumann's method, it can be shown that the corresponding FTCS finite difference equation is stable if

$$s_x + s_y + s_z \leqslant \tfrac{1}{2}. \tag{6.1.11}$$

If the dimensionless diffusion coefficients α_x, α_y, α_z and the grid-spacings Δx, Δy, Δz are such that $s_x = s_y = s_z = s$, then (6.1.11) becomes $s \leqslant 1/6$, which is more restrictive than either the one- or two-dimensional cases.

The Pure Convection Equation

The equation which describes the pure convection problem in two space dimensions is obtained from Equation (6.1.1) by setting $\alpha_x = \alpha_y = 0$; this gives

$$\frac{\partial \bar{\tau}}{\partial t} + u \frac{\partial \bar{\tau}}{\partial x} + v \frac{\partial \bar{\tau}}{\partial y} = 0. \tag{6.1.12}$$

If u and v are both positive, then using upwind differencing for the spatial derivatives and forward differencing for the time derivative, the finite difference form of (6.1.12) at the (j,k,n) grid-point is

$$\frac{\bar{\tau}^{n+1}_{j,k} - \bar{\tau}^{n}_{j,k}}{\Delta t} + O\{\Delta t\} + u\{\frac{\bar{\tau}^{n}_{j,k} - \bar{\tau}^{n}_{j-1,k}}{\Delta x} + O\{\Delta x\}\}$$

$$+ v\{\frac{\bar{\tau}^{n}_{j,k} - \bar{\tau}^{n}_{j,k-1}}{\Delta y} + O\{\Delta y\}\} = 0. \tag{6.1.13}$$

Rearranging and dropping terms of $O\{(\Delta t)^2, \Delta t\Delta x, \Delta t\Delta y\}$ gives an explicit two-level finite difference equation for finding an approximate solution to (6.1.12), namely

$$\tau_{j,k}^{n+1} = c_y \tau_{j,k-1}^{n} + c_x \tau_{j-1,k}^{n} + (1-c_x-c_y)\tau_{j,k}^{n}, \; j=1(1)J-1, \; k=1(1)K-1, \quad (6.1.14)$$

where $c_x = u\Delta t/\Delta x$ and $c_y = v\Delta t/\Delta y$.

Application of the von Neumann stability analysis shows that the gain $|G|$ for the (m_x, m_y) Fourier mode of the error distribution at any time level is given by the equation

$$|G|^2 = (1-c_x+c_x\cos\beta_x-c_y+c_y\cos\beta_y)^2 + (c_x\sin\beta_x+c_y\sin\beta_y)^2. \quad (6.1.15)$$

It can be shown that $|G|^2 \leqslant 1$ for all values of β_x and β_y, and therefore (6.1.14) is stable, if

$$c_x + c_y \leqslant 1. \quad (6.1.16)$$

If u, v and Δx, Δy are such that $c_x = c_y = c$, Equation (6.1.16) becomes $c \leqslant \frac{1}{2}$, a condition which is twice as restrictive as it is when the corresponding numerical method is used to solve the one-dimensional convection equation.

This method can also be used to solve approximately the three-dimensional convection problem, subject to the stability condition

$$c_x + c_y + c_z \leqslant 1, \quad (6.1.17)$$

or, if $c_x = c_y = c_z = c$, that $c \leqslant 1/3$.

The Transport Equation

Using central difference forms for the spatial derivatives with forward differencing in time in Equation (6.1.1) at the (j,k,n) grid-point gives, on rearranging and neglecting terms of $O\{(\Delta t)^2, \Delta t(\Delta x)^2, \Delta t(\Delta y)^2\}$, the explicit two-level finite difference equation

$$\tau_{j,k}^{n+1} = (s_x+\tfrac{1}{2}c_x)\tau_{j-1,k}^{n} + (s_y+\tfrac{1}{2}c_y)\tau_{j,k-1}^{n} + (1-2s_x-2s_y)\tau_{j,k}^{n}$$

$$+ (s_y-\tfrac{1}{2}c_y)\tau_{j,k+1}^{n} + (s_x-\tfrac{1}{2}c_x)\tau_{j+1,k}^{n}, \; j=1(1)J-1, \; k=1(1)K-1. \quad (6.1.18)$$

Substitution of Equation (6.1.7) for $\xi_{j,k}^{n}$ in the error equation corresponding to (6.1.18) gives the amplification factor G of the (m_x, m_y) Fourier component of the error distribution at any time level as it propagates to the next time level, whence the gain $|G|$ is found by solving the equation

$$|G|^2 = 1 - 4\{(s_x+s_y) - (s_x\cos\beta_x+s_y\cos\beta_y)\}$$

$$+ 4\{(s_x+s_y) - (s_x\cos\beta_x+s_y\cos\beta_y)\}^2 \quad (6.1.19)$$

$$+ (c_x\sin\beta_x)^2 + 2c_xc_y\sin\beta_x\sin\beta_y + (c_y\sin\beta_y)^2.$$

Let $c_x \leqslant 2s_x$ and $c_y \leqslant 2s_y$, so that (6.1.19) becomes

$$|G|^2 \leqslant 1 - 4\{(s_x+s_y) - (s_x\cos\beta_x+s_y\cos\beta_y)\}$$
$$+ 4\{(s_x+s_y)^2 - 2(s_x+s_y)(s_x\cos\beta_x+s_y\cos\beta_y)$$
$$+ (s_x)^2 + 2s_xs_y\cos(\beta_x-\beta_y) + (s_y)^2\}. \quad (6.1.20)$$

Since the maximum value of $\cos(\beta_x-\beta_y)$ is 1, then

$$(s_x)^2 + 2s_xs_y\cos(\beta_x-\beta_y) + (s_y)^2 \leqslant (s_x+s_y)^2,$$

and

$$|G|^2 \leqslant 1 - 4\{s_x(1 - \cos\beta_x) + s_y(1 - \cos\beta_y)\}\{1 - 2(s_x+s_y)\}. \quad (6.1.21)$$

Furthermore, because $|\cos(\beta_x)| \leqslant 1$ and $|\cos(\beta_y)| \leqslant 1$, it follows that $|G|^2 \leqslant 1$ so long as

$$s_x + s_y \leqslant \tfrac{1}{2}. \quad (6.1.22a)$$

Together with

$$c_x \leqslant 2s_x, \quad c_y \leqslant 2s_y, \quad (6.1.22b)$$

the relation (6.1.22) is sufficient for the stability of Equation (6.1.18).

If $s_x = s_y = s$, condition (6.1.22a) becomes $s \leqslant \frac{1}{4}$ which is twice as restrictive as the stability criterion for this method applied to the one-dimensional transport equation. Also, if $c_x = c_y = c$, then we require $c \leqslant 2s \leqslant \frac{1}{2}$, which is again more restrictive than the corresponding one-dimensional case.

If u and v are positive in Equation (6.1.1) applied at the (j,k,n) grid-point, using the upwind difference forms for the spatial derivatives in the convective terms with the forward difference form for the time derivative gives

$$\tau^{n+1}_{j,k} = (s_x+c_x)\tau^n_{j-1,k} + (s_y+c_y)\tau^n_{j,k-1} + (1 - 2s_x - 2s_y - c_x - c_y)\tau^n_{j,k}$$
$$+ s_x\tau^n_{j+1,k} + s_y\tau^n_{j,k+1}, \; j=1(1)J-1, \; k=1(1)K-1, \quad (6.1.23)$$

on rearranging and omitting terms of $O\{(\Delta t)^2, \Delta t\Delta x, \Delta t\Delta y\}$.

Application of the von Neumann method shows that the stability requirement for (6.1.23) is

$$2s_x + 2s_y + c_x + c_y \leqslant 1. \quad (6.1.24)$$

In particular, if $c_x = c_y = c$ and $s_x = s_y = s$, it follows that this upwind difference method is stable if

$$c + 2s \leqslant \tfrac{1}{2}, \quad (6.1.25)$$

which again is twice as restrictive as the upwind method applied in the one-dimensional case. In general, for any u and v the requirement on the time step is found to be

$$\Delta t \leqslant \{\frac{2\alpha_x}{(\Delta x)^2} + \frac{2\alpha_y}{(\Delta y)^2} + \frac{|u|}{\Delta x} + \frac{|v|}{\Delta y}\}^{-1}. \qquad (6.1.26)$$

A consistency analysis of Equation (6.1.23) yields the differential equation,at the (j,k,n) grid point,

$$\frac{\partial\bar{\tau}}{\partial t} + u\frac{\partial\bar{\tau}}{\partial x} + v\frac{\partial\bar{\tau}}{\partial y} = (\alpha_x+\alpha_x')\frac{\partial^2\bar{\tau}}{\partial x^2} + (\alpha_y+\alpha_y')\frac{\partial^2\bar{\tau}}{\partial y^2} + O\{\Delta t,(\Delta x)^2,(\Delta y)^2\} \qquad (6.1.27a)$$

where

$$\left.\begin{aligned} \alpha_x' &= \tfrac{1}{2}u\Delta x(1-c_x), \\ \alpha_y' &= \tfrac{1}{2}v\Delta y(1-c_y). \end{aligned}\right\} \qquad (6.1.27b)$$

As $\Delta x \to 0$, $\Delta y \to 0$, $\Delta t \to 0$ then $\alpha_x' \to 0$, $\alpha_y' \to 0$ and the higher order terms on the right side of Equation (6.1.27a) also tend to zero. Therefore, the finite difference equation (6.1.23) is consistent with the two-dimensional transport equation.

Even though the upwind differencing method leads to convergence of the solution of (6.1.23) to the solution of the two-dimensional transport equation if $c_x + c_y + 2s_x + 2s_y \leqslant 1$, the accuracy of the solution for a given Δx, Δy, Δt is strongly influenced by the false diffusion introduced by the terms $\alpha_x' \, \partial^2\bar{\tau}/\partial x^2$ and $\alpha_y' \, \partial^2\bar{\tau}/\partial y^2$. Unless $\alpha_x' \ll \alpha_x$ and $\alpha_y' \ll \alpha_y$, the presence of this artificial diffusion may produce a numerical solution which differs greatly from the true solution.

The appearance of the coefficients α_x' and α_y' is important because with small values of α_x and α_y, and large values of u and v, the false diffusion introduced will dominate the true diffusion unless Δx and Δy are sufficiently small. de Vahl Davis and Mallinson (1974) give examples from their work on forced convection in two directions which illustrate this point. They showed that the effect of false diffusion was greater in regions where the fluid was moving at 45 degrees to the two coordinate directions. Also, for a nominal Reynolds number of 1000 they found that the effective Reynolds number was as low as 240 in some regions, due to the effective increase in the diffusion coefficients α_x, α_y by the amounts α_x' and α_y'.

Although the use of upwind differences has a beneficial effect on the stability of the solution, the method only achieves this at the cost of reduced accuracy, which may be dramatic at large Reynolds numbers and may make the method almost useless. Whenever the upwind differencing method is used, the real diffusion coefficient α must be large enough to ensure that the numerical diffusion coefficient α' is negligible in comparison.

6.2 Implicit Finite Difference Methods

The stability restrictions which applied to the explicit equations described in the previous section for solving the transport equation in multi-dimensional space are overcome if implicit methods are used, thereby permitting the use of much larger time steps when using finite difference methods. For example, the Crank-Nicolson method for solving the two-dimensional *diffusion equation* (6.1.3) is based on evaluating the differential equation (6.1.3) at the point $(j\Delta x,k\Delta y,n\Delta t+\tfrac{1}{2}\Delta t)$, namely

$$\left.\frac{\partial\bar{\tau}}{\partial t}\right|_{j,k}^{n+\frac{1}{2}} = \alpha_x \left.\frac{\partial^2\bar{\tau}}{\partial x^2}\right|_{j,k}^{n+\frac{1}{2}} + \alpha_y \left.\frac{\partial^2\bar{\tau}}{\partial y^2}\right|_{j,k}^{n+\frac{1}{2}}. \qquad (6.2.1)$$

Replacing the space derivatives by the average of their values at the time levels n and $(n+1)$, gives

$$\left.\frac{\partial\bar{\tau}}{\partial t}\right|_{j,k}^{n+\frac{1}{2}} = \frac{\alpha_x}{2}\left\{\left.\frac{\partial^2\bar{\tau}}{\partial x^2}\right|_{j,k}^{n} + \left.\frac{\partial^2\bar{\tau}}{\partial x^2}\right|_{j,k}^{n+1} + O\{(\Delta t)^2\}\right\}$$

$$+ \frac{\alpha_y}{2}\left\{\left.\frac{\partial^2\bar{\tau}}{\partial y^2}\right|_{j,k}^{n} + \left.\frac{\partial^2\bar{\tau}}{\partial y^2}\right|_{j,k}^{n+1} + O\{(\Delta t)^2\}\right\}. \tag{6.2.2}$$

Replacing all derivatives by their centred difference forms, then rearranging as an implicit equation with $\bar{\tau}_{j-1}^{n+1}$, $\bar{\tau}_j^{n+1}$, $\bar{\tau}_{j+1}^{n+1}$ on the left side, and neglecting terms of $O\{(\Delta t)^3, \Delta t(\Delta x)^2, \Delta t(\Delta y)^2\}$, gives

$$-\tfrac{1}{2}s_x\tau_{j+1,k}^{n+1} - \tfrac{1}{2}s_x\tau_{j-1,k}^{n+1} + (1+s_x+s_y)\tau_{j,k}^{n+1} - \tfrac{1}{2}s_y\tau_{j,k+1}^{n+1} - \tfrac{1}{2}s_y\tau_{j,k-1}^{n+1}$$

$$= \tfrac{1}{2}s_x\tau_{j+1,k}^{n} + \tfrac{1}{2}s_x\tau_{j-1,k}^{n} + (1-s_x-s_y)\tau_{j,k}^{n}$$

$$+ \tfrac{1}{2}s_y\tau_{j,k+1}^{n} + \tfrac{1}{2}s_y\tau_{j,k-1}^{n} \;. \tag{6.2.3}$$

For $j=1(1)J-1$, $k=1(1)K-1$, the set of Equations (6.2.3) can be written with the unknown values of $\tau_{j,k}^{n+1}$ on the left side, and all known terms on the right side, namely the values of τ from the nth time level plus known values from the boundaries at the $(n+1)$th time level. By solving this system of linear algebraic equations the unknown values of τ at interior grid-points at the $(n+1)$th time level may be computed.

Application of von Neumann's stability analysis shows that the system of equations (6.2.3) is unconditionally stable: that is, errors introduced at any time level due to round-off or other causes do not magnify exponentially at higher time levels. However, it does require the solution, at each time level, of a system of equations of order equal to the number of unknown values of τ, namely $(J-1)\times(K-1)$ in the case of the rectangular region in space with known boundary values. The major problem with this method is therefore the large amount of computational time needed to obtain a solution at each time level using conventional elimination methods of solution. Because this system is not tri-diagonal, the very rapid Thomas algorithm cannot be used. However, iterative methods or special direct methods may be used to compute the values of τ in an economical fashion.

The set of equations (6.2.3) can be written with the unknown values of $\tau_{j,k}^{n+1}$ occurring in the $\{(k-1)(J-1)+j\}$th place and the equations written in order such that the one with the right side $d_{j,k}^{n}$ is the $\{(k-1)(J-1)+j\}$th equation. Then the term $(1+s_x+s_y)\tau_{j,k}^{n+1}$ lies on the leading diagonal of the resulting set of $(J-1)\times(K-1)$ linear equations. The terms $-\tfrac{1}{2}s_x\tau_{j-1,k}^{n+1}$ and $-\tfrac{1}{2}s_x\tau_{j+1,k}^{n+1}$ lie each side of the term $(1+s_x+s_y)\tau_{j,k}^{n}$ and the term $-\tfrac{1}{2}s_y\tau_{j,k-1}^{n+1}$ lies $(J-1)$ places to the left and $-\tfrac{1}{2}s_y\tau_{j,k+1}^{n+1}$ lies $(J-1)$ places to the right of the leading diagonal. Therefore the coefficient matrix has zero coefficients everywhere except on the leading diagonal, one diagonal each side of it, and two diagonals $(J-1)$ places each side of the leading diagonal.

Written in this way the coefficient matrix formed from all equations of the form (6.2.3) is diagonally dominant for all s_x and s_y, because

$$|1+s_x+s_y| > |-\tfrac{1}{2}s_x| + |-\tfrac{1}{2}s_x| + |-\tfrac{1}{2}s_y| + |-\tfrac{1}{2}s_y| \ . \tag{6.2.4}$$

Since diagonal dominance of the coefficient matrix of a set of linear equations implies unconditional stability in the solution of these equations, it follows that the implicit method defined by Equation (6.2.3) is always solvable. Some direct methods of solving such systems using elimination techniques are described in Mann (1981), and various iterative techniques are given in Colgan (1981).

When the full *transport equation* is considered, the form of the implicit finite difference equations change. Using the central difference approximation to the convection terms, the finite difference approximation to Equation (6.1.1) evaluated at the point $(j\Delta x, k\Delta y, n\Delta t+\tfrac{1}{2}\Delta t)$ is

$$\frac{\tau^{n+1}_{j,k} - \tau^{n}_{j,k}}{\Delta t} = \{P^{n+1}_{j,k} + P^{n}_{j,k}\}/2, \tag{6.2.5a}$$

where

$$P^n_{j,k} = -u\frac{(\tau^n_{j+1,k} - \tau^n_{j-1,k})}{2\Delta x} - v\frac{(\tau^n_{j,k+1} - \tau^n_{j,k-1})}{2\Delta y}$$

$$+\alpha_x\frac{(\tau^n_{j+1,k} - 2\tau^n_{j,k} + \tau^n_{j-1,k})}{(\Delta x)^2} + \alpha_y\frac{(\tau^n_{j,k+1} - 2\tau^n_{j,k} + \tau^n_{j,k-1})}{(\Delta y)^2} \tag{6.2.5b}$$

and $P^{n+1}_{j,k}$ is a similar expression in terms of τ^{n+1}. Equation (6.2.5a) has a truncation error of $O\{(\Delta t)^2,(\Delta x)^2,(\Delta y)^2\}$. Rearranging it gives the implicit finite difference equation

$$(s_y+\tfrac{1}{2}c_y)\tau^{n+1}_{j,k-1} + (s_x+\tfrac{1}{2}c_x)\tau^{n+1}_{j-1,k} - 2(1+s_x+s_y)\tau^{n+1}_{j,k}$$

$$+ (s_x-\tfrac{1}{2}c_x)\tau^{n+1}_{j+1,k} + (s_y-\tfrac{1}{2}c_y)\tau^{n+1}_{j,k+1} = Q^n_{j,k}\ ,\ j=1(1)J-1,\ k=1(1)K-1, \tag{6.2.6a}$$

where $Q^n_{j,k}$ contains all the terms at time level n, namely

$$Q^n_{j,k} = -(s_y+\tfrac{1}{2}c_y)\tau^n_{j,k-1} - (s_x+\tfrac{1}{2}c_x)\tau^n_{j-1,k} - 2(1-s_x-s_y)\tau^n_{j,k}$$

$$- (s_x-\tfrac{1}{2}c_x)\tau^n_{j+1,k} - (s_y-\tfrac{1}{2}c_y)\tau^n_{j,k+1} \ . \tag{6.2.6b}$$

At each time step, the use of Equation (6.2.6a) contributes an error of $O\{(\Delta t)^3,\Delta t(\Delta x)^2,\Delta t(\Delta y)^2\}$ to the discretisation error.

Diagonal dominance and hence solvability of the set of linear algebraic equations (6.2.6a) is assured if the set of unknowns $\tau^{n+1}_{j,k}$ are ordered as described previously, and

$$2(1+s_x+s_y) \geqslant (s_y+\tfrac{1}{2}c_y) + (s_x+\tfrac{1}{2}c_x) + |s_x-\tfrac{1}{2}c_x| + |s_y-\tfrac{1}{2}c_y| . \tag{6.2.7}$$

A number of cases need to be considered, depending on the sign of $s_x-\tfrac{1}{2}c_x$ and $s_y-\tfrac{1}{2}c_y$, if (6.2.7) is to be satisfied. For simplicity, assume that $s_x = s_y = s$ and $c_x = c_y = c$. Then the stability requirement is

$$1 + 2s \geqslant s+\tfrac{1}{2}c + |s-\tfrac{1}{2}c| .$$

For $c \leqslant 2s$, solvability of the system (6.2.6a) is assured. However, if $c > 2s$, the requirement becomes

$$c \leqslant 2s + 1. \qquad (6.2.8)$$

Therefore, c may exceed $2s$, but only by 1. Clearly, if the cell Reynolds number $r = c/s = u\Delta x/\alpha$ is less than $(2+s^{-1})$, then the above condition (6.2.8) is met.

The stability of (6.2.6a) can be investigated using the von Neumann method of analysis. Considering the simplified case with $s_x = s_y = s$, $c_x = c_y = c$, then the amplification factor G is found to be

$$G = \frac{2-4s+2s(\cos\beta_x+\cos\beta_y)-ic(\sin\beta_x+\sin\beta_y)}{2+4s-2s(\cos\beta_x+\cos\beta_y)+ic(\sin\beta_x+\sin\beta_y)}. \qquad (6.2.9)$$

Therefore

$$|G|^2 - 1 = \frac{-8s\{(1-\cos\beta_x)+(1-\cos\beta_y)\}}{\{2+4s-2s(\cos\beta_x+\cos\beta_y)\}^2+c^2\{\sin\beta_x+\sin\beta_y\}^2}, \qquad (6.2.10)$$

which is non-positive for all values of β_x and β_y since s is positive. Therefore $|G| \leqslant 1$ for all s and c, so the finite difference equation is unconditionally stable.

This finite difference method is therefore both stable and solvable if (6.2.8) is satisfied.

If the convection terms in the transport equation evaluated at the point $(j\Delta x, k\Delta y, n\Delta t+\frac{1}{2}\Delta t)$ are approximated by upwind differences, the corresponding finite difference equation for $u > 0$, $v > 0$ is

$$(s_y+c_y)\tau^{n+1}_{j,k-1} + (s_x+c_x)\tau^{n+1}_{j-1,k} - 2(1+s_x+s_y+\tfrac{1}{2}c_x+\tfrac{1}{2}c_y)\tau^{n+1}_{j,k}$$

$$+ s_x\tau^{n+1}_{j+1,k} + s_y\tau^{n+1}_{j,k+1} = Q^n_{j,k}, \quad j=1(1)J-1,\ k=1(1)K-1, \qquad (6.2.11a)$$

where $Q^n_{j,k}$ contains all the terms in τ^n and is given by

$$Q^n_{j,k} = -(s_y+c_y)\tau^n_{j,k-1} - (s_x+c_x)\tau^n_{j-1,k} - 2(1-s_x-s_y-\tfrac{1}{2}c_x-\tfrac{1}{2}c_y)\tau^n_{j,k}$$

$$- s_x\tau^n_{j+1,k} - s_y\tau^n_{j,k+1}. \qquad (6.2.11b)$$

Solvability of (6.2.11a) is assured if there is diagonal dominance of the coefficient matrix; that is, with the unknowns $\tau^{n+1}_{j,k}$ ordered as described previously, the condition to be met is that

$$|-2(1+s_x+s_y+\tfrac{1}{2}c_x+\tfrac{1}{2}c_y)| \geq |s_x+c_y| + |s_x+c_y| + |s_x| + |s_y|. \qquad (6.2.12)$$

Since this is true for all positive values of s_x, s_y, c_x, c_y, the system of equations (6.2.11) is always solvable.

The stability of the finite difference equation (6.2.11a) may be determined by the von Neumann method. If $s_x = s_y = s$, $c_x = c_y = c$, then the amplification factor is

$$G = \frac{-(s+c)e^{-i\beta_y} - (s+c)e^{-i\beta_x} - 2(1-2s-c) - se^{i\beta_x} - se^{i\beta_y}}{(s+c)e^{-i\beta_y} + (s+c)e^{-i\beta_x} - 2(1+2s+c) + se^{i\beta_x} + se^{i\beta_y}}$$

$$= \frac{\{2-(2s+c)(2-\cos\beta_x - \cos\beta_y)\} - ic\{\sin\beta_x + \sin\beta_y\}}{\{2+(2s+c)(2-\cos\beta_x - \cos\beta_y)\} + ic\{\sin\beta_x + \sin\beta_y\}}. \qquad (6.2.13)$$

Therefore

$$|G|^2 - 1 = \frac{\{2-(2s+c)(2-\cos\beta_x - \cos\beta_y)\}^2 - \{2+(2s+c)(2-\cos\beta_x - \cos\beta_y)\}^2}{\{2+(2s+c)(2-\cos\beta_x - \cos\beta_y)\}^2 + c^2\{\sin\beta_x + \sin\beta_y\}^2}$$

$$= \frac{-8(2s+c)\{(1-\cos\beta_x) + (1-\cos\beta_y)\}}{\{2+(2s+c)(2-\cos\beta_x - \cos\beta_y)\}^2 + c^2\{\sin\beta_x + \sin\beta_y\}^2}. \qquad (6.2.14)$$

Since s, c are positive and $-1 \le \cos\beta_x \le 1$, $-1 \le \cos\beta_y \le 1$, then the right side of (6.2.14) is always less than, or equal to, zero. Therefore

$$|G| \le 1 \quad \text{for all } s, c, \beta_x, \beta_y,$$

and (6.2.11) is unconditionally stable.

6.3 Time-Splitting Methods

Alternating Direction Implicit Methods

Using the Crank-Nicolson method to solve the two-dimensional diffusion equation (6.1.3) gave the unconditionally stable implicit finite difference equation (6.2.3). The resulting set of linear algebraic equations is not tri-diagonal so the Thomas algorithm cannot be used in their solution.

The *alternating direction implicit* method, sometimes called the method of variable direction, was introduced by Peaceman and Rachford (1955) and Douglas (1955) in order to develop an implicit finite difference method which only required the inversion of tri-diagonal systems of equations. This method requires two sets of calculations in order to take one step forward in time; a set of intermediate values τ^* with no physical significance are computed at the mid-point of the normal time interval Δt.

Consider the procedure of stepping from time level n to time level $(n+1)$, with boundary values $\bar{\tau}$ given on $x = 0,1$ and $y = 0,1$. Firstly, values of $\tau^*_{j,k}$ are calculated at the space position $(j\Delta x, k\Delta y)$ at the time $(n+\frac{1}{2})\Delta t$ using central difference approximations for both $\partial^2\bar{\tau}/\partial x^2$ and $\partial^2\bar{\tau}/\partial y^2$, the former being written at the time level $(n+\frac{1}{2})\Delta t$ and the latter at the time level $n\Delta t$. The resulting finite difference approximation for (6.1.3) is

$$\frac{\tau^*_{j,k} - \tau^n_{j,k}}{\frac{1}{2}\Delta t} = \alpha_x\left\{\frac{\tau^*_{j+1,k} - 2\tau^*_{j,k} + \tau^*_{j-1,k}}{(\Delta x)^2}\right\} + \alpha_y\left\{\frac{\tau^n_{j,k+1} - 2\tau^n_{j,k} + \tau^n_{j,k-1}}{(\Delta y)^2}\right\}. \qquad (6.3.1)$$

Using this equation for each $k=1(1)K-1$, the unknown values of τ^* at the time level $(n+\frac{1}{2})\Delta t$ are computed along a line of points $j\Delta x$, $j=1(1)J-1$, in the x-direction using known values of τ^n which lie along the lines of points $k\Delta y$, $(k-1)\Delta y$, $(k+1)\Delta y$ and known values of $\tau^*_{j,k} = \bar{\tau}^{n+\frac{1}{2}}_{j,k}$ on the boundaries $x = 0$ and $x = 1$.

The rearranged form of Equation (6.3.1) with only values of τ^* on the left side is, for $k = 1(1)K-1$,

$$-\tfrac{1}{2}s_x\tau^*_{j-1,k} + (1+s_x)\tau^*_{j,k} - \tfrac{1}{2}s_x\tau^*_{j+1,k}$$

$$= \tfrac{1}{2}s_y\tau^n_{j,k-1} + (1-s_y)\tau^n_{j,k} + \tfrac{1}{2}s_y\tau^n_{j,k+1},\ j=1(1)J-1. \qquad (6.3.2)$$

The system of equations obtained for a given k by substituting $j=1,2,3,\ldots$ in Equation (6.3.1) and using the known values $\tau^n_{j,k}$, $\tau^n_{j,k\pm1}$, $j=0(1)J$, $\tau^*_{0,k}$ and $\tau^*_{J,k}$, is tri-diagonal, so it can be solved using the Thomas algorithm. On a rectangular grid where $j=0(1)J$, and $k=0(1)K$, then $(K-1)$ applications of the Thomas algorithm gives the intermediate values $\tau^*_{j,k}$ for all interior grid-points.

Checking the stability of the finite difference equation (6.3.2) by the von Neumann method, the error distribution at the time level $n\Delta t$ is considered to have spatial Fourier components of the form $\xi^n_{j,k} = e^{i(j\beta_x+k\beta_y)}$ while it has the component $\xi^*_{j,k} = G'e^{i(j\beta_x+k\beta_y)}$ at the time level $(n+\tfrac{1}{2})\Delta t$. The amplification factor of this error when computing the intermediate values at time level $(n+\tfrac{1}{2})\Delta t$ from those at time level $n\Delta t$ is

$$G' = \frac{1-2s_y\sin^2(\beta_y/2)}{1+2s_x\sin^2(\beta_x/2)}. \qquad (6.3.3)$$

For $\beta_x \to 0$ and $\beta_y \to \pi$, $|G'| > 1$ for sufficiently large values of $s_y > s_x$. This step is therefore only conditionally stable.

At the second step, values of $\tau^{n+1}_{j,k}$ are computed from the values of $\tau^*_{j,k}$ found by the first step, by writing the central difference approximation for $\partial^2\bar{\tau}/\partial x^2$ at the intermediate time level $(n+\tfrac{1}{2})\Delta t$ and the central difference approximation for $\partial^2\bar{\tau}/\partial y^2$ at the new time level $(n+1)\Delta t$. The resulting finite difference approximation for the two-dimensional diffusion equation is

$$\frac{\tau^{n+1}_{j,k} - \tau^*_{j,k}}{\tfrac{1}{2}\Delta t} = \alpha_x\left\{\frac{\tau^*_{j+1,k} - 2\tau^*_{j,k} + \tau^*_{j-1,k}}{(\Delta x)^2}\right\}$$

$$+ \alpha_y\left\{\frac{\tau^{n+1}_{j,k+1} - 2\tau^{n+1}_{j,k} + \tau^{n+1}_{j,k-1}}{(\Delta y)^2}\right\}, \qquad (6.3.4)$$

which can be rewritten in the implicit form for each $j=1(1)J-1$,

$$-\tfrac{1}{2}s_y\tau^{n+1}_{j,k-1} + (1+s_y)\tau^{n+1}_{j,k} - \tfrac{1}{2}s_y\tau^{n+1}_{j,k+1}$$

$$= \tfrac{1}{2}s_x\tau^*_{j-1,k} + (1-s_x)\tau^*_{j,k} + \tfrac{1}{2}s_x\tau^*_{j+1,k},\ k=1(1)K-1. \qquad (6.3.5)$$

In contrast to Equation (6.3.2), Equation (6.3.5) is implicit in the y-direction. Along each grid line $x = j\Delta x$, $j=1(1)J-1$, a tri-diagonal system of equations is obtained which can be solved by means of the Thomas algorithm. With $(J-1)$ applications of this algorithm, the values of τ can be found at all interior grid-points at the time level $(n+1)$. The values of the intermediate variable $\tau^*_{j,k}$ have no physical significance; they should be discarded as soon as the values at the next time level have been calculated.

Application of the von Neumann stability analysis shows that the second step (6.3.5) is also only conditionally stable, the amplification factor of the error propagation being

$$G'' = \frac{1-2s_x \sin^2(\beta_x/2)}{1+2s_y \sin^2(\beta_y/2)} . \tag{6.3.6}$$

However, the double process in which Equation (6.3.2) is followed by Equation (6.3.5) at each time step, has an amplification factor of

$$G = G'G'' = \left\{\frac{1-2s_y \sin^2(\beta_y/2)}{1+2s_x \sin^2(\beta_x/2)}\right\} \cdot \left\{\frac{1-2s_x \sin^2(\beta_x/2)}{1+2s_y \sin^2(\beta_y/2)}\right\} , \tag{6.3.7}$$

and $|G| \leqslant 1$ for all values of s_x, s_y, β_x, β_y. The two conditionally stable steps have combined to produce a method which is always stable.

Although it may not seem plausible that intermediate values which have no significance, computed half a time step forward, can be used to obtain correct values another half time step later, it was seen in Section 4.2 that the Crank-Nicolson equation for the one-dimensional diffusion equation can be considered the result of such a process.

This method has a truncation error of $O\{(\Delta t)^2,(\Delta x)^2,(\Delta y)^2\}$. The second order accuracy is obvious when Equations (6.3.1) and (6.3.4) are added to give

$$\frac{\tau^{n+1}_{j,k} - \tau^{n}_{j,k}}{\Delta t} = \alpha_x \left\{\frac{\tau^{*}_{j+1,k} - 2\tau^{*}_{j,k} + \tau^{*}_{j-1,k}}{(\Delta x)^2}\right\}$$

$$+ \tfrac{1}{2}\alpha_y \left\{\frac{\tau^{n}_{j,k+1} - 2\tau^{n}_{j,k} + \tau^{n}_{j,k-1}}{(\Delta y)^2} + \frac{\tau^{n+1}_{j,k+1} - 2\tau^{n+1}_{j,k} + \tau^{n+1}_{j,k-1}}{(\Delta y)^2}\right\}. \tag{6.3.8}$$

The truncation error is of order $(\Delta t)^2$ because the central difference approximation was used about the point $(j\Delta x, k\Delta y, n\Delta t+\tfrac{1}{2}\Delta t)$ for the derivative $\partial\bar{\tau}/\partial t$. The error is also second order in Δx because the central difference approximation was used for the spatial derivative $\partial^2\bar{\tau}/\partial x^2$ at the same point $(j\Delta x, k\Delta y, n\Delta t+\tfrac{1}{2}\Delta t)$. The truncation error for the approximation to $\partial^2\bar{\tau}/\partial y^2$ at that point is $O\{(\Delta t)^2,(\Delta y)^2\}$, since it is the average of central difference approximations at the time levels n and $(n+1)$. Clearly, the complete truncation error is $O\{(\Delta t)^2,(\Delta x)^2,(\Delta y)^2\}$, and the contribution to the discretisation error at each time step is $O\{(\Delta t)^3,\Delta t(\Delta x)^2,\Delta t(\Delta y)^2\}$.

The alternating direction implicit method can also be applied to solve the three-dimensional diffusion problem. The most obvious method, described by Brian (1961), uses the same process as the two-dimensional case; that is, the time interval Δt is divided up into three equal parts. Hence, three calculations are performed with two sets of artificial intermediate values being calculated at $(n+1/3)\Delta t$ and $(n+2/3)\Delta t$. This method, however, has only first-order accuracy in time and becomes unstable for $s > 3/2$.

Another method for the advancement of the 3D diffusion equation over the time interval Δt, which is second order accurate in time and is unconditionally stable, is described by Douglas (1962). This method is once again a three-step process but differs from Brian's method in that the two intermediate fictitious values are not calculated at the times $(n+1/3)\Delta t$ and $(n+2/3)\Delta t$ but rather they are used to find successive approximations for $\bar{\tau}$ at the $(n+1)$th time level. That is, this process is similar to a three-step iterative procedure for calculating τ^{n+1} in which the formula for the iteration process changes at each iteration.

This method gives the following finite difference equations where $\tau^{*}_{j,k,\ell}$ and $\tau^{**}_{j,k,\ell}$ refer to the first and second approximations respectively for the value of $\tau^{n+1}_{j,k,\ell}$ at the grid-point $(x_j, y_k, z_\ell, t_{n+1})$:

$$\frac{\tau^{*}_{j,k,\ell} - \tau^{n}_{j,k,\ell}}{\Delta t} = \frac{\alpha_x}{2}\left\{\frac{\tau^{*}_{j+1,k,\ell} - 2\tau^{*}_{j,k,\ell} + \tau^{*}_{j-1,k,\ell}}{(\Delta x)^2} + \frac{\tau^{n}_{j+1,k,\ell} - 2\tau^{n}_{j,k,\ell} + \tau^{n}_{j-1,k,\ell}}{(\Delta x)^2}\right\} + \alpha_y\left\{\frac{\tau^{n}_{j,k+1,\ell} - 2\tau^{n}_{j,k,\ell} + \tau^{n}_{j,k-1,\ell}}{(\Delta y)^2}\right\} + \alpha_z\left\{\frac{\tau^{n}_{j,k,\ell+1} - 2\tau^{n}_{j,k,\ell} + \tau^{n}_{j,k,\ell-1}}{(\Delta z)^2}\right\}, \qquad (6.3.9a)$$

which is implicit in $\tau^{*}_{j-1,k,\ell}$, $\tau^{*}_{j,k,\ell}$ and $\tau^{*}_{j+1,k,\ell}$ for each value of k and ℓ;

$$\frac{\tau^{**}_{j,k,\ell} - \tau^{n}_{j,k,\ell}}{\Delta t} = \frac{\alpha_x}{2}\left\{\frac{\tau^{*}_{j+1,k,\ell} - 2\tau^{*}_{j,k,\ell} + \tau^{*}_{j-1,k,\ell}}{(\Delta x)^2} + \frac{\tau^{n}_{j+1,k,\ell} - 2\tau^{n}_{j,k,\ell} + \tau^{n}_{j-1,k,\ell}}{(\Delta x)^2}\right\} + \frac{\alpha_y}{2}\left\{\frac{\tau^{**}_{j,k+1,\ell} - 2\tau^{**}_{j,k,\ell} + \tau^{**}_{j,k-1,\ell}}{(\Delta y)^2} + \frac{\tau^{n}_{j,k+1,\ell} - 2\tau^{n}_{j,k,\ell} + \tau^{n}_{j,k-1,\ell}}{(\Delta y)^2}\right\} + \alpha_z\left\{\frac{\tau^{n}_{j,k,\ell+1} - 2\tau^{n}_{j,k,\ell} + \tau^{n}_{j,k,\ell-1}}{(\Delta z)^2}\right\}. \qquad (6.3.9b)$$

which is implicit in $\tau^{**}_{j,k-1,\ell}$, $\tau^{**}_{j,k,\ell}$ and $\tau^{**}_{j,k+1,\ell}$ for each j and ℓ;

$$\frac{\tau^{n+1}_{j,k,\ell} - \tau^{n}_{j,k,\ell}}{\Delta t} = \frac{\alpha_x}{2}\left\{\frac{\tau^{*}_{j+1,k,\ell} - 2\tau^{*}_{j,k,\ell} + \tau^{*}_{j-1,k,\ell}}{(\Delta x)^2} + \frac{\tau^{n}_{j+1,k,\ell} - 2\tau^{n}_{j,k,\ell} + \tau^{n}_{j-1,k,\ell}}{(\Delta x)^2}\right\}$$

$$+ \frac{\alpha_y}{2}\left\{\frac{\tau^{**}_{j,k+1,\ell} - 2\tau^{**}_{j,k,\ell} + \tau^{**}_{j,k-1,\ell}}{(\Delta y)^2}\right.$$

$$\left. + \frac{\tau^{n}_{j,k+1,\ell} - 2\tau^{n}_{j,k,\ell} + \tau^{n}_{j,k-1,\ell}}{(\Delta y)^2}\right\}$$

$$+ \frac{\alpha_z}{2}\left\{\frac{\tau^{n+1}_{j,k,\ell+1} - 2\tau^{n+1}_{j,k,\ell} + \tau^{n+1}_{j,k,\ell-1}}{(\Delta z)^2}\right.$$

$$\left. + \frac{\tau^{n}_{j,k,\ell+1} - 2\tau^{n}_{j,k,\ell} + \tau^{n}_{j,k,\ell-1}}{(\Delta z)^2}\right\} \qquad (6.3.9c)$$

which is implicit in $\tau^{n+1}_{j,k,\ell-1}$, $\tau^{n+1}_{j,k,\ell}$ and $\tau^{n+1}_{j,k,\ell+1}$. This method becomes only conditionally stable if in Equation (6.3.9c) the most recent approximation τ^{**} is used instead of τ^{*} in the finite difference approximation to the derivative $\partial^2\bar{\tau}/\partial x^2$ (see Richtmeyer and Morton, 1967). The Thomas algorithm can again be used to solve the sets of implicit equations which result from applying this method.

Alternating direction implicit methods are used widely for the solution of parabolic and hyperbolic partial differential equations in both two and three dimensions. However, Birkhoff and Varga (1959) point out that the unconditional stability of the method applied to the diffusion equation (see Douglas 1955, 1957) does not necessarily apply to differential equations with variable coefficients nor to non-rectangular regions. Nevertheless, numerical experimentation indicates that they are applicable to a wider range of partial differential equations than those with constant coefficients prescribed on a rectangular domain.

When alternating direction implicit methods are applied to the complete transport equation, it is found (see Pearson, 1964) that the inclusion of the advection terms does not change the unconditional stability.

Locally One-Dimensional Methods

Recently, Russian numerical analysists such as D'Yakonov (1963), Samarskii (1964), Yanenko (1971) and Marchuk (1975) have developed a splitting of time dependent partial differential equations in two or more space variables, so that the resulting set of partial differential equations are one-dimensional in space. These methods are called *locally one-dimensional*.

By this method the two-dimensional diffusion equation (6.1.3) is split into the pair of equations

$$\tfrac{1}{2}\frac{\partial\bar{\tau}}{\partial t} = \alpha_x \frac{\partial^2\bar{\tau}}{\partial x^2}\ , \qquad (6.3.10a)$$

and

$$\tfrac{1}{2}\frac{\partial\bar{\tau}}{\partial t} = \alpha_y \frac{\partial^2\bar{\tau}}{\partial y^2}\ . \qquad (6.3.10b)$$

In advancing a calculation from the time level n to time level $(n+1)$ it is assumed that Equation (6.3.10a) holds from $t = n\Delta t$ to $t = (n+\tfrac{1}{2})\Delta t$ giving at each point in space a set of values τ^* which have no physical significance and must be discarded. Using the values of τ^*, the set of values of τ at $t = (n+1)\Delta t$ are computed using Equation (6.3.10b).

The split equations (6.3.10) can be solved by either explicit or implicit methods. Application of the FTCS method to these two equations leads to the explicit forms

$$\tau^{*}_{j,k} = s_x \tau^{n}_{j-1,k} + (1-2s_x)\tau^{n}_{j,k} + s_x \tau^{n}_{j+1,k}, \quad j=1(1)J-1, \quad k=1(1)K-1, \qquad (6.3.11a)$$

and

$$\tau^{n+1}_{j,k} = s_y \tau^{*}_{j,k-1} + (1-2s_y)\tau^{*}_{j,k} + s_y \tau^{*}_{j,k+1}, \quad j=1(1)J-1, \quad k=1(1)K-1. \qquad (6.3.11b)$$

The stability condition for the pair of equations (6.3.11) can be determined by the von Neumann method. Inserting

$$\xi^{n}_{j,k} = e^{i(j\beta_x + k\beta_y)},$$

the Fourier component of the error distribution at the nth time level, into the corresponding error equations to (6.3.11a,b) gives the amplification factor

$$G = \{1-4s_x \sin^2(\beta_x/2)\}\{1-4s_y \sin^2(\beta_y/2)\} \qquad (6.3.12)$$

for the propagation from the time level n to $(n+1)$. Therefore the stability requirement $|G| \leqslant 1$ holds if

$$|1-4s_x \sin^2(\beta_x/2)|\,|1-4s_y \sin^2(\beta_y/2)| \leqslant 1. \qquad (6.3.13)$$

This condition is satisfied if both $|1-4s_x \sin^2(\beta_x/2)| \leqslant 1$ and $|1-4s_y \sin^2(\beta_y/2)| \leqslant 1$. Using the results of Section 3.3, this is true if both $s_x \leqslant \frac{1}{2}$ and $s_y \leqslant \frac{1}{2}$.

When the non-dimensional diffusion coefficients α_x, α_y and the grid-spacings Δx, Δy are such that $s_x = s_y = s$, the stability requirement becomes $s \leqslant \frac{1}{2}$. This is an improvement on the usual explicit restriction of $s \leqslant \frac{1}{4}$ which holds for the one-step FTCS method of solving the two-dimensional diffusion equation. This is because Equations (6.3.11) relate the value of τ at one point at the time level $(n+1)$ with values at nine points at the time level n, since elimination of τ^* from these finite difference equations gives

$$\begin{aligned}\tau^{n+1}_{j,k} = {} & s_x s_y \tau^{n}_{j-1,k-1} + (1-2s_x)s_y \tau^{n}_{j,k-1} + s_x s_y \tau^{n}_{j+1,k-1} \\ & + s_x(1-s_y)\tau^{n}_{j-1,k} + (1-2s_x)(1-2s_y)\tau^{n}_{j,k} + s_x(1-s_y)\tau^{n}_{j+1,k} \qquad (6.3.14) \\ & + s_x s_y \tau^{n}_{j-1,k+1} + (1-2s_x)s_y \tau^{n}_{j,k+1} + s_x s_y \tau^{n}_{j+1,k+1}.\end{aligned}$$

This contrasts with the one-step FTCS explicit method (6.1.5) which relates a value of τ at any time level with values of τ at only five points at the previous time level.

Two other features of this locally one-dimensional method are worth noting. Firstly, (6.3.14) approximates the two-dimensional diffusion equation (6.1.3) with a truncation error of $O\{\Delta t, (\Delta x)^2, (\Delta y)^2\}$, the same as the one-step FTCS method of solution. Secondly, the minimum number of operations required for a single time step of the one-step FTCS method (6.1.5) is about 3JK, which is less than for the time-split method which requires approximately 4JK multiplications or divisions, whether (6.3.11a,b) or (6.3.14) is used. But the more restrictive stability requirement of (6.1.5) means that twice as many time steps are required to reach the same time level if the maximum time steps are taken, thereby doubling the effective number of applications of the one-step FTCS method. This method therefore requires fifty percent more computer time than the locally one-dimensional FTCS scheme.

Alternatively, implicit techniques may be used to solve (6.3.10a,b). Applying the Crank-Nicolson method gives the approximation to the Equation (6.3.10a) at the point $(j\Delta x, k\Delta y, n\Delta t+\frac{1}{4}\Delta t)$ in the solution domain,

$$\tfrac{1}{2}\left\{\frac{\tau^*_{j,k} - \tau^n_{j,k}}{\tfrac{1}{2}\Delta t}\right\} = \frac{\alpha_x}{2}\left\{\frac{\tau^*_{j+1,k} - 2\tau^*_{j,k} + \tau^*_{j-1,k}}{(\Delta x)^2} + \frac{\tau^n_{j+1,k} - 2\tau^n_{j,k} + \tau^n_{j-1,k}}{(\Delta x)^2}\right\} . \tag{6.3.15}$$

Rearrangement yields, for each $k=1(1)K-1$, the set of implicit finite difference equations

$$-\tfrac{1}{2}s_x\tau^*_{j-1,k} + (1+s_x)\tau^*_{j,k} - \tfrac{1}{2}s_x\tau^*_{j+1,k}$$
$$= \tfrac{1}{2}s_x\tau^n_{j-1,k} + (1-s_x)\tau^n_{j,k} + \tfrac{1}{2}s_x\tau^n_{j+1,k}, \quad j=1(1)J-1. \tag{6.3.16}$$

Since $\tau^*_{0,k} = \bar{\tau}^{n+\frac{1}{2}}_{0,k}$, $\tau^*_{J,k} = \bar{\tau}^{n+\frac{1}{2}}_{J,k}$ are known boundary values for each k, then transferring these values to the right side of (6.3.16) with the known values $\tau^n_{j,k}$ gives a tri-diagonal set of algebraic equations in the unknowns $\tau^*_{j,k}$, $j=1(1)J-1$. This set of equations is solved using the Thomas algorithm.

Similarly, the Crank-Nicolson approximation to Equation (6.3.10b) yields, for each $j=1(1)J-1$, the set of implicit finite difference equations

$$-\tfrac{1}{2}s_y\tau^{n+1}_{j,k-1} + (1+s_y)\tau^{n+1}_{j,k} - \tfrac{1}{2}s_y\tau^{n+1}_{j,k+1}$$
$$= \tfrac{1}{2}s_y\tau^*_{j,k-1} + (1-s_y)\tau^*_{j,k} + \tfrac{1}{2}s_y\tau^*_{j,k+1}, \quad k=1(1)K-1. \tag{6.3.17}$$

For each j, use of the known values of $\tau^*_{j,k}$, $k=0(1)K$, with the boundary values $\bar{\tau}^{n+1}_{j,0}$ and $\bar{\tau}^{n+1}_{j,K}$ produces a tri-diagonal set of algebraic equations, again solved using the Thomas algorithm.

The von Neumann stability analysis shows that the system of Equations (6.3.16), (6.3.17) is unconditionally stable. For an error distribution of the form $\xi^n_{j,k} = e^{i(j\beta_x + k\beta_y)}$ at the time level n, the amplification factor produced by application of the finite difference equation (6.3.16) is

$$G' = \frac{1-2s_x\sin^2(\beta_x/2)}{1+2s_x\sin^2(\beta_x/2)} .$$

Similarly, the amplification factor produced by application of (6.3.17) to step from the values of $\tau^*_{j,k}$ at time $(n+\frac{1}{2})\Delta t$ to $\tau^{n+1}_{j,k}$ is

$$G'' = \frac{1-2s_y\sin^2(\beta_y/2)}{1+2s_y\sin^2(\beta_y/2)} .$$

The amplification factor for the complete step from time level n to $(n+1)$ is

$$G = G'G'',$$

and as $|G'| \leq 1$ for all s_x, β_x, $|G''| \leq 1$ for all s_y, β_y, then $|G| \leq 1$ for all s_x, s_y, and β_x, β_y. Therefore the combination of (6.3.16) with (6.3.17) is stable for all s_x, s_y.

Each of the implicit steps (6.3.16) and (6.3.17) is diagonally dominant for all s_x, s_y respectively. Therefore this locally one-dimensional method is solvable for all s_x, s_y

In a similar way, the time-dependent diffusion equation (6.1.10) in three space dimensions can be split into three parts,

$$\frac{1}{3}\frac{\partial\bar{\tau}}{\partial t} = \alpha_x \frac{\partial^2\bar{\tau}}{\partial x^2}, \tag{6.3.18a}$$

$$\frac{1}{3}\frac{\partial\bar{\tau}}{\partial t} = \alpha_y \frac{\partial^2\bar{\tau}}{\partial y^2}, \tag{6.3.18b}$$

and

$$\frac{1}{3}\frac{\partial\bar{\tau}}{\partial t} = \alpha_z \frac{\partial^2\bar{\tau}}{\partial z^2}. \tag{6.3.18c}$$

Again this gives a set of one-dimensional equations to be solved by either explicit or implicit methods for each time step of length $\Delta t/3$ to give values of τ^{n+1} at the end of the final step. Thus Equation (6.3.18a) is solved over the interval $t = n\Delta t$ to $t = (n+1/3)\Delta t$ to give a fictitious set of values τ^* at the time $t = (n+1/3)\Delta t$; Equation (6.3.18b) is then used with the values τ^* at the beginning of the time interval from $(n+1/3)\Delta t$ to $(n+2/3)\Delta t$ to obtain a fictitious set of values τ^{**} at the time $t = (n+2/3)\Delta t$; these are used at the start of a further time step of $\Delta t/3$ with Equation (6.3.18c) to give an approximate value for $\bar{\tau}$ at the time $t = (n+1)\Delta t$.

Like the two-dimensional case, the implicit method is both stable and solvable for all s_x, s_y, s_z, whereas the explicit method is only conditionally stable.

Gourlay and Mitchell (1969) have pointed out an interesting connection between the alternating direction implicit method given by Equations (6.3.2) and (6.3.5) and the locally one-dimensional method described by (6.3.16), (6.3.17), for solving the two-dimensional time dependent diffusion equation.

Because they are relatively new, not many applications of locally one-dimensional methods can be found in the literature, but their use is increasing, particularly because of their rather simple form when applied to three-dimensional problems.

6.4 Incorporating Boundary Conditions at Irregular Boundaries

When the two-or three-dimensional physical region to which the partial differential equation applies has an irregular boundary, that is, when segments of the boundary cannot be represented by a constant spatial coordinate, say x equals a constant in Figure 6.1, finite differences can still be employed to approximately solve the equation. However, care must be taken if the finite difference approximations to the boundary conditions are to be as accurate as the finite difference equation applied to the interior of the space region.

In the following, we suppose that the two-dimensional transport equation (6.1.1) is to be solved in a physical region R such as that shown in Figure 6.1, bounded by the curve R^+. As seen in Section 4.4, along different parts of R^+ the boundary conditions may be classified as one of three types. Either,

(a) the dependent variable $\bar{\tau}$ is specified at each point (x,y) on the boundary of the space region R. As mentioned previously this type of boundary

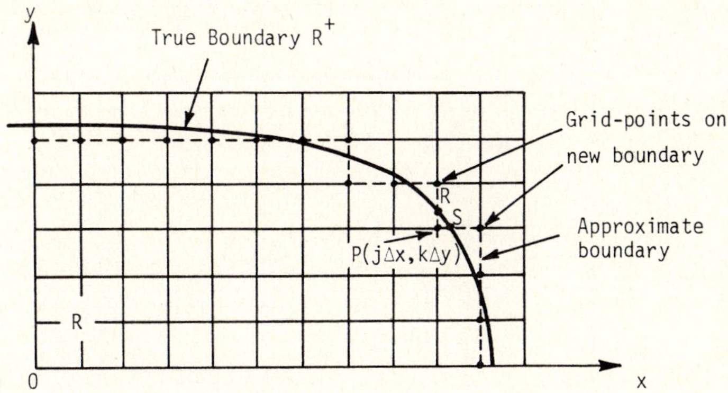

Figure 6.1: An irregular boundary in the rectangular cartesian coordinate system.

condition is said to be of <u>Dirichlet</u> type; or

(b) the normal derivative to the boundary is specified at each point on the boundary. This type of boundary condition is referred to as the <u>Neumann type</u>; or

(c) a combination of the Dirichlet and Neumann boundary conditions in which a linear function of $\bar{\tau}$ and $\partial\bar{\tau}/\partial\eta$ is specified at points on the boundary. This is called the <u>mixed boundary condition</u>.

In the process of finding a solution to the two-dimensional transport equation (6.1.1), these three types of boundary conditions must be treated separately.

THE DEPENDENT VARIABLE SPECIFIED ON THE BOUNDARY - THE DIRICHLET CONDITION

Let the value of the dependent variable be given by

$$\bar{\tau}(x,y,t) = B(x,y,t) \tag{6.4.1}$$

for all points (x,y) on the boundary R^+.

Transfer of boundary values

The most obvious way to incorporate known values on a boundary into a finite difference scheme is to choose the interior gridpoint which lies closest to the point where the boundary intersects a grid-line, and at that gridpoint let the approximation τ take the boundary value. This has the effect of reshaping the boundary so that it conforms to the discrete points of the grid.

The dashed line in Figure 6.1 shows how a reshaped boundary could be drawn through gridpoints in the (x,y) plane if this method is used. Once the boundary is altered in this way the problem can be solved by any of the methods discussed previously.

The obvious disadvantage of treating the boundary condition in the above way is the low order of accuracy which results when the finite difference equations are applied at interior gridpoints near the boundary, since the shift of value introduces a contribution of either $O\{\Delta x\}$ or $O\{\Delta y\}$ to the discretisation error.

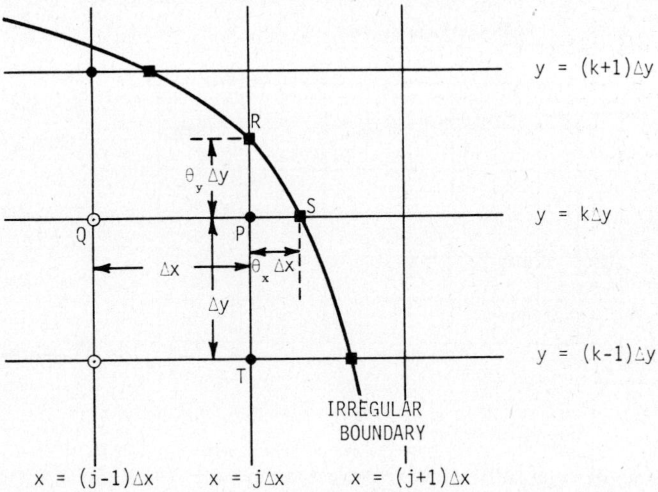

Figure 6.2: Incorporating boundary values into finite difference equations

Taylor series method

This method generally results in a higher order of accuracy than reshaping the boundary so that it conforms with segments of the grid-lines. The extra work required is generally worthwhile because, in this case, the same sized truncation error at the boundary can be obtained using a coarse grid as that obtained using a very fine grid when boundary values are transferred to nearby gridpoints.

In this method, at any time level, the gridpoints in the region R are divided into two disjoint sets. Regular interior gridpoints are those for which no neighbouring gridpoint in the direction of either coordinate axis lies outside the boundary. In Figure 6.2, these points are denoted by open circles. Interior gridpoints in R which have at least one neighbouring gridpoint in the direction of either coordinate axis lying outside the boundary are termed irregular points. These points are denoted by closed circles in Figure 6.2.

For regular points the usual finite difference forms for the derivatives apply, and equations such as those described in Sections 6.1 and 6.2 result. For irregular interior points the finite difference forms for the derivatives are altered in the following way. Discrete points are placed on the boundary at the intersection of the boundary with each grid line. These are shown as closed squares in Figure 6.2.

Consider the point $P(j\Delta x,k\Delta y)$ in the figure. Denote by $B^n_{j,k+\theta_y}$ the known value of $\bar{\tau}^n_R$ at time $t = n\Delta t$, on the boundary where it cuts the grid line between P and $(j\Delta x,k\Delta y+\Delta y)$, a distance $\theta_y\Delta y$ from the point P. Also denote by $B^n_{j+\theta_x,k}$ the known value of $\bar{\tau}^n_S$ on the boundary where it cuts the grid line $y = k\Delta y$ between $x = j\Delta x$ and $(j+1)\Delta x$, a distance $\theta_x\Delta x$ from the point P, at time $t = n\Delta t$.

To illustrate the construction of the finite difference approximations for various spatial derivatives at irregular gridpoints at time $n\Delta t$, we will obtain $\partial\bar{\tau}/\partial x$, $\partial\bar{\tau}/\partial y$, $\partial^2\bar{\tau}/\partial x^2$, $\partial^2\bar{\tau}/\partial y^2$ simultaneously at P in terms of values of $\bar{\tau}$ at P and the four points R, S, T, Q surrounding P. Now $\bar{\tau}$ is known at the boundary, so $\bar{\tau}_R^n = B_{j,k+\theta_y}$, $\bar{\tau}_S^n = B_{j+\theta_x,k}$ are known. Clearly, the usual central difference approximations at P cannot be used since $\bar{\tau}_Q^n, \bar{\tau}_T^n, \bar{\tau}_R^n, \bar{\tau}_S^n$ are not evenly spaced about P along either grid line. Thus we have

$$\begin{aligned}
\bar{\tau}_Q^n &= \bar{\tau}_{j-1,k}^n &&= \bar{\tau}(j\Delta x-\Delta x, k\Delta y, n\Delta t)\\
\bar{\tau}_T^n &= \bar{\tau}_{j,k-1}^n &&= \bar{\tau}(j\Delta x, k\Delta y-\Delta y, n\Delta t)\\
\bar{\tau}_R^n &= \bar{\tau}_{j,k+\theta_y}^n &&= \bar{\tau}(j\Delta x, k\Delta y+\theta_y\Delta y, n\Delta t)\\
\bar{\tau}_S^n &= \bar{\tau}_{j+\theta_x,k}^n &&= \bar{\tau}(j\Delta x+\theta_x\Delta x, k\Delta y, n\Delta t).
\end{aligned} \tag{6.4.2}$$

Expanding these functions of $\bar{\tau}$ in their Taylor series about $(j\Delta x, k\Delta y, n\Delta t)$ gives, up to terms of order $(\Delta x)^3$ and $(\Delta y)^3$, the four equations

$$\begin{aligned}
\bar{\tau}_{j-1,k}^n &= \bar{\tau}_{j,k}^n - \Delta x \left.\frac{\partial\bar{\tau}}{\partial x}\right|_{j,k}^n + \tfrac{1}{2}(\Delta x)^2 \left.\frac{\partial^2\bar{\tau}}{\partial x^2}\right|_{j,k}^n + O\{(\Delta x)^3\},\\
\bar{\tau}_{j,k-1}^n &= \bar{\tau}_{j,k}^n - \Delta y \left.\frac{\partial\bar{\tau}}{\partial y}\right|_{j,k}^n + \tfrac{1}{2}(\Delta y)^2 \left.\frac{\partial^2\bar{\tau}}{\partial y^2}\right|_{j,k}^n + O\{(\Delta y)^3\},\\
B_{j,k+\theta_y}^n &= \bar{\tau}_{j,k}^n + \theta_y\Delta y \left.\frac{\partial\bar{\tau}}{\partial y}\right|_{j,k}^n + \tfrac{1}{2}(\theta_y\Delta y)^2 \left.\frac{\partial^2\bar{\tau}}{\partial y^2}\right|_{j,k}^n + O\{(\theta_y\Delta y)^3\},\\
B_{j+\theta_x,k}^n &= \bar{\tau}_{j,k}^n + \theta_x\Delta x \left.\frac{\partial\bar{\tau}}{\partial x}\right|_{j,k}^n + \tfrac{1}{2}(\theta_x\Delta x)^2 \left.\frac{\partial^2\bar{\tau}}{\partial x^2}\right|_{j,k}^n + O\{(\theta_x\Delta x)^3\}.
\end{aligned} \tag{6.4.3}$$

These may be written in matrix form, as

$$\begin{bmatrix} 0 & \theta_y\Delta y & 0 & \tfrac{1}{2}(\theta_y\Delta y)^2 \\ \theta_x\Delta x & 0 & \tfrac{1}{2}(\theta_x\Delta x)^2 & 0 \\ 0 & -\Delta y & 0 & \tfrac{1}{2}(\Delta y)^2 \\ -\Delta x & 0 & \tfrac{1}{2}(\Delta x)^2 & 0 \end{bmatrix} \begin{bmatrix} \left.\frac{\partial\bar{\tau}}{\partial x}\right|_{j,k}^n \\ \left.\frac{\partial\bar{\tau}}{\partial y}\right|_{j,k}^n \\ \left.\frac{\partial^2\bar{\tau}}{\partial x^2}\right|_{j,k}^n \\ \left.\frac{\partial^2\bar{\tau}}{\partial y^2}\right|_{j,k}^n \end{bmatrix} = \begin{bmatrix} B_{j,k+\theta_y}^n - \bar{\tau}_{j,k}^n + O\{(\theta_y\Delta y)^3\} \\ B_{j+\theta_x,k}^n - \bar{\tau}_{j,k}^n + O\{(\theta_x\Delta x)^3\} \\ \bar{\tau}_{j,k-1}^n - \bar{\tau}_{j,k}^n + O\{(\Delta y)^3\} \\ \bar{\tau}_{j-1,k}^n - \bar{\tau}_{j,k}^n + O\{(\Delta x)^3\} \end{bmatrix} \tag{6.4.4}$$

This equation contains four unknowns, the solutions for which are

$$\left.\frac{\partial\bar{\tau}}{\partial x}\right|_{j,k}^n = \frac{1}{\theta_x(1+\theta_x)\Delta x}\{B_{j+\theta_x,k}^n - [1-(\theta_x)^2]\bar{\tau}_{j,k}^n - (\theta_x)^2\bar{\tau}_{j-1,k}^n\}$$

$$+ O\{\theta_x(\Delta x)^2\}, \quad 0 < \theta_x \leqslant 1, \tag{6.4.5a}$$

$$\left.\frac{\partial\bar{\tau}}{\partial y}\right|_{j,k}^{n} = \frac{1}{\theta_y(1+\theta_y)\Delta y}\{B^n_{j,k+\theta_y} - [1-(\theta_y)^2]\bar{\tau}^n_{j,k} - (\theta_y)^2\bar{\tau}^n_{j,k-1}\}$$

$$+ O\{\theta_y(\Delta y)^2\}, \quad 0 < \theta_y \leqslant 1, \tag{6.4.5b}$$

$$\left.\frac{\partial^2\bar{\tau}}{\partial x^2}\right|_{j,k}^{n} = \frac{2}{\theta_x(1+\theta_x)(\Delta x)^2}\{B^n_{j+\theta_x,k} - (1+\theta_x)\bar{\tau}^n_{j,k} + \theta_x\bar{\tau}^n_{j-1,k}\}$$

$$+ O\{(1-\theta_x)\Delta x, (\Delta x)^2\}, \quad 0 < \theta_x \leqslant 1, \tag{6.4.5c}$$

$$\left.\frac{\partial^2\bar{\tau}}{\partial y^2}\right|_{j,k}^{n} = \frac{2}{\theta_y(1+\theta_y)(\Delta y)^2}\{B^n_{j,k+\theta_y} - (1+\theta_y)\bar{\tau}^n_{j,k} + \theta_y\bar{\tau}^n_{j,k-1}\}$$

$$+ O\{(1-\theta_y)\Delta y, (\Delta y)^2\}, \quad 0 < \theta_y \leqslant 1. \tag{6.4.5d}$$

In this way, at irregular gridpoints, the first and second order spatial derivatives in the partial differential equation being discretised can always be written in terms of values of the dependent variable at that gridpoint, at neighbouring gridpoints within R, and boundary values at points where the boundary R^+ intersects the gridlines leading to exterior gridpoints.

DERIVATIVE NORMAL TO THE BOUNDARY PRESCRIBED - THE NEUMANN CONDITION

First method

Of the two methods available for handling this type of boundary condition, the simpler is that proposed by Fox (1944). The gridpoints inside the space region R are again divided into regular and irregular points. In both cases the centred finite difference forms for the spatial derivatives are of the same type as those given in (2.3.18), (2.3.19). However, in the case of the irregular points at least one of the x or y spatial derivatives will include a fictitious value of $\bar{\tau}$ from an exterior gridpoint. Thus, in Figure 6.3 the point $P(j\Delta x, k\Delta y)$ is an irregular point and the usual central difference forms for the first and second order derivatives $\partial\bar{\tau}/\partial y$ and $\partial^2\bar{\tau}/\partial y^2$ at that point include a value of $\bar{\tau}$ prescribed at the exterior point $E(j\Delta x, k\Delta y+\Delta y)$. An estimate for this value of $\bar{\tau}$ can be found in terms of the geometry of the boundary, the value of the normal derivative on the boundary and two interior gridpoints, in the following way.

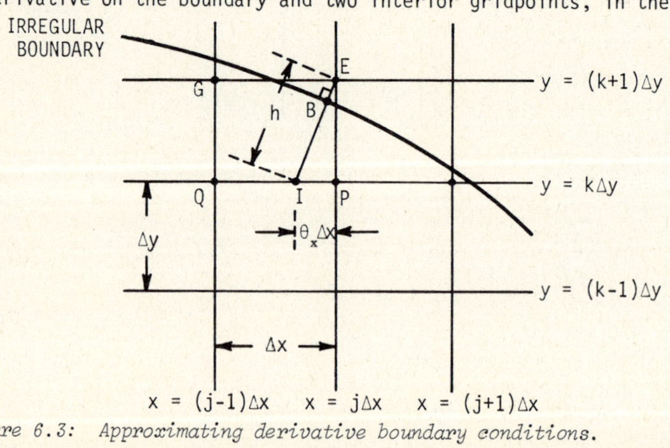

Figure 6.3: Approximating derivative boundary conditions.

From the exterior point E a line is drawn normal to the boundary to cut it at B, and extended until at I it intersects a grid line joining two adjacent interior gridpoints: see Figure 6.3. An estimate of the value of $\bar{\tau}$ at I can be obtained by linearly interpolating between the values of $\bar{\tau}$ at the two gridpoints P and Q. Thus

$$\bar{\tau}_I^n = \theta_x \bar{\tau}_Q + (1-\theta_x)\bar{\tau}_P + O\{(\Delta x)^2\}, \quad 0 < \theta_x < 1, \tag{6.4.6}$$

where the length of PI is $\overline{PI} = \theta_x \Delta x$. A value $\bar{\tau}_{j,k+1}^n = \bar{\tau}_E^n$ is then allotted to the exterior point E by relating the values of $\bar{\tau}$ at E and I through the known derivative boundary condition at B. If, at time $t_n = n\Delta t$, $\partial\tau/\partial\eta$ takes the value C_B^n at the point B, and if $\overline{EI}$ is of length h, then, unless B is the mid-point of EI,

$$\left.\frac{\partial\tau}{\partial\eta}\right|_B^n = C_B^n = \frac{\bar{\tau}_E^n - \bar{\tau}_I^n}{h} + O\{h\}. \tag{6.4.7}$$

However, if B is the midpoint of EI, the truncation error is $O\{h^2\}$. Combining (6.4.7) with (6.4.6) then leads to the relation

$$\bar{\tau}_{j,k+1}^n = \theta_x \bar{\tau}_{j-1,k}^n + (1-\theta_x)\bar{\tau}_{j,k}^n + C_B^n h + O\{h^2\} . \tag{6.4.8}$$

This value of $\bar{\tau}_{j,k+1}^n$ is then used in the central difference forms for the derivatives $\partial^2\bar{\tau}/\partial y^2$ and $\partial\bar{\tau}/\partial y$ so they can be written entirely in terms of values of $\bar{\tau}$ at interior gridpoints and the value of the normal derivative at B on the boundary. Thus we find, for example, that

$$\left.\frac{\partial\bar{\tau}}{\partial y}\right|_{j,k}^n = \frac{\theta_x \bar{\tau}_{j-1,k}^n + (1-\theta_x)\bar{\tau}_{j,k}^n - \bar{\tau}_{j,k-1}^n + C_B^n h}{2\Delta y} + O\{\Delta y\}, \tag{6.4.9}$$

since h is proportional to Δy.

The finite difference equations which result on substituting these approximations for the derivatives in the transport equation evaluated at irregular points $(j\Delta x, k\Delta y)$ and time $n\Delta t$, can be combined with the set of finite difference equations for regular gridpoints, in order to find $\bar{\tau}$ at all interior grid points.

This method can be generalized to account for irregular points of the type P in Figure 6.2, where P is adjacent to boundary gridpoints in both x and y directions. Some variations in the way of approximating $\bar{\tau}_P^n$ occur in the literature, although they are essentially the same (see, for example, Ames, 1965, and Fox, 1944).

Fox also gave a method which improves the accuracy of the approximation to the normal, but is more complicated. To illustrate his method, consider Figure 6.3. Instead of extending the normal through E to the first grid-line QP, he only extended it as far as the diagonal GP, where G, P are both interior points. He then used linear interpolation to obtain the value of $\bar{\tau}_D$ at the point of intersection D of the diagonal and the normal, in terms of values of $\bar{\tau}$ at G and P; the value of $\bar{\tau}_D$ written in terms of $\bar{\tau}_{j-1,k+1}^n$, $\bar{\tau}_{j,k}^n$, was then substituted into the value of the derivative $\partial\bar{\tau}/\partial\eta|_B^n$ expressed in terms of $\bar{\tau}_E$ and $\bar{\tau}_D$ to obtain $\bar{\tau}_{j,k+1}^n$ in terms of $\bar{\tau}_{j-1,k+1}^n$, $\bar{\tau}_{j,k}^n$.

The advantage of this method is that the normal line from E to D is shorter than the line $\overline{EI}$, so the approximation to the derivative $\partial\bar{\tau}/\partial\eta|_B^n$ is more accurate, when using the analogous equation to (6.4.7). Note that, if $G(j\Delta x-\Delta x, k\Delta y+\Delta y)$ is an external gridpoint, this alternative method is not appropriate and the normal has to be extended to I as in Figure 6.3.

Taylor series method

In this method a governing difference equation is developed for every internal point. However, at an irregular point, modifications to the forms at regular points take place by taking into account the derivative boundary condition and the non-standard grid-spacings.

The procedure for treating those boundary conditions involving the normal derivative hinge on the expression for the directional derivative,

$$\left.\frac{\partial\bar{\tau}}{\partial\eta}\right|_B = \left.\frac{\partial\bar{\tau}}{\partial x}\right|_B \cos\gamma_B + \left.\frac{\partial\bar{\tau}}{\partial y}\right|_B \sin\gamma_B\,, \tag{6.4.10}$$

where all derivatives are evaluated at the boundary point B in question and the angle γ_B specifies the direction relative to the x-axis of the normal to the boundary at B. In Figure 6.4 $\partial\bar{\tau}/\partial\eta$ is known along the unit normals $\underset{\sim}{\eta}_R$, $\underset{\sim}{\eta}_N$, $\underset{\sim}{\eta}_S$.

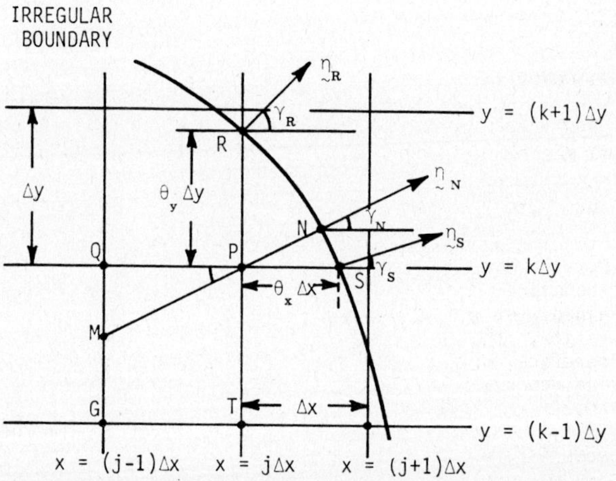

Figure 6.4: Geometry of an irregular point near the boundary

Now, as for the Dirichlet problem, consider Taylor series for $\bar{\tau}_R^n$, $\bar{\tau}_S^n$, $\bar{\tau}_N^n$ about the point $P(j\Delta x, k\Delta y)$. Thus

$$\bar{\tau}_R^n = \bar{\tau}(j\Delta x, k\Delta y + \theta_y\Delta y, n\Delta t)$$

$$= \bar{\tau}_P^n + \theta_y\Delta y \left.\frac{\partial\bar{\tau}}{\partial y}\right|_P^n + \tfrac{1}{2}(\theta_y\Delta y)^2 \left.\frac{\partial^2\bar{\tau}}{\partial y^2}\right|_P^n + \ldots$$

Replacing $\bar{\tau}$ by $\partial\bar{\tau}/\partial x$, then $\partial\bar{\tau}/\partial y$, yields the pair of equations

$$\left.\frac{\partial\bar{\tau}}{\partial x}\right|_R^n = \left.\frac{\partial\bar{\tau}}{\partial x}\right|_P^n + \theta_y\Delta y \left.\frac{\partial^2\bar{\tau}}{\partial y\partial x}\right|_P^n + O\{(\theta_y\Delta y)^2\}, \tag{6.4.11a}$$

$$\left.\frac{\partial\bar{\tau}}{\partial y}\right|_R^n = \left.\frac{\partial\bar{\tau}}{\partial y}\right|_P^n + \theta_y \Delta y \left.\frac{\partial^2\bar{\tau}}{\partial y^2}\right|_P^n + O\{(\theta_y \Delta y)^2\}. \tag{6.4.11b}$$

Since R is a boundary point we may substitute Equations (6.4.11a,b) into the right side of Equation (6.4.10) with R replacing B, giving

$$\left.\frac{\partial\bar{\tau}}{\partial\eta}\right|_R^n = \left[\left.\frac{\partial\bar{\tau}}{\partial x}\right|_P^n + \theta_y \Delta y \left.\frac{\partial^2\tau}{\partial x\partial y}\right|_P^n\right] \cos\gamma_R$$

$$+ \left[\left.\frac{\partial\bar{\tau}}{\partial y}\right|_P^n + \theta_y \Delta y \left.\frac{\partial^2\bar{\tau}}{\partial y^2}\right|_P^n\right] \sin\gamma_R + O\{(\theta_y \Delta y)^2\}. \tag{6.4.12}$$

For the boundary points N and S, two analogous equations are obtained, containing $\partial\bar{\tau}/\partial x|_P^n$, $\partial\bar{\tau}/\partial y|_P^n$, $\partial^2\bar{\tau}/\partial x^2|_P^n$, $\partial^2\bar{\tau}/\partial y^2|_P^n$ and $\partial^2\bar{\tau}/\partial x\partial y|_P^n$. Adding to these three equations the two Taylor series expansions about P for $\bar{\tau}_Q^n$ and $\bar{\tau}_T^n$ gives a total of five equations containing the five derivatives listed. The solution of this system of equations gives expressions for the various partial derivatives, up to second order, at the irregular point P in terms of the known values $\partial\bar{\tau}/\partial\eta|_R^n$, $\partial\bar{\tau}/\partial\eta|_N^n$, $\partial\bar{\tau}/\partial\eta|_S^n$, and the quantities τ_Q^n, τ_T^n.

THE MIXED BOUNDARY CONDITION

In this case $\bar{\tau}$ and $\partial\bar{\tau}/\partial\eta$ are linearly related at all points on the boundary. This condition may be represented by the relation

$$a(x,y,t)\bar{\tau} + b(x,y,t)\frac{\partial\bar{\tau}}{\partial\eta} = c(x,y,t), \tag{6.4.13}$$

for all points $B(x,y)$ on R^+. If $b(x,y,t) \equiv 0$ for all t this degenerates to the Dirichlet boundary condition and if $a(x,y,t) \equiv 0$, the Neumann boundary condition is obtained. At the point R (see Figure 6.4) where the boundary cuts the grid-line $x = j\Delta x$, the value of $\bar{\tau}_R^n$ is not known explicitly if $b(x,y,t) \neq 0$. Therefore, a linear relation is sought which involves $\bar{\tau}_R^n$ and values of $\bar{\tau}$ at neighbouring interior gridpoints and points where the boundary cuts the grid-lines.

In general, problems involving mixed boundary conditions on irregular boundaries are difficult to solve.

When considering boundary conditions on irregular boundaries in three-dimensional space, little is different from the two-dimensional case. The techniques described in this section may be extended to the three-dimensional case, but the methods are much more complicated. In general, results are usually less accurate because of the further approximations which must be made due to the incorporation of derivatives involving the extra space dimension.

7. THE WAVE EQUATION

7.1 Introduction

The one-dimensional wave equation may be written

$$\frac{\partial^2\bar{\tau}}{\partial t^2} = u^2 \frac{\partial^2\bar{\tau}}{\partial x^2}, \quad 0 \le x \le 1, \ t \ge 0. \tag{7.1.1}$$

Associated with this equation are two initial conditions, such as

$$\bar{\tau}(x,0) = f(x), \ \partial\bar{\tau}/\partial t|_{(x,0)} = h(x), \tag{7.1.2a}$$

and two boundary conditions, such as

$$\bar{\tau}(0,t) = g_0(t), \ \bar{\tau}(1,t) = g_1(t), \ t \ge 0. \tag{7.1.2b}$$

An exact solution of Equation (7.1.1) subject to initial conditions (7.1.2a), on the infinite interval $-\infty < x < \infty$, may be found using D'Alembert's method. Using the transformation $\mu = x + ut$, $\nu = x - ut$ in Equation (7.1.1) yields, if u is *constant*,

$$\frac{\partial^2\bar{\tau}(\mu,\nu)}{\partial\mu\partial\nu} = 0, \tag{7.1.3}$$

which has the general solution

$$\bar{\tau}(\mu,\nu) = F(\mu) + G(\nu), \tag{7.1.4}$$

where F and G are arbitrary functions. Thus,

$$\bar{\tau}(x,t) = F(x+ut) + G(x-ut). \tag{7.1.5}$$

Clearly, this solution consists of two parts, initially F(x) and G(x), both moving with speed u, the first in the negative x-direction and the second in the positive x-direction. The functions F and G must satisfy the initial and boundary conditions of a particular problem. With initial conditions (7.1.2a) it can be shown that

$$\bar{\tau}(x,t) = \tfrac{1}{2}\{f(x+ut) + f(x-ut)\} + \frac{1}{2u}\int_{x-ut}^{x+ut} h(\chi)d\chi. \tag{7.1.6}$$

7.2 An Explicit Finite Difference Method

One way to obtain a finite difference approximation to Equation (7.1.1) is to use central difference forms for the time and space derivatives at the (j,n) gridpoint. In this case, if u is constant,

$$\left.\frac{\partial^2\bar{\tau}}{\partial t^2}\right|_j^n = u^2 \left.\frac{\partial^2\bar{\tau}}{\partial x^2}\right|_j^n \tag{7.2.1}$$

may be written

$$\frac{\bar{\tau}_j^{n+1} - 2\bar{\tau}_j^n + \bar{\tau}_j^{n-1}}{(\Delta t)^2} + O\{(\Delta t)^2\} = u^2\{\frac{\bar{\tau}_{j+1}^n - 2\bar{\tau}_j^n + \bar{\tau}_{j-1}^n}{(\Delta x)^2} + O\{(\Delta x)^2\}\}. \tag{7.2.2}$$

A suitable finite difference approximation to Equation (7.2.2) with truncation error $O\{(\Delta t)^2,(\Delta x)^2\}$ is therefore

$$\frac{\tau_j^{n+1} - 2\tau_j^n + \tau_j^{n-1}}{(\Delta t)^2} = u^2\left\{\frac{\tau_{j+1}^n - 2\tau_j^n + \tau_{j-1}^n}{(\Delta x)^2}\right\}. \quad (7.2.3)$$

Rearranging (7.2.3) gives the three-level explicit finite difference equation

$$\tau_j^{n+1} = c^2(\tau_{j-1}^n + \tau_{j+1}^n) + 2(1-c^2)\tau_j^n - \tau_j^{n-1}, \quad (7.2.4)$$

where, as before, $c = u\Delta t/\Delta x$. At each time step the contribution to the discretisation error in this numerical method is $O\{(\Delta t)^4, (\Delta t)^2(\Delta x)^2\}$. Since this is a linear initial value problem, and the finite difference equation is consistent because the truncation error tends to zero as both Δx and Δt tend to zero, then convergence of the solution of this equation to the solution of the wave equation is ensured if (7.2.4) is stable.

If the stability is investigated by means of the von Neumann method, the amplification factor is given by the solution of

$$G^2 - 2\gamma G + 1 = 0, \quad (7.2.5)$$

where $\gamma = 1 - 2c^2 \sin^2(\beta/2)$. The solutions of (7.2.5) are

$$G_1, G_2 = \gamma \pm \sqrt{\gamma^2-1}. \quad (7.2.6)$$

If $c \leq 1$, then $|\gamma| \leq 1$ and $\sqrt{1-\gamma^2}$ is real. In this case,

$$G_1, G_2 = \gamma \pm i\sqrt{1-\gamma^2},$$

so that

$$|G_1|^2 = |G_2|^2 = \gamma^2 + (1-\gamma^2) = 1,$$

for all β and (7.2.4) is stable. If $c \geq 1$, then for some β we have $|\gamma| > 1$ and $\sqrt{\gamma^2-1}$ in (7.2.6) is real. It follows that one of G_1, G_2 has a modulus which is larger than 1 and equation (7.2.4) is unstable.

The finite difference equation (7.2.4) defines a three-level difference scheme. Therefore a special starting procedure is required in order to obtain the set of values of τ at the first time level from the set of initial values $\bar{\tau}_j^0$, $j=0(1)J$. In order to do this, the initial derivative condition is used. Given that

$$\frac{\partial\bar{\tau}(x,0)}{\partial t} = h(x),$$

the derivative may be replaced by a central difference form at the grid-point $(j,0)$, giving

$$\left.\frac{\partial\bar{\tau}}{\partial t}\right|_j^0 = \frac{\bar{\tau}_j^1 - \bar{\tau}_j^{-1}}{2\Delta t} + O\{(\Delta t)^2\} = h_j, \; j=0(1)J, \quad (7.2.7)$$

where $h_j = h(j\Delta x)$. The finite difference approximation to the initial derivative condition, with truncation error $O\{(\Delta t)^2\}$, is therefore

$$\frac{\tau_j^1 - \tau_j^{-1}}{2\Delta t} = h_j, \quad j=0(1)J, \quad (7.2.8)$$

where τ_j^{-1} is a fictitious value of τ assigned to the position $j\Delta x$ at a time Δt before the initial instant. Therefore,

$$\tau_j^{-1} = \tau_j^1 - 2\Delta t\, h_j\ ,\ j=0(1)J, \tag{7.2.9}$$

with a contribution of $O\{(\Delta t)^3\}$ to the discretisation error in τ_j^n. Substitution of this value of τ_j^{-1} in (7.2.4) with n=0, gives

$$\tau_j^1 = \tfrac{1}{2}c^2(f_{j-1} + f_{j+1}) + (1-c^2)f_j + \Delta t\, h_j\ ,\ j=1(1)J-1, \tag{7.2.10}$$

where $f_j = f(j\Delta x)$.

By means of (7.2.10) the values of τ at the first time level are computed from the known initial values $\bar{\tau}_j^0 = f_j$, j=0(1)J. Since (7.2.9) contributes a term of $O\{(\Delta t)^3\}$ to the discretisation error while at each subsequent time step (7.2.4) contributes a term of $O\{(\Delta t)^4, (\Delta t)^2(\Delta x)^2\}$, then in practical terms this method has a discretisation error of $O\{N(\Delta t)^3, N(\Delta t)^2(\Delta x)^2\}$ in the value of τ_j^N, if N is small and an error of $O\{N(\Delta t)^4, N(\Delta t)^2(\Delta x)^2\}$ if N is large.

In the special case with $h(x) \equiv 0$, so $h_j = 0$ for all j=0(1)J, and with c = 1, then (7.2.10) becomes

$$\tau_j^1 = \tfrac{1}{2}(f_{j-1} + f_{j+1}). \tag{7.2.11}$$

We can use (7.2.11) to solve a wave-equation problem and compare the numerical solution with the exact solution obtained using D'Alembert's solution (7.1.6). With u = 1, Δt = 0.1, Δx = 0.1, then c = 1. Computing the analytic and numerical solutions for $\bar{\tau}(j\Delta x, n\Delta t)$, j=1(1)9, n=0(1)10 gave the same answers to 8 significant figures.

An alternative and more accurate way of calculating the values at the first time level takes advantage of D'Alembert's solution (7.2.6). In discrete form his solution applied for $x = j\Delta x$, j=1(1)J-1, and $t = \Delta t$, is

$$\begin{aligned}
\bar{\tau}_j^1 &= \bar{\tau}(j\Delta x, \Delta t)\\
&= \tfrac{1}{2}[f(j\Delta x+u\Delta t) + f(j\Delta x-u\Delta t)] + \frac{1}{2u}\int_{j\Delta x-u\Delta t}^{j\Delta x+u\Delta t} h(\chi)d\chi\\
&= \tfrac{1}{2}[f(j\Delta x+\Delta x) + f(j\Delta x-\Delta x)] + \frac{1}{2u}\int_{j\Delta x-\Delta x}^{j\Delta x+\Delta x} h(\chi)d\chi, \text{ if } c=1,\\
&= \tfrac{1}{2}[f_{j-1} + f_{j+1}] + \frac{1}{2u}\int_{x_{j-1}}^{x_{j+1}} h(\chi)d\chi,
\end{aligned} \tag{7.2.12}$$

where the integral can be calculated using Simpson's Rule. That is

$$\tau_j^1 = \tfrac{1}{2}[f_{j-1} + f_{j+1}] + \frac{\Delta x}{6u}[h_{j-1} + 4h_j + h_{j+1}]. \tag{7.2.13}$$

If $h(x) \equiv 0$, again this results in τ_j^1 being computed by means of (7.2.11).

7.3 Implicit methods

If, in Equation (7.1.3), $\partial^2\bar{\tau}/\partial x^2|_j^n$ is replaced by the average of its values at time levels (n-1) and (n+1), then

$$\left.\frac{\partial^2\bar{\tau}}{\partial t^2}\right|_j^n = \frac{u^2}{2}\left\{\left.\frac{\partial^2\bar{\tau}}{\partial x^2}\right|_j^{n-1} + \left.\frac{\partial^2\bar{\tau}}{\partial x^2}\right|_j^{n+1} + O\{(\Delta t)^2\}\right\}. \tag{7.3.1}$$

Replacing all the derivatives in (7.3.1) by their central difference forms gives

$$\frac{\bar{\tau}_j^{n+1} - 2\bar{\tau}_j^n + \bar{\tau}_j^{n-1}}{(\Delta t)^2} + O\{(\Delta t)^2\} = \frac{u^2}{2}\left\{\frac{\bar{\tau}_{j+1}^{n-1} - 2\bar{\tau}_j^{n-1} + \bar{\tau}_{j-1}^{n-1}}{(\Delta x)^2}\right.$$

$$\left. + \frac{\bar{\tau}_{j+1}^{n+1} - 2\bar{\tau}_j^{n+1} + \bar{\tau}_{j-1}^{n+1}}{(\Delta x)^2} + O\{(\Delta t)^2, (\Delta x)^2\}\right\}. \tag{7.3.2}$$

Rearranging Equation (7.3.2) so the values at the time level (n+1) are on the left side, and terms of $O\{(\Delta t)^4, (\Delta t)^2(\Delta x)^2\}$ are dropped, the following implicit equation is obtained for $j=1(1)J-1$:

$$-\tfrac{1}{2}c^2\tau_{j-1}^{n+1} + (1+c^2)\tau_j^{n+1} - \tfrac{1}{2}c^2\tau_{j+1}^{n+1}$$

$$= 2\tau_j^n + \tfrac{1}{2}c^2\tau_{j-1}^{n-1} - (1+c^2)\tau_j^{n-1} + \tfrac{1}{2}c^2\tau_{j+1}^{n-1}. \tag{7.3.3}$$

If the values of $\bar{\tau}$ are known on the boundary, say $\tau_0^{n+1} = g_0(t_{n+1})$, $\tau_J^{n+1} = g_1(t_{n+1})$, then the set of simultaneous algebraic equations (7.3.3) in the unknowns τ_j^{n+1}, $j=1(1)J-1$, can be readily computed using the Thomas algorithm.

The stability of Equation (7.3.3) may be investigated by the von Neumann method. The amplification factor G of the m-th Fourier component of the error distribution as it propagates from one time level to the next is given by the solution of the quadratic equation

$$G^2 - 2\gamma G + 1 = 0, \tag{7.3.4}$$

where $\gamma = \{1 + 2c^2(\sin^2(\beta/2)\}^{-1} \leq 1$ for all c and β. It follows that the values of G are

$$G_1, G_2 = \gamma \pm i\sqrt{1-\gamma^2}, \tag{7.3.5}$$

where $\sqrt{1-\gamma^2}$ is real. Hence

$$|G_1|^2 = |G_2|^2 = \gamma^2 + (1-\gamma^2) = 1 \tag{7.3.6}$$

for all c and β, and the implicit scheme given by (7.3.3) is always *stable*.

Also, as

$$|1+c^2| > |-\tfrac{1}{2}c^2| + |-\tfrac{1}{2}c^2| \tag{7.3.7}$$

for all c, it follows that the coefficient matrix defined by (7.3.3) is diagonally dominant, and the system is *solvable* for all c.

The above implicit method is a special case of a general form involving a weighting factor θ in time. Since

$$\bar{\tau}_j^n = \theta\bar{\tau}_j^{n+1} + (1-2\theta)\bar{\tau}_j^n + \theta\bar{\tau}_j^{n-1} + O\{(1-2\theta)\Delta t, (\Delta t)^2\},\ 0 < \theta \leq 1,$$

it follows that the wave equation at the (j,n) gridpoint, namely Equation (7.2.1), can be written

$$\left.\frac{\partial^2\bar{\tau}}{\partial t^2}\right|_j^n = u^2\{\theta\left.\frac{\partial^2\bar{\tau}}{\partial x^2}\right|_j^{n+1} + (1-2\theta)\left.\frac{\partial^2\bar{\tau}}{\partial x^2}\right|_j^n + \theta\left.\frac{\partial^2\bar{\tau}}{\partial x^2}\right|_j^{n-1}$$

$$+ O\{(1-2\theta)\Delta t,(\Delta t)^2\}\}. \tag{7.3.8}$$

On substitution of the central difference forms for the derivatives in (7.3.8), then dropping terms of $O\{(1-2\theta)\Delta t,(\Delta t)^2,(\Delta x)^2\}$, and rearranging, the following implicit finite difference equation is obtained for $j=1(1)J-1$:

$$-\theta c^2\tau_{j-1}^{n+1} + (1+2c^2\theta)\tau_j^{n+1} - \theta c^2\tau_{j+1}^{n+1}$$

$$= c^2(1-2\theta)\tau_{j-1}^n + 2[1 - c^2(1-2\theta)]\tau_j^n + c^2(1-2\theta)\tau_{j+1}^n$$

$$+ c^2\theta\tau_{j-1}^{n-1} - (1+2c^2\theta)\tau_j^{n-1} + c^2\theta\tau_j^{n-1}. \tag{7.3.9}$$

Equation (7.3.9) has a truncation error of $O\{(1-2\theta)\Delta t,(\Delta t)^2,(\Delta x)^2\}$ and contributes terms of $O\{(1-2\theta)(\Delta t)^3,(\Delta t)^4,(\Delta t)^2(\Delta x)^2\}$ to the discretisation error at each time step.

Note that $\theta = 0$ gives the explicit equation (7.2.4) considered in Section 7.2 and $\theta = \frac{1}{2}$ gives the implicit equation (7.3.3) previously described in this section. The von Neumann stability analysis shows that this method is stable for all c if $\theta \geq \frac{1}{4}$ and stable for $c \leq (1-4\theta)^{-\frac{1}{2}}$ if $\theta < \frac{1}{4}$. Since the set of equations (7.3.9) are diagonally dominant for all θ and c the solutions may be computed with the aid of the Thomas algorithm.

Another implicit method (see Wachspress, 1966) uses central differences to approximate $\partial^2\bar{\tau}/\partial x^2$ at the (j,n+1) gridpoint, namely

$$\left.\frac{\partial^2\bar{\tau}}{\partial x^2}\right|_j^{n+1} = \frac{1}{(\Delta x)^2}\{\bar{\tau}_{j-1}^{n+1} - 2\bar{\tau}_j^{n+1} + \bar{\tau}_{j+1}^{n+1}\} + O\{(\Delta x)^2\}, \tag{7.3.10}$$

with a backward difference form for $\partial^2\bar{\tau}/\partial t^2$ at the same gridpoint, that is

$$\left.\frac{\partial^2\bar{\tau}}{\partial t^2}\right|_j^{n+1} = \frac{1}{(\Delta t)^2}\{\bar{\tau}_j^{n+1} - 2\bar{\tau}_j^n + \bar{\tau}_j^{n-1}\} + O\{\Delta t\}. \tag{7.3.11}$$

Substitution of (7.3.10) and (7.3.11) in the wave equation applied at the (j,n+1) gridpoint gives the finite difference equation

$$c^2\tau_{j-1}^{n+1} - (1+2c^2)\tau_j^{n+1} + c^2\tau_{j+1}^{n+1} = -2\tau_j^n + \tau_j^{n-1}, \quad j=1(1)J-1, \tag{7.3.12}$$

with a truncation error of $O\{\Delta t,(\Delta x)^2\}$, and a contribution of $O\{(\Delta t)^3,(\Delta t)^2(\Delta x)^2\}$ to the discretisation error at each time step. These errors are of the same order as (7.3.9) with $\theta \neq \frac{1}{2}$ but (7.3.12) is less accurate than (7.3.9) with $\theta = \frac{1}{2}$, that is with (7.3.3). Again, the tri-diagonal set of algebraic equations (7.3.12) is diagonally dominant, and the Thomas algorithm may be used to solve the system.

Application of the von Neumann method shows (7.3.12) is stable for all values of c. It should be noted that backward differences in time are used to guarantee stability for all values of c in many finite difference schemes.

7.4 Conversion to Two First Order Partial Differential Equations

If we define $p=u\partial\bar{\tau}/\partial x$ and $\bar{q} = \partial\bar{\tau}/\partial t$, Equation (7.1.1) with constant u can be expressed as two simultaneous, first order, partial differential equations,

$$\frac{\partial\bar{p}}{\partial t} = u\frac{\partial\bar{q}}{\partial x} \quad \text{and} \quad \frac{\partial\bar{q}}{\partial t} = u\frac{\partial\bar{p}}{\partial x}. \tag{7.4.1}$$

The obvious finite difference approximation which could be used to solve this pair of simultaneous partial differential equations would involve forward difference forms for $\partial\bar{p}/\partial t$ and $\partial\bar{q}/\partial t$ with central difference forms for $\partial\bar{p}/\partial x$ and $\partial\bar{q}/\partial x$. This gives, on dropping terms of $O\{\Delta t,(\Delta x)^2\}$ and rearranging, the pair of two-level explicit finite difference equations for j=1(1)J-1

$$\left.\begin{aligned} p_j^{n+1} &= p_j^n + \tfrac{c}{2}(q_{j+1}^n - q_{j-1}^n), \\ q_j^{n+1} &= q_j^n + \tfrac{c}{2}(p_{j+1}^n - p_{j-1}^n). \end{aligned}\right\} \quad n= 0,1,2,3,\ldots . \tag{7.4.2}$$

But on checking the stability of (7.4.2) it is found that it is unstable for all values of $c > 0$. The method is therefore of no practical use.

One variation of Equation (7.4.2) uses forward difference forms for $\partial\bar{p}/\partial t$ and $\partial\bar{q}/\partial x$ in the first equation discretised at the (j,n) gridpoint. Then backward difference forms for the derivatives are used to discretise the second equation at the gridpoint (j,n+1). This gives, on dropping terms of $O\{\Delta t,\Delta x\}$ and rearranging, the pair of explicit equations for j=1(1)J-1

$$\left.\begin{aligned} p_j^{n+1} &= p_j^n + c(q_{j+1}^n - q_j^n) \\ q_j^{n+1} &= q_j^n + c(p_j^{n+1} - p_{j-1}^{n+1}) \end{aligned}\right\} \quad n = 0,1,2,3,\ldots . \tag{7.4.3}$$

In general, these equations contribute errors of $O\{(\Delta t)^2,\Delta t\Delta x\}$ to the discretisation errors of p_j^{n+1}, q_j^{n+1} at each time step.

A variation of the previously described von Neumann stability analysis, so it can be applied to simultaneous partial differential equations, may be used to find the values of c for which the system of equations (7.4.3) is stable.

The errors ζ_j^n in p_j^n and η_j^n in q_j^n propagate in a way which is analogous to the calculation of p and q using Equation (7.4.3). If the m-th Fourier component of the distribution of errors in p_j^n, j=0(1)J, at time level n is given by

$$\zeta_j^n = (G_p)^n e^{i\beta j}, \tag{7.4.4a}$$

and the corresponding component of the distribution of errors in q_j^n, j=0(1)J, is given by

$$\eta_j^n = (G_q)^n e^{i\beta j}, \tag{7.4.4b}$$

where $\beta = m\pi\Delta x$, then substitution of these forms into the corresponding error equations to (7.4.3) gives, on division by $e^{i\beta J}$,

$$(G_p)^{n+1} = (G_p)^n + c(G_q)^n(e^{i\beta}-1), \tag{7.4.5a}$$

$$(G_q)^{n+1} = (G_q)^n + c(G_p)^{n+1}(1-e^{-i\beta}). \tag{7.4.5b}$$

Substitution of $(G_p)^{n+1}$ from (7.4.5a) into (7.4.5b) gives

$$(G_q)^{n+1} = c(1-e^{-i\beta})(G_p)^n + [1 + 2c^2(\cos\beta - 1)](G_q)^n. \tag{7.4.5c}$$

The pair of equations (7.4.5a,c) may be written in matrix form as

$$\underset{\sim}{G}^{n+1} = \underset{=}{A}\,\underset{\sim}{G}^n, \tag{7.4.6}$$

where

$$\underset{\sim}{G}^n = \begin{bmatrix}(G_p)^n\\(G_q)^n\end{bmatrix}, \quad \underset{=}{A} = \begin{bmatrix}1 & c(e^{i\beta}-1)\\ c(1-e^{-i\beta}) & 1-4c^2\sin^2(\beta/2)\end{bmatrix}.$$

Now, Equations (7.4.4a,b) can be written in vector form as

$$\underset{\sim}{\xi}_j^n = e^{i\beta j}\,\underset{\sim}{G}^n, \tag{7.4.7}$$

where

$$\underset{\sim}{\xi}_j^n = \begin{bmatrix}\zeta_j^n\\ \eta_j^n\end{bmatrix}.$$

For stability, we require $\underset{\sim}{\xi}_j^n \to 0$ or $\underset{\sim}{\xi}_j^n$ bounded, as $n\to\infty$, for all β; this requires $\underset{\sim}{G}^n \to \underset{\sim}{0}$ or $\underset{\sim}{G}^n$ bounded for all β, as $n\to\infty$. Using the result (3.3.13) from Noye (1978) on Equation (7.4.6), stability therefore requires the moduli of all the eigenvalues λ of the matrix $\underset{=}{A}$ to be not larger than 1, that is

$$|\lambda| \le 1.$$

The eigenvalues of $\underset{=}{A}$ are the roots of

$$|\underset{=}{A} - \lambda\underset{=}{I}| = 0,$$

where $\underset{=}{I}$ is the (2×2) unit matrix. We therefore seek the roots of

$$(1-\lambda)\{1-4c^2\sin^2(\beta/2)-\lambda\} + 4c^2\sin^2(\beta/2) = 0,$$

or of

$$\lambda^2 - 2\gamma\lambda + 1 = 0, \tag{7.4.8}$$

where $\gamma = 1 - 2c^2\sin^2(\beta/2)$. The roots of (7.4.8) are

$$\lambda_1,\ \lambda_2 = \gamma \pm \sqrt{\gamma^2-1}.$$

If $|\gamma| > 1$ then λ_1, λ_2 are real and one of $|\lambda_1|$, $|\lambda_2|$ is larger than 1. If $|\gamma| \le 1$, then λ_1 and λ_2 are the complex conjugates given by

$$\lambda_1,\ \lambda_2 = \gamma \pm i\sqrt{1-\gamma^2}.$$

Then

$$|\lambda_1| = |\lambda_2| = \sqrt{\gamma^2+(1-\gamma^2)} = 1.$$

The set of equations (7.4.3) is therefore stable if $|\gamma| \le 1$; that is, if

$$-1 \le 1 - 2c^2\sin^2(\beta/2) \le 1,$$

or

$$0 \le c^2\sin^2(\beta/2) \le 1.$$

This is true for all β if $c \le 1$.

There are other variations of the finite difference equation (7.4.2) which are also useful. For example, central difference approximations can be used for the time derivatives after the first time step, which gives

$$\left.\begin{aligned} p_j^{n+1} &= p_j^{n-1} + c(q_{j+1}^n - q_{j-1}^n) \\ q_j^{n+1} &= q_j^{n-1} + c(p_{j+1}^n - p_{j-1}^n) \end{aligned}\right\}, \quad j=1(1)J-1,\ n=1,2,3,\ldots\ . \tag{7.4.9}$$

Use of the von Neumann stability analysis shows that this method is stable for $c \le 1$.

Another variation is to replace p_j^n in (7.4.2) by $\frac{1}{2}(p_{j+1}^n + p_{j-1}^n)$ and q_j^n by a similar expression which yields the following set of difference equations

$$\left.\begin{aligned} p_j^{n+1} &= \tfrac{1}{2}(p_{j+1}^n + p_{j-1}^n) + \frac{c}{2}(q_{j+1}^n - q_{j-1}^n) \\ q_j^{n+1} &= \tfrac{1}{2}(q_{j+1}^n + q_{j-1}^n) + \frac{c}{2}(p_{j+1}^n - p_{j-1}^n) \end{aligned}\right\}, \quad j=1(1)J-1,\ n=0,1,2,.. \tag{7.4.10}$$

which is stable for $c \le 1$.

Yet another finite difference approximation used to solve the system (7.4.1) is to replace the approximation to the derivative $\partial\bar{p}/\partial x|_j^n$ in the second equation by $\partial p/\partial x|_j^{n+1}$ which introduces an error of $O\{\Delta t\}$. Then

$$\left.\begin{aligned} p_j^{n+1} &= p_j^n + \frac{c}{2}(q_{j+1}^n - q_{j-1}^n) \\ q_j^{n+1} &= q_j^n + \frac{c}{2}(p_{j+1}^{n+1} - p_{j-1}^{n+1}) \end{aligned}\right\}, \quad j=1(1)J-1,\ n=0,1,2,\ldots\ . \tag{7.4.11}$$

The stability requirement for this scheme is $c \le 2$.

7.5 Shock Waves

When dealing with shocks or discontinuities we may not know where the shock is, so we are unable to put the usual Eulerian grid over the space we are considering and still consider the shock; that is, the computation would go on as though the shock were not there.

The best method of handling the shock wave is to make one of the grid lines coincide with the shock at all time steps and calculate the effect of the shock on the neighbouring spatial grid points by using the Rankine-Hugoniot shock conditions. This can be done if the Lagrangian co-ordinate system, which follows the motion of the particles, is used.

Another method for treating shocks is to introduce a pseudo viscosity term. The idea behind this is to add an artificial viscosity term into the equation we are looking at, in such a way that there results a smooth shock transition extending over a small number of spatial grid spacings (see Richtmyer and Morton, 1967).

8. DIFFERENT GRID AND COORDINATE SYSTEMS

In the previous sections the transport equation was differenced on a uniformly spaced grid in the rectangular cartesian coordinate system. However, this may not be the most appropriate system to use.

While a finite difference approximation developed on a uniform grid may be the simplest and formally the most accurate scheme to use, it can prove deficient in some respects. For example, when computing the flow around solid bodies or solving other boundary layer problems, there are quite large velocity gradients in certain regions of the flow near the body. If there are not enough grid points across the boundary layer to resolve the velocity, the numerical solution may be a very poor approximation to the true solution in this layer. To increase the resolution in this region we could still use a uniform grid but with a much smaller grid spacing over the complete area of flow. However, this may lead to unacceptably large computational times.

To avoid excessive numbers of calculations, while still obtaining a high degree of resolution in particular regions, a non-uniform grid system may be used with a fine grid in the area of interest and a coarse grid over the rest of the flow field. Different types of non-uniform grids may be used. For instance, it is possible to have a coarse grid which changes abruptly to a fine grid in the region of interest, as described in Section 8.1, or it is possible to have a grid in which there is a gradual reduction from a large to a small grid size. This case is dealt with in Section 8.2. Grids need not be rectangular or even regular; specially shaped grids of variable size may be used in problems involving irregular boundaries (for example, see Thoman and Szewczyk, 1969).

While there are considerable benefits to be gained by having a non-uniform grid, these may be offset by other problems introduced by the change in grid size. There may be wave distortion due to a phase change at the interface of two grids of different size or shape or a wave reflection off the interface. It may also be difficult to match the finite difference approximations at the interface of different grids. The introduction of a change in grid spacing also may adversely affect the formal truncation error and the stability of the system.

Because of these difficulties, a coordinate transformation may be preferred to a change in grid shape or size in order to obtain higher resolution in a particular area of flow. It is often desirable to make a change of coordinate system if a problem can be handled more easily in another system. For example, by aligning a coordinate surface along a boundary, it is easier to apply boundary conditions. However, changing the coordinate system or making a coordinate transformation can also lead to complications which are outlined later.

8.1 Abrupt Changes in Grid Spacing

The simplest non-uniform grid scheme has a large constant grid spacing over most of the flow field with an abrupt change to a smaller constant grid spacing to cover the region of particular interest. In making this change of grid size it is desirable that the second order spatial accuracy which occurs with central differences on a uniform grid should be retained; a decrease in computational accuracy due to the change in grid size may more than offset the gain in accuracy due to the use of a finer grid.

Crowder and Dalton (1971) considered the case where the grid spacing is decreased from a constant value $\Delta x_{J-1} = x_J - x_{J-1}$ to the smaller value $\Delta x_J = x_{J+1} - x_J$ at a point J, as in Figure 8.1, in order to see how the order of the truncation error of the approximations to the spatial derivatives $\partial\bar{\tau}/\partial x$ and $\partial^2\bar{\tau}/\partial x^2$ of a function $\bar{\tau}$ was affected by this change. Expanding the functions $\bar{\tau}_{J-1}$, $\bar{\tau}_{J+1}$ in a Taylor series about $x = x_J$ at the J-th grid point, at a particular time level, the superscript for which is omitted in the following, gives

$$\bar{\tau}_{J+1} = \bar{\tau}_J + \Delta x_J \left.\frac{\partial\bar{\tau}}{\partial x}\right|_J + \frac{(\Delta x_J)^2}{2!}\left.\frac{\partial^2\bar{\tau}}{\partial x^2}\right|_J + \frac{(\Delta x_J)^3}{3!}\left.\frac{\partial^3\bar{\tau}}{\partial x^3}\right|_J + O\{(\Delta x_J)^4\}, \quad (8.1.1)$$

$$\bar{\tau}_{J-1} = \bar{\tau}_J - \Delta x_{J-1} \left.\frac{\partial\bar{\tau}}{\partial x}\right|_J + \frac{(\Delta x_{J-1})^2}{2!}\left.\frac{\partial^2\bar{\tau}}{\partial x^2}\right|_J - \frac{(\Delta x_{J-1})^3}{3!}\left.\frac{\partial^3\bar{\tau}}{\partial x^3}\right|_J + O\{(\Delta x_{J-1})^4\}. \quad (8.1.2)$$

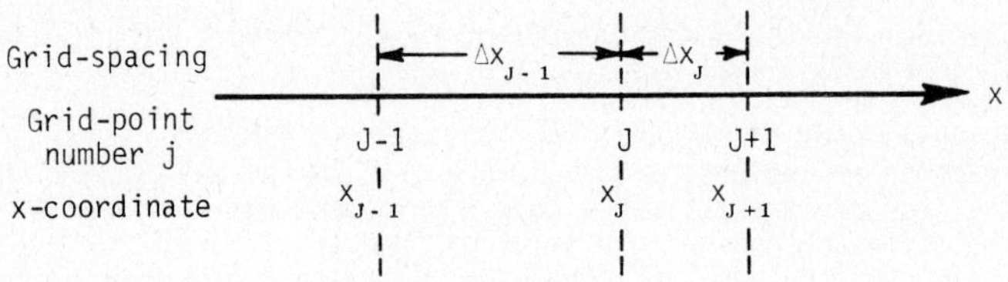

Figure 8.1: Sudden change in grid spacing at the J-th gridpoint.

Eliminating the terms containing $\partial^2\bar{\tau}/\partial x^2|_J$, by multiplying Equation (8.1.1) by $(\Delta x_{J-1})^2$ and subtracting Equation (8.1.2) multiplied by $(\Delta x_J)^2$, yields

$$\left.\frac{\partial\bar{\tau}}{\partial x}\right|_J = \frac{\bar{\tau}_{J+1}+[(\nu_J)^2-1]\bar{\tau}_J-(\nu_J)^2\bar{\tau}_{J-1}}{\Delta x_J(\nu_J+1)} - \frac{\Delta x_{J-1}\Delta x_J}{6}\left.\frac{\partial^3\bar{\tau}}{\partial x^3}\right|_J + O\{(\Delta x_{J-1})^3\}, \quad (8.1.3)$$

where $\nu_J = \Delta x_J/\Delta x_{J-1}$. The quasi-central difference approximation to the first derivative at the point where the grid spacing changes is given by the first term on the right hand side of (8.1.3), and has a truncation error of $O\{\Delta x_{J-1}\Delta x_J,(\Delta x_{J-1})^3\}$.

At all other gridpoints the standard central finite difference form (2.3.18), with $\Delta x = \Delta x_{J-1}$ if $j < J$ or $\Delta x = \Delta x_J$ if $j > J$, is used.

Similarly, by eliminating the terms with $\partial\bar{\tau}/\partial x|_J$ from Equations (8.1.1) and (8.1.2) gives

$$\left.\frac{\partial^2\bar{\tau}}{\partial x^2}\right|_J = \frac{\bar{\tau}_{J+1}-(\nu_J+1)\bar{\tau}_J+\nu_J\bar{\tau}_{J-1}}{\frac{1}{2}\Delta x_{J-1}\Delta x_J(\nu_J+1)} + \frac{(\Delta x_{J-1}-\Delta x_J)}{3}\left.\frac{\partial^3\bar{\tau}}{\partial x^3}\right|_J + O\{(\Delta x_{J-1})^2\}. \quad (8.1.4)$$

Therefore the quasi-central difference approximation to the second derivative at x_J is given by the first term on the right side of (8.1.4), and has a truncation error of $O\{\Delta x_{J-1}-\Delta x_J,(\Delta x_{J-1})^2\}$. At the other gridpoints the standard finite difference form (2.3.19) is used.

Note that the choice $\Delta x_{J-1} = \Delta x_J = \Delta x$ in (8.1.3) and (8.1.4) yield the usual central difference forms (2.3.18) and (2.3.19) for a uniform grid spacing.

Because the finite difference approximation based on Equation (8.1.4) has a truncation error of $O\{\Delta x_{J-1}-\Delta x_J,(\Delta x_{J-1})^2\}$, any finite difference methods for second order partial differential equations based upon this approximation will involve a truncation error of $O\{(\Delta x_{J-1})^2\}$ at the interface where the grid size changes abruptly, only if

$$\nu_J = 1 - \kappa\Delta x_{J-1}, \quad \kappa = O\{1\}, \tag{8.1.5}$$

as then

$$\begin{aligned}\Delta x_{J-1}-\Delta x_J &= \Delta x_{J-1}(1-\nu_J) \\ &= \kappa(\Delta x_{J-1})^2.\end{aligned} \tag{8.1.6}$$

Therefore, the truncation error may be worse than that with a uniform grid, rather than an improvement, unless the change in grid spacing is small. Crowder and Dalton (1971) compared results obtained using six different grids and concluded that a uniform square mesh was the best grid to use for the Poiseuille flow problem they were studying. This was contrary to their original hypothesis that using non-uniform, non-square grids would give a minimum overall error with their finite difference methods. Commenting on these findings, Blottner and Roache (1971) pointed out that such results are not generally true and that they probably occurred because change of grid spacing from 0.05 to 0.10 by Crowder and Dalton did not satisfy condition (8.1.5). Some applications where variable grids give better results than a constant grid are described by Moretti and Salas (1970) and Anthes (1970). MacCormack (1971) and Chavez and Richards (1970) also indicate that the loss of accuracy of the solution in a finite difference approximation in which grid size varies is not usually as bad as may be indicated by the formal truncation error, particularly if an isolated change in mesh size is used.

When a variable grid spacing is used,one of the main problems is to obtain consistent difference equations at interfaces in the grid without introducing instability. For explicit finite difference approximations, in a one-dimensional problem, stability conditions such as

$$s = \alpha\Delta t/(\Delta x)^2 \leqslant \tfrac{1}{2} \Rightarrow \Delta t \leqslant (\Delta x)^2/2\alpha \tag{8.1.7}$$

for the diffusion equation, and the Courant-Friedrichs-Lewy condition

$$c = u\Delta t/\Delta x \leqslant 1 \Rightarrow \Delta t \leqslant \Delta x/u \tag{8.1.8}$$

for the convection equation, are needed.

From Equations (8.1.7) and (8.1.8) we can see that if the grid size Δx is reduced, then the time step Δt may also have to be reduced if the method is to remain stable. For instance, if Δx_J is the smallest grid-spacing used then either $\Delta t \leqslant (\Delta x_J)^2/2\alpha$ or $\Delta t \leqslant \Delta x_J/u$ are required. Consequently, the number of computations required may increase significantly. Since such restrictive conditions do not apply to implicit methods, these are more useful in such cases, although it must be remembered that truncation errors do increase as Δt increases, thereby reducing accuracy even with implicit methods. Osher (1970) considered implicit difference equations appropriate for the solution of the diffusion equation (2.1.1) on a grid with an abrupt change in grid size and showed that they were stable over the entire spatial domain, especially at the interface.

A further complication arising from the use of a non-uniform grid is the possibility of wave reflection off the interface between the regions of different grid size. Browning et al (1973) state this need not be a problem but they point out another phenomenon, causing wave distortion, which cannot be avoided. If a wave is already well represented in the coarse grid it should propagate through the interface without any difficulty. But any wave which is poorly represented in the coarser grid will change speed when it passes through the interface into a finer grid region. If this wave passes back into the coarse grid at some other place there can be a serious interaction of the part which propagated through the fine grid with the part of the wave which remained in the coarse grid.

8.2 Variable Grid Spacing

One way of avoiding some of the problems caused by a sudden change in grid size is to use a gradual change of grid spacing in which the intervals are varied continuously from a coarse grid covering most of the flow field to a fine grid in the region of special interest. One advantage of this gradual reduction in grid size is the decrease in both the computer memory necessary and computational time required to obtain improved solutions over the same area as using a small grid spacing everywhere.

Sundqvist and Veronis (1970) have constructed a simple finite difference grid with non-uniform intervals that gives the same accuracy as the uniform grid when derivatives are represented by centred differences.

Consider the interval $0\le x\le 1$ divided into sub-intervals of decreasing length. The grid points are numbered $j=0,1,2,3,\ldots,$ the value of x at the j-th grid point being denoted x_j and the interval between x_j and x_{j+1} denoted Δx_j. Now consider $\bar{\tau}(x)$ on the domain $<0,1>$ with $\bar{\tau}_j = \bar{\tau}(x_j)$. From the Taylor series expansion of $\bar{\tau}_{j-1}$ and $\bar{\tau}_{j+1}$ about the point x_j the following finite difference forms for the first and second order space derivatives are obtained by eliminating the terms in $\partial^2\bar{\tau}/\partial x^2|_j$ and $\partial\bar{\tau}/\partial x|_j$, respectively, in the same way as (8.1.3) and (8.1.4) were derived for an abrupt change in grid spacing:

$$\left.\frac{\partial\bar{\tau}}{\partial x}\right|_j = \frac{\bar{\tau}_{j+1}+[(\nu_j)^2-1]\bar{\tau}_j-(\nu_j)^2\bar{\tau}_{j-1}}{\Delta x_j(\nu_j+1)} - \frac{\Delta x_j\,\Delta x_{j-1}}{6}\left.\frac{\partial^3\bar{\tau}}{\partial x^3}\right|_j + O\{(\Delta x_{j-1})^3\}, \qquad (8.2.1)$$

$$\left.\frac{\partial^2\bar{\tau}}{\partial x^2}\right|_j = \frac{\bar{\tau}_{j+1}-(\nu_j+1)\bar{\tau}_j+\nu_j\bar{\tau}_{j-1}}{\tfrac{1}{2}\Delta x_j\,\Delta x_{j-1}(\nu_j+1)} + \frac{(\Delta x_{j-1}-\Delta x_j)}{3}\left.\frac{\partial^3\bar{\tau}}{\partial x^3}\right|_j + O\{(\Delta x_{j-1})^2\}, \qquad (8.2.2)$$

where $\nu_j = \Delta x_j/\Delta x_{j-1}$. By choosing $\Delta x_{j-1} - \Delta x_j = O\{(\Delta x_{j-1})^2\}$ the first terms on the right hand side of Equations (8.2.1) and (8.2.2) approximate the derivatives on the left hand side with a truncation error of order $O\{(\Delta x_{j-1})^2\}$ - compare Equation (8.1.6).

In this way Sundqvist and Veronis (1970) solved the boundary layer problem proposed by Stommel (1948), using a non-uniform grid with spacings

$$\Delta x_0 = h,\ \Delta x_j = \Delta x_{j-1}(1-\kappa\Delta x_{j-1}), \qquad (8.2.3)$$

where κ is a constant of $O\{1\}$ - compare Equation (8.1.5). They compared their results with those obtained using a uniform grid, for which the solution exhibited large variations from the exact solution with an oscillatory behaviour near the boundary. It was found that the non-uniform grid solution was much more accurate near the boundary and no oscillations were evident. While this method of choosing the grid improves resolution near the boundary, overall it still requires a large number of grid points to significantly reduce the grid spacing there if κ is small.

In their analysis of the formation of a shock in a viscous fluid Moretti and Salas (1970) used both constant and variable size grid spacings. With the constant grid size their solutions exhibited oscillations while with a variable grid these oscillations were no longer present and the solution was more accurate.

There are many geophysical situations in which variable grids are used in order to get a higher resolution in certain areas. For example, ocean models may require higher resolution over continental boundary regions than over other areas of the ocean because of the rapidly changing shore line and associated shallow waters. Variable grids are also useful in modelling atmospheric phenomena. Numerical experiments by Anthes (1970) and Harrison (1973) indicate that a variable grid may give superior results to a constant grid when forecasting the movement and development of tropical storms.

8.3 Stretched Coordinates

Increasing resolution in certain areas may be achieved with a coordinate stretching transformation, such as an exponential stretch. For example, Pao and Daugherty (1969) transformed the cartesian coordinates (x,y) to the coordinates (χ,γ) by means of the relations

$$\begin{aligned} x &= \chi \\ y &= b\{\exp(a\gamma)-1\}, \end{aligned} \tag{8.3.1}$$

where a and b are arbitrary constants.

While such stretching transformations have the same purpose as that of an expanding grid, in that both reduce the number of grid points in certain regions, the two methods are fundamentally different. When the original equations are differenced on an expanding grid, there is a deterioration in the formal truncation error, whereas when the transformed equations are differenced on a uniform grid there is no such loss in formal accuracy.

However, while the formal truncation error is not degraded by coordinate stretching other problems do arise. Stability and convergence may be affected by the introduction of new terms in the transformed equations which can lead to the computation time per step being increased. There is also the possibility of wave distortion due to phase changes and damping.

Kalnay de Rivas (1972) investigated coordinate stretching and found the following.

Consider a function $\bar{\tau}(x)$ defined on a non-uniform grid in the x-direction as in Figure 8.2. Suppose grid intervals are varied by defining a stretched coordinate χ where $x = x(\chi)$ in such a way that grid intervals $\Delta\chi$ are constant. If the function being studied is defined on the region $0 \leqslant x \leqslant 1$ with a fine grid required near $x = 0$, then $x(\chi)$ should have the properties:

(i) $dx/d\chi$ should be finite over the whole interval - if it becomes infinite at some point, then there is poor resolution near that point;

(ii) $dx/d\chi = 0$ at $x = 0$, $dx/d\chi \neq 0$ if $x \neq 0$, which ensures high resolution near the point $x = 0$.

By taking Taylor series expansions about the point x_j to obtain the approximations to the first and second order derivatives, it was found that the truncation error is second order in $\Delta\chi$ even near the boundary $x = 0$, where the errors may be first order with respect to Δx_j. Because the intervals Δx_j are very small here, the overall error remains small.

A Taylor series analysis suggests that a convenient form which gives a fine resolution near $x = 0$ is

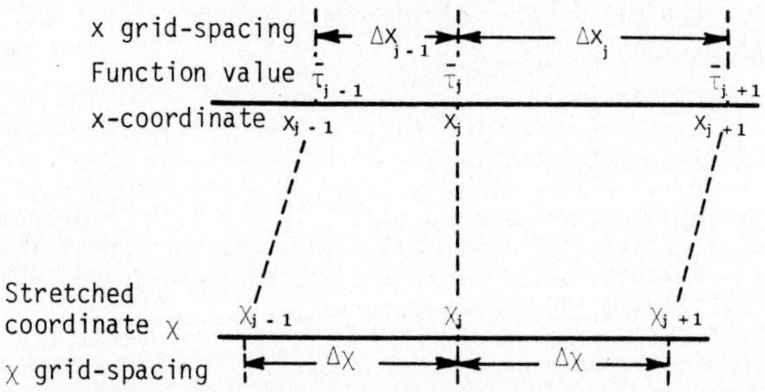

Figure 8.2: Non-uniform grid defined through the use of a stretched coordinate.

$$x = \chi^2. \tag{8.3.2}$$

If a fine grid is required at both $x = 0$ and $x = 1$, a suitable choice of stretched coordinate is

$$x = \sin^2(\pi\chi/2). \tag{8.3.3}$$

Both types of stretched coordinates defined by (8.3.2) and (8.3.3) have been successfully used in two-dimensional numerical models of the atmosphere of Venus (Kalnay de Rivas, 1971). Comparing the percentage errors introduced by the Sundqvist-Veronis grid for Stommel's problem with those introduced by using the stretched coordinate $x = \chi^2$, Kalnay de Rivas found that the latter is more accurate and that there is no tendency for relative errors to grow as $x \to 0$.

Blottner and Roache (1971) noted that, for the problem considered by Crowder and Dalton (1971), a large improvement in accuracy of the solution is achieved by transforming the independent variables, rather than by changing the grid spacing as in Sections 8.1 and 8.2. In fact, the exact solution to their problem can be obtained by means of a parabolic transformation.

8.4 Change of Coordinate System

While grid refinement has proved useful for improving the solution of a flow in special regions such as boundary layers, a coordinate transformation is generally preferred for increasing resolution in a particular area. Similarly the preferred method for treating non-rectangular boundaries is to express the governing partial differential equation in a non-rectangular coordinate system in which boundaries are coordinate surfaces. For example, flow over a cylinder of elliptic cross-section is naturally calculated in elliptic cylinder coordinates.

Thus, problems involving circular boundaries are usually solved more easily in polar coordinates because they avoid the use of complicated finite difference formulae near the curved boundary. Consider the two-dimensional diffusion equation (6.1.3) with $\alpha_x = \alpha_y = \alpha$; in polar coordinates in two-dimensional space this becomes

$$\frac{\partial \bar{\tau}}{\partial t} = \alpha\{\frac{\partial^2 \bar{\tau}}{\partial r^2} + \frac{1}{r}\frac{\partial \bar{\tau}}{\partial r} + \frac{1}{r^2}\frac{\partial^2 \tau}{\partial \theta^2}\}. \tag{8.4.1}$$

If Equation (8.4.1) is differenced on a uniform rectangular grid in the r-θ plane this corresponds to differencing (6.1.3) on a non-uniform grid consisting of radial and circular lines in the x-y plane.

However, such a coordinate transformation may pose problems. Thus, the origin of the polar coordinate system, $r=0$, becomes a line $r=0$ in the r-θ plane. Also, a singularity may be introduced at the origin because of the nature of the second and third terms in (8.4.1).

The numerical solution of certain equations, such as the Navier-Stokes equations, is often hindered by the fact that with the cartesian coordinate system and the standard finite difference approximation, we have to deal with complicated boundaries and so it is difficult to use the correct boundary conditions. Gal-Chen (1975) used a special coordinate transformation to change a domain with an irregular boundary into a regular domain; the transformed system had flat boundaries and homogeneous boundary conditions. Again, while there are considerable advantages to be obtained by this transformation, difficulties may arise due to extra terms being obtained in the transformed governing equations. A more serious problem can occur if the transformation causes new singularities to appear in the transformed equations. Since these singularities are not part of the true solution they may considerably affect numerical stability. Nevertheless, the transformation approach to flow over irregular bodies, such as atmospheric flows over the earth's surface, seems preferable to that of retaining the rectangular cartesian system and using special techniques to cope with the lower flow boundary.

8.5 Non-Rectangular Grids

Geometrical shapes other than the rectangular grid may have advantages for matching irregular boundaries. In particular triangular meshes have been successfully used in problems by Winslow (1966) and Williamson (1969). A hexagonal grid was used by Sadourny and Morel (1969) for a spherical surface.

9. IMPROVING THE ACCURACY OF SOLUTIONS

Techniques for obtaining more accurate solutions using finite difference methods which have already been discussed include reducing the truncation error by using central differences, using methods which minimise numerical diffusion, damping and dispersion, and using more accurate approximations to derivative boundary conditions. Other methods include:

(1) reducing the size of the grid spacing,
(2) using higher order difference approximations, and
(3) reducing the discretisation error by extrapolation.

9.1 Reducing the Grid Spacing

No very useful results which connect the magnitude of the discretisation error to the size of the grid spacing have yet been found. However, it is clear that the errors decrease as the grid spacing is reduced (for example, see Table 3.1). One therefore expects that using smaller and smaller grid spacings will eventually produce successive finite difference solutions which differ from the true solution by less than some required amount.

However, this approach can be very uneconomical. For instance, when solving the one-dimensional diffusion equation using the FTCS method with a fixed value of the parameter s, a reduction of the grid spacing from $\Delta x = 0.1$ (the second row in Table 3.1) to $\Delta x = 0.01$ (the last row) increased the required number of calculations 1000-fold in order to compute τ at a given position and time.

9.2 Higher Order Finite Difference Approximations

More accurate solutions to a problem can be found by representing the derivatives in the given partial differential equation by finite difference approximations of a higher order of accuracy. In this way, the truncation error of the finite difference approximation to the partial differential equation is reduced. For example, in Section 2.3 it was seen that the three-point central difference approximation to a spatial derivative on a grid with spacing h had a truncation error of $O\{h^2\}$, whereas the two-point forward and backward difference approximations had a truncation error of $O\{h\}$. From the point of view of accuracy it is preferable to use central difference approximations for spatial derivatives.

Finite difference equations of higher order of accuracy may contain a larger number of different values of the dependent variable τ, but for many differential operators, such as the Laplacian, the increase is not large.

There are many finite difference representations for the Laplacian in two dimensions,

$$\nabla^2\bar{\tau} = \partial^2\bar{\tau}/\partial x^2 + \partial^2\bar{\tau}/\partial y^2, \tag{9.2.1}$$

a number of these being listed in Bickley (1948) and Thom and Apelt (1961). The Laplacian occurs in the two-dimensional diffusion equation (6.1.3) with $\alpha_x = \alpha_y = \alpha$, which may be written in terms of the Laplacian operator ∇^2 as

$$\partial\bar{\tau}/\partial t = \alpha\nabla^2\bar{\tau} . \tag{9.2.2}$$

Consider a square grid, that is, one with equal spacings in each of the coordinate directions so $\Delta x = \Delta y = h$. Using the central difference forms to both space derivatives in (9.2.1) at the grid point (j,k,n) gives

$$\nabla^2\bar{\tau}\Big|^n_{j,k} = \frac{\bar{\tau}^n_{j+1,k} - 2\bar{\tau}^n_{j,k} + \bar{\tau}^n_{j-1,k}}{h^2} + \frac{\bar{\tau}^n_{j,k+1} - 2\bar{\tau}^n_{j,k} + \bar{\tau}^n_{j,k-1}}{h^2} + O\{h^2\},$$

or

$$\nabla^2\bar{\tau}\Big|_{j,k}^{n} = \frac{1}{h^2}\begin{bmatrix} & +\bar{\tau}^{n}_{j,k+1} & \\ +\bar{\tau}^{n}_{j-1,k} & -4\bar{\tau}^{n}_{j,k} & +\bar{\tau}^{n}_{j+1,k} \\ & +\bar{\tau}^{n}_{j,k-1} & \end{bmatrix} + O\{h^2\}. \qquad (9.2.3)$$

The five-point finite difference approximation to the Laplacian,obtained by dropping terms of $O\{h^2\}$ from (9.2.3), is often written in the stencil form

$$\nabla^2\bar{\tau} \simeq \frac{1}{h^2}\begin{bmatrix} & 1 & \\ 1 & -4 & 1 \\ & 1 & \end{bmatrix}\bar{\tau}, \qquad (9.2.4)$$

which has a second order truncation error.

A nine-point finite difference approximation to the Laplacian, with a fourth order truncation error, can be obtained by using difference approximations which use τ values two grid-points either side of the point (jh,kh). For example, using Taylor expansions about (jh,kh) it can be shown that the spatial derivative $\partial^2\bar{\tau}/\partial x^2|_{j,k}$ can be written in the form

$$\frac{\partial^2\bar{\tau}}{\partial x^2}\Big|_{j,k} = \frac{-\bar{\tau}_{j+2,k}+16\bar{\tau}_{j+1,k}-30\bar{\tau}_{j,k}+16\bar{\tau}_{j-1,k}-\bar{\tau}_{j-2,k}}{12(\Delta x)^2} + O\{(\Delta x)^4\}, \qquad (9.2.5)$$

and there is a similar expression for $\partial^2\bar{\tau}/\partial y^2|_{j,k}$. For the Laplacian, the stencil with fourth order truncation error on a square grid is therefore

$$\nabla^2\bar{\tau} \simeq \frac{1}{12h^2}\begin{bmatrix} & & -1 & & \\ & & 16 & & \\ -1 & 16 & -60 & 16 & -1 \\ & & 16 & & \\ & & -1 & & \end{bmatrix}\bar{\tau}. \qquad (9.2.6)$$

When methods of improved accuracy are developed for grid-points in the interior of the solution domain, it should be kept in mind that the formulation of the boundary conditions should be of comparable accuracy. There is little point in using finite difference approximations to the space derivatives in a partial differential equation with truncation error of fourth order, if the finite difference approximations to the space derivatives in the boundary conditions have truncation errors of first or second order (see Section 4.4).

It should be noted that even when the boundary values are known, as in the Dirichlet problem, special treatment is required at interior gridpoints next to the boundary when derivatives are written in the forms like (9.2.5) with fourth order truncation errors. When the boundary values $\bar{\tau}_{0,k}$, $\bar{\tau}_{J,k}$, $k=0(1)K$ are known, the form (9.2.5) can be used for values of $x = j\Delta x$, $j=2(1)J-2$, resulting in fourth order truncation errors. With $j=1$, the use of (9.2.5) requires knowledge about $\bar{\tau}_{-1,k}$ and with $j=J-1$, $\bar{\tau}_{J+1,k}$ is needed, and both are values at gridpoints exterior to the region considered. Equation (2.3.19) could be used, since with $j=1$ for example, it gives

$$\left.\frac{\partial^2\bar{\tau}}{\partial x^2}\right|_{1,k} = \frac{\bar{\tau}_{2,k} - 2\bar{\tau}_{1,k} + \bar{\tau}_{0,k}}{(\Delta x)^2} + O\{(\Delta x)^2\}.$$

Unfortunately, the resulting finite difference approximation to the original partial differential equation discretised at $j=1$ then has a truncation error of $O\{(\Delta x)^2\}$, which is much less accurate than the form for $j=2(1)J-2$.

The truncation error of $\partial^2\bar{\tau}/\partial x^2|_{1,k}$ can be made $O\{(\Delta x)^4\}$ if the derivative is expressed as a linear combination of $\bar{\tau}_{j,k}$, $j=0(1)5$. Use of Taylor series about $(\Delta x, k\Delta y)$ then gives the form

$$\left.\frac{\partial^2\bar{\tau}}{\partial x^2}\right|_{1,k} = \frac{10\bar{\tau}_{0,k} - 15\bar{\tau}_{1,k} - 4\bar{\tau}_{2,k} + 14\bar{\tau}_{3,k} - 6\bar{\tau}_{4,k} + \bar{\tau}_{5,k}}{12(\Delta x)^2} + O\{(\Delta x)^4\}. \quad (9.2.7)$$

With a similar expression for $\partial^2\bar{\tau}/\partial x^2|_{J-1,k}$ written in terms of $\bar{\tau}_{j,k}$, $j=J-5(1)J$, all derivatives $\partial^2\bar{\tau}/\partial x^2|_{j,k}$, $j=1(1)J-1$, can be written in terms of $\bar{\tau}_{j,k}$, $j=0(1)J$, with a truncation error of $O\{(\Delta x)^4\}$.

However, the use of finite difference equations based on derivative approximations of higher order accuracy does not necessarily mean that the finite difference solution obtained will be more accurate. Examples given by Hamming (1962), Cyrus and Fulton (1967) and Cheng (1968) clearly show that some lower order methods produce results at least as accurate as corresponding methods of higher order. The lack of success with the latter may have been due to poor treatment of the boundary conditions. The order of the discretisation errors of values computed at the boundary and in the interior of the space region concerned should be consistent.

One of the reasons for the lack of improvement when higher order methods are used to solve partial differential equations is that the order of the method is significant only as $\Delta x \to 0$, $\Delta y \to 0$, $\Delta z \to 0$, $\Delta t \to 0$, because the order of the truncation error is only that of the leading term which has been omitted. This order is more significant to ordinary differential equations where much smaller grid spacings, Δx say, can be used than the Δx, Δy, Δz, Δt for partial differential equations. This is because finite difference solutions in the former case only involve the storage of arrays of dimension J, whereas difficulty is often encountered when arrays of dimension $J \times K \times L$, the product of the number of grid points in the x-direction with the numbers in the y- and z-directions, are required to be stored in the solution of three-dimensional partial differential equations.

However, the possibility of improved accuracy is not the only fact to be considered when higher order methods are evaluated. In practice, stability, rate of convergence of the solution, and special procedures required near the boundaries must be considered.

9.3 Reduction of the Discretisation Error by Extrapolation

This method, introduced by Richardson (1910) and called by him the *deferred approach to the limit*, may lead to considerable improvements of numerical results when solving partial differential equations by finite difference methods. Applications of this method to equations solved on rectangular regions in space are described in Richardson and Gaunt (1927) and Salvadori (1951).

The reduction of the error depends upon there being a reliable estimate of the discretisation error as a function of the grid spacing. Let $\bar{\tau}$ be the exact solution of the differential equation and $\tau^{(1)}$, $\tau^{(2)}$ the finite difference solutions, at a particular point in the solution domain computed using grid spacings h_1 and h_2 respectively. If the leading term in the discretisation error is of $O\{h^m\}$ for a grid spacing of h, then as

$$\bar{\tau} \simeq \tau + K(h)^m, \tag{9.3.1}$$

where τ is the finite difference approximation to $\bar{\tau}$, it follows that

$$\bar{\tau} - \tau^{(1)} \simeq K(h_1)^m, \tag{9.3.2a}$$

and

$$\bar{\tau} - \tau^{(2)} \simeq K(h_2)^m. \tag{9.3.2b}$$

Eliminating K and solving for $\bar{\tau}$ gives a more accurate estimate, τ_E of the solution in terms of the two approximate solutions $\tau^{(1)}$ and $\tau^{(2)}$, namely

$$\tau_E = \frac{(h_2)^m\tau^{(1)} - (h_1)^m\tau^{(2)}}{(h_2)^m - (h_1)^m} . \tag{9.3.3}$$

Thus $\bar{\tau}$ is found more accurately than $\bar{\tau}^{(1)}$ or $\bar{\tau}^{(2)}$, provided a reliable estimate of m is available.

For instance, the FTCS equation (2.4.4) used to solve the one-dimensional diffusion equation (2.1.1), with the initial and boundary conditions given in Section 2.1, contributes a term of $O\{(\Delta t)^2, \Delta t(\Delta x)^2\}$ to the discretisation error at each time step if $s \neq 1/6$. Consider the approximations to the value of $\bar{\tau}(0.4,8)$ if $s = \frac{1}{2}$. As the number of time steps required to reach $t_N = N\Delta t = 8$ is inversely proportional to Δt, then the total discretisation error in the approximations to $\bar{\tau}(0.4,8)$ are

$$e = O\{N(\Delta t)^2, N\Delta t(\Delta x)^2\} = O\{\Delta t, (\Delta x)^2\}.$$

Also, as Δt is proportional to $(\Delta x)^2$ for a fixed value of s, then the total discretisation error may be written as

$$e = O\{(\Delta x)^2\}$$

Clearly we may take $m = 2$ in this example, so long as s is held constant.

From Table 3.1, it is seen that with $J=10$, the approximation to $\bar{\tau}(0.4,8) = 0.4503963$ is $\tau^{(1)} = 0.4754791$, and with $J=20$, $\tau^{(2)} = 0.4566436$. Substitution of $\tau^{(1)}$ and $\tau^{(2)}$ into (9.3.3) with $h_1 = 0.1$, $h_2 = 0.05$ gives the improved approximation $\tau_E = 0.4503651$. Note that, while $\tau^{(1)}$ is correct to only one significant figure and $\tau^{(2)}$ to only two figures, the extrapolated value τ_E is correct to four figures.

When the value of m is not known at particular gridpoints, it can be determined from three finite difference solutions obtained for the same point using grid spacings which have been successively reduced by multiplying by some faction $f < 1$ so that

$$h_3 = fh_2 = (f)^2 h_1. \tag{9.3.4}$$

Subtracting (9.3.2a,b) gives

$$\tau^{(2)} - \tau^{(1)} \simeq K\{(h_1)^m - (h_2)^m\} = K(h_1)^m\{1-(f)^m\}, \tag{9.3.5a}$$

and, by subtracting (9.3.2b) from the equivalent form with $h = h_3$, we find

$$\tau^{(3)} - \tau^{(2)} \simeq K\{(h_2)^m-(h_3)^m\} = (f)^m K(h_1)^m\{1-(f)^m\}. \qquad (9.3.5b)$$

Dividing (9.3.5b) by (9.3.5a) gives

$$\frac{\tau^{(3)} - \tau^{(2)}}{\tau^{(2)} - \tau^{(1)}} = (f)^m, \qquad (9.3.6)$$

so

$$m = \log_f \left\{\frac{\tau^{(3)} - \tau^{(2)}}{\tau^{(2)} - \tau^{(1)}}\right\}. \qquad (9.3.7)$$

Application of Equation (9.3.3) with $h_2 = fh_1$ gives

$$\tau_E = \frac{\tau^{(2)} - (f)^m \tau^{(1)}}{1-(f)^m}. \qquad (9.3.8)$$

Substitution of (9.3.6) for $(f)^m$ in this equation yields

$$\tau_E = \frac{\tau^{(1)}\tau^{(3)} - (\tau^{(2)})^2}{\tau^{(1)} - 2\tau^{(2)} + \tau^{(3)}}, \qquad (9.3.9)$$

which is a better estimate of the exact solution at the particular grid-point than $\tau^{(1)}$, $\tau^{(2)}$ or $\tau^{(3)}$.

This method of reducing the discretisation error has been applied to the one-dimensional diffusion problem presented in Section 2.1, which was solved using the FTCS explicit method in Section 2.4. From the results presented in Table 3.1 it is seen that, when $s = \frac{1}{2}$ and J=10,20,40 the approximations to $\bar{\tau}(0.4,8)$ obtained were $\tau^{(1)} = 0.4754791$, $\tau^{(2)} = 0.4566436$, $\tau^{(3)} = 0.4519564$, respectively. The first two approximations to the exact value $\bar{\tau} = 0.4503963$ have relative errors of 5.6% and 1.4% respectively, and the third has a relative error of 0.35%. Substitution of these values into Equation (9.3.9) gives the value $\tau_E = 0.4504036$ which has a relative error of 0.0016%. While $\tau^{(1)}$ is correct to only one figure, $\tau^{(2)}$ to two figures and $\tau^{(3)}$ to three figures, the extrapolated value τ_E is correct to five figures when rounded off.

Clearly, this is a useful way of increasing the accuracy of the result at a given gridpoint without having to excessively increase the number of calculations required. To obtain the result $\tau_E = 0.4504036$ with $s = \frac{1}{2}$ required $16 + 64 + 256 = 336$ applications of the explicit finite difference equation (2.4.4). This result is more accurate than that obtained in Table 3.1 using the FTCS method $J = 100$ and $s = 1/6$, which required 4800 applications of the Equation (2.4.4). However, note that the more accurate values are only obtained on the coarsest grid used, that is at $x = j/10$, $j=1(1)9$, in the above example.

This method may be of uncertain value near boundaries which are not straight (since boundary value interpolation is then required), near corners with interior angles greater than 180°, and near boundaries on which the given values of the dependent variable or its normal derivative are not spatially smooth. For details about these restrictions see Wasow (1955) and Forsythe and Wasow (1960).

10. SUMMARY

Analytical methods of solving partial differential equations are usually restricted to linear cases with simple geometries and boundary conditions. The increasing availability of more and more powerful digital computers has made more common the use of numerical methods for solving such equations, in addition to non-linear equations with more complicated boundaries and boundary conditions.

In the previous sections, one particular numerical method, the method of finite differences, has been described. This method is based on the representation of the continuously defined function $\bar{\tau}(x,y,z,t)$, and its derivatives, in terms of values of an approximation τ defined at particular, discrete points called gridpoints. These are generally the points of intersection of a grid of straight lines $x = x_j$, $y = y_k$, $z = z_\ell$, $t = t_n$, $j,k,\ell,n=1,2,3,\ldots$, that cover the domain of $\bar{\tau}$. From the appropriate Taylor series expansions of $\bar{\tau}$ about such gridpoints, forward, backward and central difference approximations to derivatives of $\bar{\tau}$ can be developed to convert the given partial differential equation and its initial and boundary conditions to a set of linear algebraic equations linking the approximations τ defined at the gridpoints.

The resulting numerical model is composed of the finite difference analogues (the difference equations) of the mathematical model (the partial differential equation and the initial and boundary conditions) that describes the original problem. While the mathematical model represents the problem in terms of the dependent variable $\bar{\tau}$ defined continuously in the domain of the independent variables x,y,z and t, the finite difference numerical model defines the dependent variable pointwise at the discrete gridpoints (x_j, y_k, z_ℓ, t_n) within the domain of the mathematical model. The resulting numerical solution of the finite difference equation is a table of numbers, each one being associated with one point in the grid. Distinct finite difference numerical models of the same mathematical model are obtained by using different grids and different finite difference approximations to the continuous derivatives involved in the mathematical model.

The finite difference method can be used to solve even the most complicated mathematical models where they consist of partial differential equations and their associated auxiliary conditions. More information about these techniques may be found in books by Ames (1969), Forsythe and Wasow (1960), Fox (1962), Richtmeyer and Morton (1967) and Roache (1974), as well as in scientific and engineering journals such as the Journal of Computational Physics.

The formulation of a well posed mathematical model of a continuum problem requires that the problem be reduced to a set of differential equations and associated boundary conditions that accurately describe the problem. The formulation of a well posed numerical model requires both the derivation of a set of difference equations based on the mathematical model and a demonstration of the adequacy of these difference equations. The latter involves an analysis of the consistency, convergence and stability of the numerical model.

The difference equations and auxiliary conditions of the numerical model must be *consistent* with the differential equations and initial and boundary conditions of the mathematical model. The numerical model is consistent if the truncation error, that is, the discrepancy between the finite difference approximations and the continuous derivatives, tends to zero as the grid spacings get smaller and smaller. Second, the solution to the finite difference equation must *converge* to the solution of the differential equation as the grid spacings get smaller and smaller. That is, the difference between the exact solutions of the numerical and mathematical models should vanish as the grid spacings tend to zero. Finally, it must be possible to solve the finite difference equation on a computer. As computers can store only a finite number of digits to represent each number (between 7 and 28 decimal digits) at each step of any calculation, round-off errors may occur. The computation is *stable* if the growth of these errors is reasonable or controlled.

A numerical model with consistent equations, convergent solutions and stable error propagation forms a computationally stable scheme which gives results which closely approximate the exact solution of the mathematical model. This was illustrated in Sections 2, 4, 5, 6 and 7, in which finite difference solutions obtained from various numerical models were compared with exact solutions of the original mathematical models.

Convergence of the numerical solution of particular finite difference schemes is often difficult to prove. However, for mathematical models which are well posed, linear, initial-value problems, Lax's Equivalence Theorem can be used. This theorem states that the solution of a numerical model which is consistent with such a mathematical model, converges if and only if the finite difference equation is stable.

In principle, the method of finite differences can be used on non-linear mathematical models, but consistency, stability and convergence are more difficult to prove. For instance, von Neumann's method of stability analysis cannot be used other than locally, since it only applies to linear finite difference equations. In many cases, numerical experimentation, such as solving the finite difference equation using progressively smaller grid spacings and examining the behaviour of the sequence of values of τ obtained at a given point (x,y,z,t), is the only method available with which to assess the numerical model.

The various methods of obtaining a finite difference numerical model corresponding to a particular mathematical model may result in either explicit or implicit finite difference equations. The former may be solved directly with a minimum of programming effort and computation time in order to advance from one time level to another in an initial-value problem, but are generally only conditionally stable.

Generally, implicit finite difference schemes are unconditionally stable. This implies computational stability for any size of the time increment. However, the size of Δt is still limited by the accuracy required in the answers; use of very large values of the time step results in poor answers because of unacceptably large truncation errors produced.

Implicit finite difference schemes depend on the solution of a large number of linear algebraic equations. These may be solved by elimination processes, such as the Thomas algorithm used for the tri-diagonal system of linear algebraic equations obtained when the Crank-Nicolson finite difference method is used to solve the one-dimensional diffusion equation. Alternatively, iterative methods such as successive over-relaxation may be used. Iterative methods may be used at each time level when solving numerical models of multi-dimensional equations in space. Recently, alternating direction implicit methods and locally one-dimensional methods have been increasingly used to solve initial-value, boundary-value problems involving more than one space dimension.

A similar numerical technique in vogue for solving equilibrium type problems is the finite element method (see Zienkiewicz, 1967). This method is not classified as a finite difference technique since it differs from finite difference methods in two ways. First, instead of involving a set of grid points at the intersection of a set of parallel lines, it employs a series of "finite elements" connected to discrete points on the boundary of the space region and covering the region. Second, the left side of the partial differential equation, rearranged so it is equated to zero, is replaced over the region by its equivalent integral form, the value of which is minimized using the calculus of variations. The resulting set of linear algebraic equations connecting values of the dependent variable at various points over the space region involved, must then be solved. Unlike the well-structured sets of equations which result from implicit finite differencing, which are usually strongly banded and can be solved using special algorithms, the form of the equations resulting from the finite element method are more random in nature. They usually require reordering to develop any inherent banding. For time-dependent initial-value problems in two- and three-dimensions, the finite element method in space is usually linked with finite differencing in time. The finite element method is seldom extended to include the independent variable, time.

APPENDIX : The Thomas Algorithm

The system

$$
\begin{aligned}
b_1\tau_1 + c_1\tau_2 \qquad\qquad\qquad\qquad\qquad\qquad &= d_1\\
a_2\tau_1 + b_2\tau_2 + c_2\tau_3 \qquad\qquad\qquad\qquad &= d_2\\
a_3\tau_2 + b_3\tau_3 + c_3\tau_4 \qquad\qquad\qquad &= d_3\\
\vdots \qquad\qquad\qquad\qquad &\quad \vdots\\
a_j\tau_{j-1} + b_j\tau_j + c_j\tau_{j+1} \qquad &= d_j\\
\vdots \qquad\qquad &\quad \vdots\\
a_{M-1}\tau_{M-2} + b_{M-1}\tau_{M-1} + c_{M-1}\tau_M &= d_{M-1}\\
a_M\tau_{M-1} + b_M\tau_M &= d_M
\end{aligned}
\tag{A1.1}
$$

in which the a's, b's, c's and d's are known, is usually solved for the unknown values τ_j, j=1,2,..,M, by an elimination procedure attributed to Thomas (1949). The first equation listed in (A1.1) is used to eliminate τ_1 from the second equation, the new second equation is used to eliminate τ_2 from the third equation and so on until finally, the new second-last equation is used to eliminate τ_{M-1} from the last equation giving one equation with only one unknown, τ_M. The unknowns $\tau_{M-1},\tau_{M-2},..,\tau_2,\tau_1$ are then found in turn by back-substitution. This method is therefore a particular example of the Gauss elimination method with back substitution

Suppose τ_{j-2} has been eliminated from the (j-1)th equation in the set (A1.1) which has become

$$
\begin{aligned}
\beta_1\tau_1 + \gamma_1\tau_2 \qquad\qquad\qquad\qquad\qquad &= \delta_1\\
\beta_2\tau_2 + \gamma_2\tau_3 \qquad\qquad\qquad\qquad &= \delta_2\\
\beta_3\tau_3 + \gamma_3\tau_4 \qquad\qquad\qquad &= \delta_3\\
\vdots \qquad\qquad\qquad &\quad \vdots\\
\beta_{j-1}\tau_{j-1} + \gamma_{j-1}\tau_j \qquad\qquad &= \delta_{j-1}\\
a_j\tau_{j-1} + b_j\tau_j + c_j\tau_{j+1} \qquad &= d_j\\
\vdots \qquad &\quad \vdots\\
a_M\tau_{M-1} + b_M\tau_M &= d_M
\end{aligned}
\tag{A1.2}
$$

where $\beta_1 = b_1, \gamma_1 = c_1, \delta_1 = d_1$. Using the new (j-1)th equation to eliminate τ_{j-1} from the jth equation in (A1.2) gives

$$
\left(b_j - \frac{a_j\gamma_{j-1}}{\beta_{j-1}}\right)\tau_j + c_j\tau_{j+1} = d_j - \frac{a_j\delta_{j-1}}{\beta_{j-1}}, \tag{A1.3a}
$$

or

$$
\beta_j\tau_j + \gamma_j\tau_{j+1} = \delta_j . \tag{A1.3b}
$$

A comparison of Equations (A1.3a) and (A1.3b) shows that the coefficients β_j, γ_j, δ_j are given in terms of β_{j-1}, γ_{j-1}, δ_{j-1} and a_j, b_j, c_j, d_j by the relations

$$\varepsilon_j = a_j / \beta_{j-1}, \quad \text{(A1.4a)}$$

$$\beta_j = b_j - \varepsilon_j \gamma_{j-1}, \quad \text{(A1.4b)}$$

$$\gamma_j = c_j, \quad \text{(A1.4c)}$$

$$\delta_j = d_j - \varepsilon_j \delta_{j-1}. \quad \text{(A1.4d)}$$

Commencing with $\beta_1 = b_1$, $\gamma_1 = c_1$ and $\delta_1 = d_1$, Equations (A1.4) provide a method of finding ε_j, β_j, γ_j, δ_j for j=2 (1)M.

When j=M, we obtain

$$\beta_M \tau_M = \delta_M$$

since $\gamma_M = c_M = 0$. Thus

$$\tau_M = \delta_M / \beta_M ,$$

and the remaining values of τ are found by substitution into a rearranged form of Equation (A1.3b), namely

$$\tau_j = (\delta_j - \gamma_j \tau_{j+1}) / \beta_j , \ j=M-1(-1)1. \quad \text{(A1.5)}$$

The complete algorithm, with c_j substituted for γ_j in the previous results, is:

Set $\beta_1 = b_1$, $\delta_1 = d_1$.

For j=2 (1)M *compute*

$$\varepsilon_j = a_j / \beta_{j-1}$$

$$\beta_j = b_j - \varepsilon_j c_{j-1}$$

$$\delta_j = d_j - \varepsilon_j \delta_{j-1}.$$

Then

$$\tau_M = \delta_M / \beta_M .$$

For j=M-1(-1)1 *compute*

$$\tau_j = (\delta_j - c_j \tau_{j+1}) / \beta_j .$$

This method is stable with regard to propagation of round-off error as long as $|b_j| \geq |a_j| + |c_j|$ for all j, with strict inequality for one value of j.

The normal Gauss elimination methods require about $M^3/3$ multiplications and divisions (about 9,000 if M=30) to solve a system of M linear algebraic equations in M unknowns. The Thomas method, which takes advantage of the tridiagonal nature of the system, requires only approximately 5M multiplications and divisions (150 if M=30) since the algorithm eliminates all multiplications by zero coefficients of τ which take place when the general Gauss elimination method is applied.

If there are N such systems as (A1.1) to be solved, the quantities ε_j and β_j need be computed only once because the values of a_j , b_j , c_j are fixed. Equations (A1.4d) and (A1.5) must each be computed for N systems, for a total of approximately 3MN arithmetic operations.

Even greater economy can be achieved if the values of the a_j , b_j , c_j are the same for each j in most of the equations (A1.1). This is the case for the equations (4.2.3) of the classical implicit method and the equations (4.2.12) of the Crank-Nicolson method for solving the diffusion problem.

REFERENCES

Ames, W.F. (1965), *Nonlinear Partial Differential Equations in Engineering*, Academic Press, New York.

Ames, W.F. (1969), *Numerical Methods for Partial Differential Equations*, Barnes and Noble, Inc., New York, New York.

Amsden, A.A., and Harlow, F.H. (1964), "Slip Instability", *Physics of Fluids*, Vol. 7, pp. 327-334.

Anthes, R.A. (1970), "Numerical Experiments with a Two Dimensional Horizontal Variable Grid", *Monthly Weather Review*, Vol. 98, No. 11, pp. 810-822.

Bickley, W.G. (1948), "Finite-Difference Formulae for the Square Lattice", *Quarterly Journal of Mechanics and Applied Mathematics*, Vol. 1, pp. 35-42.

Birkhoff, G., and Varga, R.S. (1959), "Implicit Alternating Direction Methods", *Transactions American Mathematical Society*, Vol. 92, pp. 13-24.

Blottner, F.G., and Roache, P.J. (1971), "Nonuniform Mesh Systems", *Journal of Computational Physics*, Vol. 8, pp. 498-499.

Brian, P.L.T. (1961), "A Finite-Difference Method of High-Order Accuracy for the Solution of Three-Dimensional Transient Heat Conduction Problems", *American Institute of Chemical Engineering Journal*, Vol. 7, No. 3, pp. 367-370.

Browning, G., Kreiss, H.O., and Oliger, J. (1973) "Mesh Refinement", *Mathematics of Computation*, Vol. 27, No. 121, pp. 29-39.

Carnahan, B., Luther, H.A., and Wilkes, J.O. (1969), *Applied Numerical Methods*, John Wiley, New York.

Carre, B.A. (1961), "The Determination of the Optimum Accelerating Factor for Successive Over-Relaxation", *Computer Journal*, Vol. 4, No. 1, pp. 73-78.

Carslaw, H.S., and Jaeger, J.C. (1962), *Conduction of Heat in Solids*, Oxford University Press, London.

Chavez, S.P., and Richards, C.G. (1970), *A Numerical Study of the Coanda Effect*, 70-FIcs-12, The American Society of Mechanical Engineers, United Engineering Center, New York, New York.

Cheng, S.I. (1968), *Accuracy of Difference Formulation of Navier-Stokes Equations*, A.M.S. Department, Princeton University, Princeton, New Jersey.

Colgan, L. (1981), "Iterative Methods for Solving Large Sparse Linear Systems", *Numerical Solution of Partial Differential Equations*, ed. John Noye, North Holland, Amsterdam, pp. 367-396.

Courant, R., Isaacson, E., and Rees, M. (1952), "On the Solution of Nonlinear Hyperbolic Differential Equations by Finite Differences", *Communications on Pure and Applied Mathematics*, Vol. 5, pp. 243-255.

Crank, J., and Nicolson, P. (1947), "A Practical Method for Numerical Evaluation of Solutions of Partial Differential Equations of the Heat-Conduction Type", *Proceedings of the Cambridge Philosophical Society*, Vol. 43, No. 50, pp. 50-67.

Crowder, H.J., and Dalton, C. (1971), "Errors in the Use of Nonuniform Mesh Systems", *Journal of Computational Physics*, Vol. 7, pp. 32-45.

Cyrus, N.J., and Fulton, R.E. (1967), *Accuracy Study of Finite Difference Methods*, NASA TN D-4372, National Aeronautics and Space Administration, Langley Research Center, Langley Station, Hampton, Virginia.

de Vahl Davis, G., and Mallinson, G.D. (1976), "An Evaluation of Upwind and Central Difference Approximations by a Study of Recirculating Flow", *Computers and Fluids*, Vol. 4, pp. 29-43.

Douglas, J., Jr. (1955), "On the Numerical Integration of $\partial^2u/\partial x^2 + \partial^2u/\partial y^2 = \partial u/\partial t$ by Implicit Methods", *Journal Society of Industrial Applied Mathematics*, Vol. 3, No. 1, pp. 42-65.

Douglas, J., Jr. (1957), "A Note on the Alternating Direction Implicit Method for the Numerical Solution of Heat Flow Problems", *Proceedings of the American Mathematical Society*, Vol. 8, pp. 409-412.

Douglas, J., Jr. (1962), "Alternating Direction Methods for Three Space Variables", *Numerische Mathematik*, Vol. 4, pp. 41-63.

DuFort, E.C., and Frankel, S.P. (1953), "Stability Conditions in the Numerical Treatment of Parabolic Differential Equations", *Mathematical Tables and Other Aids to Computation*, Vol. 7, pp. 135-152.

D'Yakonov, E.G. (1963), "Difference Schemes with Split Operators for Multi-Dimensional Unsteady Problems", *U.S.S.R. Comp. Math.*, Vol. 4, No. 2, pp. 92-110.

Eddy, E.P. (1949), *Stability in the Numerical Solution of Initial Value Problems in Partial Differential Equations*, NOLM 10232, Naval Ordinance Laboratory, White Oak, Silver Spring, Maryland.

Forsythe, G.E., and Wasow, W. (1960), *Finite Difference Methods for Partial Differential Equations*, Wiley, New York.

Fox, L. (1944), "Solution by Relaxation Methods of Plane Potential Problems with Mixed Boundary Conditions", *Quarterly of Applied Mathematics*, Vol. 2, pp. 251-257.

Fox, L. (1962), *Numerical Solution of Ordinary and Partial Differential Equations*, Addison-Wesley Publishing Company, Inc., Reading, Massachusetts.

Frankel, S.P. (1950), "Convergence Rates of Iterative Treatments of Partial Differential Equations", *Mathematical Tables and Other Aids to Computation*, Vol. 4, pp. 65-75.

Frankel, S.P. (1956), "Some Qualitative Comments on Stability Considerations in Partial Differential Equations", *Proceedings Sixth Symposia in Applied Mathematics, AMS, Vol. 6: Numerical Analysis*, pp. 73-75.

Fromm, J.E., and Harlow, F.H. (1963), "Numerical Solution of the Problem of Vortex Sheet Development", *Physics of Fluids*, Vol. 6, No. 7, pp. 975-982.

Gal-Chen, T. (1975), "On the Use of a Coordinate Transformation for the Solution of the Navier Stokes Equations", *Journal of Computational Physics*, Vol. 17, pp. 209-228.

Gourlay, A.R., and Mitchell, A.R. (1969), "The Equivalence of Certain Alternating Direction and Locally One-Dimensional Difference Methods", *SIAM Journal on Numerical Analysis*, Vol. 6, No. 1, pp. 37-46.

Hahn, S.G. (1958), "Stability Criteria for Difference Schemes", *Communications on Pure and Applied Mathematics*, Vol. 11, pp. 243-255.

Hamming, R.W. (1962), *Numerical Methods for Scientists and Engineers*, McGraw-Hill Book Company, Inc., New York.

Harrison, E.J. (1973), "Three-Dimensional Numerical Simulations of Tropical Systems Utilizing Nested Finite Grids", *Journal of the Atmospheric Sciences*, Vol. 30,pp. 1528 - 1543.

Hirt, C.W. (1968), "Heuristic Stability Theory for Finite-Difference Equations", *Journal of Computational Physics*, Vol. 2, pp. 339-355.

Hung, T.K., and Macagno, E.O. (1966), "Laminar Eddies in a Two-Dimensional Conduit Expansion", *La Houille Blanche*, Vol. 21, No. 4, pp. 391-400.

Isenberg, J., and de Vahl Davis, G. (1975), "Finite Difference Methods in Heat and Mass Transfer", in *Topics in Transport Phenomena*, ed. C. Gutfinger, Hemisphere Publishing Company, Washington, D.C.

John, F. (1952), "On the Integration of Parabolic Equations by Difference Methods", *Communications on Pure and Applied Mathematics,* Vol. 5, pp. 155-211.

Kalnay de Rivas, E. (1971), *The Circulation of the Atmosphere of Venus*, Ph.D. Thesis, Department of Meteorology, M.I.T., Massachusetts, U.S.A.

Kalnay de Rivas, E. (1972), "On the Use of Non-Uniform Grids in Finite-Difference Equations", *Journal of Computational Physics,* Vol. 10, pp. 202-210.

Kreyszig, E. (1972), *Advanced Engineering Mathematics,* 3rd Edition, John Wiley & Sons, Inc., New York.

Kusic, G.L., and Lavi, A. (1968), "Stability of Difference Methods for Initial-Value Type Partial Differential Equations", *Journal of Computational Physics,* Vol. 3, pp. 358-378.

Kusic, G.L. (1969), "On Stability of Numerical Methods for Systems of Initial-Value Partial Differential Equations", *Journal of Computational Physics,* Vol. 4, pp. 272-275.

Laasonen, P. (1949), "Über eine Methode zur Lösung der Warmeleitungsgleichung", *Acta Mathematica*, Vol. 81, p. 309.

Lax, P.D., and Richtmyer, R.D. (1956), "Survey of the Stability of Linear Finite Difference Equations", *Communications on Pure and Applied Mathematics,* Vol. 9, pp. 267-293.

MacCormack, R.W. (1971), "Numerical Solution of the Interaction of a Shock Wave with a Laminar Boundary Layer", *Proceedings of Second International Conference on Numerical Methods in Fluid Dynamics,* ed. M. Holt, Lecture Notes in Physics, Vol. 8, Springer-Verlag, New York.

Mann, K. (1981),"Inversion of Large Sparse Matrices - Direct Methods", *Numerical Solution of Partial Differential Equations,* ed. John Noye, North-Holland, Amsterdam, pp. 311-355.

Marchuk, G.I. (1975), *Methods of Numerical Mathematics,* Springer-Verlag, New York.

Moretti, G., and Salas, M.D. (1970), "Numerical Analysis of Viscous One-Dimensional Flows", *Journal of Computational Physics,* Vol. 5, pp. 487-506.

Myers, G.E. (1971), *Analytical Methods in Conduction Heat Transfer,* McGraw-Hill Book Company.

Noye, B.J. (1978), "Finite Difference Techniques", *Numerical Simulation of Fluid Motion,* ed. John Noye, North-Holland, Amsterdam, pp. 1-112.

O'Brien, G.G., Hyman, M.A., and Kaplan, S. (1950), "A Study of the Numerical Solution of Partial Differential Equations", *Journal of Mathematics and Physics,* Vol. 29, pp. 223-251.

Osher, S.J. (1970), "Mesh Refinements for the Heat Equation", *S.I.A.M. Journal of Numerical Analysis*, Vol. 7, pp. 219-227.

Pao, Y.-H., and Daugherty, R.J. (1969), *Time-Dependent Viscous Incompressible Flow Past a Finite Flat Plate*, Boeing Scientific Research Laboratories, D1-82-0822, January, 1969.

Payne, R.B. (1958), "Calculation of Unsteady Viscous Flow Past a Circular Cylinder", *Journal of Fluid Mechanics,* Vol. 4, p. 81.

Peaceman, D.W., and Rachford, H.H., Jr. (1955), "The Numerical Solution of Parabolic and Elliptical Differential Equations", *Journal Society Indust. Applied Mathematics,* Vol. 1, No. 1, March 1955, pp. 28-41.

Pearson, C.E. (1964), *A Computational Method for Time-Dependent Two-Dimensional Incompressible Viscous Flow Problems,* Sperry Rand Research Centre Report, SRRC-RR-64-17, February, 1964.

Richardson, L.F. (1910), "The Approximate Arithmetical Solution by Finite Differences of Physical Problems Involving Differential Equations, with an Application to the Stresses in a Masonry Dam", *Transactions of the Royal Society of London*, Series A., Vol. 210, pp. 307-357.

Richardson, L.F., and Gaunt, J.A. (1927), "The Deferred Approach to the Limit", *Philosophical Transactions of the Royal Society, London*, Series A, Vol. 226, pp. 299-361.

Richtmyer, R.D. (1957), *Difference Methods for Initial-Value Problems*, Interscience Publishers, Inc., New York, New York.

Richtmyer, R.D., and Morton, K.W. (1967), *Difference Methods for Initial-Value Problems*, Second Edition, Interscience Publishers, J. Wiley and Sons, New York, New York.

Roache, P.J. (1974), *Computational Fluid Dynamics*, Hermosa Publishers, Albuquerque.

Sadourny, R., and Morel, P. (1969), "A Finite-Difference Approximation of the Primitive Equations for a Hexagonal Grid on a Plane", *U.S. Monthly Weather Review*, Vol. 97, No. 6 , pp. 439-445.

Salvadori, M.G. (1951), "Extrapolation Formulas in Linear Difference Operators", *Proceedings 1st Congress Applied Mechanics*, New York, pp. 15-81.

Samarskii, A.A. (1964), "Local One-Dimensional Difference Schemes for Multi-Dimensional Hyperbolic Equations in an Arbitrary Region", *Z. Vycisl. Mat. i Mat. Fiz.*, Vol. 4, pp. 21-35.

Smith, G.D. (1969), *Numerical Solution of Partial Differential Equations*, Oxford University Press, London.

Stommel, H. (1948), "The Westward Intensification of Wind-Driven Ocean Currents", *Transactions of American Geophysical Union*, Vol. 29, pp. 202-206.

Sundqvist, H., and Veronis, G. (1970), "A Simple Finite-Difference Grid with Non-Constant Intervals", *Tellus*, Vol. 22, pp. 26-31.

Thom, A., and Apelt, C.J. (1961), *Field Computations in Engineering and Physics*, D. Van Nostrand Company, Ltd., London.

Thoman, D., and Szewczyk, A.A. (1969), "Time Dependent Viscous Flow over a Circular Cylinder", *The Physics of Fluids Supplement II*, pp. 76-87.

Thomas, G.B. (1968), *Calculus and Analytic Geometry*, 4th Edition, Addison-Wesley Publishing Co., Reading, Massachusetts.

Thomas, L.H. (1949), *Elliptic Problems in Linear Difference Equations over a Network*, Watson Scientific Computing Laboratory, Columbia University, New York.

Torrance, K.E. (1968), "Comparison of Finite-Difference Computations of Natural Convection", *Journal of Research of the National Bureau of Standards*, Vol. 72B, No. 4, pp. 281-301.

Trapp, J.A., and Ramshaw, J.D. (1976), "A Simple Heuristic Method for Analyzing the Effect of Boundary Conditions on Numerical Stability", *Journal of Computational Physics*, Vol. 20, pp. 238-242.

Wachspress, E.L.(1966), *Iterative Solution of Elliptic Systems and Applications to the Neutron Diffusion Equations of Reactor Physics*, Prentice-Hall, Englewood Cliffs, New Jersey.

Wasow, W. (1955), "Discrete Approximations to Elliptic Differential Equations", *Zeitschrift Angew. Mathematical Physics*, Vol. 6, pp. 81-97.

Williamson, D. (1969), "Numerical Integration of Fluid Flow Over Triangular Grids", *Monthly Weather Review*, Vol. 97, No. 12, pp. 885-895.

Winslow, A.M. (1966), "Numerical Solution of the Quasilinear Poisson Equation in a Nonuniform Triangle Mesh", *Journal of Computational Physics*, Vol. 1, No. 2, p. 149.

Yanenko, N.N. (1971), *The Method of Fractional Steps: The Solution of Problems of Mathematical Physics in Several Variables*, English Translation edited M. Holt, Springer-Verlag, New York.

Young, D. (1954), "Iterative Methods for Solving Partial Differential Equations of Elliptic Type", *Transactions American Mathematical Society*, Vol. 76, pp. 92-111.

Zienkiewicz, O.C. (1967), *The Finite Element Method in Structural and Continuum Mechanics*, McGraw-Hill, London.

NUMERICAL SOLUTIONS OF PARTIAL DIFFERENTIAL EQUATIONS
J. Noye (Editor)

BURGERS' EQUATION: A MODEL FOR ALL REASONS

Clive A. J. Fletcher

University of Sydney, New South Wales, Australia

The features that make Burgers' equation a suitable model for convection diffusion problems are discussed. Some of the diverse physical phenomena which are modelled accurately by Burgers' equation are described. Burgers' equation has an exact solution for many combinations of boundary and initial conditions. Examples are given where this feature of Burgers' equation has been exploited to test different computational techniques. In particular, a detailed examination is made of the various Galerkin methods: the traditional Galerkin method, the spectral method and the Galerkin finite element method. Each Galerkin method is described and applied to the propagating 'shock' problem governed by Burgers' equation to permit a quantitative comparison of the various methods to be made.

TABLE OF CONTENTS

0 INTRODUCTION

The purpose of this paper is twofold. Firstly, attention is drawn to Burgers' equation as a useful model for many physically interesting problems, particularly those of a fluid-flow nature, in which either shocks or viscous dissipation is significant in part of the region. For many combinations of initial and boundary conditions, an exact solution of Burgers' equation is readily available. Consequently, approximations to physically more complex problems are often sought so that the exact solution of Burgers' equation can be exploited.

Burgers' equation is probably one of the simplest nonlinear partial differential equations for which it is possible to obtain an exact solution. Also, depending on the magnitude of the various terms in Burgers' equation, it behaves as an elliptic, parabolic or hyperbolic partial differential equation. Therefore, Burgers' equation has been used widely as a model equation for testing and comparing computational techniques.

The second purpose of this paper is to use Burgers' equation as a convenient model equation to illustrate various Galerkin methods. The order of presentation will be, in a sense, historical. The traditional Galerkin method will be described, mainly through the medium of specific, simple examples. Some typical applications will be given and the strengths and inherent weaknesses of the method will be highlighted. Thereafter, the spectral and Galerkin finite element methods will be described, again, mainly via specific examples. Comparisons of the computed solutions of Burgers' equation for the different categories of Galerkin method will be made.

The layout of the rest of the paper is as follows. Burgers' equation is described in Chapter 1 along with some background material of a mathematical nature. This will include the Cole-Hopf transformation and some common exact solutions. Various physical applications that are governed (sometimes approximately) by Burgers' equation are presented in Chapter 2. In Chapter 3 the use of Burgers' equation as a model problem for assessing computational techniques is described. In Chapters 4, 5 and 6 the traditional Galerkin, spectral and Galerkin finite element methods, respectively, are discussed in relation to Burgers' equation.

1. BACKGROUND

Burgers' equation is

$$u_t + uu_x - \nu u_{xx} = 0 \tag{1.1}$$

where, in a typical application, u is a velocity-like dependent variable and ν is a viscosity-like parameter. The subscript t indicates a time derivative and the subscript x a spatial derivative. If Eq. (1.1) is interpreted as a non-dimensional equation, ν is the inverse of the Reynolds number, typically.

The first solutions (steady) to Eq. (1.1) were given by Bateman (1915). However, the equation takes its name from the extensive research of Burgers (e.g., 1939, 1948,1964,1972,1974) in modelling turbulence. The equation is a useful mathematical model for such diverse physical problems as shock flows, traffic flow, acoustic transmission in fogs, etc.; in fact any nonlinear wave propagation problem subject to dissipation. The dissipation may arise from viscosity, heat conduction, chemical reaction, mass diffusion, thermal radiation, etc.

Lighthill (1956) formally derived Eq. (1.1) as a second-order approximation to the one-dimensional unsteady Navier-Stokes equations governing the flow through a propagating shock wave. Moran and Shen (1966) obtained a similar result via the method of matched asymptotic expansions. Namely, that for the shock formation ahead of an impulsively-started piston, Burgers' equation governs the flowfield behavior for times large compared with the initial piston motion. Comparable results, using the method of matched asymptotic expansions, for the situation where the diffusive (dissipative) mechanism could be due to viscosity, thermal conductivity, chemical reaction, radiation or a combination have been obtained and described by Lick (1967,1970).

As it stands, Eq. (1.1) is a quasi-linear parabolic partial differential equation that describes the evolution of some dependent variable, u, with time. The equation is linear in the second x and t derivatives of u but nonlinear in u. An important feature of the equation is that it is a prototype equation for the balance between the nonlinear convective term, uu_x, and the diffusive term, νu_{xx}. The ability to handle this balance efficiently is probably the single most difficult aspect of computing fluid dynamic problems.

The interaction of these two effects can be appreciated by simplifying Eq. (1.1). Firstly, if the diffusive term is dropped, Eq. (1.1) becomes

$$u_t + uu_x = 0 \quad . \tag{1.2}$$

Equation (1.2) is hyperbolic and is a model for the convection of disturbances in inviscid flow. A typical situation is shown in Figure 1.1. A wave is convecting from left to right and solutions for successive times, $t = 0$, t_1, t_2 are illustrated. Points on the wave with larger u convect faster and consequently overtake

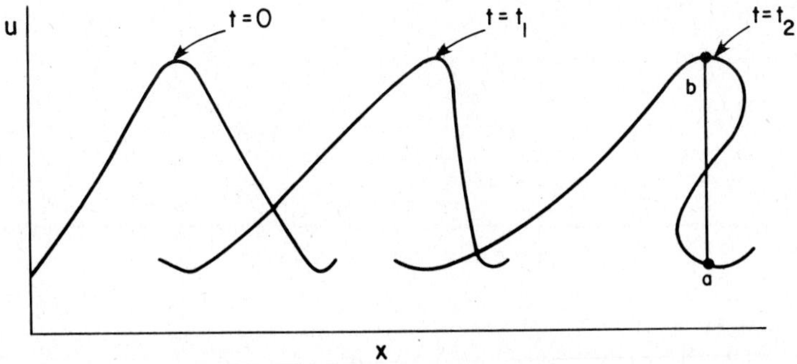

Figure 1.1 Formation of a multivalued solution of $u_t + uu_x = 0$

parts of the wave convecting with smaller u. At $t = t_2$ u has three values at a given x. For Eq. (1.2) to have a unique solution (and a physically sensible result) it is necessary to postulate a shock (ab in Figure 1.1) across which u changes discontinuously.

If in Eq. (1.1) the nonlinear convective term is dropped, the result is

$$u_t = \nu u_{xx} \tag{1.3}$$

which is the classical diffusion or heat conduction equation. If u is interpreted as the temperature of a bar of length, L, and ν as the thermal conductivity, the solution would be represented schematically as in Figure 1.2. An initial temperature profile is shown at $t = 0$. Due to heat transfer from the ends of the bar the 'temperature' falls everywhere and eventually, at $t = \infty$, the temperature will be

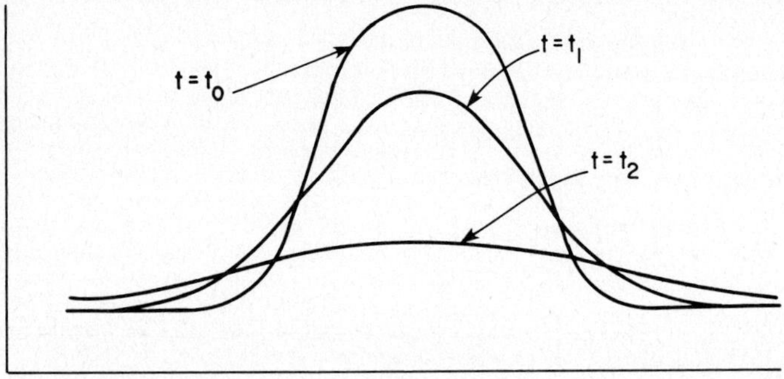

Figure 1.2 Evolution of the solution of $u_t = u_{xx}$

constant at all points along the bar.

If the two processes of convection and diffusion are considered together, i.e., Eq. (1.1), the evolution of the solution can be shown schematically as in Figure 1.3. Clearly the maximum amplitude of u is diminishing and the leading

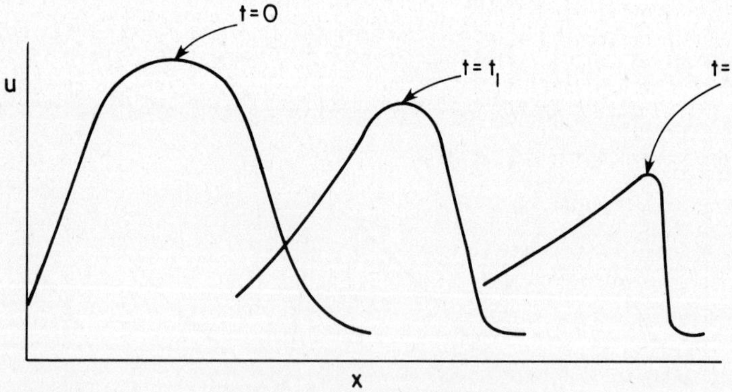

Figure 1.3 Evolution of the solution of $u_t + uu_x = \nu u_{xx}$

profile (right hand side) becomes steeper. However, the dissipative term, νu_{xx}, always prevents a multivalued solution from developing, since its contribution becomes larger where the steepening occurs (shock formation).

Burgers' equation can also represent the balance between convection and dissipation in the steady-state sense. If u_t is dropped from Eq. (1.1), the result is

$$uu_x = \nu u_{xx} \quad . \tag{1.4}$$

Equation (1.4) would be an elliptic partial differential equation in more than one spatial direction. If $u = u_1$ at $x = -\infty$ and $u = u_2$ at $x = +\infty$ the solution of (1.4) can be shown schematically in Figure 1.4. In the limit $\nu \to 0$ the solution changes discontinuously.

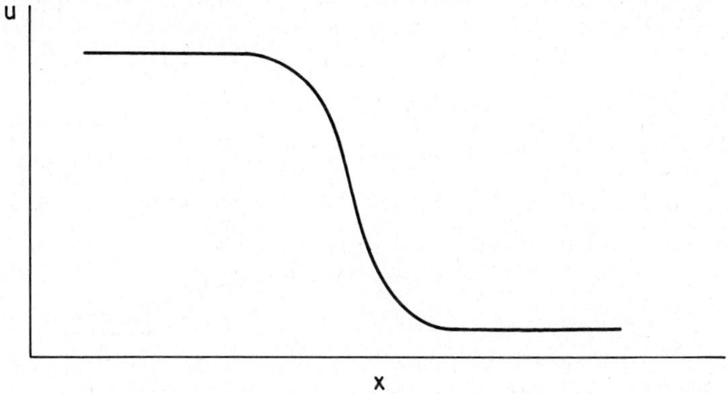

Figure 1.4 Typical solution of $uu_x = \nu u_{xx}$

Thus it is apparent that, for different conditions, Burgers' equation can be classified as an elliptic, hyperbolic or parabolic partial differential equation and that it is the simplest nonlinear model for considering convection diffusion interactions.

As a medium for testing new computational schemes Burgers' equation has one other very important attribute. For a wide range of initial and boundary conditions exact solutions exist and can be obtained easily. Independently Hopf (1950) and Cole (1951) discovered a transformation that would reduce Eq. (1.1) to the linear diffusion equation, (1.3). The relevance of this transformation is that the large number of known exact solutions of Eq. (1.3), e.g. see Carslaw and Jaegar (1959), become, after transformation, solutions of Eq. (1.1).

The Cole-Hopf transformation, after Whitham (1974), can be introduced in two stages. First let

$$u = \psi_x \tag{1.5}$$

so that Eq. (1.1) becomes, after integration,

$$\psi_t + \tfrac{1}{2}\psi_x^2 = \nu\psi_{xx} \quad . \tag{1.6}$$

Then if

$$\psi = -2\nu \log \phi \quad , \tag{1.7}$$

Eq. (1.6) becomes

$$\phi_t = \nu\phi_{xx} \tag{1.8}$$

which is just Eq. (1.3). The Cole-Hopf transformation can be generalised (Ames, 1965) as follows. Starting from the linear heat conduction equation (1.8) let

$$\phi = F(\psi) \quad , \tag{1.9}$$

so that Eq. (1.8) becomes

$$\psi_t - \nu(\psi_x)^2 \cdot \frac{F''}{F'} = \nu\psi_{xx} \quad . \tag{1.10}$$

Differentiating with respect to x and setting $u = \psi_x$ gives

$$u_t - 2\nu \cdot uu_x \cdot \frac{F''}{F'} - \nu u^3 \{\frac{F'''}{F'} - (\frac{F''}{F'})^2\} = \nu u_{xx} \quad , \tag{1.11}$$

and Burgers' equation is the special case

$$F(\psi) = \exp(-\psi/2\nu) \quad . \tag{1.12}$$

Initial conditions can be readily transformed. However, boundary conditions are not handled so easily and many exact solutions are for the interval, $-\infty < x < \infty$. The known exact solutions of Eq. (1.1) have been tabulated by Benton and Platzman (1972). For instance, the exact solution for the situation shown in Figure 1.4 is

$$u = 0.5(u_1 + u_2) - 0.5(u_1 - u_2)\tanh((u_1 - u_2)(x - 0.5(u_1 + u_2)t)/4\nu) \tag{1.13}$$

The Cole-Hopf transformation, Eqs. (1.5) and (1.7), can be applied to the initial conditions for Eq. (1.1), i.e.,

$$u(x,0) = u_o(x) \quad , \tag{1.14}$$

to give

$$\phi(x,0) = \phi_o(x) = \exp[-\{\int_0^x u_o(\xi)d\xi\}/2\nu] \quad . \tag{1.15}$$

For the infinite domain $(-\infty < x < \infty)$ the solution of Eq. (1.8) subject to the initial condition, Eq. (1.15), is particularly concise,

$$\phi(x,t) = \frac{1}{\sqrt{4\pi\nu t}} \int_{-\infty}^{\infty} \exp\left[\frac{-(x-\xi)^2}{4\nu t}\right] \cdot \phi_o(\xi)d\xi \quad . \tag{1.16}$$

Consequently the corresponding solution to Burgers' equation, (1.1), is, through the Cole-Hopf transformation,

$$u(x,t) = \int_{-\infty}^{\infty} \{\frac{x-\xi}{t}\}\exp\{-F/2\nu\}d\xi \Big/ \int_{-\infty}^{\infty} \exp\{-F/2\nu\}d\xi \tag{1.17}$$

where

$$F(\xi\,;x,t) = \int_0^{\xi} u_o(\xi')d\xi' + (x-\xi)^2/2t \quad . \tag{1.18}$$

As soon as a boundary condition must be applied at a finite x location, a difficulty arises. Thus consider the half-plane problem with initial and boundary conditions

$$u(x,0) = u_o(x) \quad , \quad 0 \leqslant x \leqslant \infty$$

and

$$u(0,t) = g(t) \quad , \quad t > 0 \quad . \tag{1.19}$$

To solve Eq. (1.1) subject to Eq. (1.19) requires the solution of an integral equation for h(t),

$$g(t) \cdot h(t) = 2\nu \int_0^t \frac{h'(t)}{\sqrt{\pi(t-\tau)}} d\tau + P(t) \quad , \tag{1.20}$$

where P(t) is a function of $u_0(x)$. After applying the Cole-Hopf transformation, h(t) becomes the boundary condition for ϕ,

$$\phi(0,t) = h(t) \tag{1.21}$$

Rodin (1970a) concludes that the appearance of Eq. (1.20) precludes the possibility of obtaining an exact solution and that it is more efficient to postulate a series form of solution from the start. Rodin considers the motion induced by a piston in the half-plane, $0 \leqslant x < \infty$, with initial motion of the piston, H(t). Thus the boundary condition at $x = 0$ becomes

$$u(H(t),t) = H'(t) \quad . \tag{1.22}$$

Expanding as a Taylor series and assuming that the displacement and velocity of the piston are large, he obtains

$$\begin{aligned} u_x(0,t) &= H'(t)/H(t) \quad , \\ u(0,t) &= 0 \quad . \end{aligned} \tag{1.23}$$

Rodin then obtains a series solution of the form

$$u(x,t) = -2\nu \sum_{n=0}^{\infty} [H(t)^{-0.5}]^{(n+1)} \frac{x^{2n+1}}{\nu(2n+1)!} \Bigg/ \sum_{n=0}^{\infty} [H(t)^{-0.5}]^{(n)} \frac{x^{2n}}{\nu(2n)!} \ . \tag{1.24}$$

For the bounded region, $0 \leqslant x \leqslant a$, a separation of variables solution can be obtained (Dennemeyer, 1968). Thus let

$$\phi = X(x) \cdot T(t) \quad . \tag{1.25}$$

For homogeneous boundary conditions an eigenvalue problem results so that Eq. (1.25) must be replaced by a series expansion which takes the particular form,

$$\phi(x,t) = \sum_{n=1}^{\infty} A_n \exp[-(\tfrac{n\pi}{a})^2 \cdot \nu t] \sin \tfrac{n\pi}{a} x \tag{1.26}$$

where

$$A_n = \frac{2}{a} \int_0^a \sin \frac{n\pi}{a} x \cdot \phi_0(x) dx \tag{1.27}$$

and

$$u(x,t) = \frac{\sum_{n=1}^{\infty} \frac{n\pi}{a} A_n \exp[-(\frac{n\pi}{a})^2 \nu t] \cos \frac{n\pi}{a} x}{\sum_{n=1}^{\infty} \frac{n\pi}{a} A_n \exp[-(\frac{n\pi}{a})^2 \nu t] \sin \frac{n\pi}{a} x} \quad . \tag{1.28}$$

Boldrighini (1977) considers Burgers' equation with stationary non-zero boundary conditions on a finite domain and obtains a solution via a separation of variables approach. Boldrighini also demonstrates that the corresponding steady-state solution is unique.

Equation (1.1) can be manipulated to resemble a Riccati equation

$$v_x = f(x) + g(x) \cdot v + h(x) \cdot v^2 \quad . \tag{1.29}$$

As with the Cole-Hopf transformation, the first step is to introduce

$$u = \psi_x \quad . \tag{1.30}$$

A similarity variable, $z = x^2/t$, is introduced and if

$$\psi(x,t) = f(z) \quad , \tag{1.31}$$

then Eq. (1.1) becomes

$$f_{zz} = -[(z + 2\nu)/4\nu z] \cdot f_z + (f_z)^2/2\nu \quad , \tag{1.32}$$

i.e., a Riccati equation for f_z. Rodin (1970b) uses this formulation to obtain some exact solutions to the inhomogeneous Burgers' equation

$$u_t + uu_x - \nu u_{xx} = F(x,t) \tag{1.33}$$

for particular choices of F.

An interesting problem is whether the solution of Burgers' equation as $\nu \to 0$ converges to the solution of the inviscid Burgers' equation, (1.2). In fact, it does and this has been established for the infinite x domain by Hopf (1950) and for the finite x domain by Ton (1975). Hopf established that Burgers' equation as an initial value problem is well posed and that solutions to Eq. (1.1) are unique if $\mu > 0$. Hopf found that when $\mu = 0$ discontinuous solutions are possible and that it is necessary to consider the equivalent integral formulation to establish uniqueness. Thus instead of Eq. (1.1) the following is considered,

$$\iint f(u_t + uu_x - \nu u_{xx})dx\ dt = 0 \quad , \tag{1.34}$$

where f(x,t) is any function of class C^2 in $t > 0$ and vanishes outside of some circle contained in $t > 0$. Then

$$\iint (uf_t + \frac{u^2}{2} f_x + \nu u \cdot f_{xx})dx\ dt = 0 \tag{1.35}$$

will have a unique solution for u with $\nu \geqslant 0$. In order to model certain aspects of turbulence Hopf (1950) demonstrates that the double limit should be applied in the order $t \to \infty$, $\mu \to 0$. It is shown that, reversing the limit, produces a different solution.

Although Burgers' equation is normally thought of as a one-dimensional (spatial) equation, it is possible to obtain higher-dimensional Burgers' equations. For example, in two dimensions, one obtains

$$u_t + uu_x + vu_y = \nu(u_{xx} + u_{yy}) \tag{1.36}$$

and

$$v_t + uv_x + vv_y = \nu(v_{xx} + v_{yy}) \tag{1.37}$$

These equations resemble the incompressible Navier-Stokes equations without pressure gradient terms. Jain and Holla (1978) have solved these equations numerically by adopting a fractional step, cubic spline approach. Unlike the one-dimensional Burgers' equation, no exact solution for the two dimensional Burgers' equation appears to have been presented in the literature.

But, as noted by Cole (1951) and Ames (1965), the Cole-Hopf transformation generalises to multi-dimensions. Thus, in two dimensions, the transformations

$$u = -2\nu \frac{\partial}{\partial x}(\ln \phi) = -2\nu\, \phi_x/\phi \tag{1.38}$$

and

$$v = -2\nu \frac{\partial}{\partial y}(\ln \phi) = -2\nu\, \phi_y/\phi \tag{1.39}$$

transform the two dimensional heat conduction equation

$$\phi_t = \phi_{xx} + \phi_{yy} \tag{1.40}$$

into Eqs. (1.36) and (1.37). For the pure initial value problem (i.e., $-\infty \leqslant x < \infty$, $-\infty < y < \infty$) with

$$\phi(x,y,0) = \phi_0(x,y) \tag{1.41}$$

the solution is (extending Eq. (1.16))

$$\phi(x,y,t) = \frac{1}{4\pi\nu t}\int_{-\infty}^{\infty}\int_{-\infty}^{\infty} \exp[-\{(x-\xi)^2 + (y-\eta)^2\}/4\nu t]\cdot\phi_0(\xi,\eta)d\xi d\eta \tag{1.42}$$

The initial conditions u_0, v_0 can be obtained from Eq. (1.41) by applying Eqs. (1.38) and (1.39), and similarly the solutions for u and v are

$$u(x,y,t) = \frac{\displaystyle\int_{-\infty}^{\infty}\int_{-\infty}^{\infty} \frac{(x-\xi)}{t}\exp[-\{(x-\xi)^2 + (y-\eta)^2\}/4\nu t]\cdot\phi_0(\xi,\eta)d\xi d\eta}{\displaystyle\int_{-\infty}^{\infty}\int_{-\infty}^{\infty} \exp[-\{(x-\xi)^2 + (y-\eta)^2\}/4\nu t]\phi_0(\xi,\eta)d\xi d\eta} \tag{1.43}$$

and

$$v(x,y,t) = \frac{\displaystyle\int_{-\infty}^{\infty}\int_{-\infty}^{\infty} \frac{(y-\eta)}{t}\exp[-\{(x-\xi)^2 + (y-\eta)^2\}/4\nu t]\cdot\phi_0(\xi,\eta)d\xi d\eta}{\displaystyle\int_{-\infty}^{\infty}\int_{-\infty}^{\infty} \exp[-\{(x-\xi)^2 + (y-\eta)^2\}/4\nu t]\phi_0(\xi,\eta)d\xi d\eta} \tag{1.44}$$

The addition of a third derivative term to Burgers' equation produces the Korteweg-deVries-Burgers' equation,

$$u_t + uu_x - \nu u_{xx} + \alpha u_{xxx} = 0 \quad . \tag{1.45}$$

This equation governs a form of wave motion in which the waves can not only be dissipated (by νu_{xx}) but also be dispersed (by αu_{xxx}). Like Burgers' equation the Korteweg-deVries equation ($\nu = 0$) also possesses an exact solution for many particular boundary condition (Miura, 1976). Equation (1.45) is important in the analysis of computational schemes since Taylor expansions of typical convection diffusion schemes have as the leading terms the Korteweg-deVries-Burgers equation, (1.45).

It is instructive to follow the analysis of Eq. (1.45) due to Jeffrey and Kakutani (1972) to establish the character of the solution. They introduce the variable $\zeta = x - \lambda t$ and integrate Eq. (1.45) with respect to ζ, to give

$$\alpha\cdot u_{\zeta\zeta} = -0.5\,u^2 + \lambda u + A + \nu\cdot u_\zeta \tag{1.46}$$

where A is the integration constant. Equation (1.46) can be split into two, simultaneous first-order equations,

$$u_\zeta = v \tag{1.47}$$

and

$$v_\zeta = \frac{\nu}{\alpha} \cdot v - \frac{1}{2\alpha}(u - u_1)(u - u_2) \tag{1.48}$$

where

$$u_{1,2} = \lambda \pm \sqrt{\lambda^2 + 2A} \; . \tag{1.49}$$

Two singular points exist at which $v = 0$ and $u = u_{1,2}$. It then follows that:

if $\nu^2 \geqslant 4\alpha\sqrt{\lambda^2 + 2A}$	$(u_1,0)$ is a nodal point: case A
if $0 < \nu^2 < 4\alpha\sqrt{\lambda^2 + 2A}$	$(u_1,0)$ is a spiral point: case B
if $\nu = 0$	$(u_1,0)$ is a centre: case C

$(u_2,0)$ is always a saddle point.

Typical solutions for u (shock profiles) are shown in Figure 1.5. If dissipation is dominant (case A), the shock has a monotonic profile. If dispersion is dominant (case B), the shock profile is oscillatory. This is the typical case

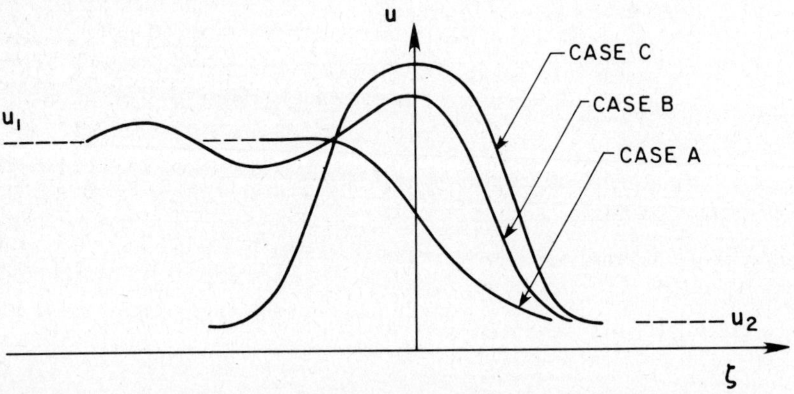

Figure 1.5 Typical solutions to Korteweg-deVries-Burgers' equation

in computing shock profiles numerically. If there is no dissipation, a solitary wave results. Clearly in generating computational schemes to model shock waves it is important to minimise terms associated with dispersion.

2. PHYSICAL MODELS

Many apparently unconnected problems can be modelled, approximately or exactly, by Burgers' equation. In this chapter we consider some of the more illuminating examples that are available. By the nature of Burgers' equation (1.1) all the problems can be associated with an evolutionary process in which a convective phenomenon is in conflict with a diffusive phenomenon.

2.1 Shock waves

Figure 2.1 shows an impulsively-started piston moving at a steady velocity, V_p, into a tube containing a compressible fluid, e.g., air, initially at rest. The equations that govern the one-dimensional unsteady motion of the fluid in the tube will be reduced to Burgers' equation.

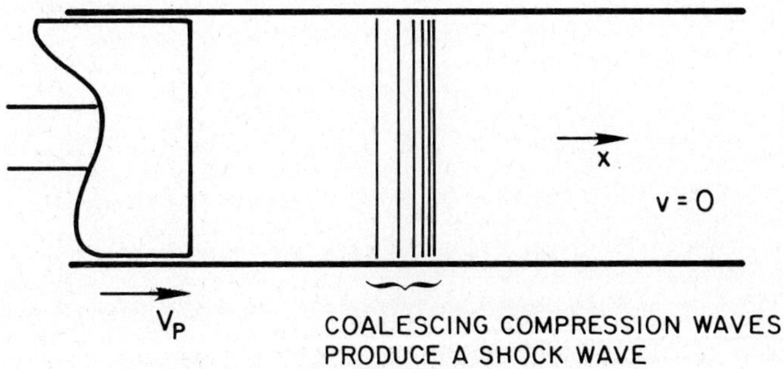

Figure 2.1 Incipient shock formation

The motion of the piston creates compression waves that propagate to the right at a velocity greater than V_p. The compression waves eventually coalesce, due to the nonlinear nature of the convection, to form a single shock wave. Ahead (to the right) of the shock wave the fluid is undisturbed, $v = 0$. Behind the shock (i.e., between the shock wave and the piston) the fluid moves with the speed of the piston, V_p. As the velocity profile, associated with the coalescence of the compression waves, steepens due to the nonlinear convective effect, viscous forces come into play tending to smooth the velocity profile through the shock wave and, by implication, spread the shock thickness over a finite width. Thus the velocity undergoes a rapid increase through the shock to reach a steady value adjacent to the piston. However, the shock location is a function of time, since the shock travels at a speed greater than V_p.

This problem is governed by the continuity equation

$$\rho_t + \rho v_x + v\rho_x = 0 \quad , \tag{2.1}$$

where ρ is the density and v is the velocity. The x-momentum equation can be written

$$v_t + vv_x + p_x/\rho = \delta \cdot v_{xx} \tag{2.2}$$

where δ is the "diffusivity of sound" (after Lighthill, 1956). It is assumed that entropy changes are small. It is convenient to replace the density by the sound speed, a, via

$$a/a_o = (\rho/\rho_o)^{0.5(\gamma-1)} \tag{2.3}$$

where γ is the specific heats ratio and the subscript o refers to the undisturbed (constant) values. Equations (2.1) and (2.2) become

$$a_t + va_x + \frac{(\gamma-1)}{2} a \cdot v_x = 0 \tag{2.4}$$

and

$$v_t + vv_x + \frac{2}{\gamma-1} \cdot aa_x = \delta v_{xx} \quad , \tag{2.5}$$

where δ, the "diffusivity of sound," is a function of the undisturbed (to the right of the shock) viscosity, density, specific heat and thermal conductivity of the medium.

Equations (2.4) and (2.5) can be simplified by introducing the Riemann invariants,

$$r = a/(\gamma-1) + v/2$$

and (2.6)

$$s = a/(\gamma-1) - v/2$$

to give

$$r_t + (a+v)r_x = \tfrac{1}{2}\delta(r_{xx} - s_{xx}) \quad (2.7)$$

$$s_t - (a-v)s_x = \tfrac{1}{2}\delta(s_{xx} - r_{xx}) \quad (2.8)$$

For the particular problem under consideration, the propagation of a disturbance into an initially undisturbed region, $s = s_o$ where $s_o = a_o/(\gamma-1)$ (from Eq. (2.6)). The problem is governed by Eq. (2.7). But from Eq. (2.6),

$$(a+v) = \frac{\gamma+1}{2}\, r + \tfrac{1}{2}(\gamma-3)s_o \quad , \quad (2.9)$$

so that Eq. (2.7) becomes

$$r_t + \{0.5(\gamma+1)r + 0.5(\gamma-3)s_o\}r_x = \frac{\delta}{2}\, r_{xx} \quad . \quad (2.10)$$

As the final step we introduce

$$u = 0.5(\gamma+1)(r-r_o) = 0.5(\gamma+1)r + 0.5(\gamma-3)s_o - a_o \quad (2.11)$$

and

$$\xi = x - a_o t \quad (2.12)$$

to give Burgers' equation

$$u_t + u \cdot u_\xi = \frac{\delta}{2} \cdot u_{\xi\xi} \quad . \quad (2.13)$$

From Eq. (2.11) $u \approx a + v - a_o$, i.e., an excess wavelet velocity. The coordinate ξ is measured relative to a frame of reference moving with the undisturbed speed of sound, a_o.

The general solution to (2.13) is given by Eq. (1.17). For the specific initial conditions

$$u = u_1 = 0.5(\gamma+1)V_p \quad \text{for} \quad \xi < 0,\ t = 0$$

and (2.14)

$$u = 0 \quad \text{for} \quad \xi \geqslant 0,\ t = 0 \quad ,$$

the solution is

$$u(\xi,t)/u_1 = I_A/(I_A + I_B \cdot \exp\{u_1(\xi - 0.5u_1 t)/\delta\}) \quad (2.15)$$

where

$$I_A = \int_{\xi-u_1 t}^{\infty} \exp\{-\zeta^2/2\delta t\}d\zeta \quad (2.16)$$

and

$$I_B = \int_{-\xi}^{\infty} \exp\{-\zeta^2/2\delta t\}d\zeta \quad . \quad (2.17)$$

Velocity solutions for increasing time and diffusivity are shown in Figures 2.2

and 2.3. The effective propagation velocity in the (ξ,t) plane is $0.5u_1$.

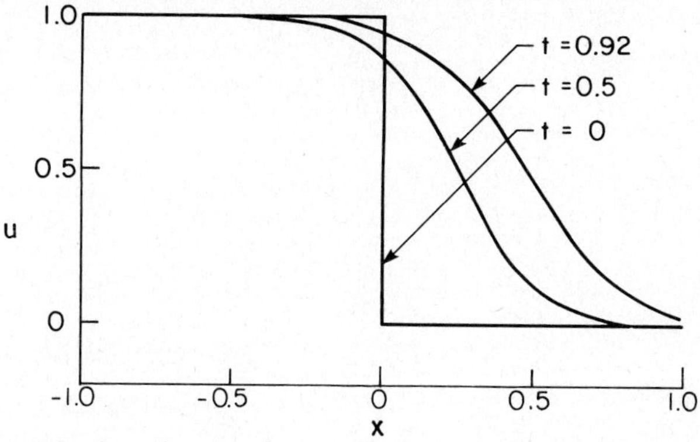

Figure 2.2 Velocity distributions through shock for various times, $\delta = 0.2$

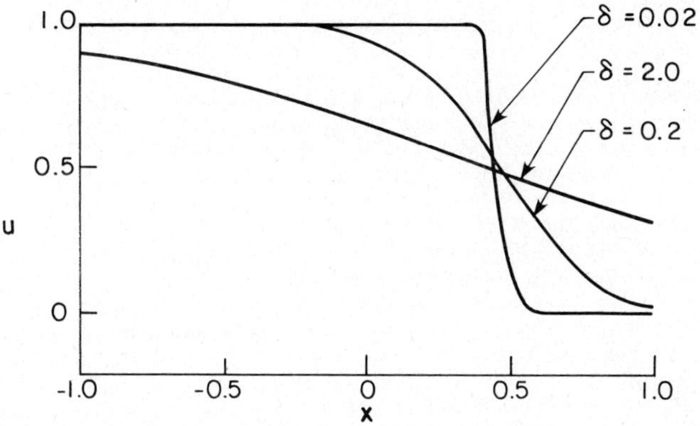

Figure 2.3 Velocity distributions through shock for different diffusivities, $t = 0.92$

The use of Burgers' equation is also appropriate to other gas dynamic phenomena. For instance, the convection and decay of a compression pulse, i.e., a Dirac delta function at zero time, is governed by Burgers' equation. For this problem a Reynolds number (expressing the balance between convection and viscous diffusion) can be defined by

$$R_e = \frac{1}{\delta}\int_{-\infty}^{\infty} u \, d\xi \quad . \tag{2.18}$$

For this situation, $u(t/\delta)^{1/2}$ is a function of the similarity variable, $\eta = \xi/(\delta t)^{1/2}$, only and the asymptotic solution is given by

$$u = (\delta/t)^{1/2} \cdot \exp\{-0.5\eta^2\}/[P + I_C] \tag{2.19}$$

where

$$P = \sqrt{2\pi}/(\exp(R_e) - 1)$$

and

$$I_C = \int_{\eta}^{\infty} \exp(-0.5\zeta^2)d\zeta \quad . \tag{2.20}$$

Some typical results are presented in Figure 2.4. At small Reynolds number the

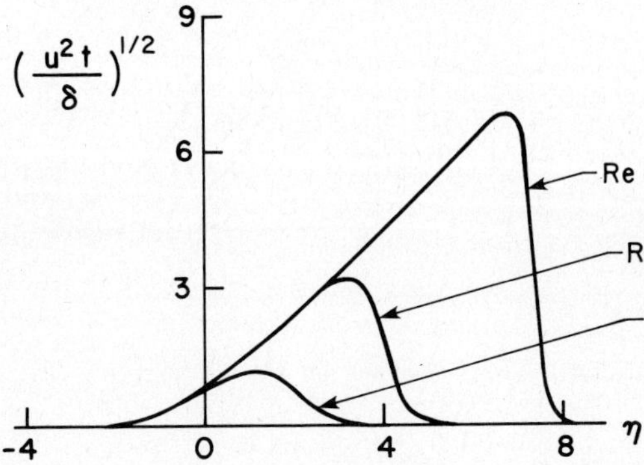

Figure 2.4 Compression pulse solutions for various Reynolds numbers

compression wave diffuses rapidly before significant convection can take place (in the ξ,t plane). However, at large Reynolds number, the compression wave develops a steep leading face and convects over a considerable range of ξ values before dissipation is effective.

If the just described compression wave is followed by an equal (and opposite) expansion wave, the result is the N-wave. In this case a Reynolds number can be defined by considering half of the wave, i.e.,

$$R_e = \frac{1}{\delta}\int_{\xi_u=0}^{\infty} u \, d\zeta \quad , \tag{2.21}$$

and the solution is

$$u = \xi/t/\{1 + \exp(0.5\eta^2)/(\exp(R_e) - 1.)\} \quad . \tag{2.22}$$

Typical solutions at various Reynolds numbers are shown in Figure 2.5. The steepening of the wave form with increasing Reynolds number is apparent. The corresponding pressure solution provides a useful model for sonic boom phenomena (Seebass, 1969).

Sachdev and Seebass (1973) have looked at the more general N-wave problem which is governed by the equation

$$u_t + uu_\xi + 0.5\alpha \, u/t = 0.5\delta \, u_{xx} \tag{2.23}$$

where $\alpha = 0,1$ and 2 for plane, cylindrical or spherical symmetry, respectively. For $\alpha=1$ or 2 the Cole-Hopf transformation does not give the linear heat conduction equation and no exact solutions are available. Therefore Sachdev and Seebass obtained numerical solutions using an implicit predictor-corrector finite difference scheme. The accuracy of the scheme was established for the plane case ($\alpha = 0$).

It was found by Sachdev and Seebass that, for Re (defined by Eq. (2.21)) > 0.5, the expression for the velocity profile in a plane shock (after Taylor, 1910)

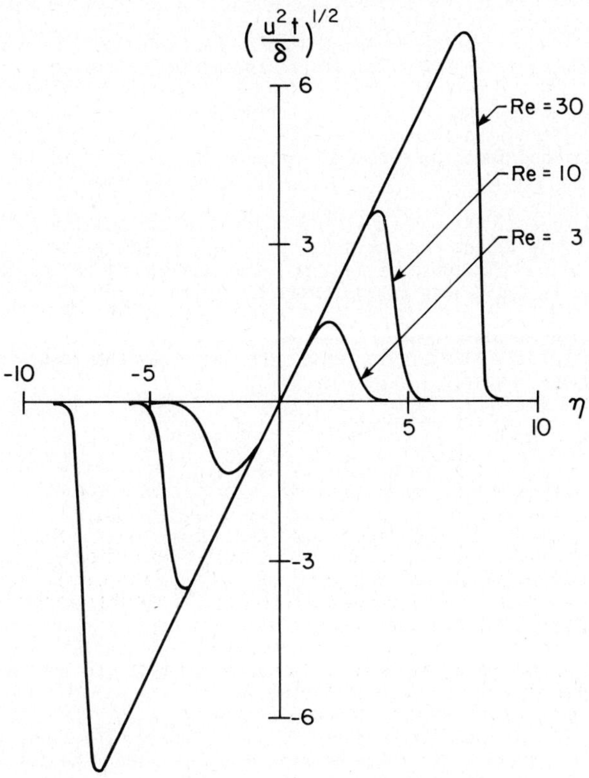

Figure 2.5 N-wave solutions for various Reynolds numbers

i.e.

$$u = u_{max}/[1 + \exp\{u_{max}(\xi-\xi_s)/\delta\}] \qquad (2.24)$$

is also accurate at the shock for the cylindrical and spherical cases. Solutions for cylindrical and spherical N-waves are relevant to the study of sonic booms from aircraft and explosions.

Lighthill (1956) also presents solutions for the problem of one shock overtaking and coalescing with another. In addition, as with N-waves, the formation and decay of shocks in nonplanar geometry are also shown to be governed by Burgers' equation.

The derivation of Burgers' equation (2.13) as the governing equation for many gas dynamic phenomenon has followed Lighthill (1956). However, using the method of matched asymptotic expansions it is also possible to derive Burgers' equation for any time after the initiation of the motion (e.g. Moran and Shen, 1966).

2.2 Acoustic transmission

Acoustic transmission in fogs has been studied by Davidson (1976). A fog can be idealised as a mixture of an inert (perfect) gas and the saturated liquid vapour phases of another substance (e.g. water). The liquid phase is assumed to be uniformly distributed throughout the mixture as liquid droplets. The behaviour of the fog arises from the mass transfer between the vapour and liquid (droplet) phases and the heat and momentum exchanges between the gas mixture and the droplets.

By considering the one-dimensional unsteady mass, momentum and energy balances for the inert gas, the vapour and the droplets, the droplet-gas transport phenomena, certain global relationships and the ideal gas law it is possible to reduce the problem to ten governing equations (Davidson, 1976).

By considering perturbations about the quiescent values and only retaining second order terms, it is possible to reduce the ten equations to a single inhomogeneous Burgers' equation,

$$u'_\ell - a'u'_g - 0.5(\gamma+1)u'u'_g - \delta u_{gg} = F \tag{2.25}$$

where u' is the dimensionless perturbation velocity in the x-direction,
ℓ is a dimensionless distance, $\ell = \omega x/a_0$,
g is a dimensionless retarded time, $g = \omega t - (1-a')\omega x/a_0$,
$a' = 1 - a/a_0$, where a is the sound speed and subscript o indicates the quiescent values,
and δ is an acoustic diffusion, $\delta = (\mu\omega/2\rho_0 a_0^2)[\frac{4}{3} + \frac{\gamma-1}{P_r}]$.

F is a forcing function that is dependent on the various relaxation times controlling the mass, momentum and heat transfer between the droplets and the mixture and on the density ratios of the droplet and vapour to the mixture. The explicit form of F is given by Davidson (1976).

Solutions to Eq. (2.25) have been obtained for the problem of a plane wave propagating down a semi-infinite tube containing a fog. The tube has a sinusoidally oscillating piston at one end (Figure 2.6).

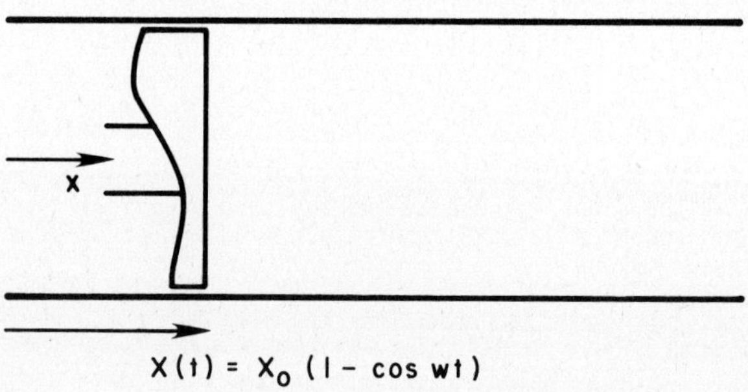

Figure 2.6 Flowfield for acoustic transmission in fog problem

The boundary conditions are

$$\begin{aligned} u(X,t) &= \omega X_0 \sin \omega t \\ u(\infty,t) &= 0 \ . \end{aligned} \tag{2.26}$$

After non-dimensionalisation, an acoustic Mach number, ε, can be defined by $\varepsilon = u_{max}/a_o$ and the boundary conditions become

$$u'(X,s) = \varepsilon \sin s$$

and

$$u'(\infty,s) = 0 \quad , \tag{2.27}$$

where $s = \omega t$. The problem, Eq. (2.25), can be linearized by expanding u' as a function of ε to give

$$u' = \varepsilon u^{(1)} + \varepsilon^2 u^{(2)} \ldots \tag{2.28}$$

Substitution into Eq. (2.25) and making use of Eq. (2.27) gives

$$u_\ell^{(1)} - a^{(1)} \cdot u_g^{(1)} - \delta u_{gg}^{(1)} = F(u^{(1)}) \tag{2.29}$$

with

$$u^{(1)}(0,g) = \sin g^{(1)} \quad \text{and} \quad u^{(1)}(\ell,g) \to 0 \text{ as } \ell \to \infty \quad . \tag{2.30}$$

Solutions of the form

$$u^{(1)} = I_m[\exp\{i(g^{(1)} + ik^{(1)}\ell\}] \tag{2.31}$$

produce results for $g^{(1)}$ and $k^{(1)}$ that are very frequency dependent. The variation of $k^{(1)}$, the attenuation coefficient, with non-dimensional frequency is shown in Figure 2.7. The peak at $\omega\tau \approx 0.01$ is associated with the droplet mass

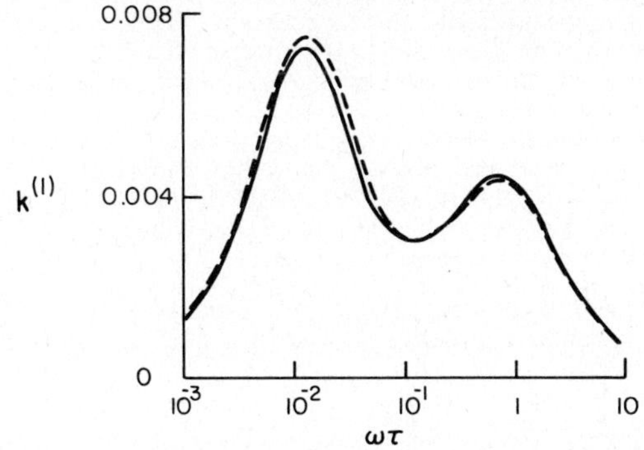

Figure 2.7 Variation of attenuation coefficient, $k^{(1)}$, with frequency, $\omega\tau$. Solid line (Eq. (2.29)), dashed line (numerical, Davidson, 1975).

transfer relaxation time. The peak at $\omega\tau \approx 1.0$ is associated with the droplet momentum and heat transfer relaxation times. The dashed line solution (Davidson, 1975) is obtained from numerical integration of the ten governing equations from which Burgers' equation was derived.

The corresponding variation of the dispersion parameter, $a^{(1)}$, with non-dimensional frequency is shown in Figure 2.8. It is clear that the dispersion attenuates rapidly with increasing frequency. It is apparent, from both Figures 2.7 and 2.8, that the agreement between the numerical solutions (dashed line) and the solution of Eq. (2.29) is good. The agreement is better where the magnitudes of the parameters, $k^{(1)}$ and $a^{(1)}$, is smaller. This is consistent with the assumptions about dissipation and nonlinearities not being too severe in deriving Burgers' equation.

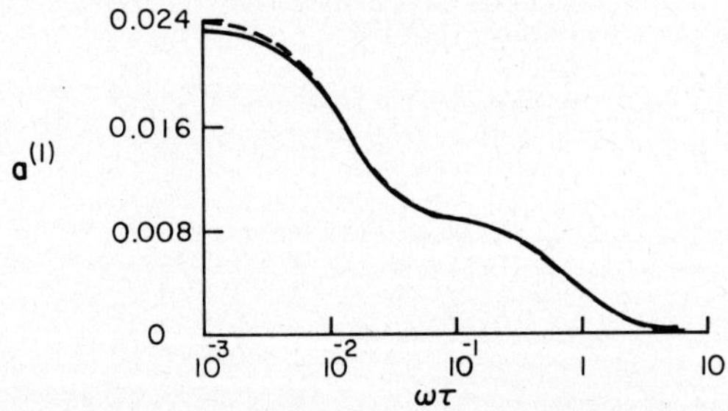

Figure 2.8 Variation of the dispersion parameter, $a^{(1)}$, with frequency, $\omega\tau$, solid line (Eq. (2.29)), dashed line (numerical, Davidson, 1975).

The following problem in underwater acoustics has been considered by Cary (1973,1975). If two primary acoustic signals are projected along the same path but at different frequencies, a subharmonic acoustic signal at the difference frequency can be generated. This problem is governed by Burgers' equation with parametric excitation of the boundary conditions. If the frequency ratio of the two primary signals, n, is in the range $1.5 \leqslant n < 2$, Cary found that the amplitude of the difference signal will increase above its initial value due to the non-linear nature of Burgers' equation. This occurs if the frequency regimes are well away from the relaxation frequency (associated with high absorption) and if the intensities are not so high as to cause shock waves to appear.

The propagation of finite-amplitude, low frequency sound waves in a media with high thermal conductivity has been shown, by Lerner and Fridman (1978), to be governed by Burgers' equation. Typical solutions are presented and discussed.

2.3 Traffic flow

A flux or current, J, and a related density or concentration, n, can, in general, be connected through a continuity equation (after Whitham, 1974). For a one-dimensional problem, we have

$$n_t + J_x = 0 \quad . \tag{2.32}$$

If another relationship can be prescribed between J and n, then closure may be expected.

For the traffic flow problem J may be interpreted as the volume flow rate of cars at a particular location and time. n may be interpreted as the number of cars per unit length of highway. Although a pedestrian view of the traffic flow problem might focus on the discrete or microscopic aspects, here we are constructing a continuum or macroscopic model. If v is the local mean velocity, a first approximation to the dependence of J on n might be

$$J = n \cdot v \quad . \tag{2.33}$$

Clearly in the limit $n \to 0$, $J \to 0$ however fast individual cars travel. As the concentration reaches saturation, i.e., the cars become nose-to-tail, we would expect v to fall to zero. A simple representation of this would be

$$v = v_o(1 - n/n_s) \tag{2.34}$$

i.e. v_o is the speed when the concentration is zero and n_s is the saturation concentration, at which $v = 0$. Combining (2.33) and (2.34) produces the relationship shown in Figure 2.9.

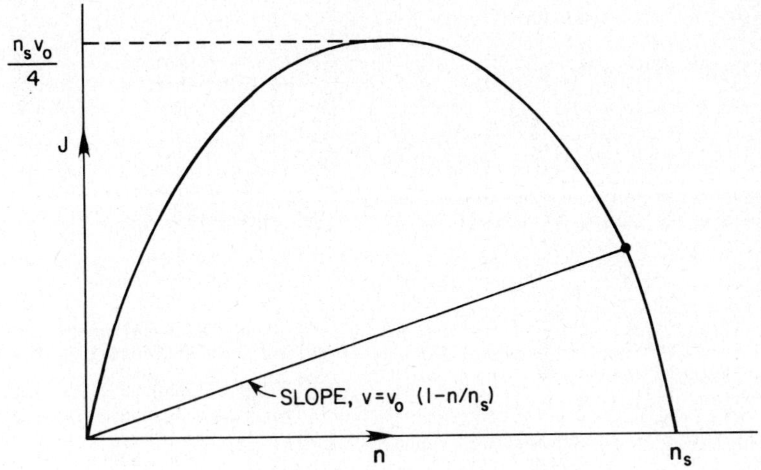

Figure 2.9 Idealised flux density relationship

In practice drivers may be expected to look ahead and to modify their speed accordingly. To represent this, Eq. (2.33) is replaced by

$$J = nv - Dn_x \tag{2.35}$$

where D is a diffusion coefficient given by (Musha and Higuchi, 1978)

$$D = \tau \cdot v_r^2 \tag{2.36}$$

where v_r is a random velocity and τ is the mean collision time for the cars. Combining equations (2.32), (2.34) and (2.35) gives

$$(n_t + v_o n_x) - 2(\frac{n}{n_s}) \cdot v_o \cdot n_x - D \cdot n_{xx} = 0 \quad . \tag{2.37}$$

It is convenient to introduce a moving reference frame by

$$\xi = -x + v_o t \quad . \tag{2.38}$$

If, in addition, n is nondimensionalised with $n_s/2$ and the time by t_o the result is

$$n_t + nn_\xi - \frac{1}{Re} \cdot n_{\xi\xi} = 0 \tag{2.39}$$

i.e., a Burgers' equation for the car concentration. Re is a dimensionless constant, analogous to the Reynolds number, defined by

$$Re = (\frac{v_o}{v_r})^2 \cdot \frac{t_o}{\tau} \quad , \tag{2.40}$$

where the characteristic time, t_o, is the mean time between successive cars passing a stationary observer.

In order to use Eq. (2.39) as a model for real traffic flows it is necessary to consider the power spectrum of Burgers' flow. For transient Burgers' flow assuming an initial Gaussian concentration distribution it is found that the power spectral density depends on the normalised wave number K, like K^{-q}, where initially $q \approx 4$ and asymptotically $q \approx 1.0$.

To obtain the power spectrum for steady Burgers' flow a random force f is added to Eq. (2.39). Taking the ensemble average and requiring that $\langle n \rangle$ is constant and $\langle f \rangle = 0$ gives

$$\tfrac{1}{2} \langle n^2 \rangle_x = 0 \quad . \tag{2.41}$$

If Burgers' equation plus the random force, f, are multiplied by n and an ensemble average taken, the result is, if $\langle n^2 \rangle$ is constant in space (from Eq. (2.41))

$$\tfrac{1}{2} \langle n^2 \rangle_t + \frac{1}{Re} \langle (n_x)^2 \rangle = \langle nf \rangle \quad . \tag{2.42}$$

But for steady flow, $\langle n^2 \rangle_t = 0$, so that Eq. (2.42) gives the requirement that a random force will lead to a steady Burgers' equation. It is found that the power spectral density for the steady Burgers' equation leads to a dependence on wave number like $K^{-1.4}$ which is not very different from the transient case. For the traffic flow problem it is convenient to express the power spectral density, S, as a function of the frequency, f. Then

$$S(f) = \frac{10^{-2}}{(2\pi)^{0.4}} (n_s v_o)^2 \cdot \frac{1}{t_o^{0.4}} \cdot \frac{1}{f^{1.4}} + 2(n v_o) \quad . \tag{2.43}$$

A comparison of this equation with experimental observations is shown in Figure 2.10. Musha and Higuchi (1978) consider the agreement for the steady Burgers flow satisfactory.

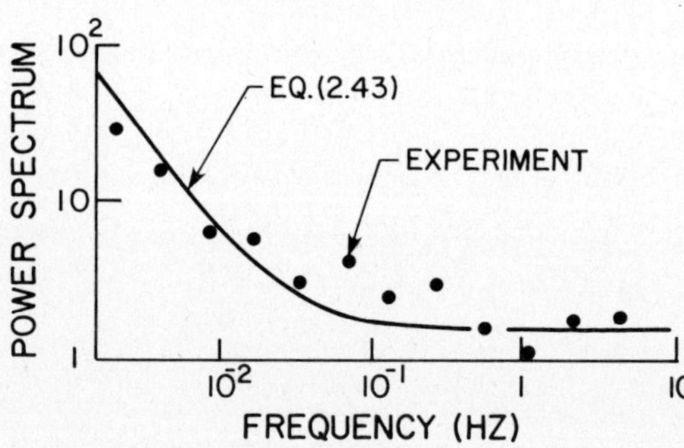

Figure 2.10 Comparison of Eq. (2.43) with experimental power spectrum

2.4 Turbulence

In an attempt to formulate a simple mathematical model that would exhibit the essential features demonstrated by turbulence in hydrodynamic flows, Burgers (e.g. 1948) introduced the equation that has now come to be known as Burgers' equation.

In considering turbulent flow in a channel, Burgers introduced two variables, U and v. U was a function of time only and was equivalent to a mean flow. v was a function of y (measured across the channel) and time, t, and was equivalent to a fluctuating or turbulent motion. Burgers took as governing equations

$$bU_t = P - \nu U/b - [\int_0^b v^2 dy]/b \tag{2.44}$$

and

$$v_t + 2vv_y - \nu v_{yy} = \nu U/b \quad . \tag{2.45}$$

In Eq. (2.44) the term P represents the external forces and the last term on the right-hand side is associated with the Reynolds stress. The viscous term, $\nu U/b$, is introduced into both equations as a mechanism by which energy can be transferred from the primary motion to the secondary motion. To investigate turbulence in an infinite domain, Burgers set $U = 0$ giving the homogeneous (and more familiar) form of Burgers' equation. Burgers (1948) obtained the result for steady Burgers flow that the energy spectrum function, E(k), varied with the wavenumber, k, like

$$E(k) \sim k^{-2} \qquad \text{for} \quad k \ll Re \tag{2.46}$$

$$E(k) \sim \exp(-k) \quad \text{for} \quad k > Re \quad . \tag{2.47}$$

Comparable results have been obtained by subsequent investigators (e.g. Tatsumi, 1980) if Eq. (2.46) is identified with the inertial range and Eq. (2.47) is identified with the viscous dissipation range of wavenumbers.

To analyse homogeneous isotropic turbulence as well as Burgers turbulence, Tatsumi (1980) introduces a modified zero cumulant approximation in wave number space. It is convenient to define the Fourier transform of the velocity field, v(x,t), by

$$u(k,t) = \frac{1}{2\pi}\int v(x,t)\exp(-ikx)dx \quad . \tag{2.48}$$

The equation for the energy spectrum function, E(k,t), is

$$E_t(k,t) + 2\nu k^2 E(k,t) = T(k,t) \tag{2.49}$$

where

$$T(k,t) = k\int_{-\infty}^{\infty} [1 - \exp\{-2\nu(k^2 + k'^2 + kk')t\}] \cdot B/[2\nu(k^2 + k'^2 + kk')]dk'$$

and

$$B = \{k \cdot E(k',t) + k'E(k,t)\}E(k+k',t) - (k+k')E(k,t)E(k',t) \quad .$$

For initial conditions

$$\begin{aligned} E(k,0)/E_0 &= \exp(-(k/k_0)^2) && \text{case A} \\ E(k,0)/E_0 &= (k/k_0)^2\exp(-(k/k_0)^2) && \text{case B} \quad , \end{aligned} \tag{2.50}$$

numerical solutions to Eq. (2.49) have been obtained for Reynolds numbers, $Re = u_0/\nu k_0$, up to 10^4. For high wave numbers in the energy-containing range of wave numbers the solution has the form

$$\begin{aligned} E(k,t)/E_0 &\sim \tau^{-4/3}k^{-2} && \text{case A} \\ E(k,t)/E_0 &\sim \tau^{-1.51}k^{-2} && \text{case B} \quad , \end{aligned} \tag{2.51}$$

where τ is a nondimensional time, $\tau = E_0^{1/2} \cdot k_0^{3/2} \cdot t$. A similar form of solution was obtained by Kida (1979) who considers the initial profile as a collection of sawtooth waves, each of which satisfies Burgers' equation. In the limit of $\nu \to 0$ and t large (i.e. the turbulence problem) any initial profile subjected to Burgers' equation will eventually look like a distribution of triangular waves (Burgers, 1974).

For the energy-dissipation range of wave numbers, i.e. for which $(\nu t)^{\frac{1}{2}} k >> 1$, a similarity form of the solution of Eq. (2.49) is possible, i.e.

$$E(k,t)/E_0 = Re^{\alpha} \cdot \tau^{\beta} \cdot F[k/(Re^{\gamma}\tau^{\delta})] \quad . \tag{2.52}$$

In addition the energy dissipation, $\varepsilon(t)$, is given by

$$\varepsilon(t) = -\frac{d}{dt}\int_0^{\infty} E(k,t)dk = 2\nu\int_0^{\infty} k^2 \cdot E(k,t)dk \quad . \tag{2.53}$$

Substitution of Eq. (2.52) into (2.53) gives

$$\begin{aligned} &\alpha = -5/4,\ \gamma = 3/4,\ \beta = \delta = -5/12 \quad \text{case A} \\ &\alpha = -5/4,\ \gamma = 3/4,\ \beta = \delta \quad \text{case B} \quad . \end{aligned} \tag{2.54}$$

Tatsumi (1980) notes that the values of α and γ in Eq. (2.54) are identical with those obtained from considering 'real' turbulence, i.e. homogeneous, isotropic turbulence governed by the three-dimensional Navier-Stokes equations. In addition, the Reynolds number dependence in the energy-containing range is the same for both types of turbulence. However, the dependence of the energy spectrum function on the wave number is different. From Eq. (2.51) this is like k^{-2} for Burgers turbulence but like $k^{-5/3}$, on average, for Navier-Stokes turbulence. But for high Reynolds number flows ($Re \approx 10^4$ to 10^6) there is a large range of wave numbers for which the dependence is also like k^{-2} for Navier-Stokes turbulence.

However, significant differences do occur between Burgers and Navier-Stokes turbulence that can be related to the underlying governing equations. Firstly the presence of the pressure gradient terms in the Navier-Stokes equations is a continuing source of new random motion. In contrast, the randomness in Burgers turbulence comes from the initial data so that the randomness slowly reduces as the solution evolves (in time). This corresponds to the coalescence of the triangular waves or shocks. Thus for large time Burgers turbulence is characterised by large regions of the flow where the solution is changing slowly and sharply defined shock regions where the viscous dissipation is occurring.

The simple nature of Burgers' equation and the availability of an exact solution have been exploited by Love (1980) to test subgrid scale turbulence models. Solutions to the inhomogeneous Burgers' equation (1.33), with F as a stochastic forcing function, were obtained computationally via a finite difference formulation. Two cases were considered, a freely decaying ensemble of shocks and a stochastically forced ensemble of shocks.

For periodic boundary conditions solutions have been obtained on a grid of sufficient refinement that all length scales present can be properly resolved. Love takes these solutions to be 'exact.' The results indicate that large and small eddies are separated in wave number space by a large inertial range (as was found by Tatsumi, 1980) and that the dependence of the energy spectrum function on wave number is like k^{-2} (as found by Burgers, 1974, Tatsumi, 1980, etc.). However, Love found that for small wave numbers (corresponding to the regions between the 'shocks') there is a downward cascade of energy for Burgers turbulence that does not occur for Navier-Stokes turbulence.

The velocity fields obtained from the 'exact' (fine grid) solutions have been filtered (Gaussian) to attenuate the viscous dissipation spectrum. The evolution of the filtered velocity field has a slightly different spectrum due to the loss of energy via the filter. The inclusion of subgrid models for the energy dissipation produces solutions in reasonable agreement with the exact solution.

The effectiveness of the subgrid modelling was assessed by obtaining solutions on a coarse grid for the stochastically forced ensemble. The coarse grid solutions produce too large a loss of energy for early time but the steady-state wave number distribution is in good agreement with the fine grid (filtered) solutions.

Since Burgers' equation describes the formation and decay of weak shock waves, it does find direct application in modelling compression turbulence (Tatsumi and Tokunaga, 1974). Compression turbulence is associated with significant magnitudes of the longitudinal velocity, density and temperature fluctuations resulting from the compressible nature of the flow.

Compression turbulence can be thought of as a random assembly of sound waves which, for a Mach number, $M > 1$ and Re large, form essentially triangular shock waves. The viscous dissipation is associated almost entirely with the shock fronts where the velocity gradient is very large. By considering a compressible flow with superimposed random disturbances having triangular wave forms, each of which is a solution of the Burgers' equation, Tatsumi and Tokunaga (1974) demonstrate that the law of decay of turbulent energy is the same for compression turbulence as it is for Burgers turbulence, namely like $t^{2(\alpha-1)}$, $0 \leqslant \alpha < 1$ after Tatsumi and Kida (1972).

2.5 Supersonic flow about aerofoils

For slender wings or aerofoils subjected to a supersonic inviscid flow, it is convenient to think of the wing as a perturbation of an, otherwise, uniform free stream. If the governing partial differential equation is expressed as a velocity potential (ϕ) as a function of position (x,y) it is possible to develop a hierarchy of solution, the lowest order of which is linear.

In fact the first-order problem can be thought of as a body which communicates its disturbing effect on the flow by passing information along straight characteristics (or Mach lines) (the governing partial differential equation is hyperbolic) which emanate from the body. The governing equation is

$$(1 - M_\infty^2)\phi_{xx} + \phi_{yy} = 0 \quad , \tag{2.55}$$

where M_∞ is the Mach number of the freestream. Although solutions to Eq. (2.55) give an accurate description of the flow at an intermediate distance from the body, in the farfield a nonlinear second-order correction must be added to account for the merging of the characteristics. In this case, Eq. (2.55) is replaced by

$$(1 - M_\infty^2 - (\gamma+1)M_\infty^4\phi_x)\phi_{xx} + \phi_{yy} = 0 \quad . \tag{2.56}$$

However, when characteristics merge the velocity potential description is inappropriate locally, since a shock wave will occur. One of the current computational research problems is to use a potential description of the flow, like Eq. (2.56), that strictly excludes the appearance of shocks, to describe physical situations, like the transonic flow around an aerofoil or through a nozzle where shocks may occur. The attraction of equations like (2.56) is that through use of the Prandtl-Glauert transformation, perturbations of the Laplace or wave equation can be obtained.

One way to permit shocks to appear in the solution is to add a diffusive term to Eq. (2.56). This idea has been used in transonic flow by Sichel (1963) and for supersonic flow by Chin (1978). In this case Eq. (2.56) is replaced by

$$\delta\phi_{xxx} + (1 - M_\infty^2 - (\gamma+1)M_\infty^4\phi_x)\phi_{xx} + \phi_{yy} = 0 \quad , \tag{2.57}$$

where δ is a small parameter. The term $\delta\phi_{xxx}$ will only contribute to the solution in regions where shocks occur. In the limit $\delta \to 0$ the inviscid solution is recovered. The effect of including the term $\delta\phi_{xxx}$ permits a continuous solution to be obtained in the vicinity of the shock so that the gross effect of the shock on, say, the pressure distribution on the aerofoil can be obtained. Typically, boundary conditions to be used with Eq. (2.57) are

$$\phi_y(x,0) = \varepsilon T'(x) \quad , \quad \phi = 0 \quad \text{at} \quad x = -\infty \text{ (upstream)} \tag{2.58}$$

where ε is the thickness ratio of the body and T' is the local, normalised body slope.

The relevance of Eq. (2.57) is that with a suitable transformation (Chin (1978)) Burgers' equation can be obtained and a closed form solution follows. Physically the problem has two obvious co-ordinate directions, the freestream and the characteristics. This suggests the transformation

$$\xi = x - (M_\infty^2 - 1)^{1/2} y \quad \text{and} \quad \eta = (M_\infty^2 - 1)^{1/2} \cdot y \quad . \tag{2.59}$$

Substitution into Eq. (2.57) generates certain second-order terms in $\phi_\eta, \phi_{\eta\eta}$ which can be dropped (Hayes (1954)). Additional transformations are introduced by

$$\zeta = (2(M_\infty^2 - 1))^{1/2} \cdot \xi$$

and

$$u = \phi_x \cdot (\gamma+1)M_\infty^4/(2(M_\infty^2 - 1))^{1/2} \quad , \tag{2.60}$$

where u is a modified velocity component in the freestream direction. Equations (2.57) and (2.58) become

$$u_\eta + u \cdot u_\zeta = \delta u_{\zeta\zeta} \tag{2.61}$$

and

$$u = -\varepsilon \cdot \frac{\gamma+1}{\sqrt{2}} \cdot \frac{M_\infty^4}{M_\infty^2 - 1} \cdot T' \cdot \xi \quad \text{at} \quad \eta = 0 \quad ,$$

and

$$u = 0 \quad \text{at} \quad \zeta = -\infty \text{ (upstream)} \quad . \tag{2.62}$$

The solution to these equations is just Eq. (1.13).

All techniques for solving the transonic full potential equation are iterative. The solutions to Eqs. (2.61) and (2.62) would form an appropriate starting point for the iterative scheme. To the extent that the solution to Eqs. (2.61) and (2.62) are close to the desired solution in some sense one would expect the subsequent iteration to converge rapidly. To the author's knowledge, the solution of Burgers' equation for this problem has never been used in such a way.

2.6 Other physical problems

Gurbatov et al. (1977) have investigated the statistical characteristics of the low frequency components of a signal modulated by a quasimonochromatic wave. The dependence of the spectra and probability distribution function on the amplitude modulation index and acoustic Reynolds number is indicated. Rizun and Engel'brekht (1975) have obtained solutions to cylindrical and spherical Burgers' equations in order to study wave propagation in a thermoelastic medium. Kuznetsov et al. (1977) demonstrate that the dynamics of a liquid containing gas bubbles is determined by the Korteweg-deVries-Burgers equation. By a proper choice of dimensionless groups it is possible to interpret the propagation of signals in terms of monotonic and oscillating shock waves, solitons and wave pockets. Kuo and Tanner (1971) examine the decay of turbulence in a polymer solution by substituting a non-Newtonian viscosity into Burgers' equation. Numerical experiments compared various spectral properties with the corresponding Newtonian fluid results.

3. COMPUTATIONAL EXAMPLES

In this chapter the use of Burgers' equation as a model problem to compare different computational algorithms is described. The prime feature of Burgers' equation that makes this attractive is the availability of an exact solution. However, since Burgers' equation also serves as a model problem for convective and diffusive fluid flows it is particularly appropriate for situations where the computational techniques are being developed for more complex fluid flow applications.

3.1 Comparison of finite difference schemes

Taylor et al. (1972) have used Burgers' equation to test a number of finite difference schemes for viscous flows in which shocks are expected. Thus the appropriate model problem is that of a propagating 'shock' (section 2.1), which is governed by

$$u_t + 0.5(u^2)_x - \frac{1}{Re} \cdot u_{xx} = 0 \quad , \tag{3.1}$$

with initial conditions

$$\begin{aligned} u(x,0) &= 1 \qquad x \leqslant 0 \\ &= 0 \qquad x > 0 \end{aligned} \tag{3.2}$$

Taylor et al. modelled the convective term with first-order and third-order Rusanov schemes and the viscous term by explicit, implicit and Dufort-Frankel second-order centered difference schemes.

The first-order Rusanov scheme for $u_t + f_x = 0$ is $(f \equiv 0.5u^2)$

$$u_i^{n+1} - u_i^n + 0.5\,\frac{\Delta t}{\Delta x}\,(f_{i+1}^n - f_{i-1}^n) - 0.5(\phi_{i+1/2}^n - \phi_{i-1/2}^n) = 0 \tag{3.3}$$

where

$$\phi_{i+1/2}^n = 0.5(\alpha_i^n + \alpha_{i+1}^n)(u_{i+1}^n - u_i^n)$$

and

$$\alpha_i^n = w(\Delta t/\Delta x) \cdot u_i^n \quad .$$

It can be seen that w is a parameter that controls the amount of artificial viscosity. The third-order Rusanov scheme is

$$\begin{aligned} u_{i+1/2}^{*} &= 0.5(u_i^n + u_{i+1}^n) + 1/3 \cdot (\Delta t/\Delta x)(f_{i+1}^n - f_i^n) \quad , \\ u_i^{**} &= u_i^n + 2/3 \cdot (\Delta t/\Delta x)(f_{i+1/2}^{*} - f_{i-1/2}^{*}) \quad , \end{aligned} \tag{3.4}$$

$$u_i^{n+1} - u_i^n - (\Delta t/\Delta x)(2f_{i-2}^n - 7f_{i-1}^n + 7f_{i+1}^n - 2f_{i+2}^n)/24$$
$$+ 3(\Delta t/\Delta x)(f_{i+1}^{**} - f_{i-1}^{**})/8$$
$$+ w(u_{i-2}^n - 4u_{i-1}^n + 6u_i^n - 4u_{i+1}^n + u_{i+2}^n)/24 = 0 \quad . \qquad (3.5)$$

The Dufort-Frankel representation of u_{xx} is

$$(u_{i-1}^* - (4/3)u_i^{n+1} - (2/3)u_i^n + u_{i+1}^*)/\Delta_x^2 \qquad (3.6)$$

The time level * is between n (the current time level) and n+1 (the advanced time level). Thus a predictor is needed to obtain the solution at *, typically the explicit scheme.

A numerical stability study indicates that all schemes are restricted in the allowable time step, Δt, as the Reynolds number is increased. This is to be expected since in the limit of $Re \to \infty$ the schemes revert to the purely explicit Rusanov schemes. However, the results presented indicate that for large Reynolds number ($Re \sim 1000$) a stable scheme is possible up to a Courant number = 1 with no artifical viscosity ($w = 0$ in Eq. (3.5)).

Results are presented for the L_2 error in the computed solutions at various Reynolds numbers in Table 3.1. These results were obtained after integrating the difference schemes for 200 time steps. Taylor et al. (1972) concluded that the third-order Rusanov scheme for the inviscid terms and the second-order implicit scheme for the viscous terms gives the best results of the schemes considered. However, since the tridiagonal algorithm requires a larger execution time, it is not clear that this combination is computationally the most efficient. It is apparent that for large Re there is no difference in accuracy between the explicit and implicit viscous schemes.

Scheme	Re = 100 L_2 error	Re = 200 L_2 error	Re = 500 L_2 error
3rd-order Rusanov/implicit	0.0005	0.0023	0.0103
3rd-order Rusanov/explicit	0.0011	0.0026	0.0103
3rd-order Rusanov/Dufort-Frankel	0.0093	0.0028	0.0099
1st-order Rusanov/implicit	0.0100	≈ 0.0138	≈ 0.02

Table 3.1 Accuracy of various finite difference schemes ($\Delta x = 0.01$, $\Delta t = 0.002$, $w = 1.$, after 200 time steps)

In most meteorological problems the governing equations are only weakly dissipative and the length of time over which an integration can be made is governed by nonlinear instabilities associated with the convective terms. It is found that this instability can be suppressed by representing the convective term in divergence form. E.g. for Burgers' equation this is just $0.5(u^2)x$.

Piacsek and Williams (1970) differenced $0.5(u^2)x$ by the representation

$$[u_{i+1/2} \cdot u_{i+1} - u_{i-1/2}u_{i-1} + u_i(u_{i+1/2} - u_{i-1/2})]/4\Delta x \equiv C_1 \qquad (3.7)$$

However, it was argued that where continuity is not enforced, Eq. (3.7) permits the equivalence of a mass build-up. Therefore Piacsek and Williams preferred to represent $0.5(u^2)x$ by

$$[u_{i+1/2}u_{i+1} - u_{i-1/2}\ u_{i-1}]/2\Delta x \equiv C_3 \qquad (3.8)$$

which satisfies global continuity and conserves u^2 globally. A Taylor expansion about the node i indicates that Eq. (3.8) is not consistent with $0.5(u^2)_x$. However, Piacsek and Williams presented numerical results for the two dimensional, inviscid Burgers' equation

$$u_t + uu_x + vu_y = 0$$

and

$$v_t + uv_x + vv_y = 0 \quad . \qquad (3.9)$$

This system was solved in a square cavity with the normal derivative zero on each wall. The initial velocity distribution is

$$u = \sin^2 \pi x \cdot \sin 2\pi y + r_1(x,y)$$
$$v = -\sin 2\pi x \cdot \sin^2 \pi y + r_2(x,y) \quad . \qquad (3.10)$$

r_1, r_2 are random components chosen to satisfy

$$D = \frac{\partial u}{\partial x} + \frac{\partial v}{\partial y} = 0 \quad . \qquad (3.11)$$

Results have been obtained on a 40×40 mesh with the following finite difference representation of Eq. (3.9)

$$\frac{u_{ij}^{n+1} - u_{ij}^{n-1}}{2\Delta t} + C_{3x}(u,u) + C_{3y}(v,u) = 0$$

and

$$\frac{v_{ij}^{n+1} - v_{ij}^{n-1}}{2\Delta t} + C_{3x}(u,v) + C_{3y}(v,v) = 0 \quad . \qquad (3.12)$$

The time integration was made over a 25 sec period and some results are shown in Table 3.2. E^α is the ratio of the final to initial kinetic energy. The C_1

Scheme	$\Delta t = 0.0005$ (50 K steps) E^α	$\Delta t = 0.0025$ (10 K steps) E^α	$\Delta t = 0.0125$ (2 K steps) E^α
C_3	1.029	1.304	∞
C_1	∞	∞	∞

Table 3.2 Change in kinetic energy after 23 secs

schemes blew up between $t = 0.5$ and 1 secs and were accompanied by a catastrophic growth in D. However, a large growth in D had a very small effect on the C_3 schemes. Clearly the growth of E^α with Δt indicates that a time integration scheme that is also conservative is required.

Cheng and Shubin (1978) have obtained solutions to the steady Burgers' equation

$$uu_x - \frac{1}{Re} u_{xx} = 0 \quad , \qquad (3.13)$$

in order to investigate the influence of cell Reynolds number, Re Δx, on accuracy. The viscous term was represented by a conventional, second-order centered difference formula. The convective term was represented by

$$\frac{\gamma}{(2+\gamma)}\, u_i \,\frac{(u_{i+1} - u_{i-1})}{2\Delta x} + \frac{2}{(2+\gamma)}\,\frac{(u^2_{i+1} - u^2_{i-1})}{4\Delta x} \quad . \tag{3.14}$$

γ is a free parameter which permits a balance to be struck between the convective and divergence forms. In particular for

$$Re^* \Delta x = 2(2+\gamma)/(1+\gamma) \tag{3.15}$$

the solution of the difference equations is exact and oscillation free. Clearly γ can be chosen so that $Re^* \Delta x$ is arbitrarily large.

Cheng and Shubin indicated for large Re Δx (say ≈ 10) solutions can be obtained (with appropriate choice of γ) that are considerably more accurate than the second-order accuracy suggested by a Taylor expansion. It is argued that the Taylor expansion result is only relevant for small Re Δx (say < 2).

Khosla and Rubin (1979) obtained the result Eq. (3.15) by introducing a filter to prevent nonlinear instabilities from developing when solving the modified Burgers' equation, Eq. (3.21), with cell Reynolds numbers much greater than two. They considered the general implicit finite difference scheme,

$$\frac{u_i^{n+1} - u_i^n}{\Delta t} + \left(\frac{u_{i-1}^{n+1} + \gamma u_i^{n+1} + u_{i+1}^{n+1}}{2+\gamma} - \alpha\right)\left(\frac{u_{i+1}^{n+1} - u_{i-1}^{n+1}}{2\Delta x}\right) - \frac{\nu}{\Delta_x^2}\left(u_{i-1}^{n+1} - 2u_i^{n+1} + u_{i+1}^{n+1}\right) = 0$$

Different values of γ correspond to different degrees of smoothing or filtering. Thus $\gamma = \infty$ corresponds to no filtering, $\gamma = 0$ to the conservation form and γ given by Eq. (3.15) gives an oscillation-free solution. It is interesting that the linear Galerkin finite element treatment of uu_x, given by Eq. (6.15), corresponds to $\gamma = 1$. However, Khosla and Rubin found that when the cell Reynolds number is greater than two and the Courant number much greater than one, two solutions are possible. This non-uniqueness has also been investigated by Kellogg et al. (1980).

3.2 Compact implicit methods

Although the centered difference representation of u_x is second order accurate, a fourth-order accurate representation is possible from the scheme

$$\frac{1}{6}(u_x)_{i-1} + \frac{2}{3}(u_x)_i + \frac{1}{6}(u_x)_{i+1} = \frac{u_{i+1} - u_{i-1}}{2\Delta x} \tag{3.16}$$

However, the derivatives now appear implicitly and, typically, auxiliary variables $(u_x)_i$, etc., need to be introduced to take advantage of the higher accuracy.

Thus if second derivatives occur the solution of a 3×3 block tridiagonal system of equations will be required (Hirsch, 1975). However, Adam (1975,1977) showed how to treat the second derivative implicitly so that only a 2×2 block tridiagonal system need be solved. Ciment et al. (1978) demonstrated that, by considering the complete spatial operator rather than individual terms, it is possible to generate a compact implicit formulation that only requires a scalar tridiagonal system of equations to be solved. However, the algebraic complexity is somewhat greater.

Burgers' equation, with Re = 1, was represented by Adam (1975) as

$$H\left\{\frac{u_i^{n+1} - u_i^n}{\Delta t}\right\} + \frac{H}{2}\{u_i^{n+1}\cdot u_{xi}^n\} + \frac{H}{2}\{u_i^n u_{xi}^{n+1}\} - \frac{\delta^2}{2\Delta_x^2}\{u_i^{n+1} + u_i^n\} = 0 \quad , \tag{3.17}$$

where H and δ^2 are operators defined by

and

$$Hu_i = u_{i+1} + 10u_i + u_{i-1} \quad , \qquad (3.18)$$
$$\delta^2 u_i = u_{i+1} - 2u_i + u_{i-1} \quad .$$

The second equation is

$$\frac{u_{xi+1}^{n+1}}{6} + \frac{2}{3} + \frac{u_{xi-1}^{n+1}}{6} = \frac{u_{i+1} - u_{i-1}}{2\Delta x} \qquad (3.19)$$

Solutions to Burgers' equations have been obtained for initial conditions

$$u(x,0) = R \sin x \qquad (3.20)$$

and boundary conditions, $u = 0$ at $x = -\pi$ and 0. Results for the above scheme (Eqs. (3.17) to (3.19)) are compared with a conventional second-order Crank-Nicolson scheme for various times in Table 3.3. ε is the absolute average error and m is

t	0.1	0.2	0.3	0.4	0.5	1.0	m	ΔT
ε/u_{max} for compact implicit	0.0049	0.0020	0.0012	0.0006	0.0004	0.7×10^{-4}	21	0.01
ε/u_{max} for Crank-Nicolson	0.083	0.066	0.037	0.024	0.019	0.010	21	0.01
ε/u_{max} for Crank-Nicolson (refined)	0.0029	0.0017	0.0011	0.0008	0.0006	0.0004	101	0.005

Table 3.3 Comparison of compact implicit and Crank-Nicolson methods

the number of spatial nodal points. According to Adam (1975) the compact implicit method is 2.3 times more expensive than the Crank-Nicolson method per step but for comparable accuracy after the same time the Crank-Nicolson method is four times more expensive. Adam also suggested the loss of accuracy with a variable grid is far less for the compact implicit method than for conventional methods.

Ciment et al. (1978) have sought solutions of a modified Burgers' equation written in the form

$$u_t + (u-\alpha)u_x - \nu u_{xx} = 0 \qquad (3.21)$$

using a compact implicit scheme in which only the function values, u, need be solved for explicitly. Equation (3.21) was solved in the region $-5 \leqslant x \leqslant 5$ subject to the initial conditions $u = 1$ for $x < 0$, $u = 0.5$ for $x = 0$ and $u = 0$ for $x > 0$. Equation (3.21) was integrated until the steady state is reached using a Crank-Nicolson time integrator. The L_2 error is compared with corresponding errors for a second-order Crank-Nicolson scheme in Table 3.4.

Scheme	N	Δx	ν	L_2 error
compact implicit	60	0.167	0.031	0.0346
(after Ciment et al., 1978)	100	0.10	0.031	0.0039
second-order	100	0.10	0.031	0.0334
Crank-Nicolson	200	0.05	0.031	0.0067

Table 3.4 Comparison of compact implicit and second-order Crank-Nicolson for steady-state Burgers' equation, $\alpha = 0.5$

Clearly the compact implicit scheme is an order of magnitude more accurate for the same grid resolution. Ciment et al. indicated that the following cell Reynolds number restriction is necessary to achieve stable results

$$R_e = (u-\alpha)\,\frac{\Delta x}{\nu} < 2.55 \tag{3.22}$$

Rubin and Graves (1975) also solved the propagating shock problem, but using a cubic spline formulation. For low to moderate Reynolds numbers (i.e., relatively diffuse 'shocks') the technique gives solutions of high accuracy. The formulation appears to be comparable in accuracy to the compact implicit schemes considered above. For higher Reynolds number the solution is oscillatory adjacent to the shock unless local mesh refinement is introduced. An interesting result, obtained by Rubin and Graves, is that the non-divergence form, uu_x, generally gives more accurate results than the divergence form, $0.5(u_x)^2$; this is opposite to the situation for finite difference schemes.

Rubin and Khosla (1976) developed comparable schemes to Adam by starting from high-order Hermitian splines and obtained solutions to Burgers' equation.

3.3 Finite element methods

Arminjon and Beauchamp (1978, 1979 and 1981) have obtained solutions to the one- and two-dimensional, inviscid and continuous Burgers' equations.

For the continuous, one-dimensional problem, Arminjon and Beauchamp (1978), a variational formulation is written, for $0 \leqslant x \leqslant 1$ and $\tau_1 < t < \tau_2$,

$$\iint u\phi_t \,dtdx + \iint \left(\frac{u^2}{2} - \nu u_x\right)\phi_x \,dxdt = \int_0^1 u(x,\tau_2)\cdot\phi(x,\tau_2)dx - \int_0^1 u(x,\tau_1)\cdot\phi(x,\tau_1)dx \tag{3.23}$$

where u is governed by Burgers' equation and ϕ is an analytic function of x and t. A simple form of ϕ would be

$$\phi(x,t) = a_1 + a_2 x + a_3 t + a_4 xt \quad . \tag{3.24}$$

Clearly ϕ can be interpreted as a bilinear element in the (x,t) plane. With some simplifying assumptions on how the integrals are evaluated the following algebraic scheme is obtained

$$\frac{u_i^{n+1} - u_i^n}{\Delta t} + \frac{1}{4}\left[\frac{u_{i+1}^{2n+1} - u_{i-1}^{2n+1}}{2\Delta x} + \frac{u_{i+1}^{2n} - u_{i-1}^{2n}}{2\Delta x}\right]$$

$$- \frac{\nu}{2}\left[\frac{(u_{i-1}^{n+1} - 2u_i^{n+1} + u_{i+1}^{n+1})}{2\Delta x} + \frac{(u_{i-1}^n - 2u_i^n + u_{i+1}^n)}{2\Delta x}\right] \tag{3.25}$$

Clearly the scheme is equivalent to a second-order Crank-Nicolson scheme in which the convective term is represented in divergence form, $0.5(u^2)_x$. Arminjon and Beauchamp (1978) presented results for $\nu = 1$ with a starting solution

$$u(x,0) = 2\pi\nu \frac{(0.25 \sin \pi x + \sin 2\pi x)}{(1 + 0.25 \cos \pi x + 0.5 \cos 2\pi x)} \quad . \tag{3.26}$$

The results confirmed that the scheme, Eq. (3.25), is second-order accurate in space.

To solve the inviscid Burgers' equation,

$$u_t + uu_x = 0 \tag{3.27}$$

Arminjon and Beauchamp (1981) developed two discontinuous finite element methods, i.e., at each node u_i can take two values. This permits shocks to appear in the solution in such a way that the correct change in solution across the shock and the correct shock speed is obtained.

Arminjon and Beauchamp (1981) followed Fortin (1979) by replacing u_t with $(u^n - u^{n-1})/\Delta t$ and by constructing the following variational relationship

$$\int_0^1 \left(\frac{u^n - u^{n-1}}{t}\right)\psi \, dx = 0.5 \int_0^1 (u^n)^2 \, \psi_x \, dx \quad . \tag{3.28}$$

The test function, ψ, is a cubic polynomial and the trial function for u is a linear polynomial. This leads to the following system of equations (two at each node) at internal nodes,

$$\left[\tfrac{3}{2}(\underset{+}{u_i^n})^2 + 2\underset{+}{u_i^n}\cdot\underset{-}{u_{i+1}^n} + \tfrac{3}{2}(\underset{-}{u_{i+1}^n})^2 - \tfrac{3}{2}(\underset{+}{u_{i-1}^n})^2 - 2\underset{+}{u_{i-1}^n}\cdot\underset{-}{u_i^n} - \tfrac{3}{2}(\underset{+}{u_{i-1}^n})^2\right]$$

$$+\frac{\Delta x}{2\Delta t}\left[3\underset{+}{u_{i-1}^n} + 7\underset{-}{u_i^n} + 7\underset{+}{u_i} + 3\underset{-}{u_{i+1}^n}\right] = \frac{\Delta x}{2\Delta t}\left[3\underset{+}{u_{i-1}^{n-1}} + 7\underset{-}{u_i^{n-1}} + 7\underset{+}{u_i^{n-1}} + 3\underset{-}{u_{i+1}^{n-1}}\right] \;, \tag{3.29}$$

$$\left[(\underset{+}{u_{i-1}^n})^2 + 0.5\,\underset{+}{u_{i-1}^n}\cdot\underset{-}{u_i^n} - 1.5(\underset{-}{u_i^n})^2 + 1.5(\underset{+}{u_i^n})^2 + 0.5\,\underset{+}{u_i^n}\cdot\underset{-}{u_{i+1}^n} + (\underset{-}{u_{i+1}^n})^2\right]$$

$$+\frac{\Delta x}{\Delta t}\left[\underset{-}{u_{i+1}^n} + 1.5(\underset{+}{u_i^n} - \underset{-}{u_i^n}) - \underset{+}{u_{i-1}^n}\right] = \frac{\Delta x}{\Delta t}\left[\underset{-}{u_{i+1}^{n-1}} + 1.5(\underset{+}{u_i^{n-1}} + \underset{-}{u_i^{n-1}}) - \underset{+}{u_{i-1}^{n-1}}\right] \;.$$

Where the solution is expected to be continuous the final solution is

$$u_i^n = 0.5(\underset{-}{u_i^n} + \underset{+}{u_i^n}) \;. \tag{3.30}$$

Results are presented by Arminjon and Beauchamp (1981) for the solution of the inviscid Burgers' equation with initial conditions

$$\begin{aligned} u(x,0) &= 2x \qquad 0 \leqslant x < 0.5 \\ &= 1 - x \quad 0.5 \leqslant x \leqslant 1 \end{aligned} \tag{3.31}$$

This problem has a discontinuity at $x = 0.5$ which propagates to the right with increasing t. After 60 time steps ($\Delta t = 1/200$), with $\Delta x = 1/40$, the mean relative error is 0.28%. For other problems involving the inviscid Burgers' equation, for which continuous solutions exist the present discontinuous algorithm gave superior results to the use of the related continuous formulation (Eq. (3.25)). However, the opposite was true when considering problems governed by the full (viscous) Burgers' equation.

Nickell et al. (1979) applied a conventional Galerkin finite element method in space to Burgers' equation to reduce the equation to a system of ordinary differential equations in time. By expanding the temporal solution in terms of the eigenfunctions of the matrix of ordinary differential equations, the solution of the ordinary differential equations can be transformed to the obtaining of the coefficients in the eigenfunction expansion. If the contributions from the known initial and boundary conditions can be linearized over a time step, Δt, an exact solution for the coefficients in the eigenfunction expansion can be written down.

This method of time integration was compared with a Crank-Nicolson scheme for the propagating 'shock' problem. Three values of Reynolds number, Re = 0.02, 1, and 50, were considered. Both schemes produced identical results for a sufficiently small time step, Δt. However, for coarse time steps the eigenfunction expansion method produced more accurate answers. However, the relative computational efficiency of the two methods was not compared.

3.4 Traditional Galerkin and spectral methods

A Galerkin method with B-splines as basis functions has been used by Davies (1978) to solve Burgers' equation for the propagating 'shock' problem.

The representation for u is

$$u(x,t) = \sum_{i=1}^{m} a_i(t) \cdot M_{ni}(x) \tag{3.32}$$

where $M_{ni}(x)$ is a B-spline of order n and m is the number of unknowns $a_i(t)$. A B-spline of order n has the restricted support property

$$M_i(x) \geqslant 0 \quad x_{i-n} \leqslant x \leqslant x_i$$

$$= 0 \quad x < x_{i-n} \text{ or } x > x_i \quad .$$

Therefore 2n nodes are required outside the region of interest ($0 \leqslant x \leqslant 1$) and the number of equations required is $m = N + n$ where N is the number of nodes in the interior of (0,1). Application of the Galerkin method produces the matrix equation

$$\underline{C}\,\dot{\underline{A}} = \frac{1}{Re}\,\underline{B}\;\;\underline{A} - \underline{D}\,\underline{A} \quad , \tag{3.33}$$

where $\underline{A}$ is a vector of the unknown coefficients, a_i. A component of $\underline{C}$ is evaluated from

$$C_{ik} = \int_0^1 M_k(x) \cdot M_i(x)dx \quad . \tag{3.34}$$

Components of matrices $\underline{B}$ and $\underline{D}$ have the same form as those arising in a Galerkin finite element method (see chapter 6) except that, here, they are based on B-spline rather than polynomial shape functions.

After inverting $\underline{C}$ the system of equations are integrated with a fourth-order Runge-Kutta scheme and a time step size, Δt, sufficiently small that any errors at subsequent time are due to the spatial approximation.

Starting from a square wave at t=0, the solution at t=0.5 is compared with the exact solution for Reynolds numbers up to 267 for splines of order 2, 4 and 6. Various discretisations in the x-direction with up to 25 nodes are considered. Typical results for the maximum value of u (= 1.00, if exact) are given in Table 3.5. For the coarse grid and highest Reynolds number the maximum value of u is essentially determined by an oscillatory solution associated with cell Reynolds number effects.

The solution accuracy generally improves with grid refinement, an increase in the order of the spline or reduction in Reynolds number. However, the effectiveness of increasing the order of the spline reduces as the grid is refined at lower Reynolds number. Davies presented additional results that indicate an improvement in accuracy can be obtained by refining the grid in the region where the large gradient occurs (shock).

The inviscid Burgers' equation, with initial condition,

$$u(x,0) = -0.45x + 0.55 \quad -1 \leqslant x \leqslant 1 \tag{3.35}$$

and boundary condition $u(-1,t) = 1$ has been considered by Morchoisne (1979). With increasing time a shock develops and propagates to the right. Morchoisne used a pseudospectral method in which both the spatial and temporal behavior are represented by Chebyshev polynomials. Due to the nonlinear term an iterative process is built up from a linearisation about the nth level. Thus

Reynolds number	Order of spline	Maximum value of u with 10 nodes	Maximum value of u with 25 nodes
267	2	1.275	1.160
267	4	1.141	1.115
267	6	1.142	1.083
133	2	1.099	1.020
133	4	1.060	1.014
	6	1.063	1.007
100	2	1.091	1.003
100	4	1.026	1.004
100	6	1.017	1.003

Table 3.5 Numerical solution of Burgers' equation at $t = 0.5$ on a uniform grid for B-spline Galerkin formulation

$$L_n(u^{n+1} - u^n) = -R^{(n)} \tag{3.36}$$

where $R^{(n)}$ is the equation residual, i.e., $R^{(n)} = u_t^{(n)} + u^{(n)}u_x^{(n)}$.

The operator L is interpreted as a split operator

$$L_n \equiv L_{n,x} \cdot L_{n,t} = \frac{1}{\sigma\varepsilon}\,(1 + \varepsilon u^{(n)}\,\frac{\partial}{\partial x})(1 + \varepsilon\,\frac{\partial}{\partial t})u \quad , \tag{3.37}$$

where ε and σ are adjusted to maintain stability. Thus, Eq. (3.36) can be written,

$$u^{n+1}(x,t) = u^n(x,t) - L_{n,t}^{-1} \cdot L_{n,x}^{-1}\,R^{(n)} \quad . \tag{3.38}$$

In evaluating $u \cdot u_x$, the multiplication is made in physical space rather than in Fourier space. The solution algorithm then consists of the following steps.

1. Choose initial value $u^{(0)}(x,t)$
2. Calculate $R^{(n)}$
3. Calculate f from $f = (1 + \varepsilon u^{(n)}\,\frac{\partial}{\partial x})^{-1}[-\sigma\varepsilon R^{(n)}]$
4. Calculate $\delta_n u$ from $\delta_n u = (1 + \varepsilon\,\frac{\partial}{\partial t})^{-1}[f]$
5. Evaluate $u^{n+1} = u^n + \delta_n u$
6. If not converged go to step 2

Results for u(x) show good agreement with the exact solution for $t = 1$ and $t = 2$, when a nine term Chebyshev expansion in x and t is used (Morchoisne, 1979).

Gazdag (1973) has obtained solutions to the propagating shock problem governed by Burgers' equation using a spectral method and a Taylor series expansion time step integrator. Higher order time derivatives (up to third) were obtained from the spatial derivatives by repeated differentiation of the right-hand side of Burgers' equation. In order to use a Fourier series expansion it was necessary to construct an unrealistic upstream velocity profile so that the spatial solution could be considered periodic. Results for $\Delta x = 0.01$, $\Delta t = 0.001$ and a Reynolds number up to 500 were presented. Solutions were oscillatory when $Re \cdot \Delta x > 2$.

3.5 Other computational examples

Jain and Holla (1978) have obtained solutions to Burgers' equation on $0.5 \leq x \leq 1.5$ and $t \geq 0.5$ for an initial condition

$$u(x,0) = \frac{1}{Re}[x + \tan(x/2)] \quad . \tag{3.39}$$

The method is based on splitting Burgers' equation into

$$\begin{aligned} &0.5\, u_t = -uu_x \\ \text{and} \quad &0.5\, u_t = \frac{1}{Re} u_{xx} \end{aligned} \tag{3.40}$$

and then using cubic splines in space, and a forward difference representation in time to construct the following finite difference scheme. For the first half step

$$\{1 + \frac{u_i^n}{6} \delta_x^2 (u_i^n)^{-1} + \frac{r}{2} \Delta x\, \theta_1\, u_i^n\, \delta_x\}[u_i^{n+\frac{1}{2}} - u_i^n] = -\frac{r}{2} \Delta x\, u_i^n\, \delta_x\, u_i^n \tag{3.41}$$

where $\delta_x u_i^n = u_{i+1}^n - u_{i-1}^n$, $\delta_x^2 u_i^n = u_{i+1}^n - 2u_i^n + u_{i-1}^n$, $r = \Delta t/\Delta_x^2$ and θ, is a parameter to control the implicitness of the scheme. For the second half step

$$\{1 + (\frac{1}{6} - \frac{r\theta_2}{Re})\delta_x^2\}(u_i^{n+1} - u_i^{n+\frac{1}{2}}) = \frac{r}{Re}\delta_x^2\, u_i^{n+\frac{1}{2}} \quad . \tag{3.42}$$

The scheme requires the alternate solution of a tridiagonal system of equations and is stable if $\frac{1}{2} < \theta_1$, $\theta_2 < 1$. The truncation error is $O(\Delta t + \Delta x^2)$.

Jain and Holla developed a comparable scheme for the two dimensional Burgers' equation, in which four separate integrations are made. For the problem given by Eq. (3.39), solutions are presented with $\Delta x = 0.05$, $r = 1.5$, $Re = 100$ and 1000 for $t = 1.125$. Very close agreement with the exact solution is obtained. Solutions for a two dimensional problem are presented for Reynolds numbers up to 1000.

Bellman et al. (1965) expanded u as a Fourier series and sought solutions for u periodic in x starting from $u(x,0) = 1$. The resulting ordinary differential equations were integrated using a Runge-Kutta scheme and a variable-step Adams-Moulton technique. The corresponding two-dimensional problem was formulated but not solved.

Wadia and Payne (1979) obtained solutions to Burgers' equation with initial conditions

$$u(x,0) = \sin \pi x \qquad 0 \leq x \leq 1 \tag{3.43}$$

and boundary conditions $u(0,t) = u(1,t) = 0$, $t \geq 0$. As time increases this problem develops a 'shock' located at the right-hand boundary. The formulation consists of a subdomain weighted residuals method in time followed by a finite difference discretisation in space. The subdomain method assumes a unit weight function between successive time levels n and n+1 and that u_i is a piecewise linear function in time. Second order centered spatial differences are used and the nonlinear equations are solved with a Gauss-Seidel scheme. The results were compared with a Galerkin finite element method in time, for $Re = 100$, and the subdomain method was found to be more economical but less accurate.

Gey and Lesser (1969) obtained solutions to the same problem, by assuming a series representation in time

$$u = \sum_i a_i \cdot t^i \tag{3.44}$$

to obtain an ordinary differential equation for $a_i(x)$,

$$a_{i+1}(x) = \frac{1}{i+1}\left\{\frac{\partial^2 a_i}{\partial x^2} - \sum_{j+k=i} a_j(x) \cdot \frac{\partial a_k}{\partial x}\right\}$$

$$a_o(x) = \sin x \quad . \tag{3.45}$$

By using a program capable of symbolic algebraic manipulation the coefficients a_i were obtained algebraically. Solutions for u(x) were presented for various t at Re = 1.

Yee, Beam and Warming (1981) applied a finite difference scheme to the inviscid Burgers' equation in conservation form in the interval, $0.5 \leq x \leq 1$, with initial conditions,

$$u(x,0) = x \tag{3.46}$$

and boundary conditions,

$$u(0.5,t) = 0.5/(1+t) \quad . \tag{3.47}$$

This combination has the exact solution,

$$u(x,t) = x/(1+t) \quad . \tag{3.48}$$

Yee et al. used a second-order centered difference representation for the spatial derivative and a backward Euler scheme for the time derivative. Different numerical boundary conditions were applied at x = 1. It was found that using

$$u_o^{n+1} = 2u_1^{n+1} - u_2^{n+1} \quad , \tag{3.49}$$

i.e., equivalent to $u_{xx} = 0$, gave smooth results but the use of

$$u_o^{n+1} = u_1^{n+1} \quad , \tag{3.50}$$

i.e., equivalent to $u_x = 0$, gave an oscillatory solution in the interior. These results were essentially unaltered by staggering the time, e.g., using

$$u_o^{n+1} = 2u_1^n - u_2^{n-1} \quad . \tag{3.51}$$

Pierson and Kutler (1980) developed a scheme for choosing an optimal grid based on minimising, in the L_2 sense, the leading term in the truncation error. The scheme involves defining a grid transformation (to a uniform grid) as a Chebyshev polynomial expansion with unknown coefficients that are determined to minimise a 'global truncaction error parameter.' A test on the inviscid Burgers' equation and the propagating shock problem indicates that an improvement in accuracy is possible comparable to doubling the number of points on a uniform grid. No indication is given of the extra computation time associated with the optimisation. For time dependent problems it is necessary to repeat the optimal grid calculation periodically. As the authors note minimising the truncation error may not produce the optimal minimisation of solution error.

In this chapter the examples have been chosen to demonstrate the breadth of computational techniques and formulations for which Burgers' equation has been an appropriate test problem.

4. TRADITIONAL GALERKIN METHOD

The Galerkin method originated with a paper (Galerkin, 1915) concerning the elastic behavior of rods and plates. The method became more widely known (in the Western literature) through the work of Duncan (1937,1938) on the dynamic behavior of aeronautical structures.

In a modern context it is appropriate to think of the Galerkin method as a member of the class of methods of weighted residuals (Crandall, 1956; Finlayson and Scriven, 1966). Collatz (1960) considered the same idea through different "error distribution principles."

In this chapter the traditional Galerkin method is introduced as an example of the method of weighted residuals. Application of the traditional Galerkin method to a simple ordinary differential equation (after Duncan, 1937) illustrates the need to choose the weighting function from the lowest-order members of a complete set of functions. Applying the method to Burgers' equation provides baseline data with which results presented in subsequent chapters will be compared. The link between the Galerkin method and other methods such as the Rayleigh-Ritz method is emphasised. The use of the traditional Galerkin method in a wide range of disciplines is illustrated. The chapter closes with a discussion of the inherent shortcomings of the traditional Galerkin method and the way in which the spectral and Galerkin finite element methods have sought to overcome them.

4.1 Method of weighted residuals

The method of weighted residuals (MWR) can be described in the following way. In obtaining the solution of a differential equation

$$L(u) = 0 \quad , \tag{4.1}$$

with initial condition, $I(u) = 0$, and boundary conditions, $S(u) = 0$, an appropriate solution, v, is sought which satisfies both $L(v) = R$ and $S(v) = R_b$. One would like the equation residual, R, and the boundary residual, R_b, to be as small as possible.

A trial solution, of the following form, is assumed,

$$v(\bar{x},t) = v_0(\bar{x},t) + \sum_{j=1}^{N} a_j(t) \cdot \phi_j(\bar{x}) \quad , \tag{4.2}$$

in which $\phi_j(\bar{x})$ are known analytic functions and the a_j's are to be determined as the solution develops (in time dependent problems) or once and for all (in steady problems).

It would also be possible to let $a = a(\bar{x})$ and $\phi = \phi(t)$ but this is unusual (see Wadia and Payne, 1979). By forcing the form of the analytic solution on the approximate solution, the approximate solution is constrained. In a sense making N larger relaxes the constraints. However, for computational efficiency the error between u and v should be acceptably small with N as small as possible. The Galerkin method, as used by Galerkin and Duncan, etc., was a pre-computer method for which a typical value of N would be $N = 5$.

The form of v_0 (in Eq. (4.2)) is chosen to satisfy the boundary conditions and the initial conditions exactly if possible. Ames (1965) calls a method of weighted residuals (MWR) which satisfies the boundary conditions exactly and the governing equation approximately an interior method. This is opposite to the idea which motivates the method of singularities in which exact solutions of the partial differential equation are known and these are added together to satisfy the boundary conditions approximately. It is also possible to formulate a method (Shuleshko, 1959) in which both the partial differential equation and the boundary conditions are satisfied approximately.

Clearly substitution of the trial solution, Eq. (4.2), into the governing equation, (4.1), will cause a non-zero residual for all choices of a_j unless the trial solution contains the exact solution, which is unlikely to occur in real problems. The equation residual, R, is typically continuous in the spatial domain ($\bar{x}$) and with increasing N a choice of the a_j's is possible for which $\|R\|$ becomes smaller and smaller. For a given N the a_j's are chosen by requiring that an integration of the weighted residual (and hence the name) over the domain is zero. Thus

$$\int W_k \cdot R \, d\bar{x} = 0 \quad . \tag{4.3}$$

By letting $k = 1, \ldots N$ a system of equations involving only the a_j's is obtained. For unsteady problems this would be a system of ordinary differential equations; for steady problems a system of algebraic equations is obtained. Different choices of W_k give rise to the different methods within the class. Some of these methods are:

i) Subdomain method.

The domain is split up into N subdomains, D_k, which may overlap, and

$$\begin{aligned} W_k &= 1 \quad \text{in } D_k \\ &= 0 \quad \text{outside } D_k \quad . \end{aligned} \tag{4.4}$$

An example of this method is the solution of the blunt body problem in supersonic inviscid flow (Belotserkovskii and Chushkin, 1965).

ii) Collocation method

$$W_k(\bar{x}) = \delta(\bar{x}-\bar{x}_k) \tag{4.5}$$

where δ is the Dirac delta function. Thus the collocation method sets $R = 0$ at $\bar{x} = \bar{x}_k$. Typical finite difference methods are collocation methods with $\bar{x}_k$ the nodes of a uniform grid. Villadsen and Stewart (1967) choose $\bar{x}_k$ as the zeros of Jacobi polynomials which gives a method of orthogonal collocation. They have applied this method to viscous flow in a channel.

iii) Least squares

$$W_k(\bar{x}) = \partial R/\partial a_k \quad . \tag{4.6}$$

Then Eq. (4.3) is equivalent to the requirement that

$$\int R^2 \, dx \text{ is a minimum} \quad .$$

The method of least-squares has been used by Narasimha and Deshpande (1969) to examine shock structure via the Boltzmann equation. In conjunction with the finite element method, Lynn (1974) has applied a least-squares formulation to the boundary layer equations, Fletcher (1979) and Chattot et al. (1979) have computed solutions to compressible, inviscid flows and Steven and Milthorpe (1978) have considered viscous, incompressible flows.

iv) Galerkin method

$$W_k(\bar{x}) = \phi_k(\bar{x}) \tag{4.7}$$

i.e., the weighting functions are chosen from the same family as the trial functions in Eq. (4.2). If the functions form a complete set, Eq. (4.3) indicates that the residual must be orthogonal to every member of the complete set. As N tends to infinity the approximate solution, v, will converge to the exact solution, u.

The various methods of weighted residuals are considered in depth by Finlayson (1972). An interesting feature of the method (WRM) explored by Finlayson is that evaluation of $||R||$, which is straightforward, will give some indication of the solution error $||u-v||$. Thus for nonlinear problems where an iterative procedure is required monitoring $||R||$ or $R(\bar{x})$ will indicate the progress towards convergence and the regions of the domain where convergence is worst. In the rest of the paper only the Galerkin method will be considered.

4.2 Duncan's example

Consider (Duncan, 1937) the ordinary differential equation,

$$\frac{dy}{dx} - y = 0 \quad , \tag{4.8}$$

with boundary condition $y=1$ at $x=0$. The exact solution is $y=\exp(x)$. A Galerkin method starts by assuming an appropriate trial function, e.g., let

$$y_a = 1 + \sum_{j=1}^{N} a_j \cdot x^j \quad . \tag{4.9}$$

The form of Eq. (4.9) satisfies the boundary condition exactly. Substitution of Eq. (4.9) into Eq. (4.8) produces the residual

$$R = -1 + \sum_{j=1}^{N} a_j \cdot \{j \cdot x^{j-1} - x^j\} \quad . \tag{4.10}$$

For the Galerkin method,

$$W_k(x) = x^{k-1} \quad , \quad k = 1,2,\ldots N \quad , \tag{4.11}$$

and evaluation of Eq. (4.3) gives

$$\underline{P} \cdot \underline{A} = \underline{B} \tag{4.12}$$

where $\underline{A}$ is the vector of unknowns a_j. An element of $\underline{P}$ is evaluated from

$$P_{kj} = \int_0^1 (j \cdot x^{j-1} - x^j)x^{k-1} \, dx \quad , \tag{4.13}$$

and an element of $\underline{B}$ from

$$b_k = \int_0^1 1 \cdot x^{k-1} \, dx \quad ,$$

For $N=3$ the approximate solution, Eq. (4.9), is

$$y_a = 1 + 1.0141\,x + 0.4225\,x^2 + 0.2817\,x^3 \quad . \tag{4.14}$$

Solutions of increasing order are compared with the exact solution in Table 4.1. It is clear that both the solution error, $||y-y_a||_2$, and the equation residual, $||R||_2$, diminish rapidly with increasing order. To achieve high accuracy with few unknowns is an important property for a hand method. It can also be seen, from Table 4.1, that $||R||_2$ is a reasonable indicator of $||y-y_a||_2$. The importance of choosing $W_k(x)$ from a complete set of functions can be appreciated from the results shown in Table 4.2. Each approximate solution is a cubic but progressively higher-order weight functions have been used for successive solutions to the right. It can be seen, by examining $||y-y_a||_2$ and $||R||_2$, that the solution accuracy progressively deteriorates as the order of weight function is increased.

x	Approximate solution, y_a			Exact solution $y = \exp(x)$
	linear	quadratic	cubic	
0	1	1	1	1
0.2	1.4	1.2057	1.2220	1.2214
0.4	1.8	1.4800	1.4913	1.4918
0.6	2.2	1.8229	1.8214	1.8221
0.8	2.6	2.2349	2.2259	2.2251
1.0	3.0	2.7143	2.7183	2.7183
$\|\|y-y_a\|\|_2$	0.3022	0.0097	0.0006	
$\|\|R\|\|_2$	0.5774	0.0639	0.0053	

Table 4.1 Traditional Galerkin solutions for Duncan's example

x	Cubic solution; with weight function, x^k			exact $y = \exp(x)$
	k = 0,1,2	k = 1,2,3	k = 2,3,4	
0	1	1	1	1
0.2	1.2220	1.2248	1.2282	1.2214
0.4	1.4913	1.4952	1.5004	1.4918
0.6	1.8214	1.8255	1.8317	1.8221
0.8	2.2259	2.2303	2.2375	2.2251
1.0	2.7183	2.7241	2.7330	2.7183
$\|\|y-y_a\|\|_2$	0.0006	0.0039	0.0096	
$\|\|R\|\|_2$	0.0053	0.0086	0.0150	

Table 4.2 Different weight functions for Duncan's example

4.3 Burgers' equation

Applying the traditional Galerkin method to Burgers' equation is more complicated than for Duncan's example. Firstly Burgers' equation is a partial differential equation so that application of the traditional Galerkin method produces an ordinary differential equation which has to be integrated in time. In principle it would be possible to introduce a trial function that was a function of both independent variables and hence obtain a system of algebraic equations. However, for a hand calculation, this would require a greater effort to solve. A second complication is that Burgers' equation is nonlinear. If the system of ordinary equations, for the $a_j(t)$'s, is integrated explicitly this complication manifests itself as the evaluation of a double summation. The complication is greater if the partial differential equation is reduced to a system of algebraic equations.

Then an iterative method, like Newton's method, will be required.

Consider Burgers' equation in the form

$$u_t + uu_x - \frac{1}{Re} \cdot u_{xx} = 0 \ , \tag{4.15}$$

with boundary conditions

$$u(-1,t) = 1 \ , \quad u(1,t) = 0 \ , \tag{4.16}$$

and initial conditions

$$\begin{aligned} u(x,0) &= 1 \ , \quad -1 \leqslant x \leqslant 0 \\ u(x,0) &= 0 \ , \quad 0 < x \leqslant 1 \ . \end{aligned} \tag{4.17}$$

Equations (4.15) to (4.17) govern a physical situation in which an initially discontinuous shock wave propagates to the right and whose profile is slowly smoothed.

Although it would be possible to introduce a trial solution similar to Eq. (4.9) it is preferable to develop the trial solution as an ascending series of Chebyshev polynomials.

The reason for introducing Chebyshev polynomials is that by interpolating the solution more accurately, particularly close to the boundaries $x = \pm 1$, the resulting equations are better conditioned than if polynomial trial functions (as in Eq. (4.9)) were used. Chebyshev polynomials, $T_j(x)$, are orthogonal over the interval (-1,1) with respect to the weight, $(1-x^2)^{-\frac{1}{2}}$, i.e.,

if

$$(f,g) \equiv \int_{-1}^{1} f(x) \cdot g(x) \cdot (1-x^2)^{-\frac{1}{2}} \, dx \tag{4.18}$$

then

$$\begin{aligned} (T_i,T_j) &= 0 \qquad \text{if} \quad i \neq j \\ &= \pi/2 \quad \text{if} \quad i = j \neq 0 \\ &= \pi \qquad \text{if} \quad i = j = 0 \end{aligned} \tag{4.19}$$

It can be seen, from the nature of weight function, that the use of Chebyshev polynomials gives more significance to points adjacent to $x = \pm 1$. In contrast the use of conventional polynomials for interpolation can be appreciated by considering Runge's example (Forsythe, Malcolm and Moler, 1977), in which the function $f(x) = 1/(1+25x^2)$ is fitted with a high order polynomial over the interval (-1,1). It is found that although the interpolating polynomial matches f(x) at the interpolation points (equally spaced) the interpolating functions display gross oscillations between the interpolation points near the boundary of the region.

The relevance of considering the interpolatory nature of the trial solution, is that the initial values of the a_j's are determined by fitting the initial data and the subsequent growth of the a_j's, according to the Galerkin equation, is such as to keep the trial solution a reasonable interpolatory fit of the exact solution.

The lower Chebyshev polynomials are (Dahlquist et al., 1974)

$$T_0(x) = 1 \ , \quad T_1(x) = x \ , \quad T_2(x) = 2x^2 - 1 \tag{4.20}$$

and the following recurrence relationship is available for higher order polynomials

$$T_{n+1}(x) = 2x \cdot T_n(x) - T_{n-1}(x) \quad . \tag{4.21}$$

Returning to Burgers' equation, the following trial solution is assumed

$$u(x,t) = \sum_{j=0}^{N} a_j(t) \cdot T_j(x) \tag{4.22}$$

where two of the coefficients, a_j, are related to the other unknown coefficients so that the boundary conditions, Eqs. (4.16), are satisfied. Substituting Eq. (4.22) into Eq. (4.15) produces the residual

$$R = \sum_j \dot{a}_j T_j(x) + \sum_j a_j \sum_i a_i T_j(x) \frac{dT_i(x)}{dx} - \frac{1}{Re} \sum_j a_j \frac{d^2T_j(x)}{dx^2} \quad . \tag{4.23}$$

Application of the Galerkin method gives

$$\int_{-1}^{1} R \cdot T_k(x)dx = 0 \quad , \quad k = 0,\ldots N \quad . \tag{4.24}$$

So that only first derivatives of T_j will appear it is necessary to apply Green's theorem to the viscous term. The overall result is

$$\underline{M} \ \dot{\underline{A}} + (\underline{B}+\underline{C})\underline{A} = 0 \quad , \tag{4.25}$$

where an element of $\dot{\underline{A}}$ is $\dot{a}_j$, an element of $\underline{M}$ is given by

$$m_{k,j} = \int_{-1}^{1} T_j(x) \cdot T_k(x)dx \quad , \tag{4.26}$$

an element of $\underline{B}$ is given by

$$b_{k,j} = \sum_i a_i \int_{-1}^{1} \frac{dT_i(x)}{dx} \cdot T_j(x) \cdot T_k(x)dx \tag{4.27}$$

and an element of $\underline{C}$ by

$$c_{k,j} = \frac{1}{Re} \int_{-1}^{1} \frac{dT_j(x)}{dx} \cdot \frac{dT_k(x)}{dx} \ dx \quad . \tag{4.28}$$

The initial profile (4.17) gives the initial values for the a_j's, after solving the system of algebraic equations

$$\underline{M} \ \underline{A} = \underline{D} \tag{4.29}$$

where an element of $\underline{D}$ is given by

$$d_k = \int_{-1}^{0} T_k(x) \cdot 1 \cdot dx \quad . \tag{4.30}$$

The system of equations (4.25) is integrated using a fourth order Runge-Kutta scheme. The scheme uses a variable stepsize in time that is adjusted to be as large as possible without causing instability. After a fixed time, typically $t = 0.92$, the solution is compared with that of the exact solution, Eq. (1.17). Results for $Re = 10$ for various orders of representation are shown in Table 4.3 and Figure 4.1.

For the solutions shown in Figure 4.1 the last two unknown coefficients (a_j's) in the trial solution have been determined as functions of the other unknown coefficients to satisfy the boundary conditions at $x = -1$ and 1. At the Reynolds number considered ($Re = 10$) and at the time of the comparison, the boundary conditions at $x = -1$ and 1 are no longer 1 and 0 precisely. For the solutions presented the values of u at $x = \pm 1$ have been set equal to the exact solution there every five steps. It can be seen that the largest errors in the approximate solutions occur at the point in the flowfield where the exact solution is changing most rapidly. This, of course, occurs where the shock is situated. A consideration of $||u-u_e||$ and Figure 4.1 indicates that the increase in accuracy with increasing order is substantial.

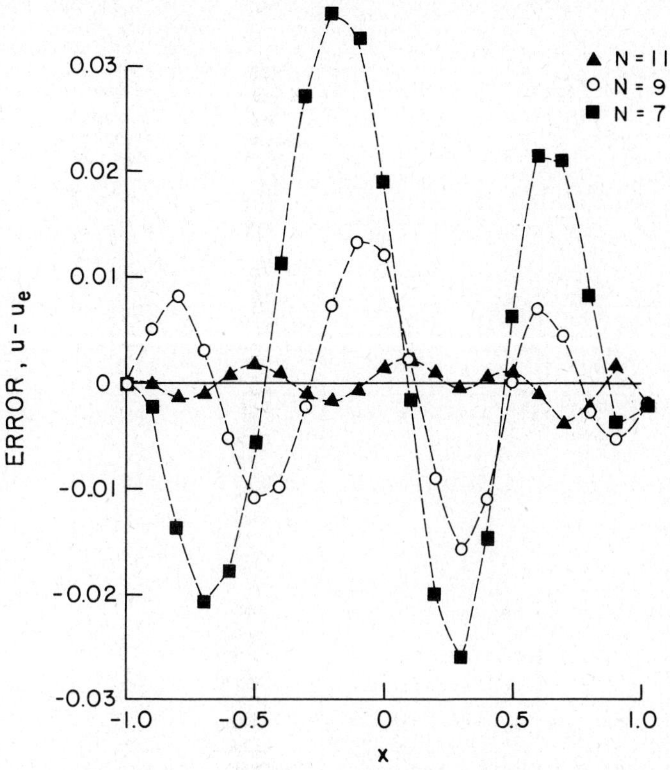

Figure 4.1 Solution of Burgers' equation by traditional Galerkin method at Re = 10

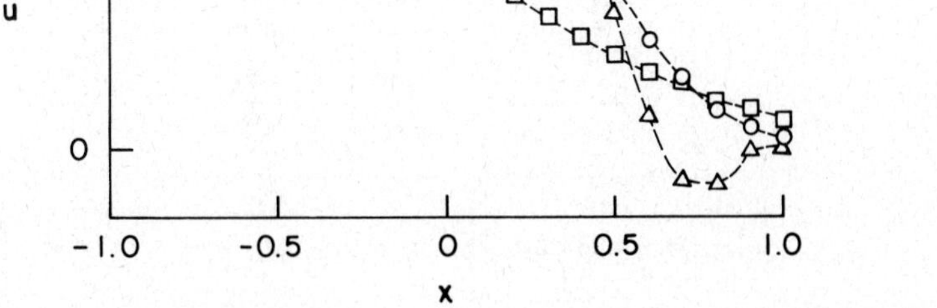

Figure 4.2 Solution of Burger's equation for increasing Reynolds number

For the same order of trial function (N = 9) a comparison of exact and approximate solutions are presented for various Reynolds numbers in Table 4.4 and Figure 4.2. Each solution has been obtained at a different time. The effect of increasing the Reynolds number is to increase the sharpness of the shock and this can be seen, most readily, by comparing the exact solutions. At a high Reynolds number the approximate solution demonstrates its inability to follow the exact profile by developing substantial oscillations in space adjacent to the location of the shock (Figure 4.2). In turn this produces a substantial inaccuracy as is apparent by considering $||u-u_e||_2$. In principle the oscillations could be reduced, and the accuracy increased, by increasing N. However this would increase the evaluation time and the influence on computational efficiency is not obvious.

x	Approximate Solution			Exact Solution
	N = 7	N = 9	N = 11	
-1.0	1.0000	1.0000	1.0000	1.0000
-0.9	0.9978	1.0054	1.0003	1.0000
-0.8	0.9863	1.0082	0.9987	0.9999
-0.7	0.9792	1.0029	0.9987	0.9998
-0.6	0.9818	0.9942	1.0002	0.9996
-0.5	0.9932	0.9881	1.0007	0.9990
-0.4	1.0088	0.9879	0.9985	0.9977
-0.3	1.0216	0.9924	0.9969	0.9948
-0.2	1.0241	0.9960	0.9869	0.9888
-0.1	1.0096	0.9903	0.9759	0.9768
0	0.9730	0.9661	0.9551	0.9539
0.1	0.9115	0.9159	0.9151	0.9131
0.2	0.8251	0.8359	0.8461	0.8451
0.3	0.7169	0.7272	0.7426	0.7428
0.4	0.5927	0.5962	0.6077	0.6074
0.5	0.4606	0.4543	0.4554	0.4545
0.6	0.3306	0.3162	0.3080	0.3092
0.7	0.2134	0.1969	0.1885	0.1926
0.8	0.1192	0.1084	0.1088	0.1110
0.9	0.0562	0.0550	0.0618	0.0602
1.0	0.0288	0.0288	0.0288	0.0309
$\|\|u-u_e\|\|_2$	0.0189	0.0083	0.0016	

Table 4.3 Solution of Burgers' equation by traditional Galerkin method. T = 0.92, Re = 10.

No comment is made on the computational efficiency of the traditional Galerkin method applied to Burgers' equation. A high-order time integrator, e.g., the fourth-order Runge-Kutta scheme, has been used with a step-size small enough to ensure stability. Also the error associated with the temporal integration is negligible compared with the error in spatial representation. This has been done to facilitate comparison with the spectral and Galerkin finite element methods but is not necessarily optimal for computational efficiency.

x	Re = 1.0, t = 0.23		Re = 10, t = 0.92		Re = 100, t = 0.92	
	Approx	Exact	Approx	Exact	Approx	Exact
-1.0	0.9609	0.9599	1.0000	1.0000	1.0000	1.0000
-0.9	0.9443	0.9447	1.0054	1.0000	0.9956	1.0000
-0.8	0.9263	0.9253	1.0082	0.9999	1.0456	1.0000
-0.7	0.9029	0.9008	1.0029	0.9998	1.0672	1.0000
-0.6	0.8731	0.8709	0.9942	0.9996	1.0402	1.0000
-0.5	0.8369	0.8350	0.9881	0.9990	0.9831	1.0000
-0.4	0.7947	0.7931	0.9879	0.9977	0.9303	1.0000
-0.3	0.7469	0.7452	0.9924	0.9948	0.9128	1.0000
-0.2	0.6938	0.6919	0.9960	0.9889	0.9444	1.0000
-0.1	0.6359	0.6340	0.9903	0.9768	1.0159	1.0000
0	0.5742	0.5728	0.9661	0.9539	1.0963	1.0000
0.1	0.5099	0.5097	0.9159	0.9131	1.1411	1.0000
0.2	0.4449	0.4464	0.8359	0.8451	1.1057	1.0000
0.3	0.3814	0.3844	0.7272	0.7428	0.9613	0.9998
0.4	0.3213	0.3254	0.5962	0.6074	0.7099	0.9714
0.5	0.2665	0.2706	0.4543	0.4545	0.3933	0.1861
0.6	0.2179	0.2210	0.3162	0.3092	0.0905	0.0015
0.7	0.1756	0.1772	0.1969	0.1925	-0.1017	0.0000
0.8	0.1388	0.1395	0.1084	0.1110	-0.1154	0.0000
0.9	0.1066	0.1077	0.0550	0.0601	0.0091	0.0000
1.0	0.0793	0.0817	0.0288	0.0309	0.0000	0.0000
$\|u-u_e\|_2$	0.0022		0.0083		0.1049	

Table 4.4 Solution of Burgers' equation by traditional method, N = 9, various Reynolds numbers

4.4 Connection with other methods

The relation of the Galerkin method to the other methods of weighted residuals has already been indicated. Here we are interested in situations where the Galerkin method is equivalent to other methods in the sense of leading to the same approximate equations and solutions.

The first situation is relatively trivial. Suppose a given problem can be solved by the method of separation of variables. If the same problem is solved by the Galerkin method with the trial solution based on the eigenfunctions of the problem (e.g. as obtained by the separation of variables approach) then the same solution will be obtained. Thus, if the trial solution contains the exact solution the Galerkin method will capture it.

There is an equivalence between the Galerkin method and the orthogonal collocation method (Villadsen and Stewart, 1967) if Gauss quadrature is used to evaluate the integrals in equations like (4.26) and certain other conditions are met.

The most important link is between the Galerkin method and variational methods, in particular the Rayleigh-Ritz method. Under many circumstances both methods produce the same approximate equations and solutions. The importance is that if a variational formulation for the problem exists then the approximate solution will be an upper or a lower bound on the exact solution. Clearly the Galerkin solution will also be the same bound on the exact solution. Of course the Galerkin method has wider application in that it can be used when a variational formulation does not exist.

In two dimensions it is assumed that a variational formulation can be written,

$$I(u) = \iint F(x,y,u_x',u_y')dx\,dy \tag{4.31}$$

with appropriate boundary conditions, and that the exact solution, $u = u_e(x,y)$ corresponds to $I(u_e)$ having a minimum. The Rayleigh-Ritz method assumes a trial solution for u e.g.

$$u_a = \sum_{j=1}^{N} a_j \cdot \phi_j(x,y) \quad . \tag{4.32}$$

It is expected that $I(u_a)$ will be close to $I(u_e)$ in some sense if N is large. The coefficients a_j can be obtained by making $I(u_a)$ a minimum, i.e., by setting

$$\frac{\partial I(u_a)}{\partial a_k} = 0 \quad , \quad k = 1,2,\ldots N \quad . \tag{4.33}$$

The equivalence with the Galerkin method can be illustrated by considering the Poisson equation,

$$u_{xx} + u_{yy} = f \quad . \tag{4.34}$$

with boundary condition $u = 0$ on the boundary. The equivalent variational problem is that

$$I(u) \equiv \iint (u_x^2 + u_y^2 + 2fu)dx\,dy \tag{4.35}$$

has a minimum corresponding to the exact solution, with the boundary condition $u = 0$ on the boundary. Substitution of Eq. (4.32) into Eq. (4.35) and evaluation of Eq. (4.33) gives

$$\frac{\partial I(u_a)}{\partial a_k} = 2 \iint \{\frac{\partial \phi_k}{\partial x}[\sum_{j=1}^{N} a_j \cdot \frac{\partial \phi_j}{\partial x}] + \frac{\partial \phi_k}{\partial y}[\sum_{j=1}^{N} a_j \cdot \frac{\partial \phi_j}{\partial y}] + f \cdot \phi_k\}dx\,dy = 0 . \tag{4.36}$$

Application of the Galerkin method to Eq. (4.34) gives

$$\iint \phi_k[u_{xx} + u_{yy} - f]dx\,dy = 0 \quad . \tag{4.37}$$

Application of Green's theorem and noting the homogeneous boundary condition gives

$$- \iint \{ \frac{\partial \phi_k}{\partial x} \frac{\partial u_a}{\partial x} + \frac{\partial \phi_k}{\partial y} \cdot \frac{\partial u_a}{\partial y} + f \cdot \phi_k \} = 0 \quad . \tag{4.38}$$

Substitution of Eq. (4.32) into (4.38) gives Eq. (4.36) demonstrating that the two methods are equivalent for this particular problem.

4.5 Applications

In the past the traditional Galerkin method was widely used but now the method has been substantially replaced by the Galerkin finite element methods and to a lesser extent by spectral methods. However applications of the traditional method still appear in the literature occasionally.

The early work of Galerkin (1917) and Duncan (1937,1938) has already been mentioned. Bickley (1941) applied the Galerkin method to the partial differential equation,

$$\phi_t - a\phi_{xx} = 0 \quad , \tag{4.39}$$

to investigate problems of electrical circuit design. Bickley also presented some comparisons with the methods of collocation and least-squares.

An early application in fluid dynamics is due to Kawaguti (1955) who studied low speed viscous flow past a sphere to determine the critical Reynolds above which a previously steady flow would become unsteady. This was done by perturbing an initially steady solution. The Galerkin method was used to provide the steady solution. The governing equation for steady, axisymmetric viscous, incompressible flow is

$$\psi_r \{ \frac{1}{r^2 \sin^2 \theta} \cdot D^2 \psi \}_\theta - \psi_\theta \{ \frac{1}{r^2 \sin^2 \theta} \cdot D^2 \psi \} - \frac{2}{Re \cdot \sin \theta} \cdot D^4 \psi = 0 \tag{4.40}$$

where

$$D^2 \equiv \frac{\partial^2}{\partial r^2} + \frac{\sin \theta}{r^2} \frac{\partial}{\partial \theta} (\frac{1}{\sin \theta} \cdot \frac{\partial}{\partial \theta}) \quad .$$

In Eq. (4.40) ψ is the stream function, (r,θ,ϕ) are spherical polar coordinates and velocity components (u,v,w) are related to the stream function by

$$u = \frac{1}{r^2 \sin \theta} \frac{\partial \psi}{\partial \theta} \quad , \quad v = \frac{-1}{r \sin \theta} \cdot \frac{\partial \psi}{\partial r} \quad \text{and} \quad w = 0 \quad . \tag{4.41}$$

In Eq. (4.40) Re is the Reynolds number based on the diameter of the sphere. Boundary conditions are required at the surface of the sphere and far from the sphere. At the surface,

$$\psi = \frac{\partial \psi}{\partial r} = 0 \quad \text{at} \quad r = 1 \quad . \tag{4.42}$$

In the farfield,

$$\psi = \frac{r^2}{2} \sin^2 \theta \quad \text{at} \quad r \to \infty \tag{4.43}$$

Kawaguti introduced a trial solution in r and θ. For larger Reynolds numbers this was

$$\psi = \{ \frac{r^2}{2} + \frac{a_1}{r} + \frac{a_2}{r^2} + \frac{a_3}{r^3} + \frac{a_4}{r^4} \} \sin^2 \theta$$
$$+ \{ \frac{b_1}{r} + \frac{b_2}{r^2} + \frac{b_3}{r^3} + \frac{b_4}{r^4} \} \sin^2 \theta \cdot \cos \theta \quad . \tag{4.44}$$

The form of Eq. (4.44) automatically satisfies the farfield boundary condition, Eq. (4.43). To satisfy the surface boundary conditions for all θ, four of the unknowns in Eq. (4.44) can be eliminated. If Eq. (4.44) is substituted into Eq. (4.40) some residual, R, results. By forcing the residual to be zero on the surface of the body, i.e. by demanding that Eq. (4.40) is satisfied exactly, Kawaguti obtains two additional conditions for the four remaining unknown coefficients in Eq. (4.44). The remaining two equations are obtained by imposing the following weighted residual relationships,

$$\int_1^\infty\int_{-1}^1 \frac{1}{r}\{P_o(z) - P_2(z)\}\cdot R\cdot dz\cdot dr = 0 \tag{4.45}$$

and

$$\int_1^\infty\int_{-1}^1 \frac{1}{r^2}\{P_1(z) - P_3(z)\}\cdot R\cdot dz\cdot dr = 0 \quad . \tag{4.46}$$

In Eqs. (4.45) and (4.46), $P_n(z)$ is a Legendre function of order n and $z = \cos\theta$.

The results, that Kawaguti obtained, predicted the drag coefficient quite closely but were less accurate for the pressure distribution and vorticity distribution at the surface.

Snyder and Stewart (1966) have considered an important problem in chemical engineering. I.e. very slow viscous flow through a densely packed bed of spheres. For very slow (creeping) flow the convective (inertial) terms can be ignored, in comparison with the viscous and pressure terms; this renders the equations linear. Imposition of symmetry provides many of the boundary conditions. After applying the traditional Galerkin formulation it was found that not enough equations were available, therefore additional orthogonal relations had to be imposed. The evaluations of the integrals, like Eq. (4.28), was undertaken with Gaussian quadrature. An indication of the accuracy can be obtained from the fact that the predicted flow rate was within 5% of the measured flow rate with 28 unknowns in the expansions for the three velocity components and pressure.

Williams (1975) has used a traditional Galerkin method to investigate the motion of a mooring cable which is given a harmonic displacement at its upper end. Williams considered a two dimensional configuration in which the applied force was in the plane of the cable and tangential to the cable at the upper end. The cable was assumed to be elastic and under the external loading of its own weight and a hydrodynamic drag force.

Williams develops the general equations for the tension and instantaneous displacement by considering the small perturbations from the catenary shape that is the appropriate static orientation of the cable in the absence of hydrodynamic drag. In addition, the equations are simplified by assuming small strain and rotation. If the displacement of the disturbing force is made small compared with the length of the cable and if the motion away from the ends of the cables is considered, the following system of partial differential equations can be obtained,

$$u_{ss} - \frac{N_F}{\overline{E}} u_{tt} = (\tfrac{\ell}{R}\, v)_s \tag{4.47}$$

and

$$\frac{N_F}{\overline{E}} v_{tt} + \frac{C_D\cdot N_D}{2\overline{E}} v_t\cdot|v_t| + (\tfrac{\ell}{R})^2\cdot v = \tfrac{\ell}{R}\, u_s \quad , \tag{4.48}$$

where u and v are the displacements due to the dynamic motion tangential to and normal to the cable. s is the nondimensional distance along the cable. ℓ is the length and R the radius of curvature of the cable. $N_F = Aw^2/g$ where A is the amplitude of the disturbing force and w is the frequency of the disturbing force. $\bar{E} = A/\ell\varepsilon_1$ where $\varepsilon_1 = 4\mu g\ell/\pi d^2 E$ and μ is the mass per unit length, d the cable diameter and E the Young's modulus for the cable material. C_D is the drag coefficient and $N_D = \rho_w A^2 w^2 d/\mu g$ where ρ_w is the density of the fluid surrounding the cable. The parameters in the problem can be rearranged to give

$$N_D = \frac{4}{\pi} \left(\frac{\rho_w}{\rho_s}\right) \cdot \left(\frac{A}{d}\right) \cdot N_F \quad , \tag{4.49}$$

where ρ_s is the density of the cable. A considerable simplification can be made to Eqs. (4.47) and (4.48) if $N_F \to 0$ as N_D remains fixed. This implies a very low frequency disturbing force whose amplitude is large compared with the cable diameter. Then one obtains

$$v = \left(\frac{\ell}{R}\right)^{\frac{1}{2}} \cdot F(t) \tag{4.50}$$

$$T = 0.5\, C_D \cdot N_D \cdot F_t \cdot |F_t| \tag{4.51}$$

and

$$u = F \int_0^s \left(\frac{\ell}{R}\right)^{3/2} dx + 0.5\, \frac{C_D \cdot N_D}{\bar{E}}\, F_t \cdot |F_t| \cdot s \tag{4.52}$$

F(t) is a function to be determined and T is the nondimensional tension associated with the dynamic motion. F is obtained from the boundary condition

$$u(1,t) = \lambda F + 0.5\, \frac{C_D \cdot N_D}{\bar{E}}\, F_t \cdot |F_t| = \cos t \quad , \tag{4.53}$$

where λ is related to the cable configuration. Equation (4.53) is solved by the Galerkin method by assuming a trial solution for F of the form,

$$F = A \sin t + B \cos t \quad . \tag{4.54}$$

Imposing the orthogonality condition (weighted integral of the residual) allows Eq. (4.54) to be written

$$F = \frac{\cos \alpha_o}{\lambda} \cos(t - \alpha_o) \tag{4.55}$$

where

$$\cos \alpha_o = [(1 + 4C_o^2)^{\frac{1}{2}} - 1]^{\frac{1}{2}}/(\sqrt{2}C_o) \tag{4.56}$$

and

$$C_o = 4C_D \cdot N_D/3\pi\lambda^2\bar{E} \quad .$$

The corresponding maximum dynamic tension is given by

$$T_{dyn,max} = 0.5\, C_D \cdot (\cos \alpha_o/\lambda)^2 \cdot \rho_w \ell \cdot d \cdot (Aw)^2 \quad . \tag{4.57}$$

It is also possible to obtain F by direct numerical integration of Eq. (4.53) but this does not indicate the nature of the dependence of F on t, as does Eq. (4.55). Some indication of the accuracy of the Galerkin formulation can be obtained from Table 4.5 (after Williams, 1975).

Durvasula (1971) has applied the traditional Galerkin method to the problem of determining the critical flutter frequency for clamped skew panels. It is assumed that the panel is subjected to a supersonic flow, that structural damping can be ignored, that the panel is thin and that only small deflections, v, are permitted. The governing equation is then

$$D\, \nabla^4 v + N_x v_{xx} + 2N_{xy} v_{xy} + N_y v_{yy} + \rho h\, v_{tt} = 0 \quad , \tag{4.58}$$

where

$$\nabla^4 \equiv \partial^4/\partial x^4 + 2\partial^4/\partial x^2 \partial y^2 + \partial^4/\partial y^4 \quad ,$$

λ	N_D	α_o (deg.)	$F(t)_{max}$	
			Galerkin	Numerical
0.5431	0.2137	8.95	1.819	1.839
	0.4808	18.8	1.743	1.800
	0.8547	29.2	1.607	1.689
0.1664	0.2137	48.5	3.979	4.305
	0.4808	61.5	2.872	3.149
	0.8547	68.4	2.216	2.431

Table 4.5 Galerkin and numerical solutions of Eq. (4.53)

D is the plate rigidity and N_x etc. are in-plane forces. To solve Eq. (4.58) for the critical flutter condition using the Galerkin method it is assumed that the motion is purely harmonic, i.e. that

$$v = \{W \cdot \exp(i\omega t)\}_{real} , \tag{4.59}$$

and that a trial function for W can be written as

$$W = \sum_{r=1}^{M} \sum_{s=1}^{N} C_{rs} \cdot X_r(\xi) \cdot Y_s(\eta) \tag{4.60}$$

where $X_r(\xi)$, $Y_s(\eta)$ are beam characteristic functions (Durvasula, 1969) associated with the normalised mode shapes of a uniform clamped beam. ξ,η are normalised coordinates within the plane of the panel. Application of the Galerkin method produces the eigenvalue problem

$$E_{mnrs} \cdot C_{rs} = k^{*2} C_{rs} \tag{4.61}$$

where k^{*2} is a frequency parameter. One can relate the dynamic pressure, q, to the plate rigidity D by

$$Q = 2qa^3/\beta D\pi^4 \tag{4.62}$$

where $\beta = (M_\infty^2 - 1)^{\frac{1}{2}}$. The parameter Q essentially controls the nature of the oscillatory behaviour. For $Q = 0$, the problem is a free vibration problem. As Q is increased some of the eigenvalues approach each other. The point at which two eigenvalues coalesce will be called Q_{cr}, the critical value of Q. If $Q > Q_{cr}$ the pair of eigenvalues will be complex and the oscillation divergent (unstable). Some typical values of Q_{cr} are given in Table 4.6. a/b is the aspect ratio of the plate, ψ is the skew angle of the plate, λ is the yaw angle of the plate and M,N are defined by Eq. (4.60). It is apparent that a minimum number of terms in Eq. (4.60) is required to establish that Q_{cr} no longer varies. Durvasula indicates that $M = 4$, $N = 4$ is the lowest order representation that will give an accurate estimate of Q_{cr}. Durvasula found that Q_{cr} increases monotonically with for $a/b = 1$ but initially decreases, increasing thereafter, for $a/b = 0.5$.

4.5 Disadvantages of the traditional Galerkin method

For the two model problems given in sections 4.2 and 4.3 high accuracy is indicated for a relatively low order trial solution (and hence small effort). However, for more complex problems, an error of less than five to ten per cent when

a/b	ψ	λ	M	N	Q_{cr}
1	0	0	2	2	6.35
			3	3	10.45
			4	3	8.60
			4	4	8.60
1	0	15	2	2	6.55
			4	3	8.70
			4	4	8.75
1	15	0	2	2	6.95
			4	3	9.14
			4	4	9.11

Table 4.6 Critical dynamic pressure parameter, Q_{cr}, dependence on geometry and order of representation

using the traditional Galerkin method is difficult to achieve without substantially increasing the order of the trial solution.

The ready availability of computers has created a demand for solutions of higher accuracy. However, a direct increase in the order of the analytic representation in the Galerkin method creates an immediate problem. That is, there is little difference, typically, between the nth and $(n-1)^{th}$ weight functions. Thus, the algebraic equations that result from the application of the traditional Galerkin method would be almost linearly dependent for large n.

If Duncan's example is considered with the weight function, x^{99} and x^{100}, the resulting equations are

$$0.000099\ a_1 + 0.009998\ a_2 + 0.019703\ a_3 \ldots 0.494975\ a_{99} + 0.497513\ a_{100} = 0.01 \quad (4.63)$$

and

$$0.000097\ a_1 + 0.009899\ a_2 + 0.019511\ a_3 \ldots 0.492487\ a_{99} + 0.495025\ a_{100} = 0.0099 \quad (4.64)$$

Clearly the resulting matrix of equations will be very ill-conditioned. In addition, the matrix will be dense and consequently will have a high operation count. In particular if any of the matrices have to be factorised, e.g. the matrix M in Eq. (4.25), the operation count will be of $O(N^3)$ where N is the order of the matrix.

A second disadvantage is that imposing the weighted residual integral has required that the limits of integration coincide with coordinate lines. In principle this problem can be avoided for distorted but simple geometries by mapping the distorted physical region into a regular region. However this not only increases the algebraic complexity but one would also expect the accuracy to be degraded since the grid is being severely distorted in the physical plane.

A third disadvantage concerning products of series arises with nonlinear equations, e.g. the convective term in Eq. (4.15). In typical nonlinear problems perhaps three or more series will be multiplied together and, since an iterative solution will probably be required, the coefficients in the series will need to be upgraded at every step of the iteration. Clearly this will impose a severe operational overhead if the number of terms in each series is large.

Improvements, associated with computer applications, have gone in two directions. Firstly the problem of matrix ill-conditioning can be avoided by using orthogonal basis functions so that a matrix like M in Eq. (4.25) has only diagonal entries and is trivial to invert. Spectral methods are of this type and will be considered in Chapter 5.

The second direction is to give the unknowns a direct physical significance by making them nodal unknowns of the problem. The related basis functions then become interpolatory in nature. In addition the basis functions are deliberately restricted to low order and local support. Linear and quadratic polynomials are typical. As a consequence matrices like M, B or C in Section 4.3, although not diagonal, are very sparse and the non-zero entries in the matrix occur close to the main diagonal. If the trial solutions are introduced in discrete subregions or 'finite elements' the treatment of arbitrary domain boundaires is greatly facilitated. The Galerkin finite element method incorporates the above features and it will be described in Chapter 6.

5. SPECTRAL METHODS

Like the traditional Galerkin methods, spectral methods are global in the sense that the trial solution spans the whole region. Thus altering a coefficient in the trial solution immediately influences the whole region. The essential difference from the traditional Galerkin method is in the use of orthogonal functions and the beneficial effect that this has on matrix conditioning. The orthogonal functions are typically Chebyshev or Legendre polynomials, Fourier series or special functions suggested by eigenfunctions of a related problem. In this way the method is related to the method of separation of variables.

The efficiency of the solution (i.e. achieving acceptable accuracy with as few terms in the series as possible) depends on the choice of the orthogonal functions. Orszag (1980) suggests that if the boundary conditions are compatible an eigenfunction expansion should be used. Alternatively if u is infinitely differentiable and satisfies the boundary conditions

$$\partial^{2n}u/\partial x^{2n} = 0 \quad \text{for} \quad x = 0,\pi \quad , \tag{5.1}$$

for all nonnegative n then a Fourier expansion is appropriate. In this case the error will go to zero uniformly faster than any power of N^{-1} as $N\to\infty$. If no such special conditions apply, then orthogonal polynomials are to be preferred. Since spectral methods require more computational effort per degree of freedom, N, than do local schemes, such as the finite difference or finite element methods, it is important that the convergence rate with N is significantly higher than for local methods.

In this chapter the spectral method is described by way of specific examples; firstly it is applied to the Poisson equation and the heat conduction equation. In applying the method to Burgers' equation the problem of handling a nonlinear term arises. Difficulties in handling the nonlinear term are discussed in relation to transform methods, aliasing and the pseudo-spectral method. Subsequently the concept of developing special trial functions for a particular problem is considered and finally some applications are described.

5.1 Poisson equation

A solution to the Poisson equation,

$$u_{xx} + u_{yy} - f = 0 \quad , \tag{5.2}$$

will be sought using a spectral method in the rectangular region $0 \leq x \leq 1$, $0 \leq y \leq 1$. A trial solution based on trigonometric functions is introduced as

$$u_a = u_o(x,y) + \sum_{i=1}^{N} \sum_{j=1}^{N} a_{ij} \sin i\pi x \cdot \sin j\pi y \quad . \tag{5.3}$$

It is assumed that $u_o(x,y)$ satisfies the boundary conditions so that the trial functions can satisfy homogeneous boundary conditions. The trial functions are eigenfunctions of the related equation

$$u_{xx} + u_{yy} = \lambda \cdot u \quad . \tag{5.4}$$

To make the presentation more uniform it is convenient to write Eq. (5.3) as a single expansion

$$u_a = u_o + \sum_{j=1}^{N^2} b_j \cdot \phi_j(x,y) \tag{5.5}$$

where

$$\phi_1 = \sin \pi x \cdot \sin \pi y$$
$$\phi_2 = \sin 2\pi x \cdot \sin \pi y$$
$$\phi_3 = \sin \pi x \cdot \sin 2\pi y$$
etc.

Applying the conventional Galerkin procedure one obtains the result

$$\int_0^1\int_0^1 [u_{o_{xx}} + u_{o_{yy}}]\phi_k \, dx \, dy + \sum_{j=1}^{N^2} b_j \int_0^1\int_0^1 \{\phi_{j_{xx}} + \phi_{j_{yy}}\}\phi_k \, dx \, dy$$
$$- \int_0^1\int_0^1 f \cdot \phi_k \, dx \, dy = 0 \quad , \quad k = 1,\ldots N^2 \tag{5.6}$$

But since the ϕ_j's are orthogonal functions there is only a contribution from the summation, namely when $k = j$. Thus Eq. (5.6) can be written

$$b_k[\int_0^1\int_0^1 \phi_{k_{xx}} + \phi_{k_{yy}}\}\phi_k \, dx \, dy] = \int_0^1\int_0^1 f - (u_{o_{xx}} + u_{o_{yy}})\} \cdot \phi_k \, dx \, dy$$
$$k = 1,\ldots N^2 \quad . \tag{5.7}$$

Then the solution for b_k follows directly when the integrals have been evaluated. This may be contrasted with the situation for the traditional Galerkin method where matrices like $\underline{M}$ in Eq. (4.25) must be inverted before the solution can be obtained. Once the $\bar{b}_k$'s have been obtained from Eq. (5.7) the solution follows directly from Eq. (5.5).

5.2 Heat conduction equation

This is a linear problem to which the spectral method will be applied to reduce the original partial differential equation to an ordinary differential equation. The governing equation is

$$\psi_x - \psi_{yy} = 0 \quad , \tag{5.8}$$

with boundary conditions

$$\psi(0,y) = \sin \pi y + y \quad , \quad \psi(x,0) = 0 \quad , \quad \psi(x,1) = 1 \quad . \tag{5.9}$$

In this problem ψ is a nondimensional velocity of the viscous flow between two parallel plates (Fletcher, 1978a). A trial solution can be introduced that incorporates the boundary condition at $x = 0$

$$\psi_a = \sin \pi y + y + \sum_{j=1}^{N} a_j(x) \cdot \sin j\pi y \quad . \tag{5.10}$$

Application of the spectral method produces the following ordinary differential equation,

$$\frac{da_k}{dx} + (k\pi)^2 \cdot a_k + r_k = 0 \quad , \quad k = 1,\ldots N \quad , \tag{5.11}$$

where

$$r_k = \pi^2 \quad \text{if} \quad k = 1$$
$$= 0 \quad \text{if} \quad k \neq 1 \quad . \tag{5.12}$$

The solution of Eqs. (5.11), (5.12) can be written down directly as

$$a_1 = \exp(-\pi^2 x) - 1$$

and

$$a_k = 0 \quad , \quad k = 2,\ldots N \tag{5.13}$$

so that Eq. (5.10) becomes

$$\psi_a = \sin \pi y \cdot \exp(-\pi^2 x) + y \quad , \tag{5.14}$$

which is, in fact, the exact solution of Eqs. (5.8) and (5.9). This has happened because the trial solution, fortuitously, includes the exact solution. One would expect this to be a rather rare occurrence. A more realistic solution can be achieved if the entry velocity distribution at $x = 0$ is given by

$$\psi(0,y) = 5y - 4y^2 \quad . \tag{5.15}$$

The development is similar to the previous case except that r_k in Eq. (5.12) is replaced by

$$r_k = 16 \int_0^1 \sin k\pi y \cdot dy \quad . \tag{5.16}$$

For this case Eq. (5.11) has been integrated with an Euler scheme and $\Delta x = 0.001$. The results for different N, in Eq. (5.10), are shown in Table 5.1. Also shown in Table 5.1 are the exact solution and a solution ψ_b which only has errors associated with the numerical integration scheme. The exact solution, shown in Table 5.1, can be obtained by the separation of variables technique as

$$\psi = y + 4 \sum_{j=1}^{\infty} \left[\frac{2}{(2j-1)\pi}\right]^3 \sin(2j-1)\pi y \cdot \exp[-\{(2j-1)\pi\}^2 x] \quad . \tag{5.17}$$

The approximate solution, ψ_b, is obtained from

$$\psi_b = y + 4 \sum_{j=1}^{\infty} \left[\frac{2}{(2j-1)\pi}\right]^3 \sin(2j-1)\pi y \cdot X_j(x) \tag{5.18}$$

where $X_j(x)$ is obtained from the numerical solution, using the same Euler scheme as for ψ_a, of

$$\frac{\partial X_j}{\partial x} + \{(2j-1)\pi\}^2 X_j = 0 \quad . \tag{5.19}$$

x (at y = 0.5)	Approximate solution, ψ_a			Exact Solution, ψ	Approximate Solution, ψ_b
	N = 1	N = 3	N = 5		
0	1.5000	1.5000	1.5000	1.5000	1.5000
0.02	1.3143	1.3466	1.3384	1.3408	1.3404
0.04	1.1620	1.1993	1.1911	1.1943	1.1931
0.06	1.0371	1.0752	1.0670	1.0707	1.0690
0.08	0.9347	0.9729	0.9647	0.9686	0.9667
0.10	0.8507	0.8889	0.8807	0.8847	0.8828
0.12	0.7819	0.8201	0.8118	0.8158	0.8139
0.14	0.7254	0.7636	0.7553	0.7592	0.7574
0.16	0.6791	0.7173	0.7090	0.7182	0.7111
0.18	0.6411	0.6793	0.6710	0.6747	0.6731
0.20	0.6099	0.6481	0.6399	0.6434	0.6420

Table 5.1 Solutions of Eq. (5.8)

Examination of Table 5.1 indicates that better agreement is achieved with increasing N. The error between ψ_a and the exact solution, ψ, comes partly from the numerical integration error and partly from the approximation inherent in the Galerkin formulation. The difference between ψ_a and ψ_b is independent of x and arises from the Galerkin formulation; in fact, from the fitting of the initial data (on x = 0) in terms of the trial functions.

5.3 Burgers' equation

The governing equation is

$$u_t + uu_x - \frac{1}{Re} u_{xx} = 0 \quad , \tag{5.20}$$

with boundary conditions

$$u(-1,t) = 1 \quad , \quad u(1,t) = 0$$

and

$$u(x,0) = 1 \quad , \quad x \leqslant 0 \tag{5.21}$$

$$u(x,0) = 0 \quad , \quad x > 0 \quad .$$

This problem is that of a propagating "shock" which was also considered in conjunction with the traditional Galerkin method in Section 4.3. The main complication compared with the heat conduction equation is the appearance of the nonlinear term. However, the method of solution develops in essentially the same way. A trial solution is introduced by

$$u(x,t) = \sum_{j=0}^{N} a_j(t) \cdot P_j(x) \quad , \tag{5.22}$$

where $P_j(x)$ are Legendre functions (Dahlquist et al., 1974) which are orthogonal polynomials over the interval $-1 \leqslant x \leqslant 1$. Following the same procedure as in Section 4.3, one obtains a set of algebraic equations which can be written,

$$\underline{M}\ \dot{\underline{A}} + (\underline{B} + \underline{C})\underline{A} = 0 \tag{5.23}$$

where an element of $\underline{M}$ is given by

$$m_{k,k} = 2/(2k+1)$$
$$m_{j,k} = 0 \quad , \quad \text{if } j \neq k \quad . \tag{5.24}$$

An element of $\underline{B}$ is given by

$$b_{k,j} = \sum_i a_i \int_{-1}^{1} \frac{dP_i(x)}{dx} \cdot P_j(x) \cdot P_k(x) dx \quad , \tag{5.25}$$

and an element of $\underline{C}$ by

$$c_{k,j} = \frac{1}{Re} \int_{-1}^{1} \frac{dP_j(x)}{dx} \cdot \frac{dP_k(x)}{dx}\, dx \quad . \tag{5.26}$$

The initial condition, Eq. (5.21), is converted into initial values of the a_j's by solving the system,

$$\underline{M}\ \underline{A} = \underline{D} \tag{5.27}$$

where an element of $\underline{D}$ is given by

$$d_k = \int_{-1}^{0} P_k(x) \cdot 1 \cdot dx \quad . \tag{5.28}$$

The system of equations, Eq. (5.23), is integrated in time with a fourth-order Runge-Kutta scheme and a variable step-size (temporal). After a certain time, $\Delta t \,\Omega\, 0.80$, the solution is compared with the exact solution. For $Re = 10$, solutions with different orders of the representation, N in Eq. (5.22), are presented in Table 5.2 and Figure 5.1. The last two coefficients in Eq. (5.22) have been determined, in terms of the other coefficients in Eq. (5.22) so that the boundary conditions at $x = \pm 1$ are satisfied.

For the solutions presented in Tables 5.2 and 5.3 the boundary conditions at $x = \pm 1$ have been set equal to the exact solution every five steps of the time integration. The results presented in Table 5.2 indicate that as the order of the representation is increased the accuracy improves rapidly. Comparing the values of $\|u-u_e\|_2$ in both Tables 4.3 and 5.2 it can be seen that the accuracy of the spectral and traditional Galerkin methods is comparable although less computational work is required with the spectral method.

Solutions at different Reynolds numbers are presented in Table 5.3 and Figure 5.2. The main problem in obtaining accurate solutions is related to the difficulty of the approximate solution in representing the exact solution in the region where the shock is situated. In this region u is changing most rapidly. This problem becomes more difficult as the shock becomes steeper, i.e. as the Reynolds number increases. This can be seen most easily by examining the values of the L_2 norm, $\|u-u_e\|_2$. The solutions can be compared with those for the traditional Galerkin method, which are presented in Table 4.4. It can be observed that the solutions are of comparable accuracy, as might be expected.

Gottlieb and Orszag (1977) have formulated a general Chebyshev spectral approach to Burgers' equation and have considered in more detail a linearised Burgers' equation, i.e. the convective term is replaced by $u_0 u_x$ where u_0 is a constant. They found that a Legendre Galerkin spectral approximation was stable and convergent for the linearised Burgers' equation used with boundary conditions

$$u(-1,0) = 0 \quad , \quad u(1,0) = 0 \tag{5.29}$$

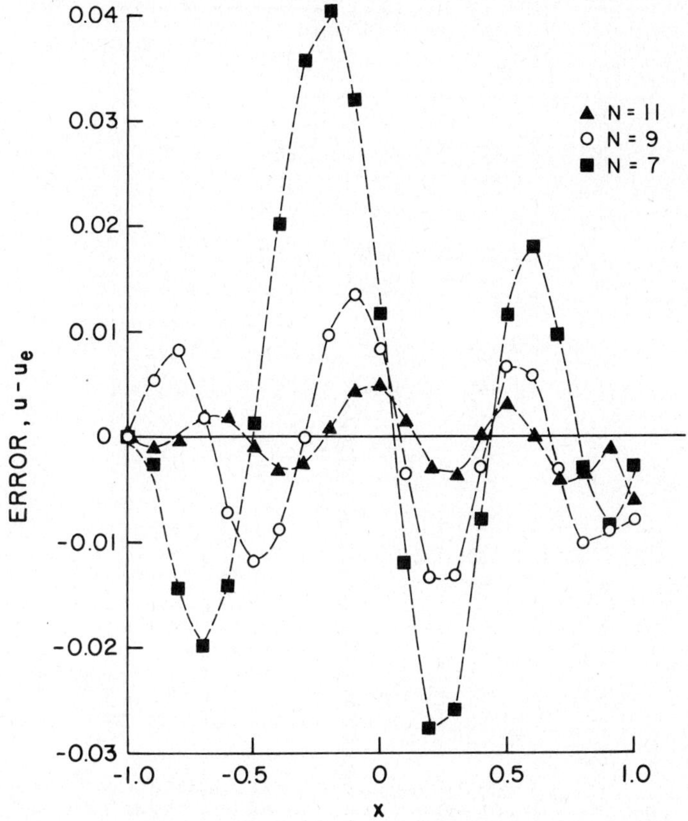

Figure 5.1 Distribution of error for the solution of Burgers' equation, Re = 10

and initial conditions

$$u(x,0) = g(x) \quad . \tag{5.30}$$

However for a Chebyshev Galerkin spectral formulation the solution was not convergent unless $N > N_{crit}$ where N_{crit} depended on the value of $Re \cdot u_0$. For $Re \cdot u_0 = 100$, $N_{crit} = 15$, for $Re \cdot u_0 = 2500$, $N_{crit} = 101$. It turns out that for the boundary conditions considered a narrow boundary layer develops at the outflow $(x = 1)$ for large $Re \cdot u_0$ and satisfying the outflow boundary condition is the cause of the instability.

5.4 Treatment of nonlinear terms

It can be seen that in obtaining the solution to Burgers' equation the evaluation of the nonlinear term requires an N^3 operation at every timestep, where N is the

x	Approximate Solution			Exact Solution
	N = 7	N = 9	N = 11	
-1.0	1.0000	1.0000	1.0000	1.0000
-0.9	0.9974	1.0054	0.9986	1.0000
-0.8	0.9856	1.0082	0.9997	0.9999
-0.7	0.9799	1.0020	1.0017	0.9998
-0.6	0.9855	0.9926	1.0012	0.9996
-0.5	1.0000	0.9870	0.9978	0.9988
-0.4	1.0176	0.9882	0.9937	0.9972
-0.3	1.0290	0.9934	0.9904	0.9934
-0.2	1.0258	0.9947	0.9857	0.9852
-0.1	1.0007	0.9816	0.9730	0.9687
0	0.9491	0.9442	0.9422	0.9372
0.1	0.8696	0.8758	0.8829	0.8816
0.2	0.7643	0.7748	0.7887	0.7920
0.3	0.6390	0.6463	0.6611	0.6648
0.4	0.5025	0.5011	0.5109	0.5104
0.5	0.3657	0.3546	0.3573	0.3542
0.6	0.2405	0.2233	0.2224	0.2225
0.7	0.1377	0.1213	0.1236	0.1280
0.8	0.0653	0.0561	0.0646	0.0684
0.9	0.0259	0.0241	0.0330	0.0344
1.0	0.0136	0.0078	0.0102	0.0164
$\|u-u_e\|_2$	0.0200	0.0086	0.0031	

Table 5.2 Solution of Burgers' equation by spectral method. T = 0.80 and Re = 10

number of unknowns in the trial solution. If N is relatively small this is not a major problem but to use a spectral method in, say, global weather forecasting, a large value of N is required. The treatment of the nonlinear terms, as described above, then becomes prohibitively expensive.

This problem has been overcome by making use of the fast Fourier transform (if the trial solution is appropriate) to evaluate the nonlinear terms in physical space. Typically the fast Fourier transform has an operation count of N log N. However the need to evaluate the nonlinear terms locally (in physical space) suggests replacing the Galerkin technique with a collocation (pseudospectral) technique. The pseudospectral technique is approximately twice as efficient as the Galerkin spectral technique in handling nonlinear terms.

The philosophy of the transform technique is expressed by Orszag (1980) as follows: "Transform freely between the physical (x_i) and the spectral (a_j)

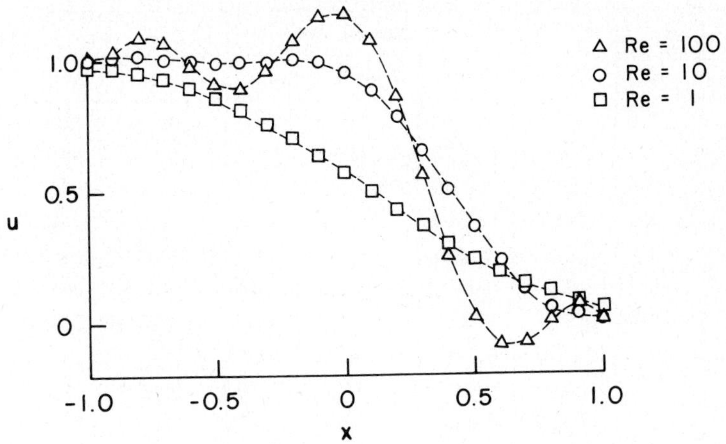

Figure 5.2 Solution of Burgers' equation for increasing Reynolds number

representations, evaluating each term in whatever representation that term is most accurately, and simply given."

Thus in applying the pseudospectral transform technique to Burgers' equation the derivative terms are evaluated via the spectral representation, e.g.

$$u_{xx}(x_i,t) = \Sigma\, a_j(t)\; d^2P/dx^2(x_i) \quad , \tag{5.31}$$

whereas u is upgraded by enforcing the partial differential equation exactly at specific locations, x_i (collocation),

$$\dot{u}(x_i,t) = -\{uu_x - \frac{1}{Re}\, u_{xx}\}_{x_i} \tag{5.32}$$

Finally the fast Fourier transform is used to recover a_j's from the x_i's. Although part of the computational efficiency of using the transform technique comes from the fast Fourier transform itself a much larger contribution comes from being able to separate a multi-dimensional transformation into a succession of one-dimensional transforms. Orszag (1980) cites the example of a three dimensional computation of the incompressible Navier-Stokes equations, which would require about 5×10^5 secs per time step if the spectral method were applied conventionally. However by performing a sequence of one-dimensional fast Fourier transforms the computation time is about 20 secs per time step. However of the improvement, a factor of roughly 10^4 comes from being able to apply the sequence of one-dimensional transforms.

The products that occur in nonlinear terms cause higher frequencies to appear. This is the phenomenon of aliasing. The Galerkin spectral method handles these terms correctly. But if the number of collocation points is equal to N (to obtain the efficiency advantage) the aliasing problem introduces an error. Traditionally this error has been thought to contribute to nonlinear instability associated with integrating hyperbolic systems of equations to large time, e.g. in weather forecasting. However Orszag (1972) argues that most physical problems have sufficient damping, e.g. viscous effects, to prevent aliasing instability. Although this is probably true, more recently Schamel and Elsasser (1976) have applied both the Galerkin spectral and the pseudospectral methods to Burgers'

	Re = 1.0, t = 0.19		Re = 10, t = 0.80		Re = 100, t = 0.66	
x	Approx.	Exact	Approx.	Exact	Approx.	Exact
-1.0	0.9718	0.9706	1.0000	1.0000	1.0000	1.0000
-0.9	0.9593	0.9574	1.0054	1.0000	1.0052	1.0000
-0.8	0.9426	0.9395	1.0082	0.9999	1.0830	1.0000
-0.7	0.9198	0.9162	1.0020	0.9998	1.0646	1.0000
-0.6	0.8898	0.8865	0.9926	0.9996	0.9760	1.0000
-0.5	0.8521	0.8499	0.9870	0.9988	0.8968	1.0000
-0.4	0.8068	0.8058	0.9882	0.9971	0.8853	1.0000
-0.3	0.7544	0.7545	0.9934	0.9933	0.9515	1.0000
-0.2	0.6957	0.6965	0.9947	0.9849	1.0603	1.0000
-0.1	0.6321	0.6328	0.9816	0.9681	1.1509	1.0000
0	0.5648	0.5651	0.9442	0.9360	1.1634	1.0000
0.1	0.4956	0.4954	0.8758	0.8793	1.0609	1.0000
0.2	0.4262	0.4258	0.7748	0.7882	0.8436	0.9986
0.3	0.3585	0.3585	0.6463	0.6514	0.5508	0.8218
0.4	0.2943	0.2955	0.5011	0.5040	0.2503	0.0297
0.5	0.2355	0.2383	0.3546	0.3479	0.0165	0.0002
0.6	0.1834	0.1879	0.2233	0.2174	-0.0967	0.0000
0.7	0.1393	0.1449	0.1213	0.1244	-0.0813	0.0000
0.8	0.1037	0.1092	0.0561	0.0662	0.0095	0.0000
0.9	0.0762	0.0804	0.0241	0.0331	0.0718	0.0000
1.0	0.0555	0.0578	0.0078	0.0157	0.0000	0.0000
$\|u-u_e\|_2$	0.0029		0.0086		0.1175	

Table 5.3 Solution of Burgers' equation by spectral method, N = 9, various Reynolds number.

equation and show that although the Galerkin method produces solutions after large time which coincide with the exact solution, the aliasing associated with the pseudospectral method introduces an error which counteracts the nonlinear steepening effect, i.e. an artificial dissipation.

5.5 Orthonormal methods of integral relations

In this section it is shown how the method of integral relations, which is a member of the class of methods of weighted residuals, can be upgraded to a spectral method, the orthonormal method of integral relations. In so doing the high accuracy of spectral methods is combined with the underlying economy of the method of integral relations.

The method of integral relations was applied to the boundary layer equations

$$u_x + v_y = 0 \tag{5.33}$$

$$uu_x + vu_y = u_e \cdot u_{e_x} + \nu u_{yy} \quad , \tag{5.34}$$

by Dorodnitsyn (1960). The method consists of eliminating v from Eqs. (5.33) and (5.34) by forming the product, $f_k \times$ Eq. (5.33) + $df_k/du \times$ Eq. (5.34) and integrating the resulting equation across the boundary layer with respect to y. Subsequently the independent variables are changed from (x,y) to (x,u). The result is an ordinary differential equation,

$$\frac{d}{dx}\int_0^1 \theta \cdot u \cdot f_k \, du = \frac{u_{e_x}}{u_e}\int_0^1 \frac{df_k}{du} \cdot \theta \cdot (1-u^2)du - [\frac{df_k}{du} \cdot \tau]_{wall} - \int_0^1 \frac{d^2 f_k}{du^2} \cdot \tau \cdot du \; . \tag{5.35}$$

In Eq. (5.35), $\theta = 1/\tau$ and τ is the shear stress, $f_k(u)$ is a weight function to be prescribed and u_e is the known velocity at the outer edge of the boundary layer. The method proceeds by prescribing the dependence of θ on u as

$$\theta = \frac{1}{1-u}\,(a_0 + \sum_j a_j \, u^j) \quad . \tag{5.36}$$

The factor (1-u) is required so that τ goes to zero at the outer edge of the boundary layer ($u = 1$). For the weight function, f_k, Dorodnitsyn used

$$f_k(u) = (1-u)^k \quad . \tag{5.37}$$

Substitution of Eqs. (5.36) and (5.37) into Eq. (5.35) and evaluation of the integrals produces a system of ordinary differential equations for the a_j's that can be written,

$$\underline{B} \;\; [da_j/dx] = \underline{C} \tag{5.38}$$

where $\underline{B}$ and $\underline{C}$ are matrices that must be re-evaluated at every step since they depend on the solution. It is apparent, from the form of f_k in Eq. (5.37), that for large k the difference between f_k and f_{k-1} is not very great and the resulting equation in the system (5.38) will be almost linearly dependent. Consequently the equations $\underline{B}$ and $\underline{C}$ become progressively more ill-conditioned and in the case of $\underline{B}$, difficult to invert accurately.

This problem is overcome in the orthonormal method of integral relations by generating a set of orthonormal functions, g_j, from the Dorodnitsyn weighting functions, f_k. The functions g_j replace f_k in Eq. (5.35). In addition, the analytic representation for θ, Eq. (5.36), is replaced by an analytical representation in terms of the g_j's. Thus a Galerkin spectral formulation is obtained in which matrix $\underline{B}$ is diagonal, and hence trivial to invert, and matrix $\underline{C}$ is well-conditions. This interpretation of the orthonormal method of integral relations as a Galerkin spectral method is described in more detail in Fletcher (1978b).

The orthonormal functions, g_j, are generated as follows. Firstly, a general dependence on f_k is introduced by

$$g_j = \sum_{k=1}^{j} e_{kj} \cdot f_k \quad , \tag{5.39}$$

where the coefficients e_{kj} are evaluated using the Gram-Schmidt orthonormalisation process (e.g. Isaacson and Keller, 1966). To obtain a diagonal form for B, the functions g_j are made orthonormal with respect to a given weighting function, w(u), according to the following inner product

$$(g_i, g_j) = \int_0^1 g_i(u) \cdot g_j(u) \cdot w(u) du \tag{5.40}$$

and

$$(g_i, g_j) = 0 \quad \text{if} \quad i \neq j$$
$$= 1 \quad \text{if} \quad i = j \quad .$$

The following analytic representation for θ is assumed (to replace Eq. (5.36)),

$$\theta = \frac{1}{1-u}\,(b_o + \sum_j b_j \cdot g_j) \quad . \tag{5.41}$$

The non-orthogonal leading term, b_o, is retained so that θ will behave correctly at $u = 1$. Substitution of Eq. (5.41) into Eq. (5.35) and replacement of f_k with g_k gives

$$\frac{d}{dx}\int_0^1 (b_o + \sum_j b_j g_j) g_k \cdot \frac{u}{1-u} \cdot du = C_k' \quad , \quad k = 1,\ldots N \tag{5.42}$$

where

$$C_k' = \frac{u_{e_x}}{u_e}\int_0^1 \theta\,\frac{dg_k}{du} \cdot (1-u^2)du - [\frac{dg_k}{du} \cdot \tau]_{wall} - \int_0^1 \frac{d^2 g_k}{du^2} \cdot \tau\, du \quad . \tag{5.43}$$

Comparing Eqs. (5.40) and (5.43), it is clear that if B is to be diagonal then

$$w(u) = u/(1-u) \quad . \tag{5.44}$$

Thus Eq. (5.42) can be written

$$\frac{db_o}{dx} \cdot V_k + \frac{db_k}{dx} = C_k' \quad , \quad k = 1,\ldots N\text{-}1 \tag{5.45}$$

where

$$V_k = \int_0^1 g_k \cdot w(u)du \tag{5.46}$$

when $k = N$,

$$\frac{db_o}{dx} = C_N'/V_N \quad . \tag{5.47}$$

Thus Eq. (5.45) can be replaced by

$$\frac{db_k}{dx} = C_k' - C_N' \cdot V_k/V_N \quad . \tag{5.48}$$

From the form of Eq. (5.46), P(k) can be evaluated once and for all; however C'(k) must be re-evaluated at every step. Results are presented by Fletcher and Holt (1975) indicating that the orthonormal method of integral relations is considerably more accurate and much more economical, for the same value of N, than the original method of integral relations. Fletcher and Holt (1976) have used the orthonormal method of integral relations to study the boundary layer development on inclined cones in supersonic flow.

5.6 Applications

Two areas have been especially fruitful for spectral methods. Firstly global atmospheric modelling has a natural spherical polar coordinate system and a vorticity formulation can be solved economically by introducing spherical harmonics as trial functions. Platzman (1960,1962) has presented such a formulation. Bourke (1972,1977) has taken a Galerkin spectral formulation based on expansion in spherical harmonics, introduced a transform technique to handle the nonlinear terms and a time splitting technique to efficiently integrate the governing equation, and obtained solutions for the various mass, momentum and energy balance equations which control the large scale motion in the atmosphere. In Bourke et al. (1977) the use of the method for numerical weather prediction is described.

A second significant area has been the study of incompressible, three dimensional turbulent flows. The underlying motivation is to better understand turbulent motion and to build and test accurate turbulence closure models. Orszag and Kruskal (1968) considered unsteady, incompressible flow in a tube with periodic boundary conditions. An investigation of homogeneous turbulence was made by Orszag and Patterson (1972). Turbulent channel flow for Reynolds numbers up to 5000 has been directly simulated using a pseudospectral method by Patera and Orszag (1981).

Most of the spectral methods associated with turbulence studies have utilised a spectral or pseudospectral representation in space but a simple finite difference scheme to integrate in time. However it is also possible to introduce a spectral representation in time. This has been done by Morchoisne (1979) who obtained solutions, using a pseudospectral method, for the shock development as controlled by Burgers' equation. Results for the driven cavity problem, with Reynolds numbers up to 400, were also presented. The flow behavior for the driven cavity problem is governed by the incompressible Navier-Stokes equations. Results produced by the spectral methods are shown to be similar to those obtained with a finite difference and a finite element method.

For two test problems, passive scalar convection and the Taylor-Green vortex-decay, Orszag (1971) has compared the performance of spectral methods and finite difference methods. To achieve comparable accuracy Orszag concludes that a finite difference method requires 2^p times as many grid points, where p is the number of spatial dimensions. Haidvogel et al. (1980) have considered the inviscid barotropic vorticity equation with a variety of boundary conditions to compare the pseudospectral method with a second order finite difference scheme and a linear finite element scheme. They conclude that for mildly nonlinear problems the pseudospectral method is fifteen times more accurate than the finite difference method and almost four times more accurate than the finite element method.

The spectral method has been used to analyse the Orr-Sommerfeld equations (Orszag, 1971) and also to look at boundary layer stability (Murdock, 1977). A Galerkin spectral method based on orthonormal eigenfunctions has been used to obtain solutions for the incompressible laminar boundary flow equations by Hsu (1975).

A mixed finite difference-Legendre spectral method has been developed by Davies and Owen (1979). They considered the hydrodynamic equations that control the motion of the sea under the influence of an applied surface stress, e.g. due to the wind. The governing equations are

$$\frac{\partial}{\partial x}\int_0^h u\,dz + \frac{\partial}{\partial y}\int_0^h v\,dz + \frac{\partial\xi}{\partial t} = 0 \tag{5.49}$$

$$\frac{\partial u}{\partial t} - \gamma\cdot v = -g\frac{\partial\xi}{\partial x} + \frac{\partial}{\partial z}\left(N\frac{\partial u}{\partial z}\right) \tag{5.50}$$

$$\frac{\partial v}{\partial t} + \gamma\cdot u = -g\frac{\partial\xi}{\partial y} + \frac{\partial}{\partial z}\left(N\frac{\partial v}{\partial z}\right) \quad .$$

The unknowns are the two velocity components u and v and the elevation ξ. N is an eddy viscosity and γ is a geostrophic coefficient. It was found that high accuracy could be achieved with only six Legendre polynomials in the vertical direction.

6. FINITE ELEMENT METHODS

It may be recalled that the traditional Galerkin and spectral methods are global in character. This arises from the nature of the trial solution, e.g.

$$u(\bar{x},t) = \sum_i a_i(t) \cdot \phi_i(\bar{x}) \quad . \tag{6.1}$$

A small perturbation in $a_i(t)$ would alter the solution, $u(\bar{x},t)$, throughout the spatial domain.

This situation may be contrasted with that for the finite element method where the trial solution is expressed as a function of the local nodal unknowns, e.g.

$$u(\bar{x},t) = \sum_i \bar{u}_i(t) \cdot \phi_i(\bar{x}) \quad . \tag{6.2}$$

The trial functions, ϕ_i, are interpolatory in nature and of limited local extent. Thus it may be expected that global matrices arising from the finite element formulation will be sparse with most of the non-zero entries clustered close to the main diagonal. An advantage of the nodal unknown formulation is that the final solution is solved for directly, i.e. evaluations of equations like (6.1) are not required.

In principle it is possible to define interpolating functions in time as well, so that algebraic equations are obtained for the nodal values, $\bar{u}_i^n$. However for time dependent problems it is more usual to adopt the semi-discrete procedure indicated by Eq. (6.2); so that application of the Galerkin formulation produces a system of ordinary differential equations in time. This is solved after introducing finite difference expressions for the time derivatives.

The interpolating or shape functions, $\phi_i(x)$, are defined on elements. It follows that the equation 'centered' on the ith node would have contributions from a number of neighbouring elements. Typically, in the interior of the domain, there would be contributions from four rectangular elements or six triangular elements.

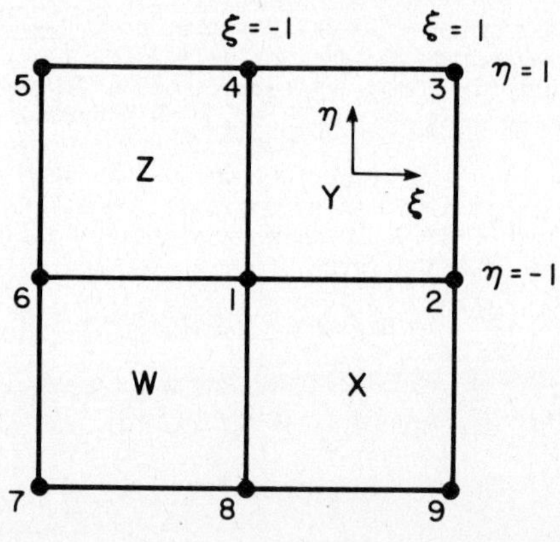

Figure 6.1 Typical rectangular element geometry

Thus an equation associated with node 1 in Figure 6.1 would have contributions from elements W,X,Y and Z. The interpolating functions, ϕ_i, are normally low-order polynomials defined in local coordinates (ξ,η). However, the use of any other convenient function, e.g. trigonometric or exponential functions, is possible and those have been used where prior knowledge of the expected solution indicated the appropriateness of a particular function.

For the linear rectangular element, Y, shown in Fig. 6.1, the form of $\phi_i(\xi,\eta)$ is

$$\phi_i(\xi,\eta) = 0.25(1+\xi_i\cdot\xi)(1+\eta_i\cdot\eta) \quad . \tag{6.3}$$

It can be seen from combining Eqs. (6.2) and (6.3) that u is interpolated bilinearly over element Y. By considering the same local coordinate system for each of the other elements, it is apparent that u is bilinearly interpolated in all four elements and that it is continuous in passing from one element to another. However first derivatives of u are discontinuous in passing from one element to another and higher derivatives are not defined. The use of a higher order (quadratic) polynomial is illustrated in section 6.3. By a special choice of the interpolating functions, i.e., hermite interpolation, it is possible to represent u in such a way that permits continuity of u and certain derivatives of u in passing from one element to another.

Although the elements shown in Fig. 6.1 are square it is possible to relate elements that are square in the (ξ,η) plane to elements that are distorted in the physical plane. This can be done by using an isoparametric formulation. A description of this is given in section 6.3 in relation to the flow around a circular cylinder. For this case the elements are defined on a polar grid but the governing equations are considered in Cartesian form.

By interpolating at the element level and by making use of the isoparametric formulation it is possible to closely represent complicated domain shapes using the finite element method. This is an important feature of the method. However the accuracy and possibly the stability of the method deteriorate if the elements are severely distorted.

The formulation of the Galerkin method, in a finite element context, follows that of the traditional Galerkin method. This will be illustrated by considering specific examples: the Poisson equation in section 6.1, Burgers' equation in section 6.2 and inviscid, incompressible flow in section 6.3. For relatively simple equations it is possible to obtain some fairly precise error estimates for the Galerkin finite element method; some of these are indicated in section 6.4. This chapter closes with some of the more interesting applications.

6.1 Poisson equation

A solution, via the Galerkin finite element formulation, is sought for the Poisson equation

$$u_{xx} + u_{yy} - f = 0 \quad , \tag{6.4}$$

in a domain D with appropriate boundary conditions given on the boundary S. Introducing a trial solution, u_a, by

$$u_a = \sum_{j=1}^{N} \phi_j(x,y)\bar{u}_{aj} \tag{6.5}$$

and applying the Galerkin formulation gives the result

$$\iint_D [\phi_k\{u_{a_{xx}} + u_{a_{yy}} - f\}]dx\, dy = 0 \quad , \quad k = 1,\ldots N \quad . \tag{6.6}$$

In Eq. (6.6) second derivatives appear which cannot be evaluated with a linear element. However it is possible to apply Green's theorem to the Laplacian in (6.6). Then one obtains

$$\int_S \phi_k\{u_{a_x} \cdot \ell_x + u_{a_y} \cdot \ell_y\}ds - \iint_D \{\phi_{k_x} \cdot u_{a_x} + \phi_{k_y} \cdot u_{a_y}\}dx\, dy$$

$$- \iint_D \phi_k \cdot f\, dx\, dy = 0 \quad , \quad k = 1,\ldots N \tag{6.7}$$

where ℓ_x, ℓ_y are the direction cosines. Substituting Eq. (6.5) into Eq. (6.7) gives the system of equations

$$\underline{C}\ \underline{U}_a = \underline{B} \quad , \tag{6.8}$$

where $\bar{u}_{a_j}$ is an element of $\underline{U}_a$. An element of $\underline{C}$ is given by

$$c_{kj} = \int_S N_k \cdot \{\frac{\partial N_j}{\partial x} \cdot \ell_x + \frac{\partial N_j}{\partial y} \cdot \ell_y\}ds - \iint_D \{\frac{\partial N_k}{\partial x} \cdot \frac{\partial N_j}{\partial x} + \frac{\partial N_k}{\partial y} \cdot \frac{\partial N_j}{\partial y}\}dx\, dy \quad , \tag{6.9}$$

and an element of $\underline{B}$ by

$$b_j = \iint_D N_k\ f\, dx\, dy \quad . \tag{6.10}$$

The form of the weighting function, N_k, in Eq. (6.9) and (6.10), is such that only the elements adjacent to the kth node (W,X,Y,Z in Fig. 6.1), contribute to the integrals over the domain D. Thus the matrix $\underline{C}$ will be relatively sparse. For linear elements the maximum number of non-zero entries in a row is nine (see Fig. 6.1). If one considers the contribution from each element in turn it is clear that contributions to $\int_S ds$ from adjacent elements cancel unless k lies on the boundary of D. This is more obvious from Eq. (6.7).

The form of Eq. (6.8) is typical of elliptic problems. For this particular problem $\underline{C}$ will be positive definite and this can be taken advantage of to solve the linear system. In practice some of the nodal values, $\bar{u}_{aj}$, will be known from the boundary conditions. Such terms and related coefficients, c_{kj}, are carried over to the right-hand side of Eq. (6.8) and the reduced system solved.

6.2 Burgers' equation

This introduces the complication of a nonlinear term and illustrates a semi-discrete Galerkin formulation. Since this equation and particular boundary condition have an exact solution it forms a convenient means of comparing solutions with those from the traditional Galerkin and spectral methods.

The same problem of the propagating 'shock,' which was solved in section 4.3 (Eqs. (4.15) to (4.17)), will be considered here. A trial solution, Eq. (6.2), is introduced into the governing equations and following the Galerkin procedure (i.e., Eqs. (4.23) and (4.24)), a set of ordinary differential equations is obtained,

$$\underline{M}\ \dot{\underline{U}}_a + (\underline{B} + \underline{C})\underline{U}_a = 0 \quad . \tag{6.11}$$

The similarity to Eq. (4.25) may be noted. An element of $\underline{U}_a$ is a nodal unknown $\bar{u}_{aj}(t)$. An element of the mass matrix, $\underline{M}$, is

$$m_{kj} = \int_{-1}^{1} \phi_k(x) \cdot \phi_j(x)dx \quad . \tag{6.12}$$

Since only one space dimension is present, only two elements will contribute to Eq. (6.12) at interior nodes. An element of $\underline{B}$ is

$$b_{k_j} = \sum_i \bar{u}_{a_i} \int_{-1}^{1} \phi_k(x)\ \ \phi_i(x)\ \ \frac{d\phi_j(x)}{dx}\ dx \quad . \tag{6.13}$$

It can be seen that $\underline{B}$ is a function of the solution. After applying Green's theorem an element of $\underline{C}$ is given by

$$c_{kj} = \frac{1}{Re}\int_{-1}^{1} \frac{d\phi_k}{dx}\cdot\frac{d\phi_j}{dx}\,dx \quad . \tag{6.14}$$

Since u at $x=\pm 1$ is given there is no contribution from the boundaries.

The structure of Eq. (6.11) can be readily appreciated for the special case of linear elements on a uniform grid. Then Eq. (6.11) becomes

$$\frac{1}{6}\dot{\bar{u}}_{a_{k-1}} + \frac{2}{3}\dot{\bar{u}}_{a_k} + \frac{1}{6}\dot{\bar{u}}_{a_{k+1}} + \frac{(\bar{u}_{a_{k-1}} + \bar{u}_{a_k} + \bar{u}_{a_{k+1}})}{3}\,\frac{(\bar{u}_{a_{k+1}} - \bar{u}_{a_{k-1}})}{2\Delta x}$$

$$- \frac{1}{Re}\,\frac{(\bar{u}_{a_{k-1}} - 2\bar{u}_{a_k} + \bar{u}_{a_{k+1}})}{\Delta^2 x} = 0 \quad . \tag{6.15}$$

This form of Eq. (6.15) is very similar to that of a centered finite difference scheme. The only difference being the distribution of u and $\dot{u}$ over adjacent nodes in the first two terms.

To obtain solutions to the system of equations like (6.15) a fourth-order Runge-Kutta scheme has been utilised to integrate the equations in time from $t=0$ up to $t=0.80$, typically. The integration scheme is the same as for the traditional Galerkin and spectral methods. At the final time the solution is compared with the exact solution. Some typical results for $Re=10$ are shown in Table 6.1 and Figure 6.2. Results for a three point finite difference scheme,

$$\dot{u}_k + u_k\,\frac{(u_{k+1} - u_{k-1})}{2\Delta x} - \frac{1}{Re}\,\frac{(u_{k-1} - 2u_k + u_{k+1})}{\Delta 2x} = 0 \tag{6.16}$$

are shown in the second column for comparison. However, it should be noted that, due to the step-size control on the Runge-Kutta scheme, the solutions correspond to different times. Thus $\| u_a - u_e \|_2$ gives a better means of comparing the various results; this has been computed from the exact solution corresponding to the approximate solution at the specified time rather than from the exact solution shown in the final column which is only applicable to $T=0.50$.

It is apparent that the linear element scheme, Eq. (6.15), is producing results of considerably greater accuracy than the comparable finite difference scheme, Eq. (6.16). Also an increase in the number of nodal points for the finite element scheme causes a substantial increase in the accuracy as does an increase in the order of the representation.

The results for a linear element, $N=11$, can be compared with the results with $N=11$ for the traditional Galerkin method, Table 4.3. It can be seen that for the same number of unknowns, nine in both situations, the traditional Galerkin method gives answers of higher accuracy. Since the trial functions are of a higher order for the traditional Galerkin formulation this higher accuracy is to be expected. However if the comparison were made on the basis of computational efficiency, i.e. the achieved accuracy per unit of execution time, one would expect the finite element method to be more efficient.

For one finite element scheme, a linear element with $N=21$, comparative results for increasing Reynolds number are shown in Table 6.2 and Figure 6.3. As with the traditional Galerkin method and the spectral method, the difficulty of following the rapid change in u with increasing Reynolds number produces a corresponding reduction in accuracy. However with the finite element method it is relatively straightforward to introduce a variable grid for which more points are grouped where the solution is changing most rapidly. Thus higher accuracy is possible with little or no additional computation. Such a situation is not available,

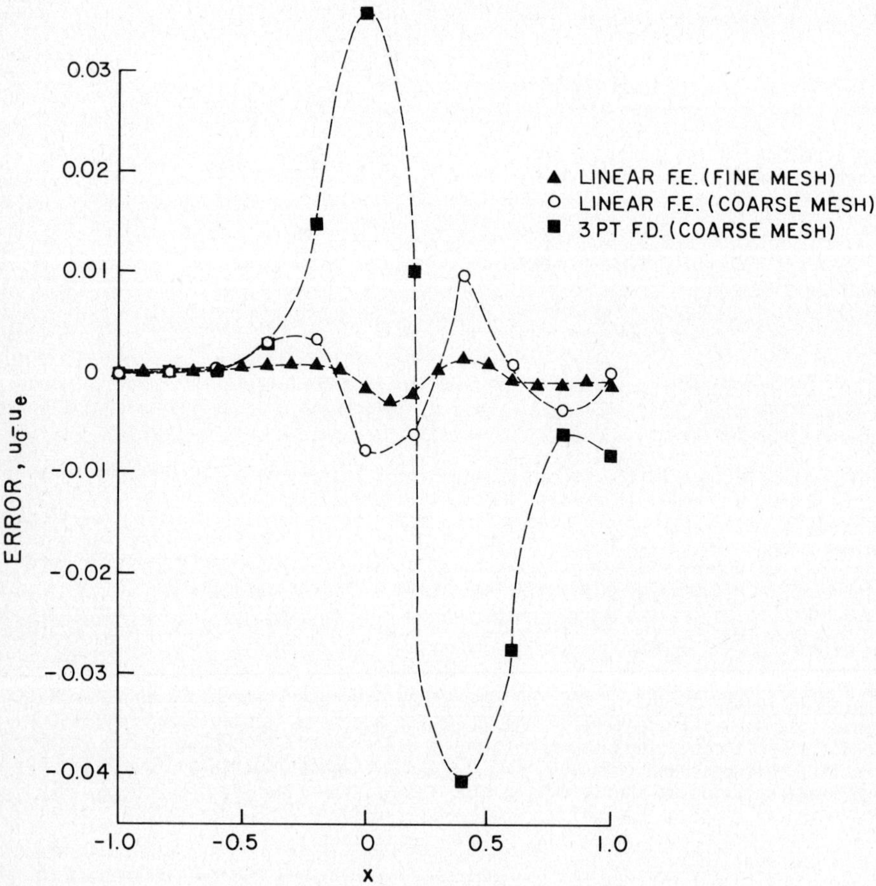

Figure 6.2 Distribution of error for the solution of Burgers' equation, Re = 10

conveniently, to the traditional Galerkin method.

6.3 Inviscid, incompressible flow

This example is considered to illustrate the use of the isoparametric formulation and the need to introduce numerical integration. This example also demonstrates the ability of achieving significantly higher accuracy by a small modification to the Galerkin method. As with Burgers' equation this problem has an exact solution.

The flow geometry for the flow past a circular cylinder is shown in Figure 6.4. Because of the symmetry only a quarter-plane need be considered. Also shown in Figure 6.4 are the distribution of elements used, in this case eight-noded

X	3PT F.D. N = 11 T = 0.81	Lin. F.E. N = 11 T = 0.47	Lin. F.E. N = 21 T = 0.50	Quad. F.E. N = 21 T = 50	Exact T = 0.50
-1.0	1.0000	1.0000	1.0000	1.0000	1.0000
-0.9			1.0000	1.0000	1.0000
-0.8	1.0000	1.0000	1.0000	1.0000	1.0000
-0.7			1.0000	0.9999	0.9999
-0.6	1.0000	1.0003	0.9998	0.9995	0.9996
-0.5			0.9991	0.9987	0.9987
-0.4	1.0000	0.9989	0.9967	0.9959	0.9960
-0.3			0.9907	0.9893	0.9891
-0.2	1.0000	0.9746	0.9734	0.9724	0.9725
-0.1			0.9367	0.9370	0.9366
0	0.9731	0.8514	0.8657	0.8671	0.8672
0.1			0.7482	0.7501	0.7509
0.2	0.8021	0.5608	0.5864	0.5882	0.5884
0.3			0.4065	0.4053	0.4062
0.4	0.4696	0.2339	0.2463	0.2440	0.2448
0.5			0.1307	0.1293	0.1301
0.6	0.1948	0.0540	0.0613	0.0611	0.0619
0.7			0.0256	0.0263	0.0268
0.8	0.0622	0.0049	0.0094	0.0103	0.0106
0.9			0.0029	0.0036	0.0039
1.0	0.0082	0.0005	0.0000	0.0011	0.0013
$\lVert u_a - u_e \rVert_2$	0.0215	0.0050	0.0011	0.0005	

Table 6.1 Solution of Burgers' equation by Galerkin finite element method, Re = 10

quadratic (serendipity) rectangular elements. The governing equations are

$$\frac{\partial u}{\partial x} + \frac{\partial v}{\partial y} = 0 \tag{6.17}$$

$$\frac{\partial u}{\partial y} - \frac{\partial v}{\partial x} = 0 \quad , \tag{6.18}$$

with boundary conditions

$$\begin{aligned} &u = 1, v = 0 \text{ at the farfield boundary} \\ \text{and} \quad &u \cos\theta - v \sin\theta = 0 \text{ at the body surface} \quad . \end{aligned} \tag{6.19}$$

Trial solutions for u and v are introduced by

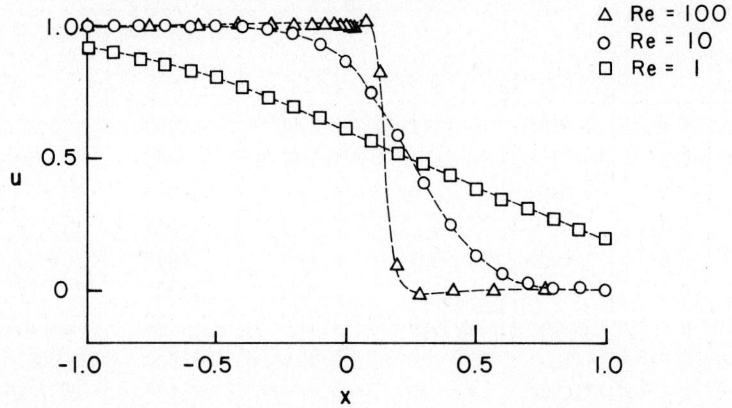

Figure 6.3 Solution of Burgers' equation for increasing Reynolds number

$$u = \sum_j \phi_j(x,y)\bar{u}_{aj} \tag{6.20}$$

and

$$v = \sum_j \phi_j(x,y)\bar{v}_{aj} \quad , \tag{6.21}$$

where the ϕ_j's are quadratic, rectangular shape functions. Execution of the Galerkin formulation produces the following algebraic equations,

$$\sum_j a_{kj} \cdot \bar{u}_{aj} + \sum_j b_{kj} \cdot \bar{v}_{aj} = 0 \tag{6.22}$$

and

$$\sum_j b_{kj} \cdot \bar{u}_{aj} - \sum_j a_{kj} \cdot \bar{v}_{aj} = 0 \tag{6.23}$$

where

$$a_{kj} = \sum_e \iint \{\phi_k \cdot \partial\phi_j/\partial x\} dx\, dy \tag{6.24}$$

and

$$b_{kj} = \sum_e \iint \{\phi_k \cdot \partial\phi_j/\partial y\} dx\, dy \quad . \tag{6.25}$$

$\sum_e$ denotes the summation over the elements adjacent to the kth node, i.e., those elements for which ϕ_k is non-zero. Since the elements are defined on a polar grid it is difficult to evaluate Eqs. (6.24) and (6.25) in the physical plane. However it is straightforward to introduce an isoparametric transformation which transforms a distorted element in the physical (x,y) plane into a regular shape in the (ξ,η) plane. This is accomplished by defining

$$x = \sum_i \phi_i(\xi,\eta)\bar{x}_i \tag{6.26}$$

$$y = \sum_i \phi_i(\xi,\eta)\bar{y}_i \quad , \tag{6.27}$$

where the ϕ_i's are just the shape functions defined in the (ξ,η) plane and $(\bar{x}_i,\bar{y}_i)$ are the physical coordinates of the ith node. After some algebra, Eqs. (6.24) and

Re = 1.0			Re = 10, t = 0.50			Re = 100, t = 0.32		
X	Approx	Exact	X	Approx	Exact	X	Approx	Exact
-1.0	0.9179	0.9174	-1.0	1.0000	1.0000	-1.00	1.0000	1.0000
-0.9	0.8998	0.8996	-0.9	1.0000	1.0000	-0.762	1.0000	1.0000
-0.8	0.8792	0.8789	-0.8	1.0000	1.0000	-0.567	1.0000	1.0000
-0.7	0.8556	0.8554	-0.7	1.0000	0.9999	-0.412	1.0000	1.0000
-0.6	0.8289	0.8288	-0.6	0.9998	0.9996	-0.290	1.0000	1.0000
-0.5	0.7991	0.7991	-0.5	0.9991	0.9987	-0.197	1.0000	1.0000
-0.4	0.7664	0.7664	-0.4	0.9967	0.9960	-0.128	1.0000	1.0000
-0.3	0.7307	0.7308	-0.3	0.9901	0.9891	-0.079	1.0000	1.0000
-0.2	0.6923	0.6926	-0.2	0.9734	0.9725	-0.045	1.0000	1.0000
-0.1	0.6517	0.6520	-0.1	0.9367	0.9366	-0.020	1.0000	1.0000
0	0.6091	0.6095	0	0.8657	0.8672	0	1.0000	0.9998
0.1	0.5651	0.5656	0.1	0.7482	0.7509	0.020	1.0003	0.9994
0.2	0.5202	0.5207	0.2	0.5864	0.5884	0.045	0.9924	0.9976
0.3	0.4750	0.4757	0.3	0.4065	0.4062	0.079	1.0195	0.9845
0.4	0.4302	0.4309	0.4	0.2463	0.2448	0.128	0.8190	0.8281
0.5	0.3862	0.3871	0.5	0.1307	0.1301	0.197	0.0942	0.1236
0.6	0.3438	0.3447	0.6	0.0613	0.0619	0.290	-0.0212	0.0009
0.7	0.3032	0.3043	0.7	0.0256	0.0268	0.412	0.0045	0.
0.8	0.2651	0.2662	0.8	0.0095	0.0106	0.567	-0.0010	0.
0.9	0.2299	0.2309	0.9	0.0029	0.0039	0.762	0.0002	0.
1.0	0.1969	0.1984	1.0	0.0000	0.0013	1.000	0.	0.
$\|u_a-u_e\|_2$	0.0007			0.0011			0.0119	

Table 6.2 Solution of Burgers' equation by Galerkin finite element method, linear element, N = 21, various Reynolds number.

(6.25) can be replaced by

$$a_{kj} = \sum_i c_{kji} \cdot \bar{y}_i \tag{6.28}$$

and

$$b_{kj} = \sum_i c_{kji} \cdot \bar{x}_i \tag{6.29}$$

where

$$c_{kji} = \int_{-1}^{1}\int_{-1}^{1} \phi_k \left\{\frac{\partial\phi_j}{\partial\xi} \cdot \frac{\partial\phi_i}{\partial\eta} - \frac{\partial\phi_i}{\partial\xi} \cdot \frac{\partial\phi_j}{\partial\eta}\right\} d\xi \, d\eta \quad . \tag{6.30}$$

It may be noted that c_{kji} is independent of the physical plane and may be evaluated, once and for all, in the (ξ,η) plane. That this is possible is a direct result of applying the Galerkin formulation to a first derivative. Thus the Jacobian, which is a function of the physical location, associated with the derivative cancels with the Jacobian associated with the integration. Clearly this situation produces a considerable economy. Typically Eq. (6.30) is sufficiently complex that numerical integration is required. With quadratic elements a 3×3 Gauss quadrature scheme produces exact integration.

Solutions have been obtained with 25 elements and 149 nodal unknowns in the quarter-plane, ABCD (Figure 6.4). To facilitate a comparison with the exact solution, the rms error between the finite element solution q_{Ta}, and the exact solution, q_{Te}, for the tangential velocity at the body

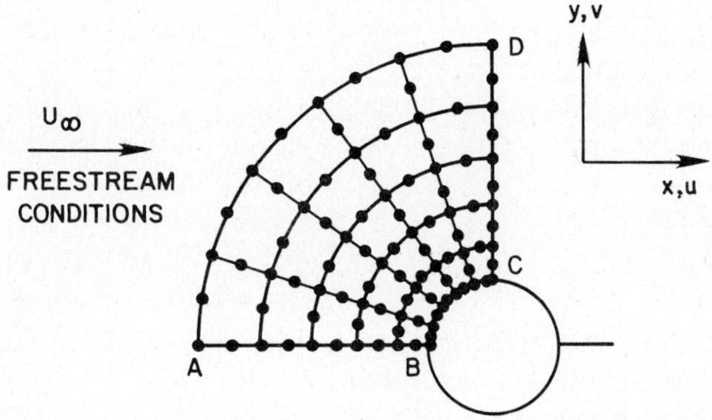

Figure 6.4 Flowfield geometry for inviscid, incompressible flow

surface has been computed by

$$\sigma = \{ \sum_{i=1}^{N} (q_{Ta} - q_{Te})^2 \}/N\}^{\frac{1}{2}} \quad . \tag{6.31}$$

A result for exact numerical integration is shown in Table 6.3.

Fletcher (1978c,1979,1980a) has considered a modification to the Galerkin method in which the weighted integration of the residual is written as

$$\iint \phi_k (R)_{\ell s} \, dx \, dy = 0 \quad , \tag{6.32}$$

where $(R)_{\ell s}$ is a least-squares fit of the residual over an element. The idea here is tnat Eq. (6.32) will constrain the approximate solution less than the conventional Galerkin formulation. If one evaluates Eq. (6.30) using a 2×2 Guass integration formula (reduced integration), this is equivalent to replacing the weighted integral of the residual with

$$\iint (\phi_k)_{\ell s} (R)_{\ell s} \, dx \, dy = 0 \quad . \tag{6.33}$$

Results for σ using Eqs. (6.32) and (6.33) are also shown in Table 6.3. It can be seen that a significant improvement in the accuracy of the solution is possible. In applying Eq. (6.32), Eq. (6.30) is written

$$C_{kji} = \int_{-1}^{1}\int_{-1}^{1} \phi_k \{ (\frac{\partial \phi_j}{\partial \xi})_{\ell s} \frac{\partial \phi_i}{\partial \eta} - (\frac{\partial \phi_j}{\partial \eta})_{\ell s} \frac{\partial \phi_i}{\partial \xi} \} d\xi \, d\eta \tag{6.34}$$

and $(\frac{\partial \phi_j}{\partial \xi})_{\ell s}$ etc. are represented by

$$(\partial \phi_j / \partial \xi)_{\ell s} = a_o + a_1 \xi + a_2 \eta + a_3 \xi \cdot \eta \tag{6.35}$$

Case	Comments	σ
1	exact numerical integration, Eq. (6.30)	0.034
2	exact numerical integration, Eq. (6.35)	0.017
3	reduced numerical integration, Eq. (6.30)	0.005

Table 6.3 Modified Galerkin finite element formulations

It was found by Fletcher (1980a) that these advantages were restricted to rectangular elements. For cases 2 and 3 with triangular elements inferior results were obtained.

6.4 Error estimates

The Galerkin finite element method has been subjected to considerable theoretical analysis with a view to providing error estimates and proving convergence. Much of the work prior to 1978 has been gathered together in Fairweather (1978). Here we consider one problem, a two-point boundary value problem in one independent variable,

$$(a(x)\cdot u_x)_x - b(x)\cdot u - f(x) = 0 \tag{6.36}$$

with boundary conditions $u(0) = u(1) = 0$. If the inner product is defined by

$$(f,g) = \int_0^1 f(x)\cdot g(x)dx \quad , \tag{6.37}$$

then the Galerkin method applied to Eq. (6.36) gives the corresponding weak formulation

$$(au_x,v_x) + (bu,v) + (f,v) = 0 \text{ for all } v \in M \quad . \tag{6.38}$$

v includes any of the trial functions used to approximate u and M is a finite dimensional subspace of H_0^1. It is assumed that subspace M is taken from a family of spaces of class $S_{k,r}$. This implies that if $M \subset H^k$ and $u \in H^r$ there exists a $\bar{u}_a \in M$ and a constant C, independent of h and u, such that

$$\| u-u_a\|_k \leqslant Ch^{r-k} \|u\|_r \tag{6.39}$$

where the norm

$$\|u\|_r = [\sum_{|r|\leqslant r} \|D^r u\|^2]^{1/2} \quad . \tag{6.40}$$

An estimate of the accuracy of the Galerkin method has been made (Fairweather, 1978) for the case $k=1$, $r \geqslant 2$ and $1 \leqslant t \leqslant r$. In this case the following estimate is available

$$\|u-u_a\| + h\|(u-u_a)_x\| \leqslant C\cdot h^t \cdot \|u\|_t \tag{6.41}$$

where $\bar{u}_a$ is the solution of Eq. (6.38). If the subspace M is based on polynomials of maximum degree s and with up to the kth derivative continuous the following L_∞ estimate can be obtained

$$\|u-u_a\|_{L_\infty} \leqslant C\cdot h^t \cdot \|u\|_{W_\infty^t} \quad , \quad 1 \leqslant t \leqslant s+1 \quad , \tag{6.42}$$

where $\| \quad \|_{W^t_\infty}$ is the equivalent L_∞ norm to Eq. (6.40).

The optimal estimate Eq. (6.41) and (6.42) are global estimates. However it is well known (Douglas and Dupont, 1974) that at certain points, e.g. the nodal points, a rate of convergence substantially greater than the above optimal estimates, can be obtained. In particular, at nodal points, x_i,

$$|(u-u_a)(x_i)| \leqslant C \cdot h^{s+t} \cdot \|u\|_{t+1} \quad , \quad 0 \leqslant t \leqslant s \quad . \tag{6.43}$$

Douglas and Dupont (1974) established that (6.43) is optimised if $t = s$.

If in addition, the Galerkin solution, $\bar{u}_a$, is post processed locally by interpolating with a polynomial of degree $2s - 1$, the following estimate is obtained

$$\|u-\hat{u}_a\|_{L_\infty} \leqslant C \cdot h^{2s} \quad , \tag{6.44}$$

where $\hat{u}_a$ is the interpolated (and hence continuous) solution of $\bar{u}_a$. Fairweather (1978) implies that comparable estimates to the above would be applicable to the related equation

$$(a(x)u_x)_x - b(x) \cdot u_x - c(x) \cdot u - f = 0 \quad . \tag{6.45}$$

with $u(0) = u(1) = 0$.

6.5 Applications

The number of applications of the Galerkin finite element method is sufficiently large that any choice will be somewhat arbitrary. Here a number of areas are mentioned that all have fluid flow connections but within that category are as diverse as possible.

The first example is chosen to illustrate the possibility of combining the Galerkin finite element method with techniques for the efficient solution of the resulting algebraic equations. In this case the alternating directions implicit (ADI) technique for finding efficient solutions in two or more directions for problems posed in an unsteady formulation. Here the unsteady formulation is used as a convenient algorithm for obtaining the steady-state solution.

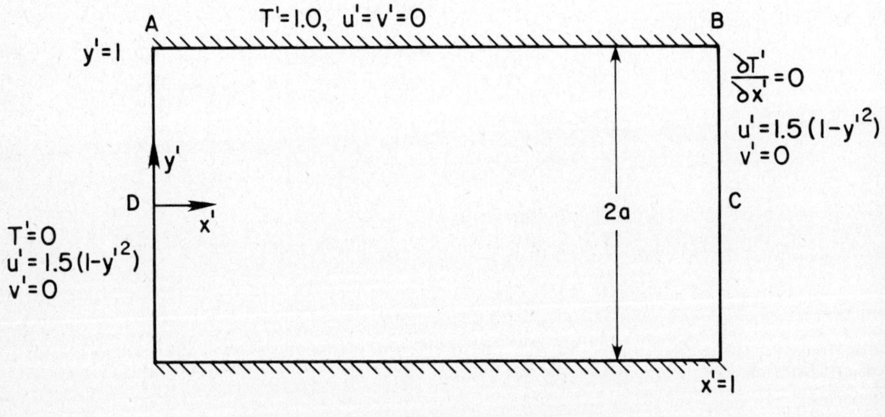

Figure 6.5 Thermal entry problem

The first example is that of the temperature distribution in a two-dimensional duct (Figure 6.5). The governing equation can be written,

$$\frac{\partial T}{\partial t} + \frac{\partial}{\partial x}(uT) + \frac{\partial}{\partial y}(vT) = \alpha_x \frac{\partial^2 T}{\partial x^2} + \alpha_y \frac{\partial^2 T}{\partial y^2} \quad . \tag{6.46}$$

It is assumed that the velocity field (u,v) is already available from the solution of the hydrodynamic entry problem. Thus the problem is linear in the temperature, T. The temperature (and velocity components) have been nondimensionalised in such a way that the nondimensional temperature will have reached the fully developed value of $T = 1$ at the boundary, BC.

To obtain algebraic equations linear rectangular elements on a uniform mesh have been used to represent T and the groups (uT) and (vT). Representing groups in this way produces an important economy in the number of algebraic terms that must be manipulated. This idea has been used previously with a least-squares finite element formulation for subsonic compressible flow (Fletcher, 1979) and with the orthonormal method of integral relations (Fletcher and Holt, 1976).

Applying the Galerkin formulation as in section 6.2 produces the following equations

$$\sum_{k,\ell} [a_{k,\ell} \cdot \dot{T}_{k,\ell} + f_{k,\ell} \cdot (uT)_{k,\ell} + g_{k,\ell} \cdot (vT)_{k,\ell} - r_{k,\ell} \cdot T_{k,\ell} - t_{k,\ell} \cdot T_{k,\ell}] = 0 \quad ,$$

$$k = i-1, i, i+1, \quad \ell = j-1, j, j+1 \tag{6.47}$$

where

$$a \equiv \frac{1}{36}\begin{bmatrix} 1 & 4 & 1 \\ 4 & 16 & 4 \\ 1 & 4 & 1 \end{bmatrix}$$

$$f \equiv \frac{1}{12\Delta x}\begin{bmatrix} -1 & 0 & 1 \\ -4 & 0 & 4 \\ -1 & 0 & 1 \end{bmatrix} , \quad g \equiv \frac{1}{12\Delta y}\begin{bmatrix} 1 & 4 & 1 \\ 0 & 0 & 0 \\ -1 & -4 & -1 \end{bmatrix} \tag{6.48}$$

$$r \equiv \frac{1}{6\Delta^2 x}\begin{bmatrix} 1 & -2 & 1 \\ 4 & -8 & 4 \\ 1 & -2 & 1 \end{bmatrix} , \quad t \equiv \frac{1}{6\Delta^2 y}\begin{bmatrix} 1 & 4 & 1 \\ -2 & -8 & -2 \\ 1 & 4 & 1 \end{bmatrix}$$

It can be seen that Eq. (6.47) is a nine-point scheme centered at the (i,j) node. Before introduction of the ADI formulation the $\dot{T}$ terms are replaced by first-order finite difference expressions. The form of the coefficients, Eq. (6.48), can be seen to be similar to those associated with the use of centered second-order finite difference expression for the spatial derivatives in Eq. (6.46). E.g.

$$f_{f.d.} \equiv \frac{1}{12\Delta x}\begin{bmatrix} 0 & 0 & 0 \\ -6 & 0 & 6 \\ 0 & 0 & 0 \end{bmatrix} , \quad t_{f.d.} \equiv \frac{1}{6\Delta^2 y}\begin{bmatrix} 0 & 6 & 0 \\ 0 & -12 & 0 \\ 0 & 6 & 0 \end{bmatrix} .$$

Thus the use of linear rectangular elements produces the same symmetry as the second-order finite difference expression except that some weight is given to adjacent grid lines.

The ADI concept is described in Mitchell and Griffiths, 1980. The basic idea is to consider all x-derivatives implicitly for the fist half time-step and all y-derivatives implicitly for the second half time-step. For the finite difference formulae shown above this produces tridiagonal systems of equations along each grid-line which can be solved very efficiently by the Thomas algorithm.

For the finite element method a difficulty arises that part of the x-derivatives (see f in Eq. (6.38)) comes from the $(i-1)^{th}$ and $(i+1)^{th}$ grid lines. To retain the availabilities of the tridiagonal algorithm it is necessary to treat these contributions explicitly. Although this degrades the accuracy of the transient solution it does not affect the accuracy of the steady-state solution. However it might be expected that the stability of the scheme will be reduced. A von Neumann stability analysis or direct calculation indicates that the scheme is, indeed, subject to a restriction on the time-step if steady-state solutions are to be obtained. The maximum time step for the scheme (1) is shown in Table 6.4. The main cause of this lack of stability is due to the coefficient, a, multiplying the time derivative term. In particular the weight associated with the (i,j) node is only 16/36 (Eq. (6.48)) compared with 1 for the finite-difference method.

Since the time derivative term has no effect on the steady-state solution it is convenient to lump all the contributions from $a_{k,\ell}$ onto the $(i,j)^{th}$ node so that $a_{i,j}=1$. This is scheme 2 in Table 6.4. As can be seen, this scheme is an improvement over scheme 1 but still subject to a time-step restriction. It is apparent that certain terms in t (Eq. (6.48)) could be treated implicitly during the first half-time-step and similarly certain terms in r during the second half-time-step. These have been treated implicitly in scheme 3 (Table 6.4) and the result is a scheme with no restriction on the time-step. Scheme 4 is the conventional ADI finite difference method. The rms difference shown in Table 6.4 is based on the difference between the computed steady-state solution and a more 'exact' solution on the centre line (DC in Figure 6.5). The solutions were obtained with a uniform 21×21 grid.

Scheme	Description	Δt_{max}	Number of time steps	Relative execution time	rms difference
1	ADIFEM	0.0003	1250	7.73	0.0502
2	ADIFEM, fully lumped	0.002	247	1.49	0.0502
3	ADIFEM, fully lumped, fully implicit	—	40	0.21	0.0502
4	ADI finite difference	—	332	1.00	0.0601

Table 6.4 Comparison of schemes for thermal entry length problem

It can be seen that the progressive reduction in the restriction on the time-step means less time-steps to reach the steady-state and consequently a shorter execution time. Although the finite difference scheme had no stability restriction it was found that too large a time-step slowed down the rate of convergence to the steady-state. Since the finite element schemes only differ in their transient behaviour the steady-state accuracy is the same for all three schemes. The entry-length problem and solution is described in more detail in Fletcher (1980b).

The general transient, convective heat-transfer problem, in which both velocity and temperature fields are obtained, has been investigated with a conventional Galerkin formulation by Gartling (1977). The above ADIFEM scheme has also been applied to the viscous, compressible flow past rectangular obstacles by Fletcher (1981a,1981b).

Kawahara et al. (1978) have solved the shallow-water equations using a Galerkin finite element formulation for problems of storm-surges in harbours and estuaries. Baker and Zelazny (1975) have applied the Galerkin finite element method to environmental pollution problems in rivers. Astley and Eversman (1981) have used the Galerkin finite element method to study the problem of acoustic transmission in non-uniform, compressible duct flow. For this particular problem they found that the Galerkin formulation gave considerably more accurate results than a least-squares formulation.

It is well known (Roache, 1972) that when solutions for convection diffusion problems are obtained using a centered finite difference scheme for the convective terms the steady-state solution will show large non-physical oscillations if the Reynolds number based on a cell is greater than 2. The same phenomenon affects the conventional Galerkin finite element method. Historically, for the finite difference method, this stability problem has been avoided by using first-order differencing (often called upwind differencing) for the convective terms. However, such a scheme is only first-order accurate. If it is interpreted as a second order scheme it is equivalent to solving the governing partial differential equation(s) with additional viscous terms added. In many situations if the grid is locally refined to achieve acceptable accuracy the cell Reynolds number will be less than two and the conventional centered difference or standard Galerkin formulae are acceptable.

Even so, considerable effort, in a finite element context, has been expended to generalise the Galerkin procedure (to what Anderssen and Mitchell (1979) call the Petrov-Galerkin method) to simulate the upwind representation of the convective term without the corresponding aritifical viscosity penalty. In this regard the work of Christie et al. (1976), Heinrich et al. (1977), Hughes (1979) and Morton and Parrott (1980) is relevant. Here we indicate some results from Kelly et al. (1980) who show that for the convection diffusion equation typified by

$$k \frac{d^2\phi}{dx^2} - u \frac{d\phi}{dx} = 0 \quad , \tag{6.49}$$

the 'upwind' scheme resulting from the Petrov-Galerkin scheme of Christie et al. (1976) can be reproduced by considering the related equation

$$(k + k') \frac{d^2\phi}{dx^2} - u \frac{d\phi}{dx} = 0 \quad , \tag{6.50}$$

where k' is referred to as a balancing dissipation and is given by

$$k' = \frac{\alpha}{2} |u| \cdot h = \frac{\alpha}{2} \cdot \gamma \cdot k \tag{6.51}$$

where h is the mesh size and α is given by

$$\alpha = \coth \frac{\gamma}{2} - \frac{2}{\gamma} \quad . \tag{6.52}$$

Kelly et al. (1980) present the following results which indicate that the introduction of artificial viscosity need not degrade the accuracy.

X	Standard Galerkin	Balanced Dissipation	Exact Solution
0	1.0000	1.0000	1.0000
0.2	1.0000	0.9997	0.9997
0.4	1.0000	0.9976	0.9976
0.6	1.0000	0.9817	0.9817
0.8	1.0000	0.8647	0.8647
1.0	0.0000	0.0000	0.0000

Table 6.5 Solutions of $-d^2\phi/dx^2 + 10d\phi/dx = 0$ with $\phi = 1$ at $x = 0$ and $\phi = 0$ at $x = 1$

7. CLOSURE

This paper has sought to achieve two objectives. Firstly Burgers' equation has been presented both as a directly useful physical representation of many flow-related problems and as a very 'simple' model for the Navier-Stokes equations. Simple in the sense that Burgers' equation is one-dimensional, neglects the pressure gradient terms and doesn't enforce continuity. But useful in the sense that Burgers' equation models the balance between convection (nonlinear) and viscous dissipation and models the development or dissipation of shocks.

Further, through the Cole-Hopf transformation, Burgers' equation has the important virtue of possessing an exact solution for many combinations of boundary conditions and initial conditions. This not only enhances the values of Burgers' equation in the direct representation of specific flow-related phenomena but also makes Burgers' equation an appropriate vehicle for testing and comparing various computational schemes.

The second objective has been to use Burgers' equation as a convenient example for illustrating the various computational Galerkin formulations: traditional Galerkin, spectral Galerkin and the Galerkin finite element method.

It is clear that the traditional Galerkin and spectral methods are global methods. That is a small change in the unknown coefficients of the trial solution, e.g. Eq. (4.17), will influence the whole domain. For the finite element method, where the unknowns in the trial solution are the nodal values themselves, Eq. (6.2), a small change in a particular unknown (nodal value) has only a local effect (immediately). In this regard the finite element method is similar to the finite difference method, i.e., they are both local methods.

There is evidence (Orszag, 1971, Haidvogel et al., 1980) to suggest that for the same number of unknowns a traditional Galerkin or spectral method will give a more accurate solution than a finite element solution. This is supported by the comparisons between Tables 4.3, 5.2 and 6.1. However, a larger execution time would be expected so that the global techniques are not necessarily more efficient, although Orszag (1980) claims that the pseudospectral method is more efficient.

In obtaining a solution of high order (the typical global situation) there is an implication of continuity of derivatives up to a corresponding order. For problems where locally discontinuous solutions (e.g. shocks) are anticipated it is reasonable to expect that a refined grid with a local low-order representation will be more successful. Clearly this type of situation favours the finite element (see Figure 6.3) or finite difference techniques. This idea is partly supported by the Rubin and Graves (1975) results for Burgers' equation.

The results of Table 6.1 indicate that a Galerkin finite element method is typically more accurate than a finite difference method of comparable order (as implied by the local truncation error). This greater accuracy is also to be anticipated from the error estimates presented in Fairweather (1978), particularly if advantage is taken of super-convergence properties. However the sparsity of a typical finite difference formulation usually implies a more economical solution and possibly a greater computational efficiency. But the results of section 6.5 for ADIFEM indicates that it is possible to achieve very efficient finite element codes.

Finite difference programs do have the advantage of being easier to write than finite element programs. In turn it is probably reasonable to suggest that finite element computer programs are easier to write than spectral computer programs. However if the subsequent usage of the program (Bourke et al., 1977) is sufficiently great, the coding effort becomes less significant.

In the past finite element methods have had the significant advantage of handling arbitrarily-shaped domains with relative ease. More recently finite difference methods for non-uniform domains have appeared, after the impetus of the grid generation techniques of Thompson et al. (1974). Even more recently, (Orszag, 1980), results have been presented to indicate that the spectral method can also handle difficult geometric situations.

In closing one is tempted to speculate about the future. It seems likely that a rapprochement between the finite element and spectral methods is possible. Thus one can postulate the solution of geometrically complicated domains with super-elements, inside of which the solution is essentially global or spectral in nature. This trend in the spectral method is apparent in Orszag (1980) and in the finite element method is apparent in the higher-order element schemes of Babuska et al. (1981). However, the problem of inter-element continuity then becomes more important.

The difficulty of handling local discontinuities and the need to compete, in terms of computational efficiency, with the finite difference methods suggests the need for an adaptive scheme in which the size of the superelements and the order of the trial solutions within them is dependent on the character of the underlying exact solution. Thus we may anticipate an arbitrary order and element scheme that will do for partial differential equations what the DIFSUB scheme of Gear (1971), etc. has done for the integration of ordinary differential equations.

Acknowledgment. The author is grateful to Professor M. Holt of the University of Calfiornia, Berkeley and to Professor R. Temam of Université de Paris-Sud, Orsay who provided an appropriate environment in 1981 for the writing of this paper. The author is also grateful to Sheila Slavin who typed this paper with her usual cheerful competence.

REFERENCES

Y. Adam (1975), "A Hermitian finite difference method for the solution of parabolic equations," Comp. & Maths. with Applics., 1, 393-406.

Y. Adam (1977), "Highly accurate compact implicit methods and boundary conditions," Journal of Computational Physics, 24, 10-22.

W. F. Ames (1965), "Nonlinear partial differential equations in engineering," Academic Press, New York, 23.

R. S. Anderssen and A. R. Mitchell (1979), "Analysis of generlised Galerkin methods in the numerical solution of elliptic equations," Math. Meth. Appl. Sci., 1, 3-15.

P. Arminjon and C. Beauchamp (1978), "A finite element method for Burgers' equation in hydrodynamics," Int. Journal for Numerical Methods in Engineering, 12, 415-428.

P. Arminjon and C. Beauchamp (1979), "Numerical solution of Burgers' equation in two space dimensions," Comp. Meths. Appl. Mech. Eng., 19, 351-365.

P. Arminjon and C. Beauchamp (1981), "Continuous and discontinuous finite element methods for Burgers' equation," Comp. Meths. Appl. Mech. Eng., 25, 65-84.

R. J. Astley and W. Eversman (1981), "Acoustic transmission in non-uniform ducts with mean flow, part II: the finite element method," J. Sound Vibration, 74, 103-121.

I. Babuska, I. N. Katz and B. A. Szabo (1981), "The p-version of the finite element method," SIAM J. Num. Anal., 18, 515-545.

A. J. Baker and S. W. Zelazny (1975), "Predictions of environmental hydrodynamics using the finite element method - applications," AIAA J., 13, 43-46.

H. Bateman (1915), "Some recent researches on the motion of fluids," Mon. Weather Rev., 43, 163-170.

R. Bellman, S. P. Azen and J. M. Richardson (1965), "On new and direct computational approaches to some mathematical models of turbulence," Quart. Appl. Math., 23, 55-67.

O. M. Belotserkovskii and P. I. Chushkin (1965), "The numerical solution of problems in gas dynamics," in "Basic developments in fluid dynamics," M. Holt (ed.), Academic Press, New York, 1-126.

E. R. Benton and G. W. Platzman (1972), "A table of solutions of the one-dimensional Burgers' equation," Quart. Appl. Math., 30, 195-212.

W. E. Bickley (1941), "Experiments in approximating to solutions of a partial differential equation," Phil. Mag., 32, 50-66.

C. Boldrighini (1977), "Absence of turbulence in a uni-dimensional model of fluid motion (Burgers model)," Meccanica, 12, 15-18.

W. Bourke (1972), "An efficient, one-level, primitive equation spectral model," Mon. Weather Rev., 100, 683-689.

W. Bourke, B. McAvaney, K. Puri and R. Thurling (1977), "Global modeling of atmospheric flow by spectral methods," Methods in Comp. Phys., 17, 267-325.

J. M. Burgers (1939), "Mathematical examples illustrating relations occurring in the theory of turbulent fluid motion," Trans. Roy. Neth. Acad. Sci., Amsterdam, 17, 1-53.

J. M. Burgers (1948), "A mathematical model illustrating the theory of turbulence" in Adv. in App. Mech., 1, 171-199.

J. M. Burgers (1964), "Statistical problems connected with the solution of a nonlinear partial differential equation," in Nonlinear problems in engineering (ed. W. F. Ames), Academic Press, New York, 123-137.

J. M. Burgers (1972), "Statistical problems connected with asymptotic solution of one-dimensional nonlinear diffusion equation," in Statistical models and turbulence (eds. M. Rosenblatt and C. van Atta), Springer, Berlin, 41-60.

J. M. Burgers (1974), "The nonlinear diffusion equation," Reidel, Holland.

H. S. Carslaw and J. C. Jaeger (1959), "Conduction of heat in solids," O.U.P., Clarendon, 2nd edition.

B. B. Cary (1973), "An exact shock wave solution to Burgers' equation for parametric excitation of the boundary," Journal of Sound and Vibration, 30, 454-464.

B. B. Cary (1975), "Asymptotic Fourier analysis of a 'sawtooth-like' wave for dual frequency source excitation," Journal of Sound and Vibration, 42, 235-241.

J. J. Chattot, J. Guiu-Roux and J. Laminie (1981), "Finite element calculation of steady transonic flow in nozzles using primary variables," Lecture Notes in Physics, 141, Springer-Verlag, 107-112.

S. I. Cheng and G. Shubin (1978), "Computational accuracy and mesh Reynods number," Journal of Computational Physics, 28, 315-326.

W. C. Chin (1978), "Pseudo-transonic equation with a diffusion term," AIAA Journal, 16, 87-88.

I. Christie, D. Griffiths, A. Mitchell and O. C. Zienkiewicz (1976), "Finite element methods for second order differential equations with significant first derivatives," Int. J. Num. Meth. Engng., 10, 1389-1396.

M. Ciment, S. H. Leventhal and B. C. Weinberg (1978), "The operator compact implicit method for parabolic equations," Journal of Computational Physics, 28, 135-166.

J. D. Cole (1951), "On a quasi-linear parabolic equation occurring in aerodynamics," Quart. Appl. Math., 9, 225-236.

L. Collatz (1960), "The numerical treatment of differential equations," Springer-Verlag, Berlin, 28.

S. H. Crandall (1956), "Engineering analysis," McGraw-Hill, New York, 151.

G. Dahlquist, A. Bjorck and N. Anderson (1974), "Numerical methods," Prentice-Hall, N.J., 104.

G. A. Davidson (1975), "Sound propagation in fogs," Journal of the Atmospheric Sciences, 32, 2201-2205.

G. A. Davidson (1976), "A Burgers' equation for finite amplitude acoustics in fogs," Journal of Sound and Vibration, 45 (4), 473-485.

A. M. Davies (1977), "A numerical investigation of errors arising in applying the Galerkin method of the solution of nonlinear partial differential equations," Comp. Meth. Appl. Mech. Eng., 11, 341-350.

A. M. Davies and A. Owen (1979), "Three-dimensional numerical sea model using the Galerkin method with a polynomial basis set," Appl. Math. Mod., 3, 421-428.

R. Dennemeyer (1968), "Introduction to partial differential equations and boundary value problems," McGraw-Hill, New York, 295.

J. Douglas and T. Dupont (1973), "Superconvergence for Galerkin methods for the two-point boundary problem via local projections," Numer. Math., 21, 270-278.

W. J. Duncan (1937), "Galerkin's method in mechanics and differential equations," ARC R & M 1978.

W. J. Duncan (1938), "The principles of Galerkin's method," ARC R & M 1848.

S. Durvasula (1969), "Natural frequencies and modes of clamped skew plates," AIAA J., 7, 1164-1167.

S. Durvasula (1971), "Flutter of clamped skew panels in supersonic flow," 8, 461-466.

G. Fairweather (1978), "Finite element Galerkin methods for differential equations," Lecture notes in pure and applied mathematics, 34, Dekker, New York.

B. A. Finlayson (1972), "The method of weighted residuals and variational principles," Academic Press, New York.

B. A. Finlayson and L. E. Scriven (1966), "The method of weighted residuals — a review," Applied Mechanics Reviews, 19, 735-748.

C. A. J. Fletcher (1978a), "The Galerkin method: an introduction," in Numerical simulation of fluid motion, J. Noye (ed.), North-Holland, 113-170.

C. A. J. Fletcher (1978b), "Application of an improved method of integral relations to the supersonic boundary layer flow about cones at large angles of attack," in Numerical simulation of fluid motion, J. Noye (ed.), North-Holland, 537-550.

C. A. J. Fletcher (1978c), "An improved finite element formulation derived from the method of weighted residuals," Comp. Meth. App. Mech. Eng., 15, 207-222.

C. A. J. Fletcher (1979a), "A primitive variable finite element formulation for inviscid, compressible flow," J. Comp. Phys., 33, 301-312.

C. A. J. Fletcher (1979b), "On the application of an improved finite element formulation with isoparametric elements," Proc. 3rd Int. Conf. in Australia on Finite Element Methods, Sydney, 1979, 671-681.

C. A. J. Fletcher (1980a), "On the application of a least-squares residual fitting finite element formulation to fluid flow problems," Comp. Meth. App. Math. Eng., 24, 251-268.

C. A. J. Fletcher (1980b), "On the application of alternating direction implicit finite element methods to flow problems," Proc. 3rd Int. Conf. Finite Elements in Flow Problems, Calgary, Canada.

C. A. J. Fletcher (1981a), "An alternating direction implicit finite element method for compressible, viscous flows," Lecture Notes in Physics, 141, Springer-Verlag, 182-187.

C. A. J. Fletcher (1981b), "On an alternating direction implicit finite element method for flow problems," Comp. Meth. Appl. Mech. Eng. (to appear).

C. A. J. Fletcher and M. Holt (1975), "An improvement to the method of integral relations," J. Comp. Phys., 18, 154-164.

C. A. J. Fletcher and M. Holt (1976), "Supersonic flow over cones at large angles of attack," J. Fluid Mechanics, 74, 561-591.

G. E. Forsythe, M. A. Malcolm and C. B. Moler (1977), "Computer methods for mathematical computations," Prentice-Hall, N.J.

M. Fortin and F. Thomasset (1979), "Mixed finite element methods for incompressible flow problems," Journal of Computational Physics, 31, 113-145.

B. G. Galerkin (1915), "Rods and plates. Series occurring in some problems of elastic equilibrium of rods and plates," Vestnik Inzhenerov. Tech., 19, 897-908.

D. K. Gartling (1977), "Convective heat transfer by the finite element method," Comp. Meth. App. Mech. Eng., 12, 365-382.

J. Gazdag (1973), "Numerical convective schemes based on accurate computation of space derivatives," J. Comp. Phys., 13, 100-113.

C. W. Gear (1971), "Numerical initial value problems in ordinary differential equations," Prentice-Hall, N.J.

F. C. Gey and M. B. Lesser (1969), "Computer generation of series and rational function solutions to partial differential initial value problems," Proc. 24th Nat. Conf. of A.C.M., Publ. P-69, 559-572.

D. Gottlieb and S. A. Orszag (1977), "Numerical analysis of spectral methods: theory and applications," SIAM Publications, Philadelphia.

S. N. Gurbatov, A. A. Dubkov and N. I. Lobachevskii (1977), "Parametric excitation of low-frequency noise waves in a nonlinear medium," Soviet Physics of Acoustics, 23, 2, 146-148.

D. B. Haidvogel, A. R. Robinson and E. E. Schulman (1980), "The accuracy, efficiency and stability of three numerical models with application to open ocean problems," J. Comp. Phys., 37, 1-53.

W. D. Hayes (1954), "Pseudotransonic similitude and first-order wave structure," Journal of the Aeronautical Sciences, 21, 721-730.

J. C. Heinrich, P. Huyakorn, O. C. Zienkiewicz and A. Mitchell (1977), "An upwind finite element scheme for two dimensional convective transport equation," Int. J. Num. Meth. Engng., 11, 131-143.

R. S. Hirsh (1975), "Higher order accurate difference solutions of fluid mechanics problems by a compact differencing technique," Journal of Computational Physics, 19, 90-109.

E. Hopf (1950), "The partial differential equation $u_t + uu_x = \mu u_{xx}$." Comm. Pure Appl. Math., 3, 201-230.

C.-C. Hsu (1975), "A Galerkin method for a class of steady, two-dimensional, incompressible, laminar boundary-layer flows," J. Fluid Mechanics, 69, 783-802.

T. J. R. Hughes (1979), "A simple scheme for developing upwind finite elements," Int. J. Num. Meth. Engng., 12, 1359-1365.

E. Isaacson and H. B. Keller (1966), "Analysis of numerical methods," Wiley, New York, 199.

P. C. Jain and D. N. Holla (1978), "Numerical solution of coupled Burgers' equations," Int. J. Nonlinear Mech., 13, 213-222.

A. Jeffrey and T. Kakutani (1972), "Weak nonlinear dispersive waves: A discussion centered around the Korteweg-deVries equation," SIAM Review, 14, 4, 582-643.

M. Kawaguti (1955), "The critical Reynolds number for the flow past a sphere," J. Phys. Soc. of Japan, 10, 694-699.

M. Kawahara, N. Takenchi and T. Yoshida (1978), "Two step explicit finite element method for tsunami wave propagation analysis," Int. J. Num. Meth. Engng., 12, 331-351.

R. B. Kellogg, G. R. Shubin and A. B. Stephens (1980), "Uniqueness and the cell Reynolds number," SIAM J. Num. Anal., 17, 733-739.

D. W. Kelly, S. Nakazawa, O. C. Zienkiewicz and J. C. Heinrich (1980), "A note on upwinding and anisotropic balancing dissipation in finite element approximations to convective diffusion problems," Int. J. Num. Meth. Engng., 15, 1705-1711.

P. K. Khosla and S. G. Rubin (1979), "Filtering of nonlinear instabilities," J. Eng. Math., 13, 127-141.

S. Kida (1979), "Asymptotic properties of Burgers' turbulence," Journal of Fluid Mechanics, 93, 337-377.

Y. Kuo and R. I. Tanner (1971), "Burgers-type model of turbulent decay in a non-Newtonian fluid," ASME paper 71-W/APM-11.

V. V. Kuznetsov, V. E. Nakoryakov, B. G. Pokusaev and I. R. Shreiber (1977), "Propagation of disturbances in a gas-liquid mixture," Soviety Physics of Acoustics, 23, 2, 153-156.

A. M. Lerner and V. E. Fridman (1978), "Model equations for the nonlinear acoustics of media with a high thermal conductivity," Soviety Physics of Acoustics, 24, 2, 128-133.

W. Lick (1967), "Wave propagation in real gases," Advanced in Applied Mechanics, 10, 1-72.

W. Lick (1970), "Nonlinear wave propagation in fluids," Annual Review of Fluid Mechanics, 2, 113-136.

M. J. Lighthill (1956), "Viscosity effects in sound waves of finite amplitude," in Surveys in Mechanics (eds. G. K. Batchelor and R. M. Davies), C.U.P., Cambridge, 250-351.

M. D. Love (1980), "Subgrid modelling studies with Burgers equation," Journal of Fluid Mechanics, 100, 87-110.

P. P. Lynn (1974), "Least-squares finite element analysis of laminar boundary layer flows," Int. J. Num. Meth. Eng., 8, 865-876.

R. H. Marshall, A. E. Churches and J. A. Reizes (1979), "A hybrid integration scheme for the solution of viscous compressible flows," Appl. Math. Modelling, 3, 459-465.

A. R. Mitchell and D. F. Griffiths (1980), "The finite difference method in partial differential equations," Wiley, New York.

R. M. Miura (1976), "The Korteweg-deVries Equation: A Survey of Results," SIAM Review, 18, 3, 412-459.

Y. Morchoisne (1979), "Resolution of Navier-Stokes equations by a space-time pseudo-spectral method," La Recherche Aerospatiale, 1979-5, 11-31.

J. P. Moran and S. F. Shen (1966), "On the formation of weak plane shock waves by impulsive motion of a piston," Journa of Fluid Mechanics, 25, part 4, 705-718.

K. W. Morton and A. K. Parrott (1980), "Generalised Galerkin methods for first-order hyperbolic equations," J. Comp. Phys., 36, 249-270.

J. W. Murdock (1979), "A numerical study of nonlinear effects on boundary layer stability," AIAA J., 15, 1167-1173.

T. Musha and H. Higuchi (1978), "Traffic current fluctuation and the Burgers' equation," Japanese Journal of Applied Physics, 17, 5, 811-816.

R. Narasimha and S. M. Deshpande (1969), "Minimum-error solutions of the Boltzmann equation for shock structure," J. Fluid Mech., 36, 555-570.

R. E. Nickell, D. Gartling and G. Strang (1979), "Spectral decomposition in advection-diffusion analysis by finite element method," Comp. Meths. Appl. Mech. Eng., 17/18, 561-580.

S. A. Orszag (1971a), "Numerical simulation of incompressible flows with simple boundaries; accuracy," J. Fluid Mech., 49, 75-42.

S. A. Orszag (1971b), "Accurate solution of the Orr-Sommerfeld equation," J. Fluid Mech., 50, 375-405.

S. A. Orszag (1972), "Comparison of pseudospectal and spectral approximation," Studies in App. Math., 51, 253-259.

S. A. Orszag (1980), "Spectral methods for problems in complex geometries," J. Comp. Phys., 37, 93-112.

S. A. Orszag and M. D. Kruskal (1968), "Formulation of the theory of turbulence," Phys. Fluids, Supplement II, 12, 250-257.

S. A. Orszag and G. S. Patterson (1972), "Numerical simulation of three-dimensional homogeneous, isotropic turbulence," Phys. Rev. Lett., 28, 76-79.

A. T. Patera and S. A. Orszag (1981), "Transition and turbulence in planar channel flows," Proc. 7th Int. Conf. Num. Meth. in Fluid Dynamics, Stanford, publ. as Lecture Notes in Physics, 141, Springer-Verlag, Berlin, 329-335.

S. A. Piacsek and G. P. Williams (1970), "Conservation properties of convection difference schemes," Journal of Computational Physics, 6, 392-405.

B. L. Pierson and P. Kutler (1980), "Optimal nodal point distribution for improved accuracy computational fluid dynamics," AIAA J., 18, 49-54.

G. W. Platzman (1960), "The spectral form of the vorticity equation," J. Meteor., 17, 635-644.

G. W. Platzman (1962), "The analytical dynamics of the spectral vorticity equation," J. Meteor., 19, 313-328.

V. I. Rizun and Iu. K. Engel'brekht (1975), "Application of the Burgers equation with a variable coefficient to the study of nonplanar wave transients," Applied Mathematics and Mechanics, 39, 3, 524-528.

P. J. Roache (1972), "Computational fluid dynamics," Hermosa, Albuquerque.

E. Y. Rodin (1970a), "On some approximate and exact solutions of boundary value problems for Burgers' equation," J. Math. Anal. Appl., 30, 401-414.

E. Y. Rodin (1970b), "A Ricatti solution for Burgers' equation," Quart. Appl. Math., 27, 541-545.

S. G. Rubin and R. A. Graves (1975), "Viscous flow solutions with a cubic spline approximation," Computers and Fluids, 3, 1-36.

S. G. Rubin and P. K. Khosla (1976), "Higher-order numerical methods derived from three-point polynomial interpolation," NASA CR-2735.

P. L. Sachdev and R. Seebass (1973), "Propagation of spherical and cylindrical N-waves," Journal of Fluid Mech., 58, part 1, 197-205.

H. Schamel and K. Elsasser (1976), "The application of the spectral method to nonlinear wave propagation," Journal of Computational Physics, 22, 501-516.

R. Seebass (1969), "Sonic boom theory," Journal of Aircraft, 6, 177-184.

P. Shuleshko (1959), "A new method of solving boundary-value problems of mathematical physics," Australian J. Appl. Sci., 10, 1-16.

L. J. Snyder and W. E. Stewart (1966), "Velocity and pressure profiles for Newtonian creeping flow in regular packed beds of spheres," Am. Inst. Chem. Eng. J., 12, 167-173.

G. P. Steven and J. Milthorpe (1978), "On a least-squares approach to the integration of the Navier-Stokes equations," in Finite Elements in Fluids, 3, Wiley, 89-110.

T. Tatsumi (1980), "Theory of homogeneous turbulence," Advances in Applied Mechanics, 20, 39-133.

T. Tatsumi and S. Kida (1972), "Statistical mechanics of the Burgers model of turbulence," Journal of Fluid Mechanics, 55, 659-675.

T. Tatsumi and H. Tokunaga (1974), "One dimensional shock turbulence in a compressible fluid," Journal of Fluid Mechanics, 65, 581-601.

G. I. Taylor (1910), "The conditions necessary for discontinuous motions in gases," Proc. Royal Society, Series A, 84, p. 371-377.

T. D. Taylor, E. Ndefo and B. S. Masson (1972), "A study of numerical methods for solving viscous and inviscid flow problems," Journal of Computational Physics, 9, 99-119.

J. F. Thompson, F. C. Thames and C. M. Martin (1974), "Automatic numerical generation of body-fitted curvilinear coordinate system for field containing any number of arbitrary two-dimensional bodies," J. Comp. Phys., 15, 299-319.

B. A. Ton (1975), "On the behaviour of the solution of the Burgers equation as the viscosity goes to zero," J. Math. Anal. Appl., 49, 713-720.

J. V. Villadsen and W. E. Stewart (1967), "Solution of boundary-value problems by orthogonal collocation," Chem. Eng. Sci., 22, 1483-1501.

A. R. Wadia and F. R. Payne (1979), "Application of the method of integral relations to unsteady fluid flow problems with shocks," in Advances in Computer Methods for Partial Differential Equations, eds. R. Vichneretski and R. S. Stepleman, IMACS, 205-213.

G. B. Whitham (1974), "Linear and nonlinear waves," Wiley, New York, 97.

H. E. Williams (1975), "Motion of a cable used as a mooring," J. Hydronautics, 9, 107-118.

H. C. Yee, R. M. Beam and R. F. Warming (1981), "Stable boundary approximations for a class of implicit schemes for the one-dimensional inviscid equations of gas dynamics," AIAA paper 81-1009, AIAA 5th Computational Fluid Dynamics Conference, Palo Alto, June 22-23, 1981.

NUMERICAL SOLUTIONS OF PARTIAL DIFFERENTIAL EQUATIONS
J. Noye (Editor)

THE FINITE ELEMENT METHOD IN ENGINEERING PRACTICE AND EDUCATION

Josef A. Tomas

Special Interest Group on Finite Element Method
Royal Melbourne Institute of Technology,
Melbourne, Victoria, Australia.

The success of the finite element method in the field of structural mechanics has been astonishing. A numerical method, which was originally invented by engineers out of necessity to solve complex practical problems, later attracted the interest of mathematicians who based it on sound mathematical foundations. The number of publications appearing every year on various aspects of the method is enormous. A wide spectrum of approaches and views, however, often creates a communication gap between the practical and theoretical "camps". On one hand, proofs based on intuition and practical experience do not attract much attention from mathematicians, but they work and give reasonable results in practice. They have frequently been a seed from which a series of mathematical papers has grown up. On the other hand, many mathematical papers are truly unreadable to a man of the practical world and often seem to conceal any possible useable content with superheavy terminology. Communication between the worlds of thinking and acting is a necessity. This paper tries to contribute to this goal by demonstrating how an engineer understands the finite element method and how he uses it.

1. INTRODUCTION

For an engineer, mathematics is an instrument which helps him to solve practical problems. He encounters complex physical objects made of a variety of materials - he improves them, manufactures them, tests them. His work sometimes proceeds under severe time constraints when the deadline always seems to be yesterday. As Poincare says: "He has no time to split hairs" (1).

His relationship with mathematical methods is entirely utilitarian. He must obtain a complete mathematical training in order to be able to recognize when to apply mathematical methods and procedures and how to apply them correctly and quickly. The more subtle part of mathematical education - clear and logical thinking - is received by him rather subconsciously and is later utilized in grasping and handling non-mathematical problems of everyday life. He is fully immersed in real three-dimensional space so that his appreciation and understanding of intricacies and beauties of abstract objects in mathematical spaces is limited.

The history of the finite element method is an excellent example of contribution of both "camps" - engineers and mathematicians - to the development of a powerful numerical method, as it is known and used today. In the beginning, a few men recognized - mostly by intuition and genius - the potential use of emerging computers in well-established engineering solution techniques. Now the relationship to more general mathematical formulations has been recognized and this method has been put on a firm mathematical basis. A third group of specialists must not be forgotten : computer programmers. Computer storage and speed are limited while the systems to be solved are large and complex. Extraordinary programming skill and experience is required if the finite element method is to be made useful. Several poor finite element programs, existing and still being used, demonstrate that it is not easy to develop a reliable and practical general purpose computer program based on well-known theory and solution methods.

A practical engineer uses the finite element method as a tool. The computer program has a place in his laboratory similar to that of any other test rig. He learns by using it. He recognizes its reliability by comparing results with natural expectations or with physical tests. If the program does not work properly or is missing important features, he gets rid of it and looks for a more suitable program. He certainly tries to understand basic ideas behind the "black-box" he is using but he has no time or interest in developing better or new features. He initiates new developments in the finite element methods by his needs leaving the details of formulation and programming to experienced specialists in relevant scientific areas.

This paper, presented to applied mathematicians, aims at demonstrating how practical engineers use the finite element method. The author also had in mind his students, to whom he would like to show that there are many things they should know in detail, other things they should have heard about and be careful when using them, and many other things which are too complex to be understood so that they must believe mathematical proofs which are absolutely true - but only in one of many mathematical spaces.

2. ENGINEERING CONCEPT - DISCRETE SYSTEMS

The finite element method evolved from various origins and along different paths. The engineering concept of the method will be presented first as it corresponds to the historical development of the method. The mathematical concept will be discussed later in Section 4.

Engineers have been solving truss and beam structures for more than a hundred years. The repetition of identical elements and the "pointwise" interconnection of elements make the application of basic principles of mechanics very simple. Although matrix algebra is not essential for the formulation and solution, it is a very convenient method of book-keeping and computer programming. It will therefore be used throughout this paper.

Let us consider a two-dimensional structure from Fig. 2.1. It consists of three members which can resist tension or compression but not bending. The individual members (elements) are numbered 1,2, 3. The elements are connected with each other at node 2 and with the frame at nodes 1, 3,4. The structure is loaded by a force F in the node 2.

The displacement of node 2 and internal forces in truss elements are to be calculated.

The usual procedure in engineering analysis is to consider the deformation of each element separately and then to satisfy the equilibrium of internal and external forces at each node.

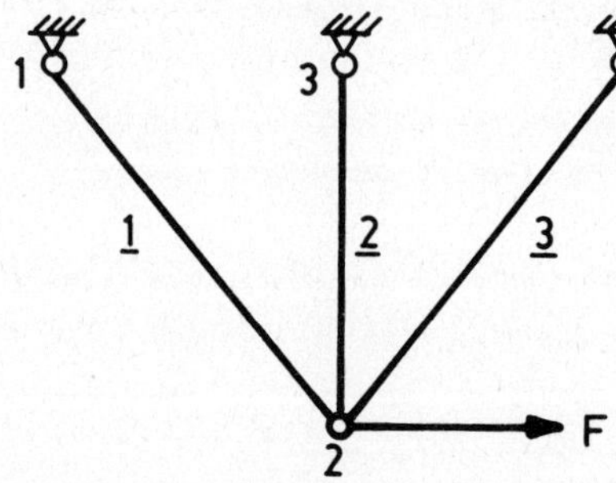

Figure 2.1

An individual element in a general position is depicted in Fig. 2.2. The node numbers 1 and 2 are local node numbers having no relation to the node numbers in Fig. 2.1. Two coordinate systems are introduced, a global system (X,Y) and a local system (x,y). The connection to other truss elements is replaced by internal forces F_1, F_2 in the local coordinate system. This is not necessary but it is convenient for computation. A linear elastic analysis with small displacements is assumed, and rotations are neglected.

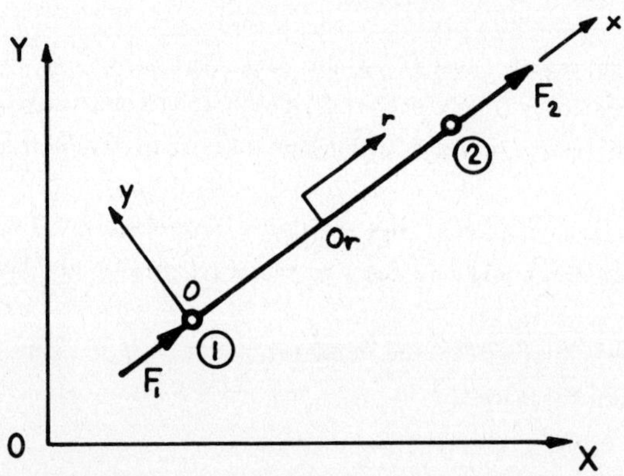

Figure 2.2

The action of internal forces F_1 and F_2 (Fig. 2.2) will result in displacements u_1 and u_2 in the direction of x.

The strain ε of the element is $(u_2 - u_1)/l$.

The stress $\sigma = \varepsilon E$, and the force in the truss is given by $F_2 = -F_1 = \sigma A$ or $F_2 = (AE/l) \cdot (u_2 - u_1)$.

This relationship can be written in a matrix form as

$$\begin{bmatrix} F_1 \\ F_2 \end{bmatrix} = \frac{A.E}{l} \begin{bmatrix} 1 & -1 \\ -1 & 1 \end{bmatrix} \cdot \begin{bmatrix} u_1 \\ u_2 \end{bmatrix} . \tag{2.1}$$

The subscript l is introduced to indicate quantities in the local system (x,y), and subscript g refers to the global system (X,Y).

Then Eq. (2.1) can be written as

$$\underset{\sim}{F}_l = \underset{\approx}{K}_l \, \underset{\sim}{u}_l . \tag{2.2}$$

The matrix $\underset{\approx}{K}$ is called the local element stiffness matrix.

Local stiffness matrices can be calculated for each element. To be able to calculate displacements of the global nodes (Fig. 2.1) the local displacements $\underset{\sim}{u}_l$ must be transformed into the global coordinate system (X,Y).

The transformation matrix is easily obtained using Fig. 2.3. The displacement u_l in the x direction is given in terms of global components u_{1x} and u_{1y} by

$$u_1 = u_{1x} \cos\alpha + u_{1y} \cos\beta .$$

A similar expression exists for u_2 - the displacement of the second node of the element. In a matrix form this can be written as

$$\begin{bmatrix} u_1 \\ u_2 \end{bmatrix} = \begin{bmatrix} \cos\alpha & \cos\beta & 0 & 0 \\ 0 & 0 & \cos\alpha & \cos\beta \end{bmatrix} \cdot \begin{bmatrix} u_{1x} \\ u_{1y} \\ u_{2x} \\ u_{2y} \end{bmatrix} \tag{2.3}$$

or

$$\underset{\sim}{u}_l = \underset{\approx}{T} \, \underset{\sim}{u}_g ,$$

where $\underset{\approx}{T}$ is the transformation matrix.

Inserting Eq.(2.3) into Eq.(2.2) results in

$$\underset{\sim}{F}_1 = \underset{\approx}{K}_1 \underset{\approx}{T} \underset{\sim}{u}_g \, . \qquad (2.4)$$

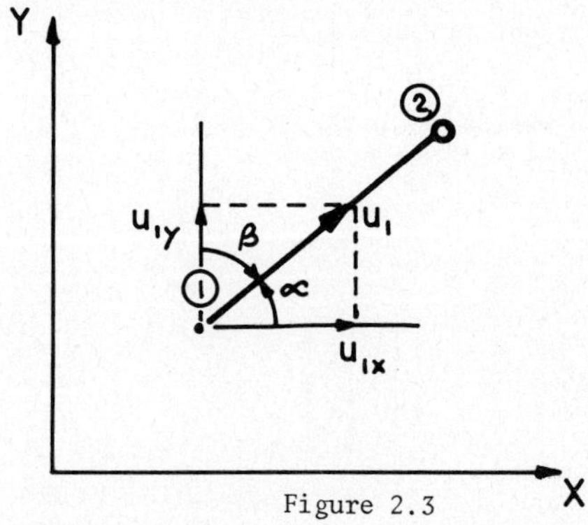

Figure 2.3

This relation is used for calculation of internal forces (they are local) after the global displacements $\underset{\sim}{u}_g$ have been calculated. Most of the computer programs precalculate the matrix $\underset{\approx}{K}$ $\underset{\approx}{T}$ for each element and store it on a tape for later use.

The equilibrium of internal and external forces in each node requires the transformation of internal forces $\underset{\sim}{F}_1$ into the global coordinate system. Fig. 2.4 shows the relationships.

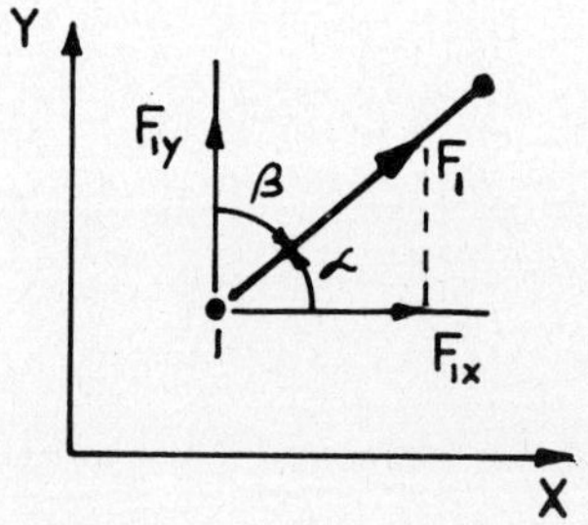

Figure 2.4

$$F_{1x} = F_1 \cos\alpha \; , \; F_{2x} = F_2 \cos\beta \; ,$$

or in a matrix form (for both nodes)

$$\begin{bmatrix} F_{1x} \\ F_{1y} \\ F_{2x} \\ F_{2y} \end{bmatrix} = \begin{bmatrix} \cos\alpha & 0 \\ \cos\beta & 0 \\ 0 & \cos\alpha \\ 0 & \cos\beta \end{bmatrix} \begin{bmatrix} F_1 \\ F_2 \end{bmatrix} ,$$

i.e.,

$$\underset{\sim}{F}_g = \underset{\approx}{T}^T \underset{\sim}{F}_1 . \qquad (2.5)$$

After multiplying Eq. (2.4) from left by the transposed transformation matrix $\underset{\approx}{T}$, the equation in the global coordinate system is obtained for each element as

$$\underset{\sim}{F}_g = \underset{\approx}{T}^T \underset{\approx}{K}_1 \underset{\approx}{T} \underset{\sim}{u}_g ,$$

or

$$\underset{\sim}{F}_g = \underset{\approx}{K}_g \underset{\sim}{u}_g \qquad (2.6)$$

where the matrix $\underset{\approx}{K}_g = \underset{\approx}{T}^T \underset{\approx}{K}_1 \underset{\approx}{T}$ is the element stiffness matrix in the global coordinate system.

Until now, the individual elements have been considered. The forces $\underset{\sim}{F}_g$ in Eq. (2.6) are internal forces at the nodes 1 and 2 in Fig. 2.2. The elements are now connected together and global points are considered (numbered 1,2,3,4 in Fig. 2.1). Each element contributes by its internal force to the global node to which it is connected. There are also external forces acting on the global nodes (some of them will normally be zero). Equilibrium requires that the sum of internal forces on each global node balance the external force on this node.

Considering Eq. (2.6) for each element, the force vector on the left side will be replaced by the force vector of external forces and the elements of $\underset{\approx}{K}_g$ matrices corresponding to the nodal point entries and displacement entries will be added together.

The result will be the equation for the complete structure.

$$\underset{\sim}{F} = \underset{\approx}{K} \; \underset{\sim}{u} \; , \qquad (2.7)$$

where $\underset{\sim}{F}$ is the vector of external forces, $\underset{\approx}{K}$ is the system stiffness matrix, $\underset{\sim}{u}$ is the vector of global displacements. Symbolically, this assembly procedure of the stiffness matrix $\underset{\approx}{K}$ is depicted in Fig. 2.5. The displacement

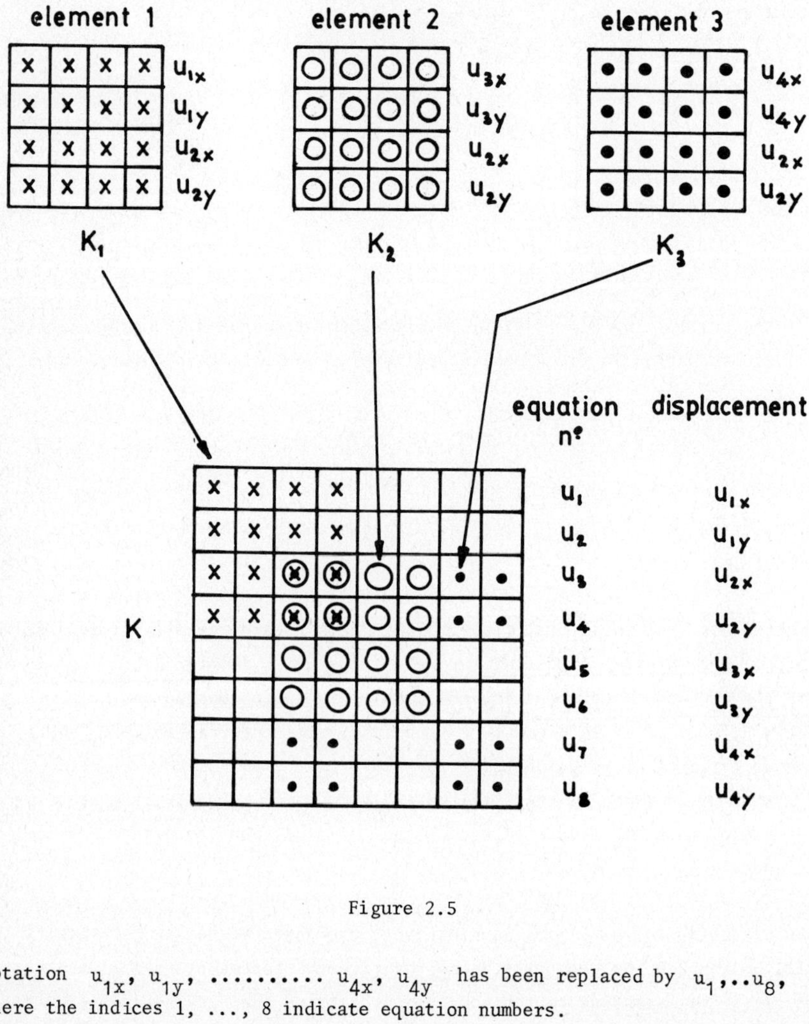

Figure 2.5

notation u_{1x}, u_{1y}, ... u_{4x}, u_{4y} has been replaced by u_1,..u_8, where the indices 1, ..., 8 indicate equation numbers.

The force vector is a vector of external forces corresponding to each displacement (equation number). Zero entry corresponds to zero external force component for the relevant equation.

The stiffness matrix $\underset{\approx}{K}$ in Eq. (2.7) is singular. This fact can be easily interpreted physically. Our assembly procedure in Fig. 2.5 assumes free nodes 1,3,4; the structure is free, so that large motions are possible without any deformations (the so called rigid body modes).

There are two ways to handle geometric (fixed) boundary conditions. The first one is to eliminate from Eq. (2.7) the rows and columns corresponding to fixed boundaries. In Fig. 2.5, this would be the rows and columns numbered 1, 2,5,6,7,8. An identification vector relating the equation numbers of the reduced system to the equation numbers of the original system has to be created and stored in order to relate displacements (now only available for free node 2) to the components of the original vector $\underset{\sim}{u}$ after solution.

The second method uses so called boundary elements. It is of interest to mention that this method is equivalent to the treatment of geometric boundary conditions suggested by Courant (2). The connections between nodes 1,3,4 and the frame are replaced by very stiff springs (Fig. 2.6).

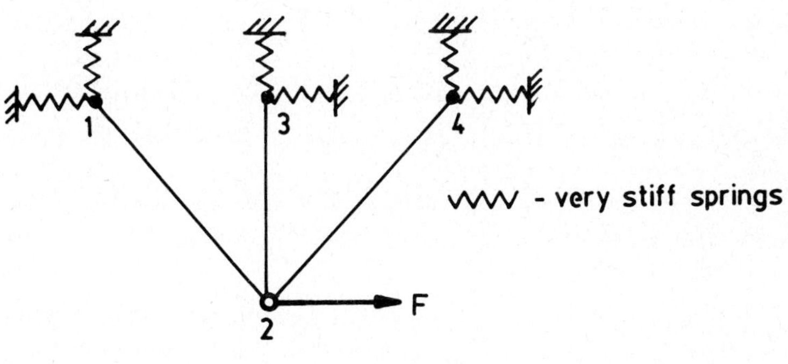

Figure 2.6

A spring is in fact identical to a truss element. The end-point connected to the frame is not taken into account. This is equivalent to eliminating additional equations corresponding to the new artificial joints introduced.

The stiffness matrix of the spring is only a single element which is added to the corresponding diagonal term of the system stiffness matrix. When Eq. (2.7) is solved by the Gaussian elimination procedure, all row matrix terms and the force component are divided by the corresponding diagonal term. This will result in the equation for the row i

$$F_i/k_{ii}=(\ldots)/k_{ii} + u_{ii} + (\ldots)/k_{ii} \tag{2.8}$$

which gives the result $u_{ii}=0$ when k_{ii} is very large.

This method can also be used to introduce prescribed displacements at some nodes. A stiff spring is added to the node with prescribed deflection d. The deflection is applied to the system through this spring. The force corresponding to the deflection d is equal to $k_s d$. Eq. (2.8) will have on the left side $F_i = k_s d$. Because the difference between k_s and k_{ii} is negligible (k_{ii} is the original k_{ii} plus k_s and $k_s >> k_{ii}$) the result will be $u_{ii} \simeq d$ as required.

Programmes CAL 78, SAP4, ADINA use the first methods; programs MARC and ABAQUS use the second method. ABAQUS uses a front-wave solution method in which the system matrix is not completely assembled so that the boundary element method does not represent any drawback and is very comfortable from the programming point of view.

After the geometric boundary conditions have been accommodated in Eq. (2.7), Eq. (2.7) is solved.

All of the programs with which the author is familiar use the Gaussian elimination method. Since the matrix $\underset{\approx}{K}$ is symmetrical and banded, an economical storage scheme is possible. SAP and MARC store the complete upper half-band, ADINA stores the upper half-band in the so-called skyline fashion (column-wise up to the last non-zero element above the diagonal), ABAQUS uses a front-wave solution method in which an element matrix is condensed, keeping only equations which are common to elements notyet included, the next element matrix is added and condensed again, and so on.

It is clear that the order of numbering nodes is important for minimizing band width, while the order of numbering elements is important for keeping the number of equations required for condensation at any time to the minimum in the front-wave solution method.

After Eq. (2.7) has been solved, the displacements of relevant nodes forming an element are extracted from the vector $\underset{\sim}{u}$ and the internal forces (stresses) are calculated from Eq. (2.4).

Although it is arguable whether the above procedure of solving the truss system should be called finite element method, nevertheless it contains all of the basic features of the true finite element method for the solution of continua. The difference is that the stiffness matrix is calculated exactly for trusses while for continua it can only be approximated. However, in both cases the forces can be applied only at nodes and the displacements obtained are the displacements of the nodes.

In a similar way, beam elements can be introduced. If elementary beam theory is assumed (normals to the neutral beam axis remain normal) the stiffness matrix can again be calculated exactly. The unknown variables will now be not only displacements of the nodes, but also their derivatives (slopes) to allow bending.

3. ENGINEERING CONCEPT - CONTINUOUS SYSTEMS

Structural problems of solid mechanics can be formulated for the finite element method directly without using variational principles or weighted residual methods. This fact is probably the main reason for the success of the method among engineers. Every third year mechanical or civil engineering student understands that the work of external forces during deformation is converted into internal energy of the body. This balance is valid for any part of the body, and so for a finite element. The finite element has a physical meaning, it can be drawn on the body, or cut-out and seen from all sides. On the other hand, formulation of the equilibrium of a point by means of a partial differential equation and subsequent numerical solution by the finite difference method represents for an engineer a double-artificial process having no connection with the reality.

The principle of virtual displacements and the principle of minimum potential energy will be used in this section to derive the stiffness matrix of a continuum. Unlike the stiffness matrix of truss or beam elements, the stiffness matrix of a continuum solid element can be evaluated only approximately.

Any real component or structure is a three-dimensional continuum. In many cases some of the dimensions can be neglected and either one-dimensional models, such as trusses or beams, or two-dimensional plates, axi-symmetric solids or shells can be constructed. The main practical gain is a substantial reduction in computer requirements.

The first attempt to solve a continuum problem by a method related to the finite element approach followed the procedure outlined in the previous section. The continuum was replaced by a framework of trusses or beams arranged in a definite pattern and having elastic properties selected in such a way that equivalent displacements and strains of the framework converged to the exact values for the continuum if the size of the individual bars was made infinitesimally small (3). This approach was abandoned when Argyris (4) and Turner et al (5) developed a more direct method of replacing a continuum by equivalent small subdivisions called later by Clough (6) finite elements. It is surprising how many newcomers to the finite element method still believe that the method replaces a continuum by a bar structure.

Let us demonstrate the method with a simple problem - a two-dimensional plate loaded in its middle plane (the so-called plane stress problem (Fig. 3.1).

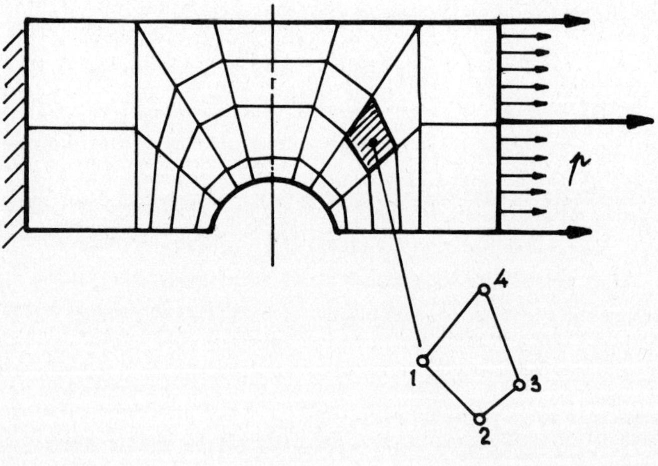

Figure 3.1

The continuum is first discretised into finite elements. There is an unlimited number of ways of constructing the element mesh. One of the possibilities is shown in Fig. 3.1. The coarseness of the mesh and the precision of the solution are directly related as will be shown later. Intuitively, the mesh should be finer in areas of large displacement gradients.

There are also many possible element shapes. Four-node quadrilateral elements have been chosen in Fig. 3.1, but other types, such as triangular elements with three or six nodes or quadrilateral elements with eight or twelve nodes could equally have been selected. The nodes have a function similar to that described earlier for trusses: their displacements are the basic unknowns of the problem, and again all loads must be replaced by equivalent forces applied at the nodes.

The displacement within each element is approximated by a shape function (also called a displacement, trial, approximation, basis or interpolation function) which has the unknown displacements of the nodes as parameters.

This displacement function allows strains within the element to be expressed in terms of the nodal displacements so that stresses throughout the element and also on its boundaries may be calculated.

The principle of virtual work is then applied to an isolated element expressing the balance between the virtual work of internal forces at the nodes and the virtual strain energy within the element for any combination of virtual displacements. The result will be a set of linear algebraic equations similar to the system of equations (2.7).

Clearly, the displacement function has to satisfy certain requirements in order to ensure continuity of displacements along boundaries between adjacent elements and to allow convergence to the exact solution with mesh refinement.

Let us cut out one element from the structure in Fig. 3.1 and number its nodes 1,2,3,4 (local nodal numbers) as shown in Fig. 3.1.

The displacement component u (in the x direction) at any point (x,y) within the element is approximated as

$$u = \sum_{i=1}^{4} N_i \, u_i \, , \tag{3.1}$$

where u_i, i = 1, ..., 4 are as yet unknown nodal displacements in the x direction. The displacement (shape) functions N_i are functions of x and y. Their proper choice will be discussed later. The same displacement functions are used for the approximation of the displacement v in the y direction.

Plane strain ε is defined in terms of displacements u and v as

$$\underset{\sim}{\varepsilon}^T = \left[\varepsilon_x \ \varepsilon_y \ \varepsilon_{xy}\right] = \left[\frac{\partial u}{\partial x} \ \frac{\partial v}{\partial y} \ \frac{\partial u}{\partial x} + \frac{\partial u}{\partial y}\right] . \tag{3.2}$$

The approximate value of the strain can be easily expressed in terms of nodal point displacements $\underset{\sim}{u}_i$, $\underset{\sim}{v}_i$ by inserting (3.1) into (3.2). We then have

$$\underset{\sim}{\varepsilon} = \underset{\approx}{B} \, \underset{\sim}{u}_i , \qquad \underset{\sim}{u}_i = \begin{bmatrix} u_i \\ v_i \end{bmatrix} , \ i=1,\ldots,4. \tag{3.3}$$

The strain displacement matrix $\underset{\approx}{B}$ is obtained by appropriate differentiation of the functions N.

The stress components can be expressed in terms of the strains. For an isotropic material, the linear elastic behaviour is described by relations (if initial strains and stresses are not considered)

$$\varepsilon_x = (\sigma_x - \nu\sigma_y)/E$$

$$\varepsilon_y = (\sigma_y - \nu\sigma_x)/E$$

$$\varepsilon_{xy} = 2(1 + \nu)\,\sigma_{xy}/E$$

which after solving for stresses gives

$$\underset{\sim}{\sigma} = \underset{\approx}{D} \, \underset{\sim}{\varepsilon} . \tag{3.4}$$

The matrix $\underset{\approx}{D}$ is called the material matrix. The action of adjacent elements on the element under consideration is replaced by equivalent nodal forces. These forces deform the element by shifting the nodal points by displacements u_i and thus strain the element throughout.

The natural procedure for deriving the conditions of equilibrium is the principle of virtual work.

A virtual displacement at the nodes $\delta \underset{\sim}{u}_i$ results in virtual strains within the element equal to

$$\delta \underset{\sim}{\varepsilon} = \underset{\approx}{B} \ \delta \underset{\sim}{u}_i . \tag{3.5}$$

The internal work per unit volume done by the stresses is

$$W_i = \delta \underset{\sim}{\varepsilon}^T \underset{\sim}{\sigma} = \delta \underset{\sim}{u}_i^T \underset{\approx}{B}^T \underset{\sim}{\sigma} . \tag{3.6}$$

The external work of nodal forces (external with respect to the element, internal with respect to the structure) is

$$W_e = \delta \underset{\sim}{u}_i^T \underset{\sim}{f}_i . \tag{3.7}$$

Integrating the internal work (3.6) over the volume of the element gives the total internal work, and this must be equal to the virtual work of external forces:

$$\delta \underset{\sim}{u}_i^T \cdot \underset{\sim}{f}_i = \delta \underset{\sim}{u}^T \int_V \underset{\approx}{B}^T \underset{\sim}{\sigma} \, dV . \tag{3.8}$$

Eq. (3.8) must be satisfied for any values of virtual displacements u_i, so that

$$\underset{\sim}{f}_i = \int_V \underset{\approx}{B}^T \underset{\sim}{\sigma} \, dV . \tag{3.9}$$

This is valid for any stress-strain relationship.

The linear elastic behaviour of the material, as expressed by the relationship (3.4), gives

$$\underset{\sim}{f}_i = \int_V \underset{\approx}{B}^T \underset{\approx}{D} \underset{\sim}{\varepsilon} \, dV = \int_V \underset{\approx}{B}^T \underset{\approx}{D} \underset{\approx}{B} \underset{\sim}{u}_i \, dV \tag{3.10}$$

or

$$\underset{\sim}{f}_i = \underset{\approx}{K}_e \underset{\sim}{u}_i$$

where the element stiffness matrix is given by

$$\underset{\approx}{K}_e = \int_V \underset{\approx}{B}^T \underset{\approx}{D} \underset{\approx}{B} \, dV. \tag{3.11}$$

Eq. (3.10) is formally identical to Eq. (2.6) derived for the truss structure in the previous Section.

After obtaining the element stiffness matrices, the assembly procedure is identical to that described in the previous Section for the truss structure. Formally, it can be derived by applying the principle of virtual work to the whole system of elements. Internal forces will be replaced by external forces acting at the nodes. The virtual work of a given external force is balanced by part of the strain energy only in those elements connected to the node at which the force is applied. This means that only these displacements enter the equation having the force on the left side. The system of equations (2.7) will be banded, and a substantial saving of computer storage can be achieved by proper nodal point numbering.

The principle of virtual work is not the only method available for the finite element formulation.

The same result can be achieved by the principle of minimization of total potential energy.

For a linear elastic material the strain energy U of a solid of volume V is given by

$$U = \frac{1}{2} \int_V \underset{\sim}{\varepsilon}^T \underset{\approx}{D} \underset{\sim}{\varepsilon} \, dV \quad . \tag{3.12}$$

The work W of external forces is

$$W = \underset{\sim}{u}^T \underset{\sim}{f}_i . \tag{3.13}$$

Replacing ε by the expression (3.3), the total potential energy P = U - W can be written as

$$P = \frac{1}{2} \int_V \underset{\sim}{u}^T \underset{\approx}{B}^T \underset{\approx}{D} \underset{\approx}{B} \underset{\sim}{u} \, dV - \underset{\sim}{u}^T \underset{\sim}{f} \, . \tag{3.14}$$

It is easy to verify that the system of equations (2.7) can be derived by minimization of the potential energy P.

The minimum of P is achieved by setting

$$\frac{\partial P}{\partial \underset{\sim}{u}} = 0 \, . \tag{3.15}$$

Applying this procedure to Eq. (3.14) yields

$$\left(\int_V \underset{\approx}{B}^T \underset{\approx}{D} \underset{\approx}{B} \, dV \right) \underset{\sim}{u} = \underset{\sim}{f} \, ,$$

which is identical to Eq. (2.7)

Recognition of the fact that the finite element formulation requires the total potential energy to be a minimum, was of critical importance for the further development of the finite element method as will be demonstrated in the next Section.

4. MATHEMATICAL CONCEPT

We have seen in the previous section that the system of equations (2.7) can be derived either by the principle of virtual work or by minimization of the total potential energy. Recognition of the equivalence of these approaches had a fundamental influence on the development of the finite element method. The integral of the total potential energy is called a functional. There are many other mathematical and physical problems in which we can define functionals whose stationary value represents the solution of the problem. This then means that the finite element method, originally developed for the solution of structural problems of solid mechanics, can be used for the solution of many other field problems.

Further, engineers and physicists have been using the Rayleigh-Ritz method since its discovery at the beginning of this century for a direct numerical technique for optimizing functionals. The principle of the Rayleigh-Ritz method is summarized as follows: a) a finite number of trial functions $\phi_1, \phi_2, \ldots\ldots, \phi_n$ are chosen; b) these functions are combined in the linear form $u = \Sigma a_i \phi_i$; c) the unknown coefficients a_i are determined in such a way that the functional I(u) is minimized. If this idea is compared with the minimization of the potential energy from the last Section, it is clear that nodal displacements correspond to the unknown parameters a_i and the shape functions N_i to the trial functions ϕ_i However, there is a fundamental difference between the classical Rayleigh-Ritz method and the finite element method. In the Rayleigh-Ritz method the trial functions are defined over the whole region and have to satisfy fixed boundary conditions along the whole boundary. The more complex the boundary or the greater the accuracy required, the more trial functions of greater accuracy have to be included. This limits the Rayleigh-Ritz method substantially for engineering application.

On the other hand, the finite element method subdivides the region into smaller subdivisions (elements) and within each element very simple trial (shape) functions are defined, usually polynomials of a low degree. Accurate approximation of complex boundaries is achieved by mesh refinement.

This equivalence of the finite element method and the Rayleigh-Ritz method gives the method a respectable mathematical basis. Error estimates and convergence criteria already existing for the classical Rayleigh-Ritz method can be applied to the finite element method.

It is worth mentioning a mathematical paper of Courant (2) from 1941. When solving a plane torsion problem, he realized that the application of the Rayleigh-Ritz method to the whole domain would be very cumbersome and therefore he divided the domain in a net of small triangular subdomains. He then defined linear trial functions over each subdomain separately and replaced the functional by a finite sum of subdomain contributions. The results were function values in nodal points. We can see that this procedure contains all basic features of the finite element technique. Unfortunately, its potential for numerical solution of practical problems was not recognized for more than two decades. This can be seen in mathematical textbooks on numerical methods for the solution of partial differential equations which only now include the finite element method after thirty years later (see e.g. (7) with no reference at all, new edition (8) with a brief chapter).

The Rayleigh-Ritz method can be applied only to functionals. Unfortunately, there are many non-structural problems where the functionals do not exist or are difficult to find. The partial differential equations governing the problem are usually known. In such cases, the classical methods of weighted residuals can be introduced in a similar way to the classical Rayleigh -Ritz method and a finite element approximation evaluated.

Let us demonstrate the procedures mentioned above on the example of a steady-state heat transfer problem in two dimensions.

The problem is governed by the differential equation

$$L(u) = \frac{\partial}{\partial x}\left(k\frac{\partial u}{\partial x}\right) + \frac{\partial}{\partial y}\left(k\frac{\partial u}{\partial y}\right) + Q(u) = 0 \ . \tag{4.1}$$

The variable u corresponds to the temperature inside a body Ω .

Let us assume for simplicity that u = 0 on the boundary Γ .

If the variable k (conductivity) in Eq. (4.1) is constant, the problem of solving the differential Eq. (4.1) is equivalent to the minimization of the functional

$$I(u) = \int_{\Omega} \left[\tfrac{1}{2}k\left(\frac{\partial u}{\partial x}\right)^2 + \tfrac{1}{2}k\left(\frac{\partial u}{\partial y}\right)^2 - Qu\right] d\,\Omega \ . \tag{4.2}$$

The proof is simple. Let us calculate a small change (variation) in I due to a change in u. Following the rules of calculus of variation, the variation in I is

$$\delta I = \int \left(k \frac{\partial u}{\partial x}\left(\frac{\partial u}{\partial x}\right) + k \frac{\partial u}{\partial y}\left(\frac{\partial u}{\partial y}\right) - Q\,\delta u\right) d\Omega \ . \tag{4.3}$$

Since $\delta\,(\partial u/\partial x) = \partial(\delta u)/\partial x$, then Eq. (4.3) can be integrated by parts. We then have

$$\delta I = -\int_{\Omega} \delta u \left\{ \frac{\partial}{\partial x}\left(k\frac{\partial u}{\partial x}\right) + \frac{\partial}{\partial y}\left(k\frac{\partial u}{\partial y}\right) + Q \right\} d\Omega + \int_{\Gamma} \delta u\left(k\frac{\partial u}{\partial n}\right) d\Gamma = 0 \cdot \tag{4.4}$$

The boundary integral over Γ is identically equal zero for $u = 0$. This follows from the given boundary condition $u = 0$. Inside the domain Ω δu is arbitrary. To satisfy Eq. (4.4), where $\delta I = 0$ for any δu, the expression in brackets must be zero. But this is identical to the original Eq. (4.1).

Instead of solving the differential equation (4.1) together with necessary boundary conditions, the function (4.2) can be solved directly by approximating the unknown temperature u by the function v.

$$u \simeq v = \Sigma N_i v_i \ , \tag{4.5}$$

where N_i are shape functions and V_i are temperatures of the nodes.

Inserting (4.5) into (4.2) gives

$$I = \int_{\Omega} \tfrac{1}{2} k \left(\Sigma \frac{\partial N}{\partial x}i\ v\right)^2 d\Omega + \int_{\Omega} \tfrac{1}{2} k \left(\Sigma \frac{\partial N}{\partial y}i\ v\right)^2 d\Omega$$
$$- \int_{\Omega} Q\ \Sigma N_i v_i\ d\Omega = 0 \ . \tag{4.6}$$

The stationary value of the approximated function (4.6) can be obtained by differentiation of (4.6) with respect to the parameters v_j, $j = 1, \ldots, n$. The system of equations (4.7) can be written in matrix form as

$$\underset{\approx}{K}\ \underset{\sim}{v} = \underset{\sim}{f} \ , \tag{4.8}$$

where $K_{ij} = K_{ji} = \int_{\Omega} k \left(\frac{\partial N_i}{\partial x}\frac{\partial N_j}{\partial x} + \frac{\partial N_i}{\partial y}\frac{\partial N_j}{\partial y}\right) d\Omega$,

$$f_j = \int_{\Omega} N_j\ Q\ d\Omega \ .$$

The system (4.8) is again formally equivalent to the system of equations (2.7).

The functional (4.2) assumes constant k. If k is not constant, the Rayleigh-Ritz method cannot be used because there is no functional related to

Eq. (4.1). A Galerkin method of weighted residuals must then be applied.

The unknown u is again approximated by (4.5). Assuming that the boundary conditions are satisfied, the approximate solution v can be made to satisfy (4.1) in the mean by making the residual L(v) orthogonal to a set of weighting functions w, that is

$$\int_\Omega w \, L(v) \, d\Omega = 0. \tag{4.9}$$

If the weighting functions are chosen as trial functions N, the well-known Galerkin procedure is obtained. Applied to the heat transfer problem we obtain

$$\int_\Omega N_i \left(\frac{\partial}{\partial x}\left(k \frac{\partial v}{\partial x}\right) + \frac{\partial}{\partial y}\left(k \frac{\partial v}{\partial y}\right) - Q \right) d\Omega = 0. \tag{4.10}$$

Integrating by parts and realizing that $N_i = 0$ at Γ
we obtain

$$\int_\Omega k \left(\frac{\partial N_i}{\partial x} \frac{\partial v}{\partial x} + \frac{\partial N_i}{\partial y} \frac{\partial v}{\partial y} \right) d\Omega - \int_\Omega N_i \, Q \, d\Omega = 0. \tag{4.11}$$

Expressing the derivatives of **v** by (4.5) will result again in the equation (4.8). This is not surprising because the Rayleigh-Ritz method and Galerkin method are identical for the problems where a functional exists.

If the original differential equation (4.1) is compared with the variational formulation (4.2) or the Galerkin formulation (4.10) a very important difference regarding the continuity of v can be immediately recognized. The differential equation (4.1) requires the second derivatives of u to exist throughout the domain. The variational or Galerkin formulation contain only the first order derivatives of v. A lower order approximation v may be used for the approximation in (4.5). Also the function u has to satisfy all boundary conditions (geometric and natural) while the approximation v has to satisfy only geometric conditions because the natural boundary conditions are satisfied implicitly by the formulation.

5. CONVERGENCE

The continuum has an infinite number of degrees of freedom. It is approximated by a finite number of shape functions so that the minimum of the potential energy (functional) is only approximated. As in the classical Rayleigh-Ritz method, the approximation of the true displacement distribution can be considered as an additional constraint of the system. Thus the true minimum of energy may be reached only in the limit when the size of the elements goes to zero. The consequence is that the finite element model is always stiffer than the true mathematical model.

We would certainly like to be sure that the true displacement distribution can be approximated as closely as necessary. This requirement can be achieved only if the chosen shape functions satisfy special criteria of completeness and conformity.

Mathematical definition of conformity can be found elsewhere (e.g. (9) or (10)). We will use a physical definition applied to the displacement functions.

Completeness requires that the displacement function is able to represent the condition of constant strain including zero strain.

Conformity (also called compatibility) requires continuity of displacements between elements so that the strains at the interface between elements remain finite.

Both criteria need to be satisfied in the limit as the size of the elements tends to zero. However, an improved accuracy is achieved if these criteria are satisfied already on elements of finite size.

If we consider that in the plane-stress problems (linear elastic) the strains are defined by first derivatives of displacements, then conformity requires C_0 continuity of the displacement functions. In beam, plate or shell problems, the strains depend on the second derivative of displacements, so that C_1 continuity is required for the displacement functions.

There are many mathematical studies analysing error bounds and rate of convergence of the finite element approximations. Although they have not yet proved to be particularly useful for quantitative estimates because they contain terms depending on the unknown solution, they nevertheless allow us to assess the order of convergence.

Again, the reader is referred for mathematical discussion elsewhere. A heuristic proof of convergence is now given in the manner presented in (11).

A polynomial expansion is assumed to approximate displacements in an element. If this expansion fits the correct solution exactly, the approximation will yield the exact answer. This is the case of beam elements, for instance, where a cubic polynomial corresponds to the exact solution of the beam model based on the elementary beam theory.

Let us now assume that the exact solution is expanded in the vicinity of a point (x_i, y_i) within the element by use of the Taylor series:

$$\underset{\sim}{u} = \underset{\sim}{u}_i + \left(\frac{\partial \underset{\sim}{u}}{\partial x}\right)_i x + \left(\frac{\partial \underset{\sim}{u}}{\partial y}\right)_i y + \ldots\ldots \tag{5.1}$$

where x and y are sides of the element, both of "size" h. If within an element of "size" h a polynomial expansion of order p is employed, this will fit locally the Taylor expansion up to that order and the error in displacements will be of the order of neglected terms, i.e. $0(h^{p+1})$. The strains (and stresses) are given by m-th derivative of displacements, so that they would converge with an error of $0(h^{p+1-m})$.

In the case of the plane stress, a linear polynomial $(p = 1)$ satisfies the compatibility criteria. The rate of convergence of displacements is then $0(h^2)$. The strain is defined as the first derivative of displacement, $m = 1$. Then the rate of convergence in strains and stresses is $0(h)$. Suppose that in a given mesh the elements are halved in size. Then the error in displacements is reduced to 25% of the original error, but the error in strains (and stresses) is reduced only to 50%. This is certainly a very practical result and a user of the finite element method should be well aware of it. In fact, by solving the same problem with two different meshes, an almost exact solution in any point can be extrapolated, in the case of a monotonic convergence.

Other errors are introduced by computers. The round-off errors depend on the number of significant digits carried in the computer. In some machines double or higher precision is necessary in order to keep the errors below reasonable limits.

Evaluation of the integrals in Eq. (4.8) is performed numerically. This introduces another error. Surprisingly, this error has an influence on the minimum value of the energy functional opposite to that introduced by the finite element discretization, that is, it makes the structure softer. This fact is exploited very efficiently in many element formulations by using reduced integration to make element behaviour closer to that of the real structure. Convergence is reserved, however it will not be monotonic. This method of improvement is discussed in Section 6.

6. SHAPE FUNCTIONS AND ELEMENT MATRICES OF CONFORMING ELEMENTS

As stated in Section 3, the displacement within a finite element is approximated in terms of the displacements of the element nodes, i.e.

$$\underset{\sim}{u} = \underset{\sim}{N}^T \underset{\sim}{u}_i \; . \tag{6.1}$$

The functions N_i must be chosen in an appropriate way in order to achieve convergence with mesh refinement. As shown in Section 5, the conditions for the convergence are those of completeness and compatibility of the shape functions (6.1).

Compatibility is automatically achieved in beam and truss elements where the adjacent elements are connected in one point. The truss element requires C_o continuity of $\underset{\sim}{u}$, while the beam requires C_1 continuity (continuity in slopes and displacements).

Compatibility is also relatively easily achieved in two- and three-dimensional solids because only C_o continuity of $\underset{\sim}{u}$ is required.

Plate bending and thin shell analyses present a problem because of the C_1 compatibility requirement, and strain being defined as the second order derivative of displacement. In fact, very few fully compatible shell elements have been developed. The reader may be surprised that compatible shell elements are used less and less in modern computer programs because of their impracticability. Non-compatible elements are preferred (this will be dicussed in Section 7).

A vast number of elements had been developed for 2-D and 3-D solids. However, the user convenience has become an increasingly important consideration to program developers and almost all of the modern programs use one type of the solid element. This is the so-called isoparametric solid element. Its formulation allows any distorted configuration or degeneration into lower order elements (for example, quadrilateral into triangle by simply merging two nodes together). Its edges can be curved so that more complicated boundaries can be easily approximated.

A two-dimensional plane-stress isoparametric quadrilateral will be briefly discussed. An extension of the formulation into three-dimensions or axi-symmetric types is straightforward and can be found in the literature (e.g. (11)).

The basic idea is quite simple. The region is subdivided into elements which may be quadrilaterals (or triangles) with straight or curved edges.

Cartesian (global) coordinates are transformed to a new cartesian (r-s) system in which element edges become straight lines with r or s equal -1 or +1 (Fig. 6.1).

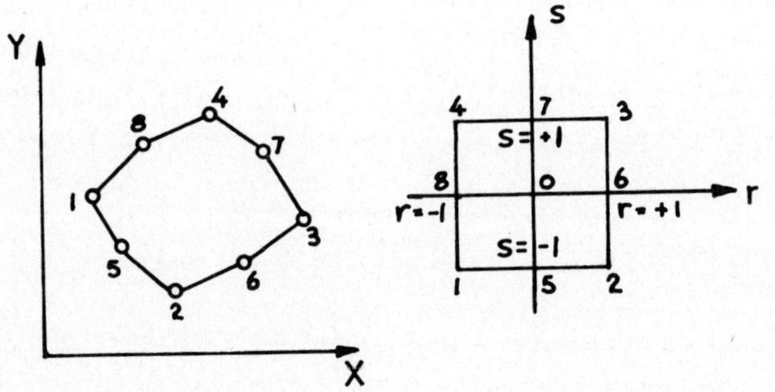

Figure 6.1

The nodes are transformed into corresponding nodes in the r-s plane. Fig. 6.1 shows this transformation for an eight node quadrilateral. The four node quadrilateral is obtained as a special case by deleting nodes 5 to 8. Higher approximations are also possible but above the cubic (4 nodes to each edge) internal nodes are required in order to satisfy compatibility, which is very impractical and so rarely used in computer programs.

The bilinear approximation of x and y will transform the r-s square into x-y quadrilateral with straight edges. The biquadratic approximation is required for the eight node element:

$$x(r,s) = a_1 + a_2 r + a_3 s + a_4 rs + a_5 r^2 + a_6 s^2 + a_7 r^2 s + a_8 rs^2. \qquad (6.2)$$

We express y(r,s) in a similar fashion.

The unknown parameters a_i are calculated from the conditions that vertices in (x,y) plane must become vertices in (r,s) plane.

The resulting transformation for the eight node element is

$$x = \sum_{i=1}^{8} N_i x_i$$

$$y = \sum_{i=1}^{8} N_i y_i \qquad (6.3)$$

where x_i, y_i are global coordinates of the nodes (known) and N_i are the shape functions. By inserting x_i (y_i) and corresponding r_i, s_i into Eq. (6.2) for all eight nodes, the eight unknown parameters a_i can be evaluated and shape functions N_i obtained for both x and y.

The result is

$$\begin{aligned} N_1 &= \tfrac{1}{4}(1-r)(1-s) - \tfrac{1}{2}N_5 - \tfrac{1}{2}N_8 \\ N_2 &= \tfrac{1}{4}(1+r)(1-s) - \tfrac{1}{2}N_5 - \tfrac{1}{2}N_6 \\ N_3 &= \tfrac{1}{4}(1+r)(1+s) - \tfrac{1}{2}N_6 - \tfrac{1}{2}N_7 \\ N_4 &= \tfrac{1}{4}(1-r)(1+s) - \tfrac{1}{2}N_7 - \tfrac{1}{2}N_8 \\ N_5 &= \tfrac{1}{2}(1-r^2)(1-s) \\ N_6 &= \tfrac{1}{2}(1-s^2)(1+r) \\ N_7 &= \tfrac{1}{2}(1-r^2)(1+s) \\ N_8 &= \tfrac{1}{2}(1-s^2)(1-r) \end{aligned} \qquad (6.4)$$

It is easy to prove that after inserting the appropriate (r,s) coordinates of a nodal point, the expansion (6.3) gives (x,y) coordinates of this node.

Also the expansion of Eq. (6.3) for an edge in (r,s) will contain only the coordinates of the nodes located on the edge (e.g. for s = -1, Eq. (6.3) will contain the nodes 1,5,2 only). This means that adjacent elements will have common edges after the transformation given by Eq. (6.3).

The isoparametric approximation is obtained by using the same shape functions (6.4) for the approximation of displacements, i.e.

$$\underset{\sim}{u} = \sum_{i=1}^{8} N_i \underset{\sim}{u}_i \quad , \qquad (6.5)$$

where $\underset{\sim}{u}_i = \begin{bmatrix} u_i \\ v_i \end{bmatrix}$ are the nodel point displacements,

$\underset{\sim}{u} = \begin{bmatrix} u \\ v \end{bmatrix}$ is the displacement of a point (r,s)

$N_i = N_i(r,s)$ is the corresponding shape function from Eq. (6.4).
Following our discussion about coordinates of points along the edges, the displacement function (6.5) will have C_0 continuity across element boundaries, that is the same displacements will be obtained along the edge of both adjacent elements from nodal point displacements because only displacements of nodal points common to both elements enter the function (6.5) along the edge. Hence the isoparametric element is compatible.

To prove completeness let us consider a linear displacement field applied to the element

$$u = a_1 + a_2 x + a_3 y \ ,$$
$$v = b_1 + b_2 x + b_3 y \ . \tag{6.6}$$

At the nodes the displacements would be

$$u_i = a_1 + a_2 x_i + a_3 y_i \ ,$$
$$v_i = b_1 + b_2 x_i + b_3 y_i \ . \tag{6.7}$$

In the isoparametric formulation the displacements are approximated as

$$u = \sum_{i=1}^{8} N_i u_i \ ,$$
$$v = \sum_{i=1}^{8} N_i u_i \ . \tag{6.8}$$

Inserting (6.7) into (6.8) gives

$$u = a_1 \Sigma N_i + a_2 \Sigma N_i x_i + a_3 \Sigma N_i y_i , \quad v= \ \ldots . \tag{6.9}$$

However, the coordinate transformation (6.3) states that $x= \Sigma N_i x_i$, $y= \Sigma N_i y_i$ so that (6.9) yields

$$u = \Sigma N_i + a_2 x + a_3 y \ ,$$
$$v = \Sigma N_i + b_2 x + b_3 y \ . \tag{6.10}$$

This is equivalent to (6.6) providing $\Sigma N_i = 1$.
It can easily be shown by inspection of expressions (6.4) that the condition $\Sigma N_i = 1$ is satisfied for four and eight node isoparametric element. Therefore isoparametric elements are complete. This is valid for any isoparametric element.

The element stiffness matrix is given by Eq. (3.11), that is

$$\underset{\approx}{K}_e = \int_V \underset{\approx}{B}^T \underset{\approx}{D} \underset{\approx}{B} \, dV. \tag{6.11}$$

The matrix $\underset{\approx}{B}$ can be written as

$$\underset{\approx}{B} = \left[\underset{\approx}{B}_1 \; \underset{\approx}{B}_2 \; \underset{\approx}{B}_3 \; \cdots \; \underset{\approx}{B}_n \right]. \tag{6.12}$$

n is the number of element nodes and submatrices B_i are given as

$$\underset{\approx}{B}_i = \begin{bmatrix} \frac{\partial N_i}{\partial x} & 0 \\ 0 & \frac{\partial N_i}{\partial y} \\ \frac{\partial N_i}{\partial y} & \frac{\partial N_i}{\partial x} \end{bmatrix}. \tag{6.13}$$

We can see that the components of the $\underset{\approx}{B}$ matrix are given by partial derivatives of the shape function N_i with respect to the global coordinates (x,y). However the shape functions N_i are specified in local coordinates r,s as defined in Eq. (6.4).

The evaluation of the integral (6.11) in local coordinates r,s requires two transformations: one for the global derivatives of the N_i's in the $\underset{\approx}{B}$ matrices, and a second for the elementary volume dV = dx dy.

Partial differentiation of N_i gives

$$\begin{bmatrix} \frac{\partial N_i}{\partial r} \\ \frac{\partial N_i}{\partial s} \end{bmatrix} = \underset{\approx}{J} \begin{bmatrix} \frac{\partial N_i}{\partial x} \\ \frac{\partial N_i}{\partial y} \end{bmatrix}, \tag{6.14}$$

where the matrix $\underset{\approx}{J}$ is the Jacobian matrix, given as

$$\underset{\approx}{J} = \begin{bmatrix} \frac{\partial x}{\partial r} & \frac{\partial y}{\partial r} \\ \frac{\partial x}{\partial s} & \frac{\partial y}{\partial s} \end{bmatrix} = \begin{bmatrix} \Sigma \frac{\partial N_i}{\partial r} \cdot x_i & \Sigma \frac{\partial N_i}{\partial r} \cdot y_i \\ \Sigma \frac{\partial N_i}{\partial s} \cdot x_i & \Sigma \frac{\partial N_i}{\partial s} \cdot y_i \end{bmatrix}. \tag{6.15}$$

Assuming that the inverse of the Jacobian matrix exists, the global derivatives of N_i's can be evaluated from (6.14).

The transformation of the elementary volume is

$$dV = dx\,dy = \det(\underset{\approx}{J})\,dr\,ds \; . \qquad (6.16)$$

The sign of the Jacobian determinant must remain unchanged throughout the element if one-to-one mapping as defined by Eq. (6.2) should exist. It can be shown that the necessary condition for isoparametric elements is that element angles are less than 180° and for eight-node elements the middle nodes are in the middle third of the side on which they are located (11).

As a result of the transformation, the integral (6.11) is written as

$$\int_{-1}^{1}\int_{-1}^{1} \underset{\approx}{F}(r,s)\,dr\,ds \; . \qquad (6.17)$$

The function $\underset{\approx}{F}(r,s)$ cannot be integrated analytically. Numerical integration has to be used for the evaluation, and this creates further interesting problems.

7. NUMERICAL INTEGRATION OF ELEMENT MATRICES

The integral (6.17) can be solved analytically for only a very few simple elements. Almost invariably, element matrices are evaluated using numerical integration technique.

Principles and details of various methods of numerical integration can be found elsewhere (for example, (12)).

Two integration methods mostly used are: a) Newton-Cotes quadrature, b) Gauss quadrature.

Newton-Cotes quadrature such as the trapezoidal or Simpson rule is mainly used for the evaluation of integrals over the cross-section of beams in large displacement analysis (13).

Gauss quadrature is used for the evaluation of matrices of isoparametric elements because such a scheme requires the least number of evaluations of functions in Eq. (6.17).

Numerical integration introduces an additional error into the calculation by replacing the integral (6.17) by a weighted sum of a few function values. It also requires a substantial amount of computer time. Therefore, it is important to determine the minimum integration order necessary for convergence. There is no reason to expect that exact integration is required in order to achieve convergence. As already shown in Section 5, a condition on the displacement (shape) function is to achieve a state or constant strain in the limit of zero element size. The energy functional (3.12) is, in case of elements with C_0 continuity, a function of first derivatives of the displacement functions N. The rate of convergence of strains is $O(h^{p-m+1})$, if the strains are given by m-th derivatives of displacements. The energy is given by the square of strains, so that the rate of convergence in energy is $O(h^{2(p-m)+1})$.

The minimum rate of convergence is O(h). This will be achieved if p = m, or p = 1 for C_0 elements. Such approximation is sufficient for convergence and will result in a constant value of F in the integral (6.17). This means that convergence will occur if the volume of the element

$$\int_V dV = \int_{-1}^{1}\int_{-1}^{1} \det(\underset{\approx}{J})\, dr\, ds$$

is correctly evaluated by numerical integration. It can easily be shown that for bilinear approximation, the det $(\underset{\approx}{J})$ is a linear function of r and s because the terms with the rs-product cancel (10). This means that one point integration is sufficient for this type of element.

Although bilinear approximation with one integration point is sufficient to assure convergence, higher order approximation will result in a higher rate of convergence. Quadratic elements have p = 2 so that the error is $O(h^3)$ if elements are not too much distorted, i.e. det $(\underset{\approx}{J})$ = const. If quadratic elements are distorted, det $(\underset{\approx}{J})$ is a polynomial of the 4th order so that p = 3 is required.

The above values are related to approximations at element level. The element stiffness matrices are then assembled and the system of linear algebraic equations

$$\underset{\approx}{K}_e \, \underset{\sim}{u}_i = \underset{\sim}{f} \tag{7.1}$$

has to be solved after the boundary conditions are inserted. With exact integration the matrix $\underset{\approx}{K}_e$ will then not be singular and the system (7.1) can be solved.

The question is how the replacement of the exact integration by the numerical integration can influence the singularity of $\underset{\approx}{K}$.

The independent relations are expressed by a strain-displacement relationship which expresses strain components in terms of a linear form of the nodal point displacements $\underset{\sim}{u}_i$. In C_o plane problems, there are three such equations (see Eq. (3.3)).

When numerical integration is used, the strains are evaluated at each of the integration points. If the number of degrees of freedom (components of $\underset{\sim}{u}_i$) is larger than the number of independent strain components, the element matrix $\underset{\approx}{K}_e$ will be singular.

Let us consider four and eight node quadrilateral elements. Fig. 7.1 shows a structure modelled by one element only. The rigid body modes are removed by appropriate supports.

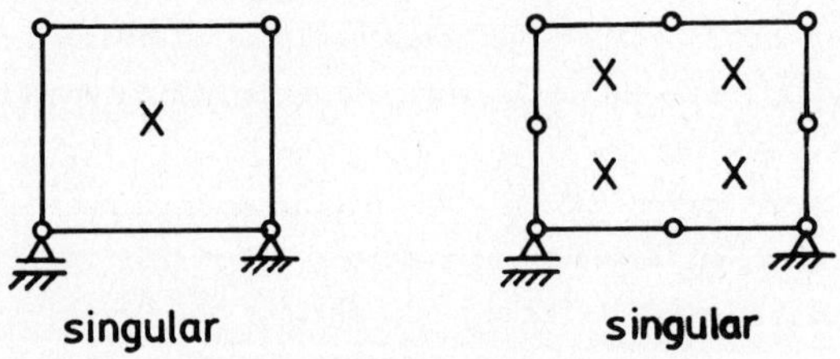

Figure 7.1

There are two degrees of freedom per node (translations in x and y direction).

In both cases, the matrix $\underset{\approx}{K}$ will be singular because the number of unknown parameters exceeds the number of independent relations.

Fig. 7.2 shows a structure modelled by two elements. The four node quadrilateral is still singular, the eight node quadrilateral is non-singular.

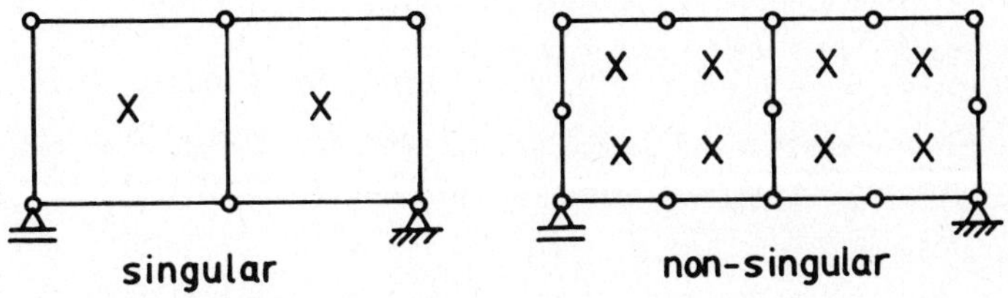

Figure 7.2

Elements with the single integration point would require more constraints in order to be non-singular. Therefore they are very rarely found in finite element programs.

The eight node quadrilateral elements with four integration points are today the most commonly used plane elements and twenty node solid elements with 2 x 2 x 2 integration points are the most commonly used three-dimensional elements.

However, as mentioned above, one integration point is sufficent for convergence with an error O(h) for any C_0 element. In fact, it was found by computer experiments that reduced integration makes elements correspond more closely to the behaviour of real solids. As already mentioned, any methods based on the Rayleigh-Ritz or Galerkin processes result in a model (in structural analysis) which is stiffer than the real object. Reduced integration has an opposite effect and the convergence is still guaranteed (at a lower rate). This is the explanation for their success of such elements in practice. However, the user of finite element programs should be aware of some problems which can result from reduced integration, such as singularity for insufficiently constrained structures.

8. NON-CONFORMING ELEMENTS

Plates and shells require C_1 continuity (displacements and their derivatives must be continuous). Determination of shape functions satisfying convergence requirements is now much more difficult. Although conforming elements for plates and shells have been derived, they are not only impractical (second order derivatives are required at some nodes; they are very difficult to combine with other elements) but their performance is very poor (a very fine mesh is required for reasonable convergence).

During the "prehistoric" period of finite element development, when the conditions of convergence and compatibility were not clearly recognised, many non-conforming elements were designed and used. Surprisingly enough, some of them gave much better results than conforming elements. It was important to explain this fact.

If strains are defined by second order derivatives of displacements (as in beams or shells), a discontinuity in first derivatives of displacements will cause infinite strains at the element interfaces. However, the convergence criteria (Section 5) need only be satisfied in the limit as element size tends to zero. Therefore, if in the limit the continuity is restored, the nonconforming element will work correctly.

A test for the acceptability of nonconforming elements has been invented as a computer test performed on an assembly of elements - a patch test (14). A patch of elements of arbitrary geometry is subject to an arbitrary constant strain condition by an appropriate deformation of the patch boundary. The patch is solved on the computer and stresses in all elements are printed. They should be identical. Strictly speaking, the patch should have infinitesimal element sizes but a fine mesh will do just as well.

The patch test was later defined mathematically, and it proved that the intuitive experimental approach as described above was correct (see, eg. (10) or (15)).

Recognition that nonconformity improves element performance suggested the idea of improving the performance of conforming elements by either adding nonconforming shape functions or by using substitute shape functions with a smoothing effect on the derivatives.

An example of improving element performance by adding nonconforming shape functions is Wilson's element (16). It is a four node isoparametric rectangular element (Section 6) to which two nonconforming shape functions are added in order to improve the element's performance in bending. Fig. 8.1(a) shows how the original element deforms if bending moments are applied, Fig. 8.1(b) demonstrates that the behaviour after two nonconforming shape functions have been added, closely corresponds to the reality.

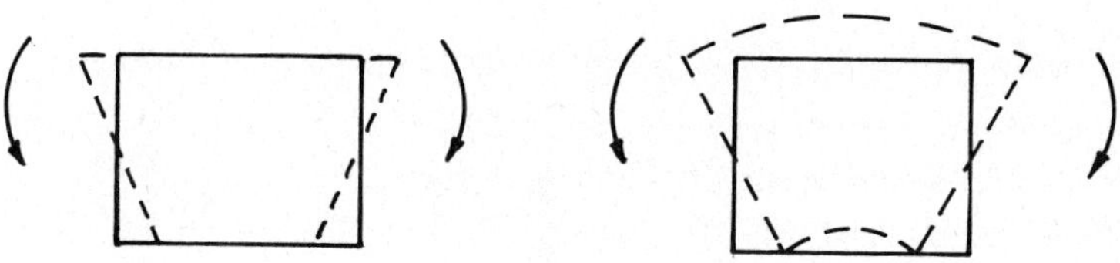

Figure 8.1

The two new shape functions are defined as

$$N_5 = (1 - r^2) ,$$

$$N_6 = (1 - s^2) . \tag{8.1}$$

They assume the values according to Eq. (8.1) inside the element. Outside of the element they are zero, therefore they are not continuous at the boundary. The two parameters u_5 and u_6 have no physical meaning. They are evaluated as functions of the remaining parameters (nodal displacements) from the condition of minimum total energy and they are then eliminated from the element stiffness matrix by static condensation.

This element passes the patch test for rectangular configuration (11).

A general isoparametric element does not pass the patch test. Taylor (17) satisfied the conditions of the test by altering the nonconforming shape functions and by modifying the numerical integration.

Substitute shape function can be demonstrated using again the example of four node isoparametric quadrilateral element. The shape function for node 1 ($r = -1$, $s = 1$) has the form $N = (1 - r)(1 - s)/4$. It has one term of the second order (rs) which provides twist but does not contribute to convergence. Fig. 8.2 (a) shows the shape function which is clearly not a plane. We can expect that the same results (or better if the overstiffening effect of the original method in general is considered) will be obtained by averaging the N surface by a linear function

$$\overline{N}_1 = a_1 + a_2 r + a_3 s \tag{8.2}$$

which would replace the derivatives $\frac{\partial N_1}{\partial r}, \frac{\partial N_1}{\partial s}$ by their constant averages (Fig. 8.2 (b)). The surface $\overline{N}_1$ (r,s) will then remain planar. This method has been successfully applied to plate bending elements (11). The performance of the nonconforming element is much better than that achieved by the original confirming elements.

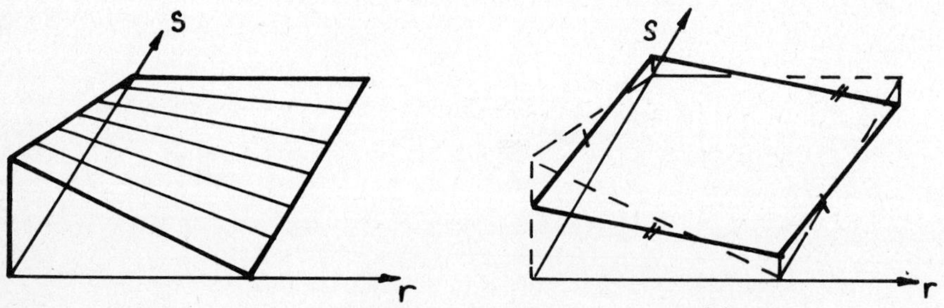

(a) (b)

Figure 8.2

One can intuitively expect that minimizing the energy functional (3.12), that is

$$U = \frac{1}{2} \int_V \underset{\sim}{\varepsilon}^T \underset{\approx}{D}\, \underset{\sim}{\varepsilon}\, dV + W \tag{8.3}$$

using the exact strains ε will be equivalent to inserting a least square fit to the exact strains. This is successful when the energy integral is evaluated exactly. Mathematical proof of this can be found in (11).

Using numerical integration, the strains are calculated in a few points only.

It is extremely important in terms of the order of the error that correct points are used. Fig. 8.3 shows exact strain (full-line) and piece-wise linear least square approximations (dotted lines). It is evident that there are points within each element where the exact and approximated strains coincide. It was shown by (18) that these points are very close to Gauss

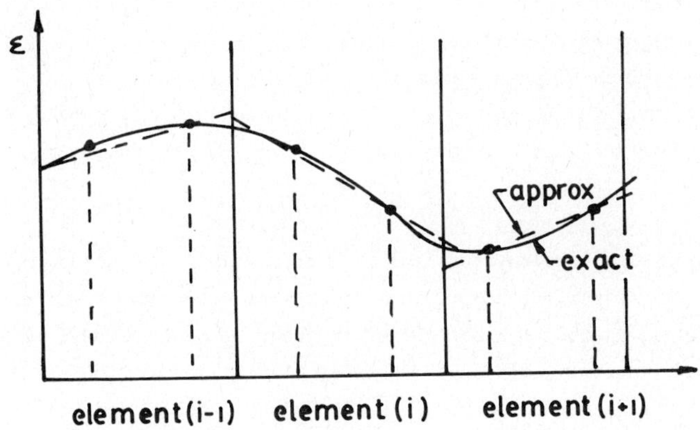

Figure 8.3

integration points (the result is exact for one-dimensional integration), which are sufficient to integrate exactly the polynomial function under the energy functional. This means that by using these points for the evaluation of strains in Eq. (8.3) the approximation will be almost one order better than sampling them elsewhere.

The practical result is that the best and most economical plane element

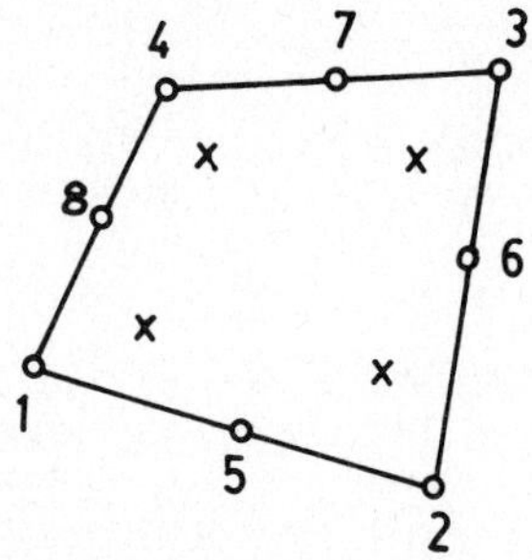

Figure 8.4

is the eight node quadrilateral (Fig. 8.4) with four Gaussian integration points. The shape functions N_i are complete polynomials of the second order (p = 2), strains are proportional to the first derivative of N_i (m = 1), that is a polynomial of order p - m = 1. The integrand in the energy integral will be a polynomial of order 2(p - m) requiring two Gaussian points (in each direction). At these points the strains are evaluated nearly one order better, that is the convergence rate of order $0(h^{p-m+1}) = 0(h^2)$ is improved to $0(h^{p-m+2}) = 0(h^3)$.

The stresses are calculated from the strains after the solution of the system (2.7). The result of the previous discussion shows that the stresses should never be calculated at nodes but at the Gauss points (1 Gauss point for a four-node quadrilateral, or 8-node 3-D solid element, 4 Gauss points for a 8-node quadrilateral, 8 Gauss points for a 20-node 3-D solid element).

Evidence that not observing this rule can lead to useless results is demonstrated in Fig. 8.5. The shaft of a seat-belt retractor was modelled by 20-node brick elements and 3 x 3 x 3 integration points were used for calculation and evaluation of stresses. The Mises stress levels plotted in Fig. 8.5 make no sense for the given loading. If the stresses are calculated at 2 x 2 x 2 = 8 integration points the results are correct.

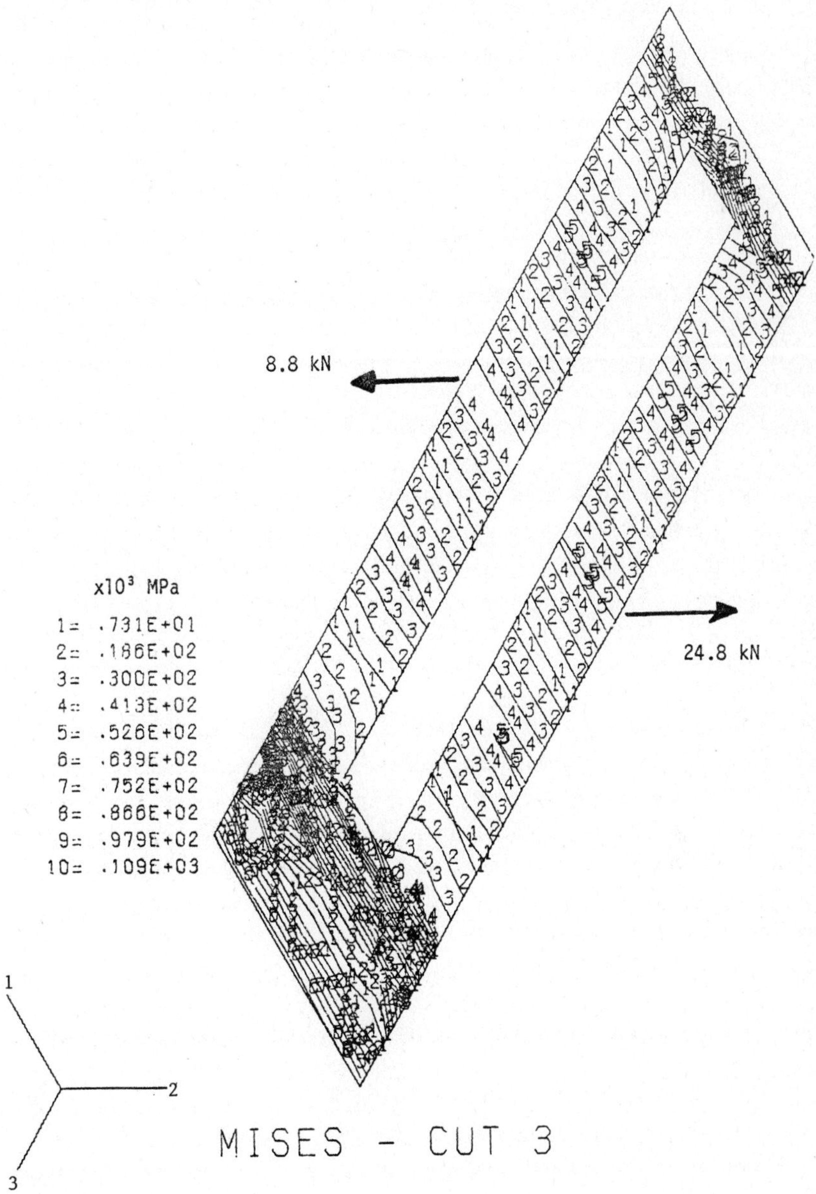
8.8 kN
24.8 kN
x10³ MPa
1= .731E+01
2= .186E+02
3= .300E+02
4= .413E+02
5= .526E+02
6= .639E+02
7= .752E+02
8= .866E+02
9= .979E+02
10= .109E+03
1
2
3
MISES - CUT 3

Figure 8.5

9. IMPOSITION OF CONSTRAINTS

Difficulties with continuity requirements in the case of beam and shell elements or difficulties in overcoming other problems such as incompressibility, have led to the formulation of various types of nonconforming elements where the coupled variables are approximated independently by separate shape functions and the relation between them is imposed as a constraint.

There are basically three methods used: a) penalty function, b) exact constraints in internal points, c) Lagrange multipliers.

The penalty function method is based on the fact that the minimization of a functional with constraints is equivalent to the minimization of the original functional with a penalty function added to it (2).

Let us consider a function $V(u)$ such as that in Eq. (4.2) and a constraint $u = g$. The solution of this problem can be achieved by minimizing the functional

$$V(u) + \alpha \int_{\Omega} (u - g)^2 \, d\Omega \tag{9.1}$$

for $\alpha \to \infty$

The Ritz method can be applied to Eq. (9.1) without any constraints. It is clear that only an approximate solution can be obtained because there is a limit on the coefficient α. For sufficiently large values of α ill-conditioned matrices will appear. However, it is reasonable to assume that there is a relationship between the error introduced by discretization (element size) and the error due to a finite value of α. Babuska (19) determined the error for geometric boundary conditions (boundary element in Section 2) where he found that

$$\alpha = c h^{-p}, \tag{9.2}$$

where h is the element "size", and p is the degree of the complete approximation polynomial.

Introducing shape functions into Eq. (9.1) and minimizing the functional with respect to the unknown parameters will result in a system of equations

$$(\underset{\approx}{K}_1 + \alpha \underset{\approx}{K}_2) \underset{\sim}{u}_i = \underset{\sim}{f}_i \; . \tag{9.3}$$

Comparing this equation with boundary element method in Section 2, we see that they are identical, $\underset{\approx}{K}_2$ being the stiffness matrix of stiff springs simulating the rigid boundary. With α very large, the $\underset{\sim}{u}_i$'s with the constraint will practically equal zero. This was required in the case of fixed boundaries.

The solution to this problem is to make the matrix $\underset{\approx}{K}_2$ singular. Then, as $\alpha \to \infty$, $\underset{\approx}{K}_2 \underset{\sim}{u}_i \to 0$, while $\underset{\sim}{u}_i \neq \underset{\sim}{0}$.

The singularity is introduced into $\underset{\approx}{K}_2$ by the use of a low order integration scheme as discussed in Section 7. The higher the degree of singularity, the greater the influence on α.

As an example let us consider the beam element (11). C_1 continuity is required (displacements and slopes). Normally Hermitian polynomials for shape functions are used to satisfy continuity requirements:

$$\underset{\sim}{u} = \begin{bmatrix} u \\ \\ \frac{du}{dx} \end{bmatrix} = \Sigma\, N_i \underset{\sim}{u}_i + \Sigma\, L_i \frac{d\underset{\sim}{u}_i}{dx} \,. \tag{9.4}$$

Using a penalty function allows us to approximate displacements and slopes independently, that is

$$u = \Sigma\, N_i\, u_i \,,$$

$$\Theta = \Sigma\, N_i\, \Theta_i \,, \tag{9.5}$$

and relate them by a constraint

$$C = \frac{du}{dx} - \Theta = 0 \,. \tag{9.6}$$

The original functional $V(u, \Theta)$ is now extended by the penalty function

$$f = \tfrac{1}{2}\alpha \int \left(\frac{du}{dx} - \Theta\right)^2 dx. \tag{9.7}$$

The influence of the singularity of the matrix $\underset{\approx}{K}_2$ is clearly demonstrated in (11) p.291. It shows the ratio of approximate to exact solution for a cantilever beam with point load at its free end. The results show that the exact integration tends to zero with large α and that remarkable results are achieved by making matrix $\underset{\approx}{K}_2$ singular by means of reduced integration.

A similar approach has been successful in applications in the field of incompressible materials (20). A constraint has been imposed on the volumetric strain which is zero for such materials.

The constraints can also be introduced in an approximate manner by imposing them at a limited number of points within the element. This introduces additional equations which allow us to eliminate some of the element variables at element level. It is clear that this procedure can violate even the C_o continuity. A patch test is necessary to assure that the element is converging.

Let us demonstrate the method again on the beam problem. The displacements and slopes are approximated independently as in Eq. (9.5). The constraint (9.6) is enforced exactly at two internal points with coordinates r_I, r_2. The beam has an additional node in the middle (Fig. 9.1).

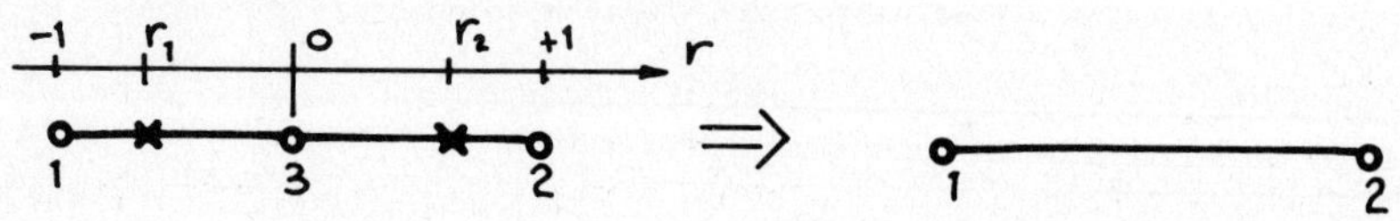

Figure 9.1

Introducing the shape functions (9.5) into the constraint (9.6) gives

$$\sum_{i=1}^{3} \frac{dN_i}{dx}(r_1)\, u_i - \sum_{i=1}^{3} N_i(r_1)\, \theta_i = 0,$$

$$\sum_{i=1}^{3} \frac{dN_i}{dx}(r_2)\, u_i - \sum_{i=1}^{3} N_i(r_2)\, \theta_i = 0. \qquad (9.8)$$

The two parameters u_3 and θ_3 of the internal node can be eliminated from the original shape functions using the relationship (9.8).

It is of interest that if the points r_1 and r_2 are Gauss points, then the stiffness matrix obtained after elimination of the internal node is precisely that obtained by Hermitian polynomials (slope and displacement continuity).

This method has been used for design of a series of plate and shell elements. Since a relatively large number of internal constraint points are now used, the elimination process is very complicated and often requires matrix inversion at element level. Clearly, the question of economy of such elements has to be considered.

An interesting example is the famous semi-loof element of Irons (14) (Fig. 9.2).

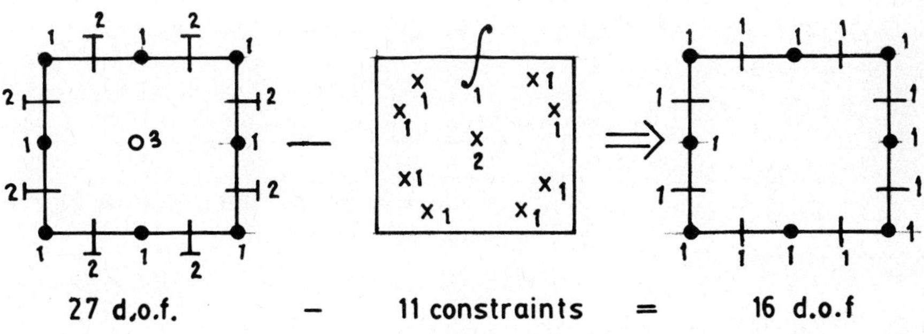

Figure 9.2

The original element has 27 degrees of freedom. This number is reduced by imposing 10 internal constraints and one integral constraint to an element with 16 degrees of freedom. The integral constraint requires that a condition in the mean on the total perimeter be satisfied.

The introduction of constraints into a variational formulation using Lagrange multipliers is a well-known method. It can also be used in the finite element formulation. The disadvantage is that additional unknowns are introduced which in the case of a large number of constraints may be prohibitive.

The Lagrangian multipliers often have physical meaning, and lead to so-called mixed variational principles and hybrid formulations. This approach allows to define compatibility inside the element and to enforce inter-element compatibility by Lagrange multipliers defined on the interelement boundary.

The interested reader can find a detailed explanation elsewhere (11).

10. NONLINEAR PROBLEMS - STATIC ANALYSIS

So far, we have applied the finite element method to linear problems. The stiffness matrix and the force vector in Eq. (3.10) were constant. In the mathematical concept, the differential equation (4.1) was linear and the functional (4.2) had the standard quadratic form.

The application of the finite element method is not limited to linear problems. It can equally well be used for the solution of problems involving very large deformations of elastic bodies, and can include nonlinear material properties (elastic-plastic analysis, creep) nonlinear field problems (diffusion, potential flow) and time dependant problems (creep, dynamic problems, impact).

Let us demonstrate the general procedure for the solution of nonlinear problems in the case of nonlinear elasticity.

The equation (3.9)

$$\underset{\sim}{f}_i = \int_V \underset{\sim}{B}^T \underset{\sim}{\sigma}\, dV \tag{10.1}$$

is still valid.

The strain-displacement relationship (3.3) is now nonlinear

$$\underset{\sim}{\varepsilon} = \underset{\sim}{\varepsilon}\,(u)\ , \tag{10.2}$$

as is also the stress-strain relationship (3.4)

$$\underset{\sim}{\sigma} = \underset{\sim}{\sigma}\,(\underset{\sim}{\varepsilon})\ . \tag{10.3}$$

Let us simplify the problem and assume that the strain ε is small and the material is linear. Then Eq. (3.4) remains in the original form (3.4)

$$\underset{\sim}{\sigma} = \underset{\approx}{D}\,\underset{\sim}{\varepsilon} \; . \tag{10.4}$$

This assumption is true for most structural problems. Large displacements occur due to rotations without causing large strains.

Let us rewrite Eq. (10.1) in the following form

$$\underset{\sim}{F}_i(u_j) = \int_V \underset{\approx}{B}^T(\underset{\sim}{u}_j)\,\underset{\sim}{\sigma}\, dV - \underset{\sim}{f}_i(\underset{\sim}{u}_j) \; . \tag{10.5}$$

There is no general method which would solve Eq. (10.5) exactly. Numerical methods must be used, and so the solution can be obtained only to a prescribed degree of accuracy. Such a solution is not unique, and the one obtained depends on the initial estimate of the solution and on the increment size used in the numerical method.

The most commonly used solution scheme in finite element programs is the combination of the incremental loading procedure and the Newton-Raphson iterative procedure (9).

<u>The Newton-Raphson method</u> is used as follows: The Eq. (10.5) is written for the $(n + 1)^{th}$ iteration step assuming that the solution for the step n is known. Then a curtailed Taylor series gives an improved solution

$$\underset{\sim}{F}_i(\underset{\sim}{u}_j^{(n+1)}) = \underset{\sim}{F}_i(\underset{\sim}{u}_j^{(n)}) + \left(\frac{\partial \underset{\sim}{F}_i}{\partial \underset{\sim}{u}_j}\right)_n \Delta \underset{\sim}{u}_j^{(n)} = 0 \; . \tag{10.6}$$

The increment $\Delta \underset{\sim}{u}_j^{(n)}$ is calculated from the linear system

$$\underset{\approx}{K}_T\, \Delta \underset{\sim}{u}_j^{(n)} = \underset{\sim}{F}_i(u_j^{(n)}) \; , \tag{10.7}$$

where $K_T = (\partial \underset{\sim}{F}_i / \partial \underset{\sim}{u}_j)_n$ is called a tangential matrix (or Jacobian matrix).

<u>The incremental loading procedure</u> is a direct iteration method. The equation (10.1) is written in the form

$$\underset{\approx}{K}(\underset{\sim}{u}_i)\underset{\sim}{u}_i = \underset{\sim}{f}_i \; . \tag{10.7}$$

If the solution at the iteration step n is $\underset{\sim}{u}_i^{(n)}$ the improved solution

$\underset{\sim}{u}_i^{(n+1)}$ can be found from Eq. (10.7), since after inserting $\underset{\sim}{u}_i^{(n)}$ into $\underset{\approx}{K}$, it becomes a linear system of equations:

$$\underset{\approx}{K}(\underset{\sim}{u}_i^{(n)})\ \underset{\sim}{u}_i^{(n+1)} = \underset{\sim}{f}_i^{(n)} \ . \tag{10.8}$$

Both the Newton-Raphson method and the incremental loading method can be used independently for the complete solution. They both have serious drawbacks. The incremental loading method often fails to converge and leads to accumulation of errors (see Fig. 10.1).

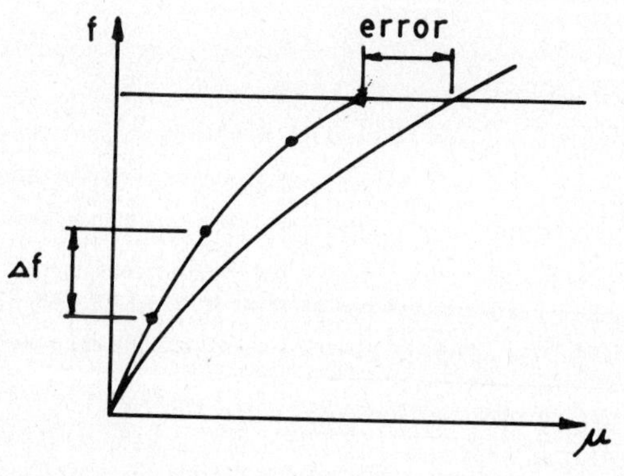

Figure 10.1

If the Newton-Raphson method is applied directly to the system (10.1) it may well diverge whenever the difference between the initial solution $u_i^{(o)}$ (found from the linearized equation) and the final solution $\underset{\sim}{u}_i^{(n)}$ is large. Also, a quite different solution than that required may be obtained because of nonlinearity.

A combination of both procedures seems to be the best and most economical way of solving Eq. (10.1).

The loads are applied in increments which are sufficiently small so that we may assume a linear solution. The result is then corrected by the Newton-Raphson method. The corrected solution is subsequently used as the.

starting point for the next incremental step (Fig. 10.2) (9).

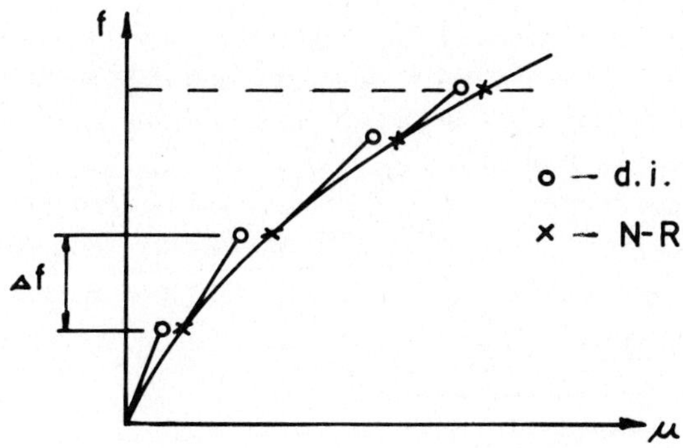

Figure 10.2

This procedure makes it possible to investigate the behaviour of the system during loading. The intermediate states can be tested for stability, bifurcation points can be found and, if necessary, multi-valued solutions may be obtained.

Calculation of the Jacobian matrix in Eq. (10.7) is computationally expensive. Most of the programs use various modifications such as the modified Newton-Raphson scheme (9) or BFGS-method (22). The trade-off is that more iteration steps are required for convergence than if the full Newton-Raphson method is used. The more non-linear the problem, the smaller the saving (13).

Let us now derive the tangential matrix $\underset{\approx}{K}_T$, which is required for the Newton-Raphson solution process (10.7) in the case of the large displacement analysis.

Eq. (10.1) is first written in an incremental form

$$\Delta F \equiv \int_V \Delta \underset{\approx}{B}^T \underset{\sim}{\sigma}\, dV + \int_V \underset{\approx}{B}^T \Delta\underset{\sim}{\sigma}\, dV - \Delta \underset{\sim}{f}_i = 0. \tag{10.9}$$

For small strain problems we have from (10.4)

$$\Delta \underset{\sim}{\sigma} = \underset{\approx}{D} \, \Delta \underset{\sim}{\varepsilon} = \underset{\approx}{D} \, \underset{\approx}{B} \, \Delta \underset{\sim}{u}_i \, , \tag{10.10}$$

so that

$$\Delta \underset{\sim}{F} = \int_V \Delta \underset{\approx}{B}^T \underset{\sim}{\sigma} \, dV + \int_V \underset{\approx}{B}^T \underset{\approx}{D} \, \underset{\approx}{B} \, dV \, \Delta \underset{\sim}{u}_i - \Delta \underset{\sim}{f}_i = 0. \tag{10.11}$$

The first matrix in Eq. (10.11) can be written as $\underset{\approx}{K}_\sigma \cdot \underset{\sim}{u}_i$ (Ref.11) and it is called the initial stress matrix. The second matrix in Eq. (10.11) can be resolved into a linear and non-linear part $\underset{\approx}{K}_L + \underset{\approx}{K}_N$. The matrix $\underset{\approx}{K}_L$ is called the small displacement matrix, and $\underset{\approx}{K}_N$ is called the initial displacement matrix. The increment in the displacement-dependant load vector $\Delta \underset{\sim}{f}_i$ can also be expressed as $\Delta \underset{\sim}{f}_i = -\underset{\approx}{R} \, \Delta \underset{\sim}{u}_i$, where the matrix $\underset{\approx}{R}$ is called the initial-load matrix. If loads do not depend on displacements, the matrix $\underset{\approx}{R}$ is zero.

The tangential matrix $\underset{\approx}{K}_T$ is then

$$\underset{\approx}{K}_T = \underset{\approx}{K}_L + \underset{\approx}{K}_\sigma + \underset{\approx}{K}_N + \underset{\approx}{R} \, . \tag{10.12}$$

The tangential stiffness matrix (10.12) can be used to predict buckling loads and modes in a modal analysis. This can be done at the beginning of the solution process or at any point during the analysis. Hibbit uses the following procedure (13) : At some point in the analysis, the stiffness matrix is stored. After one or more loading steps the stiffness of the structure is stored again. At this point, the change in stiffness associated with the added loads that would cause collapse is evaluated. The assumption is that the change in stiffness will be proportional to the change in added loads. This is a valid assumption for stiff shells or beams.

The magnification factor λ for the collapse is calculated from the eigenvalue problem.

$$(\underset{\approx}{K} + \lambda \Delta \underset{\approx}{K}) \, \underset{\sim}{v} = 0, \tag{10.13}$$

where $\underset{\approx}{K}$ is the stiffness under the "dead" load,
$\Delta \underset{\approx}{K}$ is the change in stiffness caused by the "live" loads,
λ is the "live" load magnification factor (the eigen-vector),
$\mathbf{v}$ is the buckling mode shape (the eigen-value).

Let us now demonstrate the whole process on a simple example in Fig. 10.3.

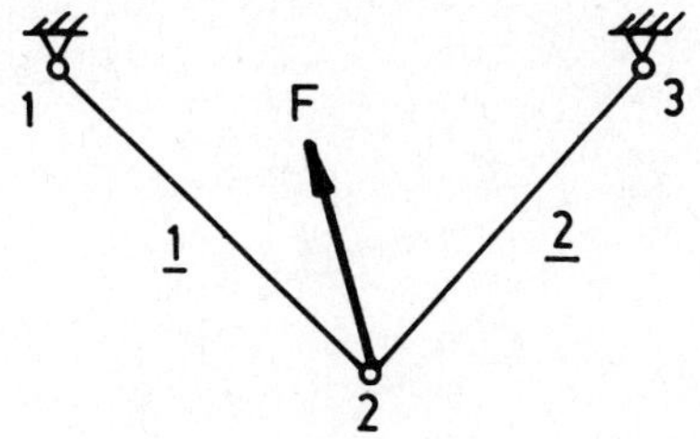

Figure 10.3

The structure consists of two truss elements 1 and 2 , fixed to the frame at points 1 and 3 and connected together at point 2. The load F is applied at point 2. The load F is large enough to cause large displacements. The stability of the structure is to be investigated at the beginning of the loading process. Following the procedure used in Section 2, the system is disassembled and the equilibrium of each element is considered in local coordinate systems (x,y). Then displacements and forces are transformed into the global coordinate system (X,Y).

Fig. 10.4 shows one of the truss elements in the original and the displaced position. u, v are the displacements of point 2 in the local coordinate system. Because the displacements are large, we have a choice of two local coordinate systems: a) in the original position (Lagrangian formulation), b) in the displaced position (Eulerian formulation). The Lagrangian formulation has been chosen in Fig. 10.4. It is generally more efficient for the solution of small strain problems.

It follows from Fig. 10.4

$$(l_o + u)^2 + v^2 = (l_o + \Delta)^2 \tag{10.14}$$

or

$$\frac{\Delta}{l_o} = \frac{u}{l_o} + \frac{u^2 + v^2}{2(l_o)^2} \tag{10.15}$$

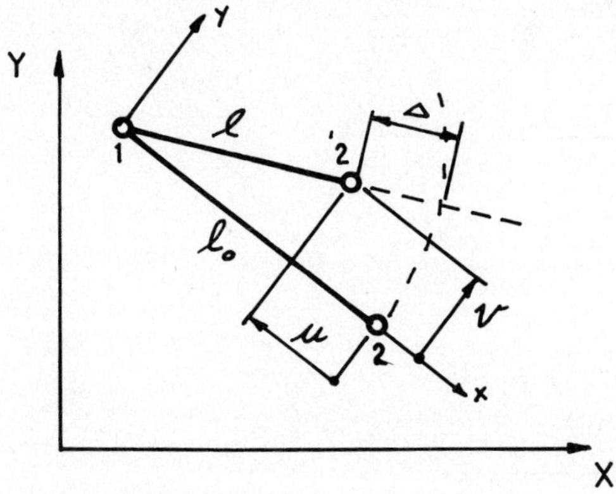

Figure 10.4

after the term $(\frac{\Delta}{l_0})^2$ has been neglected because of the assumption of small strains. Both displacements and the deformation Δ are linearly proportional to the length of the truss so that Eq. (10.15) can be written as

$$\varepsilon = \frac{\partial u}{\partial x} + \frac{1}{2}\left(\frac{\partial u}{\partial x}\right)^2 + \frac{1}{2}\left(\frac{\partial v}{\partial x}\right)^2 , \qquad (10.16)$$

where ε is the Green's strain.

To give the procedure more generality, the method of shape functions is used. An intrinsic variable r is introduced to describe the position on the element (range from -1 to 1) in the original position. The displacements and coordinates of the truss element can then be approximated (exactly in this case) as

$$\begin{bmatrix} u \\ v \\ x \end{bmatrix} = N_1 \begin{bmatrix} u_1 \\ v_1 \\ x_1 \end{bmatrix} + N_2 \begin{bmatrix} u_2 \\ v_2 \\ x_2 \end{bmatrix} . \qquad (10.17)$$

The indexed parameters are values in truss nodes, N_i are shape functions which are given by

$$N_1 = \tfrac{1}{2}(1 - r) ,$$
$$N_2 = \tfrac{1}{2}(1 + r) . \qquad (10.18)$$

The derivatives of displacements can be evaluated from Eqs. (10.17) and (10.18).

From Eq. (10.18) $\frac{\partial N_1}{\partial r} = -\frac{1}{2}$, $\frac{\partial N_2}{\partial r} = \frac{1}{2}$.

From Eq. (10.17) $\frac{\partial u}{\partial x} = \frac{\partial u}{\partial r}\frac{\partial r}{\partial x}$

and $$\frac{\partial x}{\partial r} = \frac{\partial N_1}{\partial r} x_1 + \frac{\partial N_2}{\partial r} x_2 = \frac{x_2 - x_1}{2} = \frac{l_o}{2} \ .$$

We then have
$$\frac{\partial u}{\partial x} = \left(\frac{u_2 - u_1}{2}\right)\left(\frac{2}{l_o}\right) = \frac{u_2 - u_1}{l_o} \tag{10.19a}$$
and in a similar fashion
$$\frac{\partial v}{\partial x} = \frac{v_2 - v_1}{l_o} \ . \tag{10.19b}$$

The Green's strain is then obtained from Eq. (10.16) in the form
$$\varepsilon = \frac{u_2 - u_1}{l_o} + \frac{1}{2}\left(\frac{u_2 - u_1}{l_o}\right)^2 + \frac{1}{2}\left(\frac{v_2 - v_1}{l_o}\right)^2 \ . \tag{10.20}$$
The virtual strain energy is given by
$$\int_0^l (\Delta \underset{\sim}{\varepsilon})^T \underset{\sim}{\sigma} \, dV \ . \tag{10.21}$$

In our special problem of unilateral deformation, the stress and strain used in Eq. (10.2) have the normal engineering meaning, that is they are aligned and are in the direction of the deformed element. But in more complex problems, so called conjugate pairs of stress and strain must be used (see (9) or (13)).

It follows from Eq. (10.20) that

$$\Delta \underset{\sim}{\varepsilon} = \frac{1}{l^2}\begin{bmatrix} -l-u_2+u_1 & l+u_2-u_1 & -v_2+v_1 & v_2-v_1 \end{bmatrix} \cdot \begin{bmatrix} \Delta u_1 \\ \Delta u_2 \\ \Delta v_1 \\ \Delta v_2 \end{bmatrix} \tag{10.22}$$

or
$$\Delta \underset{\sim}{\varepsilon} = \underset{\approx}{B} \cdot \Delta \underset{\sim}{u}_i \ ,$$

where the matrix B can be separated into linear and nonlinear matrix
$$\underset{\approx}{B} = \underset{\approx}{B}_L + \underset{\approx}{B}_N, \tag{10.23}$$

where $$\underset{\approx}{B}_L = 1/l \begin{bmatrix} -1 & 1 & 0 & 0 \end{bmatrix} \ ,$$

$$\underset{\approx}{B}_N = \frac{1}{l^2}\begin{bmatrix} -(u_2-u_1) & (u_2-u_1) & -(v_2-v_1) & (v_2-v_1) \end{bmatrix}$$

The initial stress matrix $\underset{\approx}{K}_\sigma$ in Eq. (10.12) requires the variation of the matrix $\underset{\approx}{B}$. From Eq. (10.22) it follows that

$$\Delta \underset{\approx}{B} = \Delta \underset{\approx}{B}_N = \frac{1}{l^2}\begin{bmatrix} 1 & -1 & 0 & 0 \\ -1 & 1 & 0 & 0 \\ 0 & 0 & 1 & -1 \\ 0 & 0 & -1 & 1 \end{bmatrix} \cdot \begin{bmatrix} \Delta u_1 \\ \Delta u_2 \\ \Delta v_1 \\ \Delta v_2 \end{bmatrix}. \tag{10.23}$$

The initial stress matrix is then evaluated by multiplying each term of the matrix $\Delta \underset{\approx}{B}_N$ in Eq. (10.23) by the value $\underset{\approx}{D}\,\underset{\sim}{\varepsilon}$, where $\underset{\sim}{\varepsilon}$ is evaluated from Eq. (10.20).

The displacement components have been written in the local coordinate system. In order to make the assembly of the global tangential matrix possible, the displacements have to be transformed into the global coordinate system in a similar fashion to the procedure used in Section 2.

$$(\Delta \underset{\sim}{u}_i)_l = \underset{\approx}{T}\,(\Delta \underset{\sim}{u}_i)_g .$$

T is the transformation matrix given as

$$\underset{\approx}{T} = \begin{bmatrix} \cos\alpha & \sin\alpha & 0 & 0 \\ -\sin\alpha & \cos\alpha & 0 & 0 \\ 0 & 0 & \cos\alpha & \sin\alpha \\ 0 & 0 & -\sin\alpha & \cos\alpha \end{bmatrix}, \tag{10.24}$$

where α is the angle between the local x and global X axes in the initial position. This matrix remains constant during iteration because the effect of geometry change is already accounted for by the expression for the Green's strain.

The displacement components in all nonlinear matrices are written in the local coordinate system. This means that after the global increments $(\Delta \underset{\sim}{u}_i)_g$ have been calculated and added to the global displacements, the local displacements have to be transformed by similar transformation to that in Eq. (10.24) before being inserted into nonlinear matrices.

There are two possible ways to proceed after each load increment.

The Total Lagrangian formulation relates the displacements throughout all load increments to the original local (or global) system. The transformation matrices remain the same, the nodal points and element geometry are not changed.

The updated Lagrangian formulation recalculates the nodal points and the element geometry after each load increment. The transformation matrix T has to be updated because of the changed angle α. The advantage of this approach lies in the decrease in the nonlinear terms. If load increments are sufficiently small, the initial displacement matrix can be neglected (23).

11. NONLINEAR PROBLEMS - FLUID FLOW

In general, the finite element formulation of fluid flow proceeds along the lines shown in Section 4. The governing differential equation is inserted into the residual integral and the Galerkin method is used to derive the matrix equation. The solution of this equation is more difficult because the system matrix is non-symmetrical, non-linear and often ill-conditioned so that instability in the iteration procedure is much more likely to occur.

Let us briefly demonstrate the formulation process on an example of two-dimensional plane flow of an incompressible viscid fluid as it was used in (24). The governing differential equations are the Navier-Stoke equations.

$$\rho u \frac{\partial u}{\partial x} + \rho v \frac{\partial u}{\partial y} - \rho f_1 - \frac{\partial \tau_{11}}{\partial x} - \frac{\partial \tau_{12}}{\partial y} = 0 ,$$

$$\rho u \frac{\partial v}{\partial x} + \rho v \frac{\partial v}{\partial y} - \rho f_2 - \frac{\partial \tau_{21}}{\partial x} - \frac{\partial \tau_{22}}{\partial y} = 0 , \tag{11.1}$$

where $\tau_{ij} = -p\,\delta_{ij} + \mu \left(\frac{\partial u}{\partial x_j} + \frac{\partial v}{\partial x_i} \right) , \; x_1 = x, \; x_2 = y,$

and u, v are the velocity components in the x and y direction, p is the pressure, ρ is the density, f is the body force, μ is the viscosity, τ_{ij} is the stress tensor and δ_{ij} the Kronecker delta.

The condition of incompressibility is given in the form

$$\frac{\partial u}{\partial x} + \frac{\partial v}{\partial y} = 0 . \tag{11.2}$$

To complete the problem a sufficient set of boundary conditions has to be specified. For example, on part of the boundary the velocity components can be prescribed, on the other part the pressure may be specified.

The region of interest is then divided into finite elements. In contrast with the formulation used in solid mechanics, the fluid elements are fixed and cannot change shape. This implies that the Eulerian description of the fluid motion must be used.

The unknown parameters are the two velocity components u_i, v_i and the pressure p_i at each of element nodal points.

Within each element, the velocity and pressure fields are approximated by shape functions with u_i, v_i, p_i; i = 1,2, ..., n at the nodal points as parameters:

$$\begin{bmatrix} u \\ v \end{bmatrix} = \sum_{i=1}^{n} N_i \begin{bmatrix} u_i \\ v_i \end{bmatrix} ,$$

$$p = \sum_{i=1}^{m} M_i \, p_i \quad . \tag{11.3}$$

The different indices n and m indicate that in general the velocity and pressure functions do not have to use the same number of nodal points for definition.

The approximations (11.3) are inserted into Eqs. (11.1) and (11.2) and subsequently made orthogonal to the interpolation functions N_i, M_i from Eq. (11.3).

After carrying out these operations and removing the nodal point values from integrals, a system of nonlinear algebraic equations is obtained

$$(\underset{\approx}{K}_L + \underset{\approx}{K}_N(\underset{\sim}{u}_i)) \, \underset{\sim}{u}_i = \underset{\sim}{f}_i , \tag{11.4}$$

where

$$\underset{\sim}{u}_i = \begin{bmatrix} u_i \\ v_i \\ p_i \end{bmatrix} .$$

The method of introduction of boundary conditions into Eq. (11.4) in terms of velocities and pressure is the same as that used in solid mechanics.

The matrices in Eq. (11.4) will be banded for a proper numbering of nodal points.

The system (11.4) is solved in (24) by the iterative method from

$$(\underset{\approx}{K}_L + \underset{\approx}{K}_N(\underset{\sim}{u}_i^{(n)})) \underset{\sim}{u}_i^{(n+1)} = \underset{\sim}{f}_i . \qquad (11.5)$$

The matrix $\underset{\approx}{K}$ is not symmetric. In (24), the transformation to a symmetric system was not successful, so that the asymmetric form has been solved. The convergence behaviour depends on the Reynolds number

$$Re = \frac{v_{ref} \cdot h}{\mu} < C. \qquad (11.6)$$

This indicates that for a particular mesh size there is a limit on viscosity for convergence. The mesh size must decrease with decreasing viscosity. The value of the constant C in expression (11.6) is unfortunately not known, so that the user must rely on computer experiments.

Other formulations of the above problem are possible, some of which use the stream function (C_1 continuity is required) or incorporate the incompressibility condition by a penalty function.

12. EDUCATION OF ENGINEERING STUDENTS AT RMIT, MELBOURNE

One important aspect of engineering education is to demonstrate to the students that mathematics cannot be applied to real objects. A mathematical model has to be constructed for a given task, and only this idealized object can be mathematically analyzed. To construct such a model sometimes requires extraordinary skill, intuition and experience. Mathematicians may be of little help because they are too busy with quite different problems.

The results from the computer analysis are valid only for the mathematical model. Extrapolation to reality is a practical problem again requiring experience, imagination and a refined criticism.

In between is the finite element program used as a tool. As in any profession, the deeper the understanding of the tool, the higher the quality of the products.

These three steps - modelling, solving, extrapolation of results - are closely interrelated as can be clearly demonstrated in two case studies where a twenty-node three-dimensional solid element was used.

A machine part - a universal joint - has been modelled using 20-node 3-D elements. One-quarter of the hardware was measured manually and nodal point coordinates and element connectivities were input into the computer. By double mirroring the geometry of the complete component was generated. The plotting program MOVIE-BYU was used to check the geometry. Fig. 12.1 shows the model.

The seemingly correct model was submitted to the finite element program. This program also has plotting facilities. A plot from the finite element program was required as a further check before the analysis. Fig. 12.2 shows the model.

We can now see that the model which seemed to be correct is worthless.

There are many edges which are completely wrong. If such model were submitted for solution, completely wrong results would be obtained.

The reason for the errors in simple. Some of the middle nodes were not in the middle of the edges. An isoparametric element approximates 3-node edges by parabolas. To obtain a close approximation to a given curve, the middle node must be at the correct position (in the middle for a circular arc). The errors could not be discovered by the graphics program because straight lines were plotted between the nodes in order to increase speed of display.

The second example is related again to the 20-node solid element. A shaft of a seat-belt retractor mechanism was modelled. This time the model was correct. The structure was analysed for given loads and Mises stresses were plotted in prescribed planes throughout the structure. The result is on Fig. 8.5.

It is clear that the results are wrong. A simple analysis by using beam elements reveals this.

What is the cause of this error? The stiffness matrices are integrated numerically. In this case, 3 x 3 x 3 integration points have been used. The stresses have been calculated in these 27 integration points and extra-

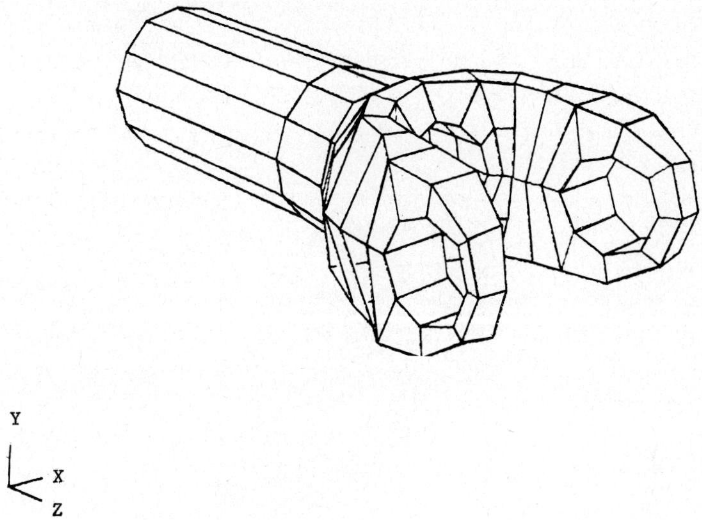

Figure 12.1

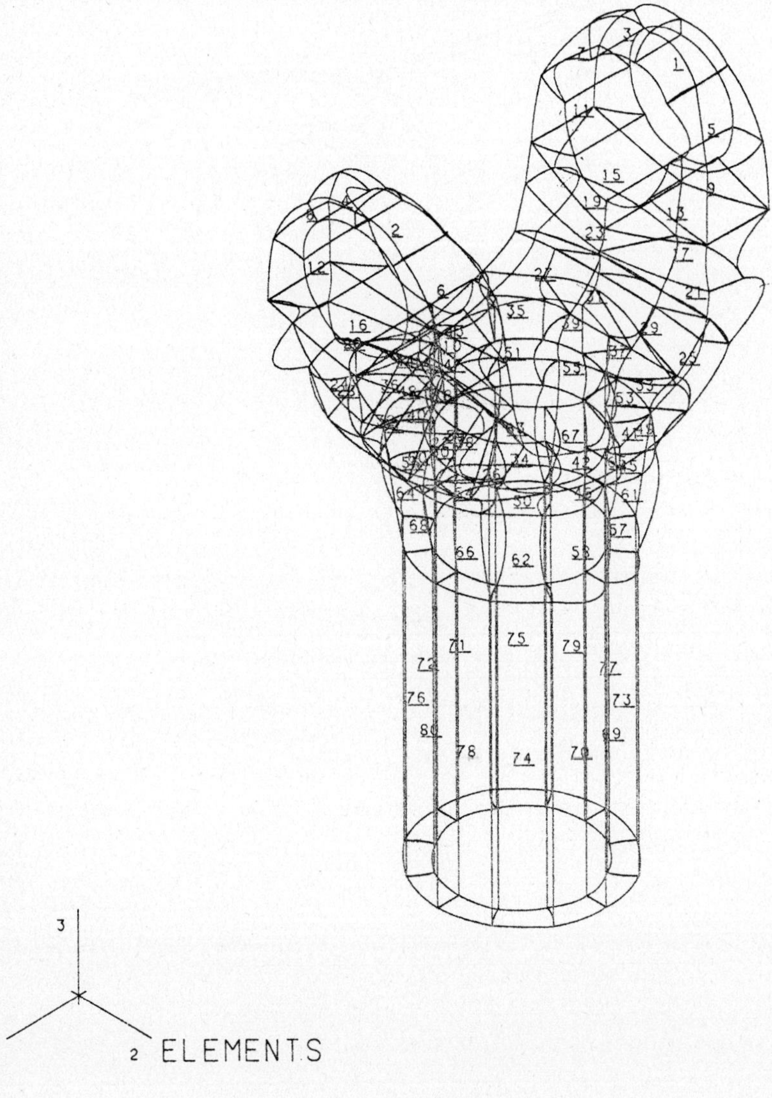

Figure 12.2

polated for contour plotting. As discussed in Section 8, the correct stresses are obtained only in 2 x 2 x 2 integration points. This means that the contours in Fig. 8.5 have been extrapolated from quite wrong values. The user again did not know enough about the tool. He had a choice of both integration schemes and selected the wrong one.

However, there are postprocessors which require 3 x 3 x 3 integration points for 20-node solids. As a consequence, the results they plot are then incorrect. The user should be interested in the quality of computer graphics software he is using.

These examples demonstrate quite clearly that the finite element method can be a dangerous weapon in the hands of under-qualified users.

At RMIT, the potential of the finite element method as a practical instrument was recognised several years ago. A group consisting of lecturers from various departments and faculties was formally established. The group objectives for the next two years have been set (25). They include hardware, software and man-power resources and their appropriate usage in graduate and postgraduate teaching and applied research.

A finite element course was designed and introduced in the second semester of 1980. The course is offered as an optional subject in mathematics to third year degree students in civil, mechanical and aeronautical engineering. It is taught by four members of the group (two applied mathematicians and two engineers - civil and mechanical).

Two finite element programs are used in the course: CAL78 (26) and TWODEPEP (27). CAL78 has three levels of user involvement. At the first level, the user uses only simple matrix operations, such as loading, adding, multiplication and solving. He has to specify all matrices including transformation matrices and solve the problem step by step. The second level allows the user to specify the order of assembly of element matrices into the global stiffness matrix by using an integer matrix. It also calculates automatically stiffness matrices of beams in global coordinates.

The third level is close to the finite element programs as used in engineering practice. Nodal point coordinates, element connectivities, material properties, loads and boundary conditions are specified in a normal way. Plane stress elements from 3 to 8 nodes are available with a different number and position of integration points. The effect of reduced integration and precision on stresses can be studied. Computer time is printed out for each operation so that the user can clearly see the cost difference in using more nodes or more integration points.

The computer graphics program M3D (28) is available for model display. The importance of careful checking of correctness of the model is emphasized again and again.

The program TWODEPEP is used in the second half of the course. It requires a quite different approach. The problem to be solved has to be described by a partial differential equation. The students are briefly informed of the method of weighted residuals on which TWODEPEP is based. Simple heat-transfer and potential fluid flow problems are formulated and solved.

The student assignments must include a discussion about the development of the model, the graphical plot of the model and validation of results by means of any suitable and available method (e.g. handbook formula, experiment, analytical solution of a simplified system).

Fourth year degree students who are working on industrial projects requiring finite element method use the nonlinear program ABAQUS (29). This program, together with mesh generation program FEMGEN and plotting program MOVIE.BYU, represent the most advanced level of finite element application. The mesh is generated on the video-screen from a basic geometric description of the component. A digitizing tablet is available for input of point coordinates from drawings. The direct input from hardware by means of laser scanning is under development by a member of the SIGFEM group.

When using FEMGEN to generate finite element mesh of a model, the basic geometric parts, such as points, lines, surfaces and bodies, are named. A data base containing the detailed information about the geometry of the component is created and saved. It contains not only coordinates of nodal points and element connectivities, but also coordinates and names of major points and tables with the definitions of the geometric parts.

In the next stage, an ABAQUS input file is created interactively in conversation mode. The communication is by means of FEMGEN point, line, surface and body names, so the user has no knowledge about nodal point or element numbers. This permits the user to generate a correct input file. This is important when such large programs as ABAQUS are used. The computer usage is limited to overnight runs for ABAQUS because of the large number of time-sharing computer users of the system. This means that one turn-around per day is the maximum.

More detailed information about student projects and postgraduate projects can be found in (30) or (31).

SIGFEM is also organising introductory and advanced courses in the finite element method for practising engineers from industry. Practical application of the method again is the main objective.

13. CONCLUSIONS

The finite element method is today used in so many areas of engineering that it is not physically possible to present all applications in a single paper.

Such diverse problems as elastoplasticity, creep, phase change, crack propagation, impact, soil mechanics, friction, and many more, can be relatively easily accommodated in the finite element formulation. Most of these features already have been incorporated in major general-purpose finite element computer programs and can be used in a routine way.

The engineer, who uses finite element codes developed by specialists at universities or professional software companies, faces two major difficulties when solving more complex problems (32).

The first difficulty is the cost of computing, including model development. This can become prohibitive, being 10 to 100 times more than for solving simple linear problems. The second is the reliability of the formulation used in the code and the proper interpretation and application of it

Many novice users of the finite element method become over-confident in the power and capabilities of modern computer, as if the computer could think for them, and solve problems before the users themselves clearly understand and formulate the problems.

The approach by some managers who are introducing the method into their companies reflects this dangerous way of thinking. They do not hesitate to spend vast amounts of money for modern computer hardware and sophisticated software, but they fight for the last penny they can save in basic training of their engineers.

The area of engineering education in modern computer methods, including the finite element method, is an exciting opportunity for co-operation between mathematicians and engineers. The use of modern methods and technology would seem to require an integrated presentation of theory and practice.

ACKNOWLEDGEMENT

The author would like to acknowledge the value of co-operation of SIGFEM members from both "camps" in helping to develop new ideas in education and applied research at RMIT. It would have been difficult to write this paper without this background.

REFERENCES

1. Poincare, H., "Science and Method", Dover 1952.

2. Courant, R., "Variational Method for the Solution of Problems of Equilibrium and Vibration", Bull. Am. Math. Soc., Vol. 49, 1-23, 1943.

3. Hrennikoff, A., "Solution of Problems in Elasticity by the Framework Method", J.Appl.Mech., Vol. 8, A169-A175, 1941.

4. Argyris, J.H., "Energy Theorems and Structural Analysis", Aircraft Eng., Vol. 26 (1954), also Butterworth 1960.

5. Turner, M.J., Clough, R.W., Martin, H.C., Topp, L.J., "Stiffness and Deflection Analysis of Complex Structures", J.Aero.Sci., Vol. 23, 805-23, 1956.

6. Clough, R.W., "The Finite Element in Plane Stress Analysis", Proc. 2nd A.S.C.E. Conf. on Electronic Computation, Pittsburgh, Pa., Sept. 1960.

7. Ames, W.F., "Numerical Methods for Partial Differential Equations", First Edition, Nelson, 1969.

8. Ames, W.F., "Numerical Methods for Partial Differential Equations", Second Edition, Academic Press, 1977.

9. Oden, J.T., "Finite Elements of Nonlinear Continua", McGraw-Hill, 1972.

10. Strang, G. , Fix, G.J., "An Analysis of the Finite Element Method, Prentice-Hall, 1973.

11. Zienkiewicz, O.C., "The Finite Element Method", McGraw-Hill, 1977.

12. Bathe, K.J., Wilson, E.L., "Numerical Methods in Finite Element Analysis", Prentice Hall, 1976.

13. Hibbit, H.D., Karlsson, B.I., Sorensen, E.P., "ABAQUS - A Finite Element Code for Nonlinear Dynamic Analysis", Theory Manual, 1980.

14. Irons, B., Ahmad, S., "Techniques of Finite Elements:, John Wiley & Sons, 1980.

15. Mitchell, A.R., Wait, R., "The Finite Element Method in Partial Differential Equations", John Wiley & Sons, 1977.

16. Wilson, E.L., Taylor, R.L., Doherty, W.P., Ghaboussi, J., "Incompatible Displacement Models", Num. & Comp. Methods in Structural Mechanics, pp. 43-51, Academic Press, 1973.

17. Taylor, R.L., Beresford, P.J., Wilson, E.L., "A Non-Conforming Element for Stress Analysis", Int. J. Num. Meth. Eng., Vol. 10, pp. 1211-1220, 1976.

18. Barlow, J., "Optional Stress Locations in Finite Element Models", Int. J. Num. Meth. Eng., Vol. 10, pp. 243-251, 1976.

19. Babuska, I., "Finite Element Method with Penalty", Rept. BN-910, Univ. Maryland.

20. Oden, J.T., Kikuchi, N., "Finite Element Methods for Constrained Problems in Elasticity", TICOM Report, The University of Texax, Austin, 1981.

21. Irons, B.M., "The Semiloof Shell Element", pp. 197-222 of "Finite Elements for Thin Shells and Curved Members", Wiley & Sons, 1976.

22. Bathe, K.J., "ADINA - A Finite Element Program for Automatic Dynamic Incremental Nonlinear Analysis", Report 82448-1, MIT, Cambridge, 1978.

23. Murray, D.W., Wilson, E.L., "Finite Element Post Buckling Analysis of Thin Elastic Plates", Proc. 2nd Conf. Matrix Methods in Struct. Mechanics, AFFDL, Oct. 1968.

24. Garthing, D.K., "Texas Fluid Analysis Program - User Manual", TICOM Report 75-2, Austin, Texas, 1974.

25. "Finite Element Methods at RMIT" - SIGFEM Report 001, RMIT, Melbourne, 1979.

26. Wilson, E.L., "CAL78 - Computer Analysis Language", University of Berkeley, 1978.

27. "TWODEPEP - User Manual", IMSL, Houston, 1980.

28. Trinder, S., "M3D - Mesh Display and Editing Program", SIGFEM Report 005, RMIT Melbourne, 1980.

29. Hibbitt, H.D., Karlsson, B.I., Sorensen, E.P., "ABAQUS - User Manual", 1981.

30. Tomas, J.A., "SIGFEM & Application Case Studies in Industry", ACADS International Symposium, ACADS Publication No. U214, pp.150-154, Melbourne, 1980.

31. Robinson, K.J., Tomas, J.A., Trinder, S.M., "Applied Research for Industry Using ADINA", ADINA Conference, MIT, 1981.

32. Oden, J.T., Bathe, K.J., "A Commentary on Computational Mechanics", Applied Mechanics Review, Vol. 31, No. 8, 1978, pp. 1053-1058.

NUMERICAL SOLUTIONS OF
PARTIAL DIFFERENTIAL EQUATIONS
J. Noye (Editor)

AN INTRODUCTION TO THE BOUNDARY ELEMENT METHOD

L.J. Wardle

Commonwealth Scientific and Industrial Research Organization
Division of Applied Geomechanics
Melbourne Australia

The problem of solving a partial differential equation within a given domain can be transformed into one solving an equivalent integral equation on the boundary of the domain. This reformulation is particularly advantageous for numerical applications as the interior of the domain does not have to be discretized. Alternative methods of formulating and solving the integral equations are described. Techniques for considering non-homogeneous and non-linear problems are discussed.

1. INTRODUCTION

This paper introduces the mathematical background to boundary integral equation methods i.e., methods by which the problem of the solution of a partial differential equation valid in a given domain is recast into the solution of an integral equation which applies only to the boundary of the domain. The principal advantage of such a reformulation is that the dimensionality of the problem is reduced by one. For example, a two-dimensional partial differential equation is replaced by a one-dimensional integral equation. In practical terms this means that two-dimensional problems involve discretization into line segments on the boundary, in contrast to finite element and finite difference procedures which require meshes over the plane domain area within the boundary (Fig. 1). For problems involving an infinite domain the boundary integral formulation is particularly advantageous because the behaviour at infinity is usually automatically included without having to discretize an artificial 'remote' boundary as with other methods.

In this paper details are given of various alternative boundary integral equation formulations. Having arrived at an integral equation there are many methods of numerical solution possible. For example, for a rectangular region it may be possible to express the unknown boundary quantities in terms of Fourier series or power series. For arbitrary geometries, however, resort must be made to numerical methods that allow the boundary geometry of the problem to discretized into 'elements' over which the boundary variables have a given polynomial order of variation, i.e. we use a finite element method on the boundary surface. To contrast this procedure with the conventional domain-type finite element method, the boundary version is now commonly called *the boundary element method*.

For many problems, the boundary element method possesses a similar level of generality to the finite element method, e.g. no limitations on the shape of the boundary or on the connectivity of the domain it encloses.

The use of integral equation formulations for partial differential equations dates back to the mid nineteenth century investigations involving the representation of harmonic functions (i.e. solutions to Laplace's equation) by single-layer and double-layer potentials. The existence of solutions to such equations was first demonstrated by Fredholm [1] using a discretization procedure. However, the use

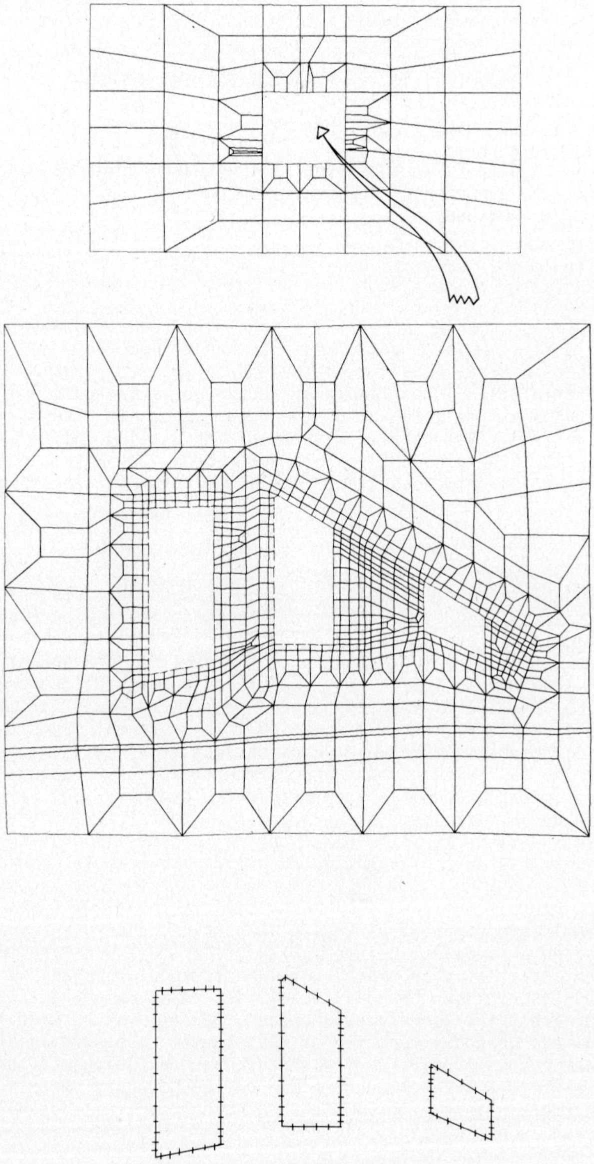

Figure 1
Boundary Element and Finite Element Meshes for Same Domain.

of a discretization procedure to actually construct solutions did not become feasible until the advent of electronic digital computers (circa 1960).

In this paper, Laplace's equation has been chosen as an example to demonstrate the basic features of boundary integral formulations and associated numerical procedures for their solution. Laplace's equation arises in a wide range of physical phenomena such as gravitational fields, electrostatic fields, steady-state heat conduction, incompressible flow, torsion and so on. For completeness, all of the essential classical results relating to surface potential representation of harmonic functions are introduced in Section 2. Using these results a variety of alternative boundary integral equations can be formulated.

In Section 3 methods used to obtain numerical solutions to boundary integral equations are illustrated by some specific examples. Section 4 gives an overview of how the basic methods can be extended to treat Poisson's equation, non-homogeneous material properties and non-linear behaviour. Section 5 discusses how the techniques have been used for the solution of other partial differential equations.

2. INTEGRAL EQUATION FORMULATIONS FOR LAPLACE'S EQUATION

To illustrate the basic features of boundary integral formulations consider Laplace's equation

$$\nabla^2\phi = 0 \tag{2.1}$$

in some domain V. (Fig. 2).

The boundary conditions on the surface S are of two types

$$\phi(P) = f(P) \qquad \textit{Dirichlet conditions} \tag{2.2}$$

or

$$\frac{\partial\phi}{\partial n}(P) = g(P) \qquad \textit{Neumann conditions} \tag{2.3}$$

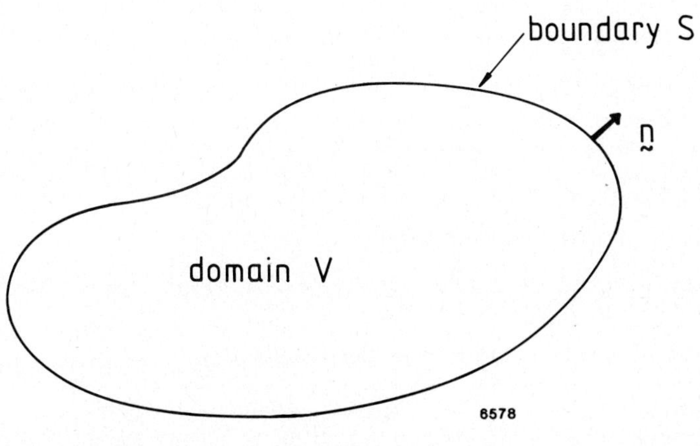

Figure 2
Definitions

2.1 Fundamental Solutions

The starting point of an equivalent integral equation formulation is to consider the appropriate fundamental singular solution for Laplace's equation, i.e. a solution of Laplace's equation that depends merely on the distance r of the point p from a fixed point q.

For such functions the Laplacian operator in three dimensions is $\partial^2/\partial r^2 + (2/r)\,\partial/\partial r$ and in two dimensions is $\partial^2/\partial r^2 + (1/r)\,\partial/\partial r$. It is easily verified by direct substitution that the functions

$$1/r(p,q) \quad \text{in three dimensions}$$

and

$$\log[1/r(p,q)] \quad \text{in two dimensions}$$

satisfy the respective Laplace's equation for $r(p,q)\neq 0$, where $r(p,q)$ is the scalar distance between points p and q.

Furthermore, if the point $r(p,q) = 0$ is included it can be shown that for three dimensions

$$\nabla^2 \frac{1}{r(p,q)} = -4\pi\delta(p,q) \tag{2.4}$$

and for two dimensions

$$\nabla^2 \log[1/r(p,q)] = -2\pi\delta(p,q) \tag{2.5}$$

where the Dirac delta function is defined to have the following properties

$$\begin{aligned} &\delta(p,q) = 0, \qquad \tilde{x}_p \neq \tilde{x}_q \\ &\int_V \rho(p)\delta(p,q)dV(q) = \rho(p) \end{aligned} \tag{2.6}$$

we infer the function

$$\psi(p,q) = \frac{1}{4\pi r(p,q)} \tag{2.7}$$

to be the fundamental solution (the factor 4π is included for convenience) for three dimensions. The fundamental solution in two dimensions can be taken to be

$$\psi(p,q) = \frac{1}{2\pi}\log\left(\frac{1}{r(p,q)}\right) \tag{2.8}$$

In the remainder of this work $\psi(p,q)$ denotes the fundamental solution defined by either equations (2.7) or (2.8) appropriate to the dimensionality of a given problem. The use of this notation in most cases permits two and three dimensions to be treated simultaneously.

2.2 Single and Double Layer Potentials

A continuous distribution of simple sources extending over the surface S and of surface density $\sigma(Q^{\dagger})$ at $Q \in S$ generates the potential

$$\phi(p) = \int_S \sigma(Q)\psi(p,Q)dS(Q) \tag{2.9}$$

and is called a single-layer (or, sometimes, simple-layer) potential.

† In all that follows the symbols p, q will represent field points in the domain V; the symbols P, Q will represent field points on the surface S of the domain.

From the single-layer distribution we can generate another distribution by considering two surfaces separated by a small distance ε. Assume each surface has a distribution of attraction of equal magnitude but of opposite sign for neighbouring points and take the magnitude of the attraction to be inversely proportional to ε. The potential associated with the point $Q(\underset{\sim}{x})$ and the point $Q(\underset{\sim}{x} + d\underset{\sim}{x})$ is given by the limit

$$\lim_{\varepsilon \to 0} \left\{ \frac{1}{\varepsilon} \left[(p,Q+\varepsilon) - (p,Q) \right] \right\} \tag{2.10}$$

which can be seen to be the derivative of the potential $\psi(p,Q)$ in the direction normal to the surface S. Integrating over all the surface points gives the double-layer potential

$$\phi(p) = \int_S \mu(Q) \frac{d\psi(p,Q)}{dn(Q)} dS(Q) \tag{2.11}$$

The function $\mu(Q)$ is the surface density of the double layer potential.

2.3 Continuity of the Surface Potentials

The single-layer and double-layer potentials obey Laplace's equation everywhere except on the surface S, in which case the integrands involve singularities when $r(p,Q) = 0$. Briefly, to investigate the behaviour of the surface potentials, the boundary is broken into two surfaces : a small disc tangent to the surface at a point $P(\underset{\sim}{x})$ and the remaining surface which will contain no singularities because $P(\underset{\sim}{x}) \neq Q(\underset{\sim}{x})$.

Provided that σ is Holder continuous at P, the single layer potential is continuous when crossing S.[2]. The double-layer potential is discontinuous when crossing S. The limiting form of equation (2.11) as $p(\underset{\sim}{x}) \to P(\underset{\sim}{x})$ from the inside is

$$\phi^+(P) = -\frac{1}{2}\mu(P) + \int_S \mu(Q) \frac{d\psi(P,Q)}{dn(Q)} dS(Q) \tag{2.12}$$

and from the outside is

$$\phi^-(P) = \frac{1}{2}\mu(P) + \int_S \mu(Q) \frac{d\psi(P,Q)}{dn(Q)} dS(Q) \tag{2.13}$$

Subtracting equation (2.12) from (2.13) shows that the double-layer potential has a discontinuity or jump of $-\mu(P)$ as the point $p(\underset{\sim}{x})$ passes from outside the region to inside the region.

2.4 Indirect Boundary Integral Equations

A variety of boundary integral equations can be obtained by considering single-layer and double-layer distributions separately.

For example, for the Drichlet boundary conditions (ϕ prescribed on S), the unknown function $\phi(p)$ may be expressed solely as a double-layer potential of unknown density $\mu(Q)$.

$$\phi(p) = \int_S \mu(Q) \frac{d\psi(p,Q)}{dn(Q)} dS(Q) \tag{2.14}$$

Using the limiting behaviour of the double-layer potential as $p(\underset{\sim}{x})$ approaches $P(\underset{\sim}{x})$ from inside S (equation 2.12) gives

$$\phi(P) = -\frac{1}{2}\mu(P) + \int_S \mu(Q) \frac{d\psi(P,Q)}{dn(Q)} dS(Q) \tag{2.15}$$

Equation (2.15) is a Fredholm equation of the second kind with $\mu(Q)$ being the only unknown.

To obtain an integral equation for Neumann boundary conditions ($d\phi/dn$ prescribed on S), the unknown function $\phi(p)$ can be expressed solely as a single-layer potential with unknown density $\sigma(Q)$

$$\phi(p) = \int_S \sigma(Q)\psi(p,Q)dS(Q) \tag{2.16}$$

For points $p(\underset{\sim}{x})\varepsilon V$ we may take the derivative of $\phi(p)$ in the ξ-direction

$$\frac{d\phi}{d\xi} = \int_S \sigma(Q)\frac{d\psi(p,Q)}{d\xi(p)}dS(Q) \tag{2.17}$$

If ξ is taken to be the direction of the outer normal to S as $p(\underset{\sim}{x})$ is taken to $P(\underset{\sim}{x})$ then as $d\phi/d\xi$ is continuous up to and including S,

$$\frac{d\phi}{dn}(P) = -\frac{1}{2}\sigma(P) + \int_S \sigma(Q)\frac{d\psi(P,Q)}{dn(P)}\,dS(Q) \tag{2.18}$$

which is a Fredholm equation of the second kind with $d\phi/dn(P)$ known and $\sigma(P)$ unknown.

The two integral equations (2.15) and (2.18) are not the only possible ones that can be formulated using the single and double layer potentials. For example, instead of using the double-layer potential for the Dirichlet boundary conditions we could have used a single-layer potential. This leads to an integral equation of the first kind. For mixed boundary value problems, the use of either single-layer or double-layer potentials alone leads to coupled first and second kind equations. An alternative is to use a double-layer distribution on the part of the boundary where ϕ is defined and single-layer on the remainder. This formulation gives an integral equation of the second kind.

2.5 Direct Boundary Integral Equation

In principle, the solution to Laplace's equation can be represented by a *combination* of single-layer and double-layer singularities,

$$\phi(p) = \int_S \{\sigma(Q)\psi(p,Q) + \mu(Q)\frac{d\psi(p,Q)}{dn(Q)}\}\,dS(Q) \tag{2.19}$$

Since there is only one boundary condition and there are two functions, there is a "degree of freedom" in the problem. Thus the singularity distributions that give rise to a particular potential are not unique. The remainder of this sub-section shows how by combining our sought after solution ϕ and the fundamental solution ψ through Green's theorem, we can arrive at a representation of the form (2.19) in which the single and double layer densities $\sigma(Q)$ and $\mu(Q)$ are expressed in terms of the actual boundary values of ϕ and its normal derivative.

If ϕ and ψ are two continuous functions with continuous first derivatives, the application of the divergence theorem to the region V with surface S gives the following Green's theorem:

$$\int_V(\phi\nabla^2\psi - \psi\nabla^2\phi)dV = \int_S(\phi\,\frac{d\psi}{dn} - \psi\,\frac{d\phi}{dn})dS \tag{2.20}$$

The normal direction to S is taken outwards from V.

We take $\phi(q)$ to be our *unknown* harmonic function (satisfies $\nabla^2\phi = 0$) and $\psi(p,q)$ as the *fundamental* solution to Laplace's equation (from Section 2.1).

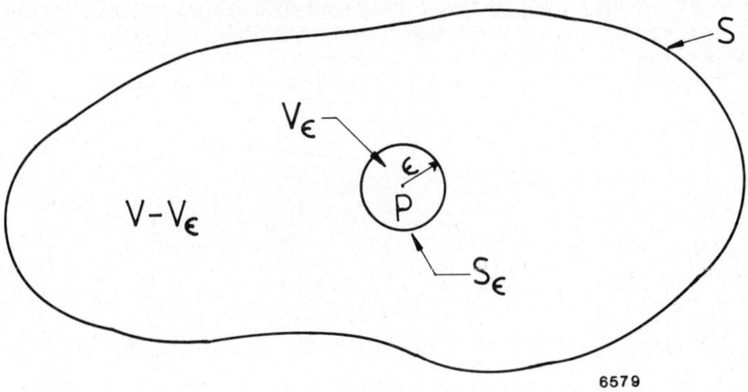

Figure 3
Domain for which Green's Theorem is written, if p is inside V.

As ψ does not obey Laplace's equation at p(x), we surround this point by a small sphere (or circle) of radius ε, denoted V_ε with surface S_ε (Fig. 3). In the region $V-V_\varepsilon$, $\nabla^2\psi = 0$. Taking Green's theorem for this region

$$\int_{V-V_\varepsilon} (\phi\nabla^2\psi - \psi\nabla^2\phi)dV = \int_{S+S_\varepsilon} (\phi \frac{d\psi}{dn} - \psi \frac{d\phi}{dn})dS \tag{2.21}$$

Because of the assumed continuity of $\phi(p)$

$$\lim_{\varepsilon\to 0} \int_{S_\varepsilon} \phi \frac{d\psi}{dn} dS = \phi(p) \cdot \lim_{\varepsilon\to 0} \int_{S_\varepsilon} \frac{d\psi}{dn} dS \tag{2.22}$$

On the surface S_ε, $d/dn = -d/dr$. Considering the three-dimensional case,

$$\frac{d\psi}{dn} = -\frac{d}{dr}\left(\frac{1}{4\pi r}\right) = \frac{1}{4\pi r^2} \tag{2.23}$$

giving

$$\int_{S_\varepsilon} \frac{d\psi}{dn} dS = \frac{1}{4\pi\varepsilon^2} \int_{S_\varepsilon} dS = 1 \tag{2.24}$$

For the two-dimensional case

$$\frac{d\psi}{dn} = -\frac{1}{2\pi} \frac{d}{dr} \ell n \left(\frac{1}{r}\right) = \frac{1}{2\pi r} \tag{2.25}$$

giving

$$\int_{S_\varepsilon} \frac{d\psi}{dn} dS = \frac{1}{2\pi\varepsilon} \int_{S_\varepsilon} dS = 1 \tag{2.26}$$

In the limit as $\varepsilon \to 0$, equation (2.21) becomes

$$\phi(p) = \int_S \{-\phi(Q)\frac{d\psi(p,Q)}{dn(Q)} + \frac{d\phi(Q)}{dn(Q)} \psi(p,Q)\} dS(Q) \tag{2.27}$$

This is the important integral identity of potential theory, that a harmonic function ϕ may be expressed as the sum of a single-layer potential with density $d\phi/dn$ and a double-layer potential with density ϕ.

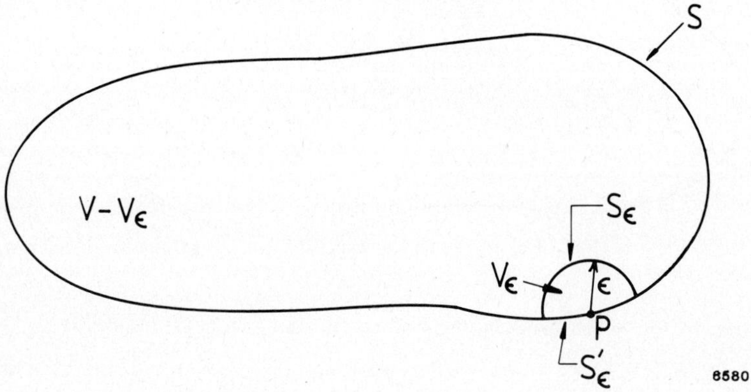

Figure 4
Domain for which Green's Theorem is written, if P is on S.

We now consider the case when the interior point p is taken to the boundary point P. The analysis is the same as for the interior case, except that instead of using the entire sphere (or circle), we only consider the part of the sphere (or circle) that is contained in V (Fig. 4). Instead of equations (2.24) or (2.26) we get

$$\lim_{\varepsilon \to 0} \int_{S_\varepsilon} \frac{d\psi}{dn} \, dS = c(P) \tag{2.28}$$

where c(P) is the proportion of surface area of S_ε contained within the domain. For a smooth boundary $c = \frac{1}{2}$.

Thus, in the limit, equation (2.27) becomes

$$c(P)\phi(P) + \int_S \phi(Q) \frac{d\psi(P,Q)}{dn(Q)} \, dS(Q) = \int_S \psi(P,Q) \frac{d\phi(Q)}{dn(Q)} \, dS(Q) \tag{2.29}$$

Equation (2.29) represents a constraint equation between the Dirichlet boundary conditions (ϕ defined) and the Neumann ($d\phi/dn$ defined) boundary conditions. For Neumann boundary conditions the right hand side of equation (2.29) is known, giving a Fredholm equation of the second kind for the unknown boundary values of the function $\phi(Q)$. For the Dirichlet problem equation (2.29) becomes a Fredholm equation of the first kind for the unknown boundary values of $d\phi/dn$. The mixed boundary value problem leads to a mixed integral equation for the unknown boundary data.

2.6 Direct vs Indirect Formulations

As the unknowns in the integral equation are physical boundary quantities (either ϕ or $d\phi/dn$), equation (2.29) is called the *direct* boundary integral equation to distinguish it from integral equations that involve a 'fictitious' source density, for example equations (2.15) and (2.18). Formulations based on the source density approach are called indirect because the 'complementary' boundary data can only be obtained after the source density distribution has been solved for.

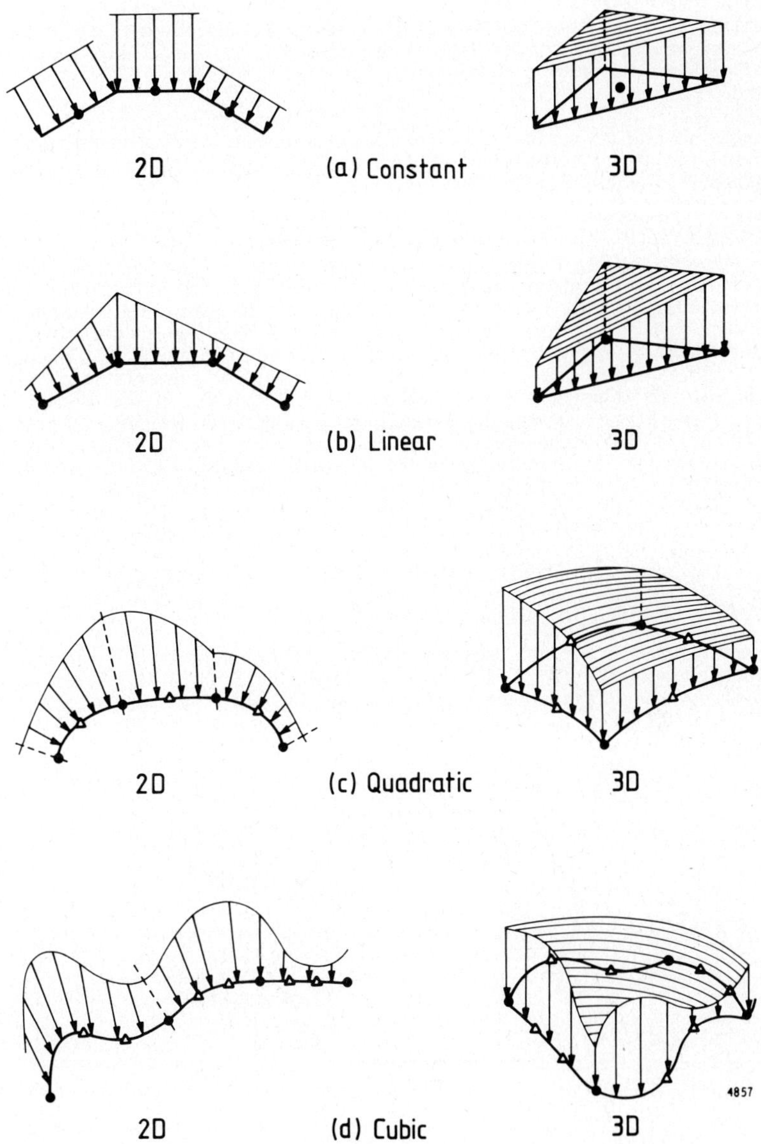

Figure 5
Common Boundary Elements

Almost without exception, authors of research papers on boundary element methods use only one of the two alternative methods of formulation.

It is fair to say that proponents of the direct method have developed the method further in terms of numerical techniques and mathematical sophistication than the supporters of the indirect methods have. However, this leaves aside the question of the numerical effectiveness of the various formulations. A proper numerical comparison of the methods still needs to be undertaken. It is possible that no one formulation is best for all problems but that each formulation has a class of problems, depending on geometry, boundary conditions etc. for which it is suited.

In the remainder of this paper, whenever a specific formulation needs to be considered, the direct formulation will be used. In general, however, the remarks apply to either method of formulation.

3. NUMERICAL SOLUTION OF BOUNDARY INTEGRAL EQUATIONS

Except for very simple geometries, analytical solutions for the boundary integral equations are unavailable and numerical methods must be used. By assuming that the boundary is divided into elements over which the data has a prescribed polynomial order of variation we can reduce the problem to that of solving a matrix system for the unknown coefficients. A variety of elements have been used (see Fig. 5). To illustrate how we can arrive at a system of linear equations we look at the so-called 'constant' element. We also discuss a number of higher-order elements of the isoparametric type (called isoparametric because the same order of polynomial variation is used for both the geometry and boundary data). These elements are particularly useful for curved geometries. Some of the factors to be considered when solving large systems of equations are pointed out. Brief mention is also made of the use of weighted residual formulations such as the Galerkin scheme.

3.1 Constant Elements

For purposes of illustration, consider the two-dimensional direct boundary integral equation (equation (2.29) with $\psi = (1/2\pi)\log(1/r)$). Divide the boundary of the body into N straight line segments (see Fig. 6). The function and its

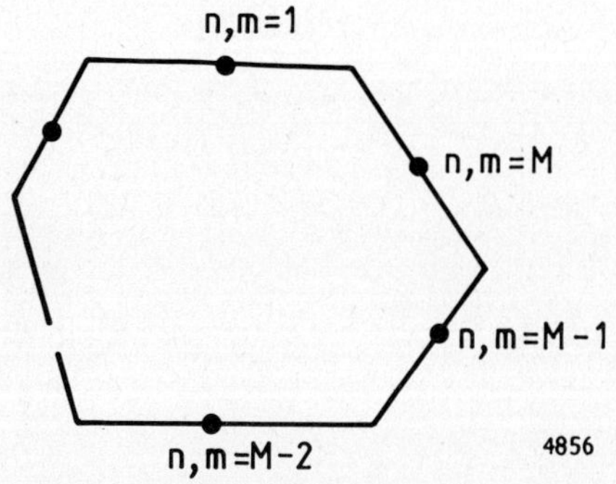

Figure 6
Constant Boundary Elements.

normal derivative are assumed constant on each segment. The boundary integral equation is written at the mid-point of each segment and the values of ϕ and $d\phi/dn$ can be extracted from the integration over each segment.

$$\pi\phi(P_n) + \sum_{m=1}^{N} \phi(Q_m) \int_{S_m} \frac{d}{dn(Q)} \log \frac{1}{r(P_n,Q)} \, dS(Q) = \sum_{m=1}^{N} \frac{d\phi}{dn}\bigg|_{Q_m} \int_{S_m} \log \frac{1}{r(P_n,Q)} \, dS(Q) \quad (n = 1,N). \tag{3.1}$$

The integrals can be expressed in closed form [3]. Care must be taken to handle the segment where the mid-point $P_n = Q_m$. This integral must be evaluated in the Cauchy Principal Value sense, i.e., by deleting a small segment of length 2ε at the mid-point and taking the limit

$$\lim_{\varepsilon \to 0} \left\{ \int_{Q_1}^{-\varepsilon} (\ ) dS(Q) + \int_{\varepsilon}^{Q_2} (\ ) dS(Q) \right\}. \tag{3.2}$$

Denoting the results of the integrals by $A^*(P_n,Q_m)$ and $B(P_n,Q_m)$, equation (3.1) becomes

$$\pi\phi(P_n) + \sum_{m=1}^{N} \phi(Q_m) A^*(P_n,Q_m) = \sum_{m=1}^{N} \frac{d\phi}{dn}\bigg|_{Q_m} B(P_n,Q_m) \qquad (n = 1,N) \tag{3.3}$$

Equation (3.3) may be rewritten in matrix form (A^* becomes A by adding π to the diagonal terms of A)

$$[A]\{\phi\} = [B]\left\{\frac{d\phi}{dn}\right\}. \tag{3.4}$$

For any well-posed boundary value problem, equation (3.4) involves just N knowns and N unknowns. This equation is normally very well conditioned and can be solved by standard matrix reduction schemes.

Values of ϕ at any interior point can be obtained using a discretized version of equation (2.27).

3.2 Ergatoudis Isoparametric Elements

One of the most efficient methods for numerical solution of boundary integral equations is based on the parametric representation of geometry and functions in terms of the Ergatoudis shape functions which are widely used in the finite element method [4]. This section introduces the quadratic element for two-dimensional problems [5]. The analogous element for three-dimensional problems is described in Ref. [6].

Assume that the boundary data at point Q within a typical element can be represented by mapping from the line $(-1<\xi<1)$ according to

$$f(Q) = \sum_{\alpha=1}^{3} M^{\alpha}(\xi) f_{\alpha} \tag{3.5}$$

where f_α is the value of the function at one of the three nodes associated with the element, (Fig. 7a). $M^{\alpha}(\xi)$ is a set of three 'shape functions' (Fig. 7b)

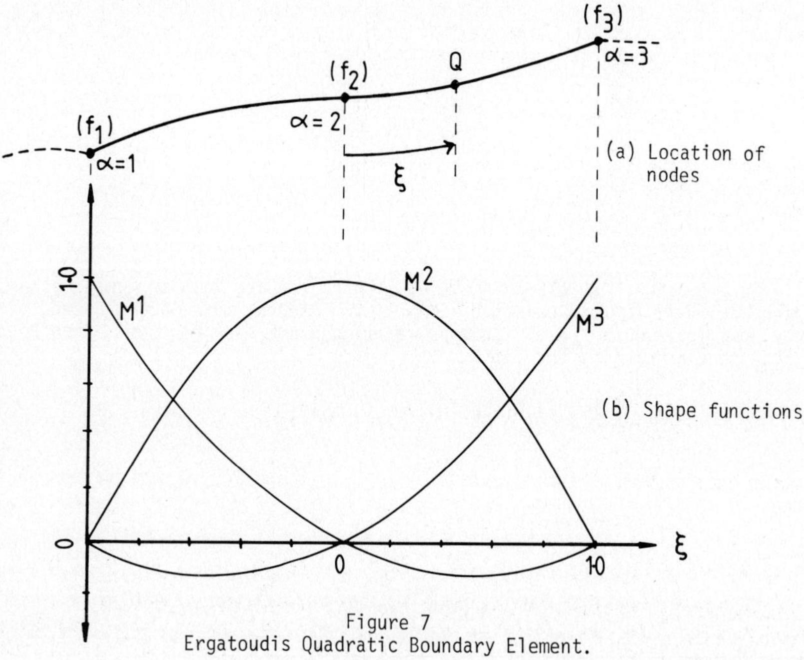

Figure 7
Ergatoudis Quadratic Boundary Element.

$$
\begin{aligned}
M^1(\xi) &= -\frac{1}{2}\xi + \frac{1}{2}\xi^2 \\
M^2(\xi) &= 1 - \xi^2 \\
M^3(\xi) &= \frac{1}{2}\xi + \frac{1}{2}\xi^2
\end{aligned}
\tag{3.6}
$$

The actual boundary curve is assumed to be described by the same form as the boundary data (hence the name 'isoparametric')

$$
\begin{aligned}
x(Q) &= \sum_{\alpha=1}^{3} M^{\alpha}(\xi)x_{\alpha} \\
y(Q) &= \sum_{\alpha=1}^{3} M^{\alpha}(\xi)y_{\alpha}
\end{aligned}
\tag{3.7}
$$

where x_α, y_α are the coordinates of the nodes.

Based on the above discretizing assumptions, equation (2.29) may be written:

$$
\begin{aligned}
&\pi c(P_n)\phi(P_n) + \sum_{m=1}^{(N/2)} \sum_{\alpha=1}^{3} \phi_m^{\alpha} \int_{S_m} M^{\alpha}(\xi) \frac{d}{dn(Q)} \log \frac{1}{r(P_n,Q(\xi))} J(\xi)d\xi \\
&= \sum_{m=1}^{N/2} \sum_{\alpha=1}^{3} \left(\frac{d\phi}{dn}\right)_m^{\alpha} \int_{S_m} M^{\alpha}(\xi) \log \frac{1}{r(P_n,Q(\xi))} J(\xi)d\xi \qquad (n = 1,N)
\end{aligned}
\tag{3.8}
$$

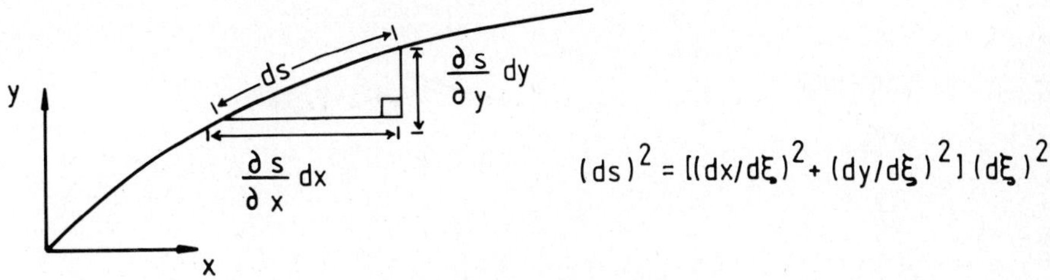

Figure 8
Notation for Curved Boundary.

In equation (3.8), $J(\xi)$ is the Jacobian given by (see Fig. 8)

$$J(\xi) = \frac{dS(Q)}{d\xi} = \left| \left(\frac{dx}{d\xi}\right)^2 + \left(\frac{dy}{d\xi}\right)^2 \right|^{\frac{1}{2}} \tag{3.9}$$

Substituting equation (3.7) gives

$$J(\xi) = \left[\left(\sum_{\alpha=1}^{3} \frac{dM^\alpha}{d\xi} x_\alpha \right)^2 + \left(\sum_{\alpha=1}^{3} \frac{dM^\alpha}{d\xi} y_\alpha \right)^2 \right]^{\frac{1}{2}} \tag{3.10}$$

The quantities ϕ_m^α, $(d\phi/dn)_m^\alpha$ represent values at local node α of element m. In general because the end nodes are associated with the elements on either side, $\phi_1^3 = \phi_2^1$, $(d\phi/dn)_1^3$ $(d\phi/dn)_2^1$ etc. The exception is at a point where discontinuous values are to be considered (say at a corner). In this case a pair of nodes can be used, each one associated with a different element.

The system of equations (3.8) becomes

$$[A]\ \{\phi\} = [B]\ \{\frac{d\phi}{dn}\} \tag{3.11}$$

The square matrices [A] and [B] contain integrals of the form

$$I_{nm}^\alpha = \int_{-1}^{1} M^\alpha(\xi)\psi(P_n,Q(\xi))J(\xi)d\xi \tag{3.12}$$

where ψ is one of the kernel functions $\log(\frac{1}{r})$ or $\frac{d}{dn}(\log(\frac{1}{r}))$.

The presence of the Jacobian (a consequence of the curved boundary) necessitates the use of numerical quadrature. For further details see Reference [5].

3.3 Hermitian Cubic Elements

Watson [7,8] has used Hermitian cubic elements for two-dimensional elasticity problems. These elements feature interelement continuity of the tangent plane and also of tangential derivatives of the boundary data. There are no midside nodes, thus simplifying data preparation.

The Hermitian cubic element has two nodes (Fig. 9a). The boundary data at point Q is representing by mapping from intrinsic coordinates ($-1<\xi<1$) according to

$$f(Q) = \sum_{\alpha=1}^{2} \{M^\alpha(\xi)f_\alpha + N^\alpha(\xi)(df/d\xi)_\alpha\} \tag{3.13}$$

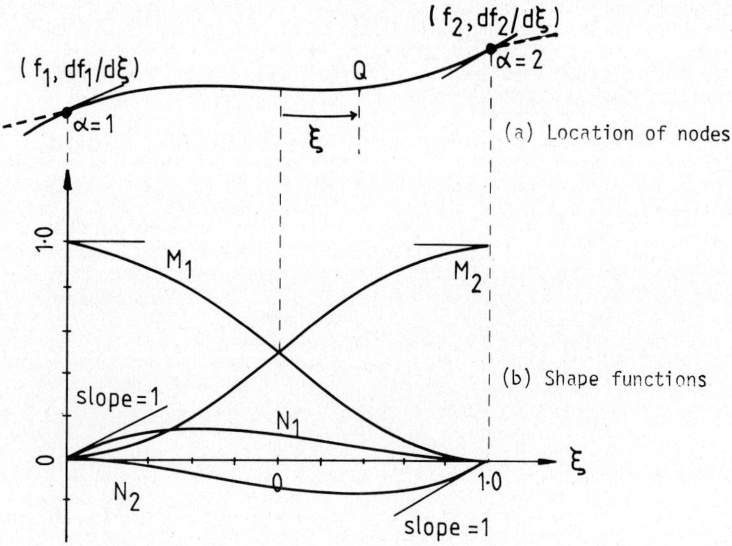

Figure 9
Hermitian Cubic Boundary Element.

where $f_\alpha(\alpha = 1,2)$ are the values of the function at the nodes and $(df/d\xi)_\alpha$ $(\alpha=1,2)$ represents the tangential derivatives of the function at the nodes.

The shape functions are given by (see Fig. 9b):

$$\begin{aligned} M^1(\xi) &= \frac{1}{4}(1+\xi)^2(2-\xi) \\ M^2(\xi) &= \frac{1}{4}(1-\xi)^2(2+\xi) \\ N^1(\xi) &= -\frac{1}{4}(1+\xi)^2(1-\xi) \\ N^2(\xi) &= \frac{1}{4}(1-\xi)^2(1+\xi) \end{aligned} \tag{3.14}$$

Again, as for the Ergatoudis elements, the Hermitian element is assumed to be isoparametric, i.e. the boundary geometry is modelled by the same shape functions.

The implementation of Hermitian cubic elements requires the derivation of an extra integral equation because of the introduction of the tangential derivatives. This is obtained by taking the tangential derivative of the basic integral equation [7,8].

3.4 Considerations of Computational Efficiency

In terms of computer effort, the two principal tasks are firstly the calculation of the coefficients and secondly the solution of the resulting linear equations. If for arguments sake we assume that a given problem is modelled by N constant elements, N^2 coefficients need to be computed. Contrasting with the symmetric banded coefficient matrices arising from the finite element method, the coefficient matrix is nonsymmetric and fully populated. The systems of equations are commonly solved by Gaussian elimination or iterative techniques. As is well known the computing effort involved in direct elimination is proportional to N^3 and is essentially independent of the number of right hand sides involved (these would arise if multiple sets of boundary conditions are considered in a single analysis). For an iterative solution the computing effort is proportional to IRN^2, where I is the number of iterations required for convergence and R is the number of right hand sides being considered.

For exterior problems (i.e. infinite domain) Hess [9] reports that iterative solutions converge in a number of iterations that is independent of N but dependent on the shape of the domain. In many applications I is of the order of 10 - 20. As a consequence iteration becomes more efficient than elimination for problems involving 100 - 300 elements. On the other hand, for interior problems (i.e. finite domain) I increases linearly with N, so in general, iteration is not competitive with direct elimination.

The computer effort involved in computing the terms in the coefficient matrix is of the order N^2. These terms involve integrating the field at the centre of one element (the 'field point') due to a prescribed source distribution within another element (the 'source element'). For flat elements these integrals are expressible analytically. The resulting expressions, which involve logarithms and inverse trignometric functions, depend on the orientation and dimensions of the source element and the relative coordinates of the field point. When a field point is far from the source element, such precision is inappropriate, and approximate expressions should be used to reduce computing costs. Hess [9] suggests that if the field point is further away than four times the maximum source element dimension, the effect of the distributed source can be replaced by that of a point source with no significant loss of accuracy. For large problems over 90% of the N^2 coefficients can be evaluated using the point source formulae, reducing the computational effort by an order of magnitude. Going further than this, for three-dimensional problems, computer time can be dramatically reduced by replacing the effects of a remote block of elements by an equivalent point source [10].

3.5 Weighted Residual Formulations

The numerical methods discussed so far are based on what is commonly known as 'nodal collocation', or point-matching. That is we have chosen a number of discrete points at which our integral equation is assumed to be satisfied so as to supply the same number of algebraic equations as we have unknowns. Alternative numerical methods can be based on assuming that the integral equation is satisfied in some sort of 'best fit' sense over the entire boundary rather than attempting to have zero error at discrete points and unknown error in between. This class of methods is commonly known as 'weighted residual methods'. Better known methods in this class are the Galerkin method, Method of moments and the Rayleigh-Ritz method. A discussion of the application of these methods applied to the integral equation for Laplace's equation is given by Reference [11]. The Galerkin scheme has been used more widely than the others. Computationally the Galerkin scheme involves double surface integrals over the boundary, as opposed to single surface integrals for nodal collocation. For a given surface discretization the Galerkin scheme will be correspondingly more costly than nodal collocation. The question is whether superior accuracy for a given discretization will allow sufficiently fewer elements to be used to offset this disadvantage.

4. EXTENSIONS TO THE BASIC BOUNDARY ELEMENT METHOD

The basic boundary integral formulations are based on the use of fundamental solutions of the governing partial differential equation over an infinite domain with homogeneous material properties. The boundary integral equations were constructed by superposition of fundamental solutions with their centres on the boundary. By distributing centres of the fundamental solution throughout the domain we show how to develop integral equation formulations for non-homogeneous partial differential equations such as Poisson's equation.

For non-homogeneous material properties in general an analytical fundamental solution does not exist. However we show how the boundary element method can be extended to solve piecewise homogeneous geometries.

Although the integral equations are based on the superposition property of linear partial differential equations, we demonstrate how various types of non-linear behaviour can be treated by iterative methods.

4.1 Non-homogeneous Partial Differential Equations

The non-homogeneous counterpart of Laplace's equation is Poisson's equation:

$$\nabla^2\phi\ (\underset{\sim}{x}) = \rho(\underset{\sim}{x}) \tag{4.1}$$

In steady flow problems $\rho(\underset{\sim}{x})$ represents the distribution of sources within the domain.

If the term $\rho(x)$ is included in Green's theorem (Equation 2.20), the boundary integral equation becomes

$$\begin{aligned} c(P)\phi(P) + \int_S \phi(Q)\, \frac{d\psi(P,Q)}{dn(Q)}\, dS(Q) &= \\ \int_S \psi(P,Q)\, \frac{d\phi(Q)}{dn(Q)}\, dS(Q) &+ \int_V \rho(q)\psi(P,q)dV(q) \end{aligned} \tag{4.2}$$

In addition to the usual surface integrals, we now have a known volume integral term involving $\rho(q)$, and for arbitrary $\rho(q)$ we must resort to a volume discretization scheme to evaluate the volume integral.

Assume that the domain is subdivided into M suitably sized 'volume elements' for which ρ can be taken as constant (Fig. 10). The volume integral then becomes

$$\int_V \rho(q)\psi(P,q)dV(q) = \sum_{\ell=1}^{M} \rho(q_\ell)\ \int_{V_\ell} \psi(P,q)dV(q) \tag{4.3}$$

The volume integrations can be performed using Gaussian quadrature.

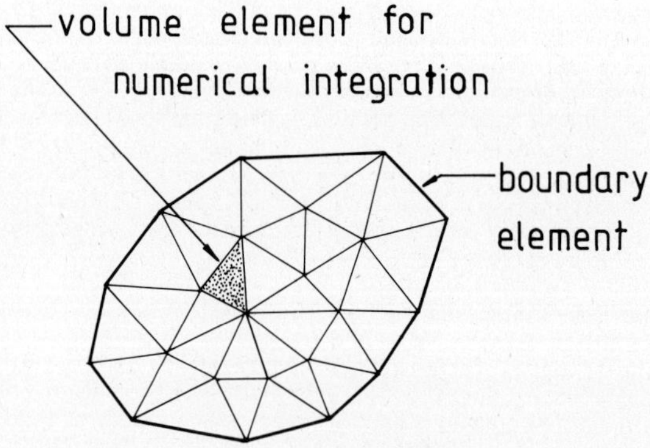

Figure 10
Volume Element Scheme for Evaluation of
Volume Integrals for Domain with Interior Source Density.

The matrix system obtained by the use of the simple constant boundary elements for Laplace's equation (Equation (3.4)) now generalizes to

$$[A]\{\phi\} = [B]\{\frac{d\phi}{dn}\} + [C]\{\rho\} \qquad (4.4)$$

where, as before, matrices A and B are NXN (N being the number of boundary elements) and matrix C is an NXM matrix used to multiply the M known values of ρ within the volume elements.

Although the volume discretization has the appearance of that used in the finite element method, no unknowns have been introduced into the interior of the domain. In section 4.3 it will be shown how the volume discretization scheme can be used to solve non-linear problems using integral formulations.

The volume discretization scheme is unnecessary if $\rho(\underset{\sim}{x})$ has certain properties. For example, if $\rho(\underset{\sim}{x})$ is harmonic, i.e. $\nabla^2\rho(\underset{\sim}{x}) = 0$, reference [5] shows that for the two-dimensional case the volume integral can be expressed as a simple surface integral.

4.2 Non-homogeneous Material Properties

An implicit assumption used in the formulation of boundary integral equations is that the material properties are uniform throughout the domain. This stems from the fact that the fundamental solutions involved are for uniform material properties. Boundary element methods can be used to analyse a piecewise homogeneous system, for example a steady-state fluid flow problem involving distinct zones, each with constant permeability.

Figure 11 shows a problem with two different permeabilities. On the interface between the two zones the potential and the normal velocity component must be continuous, i.e.

$$\begin{aligned} \phi_1 &= \phi_2 \\ C_1 \frac{\partial\phi_1}{\partial n_1} &= - C_2 \frac{\partial\phi_2}{\partial n_2} \end{aligned} \qquad (4.5)$$

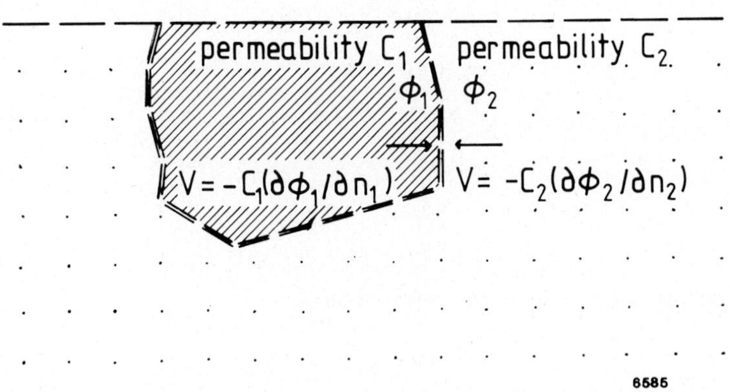

Figure 11
Boundary Element Method for Piecewise Homogeneous System

Assume that constant boundary elements are used and zone 1 has N_1 nodes, zone 2 has N_2 nodes and there are M nodes on the interface. Then there are N_1+N_2+2M unknowns of either ϕ or $d\phi/dn$ on the exterior boundaries and both ϕ and $d\phi/dn$ on the interface. Equation (3.4) applied in each zone yields N_1+N_2 equations and equations (4.5) the remaining 2M equations needed. By counting points on the interface only once, the system can be assembled into N_1+N_2 equations and unknowns.

4.3 Non-linear Problems

Non-linear effects often need to be taken into account in order to realistically model many physical phenomena. For the sake of illustration we consider steady-state flow of the type that arises in problems of heat conduction, seepage through porous media, fluid flow and so on. To solve these problems it is often necessary to obtain a numerical solution to the differential equation†

$$\frac{\partial}{\partial x_i}\left(C\,\frac{\partial\phi(\underset{\sim}{x})}{\partial x_i}\right) = q(\underset{\sim}{x}) \tag{4.6}$$

where ϕ is the potential at a point
C is the conductivity or permeability
and q is the specified source distribution in the interior of the domain

The boundary conditions are usually specified in terms of the potential ϕ or the velocity normal to the surface, V_n, given by

$$V_n = V_i\,n_i = -C\,\frac{\partial\phi}{\partial x_i}\,n_i \tag{4.7}$$

If C is constant or piecewise constant within the domain, equation (4.6) reduces to Poisson's equation or Laplace's equation (when $q = 0$) and the integral formulations discussed earlier apply.

Non-linear behaviour that may have to be taken into account in practical applications can come from one or more of the following sources:

i) part of the boundary S may not be known initially,

ii) the source q may depend on ϕ or V_i,

iii) the material property C may be dependent on ϕ or V_i.

Non-linear boundary conditions

In many flow problems such as groundwater flow, seepage through dams etc. (Fig. 12), the free surface of flow is not known *a priori*. Although on the known parts of the boundary either ϕ or V_n is known, on the unknown free surface both conditions must be satisfied simultaneously. On the free surface we require

$$\begin{aligned} \phi &= Z \\ \text{and}\quad V_n &= 0 \end{aligned} \tag{4.8}$$

where Z is the height above some reference datum.

---

† In this section the summation convention is used, i.e. repeated subscripts are summed over the number of spatial dimensions.

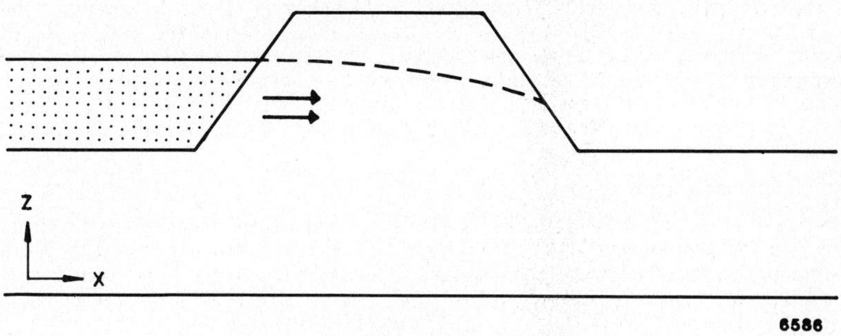

Figure 12
Example of Flow with Free Surface.

The boundary element procedure used to solve this type of problem is as follows [12]:

i) An initial guess is used for the location of the free surface.

ii) For each of the nodes on the free surface, the boundary condition $V_n = 0$ is prescribed. Of course the boundary conditions on the remainder of the boundary are prescribed unambiguously.

iii) The unspecified boundary data is solved for using the conventional boundary element method. This solution provides a set of values for ϕ on the free surface.

iv) From the relationship $\phi = Z$, a fresh estimate of the location of the free surface is obtained.

Steps (ii) and (iv) are repeated until successive locations of the free surface agree to within a prescribed tolerance. For problems solved by Liggett [12] one percent accuracy was obtained in 5 iterations.

Non-linear source

If, for example, the source $q(\underset{\sim}{x})$ is dependent on the potential ϕ, we can use the volume discretization introduced for Poisson's equation. This will lead to the final system of equations (see Equation (4.4.)).

$$[K]\ \{x\} = \{c\} + [C]\ \{q\} \tag{4.9}$$

where the vector $\{x\}$ contains the unknown boundary data and the vector $\{c\}$ represents the effects of the prescribed boundary conditions.

The computational scheme is as follows:

i) take an initial guess for $\{q\}$ (zero, say)

ii) solve equation (4.9) and determine the interior values of ϕ

iii) determine new values of $\{q\}$ from the interior values of ϕ (such a relationship has been assumed)

iv) repeat the process until the changes in $\{q\}$ between successive iterations are less than a prescribed tolerance.

Material non-linearities

In many applications the material parameter C is dependent on the values of ϕ or V_i. Assume that C is expressed in the form:

$$C = C^0 + C^d \tag{4.10}$$

where C^0 is the constant part
and C^d is the part dependent on ϕ or V_i.

The differential equation (4.6), with q = 0, becomes

$$C^0\nabla^2\phi + \frac{\partial}{\partial x_i}\left(C^d \frac{\partial\phi}{\partial x_i} \right) = 0 \tag{4.11}$$

or

$$\nabla^2\phi = \rho \tag{4.12}$$

where

$$\rho = -\frac{1}{C^0}\frac{\partial}{\partial x_i}\left(C^d \frac{\partial\phi}{\partial x_i} \right) \tag{4.13}$$

On the boundary the value of ϕ remains unaltered, but the normal velocity must be modified to

$$V_n' = V_n + C^d \frac{\partial\phi}{\partial x_i} n_i \tag{4.14}$$

The problem is reduced to the solution of equation (4.12) subject to the modified normal velocity V_n'.

We can now use the boundary integral formulation (Equation 4.2) introduced for Poisson's equation to obtain

$$c(P)\phi(P) + \int_S \phi(Q)\frac{d\psi(P,Q)}{dn(Q)}dS(Q) = \frac{1}{C^0}\int_S \psi(P,Q)\, V_n'(Q)dS(Q) - \frac{1}{C^0}\int_V \frac{\partial}{\partial x_i}\left[C^d \frac{\partial\phi(q)}{\partial x_i} \right]\psi(P,q)dV(q) \tag{4.15}$$

By applying the divergence theorem to the volume integral we obtain [13]:

$$c(P)\phi(P) + \int_S \phi(Q)\frac{d\psi(P,Q)}{dn(Q)}dS(Q) = \frac{1}{C^0}\int_S \psi(P,Q)\, V_n(Q)dS(Q) + \frac{1}{C^0}\int_V C^d \frac{\partial\phi(q)}{\partial x_i}\frac{\partial\psi(P,q)}{\partial x_i}dV(q) \tag{4.16}$$

Use of this equation does not require any modification to the normal velocity at the boundary.

The numerical solution of equation (4.16) can be obtained using the iterative procedure outlined in the previous section.

5. OTHER PARTIAL DIFFERENTIAL EQUATIONS

5.1 Steady-state Phenomena

The basic concepts of boundary element methods have been illustrated in the previous sections within the context of Laplace's equation.

Other partial differential equations arising from steady state phenomena (i.e. elliptical equations), can be transformed to boundary integrals once the appropriate fundamental solution is found. Going further than this, a direct formulation can be derived using an appropriate reciprocal relationship analogous to Green's theorem. Using these concepts all of the well known 'classical' partial differential equations, e.g. Helmholz equation, isotropic elasticity etc. have been solved by numerical techniques similar to those described in Sections 3 and 4. That is to say that to use the techniques one usually doesn't have to derive the integral equations from scratch; they are already in the literature. Some work has been done on establishing methods of formulation for quite general forms of differential equations, for example Clements and Rizzo [14] derive a direct method for two-dimensional problems governed by quite general second order elliptic systems. Such systems arise in anisotropic elasticity, for example.

It is worth pointing out that it is not guaranteed that an *analytical* fundamental solution can be found for a given system. Such examples arise in three dimensional anisotropic elasticity, which with the lowest order of symmetry requires 21 independent elastic constants. Even for cubic symmetry (3 independent constants), the resulting characteristic equation cannot be solved analytically [15]. The fundamental solution can still be derived *numerically* however [16].

5.2 Transient Phenomena

When transient phenomena are considered the complexity of the system is increased by the independent variable of time.

The most commonly used technique is to apply the Laplace transform to the partial differential equations and boundary conditions, thus temporarily eliminating time as an independent variable. Using this process, parabolic and hyperbolic systems generally reduce to elliptic systems. In the transform space the elliptic system can be reformulated in terms of a boundary integral equation which can then be solved for a sequence of values of the transform parameter. The time solution can then be obtained by numerical Laplace transform inversion. A survey of the application of this technique to transient heat conduction, viscoelasticity and wave propagation is given in Reference [17].

An alternative method is to work directly in terms of the fundamental solution for the original partial differential equation. This solution will, of course, have time dependence and have a singularity in the time domain as well as in the spatial domain. This approach has been used to solve transient heat conduction problems [18]. Integration with respect to time is performed step-wise. In contrast with the Laplace transform method, the resultant integral equations involve a volume integral.

Shaw [19] gives a survey of boundary integral techniques applied to the wave equation, and suggests that time-stepping methods are appropriate for early time and the Laplace transform method for longer time.

6. CONCLUSIONS

The boundary element method has become an important alternative to the conventional finite element method, i.e. the finite element method applied to the partial differential equations in the domain. The simplicity of the input data requirements, particularly for three-dimensional and infinite domain problems, makes boundary element computer programs easier to use than programs based on finite element or finite difference methods.

Although the basic boundary element method was developed for linear homogeneous material properties, non-homogeneous problems involving distinct homogeneous regions can also be solved by the method. Problems involving a continuous spatial variation in material properties are probably best left to the finite element method. As demonstrated in Section 4 the boundary element method can be adapted to non-linear problems, but the efficiency of the method compared to finite element or finite difference methods remains an open question. Indeed, it may be more efficient to use a combination of methods. The development of methods of coupling boundary element methods with the finite element method is an active research topic [20, 21].

The basic features of the method have been illustrated in the context of Laplace's equation. Over the last decade the method has been applied successfully to partial differential equations arising in such diverse physical applications as acoustics, electromagnetism, aerodynamics, elasticity and so on. Apart from journal papers, the interested reader is referred to the proceedings of conferences on boundary elements [22-25] and proceedings of two other conferences than contain a large proportion of papers on boundary element methods [26,27]. Also worth mentioning is a monograph of surveys of various application areas [28]. A number of books that give an introductory treatment of boundary element methods applied to potential theory and elasticity are available [29-31].

REFERENCES

[1] Fredholm, I. Solution d'un probleme fundamental de la theorie de l'elasticte Arkiv fur Mathematik Astonomie und Fysik, (1905).

[2] Smirnov, V.I., Integral equations and partial differential equations, a course of higher mathematics, Vol. IV, Pergamon, London, (1964).

[3] Cruse, T.A., Mathematical foundations of the boundary-integral equation method in solid mechanics. U.S. Air Force Office of Scientific Research, Report TR-77-1002, (1977).

[4] Zienkiewicz, O.C., Finite element methods in engineering science, 3rd ed. McGraw Hill, London, (1977).

[5] Wu, Y.S., Rizzo, R.J., Shippy, D.J. and Wagner, J.A., An advanced boundary integral equation method for two-dimensional electromagnetic field problems, Electrical Machines and Electromechanics; Vol, 1, No. 4, pp. 301-313, (1977).

[6] Lachat, J.C. and Watson, J.O., Effective numerical treatment of boundary integral equations: A formulation for three-dimensional elastostatics, Int. J. Numerical Methods in Engineering, Vol. 10, pp. 991-1005 (1976).

[7] Watson, J.O., Hermitian cubic boundary elements for plane elastostatics, Proc. 2nd Int. Symp. on Innovative Numerical Analysis in Applied Engineering Sciences, Montreal, pp. 403-412 (see Ref. [27]), (1980).

[8] Watson, J.O., Hermitian cubic boundary elements for plane problems of fracture mechanics, Res. Mechanica, (to appear)(1981).

[9] Hess, J.L., Review of integral-equation techniques for solving potential-flow problems with complicated boundaries. Proc. 2nd Int. Symp. on Innovative Numerical Analysis in Applied Engineering Sciences, Montreal, pp. 131-143 (see Ref. [27]), (1980).

[10] Diest, F.H. and Georgiadis, E., A computer system for three-dimensional elastic analysis using a boundary element approach, Chamber of Mines of South Africa, Res. Rept. No. 43/76, (December, 1976), (1976).

[11] Lean, M.H., Friedman, M. and Wexler, A., Application of the boundary element method in electrical engineering problems, Chapter 7 in Developments in Boundary Element Methods - I, (P.K. Banerjee & R. Butterfield, eds.) Applied Science Publishers, Barking, Essex, (1979).

[12] Liggett, J.A., Location of free surface in porous media, Proc. A.S.C.E. Hydraulics Division, Vol. 103, pp. 353-365, (1977).

[13] Banerjee, P.K., Non-linear problems of potential flow, Chapter 2 in Developments in Boundary Element Methods - I, (P.K. Banerjee & R. Butterfield, eds.) Applied Science Publishers, Barking, Essex, (1979).

[14] Clements, D.L. and Rizzo, F.J., Method for numerical-solution of boundary-volume problems governed by 2nd-order elliptic systems, J. Inst. Math. Appl. Vol. 22, pp. 197-202, (1978).

[15] Head, A.K., The Galois unsolvability of the sextic equation of anisotropic elasticity, J. Elasticity, Vol. 9, pp. 9-20, (1979).

[16] Wilson, R.B. and Cruse, T.A., Efficient implementation of anisotropic three-dimensional boundary-integral equation stress analysis, Int. J. for Numerical Methods in Engineering, Vol. 12, pp. 1383-1397, (1978).

[17] Shippy, D.J., Application of the boundary-integral equation method to transient phenomena in solids, Boundary Integral Equation Method : Computational applications in Applied Mechanics, (T.A. Cruse & F.J. Rizzo eds), ASME, New York, 1975, pp. 15-30, (1975).

[18] Chang, Y.P., Kang, C.S. and Chen, D.J., The use of fundamental Green's functions for the solution of problems of heat conduction in anisotropic media, Int. J. Heat & Mass Transfer, Vol. 16, pp. 1905-1918 (1973).

[19] Shaw, R.P., Boundary integral equation methods applied to wave problems, Chapter 6 in Developments in Boundary Element Methods - I. (P.K. Banerjee & R. Butterfield, eds), Applied Science Publishers, Barking, Essex, (1979).

[20] Zienkiewicz, O.C., Kelly, D.W. and Bettess, P., The coupling of the finite element method and boundary solution methods, Int. J. Numerical Methods in Engng, Vol. 11, No. 2, pp. 355-375, (1977).

[21] Kelly, D.W., Mustoe, G.G.W. and Zienkiewicz, O.C., Coupling boundary element methods with other numerical methods, Chapter 10 in Developments in Boundary Element Methods - I, (P.K. Banerjee & R. Butterfield, eds) Applied Science Publishers, Barking, Essex, (1979).

[22] Cruse, T.A. and Rizzo, F.J. (eds), Boundary-integral equation methods: computational applications in applied mechanics, AMD-11, Applied Mechanics Division, ASME, New York, 141 pp, (1975).

[23] Brebbia, C.A. (editor), Recent Advances in Boundary Element Methods, (Proc. 1st Int. Seminar, Southampton, 1978), Pentech Press, Plymouth, (1978).

[24] Brebbia, C.A. (editor), New Developments in Boundary Element Methods, (Proc. 2nd Int. Seminar, Southampton, 1980), CML Publications, Southampton, (1980).

[25] Brebbia, C.A. (editor), Further Developments in Boundary Element Methods, (Proc. 3rd Int. Seminar, California, 1981), Springer-Verlag, Berlin, (1981).

[26] Cruse, T.A., Lachat, J.C., Rizzo, F. and Shaw, R.P. (eds), Proceedings Int. Symposium on Innovative Numerical Analysis in Applied Engineering Science, (Versailles, France), C.E.T.I.M., Senlis, France (1977).

[27] Shaw, R., Pilkey, W., Pilkey, B., Wilson, R., Lakis, A., Chaudouet, A. and Marino, C. (editors), Innovative numerical analysis for the engineering sciences, (Proc. 2nd Int. Symposium on Numerical Analysis in Applied Engineering Sciences), Montreal, 1980, University Press of Virginia, Charlottesville, (1980).

[28] Banerjee, P.K. and Butterfield, R. (eds), Developments in Boundary Element Methods - I, Applied Science Publishers, Essex, (1979).

[29] Brebbia, C.A., The Boundary Element Method for Engineers, Pentech Press, Plymouth, (1978).

[30] Brebbia, C.A. and Walker, S., Boundary Element Techniques in Engineering, Newnes-Butterworth, London, (1980).

[31] Banerjee, P.K. and Butterfield, R., Boundary Element Methods in Engineering Science, McGraw-Hill, London, (1981).

NUMERICAL SOLUTIONS OF
PARTIAL DIFFERENTIAL EQUATIONS
J. Noye (Editor)

INVERSION OF LARGE SPARSE MATRICES: DIRECT METHODS

Ken J. Mann

Mathematics Department
Caulfield Institute of Technology
Victoria
Australia

Contents

1. Introduction

1.1 General Introduction

The material in this paper has been presented as a first reading for recent graduates in science and engineering to assist them in the initial stages of efficiently solving large sparse linear systems on the digital computer.

Linear systems of equations occur in many diverse areas of mathematical modelling. The analysis of experimental data by linear regression and statistical estimation techniques, and linear economic models, are some areas; others include mechanical or electrical networks, structural analysis and surveying. Many of these linear systems are large and sparse. The sparsity in management science problems is due to the fact that, although the operations of a large company might entail several thousand variables, individual constraints such as those on the production of a single factory, involve only a very few variables. In physical problems, the equation associated with an individual node will involve the variable or variables belonging to that node and the variables belonging to nodes to which it is connected in the network. Because each node is likely to be connected to only a few neighbouring nodes, the matrix is sparse.

There are numerous texts which supply various aspects of this paper in great detail with regards to derivation, error analysis, etc..., through the functional analysis approach or other procedures. However, these are not always easy to read. This paper sets out as simply as possible, using numerical examples where appropriate, some of the fundamental techniques. The reader may then approach a more rigorous analysis with a better understanding. It should be noted that even when an advanced knowledge of the area has been attained, there are many variations from the standard approaches which have been found most suitable for particular modelling problems. These would be found in the appropriate applied journals.

A survey of classical matrix theory may be found in Faddeev and Faddeeva (1963) and Householder (1964). Fox (1964) and Forsythe and Moler (1967) give an elementary approach to computational linear algebra with Forsythe (1967) presenting a basic survey. Wilkinson (1961, 1963, 1965) and Wilkinson and Reinsch (1971) present an in-depth study of numerical methods in linear algebra. The basics of direct methods for the solution of linear systems are given in Young and Gregory (1973), including an outline of several misconceptions often encountered. Reviews of direct methods for large sparse linear systems are presented by George (1977), Reid (1977 a,b) and Duff (1981) while Tewarson (1973) and Jennings (1977) are texts covering this area. Some special sparse linear systems are reviewed by Fox (1977). Many researchers are currently involved in devising efficient techniques for large sparse matrix systems.

Jennings (1977) considers that matrix programs that are unnecessarily complicated, highly inefficient or incapable of producing accurate solutions may be the result of their author's not having a sufficient background in large matrices. He states that a matrix program should attain a balance between simplicity, economy and versatility. It should also include the characteristic of being well-documented: to enable others, as well as the author, to follow the program readily and, perhaps, add to, or update, it.

It is always desirable to obtain a better understanding of the numerical algorithm by first practising some manual examples. With the understanding obtained through this, it is then easier to proceed to the coding of a FORTRAN program. To this end, some solutions are presented, and practice exercises, with answers, are then set. The appendices of this paper contain FORTRAN subroutines, well documented to enable easy usage and effective understanding of the procedures involved. Adherence to this computer language has been guided by the fact that most applied journals, texts, etc. produce their listings in this form and most practising engineers/scientists in this field know this language. In writing one's own subroutines for some of these methods it is useful to have available a selection of matrices to use for testing the programs. Westlake (1968) and Gregory and Karney (1969) have examples appropriate for this purpose, as well as for the manual exercises.

A treatment of computational linear algebra with particular reference to large sparse linear systems must include a study of the economization of computer storage, input/output procedures and central processor unit (c.p.u.) time. It also often involves a compromise between accuracy and cost.

1.2 Mathematical Introduction

A general method of representing a physical system by a mathematical model is to choose a number of points, called nodes, within the system at which equations are set up describing that aspect of the system which is of interest. We can say then, that the equations for large systems are usually associated with individual nodes and involve variables belonging to that node and to a few connected nodes, so that although the total number of variables employed may be large, the number used in any particular equation will be relatively small; i.e. each equation may contain all the variables but for all but a few the multiplying coefficients will be zero.

The system of equations considered here is a linear system which may be represented in matrix form as

$$AX = B \tag{1}$$

where A is an n×n coefficient matrix, X is the solution vector and B is the "constant" vector. If A is invertible, a solution may be found by setting

$$X = A^{-1}B \tag{2}$$

The contents of a matrix are referred to as entries rather than elements in order to avoid confusion in finite element analysis, where much discussion centres on a study of the matrices as well as on the finite elements.

Direct Methods lead to an exact solution in a finite number of steps if roundoff error is not present. Most direct methods are based on Gaussian elimination which produces a system of equations with triangular coefficient matrix which can then be easily solved. Variants arise from the different methods of intermediate storage and utilisation of special properties of certain matrices. For a system of m linear equations, Birkhoff and Fix (1974) indicate that for m less than about 2000, direct solutions are preferred, but for m greater than 2000, iterative solutions are preferred.

A matrix is considered to be sparse when it has sufficiently many zero entries to make it advantageous to employ special techniques that avoid storing or operating with the zeros. A 100 × 100 matrix with ten zero entries would simply be treated as a dense matrix. A 1000 × 1000 matrix with 4000 non-zero entries would be considered sparse.

In the analysis of a special technique for sparse matrices, it is necessary to consider "fill-ins", those zero entries that become non-zero. Direct methods lead to fill-ins being created during solution. An efficient algorithm will limit the number of fill-ins.

In considering the computer implementation of algorithms it is important to note those arithmetic processes which are more significant in the use of central processor unit time than others, whereas in accuracy considerations all the arithmetic processes are important. Multiplication and division are more time consuming than addition.

It is advantageous to the reader to be already familiar with

(i) Gaussian elimination

(ii) LU decomposition of the Crout, Doolittle, Cholesky type

(iii) positive-definite, symmetric matrices

(iv) partitioning of matrices.

Symmetric, positive-definite matrix systems arise in situations such as the application of the finite element least squares and classic variational criteria. The large systems that occur are usually sparse. A useful introduction to the fundamentals of symmetric, positive-definite matrices is given in Dahlquist and Bjorck (1974).

George (1977) states that in comparing two different approaches to solving large sparse systems of equations there are factors such as how many problems having the same structure must be solved, computer charging algorithms etc. which need be considered. He also says that because these algorithms will be applied on a computer, any comparison must involve the computer implementation of them: some data structures exact a higher price in execution and storage than others.

Quite often, the constraint of cost is real. With unlimited finance, a very good result may be obtained. With less finance, a less accurate solution results. In the computer solution of these large systems, the cost factors are essentially

(i) central processor unit time,
(ii) storage - core and backing store,
(iii) input/output time.

The many variants of Gaussian elimination that are designed for particular structures of the sparse matrix are implemented to minimize these cost factors and minimize the accumulation of round-off error. To achieve the latter, a technique is sought to

(i) keep the number of arithmetic processes to a minimum
(ii) avoid involving the zero entries
(iii) avoid fill-in.

It is desirable that the zero entries do not occupy storage space.

If the mathematical model being solved is that of a continuum problem and is not data-based and there is access to a high speed digital computer, with large core, which stores its numbers to a large number of significant digits and there is no restraint on cost, then the subject of this paper probably is of little interest. However, most engineers usually do not have the benefit of this combination.

2. Forms of Coefficient Matrices

The coefficient matrix A may be classed as dense or sparse depending on the number of non-zero elements and in each case it can be symmetric or unsymmetric. A matrix is symmetric if $a_{ij} = a_{ji}$ for all i, j, where i is the row number and j is the column number, i.e. it is symmetric about the main diagonal; $A = A^T$. It is also said to be positive-definite if $X^TAX > 0$ for all $X \neq 0$, where X^T is the transpose of X. This definition is not particularly helpful; however, it can be shown that if C is a non-singular matrix then the product $C^TC = A$ is symmetric positive-definite for any C. From this, it follows that if A is symmetric positive-definite there is a unique lower-triangular matrix L, with positive diagonal elements, such that

$$A = LL^T \tag{3}$$

This factorisation is used in the Choleski method of solution, to be described later.

Example 1

An example of $C^TC = A$ is

$$\begin{bmatrix} 3 & 4 \\ -1 & 2 \end{bmatrix} \begin{bmatrix} 3 & -1 \\ 4 & 2 \end{bmatrix} = \begin{bmatrix} 25 & 5 \\ 5 & 5 \end{bmatrix}$$

and this is then symmetric, positive-definite. It is factorised into LL^T as

$$\begin{bmatrix} 25 & 5 \\ 5 & 5 \end{bmatrix} = \begin{bmatrix} 5 & . \\ 1 & 2 \end{bmatrix} \begin{bmatrix} 5 & 1 \\ . & 2 \end{bmatrix}$$

Although there is no clear boundary at which a matrix becomes dense or sparse we will only be dealing with cases where the non-zero entries represent, at most, a few per cent of the total so that all that follows will clearly pertain to sparse matrices and in particular to large sparse matrices although, for convenience, small matrices may be shown in the examples.

A banded matrix is one in which all the non-zeros lie in a relatively narrow region about the main diagonal. The bandwidth is derived from the maximum number of non-zero entries to any one side of the diagonal. If this number is M then the bandwidth is 2M + 1. Tridiagonal and pentadiagonal matrices are two types of banded matrices for which special algorithms exist enabling solutions to be found efficiently.

$$\begin{bmatrix} 1 & 2 & & & \\ 2 & 1 & 3 & & \\ & 1 & 2 & 3 & \\ & & 3 & 1 & 2 \\ & & & 1 & 2 \end{bmatrix}$$

Figure 1
Tridiagonal Matrix

The form of the matrix may depend upon the numbering of the nodes. For the ring system shown in Figure 2, a matrix map of the nodal interconnections can be made in many ways. Of the two maps shown Figure 2(a) has a bandwidth of 6, Figure 2(b) has a bandwidth of 5.

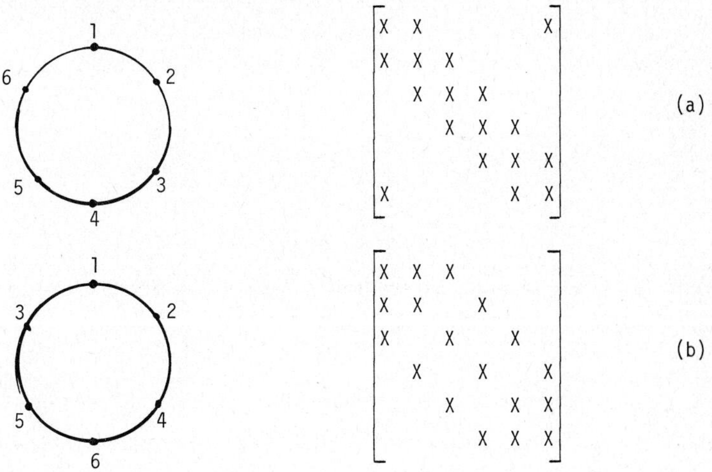

Digraph Matrix

Figure 2
Bandwidths

The X in row i, column j, denotes a coefficient a_{ij} resulting from a connection from node i to node j. The bandwidth allotted to the matrix in Figure 2(a) does not fit the formula for a banded matrix but is the maximum bandwidth of the interconnections.

In the symmetric case shown in Figure 2, the directions are not shown for the connections but in the unsymmetric case directional arrows are included in the digraph (directed graph).

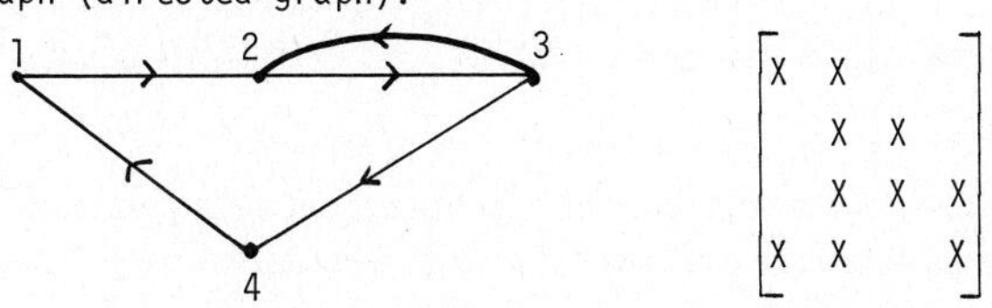

Digraph Matrix

Figure 3
Unsymmetric Matrix System

A banded matrix may have a constant bandwidth or variable bandwidth. This is, in some cases, a matter of definition, since, by including zero entries a variable bandwidth can be treated as constant. In Figure 4, the matrix is symmetrical, so that only the main diagonal and sub-diagonal entries need be shown to fully describe it. This corresponds with the fact that these are usually the only entries held in storage.

$$\begin{bmatrix}
X & & & & & & & & \\
X & X & & & & & & & \\
X & X & X & & & & & & \\
X & X & X & X & & & & & \\
0 & X & X & X & X & & & & \\
 & 0 & 0 & X & X & X & & & \\
 & & 0 & 0 & 0 & X & X & & \\
 & & & 0 & 0 & X & X & X & \\
 & & & & X & X & X & X & X
\end{bmatrix}$$

Figure 4
Banded Matrix Store

The non-zero entries form a matrix in which the bandwidth is not regular from row to row. Including the zeros shown forms a matrix with a regular bandwidth. Variable bandwidth matrices are also classifiable into those with re-entrant rows and those without. A re-entrant row is defined as one in which the column number of the first entry is less than the column number of the first entry of the previous row. In the variable bandwidth matrix shown in Figure 4, the last row is re-entrant.

When dealing with symmetric matrices, requiring only a knowledge of the diagonal and sub-diagonal entries, we often speak of the bandwidth in terms of these entries. Jennings (1977) would say that the third row of the matrix in Figure 4 had a semibandwidth of 3. George (1977) uses a term $B_i(A)$ which defines the bandwidth of the ith row and a term $B(A)$ as the maximum $B_i(A)$, $1 \leqslant i \leqslant n$. For the above matrix $B_3(A) = 3$ and $B(A) = 4$. The profile of A is then defined as $\sum_{i=1}^{n} B_i(A)$.

Given a matrix A with i rows and j columns, permutation matrices can operate on it to interchange rows or columns. The permutation PAP^T of the rows and columns of A can be used to arrange it into a form better suited to the solution process. It should be noted that, with this type of permutation, all the original entries are retained on the main diagonal.

Example 2

A permutation of a 3×3 matrix is presented as follows

$$\begin{bmatrix} 1 & 0 & 0 \\ 0 & 0 & 1 \\ 0 & 1 & 0 \end{bmatrix} \begin{bmatrix} a & b & c \\ d & e & f \\ g & h & i \end{bmatrix} \begin{bmatrix} 1 & 0 & 0 \\ 0 & 0 & 1 \\ 0 & 1 & 0 \end{bmatrix}$$

$$= \begin{bmatrix} 1 & 0 & 0 \\ 0 & 0 & 1 \\ 0 & 1 & 0 \end{bmatrix} \begin{bmatrix} a & c & b \\ d & f & e \\ g & i & h \end{bmatrix}$$

$$= \begin{bmatrix} a & c & b \\ g & i & h \\ d & f & e \end{bmatrix}$$

In this type of operation, if the ith and r^{th} columns are interchanged so also are the ith and r^{th} rows. This need not always be the case; a permutation PAQ could be used where Q is not the transpose of P.

3. Storage

3.1 Introduction

In the 2-dimensional array commonly used for A, each variable is allotted one column and each equation, corresponding to one node, is allotted one line. Hence any line of a large sparse matrix will consist mostly of zeros. This distribution may not be a static one, however. In the process of solution, additional non-zeros may be inserted. The extent of these fill-ins depends on the type of system and the manner of solution.

To carry out a solution, the elements of A are stored in computer core and accessed when required. If there is insufficient space in core, a backing (disc) store is used. Variables are transferred between main core and backing store during the solution process. For reasons of economy, and because of limitations in core space, it is useful if it can be arranged that only non-zeros be stored or that, at least, as few zeros as possible are retained. If the system can be broken up into smaller, self-consistent systems it may only be necessary to store a small part of it at any one time. As variables are eliminated during the progressive solution of the smaller sub-systems they can be transferred to disc and the storage space re-allotted to new variables.

The method of storage depends on many factors; there is no one method best for all cases. A method of storage for very sparse matrices may not be as satisfactory for matrices still classifiable as sparse but with greater numbers of non-zeros.

In the following two subsections on storage, a vector is real or integer according to the Fortran convention for the first letter of its name.

3.2 Storage for a Random Sparse Symmetric Matrix

In some applications such as finite element calculations one encounters symmetric, positive-definite matrices, where the location of the nonzero elements differs greatly from row to row. In the use of the finite element least squares and finite element classic variational criteria the resultant linear system is symmetric, positive-definite. In other large scale applications, the matrices are simply symmetric. In such cases it can be more efficient to use the "profile-storage" scheme devised by Jennings (1966).

The $n \times n$ symmetric matrix A is replaced by two vectors, E and R. In E, the entries of the lower triangle of A are stored row by row starting with the first non-zero in each row and including all subsequent entries, whether zero or non-zero, as far as the diagonal. The pointer vector, R, indicates the entries of E which are on the diagonal. R(i) points to the location in E of the diagonal entry of the ith row of A.

Example 3

$$A = \begin{bmatrix} 7 & 3 & 0 & 0 & 0 \\ 3 & 1 & -2 & 4 & 0 \\ 0 & -2 & 2 & 0 & 0 \\ 0 & 4 & 0 & 8 & -1 \\ 0 & 0 & 0 & -1 & 6 \end{bmatrix}$$

Using "profile storage", the vectors E and R are formed as follows:

$$E = (7,3,1,-2,2,4,0,8,-1,6)$$
$$R = (1,3,5,8,10)$$

Advantage or disadvantage?

In this example the lower triangular submatrix of A has 15 entries whereas the combined E and R vectors have 15 entries, so there has been no gain in storage. This is typical of small matrices in which storage is never a concern. However, for a large sparse matrix of this type the gain would be considerable. A 100×100 matrix with 400 nonzero entries has a lower triangular submatrix of 5050 entries, but, using "profile storage", vector E would have, at most, 300 entries and vector R would have exactly 100 entries, making a total of 400 in comparison to the 5050.

Exercise 1

1. Use the "profile-storage" scheme for symmetric matrices to formulate the E and R arrays from each of the following:

(a) The 100×100 matrix

$$\begin{bmatrix} 2 & -1 & & & & & \\ -1 & 2 & -1 & & & & \\ & -1 & 2 & -1 & & & \\ & & & & & & \\ & & & & -1 & 2 & -1 \\ & & & & & -1 & 1 \end{bmatrix}$$

(b) $$\begin{bmatrix} 38 & 19 & -21 \\ 19 & 21 & -12 \\ -21 & -12 & 14 \end{bmatrix}$$

(c) $$\begin{bmatrix} 5 & 3 & 0 & 1 & -6 \\ 3 & 7 & 0 & 2 & 0 \\ 0 & 0 & 2 & 0 & 4 \\ 1 & 2 & 0 & 6 & -1 \\ -6 & 0 & 4 & -1 & 11 \end{bmatrix}$$

(d) $$\begin{bmatrix} 2 & 0 & -3 & 0 & 0 \\ 0 & -4 & -1 & 2 & 0 \\ -3 & -1 & 1 & 0 & 4 \\ 0 & 2 & 0 & -6 & 0 \\ 0 & 0 & 4 & 0 & 2 \end{bmatrix}$$

<u>Answers</u>:

(a) E = (2,-1,2,-1,2,-1,2,-1,2..........2,-1,1) : 199 entries.
R = (1,3,5,7,9....................,197,199) : 100 entries.

(b) E = (38,19,21,-21,-12,14)
R = (1,3,6)

(c) E = (5,3,7,2,1,2,0,6,-6,0,4,-1,11)
R = (1,3,4,8,13)

(d) E = (2,-4,-3,-1,1,2,0,-6,4,0,2)
R = (1,2,5,8,11)

3.3 <u>Storage for a Random Sparse Unsymmetric Matrix</u>

In applications such as the computational design of electrical networks, the matrices are unsymmetric with highly irregular sparseness. In the use of the finite element Galerkin criterion for nonlinear problems, the resultant linear system is unsymmetric. Hence the "profile-storage" scheme is not appropriate.

The Gustavson (1972) method, describing the data structure of an unsymmetric matrix, A, with irregular sparseness by means of three vectors, is outlined as follows. Three vectors AH, IH, IA replace matrix A. Vector AH contains the non-zero entries of A row by row; the zero entries are not stored. Vector IH contains the column numbers of the corresponding entries of AH such that IH(j) contains the column number of the entry AH(j). Vector IA is constructed so that IA(i) gives the position in the vectors AH and IH of the first entry of the ith row of A and the last additional entry in IA is equal to the total number of elements in AH plus one.

Example 4

$$A = \begin{bmatrix} 3 & 0 & -4 & 1 & 0 \\ 0 & -2 & 3 & 0 & 0 \\ 0 & -1 & 4 & 0 & -1 \\ -2 & 0 & 0 & -3 & 0 \\ 0 & -5 & 6 & 0 & 2 \end{bmatrix}$$

Using the Gustavson scheme, vectors AH, IH and IA are formed as follows:

$$AH = (3,-4,1,-2,3,-1,4,-1,-2,-3,-5,6,2)$$
$$IH = (1,3,4,2,3,2,3,5,1,4,2,3,5)$$
$$IA = (1,4,6,9,11,14).$$

Advantage or disadvantage?

In this example, matrix A has 25 entries whereas the combined AH, IH and IA vectors have 32 entries. Therefore for this example the system has proven to be a disadvantage. It is reasonable to expect that there will be little gain in the application of this scheme to small matrices or to matrices with only a small percentage of the entries as zeros. This particular example falls into both categories. However, it can readily be seen that for a large sparse unsymmetric matrix of order 1000×1000 with a random sparseness giving, at most, 30 non-zeros in each row that, instead of storing 10^6 entries, the maximum for the total of the three vectors AH, IH and IA would be $3.10^4 + 3.10 + 1001 \doteq 6.10^4$ entries: a significant reduction. This reduction would be more pronounced for even larger systems.

Exercise 2

1. Use the Gustavson storage scheme for unsymmetric matrices to formulate arrays AH, IH and IA from each of the following:

(a) $\begin{bmatrix} 5 & 0 & -2 & 0 & 0 \\ 0 & 7 & 0 & 1 & 0 \\ 3 & 0 & 2 & 0 & 1 \\ 0 & -5 & 6 & 0 & 0 \\ 0 & 0 & 1 & 0 & 4 \end{bmatrix}$

(b) $\begin{bmatrix} 7 & 0 & -3 & 0 & -1 & 0 \\ 2 & 8 & 0 & 0 & 0 & 0 \\ 0 & 0 & 1 & 0 & 0 & 0 \\ -3 & 0 & 0 & 5 & 0 & 0 \\ 0 & -1 & 0 & 0 & 4 & 0 \\ 0 & 0 & 0 & -2 & 0 & 6 \end{bmatrix}$

(c) $\begin{bmatrix} -2.4 & 7.6 & 0 & 0 & 0 & 8.9 \\ 3.8 & -0.4 & 0 & 2.9 & 0 & 0 \\ 0 & 0 & 3.1 & -5.6 & 11.3 & 0 \\ -1.4 & 0 & 0 & 6.7 & 0 & 4.9 \\ 3.8 & 1.4 & 0 & 0 & -2.9 & 0 \\ 7.2 & 0 & 0 & 0 & -1.6 & 5.3 \end{bmatrix}$

Answers:

(a) AH = (5,-2,7,1,3,2,1,-5,6,1,4)
IH = (1,3,2,4,1,3,5,2,3,3,5)
IA = (1,3,5,8,10,12)

(b) AH = (7,-3,-1,2,8,1,-3,5,-1,4,-2,6)
IH = (1,3,5,1,2,3,1,4,2,5,4,6)
IA = (1,4,6,7,9,11,13)

(c) AH = (-2.4,7.6,8.9,3.8,-0.4,2.9,3.1,-5.6,11.3,-1.4,6.7,4.9,3.8,1.4,-2.9,7.2,-1.6,5.3)
IH = (1,2,6,1,2,4,3,4,5,1,4,6,1,2,5,1,5,6)
IA = (1,4,7,10,13,16,19)

A similar scheme has been incorporated by Fletcher (1976) in a subroutine called SPAREL which solves a system of linear algebraic equations using Gaussian elimination. A documented and ordered listing of SPAREL has been given in appendix B of this paper. Fletcher has arrays functioning identically to vectors AH and IH but, instead of the vector IA giving the position in AH and IH of the first entry of each row, he has a vector JEND such that JEND(i) indicates the number of non-zero entries in the ith-row, and a separate variable LCMAX contains the total number of non-zeros in matrix A. The overall effect of the Fletcher scheme is the same as the Gustavson scheme. There may be other similar schemes.

Example 5

For the model example 4, the Fletcher scheme forms AH, IH, and JEND and LCMAX as follows:

AH = (3,-4,1,-2,3,-1,4,-1,-2,-3,-5,6,2)
IH = (1,3,4,2,3,2,3,5,1,4,2,3,5)
JEND = (3,2,3,2,3)
LCMAX = 13

Thus the procedure is to indicate the column number of each non-zero entry in one vector, which has a one-to-one ratio with a vector containing the non-zero entries; then a further array is needed to indicate when each new row commences. This is necessary because of problems occurring in situations such as those in which the only non-zero entry in row 3 is in column 2 and the only non-zero entry in row 4 is in column 3. It would be an easy mistake to consider that they belong to the same row unless there is this further constraint.

In using SPAREL, it is necessary to set a further vector B to store the right-hand side of the system of equations and the identifier N indicating the number of equations to be solved. The solution returns in vector X.

Exercise 3

1. Arrange the system AX = B into the appropriate identifiers and arrays for solution by subroutine SPAREL (in Appendix B), where matrices A and B are given as

$$A = \begin{bmatrix} 5 & -3 & 0 & 0 & 2 \\ 0 & 2 & 4 & -1 & 0 \\ -1 & 0 & -3 & 0 & 0 \\ 2 & 0 & 0 & -5 & 4 \\ 0 & 0 & 0 & 3 & 6 \end{bmatrix} \qquad B = \begin{bmatrix} 11 \\ 21 \\ 37 \\ 14 \\ 5 \end{bmatrix}$$

Answer:

AH = (5,-3,2,2,4,-1,-1,-3,2,-5,4,3,6)
IH = (1,2,5,2,3,4,1,3,1,4,5,4,5)
JEND = (3,3,2,3,2)
B = (11,21,37,14,5)
LCMAX = 13
N = 5

3.4 Storage for Banded Matrices

A method used for storing a symmetric banded matrix is called diagonal storage. All the diagonals of the lower triangle which contain non-zero entries are stored in a rectangular array with n rows and a number of columns equal to the maximum $B_i(A) + 1$. The A_{ij} are submatrices of the symmetric matrix and $A_{ij} = A_{ji}$.

$$\begin{bmatrix} A_{11} & A_{12} & & A_{14} & & & \\ A_{21} & A_{22} & & & & & \\ & & A_{33} & & A_{35} & A_{36} & \\ A_{41} & & & A_{44} & A_{45} & & \\ & & A_{53} & & A_{55} & & A_{57} \\ & & A_{63} & & & A_{66} & A_{67} \\ & & & & A_{75} & A_{76} & A_{77} \end{bmatrix}$$

A (symmetric)

$$\begin{bmatrix} & & & A_{11} \\ & & A_{21} & A_{22} \\ & & & A_{33} \\ A_{41} & & & A_{44} \\ & A_{53} & A_{54} & A_{55} \\ A_{63} & & & A_{66} \\ & A_{75} & A_{76} & A_{77} \end{bmatrix}$$

Figure 5

Diagonal Storage of A

If there is wide variation in the value of $B_i(A)$ this method may be quite wasteful of storage space; in which case the method of profile storage, previously described, could be used.

4. Methods of Solution

4.1 Gaussian Elimination

Direct methods of solution are based on the method of Gaussian elimination.

Let A be a given square matrix of order n and B a given n-vector. The linear system of n equations

$$
\begin{aligned}
a_{11}x_1 + a_{12}x_2 + a_{13}x_3 + \dots + a_{1n}x_n &= b_1 \\
a_{21}x_1 + a_{22}x_2 + a_{23}x_3 + \dots + a_{2n}x_n &= b_2 \\
\dots\dots\dots\dots\dots\dots & \\
\dots\dots\dots\dots\dots\dots & \\
a_{n1}x_1 + a_{n2}x_2 + a_{n3}x_3 + \dots + a_{nn}x_n &= b_n
\end{aligned} \tag{4}
$$

can be put in matrix form

$$
\begin{bmatrix} a_{11} & a_{12} & a_{13} & \dots & a_{1n} \\ a_{21} & a_{22} & a_{23} & \dots & a_{2n} \\ \cdot & \cdot & \cdot & & \cdot \\ \cdot & \cdot & \cdot & & \cdot \\ a_{n1} & a_{n2} & a_{n3} & \dots & a_{nn} \end{bmatrix} \begin{bmatrix} x_1 \\ x_2 \\ \cdot \\ \cdot \\ x_n \end{bmatrix} = \begin{bmatrix} b_1 \\ b_2 \\ \cdot \\ \cdot \\ b_n \end{bmatrix} \tag{5}
$$

or, more briefly, as

$$AX = B \tag{6}$$

In order to solve the system, the matrix A is reduced to upper triangular form and the solution found by back-substitution. The method depends on the fact that the solution is unaffected by

(i) interchanging equations.

(ii) adding a scalar multiple of one equation to another.

The whole process can be outlined as

$$AX = LUX = LY = B \tag{7}$$

$$UX = Y = L^{-1}B \tag{8}$$

where L is lower triangular and U is upper triangular. X can be found by back-substitution commencing with $x_n = y_n/u_{nn}$.

In more detail, the elimination process leading to the desired upper-triangular form can be summarised as follows

$$a_{ij}^{(k+1)} = a_{ij}^{(k)} - \frac{a_{ik}^{(k)}}{a_{kk}^{(k)}} \cdot a_{kj}^{(k)} \qquad (i,j > k) \tag{9}$$

$$b_i^{(k+1)} = b_i^{(k)} - \frac{a_{ik}^{(k)}}{a_{kk}^{(k)}} \cdot b_k^{(k)} \qquad (i > k) \tag{10}$$

expressing the operations at the kth step, $k = 1,2,\dots,n-1$ commencing with $A^{(1)} = A$, $B^{(1)} = B$. This leads to the desired upper triangular system, which, on back substitution, gives

$$x_k = \frac{1}{a_{kk}^{(k)}} \left(b_k^{(k)} - \sum_{j=k+1}^{n} a_{kj}^{(k)} x_j \right) \qquad k = n, n-1, \dots, 1 \tag{11}$$

During the process each $a_{ij}^{(k+1)}$ can replace $a_{ij}^{(k)}$ in storage and the multipliers $\ell_{ik} = a_{ik}^{(k)}/a_{kk}^{(k)}$ may overwrite $a_{ik}^{(k)}$. We thus have obtained the triangular factorisation

$$A = LU \tag{12}$$

where L is unit lower triangular with off-diagonal elements ℓ_{1k} and U is the upper triangular matrix $[a_{ij}^{(i)}]$, $i \leq j$. The first few terms of L are

$$\begin{bmatrix} 1 & & & \\ (a_{21}/a_{11})^{(1)} & 1 & & \\ (a_{31}/a_{11})^{(1)} & (a_{32}/a_{22})^{(2)} & 1 & \\ \cdot\;\cdot\;\cdot & \cdot\;\cdot\;\cdot & \cdot\;\cdot\;\cdot & \cdot\;\cdot \end{bmatrix} \tag{13}$$

When a term such as $a_{ij}^{(k)}$ is a zero, $a_{ij}^{(k+1)}$ may be a new non-zero for which storage space needs to be found if, say, a Gustavson type storage method is used. This problem is dealt with by Reid (1977). Gustavson storage requires that a fresh copy be made of the row (or column) thereby temporarily wasting the space occupied by the old row (or column). A means of avoiding this is to use a linked list. The non-zeros are stored as triples (a_{ij},i,j) held contiguously, in any order, in a real array and two integer arrays. The integers indicate where the non-zero belongs in the matrix. Two more integers are held with each non-zero to give the address of the next non-zero in the row and the address of the next non-zero in the column. It is these latter two integers which expand the structure into a linked list. Pointers are also required to the first elements in the rows and columns. Thus rows and/or columns can be accessed without scanning the whole array and non-zeros can be deleted or inserted without performing a major operation involving the whole row or column.

4.2 Block-Triangular Form

The number of fill-ins can be limited by interchanging rows and columns. A simple case is illustrated below

$$\begin{bmatrix} X & X & X & X & X \\ X & X & & & \\ X & & X & & \\ X & & & X & \\ X & & & & X \end{bmatrix} \qquad \begin{bmatrix} X & & & & X \\ & X & & & X \\ & & X & & X \\ & & & X & X \\ X & X & X & X & X \end{bmatrix}$$

(a) (b)

Figure 6

Sparse Matrices

The matrix in Figure 6(b) is the matrix in Figure 6(a) permuted so that the first and last rows and the first and last columns are interchanged. Elimination applied to the former results in the creation of a totally dense matrix, whereas no new non-zeros would be created in the latter.

It is obvious that fill-in is avoided with a triangular matrix, since forward or back substitution can be commenced immediately, but a more general case is for a matrix to be block-triangular as illustrated in (14), where each diagonal block A_{kk} is a square matrix

$$A = \begin{bmatrix} A_{11} & & & & \\ A_{21} & A_{22} & & & \\ \cdot & & A_{33} & & \\ \cdot & & & \cdot & \\ \cdot & & & & \cdot \\ A_{n1} & A_{n2} & \cdot\;\cdot\;\cdot & & A_{nn} \end{bmatrix} \tag{14}$$

The equation AX = B is now broken up into a sequence of smaller problems

$$A_{kk}X_k = B_k - \sum_{j=1}^{k-1} A_{kj}X_j \qquad k = 1,2,\ldots,n \tag{15}$$

Having solved for A_{11}, for instance, all the variables in A_{21} are known so we need only solve for the unknowns in A_{22} at the next step and so on. Any fill-in occurring will always be confined to the diagonal blocks A_{kk}.

There are several algorithms for forming a block-triangular matrix, where it is possible to do so. The first step is to find a transversal i.e. set non-zeros on the diagonal. Reid (1977a) describes a method due to Hall (1956) and also outlines methods of finding the block-triangular form due to Sargent and Westerberg (1964) and Tarjan (1972). Duff and Reid (1976) have provided a Fortran program to implement the method of Tarjan.

If a matrix has t non-zeros, no more than O(t) operations are needed to place one of them on the main diagonal and it follows that no more than O(nt) operations are needed to find a transversal.

The above-mentioned algorithms will not be described here, but the following example may give some idea of what is involved in the process of block-triangularisation. Starting with a matrix A which has zeros on the main diagonal we show it after a transversal has been found, as illustrated in Figure 7. A digraph is then constructed which clearly shows that there are three composite nodes. Renumbering the ordinary nodes of the composite nodes is equivalent to a permutation of A which can result in the block-triangular form, as shown in Figure 8.

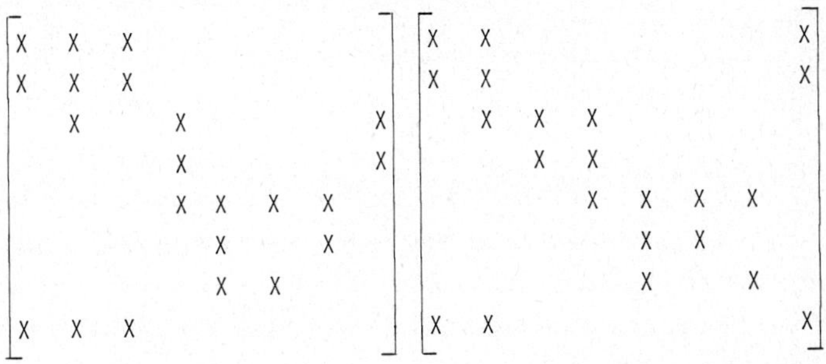

(a) Before transversal (b) After transversal

Figure 7

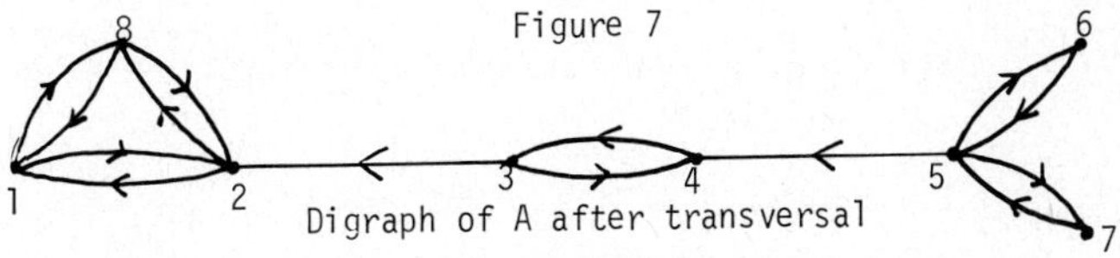

Digraph of A after transversal

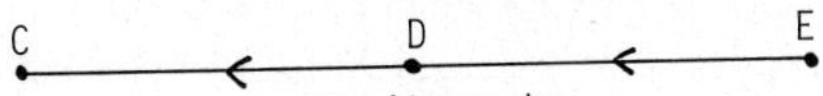

composite nodes

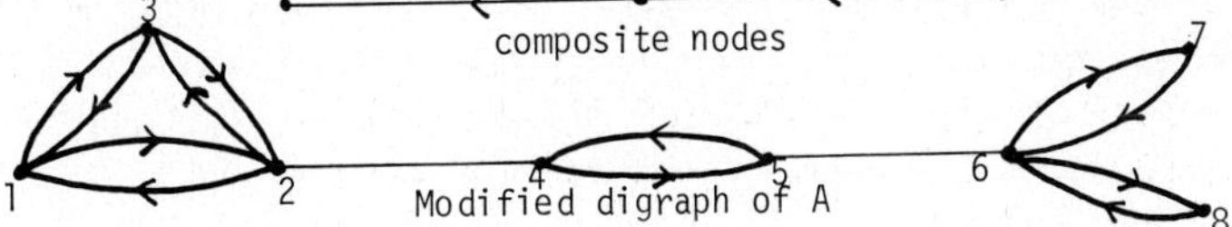

Modified digraph of A

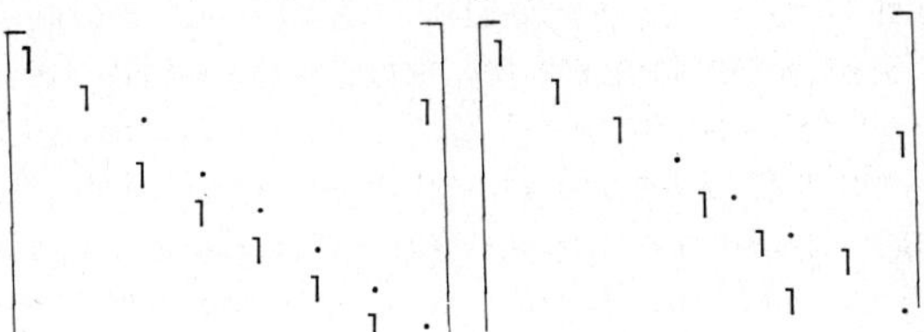

P Q

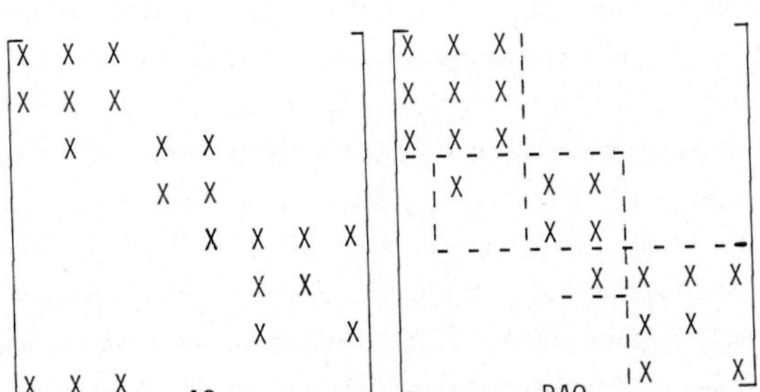

AQ PAQ

Equivalent Permutation

Figure 8

Digraph and Permutation

4.21 Pivoting

Numerical instability can arise when large numbers are added to small numbers in the elimination process as information present in the smaller number may be lost. For example in 4 decimal arithmetic

$$1.248 + 201.1 = 202.3 \tag{16}$$

and the last two digits of 1.248 have been lost. The large number could arise when one number is divided by a much smaller number. Pivoting is a strategy used to avoid this and to obtain stability in the process of solution by Gaussian elimination. Rows or columns are interchanged to ensure that the pivoting entry $a_{kk}^{(k)}$ of equations (9) and (10) satisfies one of the conditions

$$|a_{kk}^{(k)}| \geq u.\max_{k \leq i \leq n} |a_{ik}^{(k)}| \quad \text{(column search)} \tag{17}$$

$$|a_{kk}^{(k)}| \geq u.\max_{k \leq j \leq n} |a_{kj}^{(k)}| \quad \text{(row search)} \tag{18}$$

where u is a parameter chosen in the range $0 < u \leq 1$. Then, from equation (9), the inequality

$$|a_{ij}^{(k+1)}| \leq (1 + \frac{1}{u})\max_{i,j}|a_{ij}^{(k)}| \tag{19}$$

may be deduced, and we see that the growth in size of matrix entries is limited at each stage to the factor $1 + \frac{1}{u}$. Ordinary partial pivoting, as used with a dense matrix, corresponds to a choice of $u = 1$. If this were the case with a sparse matrix, the severe restriction in choice of pivots would greatly reduce the possibility of block-triangulation. Fortunately, this is not the case; u may be chosen less than 1, thus allowing exploitation of the sparsity. Curtis and Reid (1976) recommend $u = \frac{1}{4}$ on the grounds of numerical experiments. For linear programming matrices, Tomlin (1972) recommends $u = 1/100$.

From equation (9) it will be seen that the entries in a column only change if there is a non-zero in that column in the pivot row. Thus, overall growth of an entry in column j is limited by a factor

$$(1 + \frac{1}{u})^{p_j - 1} \tag{20}$$

where p_j is the total number of non-zeros, including fill-ins, in column j. For a sparse matrix, p_j may be small, so the factor may not be enormous, in contrast to the case of a dense matrix if values of $u << 1$ were used. However, even for a sparse matrix the growth factor as calculated above may be too large. In actuality, growth is likely to be much smaller than the value calculated in this manner and it may be advisable to insert a routine in the program to monitor the size of the entries. If the growth is too great, a fresh start with a new value of u is necessary.

Finally, although pivoting involves row or column interchanges, they need not be physically interchanged. Permutation vectors of length n can be used in which the interchanges can be recorded. The row permutation vector is initially $(1,2,3,...n)^T$ corresponding to the n rows. To exchange rows i and j, the ith and j^{th} entries of the permutation vector are exchanged.

In practice the criterion of Markowitz (1957) has been found generally satisfactory in choosing a pivot entry for a sparse matrix. A non-zero in $A^{(k)}$ is found which satisfies equations (17) or (18) and has the smallest product of number of other non-zeros in its row and number of other non-zeros in its column and this is then brought into the pivotal position (k,k) by interchanging rows and columns. For example, if two suitable entries are found, one of which has 5 other non-zeros in the row and 3 in the column and the other has 4 and 4 respectively, the first one will be chosen as a pivot since the product $5 \times 3 = 15$ is less than $4 \times 4 = 16$.

A fresh search is made at each stage of the elimination, requiring that the sparsity pattern be known at each stage, so it must be held in a data structure and updated at each stage. This is a disadvantage which can be overcome by ordering the columns and processing them one by one, i.e. all operations for the jth column are performed before operations on the $(j+1)^{st}$ column. This can be done if the multipliers $a_{ik}^{(k)}/a_{kk}^{(k)}$ are stored. Duff and Reid (1974) have found experimentally that it is satisfactory for the columns to be chosen in order of increasing numbers of non-zeros. The pivot $a_{kk}^{(k)}$ is chosen by searching the kth column of the rearranged matrix for an entry a_{ik} $(i \geq k)$ satisfying the stability criterion and whose row has least non-zeros in the original matrix. The sparsity pattern is not updated. This produces more fill-ins than the original Markowitz criterion but against this must be weighed the simplicity of the method. The following figures are given by Reid (1977b)

		54	57	199
Matrix order		54	57	199
Non-zeros in A		291	281	701
Non-zeros in L/U	Markowitz	381	315	1387
	a priori using A	475	408	2767

Table 1

Wilkinson (1974a) provides some basic ideas on the introduction of pivoting.

4.22 Scaling

If the ith row of the coefficient matrix and right-hand side vector is multiplied by a number k_i, the solution of the system is not affected, so we could choose k_i such that a_{ii} becomes the largest entry in the ith column and if a pivot entry is to be selected purely on the basis of it being the largest number then, for partial pivoting by rows, this would seem a suitable choice. However, multiplying the ith row by k_i does not change the error propagation as can be shown by multiplying each row by some number. From (9)

$$a_{ij}^{(k+1)} = \left[a_{ij}^{(k)} - \frac{a_{ik}^{(k)}}{k_k a_{kk}^{(k)}} . k_k a_{kj}^{(k)}\right] \quad (i,j > k) \tag{21}$$

$$= \left[a_{ij}^{(k)} - \frac{a_{ik}^{(k)}}{a_{kk}^{(k)}} . a_{kj}^{(k)}\right] \tag{22}$$

and the relative error due to the subtraction is unchanged.

In fact, leaving the equations unchanged and selecting the largest pivot is no different from the above process since we can consider each row as having already been multiplied by some constant. In order to avoid what amounts to an arbitrary choice of pivot, the matrix is scaled so that the rows are comparable in some way. Various methods of scaling have been suggested. The rows could be normalised by dividing each row by its largest entry or dividing by the average magnitude of its coefficients or by the square root of the sum of the squares of its coefficients. The scaled values should then be used in choosing the pivot entry. The scheme of Hamming (1971) is quite effective.

The process described above implies that an explicit scaling of the rows is carried out but this need not be so. Having calculated the factor by which each row should be multiplied we use this when choosing a pivot; i.e. we compare the magnitudes of the products of the entries under consideration and their associated scale factors but we leave the original matrix entries unchanged for the solution process, as scaling them could introduce an error due to the operation of division, and as we have seen scaling does not of itself reduce the error.

Scaling by columns is also a possible process but this requires re-ordering the solution entries and row pivoting is generally preferred. Full pivoting would involve interchanges of rows and columns but it is generally considered that the gain achieved is usually not sufficient to warrant the extra work involved. Again, cost is an important factor.

Whilst scaling is essential when working with a full matrix, its necessity in the case of a sparse matrix depends on the degree of sparsity. For very sparse matrices it could be omitted.

It would, however, always seem advisable to inspect the linear system to see if there is any obvious disparity in the magnitudes of the entries in the matrices.

Scaling across a row is a reasonably easy procedure to incorporate into a program. Scaling down a column requires more accounting as this then alters the value of the unknown relevant to that column in the resulting system. The following simple example illustrates this.

Example 6

$$356\ x_1 + 5.4\ x_2 = -\ 7.3$$
$$-\ 724\ x_1 + 6.7\ x_2 = 8.9$$

For this system, divide column 1 by 100 to give a system where entries are of the same order as follows

$$3.56\ x_1^* + 5.4\ x_2 = -\ 7.3$$
$$-\ 7.24\ x_1^* + 6.7\ x_2 = 8.9$$

It is necessary to note that, on solution, the x_1^* must be divided by the division factor of 100 to obtain the required x_1.

Scaling matrices automatically still remains an area of research with quite a deal of support for the scheme of Hamming (1971). This technique will not be described here. Curtis and Reid (1972) researched the area of automatic scaling of matrices for Gaussian elimination and were in favour of the method of Hamming.

4.3 Choleski Method

Symmetric positive-definite matrices usually arise as a result of the finite element method particularly with the least squares and classic-variational criteria. If

$$A^{(k)} = a_{ij}^{(k)} \qquad i \geq j,\ j \geq k \tag{23}$$

is symmetrical, then so also is

$$A^{(k+1)} = a_{ij}^{(k+1)} \qquad i \geq k+1,\ \ j \geq k+1 \tag{24}$$

and only the lower triangular part need be computed and stored. For this case it has been shown by Wilkinson (1967) that

$$\max_{i,j}|a_{ij}^{(k+1)}| \leq \max_{i,j}|a_{ij}^{(k)}| \tag{25}$$

i.e. there is no growth in the size of the matrix entries and the system is stable. Consequently, symmetric interchanges can be made without regard to the required sparsity pattern. The resulting matrix PAP^T has a triangular factorisation LL^T and L can be computed by the Cholesky algorithm. The Markowitz pivoting strategy of choosing the diagonal non-zero with least other non-zeros in its row can be used, so minimising fill-in.

If triangular decomposition is possible, a matrix, A, can be expressed as the product of two matrices

$$A = LU \tag{26}$$

where L is lower triangular and U is upper triangular. If A is symmetric positive-definite then U can be the transpose of L and

$$A = LL^T \tag{27}$$

Only the coefficients of L need be found and stored and these can be calculated by formula. A further condition on the L is that it has positive entries on the diagonal.

The method can be illustrated by means of the following 4×4 system.

$$\begin{bmatrix} a_{11} & a_{12} & a_{13} & a_{14} \\ a_{21} & a_{22} & a_{23} & a_{24} \\ a_{31} & a_{32} & a_{33} & a_{34} \\ a_{41} & a_{42} & a_{43} & a_{44} \end{bmatrix} = \begin{bmatrix} \ell_{11} & & & \\ \ell_{21} & \ell_{22} & & \\ \ell_{31} & \ell_{32} & \ell_{33} & \\ \ell_{41} & \ell_{42} & \ell_{43} & \ell_{44} \end{bmatrix} \cdot \begin{bmatrix} \ell_{11} & \ell_{12} & \ell_{13} & \ell_{14} \\ & \ell_{22} & \ell_{23} & \ell_{24} \\ & & \ell_{33} & \ell_{34} \\ & & & \ell_{44} \end{bmatrix} \tag{28}$$

where the u_{ij} have been replaced with the ℓ_{ij}, and $\ell_{ij} = \ell_{ji}$. Multiplying the two matrices on the RHS and equating the resulting coefficients to the coefficients of the LHS give

$$\ell_{11} = \sqrt{a_{11}}$$

$$\ell_{i1} = a_{i1}/\ell_{11} \qquad i = 2,3,\dots,n$$

$$\left.\begin{aligned} \ell_{jj} &= \sqrt{a_{jj} - \sum_{k=1}^{j-1} \ell_{jk}^2} \\ \ell_{ij} &= \frac{1}{\ell_{jj}}\left(a_{ij} - \sum_{k=1}^{j-1} \ell_{ik}\ell_{jk}\right) \qquad i = j+1,j+2,\dots,n \end{aligned}\right\} j = 2,3,\dots,n \tag{29}$$

$$\ell_{ij} = 0 \quad \text{for} \quad i < j$$

Then

$$LL^TX = B \tag{30}$$

and putting $L^TX = Y$ gives

$$LY = B \tag{31}$$

from which Y can be found by forward substitution and then from

$$L^T X = Y \tag{32}$$

X can be found by back substitution.

Example 7

Form A as the product $C^T C$ and then derive L, given

$$C = \begin{bmatrix} 1 & 3 & 2 \\ 2 & 2 & 1 \\ 1 & 0 & 1 \end{bmatrix}$$

Solution:

$$\begin{bmatrix} 1 & 2 & 1 \\ 3 & 2 & 0 \\ 2 & 1 & 1 \end{bmatrix}\begin{bmatrix} 1 & 3 & 2 \\ 2 & 2 & 1 \\ 1 & 0 & 1 \end{bmatrix} = \begin{bmatrix} 6 & 7 & 5 \\ 7 & 13 & 8 \\ 5 & 8 & 6 \end{bmatrix}$$

(Note that the largest number in A is always on the main diagonal)

$$\ell_{11} = \sqrt{a_{11}} = \sqrt{6} = 2.4495$$

$$\ell_{21} = \frac{a_{21}}{\ell_{11}} = \frac{7}{\sqrt{6}} = 2.8577$$

$$\ell_{31} = \frac{a_{31}}{\ell_{11}} = \frac{5}{\sqrt{6}} = 2.0412$$

$$\ell_{22} = \sqrt{a_{22} - \ell_{21}^2} = \sqrt{13 - (2.8577)^2} = 2.1985$$

$$\ell_{32} = \frac{1}{\ell_{22}}(a_{32} - \ell_{31}\ell_{21}) = \frac{1}{2.1985}(8 - 2.0412 \times 2.8577) = 0.9856$$

$$\ell_{33} = \sqrt{a_{33} - \ell_{31}^2 - \ell_{32}^2} = \sqrt{6 - (2.0412)^2 - (0.09856)^2} = 0.9285$$

$$\therefore L = \begin{bmatrix} 2.4495 & & \\ 2.8577 & 2.1985 & \\ 2.0412 & 0.9856 & 0.9285 \end{bmatrix}$$

Exercise 4

1. Find the lower-triangular matrix L with positive diagonal entries, such that

$$A = LL^T ,$$

for the following symmetric, positive-definite matrices, A.

(a) $\begin{bmatrix} 1 & 2 & 6 \\ 2 & 5 & 15 \\ 6 & 15 & 46 \end{bmatrix}$

(b) $\begin{bmatrix} 14 & 16 & 16 \\ 16 & 21 & 19 \\ 16 & 19 & 29 \end{bmatrix}$

Answers:

(a) $\begin{bmatrix} 1 & & \\ 2 & 1 & \\ 6 & 3 & 1 \end{bmatrix}$

(b) $\begin{bmatrix} 3.74 & & \\ 4.28 & 1.65 & \\ 4.28 & 0.41 & 3.24 \end{bmatrix}$

2. Solve, using Choleski factorisation, AX = B where

$$A = \begin{bmatrix} 14 & 16 & 16 \\ 16 & 21 & 19 \\ 16 & 19 & 29 \end{bmatrix} \qquad B = \begin{bmatrix} -1 \\ 2 \\ -3 \end{bmatrix}$$

given that A is symmetric, positive-definite.

Answer:
$$L = \begin{bmatrix} 3.74 & & \\ 4.28 & 1.65 & \\ 4.28 & 0.41 & 3.24 \end{bmatrix}$$

$Y = (-0.27, 1.91, -0.82)$
$X = (\ 1.38, 1.22, -0.25)$

The Choleksi method is sometimes written as

$$A = S^T S$$

where S is an upper-triangular matrix with positive entries on the diagonal. The algorithm is then written slightly differently; the effect, of course, is the same.

5. Band Matrices

5.1 Introduction

The simplest type of band matrix is one with constant bandwidth. Although they seem to be good candidates for iterative methods of solution, special algorithms exist which make direct methods preferable. The algorithms are much more efficient than Gaussian elimination on a full matrix.

Algorithms for band symmetric and unsymmetric decompositions are available in Wilkinson and Reinsch (1971) and Meis and Marcowitz (1981): the former contains Algol programs, the latter contains Fortran programs.

5.2 The Tridiagonal Matrix System

The tridiagonal algorithm, which is usually described as the Thomas Tridiagonal algorithm, may be derived in several ways. The procedure employed here is that of LU-decomposition of the Doolittle form

$$\begin{bmatrix} b_1 & c_1 & & & & \\ a_2 & b_2 & c_2 & & & \\ & a_3 & b_3 & c_3 & & \\ & & \cdots & \cdots & \cdots & \\ & & & a_{n-1} & b_{n-1} & c_{n-1} \\ & & & & a_n & b_n \end{bmatrix} \begin{bmatrix} x_1 \\ x_2 \\ x_3 \\ \cdots \\ x_{n-1} \\ x_n \end{bmatrix} = \begin{bmatrix} d_1 \\ d_2 \\ d_3 \\ \cdots \\ d_{n-1} \\ d_n \end{bmatrix} \tag{33}$$

$$\text{i.e.} \qquad AX = D \tag{34}$$

$$\text{i.e.} \qquad a_i x_{i-1} + b_i x_i + c_i x_{i+1} = d_i$$
$$\text{for} \quad 1 \leqslant i \leqslant n \tag{35}$$
$$\text{with} \quad a_1 = c_n = 0$$

If this coefficient matrix has an LU-decomposition, let it be

$$LU \equiv \begin{bmatrix} 1 & & & & \\ \alpha_2 & 1 & & & \\ & \alpha_3 & 1 & & \\ & & \cdots & \cdots & \\ & & & \alpha_n & 1 \end{bmatrix} \begin{bmatrix} \beta_1 & c_1 & & & & \\ & \beta_2 & c_2 & & & \\ & & \beta_3 & c_3 & & \\ & & & \cdots & \cdots & \\ & & & & \beta_{n-1} & c_{n-1} \\ & & & & & \beta_n \end{bmatrix} \tag{36}$$

This is an example of DOOLITTLE decomposition.

Equating entries of $A \equiv LU$ will produce the unknowns β_i, $i = 1,\ldots,n$; α_i, $i = 2,\ldots,n$.

1st row: $\quad b_1 = \beta_1 \quad \rightarrow \quad \beta_1 = b_1$

2nd row: $\quad a_2 = \alpha_2\beta_1 \quad \rightarrow \quad \alpha_2 = \dfrac{a_2}{\beta_1}$

$\quad b_2 = \alpha_2 c_1 + \beta_2 \rightarrow \quad \beta_2 = b_2 - \alpha_2 c_1$

3rd row: $\quad a_3 = \alpha_3\beta_2 \quad \rightarrow \quad \alpha_3 = \dfrac{a_3}{\beta_2}$

$\quad b_3 = \alpha_3 c_2 + \beta_3 \rightarrow \quad \beta_3 = b_3 = \alpha_3 c_2$

to lead to the generalization

$$\beta_1 = b_1 \qquad \boxed{\alpha_i = \frac{a_i}{\beta_{i-1}}, \quad \beta_i = b_i - \alpha_i c_{i-1}} \qquad i = 2,3,\ldots,n \tag{37}$$

Then $\quad AX = D$

may be written $\quad LUX = D \quad (38)$

i.e. $\begin{cases} LY = D & (39) \\ UX = Y\,. & (40) \end{cases}$

Now $LY = D$ is written as

$$\begin{bmatrix} 1 & & & & \\ \alpha_2 & 1 & & & \\ & \alpha_3 & 1 & & \\ & & \cdots\cdots & & \\ & & & \alpha_n & 1 \end{bmatrix} \begin{bmatrix} y_1 \\ y_2 \\ y_3 \\ \cdots \\ y_n \end{bmatrix} = \begin{bmatrix} \alpha_1 \\ \alpha_2 \\ \alpha_3 \\ \cdots \\ \alpha_n \end{bmatrix} \tag{41}$$

to give, on forward substitution,

$$y_1 = d_1$$
$$y_2 = d_2 - \alpha_2 y_1$$
$$y_3 = d_3 - \alpha_3 y_2$$

to lead to the generalization

$$y_1 = d_1\,, \quad y_i = d_i - \alpha_i y_{i-1} \quad \text{for } i = 2,3,\ldots,n \tag{42}$$

Hence we solve $UX = Y$ which is written as

$$\begin{bmatrix} \beta_1 & c_1 & & & & \\ & \beta_2 & c_2 & & & \\ & & \beta_3 & c_3 & & \\ & & & \cdots\cdots & & \\ & & & & \beta_{n-1} & c_{n-1} \\ & & & & & \beta_n \end{bmatrix} \begin{bmatrix} x_1 \\ x_2 \\ x_3 \\ \cdots \\ x_{n-1} \\ x_n \end{bmatrix} = \begin{bmatrix} y_1 \\ y_2 \\ y_3 \\ \cdots \\ y_{n-1} \\ y_n \end{bmatrix} \tag{43}$$

to give, on back substitution,

$$x_n = \frac{y_n}{\beta_n}$$
$$x_{n-1} = \frac{y_{n-1} - c_{n-1}x_n}{\beta_{n-1}}$$

to lead to the generalization

$$x_n = \frac{y_n}{\beta_n}\,, \quad x_i = \frac{y_i - c_i x_{i+1}}{\beta_i} \quad \text{for } i = (n-1),\ldots,2,1. \tag{44}$$

So the algorithm may be written

$$\begin{aligned}
\beta_1 &= b_1 \\
y_1 &= d_1 \\
\left.\begin{aligned}
\alpha_i &= \frac{a_i}{\beta_{i-1}} \\
\beta_i &= b_i - \alpha_i c_{i-1} \\
y_i &= d_i - \alpha_i y_{i-1}
\end{aligned}\right\} & \quad i = 2,3,\ldots,n \qquad (45) \\
x_n &= \frac{y_n}{\beta_n} \\
x_i &= \frac{y_i - c_i x_{i+1}}{\beta_i} \qquad i = (n-1),\ldots,2,1 \ .
\end{aligned}$$

Example 7

Use the tridiagonal algorithm to solve the system

$$\begin{bmatrix} 2 & -1 & & \\ -1 & 2 & -1 & \\ & -1 & 2 & -1 \\ & & -1 & 1 \end{bmatrix} \begin{bmatrix} x_1 \\ x_2 \\ x_3 \\ x_4 \end{bmatrix} = \begin{bmatrix} 10 \\ 6 \\ -8 \\ 4 \end{bmatrix}$$

Solution: Given: $b_1 = b_2 = b_3 = 2\ ,\quad b_4 = 1$

$a_2 = a_3 = a_4 = -1$

$c_1 = c_2 = c_3 = -1$

$d_1 = 10,\ d_2 = 6,\ d_3 = -8,\ d_4 = 4$

$\therefore\ \beta_1 = b_1 = 2$

$y_1 = d_1 = 10$

$\alpha_2 = a_2/\beta_1 = -1/2$

$\beta_2 = b_2 - \alpha_2 c_1 = 2 - (-\tfrac{1}{2})(-1) = 3/2$

$y_2 = d_2 - \alpha_2 y_1 = 6 - (-\tfrac{1}{2}).10 = 11$

$\alpha_3 = a_3/\beta_2 = -1/(3/2) = -2/3$

$\beta_3 = b_3 - \alpha_3 c_2 = 2 - (-2/3).(-1) = 4/3$

$y_3 = d_3 - \alpha_3 y_2 = -8 - (-2/3).11 = -2/3$

$\alpha_4 = a_4/\beta_3 = -1/(4/3) = -3/4$

$\beta_4 = b_4 - \alpha_4 c_3 = 1 - (-3/4)(-1) = \tfrac{1}{4}$

$y_4 = d_4 - \alpha_4 y_3 = 4 - (-3/4)(-2/3) = 7/2$

$\therefore\ x_4 = y_4/\beta_4 = (7/2)/(\tfrac{1}{4}) = 14$

$$x_3 = (y_3 - c_3x_4)/\beta_3 = (-2/3 - (-1)(14))/(4/3) = 10$$
$$x_2 = (y_2 - c_2x_3)/\beta_2 = (11 - (-1)(10))/(3/2) = 14$$
$$x_1 = (y_1 - c_1x_2)/\beta_1 = (10 - (-1)(14))/2 = 12$$

Answer: $(x_1,x_2,x_3,x_4) = (12,14,10,14)$.

Exercise 5

1. Use the tridiagonal algorithm to solve

$$\begin{bmatrix} 3 & -5 & & & \\ -3 & 2 & -1 & & \\ & 6 & -4 & 3 & \\ & & -2 & 5 & 4 \\ & & & 7 & -2 \end{bmatrix} \begin{bmatrix} x_1 \\ x_2 \\ x_3 \\ x_4 \\ x_5 \end{bmatrix} = \begin{bmatrix} 2 \\ -6 \\ 3 \\ -3 \\ 4 \end{bmatrix}$$

Answer: $(x_1,x_2,x_3,x_4,x_5) = (1129/486,161/162,55/54,10/27,-19/27)$.

2. Use the tridiagonal algorithm to solve, correct to 3D, the system

$$\begin{bmatrix} 4 & -1 & & & \\ -1 & 4 & -1 & & \\ & -1 & 4 & -1 & \\ & & -1 & 4 & -1 \\ & & & -2 & 1 \end{bmatrix} \begin{bmatrix} x_1 \\ x_2 \\ x_3 \\ x_4 \\ x_5 \end{bmatrix} = \begin{bmatrix} 0.4 \\ 0.8 \\ 1.2 \\ 1.6 \\ 1.6 \end{bmatrix}$$

Answer: $(x_1,x_2,x_3,x_4,x_5) = (0.223,0.491,0.940,2.070,5.740)$.

3. Use the tridiagonal algorithm to solve, correct to 3D, the system

$$\begin{bmatrix} 2 & 1 & & & \\ 3 & -2 & -3 & & \\ & 4 & -4 & 2 & \\ & & -1 & 3 & -2 \\ & & & 5 & 4 \end{bmatrix} \begin{bmatrix} x_1 \\ x_2 \\ x_3 \\ x_4 \\ x_5 \end{bmatrix} = \begin{bmatrix} 3 \\ 2 \\ 1 \\ 0 \\ -1 \end{bmatrix}$$

Answer: $(x_1,x_2,x_3,x_4,x_5) = (1.244,0.511,0.237,-0.048,-0.190)$.

Arithmetic Operations

For the whole process of LU decomposition followed by forward and backward substitution - all of which constitutes the tridiagonal algorithm - the number of arithmetic operations is represented in Table 2.

Operations	Tridiagonal	Gaussian Elimination	Gauss-Seidel for 1 iteration
Multiplication	3(n - 1)	$\frac{n^3}{3} + \frac{n^2}{2} - \frac{5n}{6}$	2n
Division	2n - 1	n(n + 1)/2	n
Addition	3(n - 1)	$\frac{n^3}{3} + \frac{n^2}{2} - \frac{5n}{6}$	2n

Table 2

From Table 2,

(i) the total number of multiplications and divisions = 5n - 4. Compare this with Gaussian elimination = $\frac{n^3}{3} + n^2 - \frac{n}{3}$; Gauss-Seidel for 1 iteration = 3n.

(ii) the total number of arithmetic operations = 8n - 7. Compare with Gaussian elimination = $\frac{2n^3}{3} + \frac{3n^2}{2} - \frac{7n}{6}$; Gauss-Seidel for 1 iteration = 5n.

Recall that for c.p.u. cost, (i) is important, while for error analysis (ii) is important.

For a tridiagonal system, Table 3 presents a comparison illustrating the effectiveness of the tridiagonal algorithm. For this table a system of 1000 equations has been chosen.

Operations	Tridiagonal	Gaussian Elimination	Gauss-Seidel for 1 iteration
Multiplication and Division	5.10^3	3.10^8	3.10^3
Arithmetic Operations	8.10^3	6.10^8	5.10^3

Table 3

Comparison for a Tridiagonal System of 1000 equations

As usual, for systems such as that for full Gaussian elimination, only the term with the highest power of n is significant.

A FORTRAN listing of the algorithm for the tridiagonal matrix system is given in Appendix A. Because the α_i are only used to obtain the corresponding β_i and y_i, they need not be stored as an array but simply occupy a variable location.

In iteration usage of this scheme, through such as time-dependent problems or degradation of nonlinear systems, there are certain terms which need be computed once only as they are independent of the particular step of the iterative scheme. This comment is mainly applicable to constant coefficient matrix problems.

It may be readily shown that the algorithm for the tridiagonal matrix may also be written as follows:

$$\left.\begin{aligned}\beta_1 &= b_1\\ \gamma_1 &= \frac{d_1}{b_1}\\ \beta_i &= b_i - \frac{a_i c_{i-1}}{\beta_{i-1}}\\ \gamma_i &= \frac{d_i - a_i\gamma_{i-1}}{\beta_i}\end{aligned}\right\} \quad i = 2,3,\ldots,n \tag{46}$$

$$x_n = \gamma_n$$

$$x_i = \gamma_i - \frac{c_i x_{i+1}}{\beta_i} \qquad i = n-1,n-2,\ldots,2,1.$$

For matrices that are symmetric, positive-definite, a unique L is formed where L is lower-triangular with positive diagonal entries

$$\text{i.e.} \qquad A \equiv LU \equiv LL^T \tag{47}$$

So for tridiagonal systems of this type, only the L need be found and so storage is conserved, realising that the $U = L^T$.

Exercise 6

In appendix A, there is a listing of a FORTRAN program for the tridiagonal algorithm. Implement this program on your computer using the following example:

$$\begin{bmatrix} 2 & -1 & & & & & \\ -1 & 2 & -1 & & & & \\ & -1 & 2 & -1 & & & \\ & & \cdots & \cdots & \cdots & & \\ & & & & -1 & 2 & -1 \\ & & & & & -1 & 1 \end{bmatrix} \begin{bmatrix} x_1 \\ x_2 \\ x_3 \\ \cdots \\ x_{n-1} \\ x_n \end{bmatrix} = \begin{bmatrix} 0 \\ 0 \\ 0 \\ \cdots \\ 0 \\ 1 \end{bmatrix}$$

Answers: $(x_1,x_2,x_3,\ldots,x_n) = (1,2,3,\ldots,n)$.

5.3 The Pentadiagonal Matrix System

The pentadiagonal algorithm may be derived through LU-decomposition similarly to that of the derivation of the tridiagonal algorithm.

$$\begin{bmatrix} c_1 & d_1 & e_1 & & & & & & \\ b_2 & c_2 & d_2 & e_2 & & & & & \\ a_3 & b_3 & c_3 & d_3 & e_3 & & & & \\ & a_4 & b_4 & c_4 & d_4 & e_4 & & & \\ & & \cdots & \cdots & \cdots & \cdots & & & \\ & & & & a_{n-2} & b_{n-2} & c_{n-2} & d_{n-2} & e_{n-2} \\ & & & & & a_{n-1} & b_{n-1} & c_{n-1} & d_{n-1} \\ & & & & & & a_n & b_n & c_n \end{bmatrix} \begin{bmatrix} x_1 \\ x_2 \\ x_3 \\ x_4 \\ \cdot\cdot \\ x_{n-2} \\ x_{n-1} \\ x_n \end{bmatrix} = \begin{bmatrix} f_1 \\ f_2 \\ f_3 \\ f_4 \\ \cdot\cdot \\ f_{n-2} \\ f_{n-1} \\ f_n \end{bmatrix} \tag{48}$$

The equations for the pentadiagonal matrix system may be written

$$a_i x_{i-2} + b_i x_{i-1} + c_i x_i + d_i x_{i+1} + e_i x_{i+2} = f_i \tag{49}$$

$$\text{for} \quad 1 \leqslant i \leqslant n$$

$$\text{with} \quad a_1 = b_1 = a_2 = e_{n-1} = d_n = e_n = 0$$

The algorithm for this system is as follows:

$$\delta_1 = \frac{d_1}{c_1}$$

$$\lambda_1 = \frac{e_1}{c_1}$$

$$\gamma_1 = \frac{f_1}{c_1}$$

$$\mu_2 = c_2 - b_2\delta_1$$

$$\delta_2 = (d_2 - b_2\lambda_1)/\mu_2$$

$$\lambda_2 = e_2/\mu_2$$

$$\gamma_2 = (f_2 - b_2\gamma_1)/\mu_2$$

$$\left.\begin{aligned} \beta_i &= b_i - a_i\delta_{i-2} \\ \mu_i &= c_i - \beta_i\delta_{i-1} - a_i\lambda_{i-2} \\ \delta_i &= (d_i - \beta_i\lambda_{i-1})/\mu_i \\ \lambda_i &= e_i/\mu_i \\ \gamma_i &= (f_i - \beta_i\gamma_{i-1} - a_i\gamma_{i-2})/\mu_i \end{aligned}\right\} \quad i = 3,4,\ldots,n-2$$

$$\beta_{n-1} = b_{n-1} - a_{n-1}\delta_{n-3}$$

$$\mu_{n-1} = c_{n-1} - \beta_{n-1}\delta_{n-2} - a_{n-1}\lambda_{n-3}$$

$$\delta_{n-1} = (d_{n-1} - \beta_{n-1}\lambda_{n-2})/\mu_{n-1}$$

$$\gamma_{n-1} = (f_{n-1} - \beta_{n-1}\gamma_{n-2} - a_{n-1}\gamma_{n-3})/\mu_{n-1}$$

$$\beta_n = b_n - a_n\delta_{n-2}$$

$$\mu_n = c_n - \beta_n\delta_{n-1} - a_n\lambda_{n-2}$$

$$\gamma_n = (f_n - \beta_n\gamma_{n-1} - a_n\gamma_{n-2})/\mu_n$$

$$x_n = \gamma_n$$

$$x_{n-1} = \gamma_{n-1} - \delta_{n-1}x_n$$

$$x_i = \gamma_i - \delta_i x_{i+1} - \lambda_i x_{i+2} \qquad i = n-2, n-3, \ldots, 2, 1 \qquad (50)$$

Storage space may be conserved during the execution of this algorithm by realising that β_i and μ_i are used only to compute δ_i, λ_i and γ_i and need not be stored once this latter group have been evaluated.

Example 8

Use the pentadiagonal algorithm to solve, correct to 2D, the system

$$\begin{bmatrix} 2 & -3 & 4 & & & \\ -5 & -2 & 1 & 2 & & \\ 1 & -3 & 3 & -4 & -1 & \\ & 2 & 4 & -1 & 3 & -2 \\ & & -1 & 3 & -2 & 5 \\ & & & -5 & 3 & -4 \end{bmatrix} \begin{bmatrix} x_1 \\ x_2 \\ x_3 \\ x_4 \\ x_5 \\ x_6 \end{bmatrix} = \begin{bmatrix} 2 \\ -3 \\ -2 \\ 4 \\ -1 \\ 5 \end{bmatrix}$$

Solution: Given: $c_1 = 2$, $c_2 = -2$, $c_3 = 3$, $c_4 = -1$, $c_5 = -2$, $c_6 = -4$.
$b_2 = -5$, $b_3 = -3$, $b_4 = 4$, $b_5 = 3$, $b_6 = 3$.
$a_3 = 1$, $a_4 = 2$, $a_5 = -1$, $a_6 = -5$.
$d_1 = -3$, $d_2 = 1$, $d_3 = -4$, $d_4 = 3$, $d_5 = 5$.
$e_1 = 4$, $e_2 = 2$, $e_3 = -1$, $e_4 = -2$.
$f_1 = 2$, $f_2 = -3$, $f_3 = -2$, $f_4 = 4$, $f_5 = -1$, $f_6 = 5$.

$$\delta_1 = d_1/c_1 = (-3)/2 = -3/2$$
$$\lambda_1 = e_1/c_1 = 4/2 = 2$$
$$\gamma_1 = f_1/c_1 = 2/2 = 1$$

$$\mu_2 = c_2 - b_2\delta_1 = -2 - (-5)(-3/2) = -9.5$$
$$\delta_2 = (d_2 - b_2\lambda_1)/\mu_2 = 1 - (-5)(2) /(-9.5) = -1.1579$$
$$\lambda_2 = e_2/\mu_2 = 2/(-9.5) = -0.2105$$

$\gamma_2 = (f_2 - b_2\gamma_1)/\mu_2 = (-3 - (-5)(1))/(-9.5) = -0.2105$

$\beta_3 = b_3 - a_3\delta_1 = -3 - (1)(-3/2) = -3/2$
$\mu_3 = c_3 - \beta_3\delta_2 - a_3\lambda_1 = 3 - (-3/2)(-1.1579) - (1)(2) = -0.7368$
$\delta_3 = (d_3 - \beta_3\lambda_2)/\mu_3 = (-4 - (-3/2)(-0.2105))/(-0.7368) = 5.8574$
$\lambda_3 = e_3/\mu_3 = (-1)/(-0.7368) = 1.3572$
$\gamma_3 = (f_3 - \beta_3\gamma_2 - a_3\gamma_1)/\mu_3 = (-2 - (-3/2)(-0.2105) - (1)(1))/(-0.7368) = 4.5094$

$\beta_4 = b_4 - a_4\delta_2 = 4 - (2)(-1.1579) = 6.3158$
$\mu_4 = c_4 - \beta_4\delta_3 - a_4\lambda_2 = -1 - (6.3158)(5.8574) - (2)(-0.2105) = -37.5732$
$\delta_4 = (d_4 - \beta_4\lambda_3)/\mu_4 = (3 - (6.3158)(1.3572))/(-37.5732) = 0.1483$
$\lambda_4 = e_4/\mu_4 = -2/(-37.5732) = 0.0532$
$\gamma_4 = (f_4 - \beta_4\gamma_3 - a_4\gamma_2)/\mu_4 = (4 - (6.3158)(4.5094) - (2)(-0.2105))/(-37.5732)$
$= 0.6403$

$\beta_5 = b_5 - a_5\delta_3 = 3 - (-1)(5.8574) = 8.8574$
$\mu_5 = c_5 - \beta_5\delta_4 - a_5\lambda_3 = -2 - (8.8574)(0.1483) - (-1)(1.3572) = -1.9564$
$\delta_5 = (d_5 - \beta_5\lambda_4)/\mu_5 = (5 - (8.8574)(0.0532))/(-1.9564) = -2.3149$
$\gamma_5 = (f_5 - \beta_5\gamma_4 - a_5\gamma_3)/\mu_5 = (-1 - (8.8574)(0.6403) - (-1)(4.5094))/$
$(-1.9564) = 1.0953$

$\beta_6 = b_6 - a_6\delta_4 = 3 - (-5)(0.1483) = 3.7415$
$\mu_6 = c_6 - \beta_6\delta_5 - a_6\lambda_4 = -4 - (3.7415)(-2.3149) - (-5)(0.0532) = -12.3952$
$\gamma_6 = (f_6 - \beta_6\gamma_5 - a_6\gamma_4)/\mu_6 = (5 - (3.7415)(1.0953) - (-5)(0.6403))/(-12.3952)$
$= -0.3311$

$x_6 = -0.3311$
$x_5 = \gamma_5 - \delta_5 x_6 = 1.0953 - (-2.3149)(-0.3311) = 0.3288$
$x_4 = \gamma_4 - \delta_4 x_5 - \lambda_4 x_6 = 0.6403 - (0.1483)(0.3288) - (0.0532)(-0.3311) = 0.6092$
$x_3 = \gamma_3 - \delta_3 x_4 - \lambda_3 x_5 = 4.5094 - (5.8574)(0.6092) - (1.3572)(0.3288) = 0.4948$
$x_2 = \gamma_2 - \delta_2 x_3 - \lambda_2 x_4 = -0.2105 - (-1.1579)(0.4948) - (-0.2105)(0.6092)$
$= 0.4907$

$x_1 = \gamma_1 - \delta_1 x_2 - \lambda_1 x_3 = 1 - (-3/2)(0.4907) - (2)(0.4948) = 0.7464$

Answer: $(x_1, x_2, x_3, x_4, x_5, x_6) = (0.75, 0.49, 0.49, 0.61, 0.33, -0.33)$.

Exercise 7

1. Use the pentadiagonal algorithm to solve, correct to 2D, the system

$$\begin{bmatrix} 3 & -4 & 2 & & & \\ -2 & 5 & -3 & 1 & & \\ 4 & 2 & -3 & 5 & 1 & \\ & -3 & 4 & 1 & -2 & 2 \\ & & -5 & 2 & -4 & 3 \\ & & & -2 & 5 & 4 \end{bmatrix} \begin{bmatrix} x_1 \\ x_2 \\ x_3 \\ x_4 \\ x_5 \\ x_6 \end{bmatrix} = \begin{bmatrix} -0.78 \\ 2.29 \\ -2.71 \\ 2.88 \\ -1.57 \\ 0.44 \end{bmatrix}$$

Answer: $(x_1,x_2,x_3,x_4,x_5,x_6) = (2.64,3.12,1.89,-2.36,-2.04,1.48)$.

There are other standard systems such as the bi-tridiagonal matrix and tri-tridiagonal matrix. The algorithms for these are listed in the appendix of Von Rosenberg (1969).

The higher order algorithms for the wider bands become progressively more complicated.

5.4 General Variable-Bandwidth Matrix System

If the nodes of a particular triangular element in a finite element mesh are numbered 5, 19 and 37 and if there is one variable at each node then the semi-bandwidth, n, of the symmetric, positive-definite matrix is $n \geq 37 - 5 = 32$. With the bandwidth given, only a triangle of coefficients is retained on core and, as elimination proceeds, this core moves diagonally downwards. When the first variable is eliminated using the first equation, all the other coefficients in the triangle are modified, and as they are modified they are shifted diagonally upwards. The second equation now takes first place in the coefficients on core. And so the system proceeds. Usually the band is sparsely filled and frequently the bandwidth varies from one position to another. Often then only the coefficients within the $n_t \times n_t$ submatrix are modified, where n_t is the temporary semi-bandwidth but this is not always satisfactory.

The above description is equally appropriate for unsymmetric systems.

5.5 Block-Tridiagonal Matrix System

Isaacson and Keller (1966), Varah (1972) and others give great detail on the block-tridiagonal matrix system.

In the finite-difference solution of partial differential equations a block tridiagonal matrix is commonly obtained. The layout is as for the one above but with sub-matrices substituted for the entries. The same formula is used but there are now matrix additions and multiplications, so that where there was a division by an entry b_i, there is now a multiplication by a matrix $[b_i]^{-1}$. This requires calculation of an inverse, which can be done by Gaussian elimination. The method is still more efficient than simple iteration or full Gaussian elimination.

For symmetric positive-definite matrices, Jennings (1977) gives a method of solution which is a type of Choleski triangular decomposition expressed in terms of submatrices and involving $7(n - 1)b^3/6$ multiplications, where the submatrices are $b \times b$. In this case the tridiagonal matrix is better shown as

$$\begin{bmatrix} A_{11} & A_{21}^T & & & \\ A_{21} & A_{22} & A_{32}^T & & \\ & A_{32} & A_{33} & A_{43}^T & \\ & & \cdots\cdots & & \\ & & & \cdot & A_{nn} \end{bmatrix} \begin{bmatrix} x_1 \\ x_2 \\ x_3 \\ \vdots \\ x_n \end{bmatrix} = \begin{bmatrix} d_1 \\ d_2 \\ d_3 \\ \vdots \\ d_n \end{bmatrix} \tag{51}$$

where the A_{ii} are square submatrices.

The lower triangular matrix L_{11} is obtained by Choleski decomposition of A_{11}, then premultiplying the first submatrix equation by L_{11}^{-1} gives

$$L_{11}^T x_1 + L_{21}^T x_2 = h_1 \tag{52}$$

The inverse L_{11}^{-1} need not be calculated since L_{21}^T and h_1 are obtained by forward substitution in the equations

$$L_{11} L_{21}^T = A_{21}^T \tag{53}$$

and

$$L_{11} h_1 = d_1 \tag{54}$$

From the first of these we obtain

$$A_{21} = L_{21} L_{11}^T \tag{55}$$

so that premultiplying the new first rows (52), by L_{21} and subtracting from row 2 gives a second line as

$$A_{22}^* x_2 + A_{32}^T x_3 = b_2^* \tag{56}$$

where

$$A_{22}^* = A_{22} - L_{21} L_{21}^T \tag{57}$$

and

$$b_2^* = d_2 - L_{21} h_1 \tag{58}$$

Putting $A_{22}^* = L_{22} L_{22}^T$, the second line, (56), is multiplied by L_{22}^{-1} and the process repeated as for the first line.

The final result is

$$\begin{bmatrix} L_{11}^T & L_{21}^T & & & \\ & L_{22}^T & L_{32}^T & & \\ & & L_{33}^T & L_{43}^T & \\ & & & \ddots & \\ & & & & L_{nn}^T \end{bmatrix} \begin{bmatrix} x_1 \\ x_2 \\ x_3 \\ \vdots \\ x_n \end{bmatrix} = \begin{bmatrix} h_1 \\ h_2 \\ h_3 \\ \vdots \\ h_n \end{bmatrix} \quad (59)$$

which can be solved by back substitution. Only a part of the matrix is held in core at any one time, the submatrices being transferred in and out as required.

6. Special Techniques

6.1 Backing Store

A simple example of the use of backing store for a symmetric band matrix is given by Jennings (1977) and a modified version is shown below for Choleski decomposition of a matrix with a semibandwidth of 3.

Referring to the formulae for the Choleski method, if only the first three rows of A are in main (core) store, the a_{ij} can be replaced with the ℓ_{ij}.

$$\begin{bmatrix} a_{11} & & & & \\ a_{21} & a_{22} & & & \\ a_{31} & a_{32} & a_{33} & & \\ \hdashline & a_{42} & a_{43} & a_{44} & \\ & & a_{53} & a_{54} & a_{55} \\ & & & \cdots & \end{bmatrix} \longrightarrow \begin{bmatrix} \ell_{11} & & & & \\ \ell_{21} & \ell_{22} & & & \\ \ell_{31} & \ell_{32} & \ell_{33} & & \\ \hdashline & a_{42} & a_{43} & a_{44} & \\ & & a_{53} & a_{54} & a_{55} \\ & & & \cdots & \end{bmatrix} \quad (60)$$

The first column can then be transferred to backing (disc) store and the fourth line of A brought into main store with all entries moved up one row and across one column. The fourth row of A (now the third in store) can then be operated on.

$$\begin{bmatrix} \ell_{22} & & \\ \ell_{32} & \ell_{33} & \\ a_{42} & a_{43} & a_{44} \\ \hdashline \end{bmatrix} \longrightarrow \begin{bmatrix} \ell_{22} & & \\ \ell_{32} & \ell_{33} & \\ \ell_{42} & \ell_{43} & \ell_{44} \\ \hdashline \end{bmatrix} \quad (61)$$

The process can be repeated until the whole matrix L is stored column by column in backing store. It can then be recalled to main store, column by column, for the operation LY = B and returned column by column to backing store. When y_i is found ($3 \leq i \leq n$), column $i - 2$ is returned to backing store. Finally L can be recalled in reverse order for the operation $L^T X = Y$, where X is found by back substitution (Y having been found by forward substitution).

The total number of multiplications and divisions in the formation of the L matrix is $\approx nb^2/2$.

With such a narrow bandwidth, there is no necessity for such frequent transfers between main and backing store. It would be preferable to hold more lines in main store and transfer several columns at a time.

The elimination scheme just described can be used with a variable bandwidth store. If there are no re-entrant rows, the extension is quite straightforward. If re-entrant rows are present the number of rows, held in main store before decomposition can commence, is increased as the diagram in Figure 9, shows.

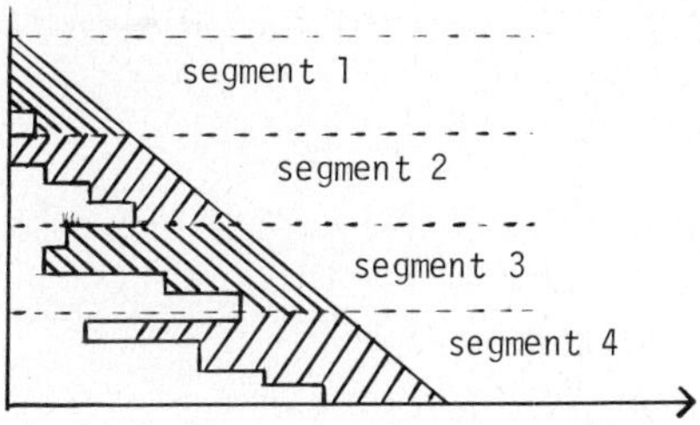

non-zero matrix elements are in the shaded area

Figure 9

For elimination of the first column segment 1 and 2 are required to be held in core. For elimination of the first columns of segment 4, all four segments are required. Jennings and Tuff (1971) have discussed this method.

6.2 Cuthill-McKee Algorithm

If a matrix has a variable bandwidth, with or without re-entrant rows, a variable band algorithm may be more efficient than a diagonal band elimination. It allows more effective use of the property that, for elimination without pivoting, zero elements before the first non-zero in any row always remain zero.

The main disadvantage is that additional storage is required for an address sequence. For very large matrices, this sequence may have to be generated by a special program segment which scans the input data. For an efficient, economical program it may be necessary to rearrange the input data by renumbering the variables. The Cuthill-McKee algorithm (1969) provides an efficient numbering scheme. Jennings (1977) gives a simple description as follows -

"(a) Choose a node to be relabelled 1. This should be located at an extremity of the graph and should have, if possible, few connections to other nodes.

(b) The nodes connected to the new node are relabelled 2, 3, etc., in the order of their increasing degree (the degree of a node is the number of nodes to which it is connected).

(c) The sequence is extended by relabelling the nodes which are directly connected to the new node 2 and which have not previously been relabelled. The nodes are again listed in the order of their increasing degree.

(d) The last operation is repeated for the new nodes 3, 4, etc., until the renumbering is complete.

The proper choice of starting node is important for the effectiveness of the algorithm, and this is often selected using the algorithm of Gibbs, Poole and Stockmeyer (1976).

The example, illustrated in Figure 10, shows how a network may be renumbered using the Cuthill-McKee algorithm.

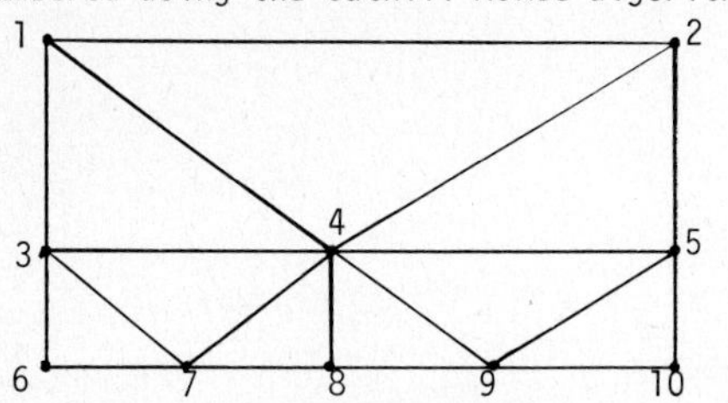

Initial numbering scheme for the network

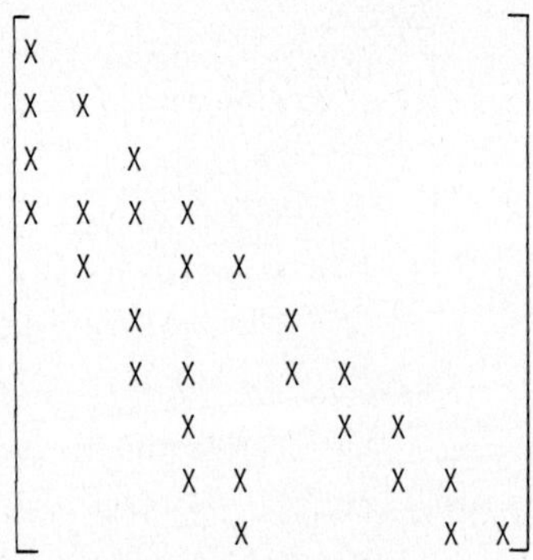

Coefficient matrix

Figure 10

The node connection list for the network of Figure 10 is given in Table 4.

Node	No. of Connections	Connection List
1	3	2,3,4
2	3	1,4,5
3	4	1,4,6,7
4	7	1,2,3,5,7,8,9
5	4	2,4,9,10
6	2	3,7
7	4	3,4,6,8
8	3	4,7,9
9	4	4,5,8,10
10	2	5,9

Table 4

Node Connection List for Network of Figure 10

Going through the list, node 6 seems to be a suitable starting node, and this is labelled 1. 3 and 7 are next renumbered as 2 and 3. The connection list for 3 now shows that 1 and 4 are next to be renumbered as 4 and 5, and since 1 has fewer connections than 4, it must be renumbered first. The process continues until all the nodes have been renumbered.

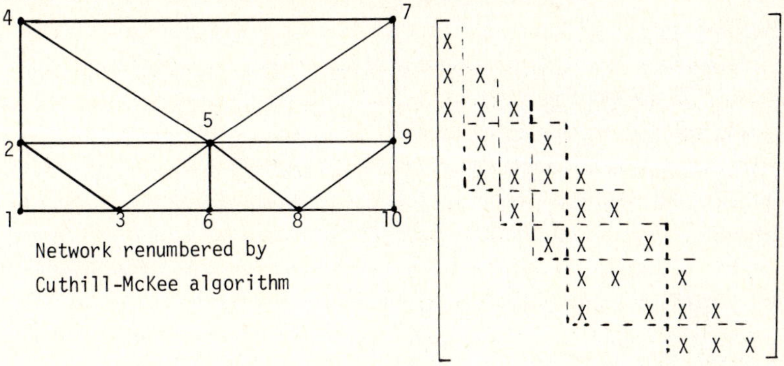

Network renumbered by Cuthill-McKee algorithm

Coefficient matrix

Figure 11

Network and matrix using Cuthill-McKee Algorithm

The Cuthill-McKee algorithm results in a variable bandwidth store which cannot have re-entrant rows. It may be considered to be made up of overlapping triangles as shown by the dashed lines in Figure 11. If the sides of the triangles span $p_1, p_2, \dots p_r$ rows and the sides of the overlaps span $q_1, q_2, \dots q_r$ rows, the number of storage locations is

$$s = \sum_{i=1}^{r} \frac{p_i(p_i + 1)}{2} - \sum_{i=1}^{r-1} \frac{q_i(q_i + 1)}{2} \tag{62}$$

Inserting figures formula (62), we obtain, for the above matrix

$$s = \frac{3.4}{2} + \frac{4.5}{2} + \frac{4.5}{2} + \frac{4.5}{2} + \frac{5.6}{2} + \frac{3.4}{2}$$

$$- \left(\frac{2.3}{2} + \frac{3.4}{2} + \frac{3.4}{2} + \frac{3.4}{2} + \frac{2.3}{2}\right)$$

$$= 57 - 24 = \underline{\underline{33}}$$

The original matrix store required 40 locations so we have reduced this by 7.

The total number of multiplications and divisions for triangular decomposition is

$$m = \sum_{i=1}^{r} \frac{p_i(p_i - 1)(p_i + 4)}{6} - \sum_{i=1}^{r-1} \frac{q_i(q_i - 1)(q_i + 4)}{6} \qquad (63)$$

6.3 The Reverse Cuthill-McKee Algorithm (RCM)

It was recognised by George (1971) that if a Cuthill-McKee numbering scheme is reversed a more efficient scheme often results. To reverse the numbers, (i,j) is replaced by (n-j+1,n-i+1). This merely reverses the triangles and the storage and computational requirements remain the same. However, the storage requirement can often be lessened by taking advantage of the re-entrant rows which may occur. It has been shown by George and also by Lin and Sherman (1976), that the Reverse Cuthill-McKee algorithm cannot give a less efficient bandwidth store than the direct method and will often be more efficient. In the example in Figure 12, use of a re-entrant row reduces the storage requirement to 32 locations, a gain of 1 over the direct method. With a larger system, the gain could be much greater.

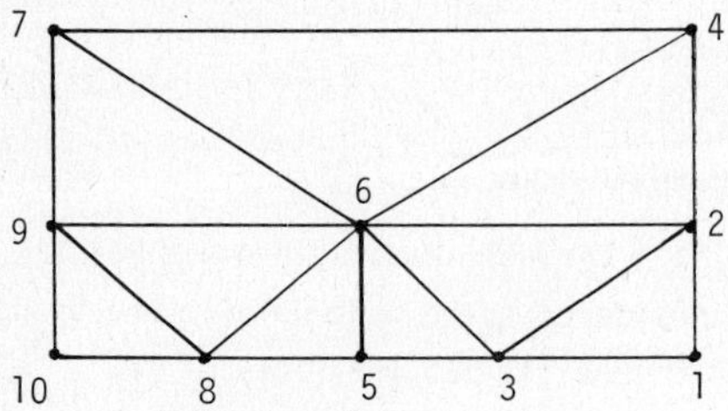

RCM numbering for the network

$$\begin{bmatrix}
X & & & & & & & & & \\
X & X & & & & & & & & \\
X & X & X & & & & & & & \\
 & X & & X & & & & & & \\
 & & X & & X & & & & & \\
 & X & X & X & X & X & & & & \\
 & & & X & & X & X & & & \\
 & & & & X & X & & X & & \\
 & & & & & X & X & X & X & \\
 & & & & & & & X & X & X
\end{bmatrix}$$

Coefficient matrix

Figure 12

Network and Matrix using Reverse Cuthill-McKee Algorithm

7. Frontal Methods

7.1 Introduction

Irons (1970) developed a Fortran program for the assembly and solution of symmetric positive-definite equations arising in the use of the finite element method, when using criteria such as the "least squares". He aimed to improve on the standard band-matrix algorithms. There is some doubt as to whether or not he developed the concept of the "frontal approach" but he was the first to develop a program based on this approach.

The frontal approach uses computer peripherals such as disc or magnetic tape to store numbers that are not immediately required for computation. As a result of this the core of the computer need not be particularly large for such programs provided this continuous Input/Output access is available.

Hood (1977a,b) developed a frontal solution program for the assembly and solution of unsymmetric matrix equations arising in certain applications of the finite element method. This program was also coded in Fortran.

The Hood program is now discussed before that of Irons because of its generality. Both programs use Gaussian elimination.

7.2 Frontal Solution for Unsymmetric Matrices

As an illustration of the Hood subroutine, a 3-element mesh with 5 nodes, as in Figure 13, will be used although, typically, a finite element system might have 5000 variables with a front width of 200 (i.e. the order of 10^6 non-zeros). All coefficients are kept in backing store, from which they are brought into core as required during assembly. With an unsymmetric matrix, the choice of pivot entry is important but a portion of the solution without pivoting will first be considered.

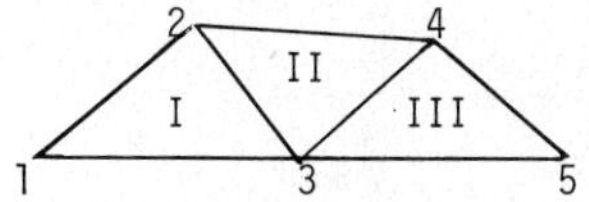

Figure 13
Three element mesh

Element I is first assembled into core. At this stage, the row and column associated with node 1 are fully summed and a_{11} may be used to eliminate the variable x_1, even though nodes 2 and 3 are not fully summed.

The total coefficient matrix for this three-element mesh is given in (64) with the submatrix, enclosed within dashed lines, being that position in core after assembly of the first element.

$$\left[\begin{array}{ccc:cc} a_{11} & a_{12} & a_{13} & & \\ a_{21} & a_{22} & a_{23} & a_{24} & \\ a_{31} & a_{32} & a_{33} & a_{34} & a_{35} \\ \hdashline & a_{42} & a_{43} & a_{44} & a_{45} \\ & & a_{53} & a_{54} & a_{55} \end{array}\right] \tag{64}$$

After assembly of the first element, the matrix equation, where superscript I denotes the element number from which the matrix entry was derived, is as follows:

$$\begin{bmatrix} a_{11}^{I} & a_{12}^{I} & a_{13}^{I} \\ a_{21}^{I} & a_{22}^{I} & a_{23}^{I} \\ a_{31}^{I} & a_{32}^{I} & a_{33}^{I} \end{bmatrix} \begin{bmatrix} x_1 \\ x_2 \\ x_3 \end{bmatrix} = \begin{bmatrix} b_1^{I} \\ b_2^{I} \\ b_3^{I} \end{bmatrix} \tag{65}$$

After eliminating x_1 and transferring row 1 to backing store the system becomes, in core,

$$\begin{bmatrix} a_{22}^{I} - \dfrac{a_{21}^{I}}{a_{11}^{I}} a_{12}^{I} & a_{23}^{I} - \dfrac{a_{21}^{I}}{a_{11}^{I}} a_{13}^{I} \\ a_{32}^{I} - \dfrac{a_{31}^{I}}{a_{11}^{I}} a_{12}^{I} & a_{33}^{I} - \dfrac{a_{31}^{I}}{a_{11}^{I}} a_{13}^{I} \end{bmatrix} \begin{bmatrix} x_2 \\ x_3 \end{bmatrix} = \begin{bmatrix} b_2^{I} - \dfrac{a_{21}^{I}}{a_{11}^{I}} b_1^{I} \\ b_3^{I} - \dfrac{a_{31}^{I}}{a_{11}^{I}} b_1^{I} \end{bmatrix} \tag{66}$$

Assembly of the next element makes node 2 fully summed. Once again there are three equations in core and the matrix system is as follows

$$\begin{bmatrix} a_{22}^{I} - \dfrac{a_{21}^{I}}{a_{11}^{I}} a_{12}^{I} + a_{22}^{II} & a_{23}^{I} - \dfrac{a_{21}^{I}}{a_{11}^{I}} a_{13}^{I} + a_{23}^{II} & a_{24}^{II} \\ a_{32}^{I} - \dfrac{a_{31}^{I}}{a_{11}^{I}} a_{12}^{I} + a_{32}^{II} & a_{33}^{I} - \dfrac{a_{31}^{I}}{a_{11}^{I}} a_{13}^{I} + a_{33}^{II} & a_{34}^{II} \\ a_{42}^{II} & a_{43}^{II} & a_{44}^{II} \end{bmatrix} \begin{bmatrix} x_2 \\ x_3 \\ x_4 \end{bmatrix}$$

$$= \begin{bmatrix} b_2^{I} - \dfrac{a_{21}^{I}}{a_{11}^{I}} b_1^{I} + b_2^{II} \\ b_3^{I} - \dfrac{a_{31}^{I}}{a_{11}^{I}} b_1^{I} + b_3^{II} \\ b_4^{II} \end{bmatrix} \tag{67}$$

Again, elimination occurs using the first entry of the first row as pivot and then the first row is transferred to backing store.

Note that the nodes of element I would have been dealt with in the same way regardless of what numbers were allotted to them.

In practice more fully summed equations would be entered to a predetermined level and a full pivotal search instituted. If a_{11}, for instance, were zero, the above solution would fail. Total pivoting, as compared with partial pivoting, involves a higher pivotal-search time and often causes a larger growth of non-zero entries with the associated increase in computation time and round-off error.

Hood indicated that in unsymmetric linear systems an entry should be used as a pivot only if its row and column are fully assembled. Therefore, assembly is continued until the front exceeds a critical size, at which stage the pivot is chosen to be the maximal entry in the fully assembled rows and columns. His reason for choosing total pivoting rather than partial pivoting for the assembled submatrix was that he had already restricted the pivotal choice by taking a small submatrix and that he felt it would be undesirable to further reduce the choice by a partial pivoting scheme.

After all the variables have been eliminated in the described manner, solution can proceed by back substitution, recalling the equations from backing store in reverse order.

The order in which the nodal solutions are evaluated is independent of their numbering but dependent on the order in which the elements are assembled to give the fully assembled equations. It is possible for a fully assembled equation to remain in core for a large part of the elimination process, simply due to its not having been chosen as a pivotal equation.

In band-solution routines the numbering of the nodes is very important. In frontal solutions, the order of the nodal numbering is completely irrelevant to the procedure; the technique is dependent on the order of the element numbering. In the assembly and solution of the equations, the only equations which must remain in core are those which are not yet fully summed. The variables represented by these equations constitute the "front". Equations behind the front have either been chosen as a pivotal equation and placed in backing store or are awaiting their selection for this. The "front width" is kept as small as possible to minimize core requirements: efficient element numbering achieves this.

Table 5 contains a selection from a comparison table constructed by Hood in support of his assertion that the frontal approach was an improvement on the band solution routines. A rectangular mesh was used for the finite elements and the nodes were numbered in the optimum manner for the band-solution routines. The equations arose in the finite element formulation of the Navier-Stokes equations for fluid flow. The results of the comparison indicate that for large scale problems there is a considerable advantage in the use of the frontal approach.

	FRONTAL SOLUTION				BAND SOLUTION		
No. of Equations	Type of Pivoting	Front Width	Computer Time(sec)	Core (words)	Band Width	Computer Time(sec)	Core (words)
876	Diagonal Total	58	156 181	6900 6900	163	344	23,000

Table 5

Comparison Between Solution Routines

Apart from its using less storage and less arithmetic than band routines, the frontal approach permits an easier refining of the mesh if it is considered too coarse in some region.

Reid (1977b), commenting on Hood's assembly and solution criteria, suggests that Hood is unreasonably cautious and that it would be preferable to follow each assembly with as many eliminations as possible subject to each pivot satisfying the relative tolerance as given in (17) and (18). Lee (1980) found that, in his use of the Hood subroutine in the study of convection, using the Navier-Stokes equations for this fluid flow, total pivoting was not necessary nor in fact was partial pivoting but purely diagonal pivoting was sufficient. This comment is really moving into variations for special cases. Mann (1981a) contains a Fortran program which includes the frontal approach for the solution of the equations resulting from a model of atmospheric convection using the finite element Galerkin criterion.

7.3 Frontal Solution for Symmetric, Positive-Definite Matrices

The Irons solution technique is similar to that described in the Hood approach. A knowledge of the basic properties of symmetric, positive-definite matrices would assist in a clear understanding of the mathematics involved.

Because of the symmetry, both core and backing store are conserved by simply using either the upper or lower triangular submatrix.

Irons (1980) states that with positive-definite matrices it makes little difference in taking the diagonal pivots in a different order whereas the use of non-diagonal pivots is highly dangerous. In his program each fully assembled equation is taken as a pivotal equation and eliminated from core to backing store as soon as it is fully assembled. There is no pivotal search. The diagonal entries are the pivots.

Because of the lack of any pivotal-search in the Irons program, it is decidedly unwise to use this program for symmetric matrices that are other than positive-definite.

Mann (1981b) contains a Fortran program which includes the frontal solution of the symmetric, positive-definite equations resulting from a model of atmospheric convection using the finite element least squares criterion.

8. Computer Packages

There are many routines for the basic techniques for dense systems available as packages. Some of these occur, in varying degree, in the International Mathematical and Statistical Libraries (IMSL), Numerical Algorithms Group (NAG) and LINPACK, with a lesser emphasis on sparse routines. From time to time computer listings of relevant packages for sparse systems are available in journals such as

International Journal for Numerical Methods in Engineering
Computers and Fluids
Computers and Structures
ACM Transactions on Mathematical Software

Usually these are written in Fortran, occasionally in Algol.

The commercial packages are usually available on magnetic tape, etc., in a form immediately applicable to specific computers. This eliminates the disadvantage of having to adapt the package to the particular computer.

The frontal approach for solving large sparse matrix systems is used in the recently available finite element package TWODEPEP.

9. Concluding Remarks

Some of the standard solution techniques, and other important considerations, involved in the solution of large sparse linear systems by direct methods have been considered in this paper. However, there are many refinements and alternatives to these approaches available for specific systems. Fox (1977) reviews some of these. The well-known applied numerical journals regularly provide these specific approaches.

As this paper is being printed, Duff (1981) is not yet available. This contains many invited papers from a recent conference on sparse matrices and its uses and would seem to be ideal for further reading from this paper.

10. References

Birkhoff, G. and Fix, G.J. (1974) Higher Order Linear Finite Element Methods, Report AD-779-341, Harvard University.

Curtis, A.R. and Reid, J.K. (1972) On the Automatic Scaling of Matrices for Gaussian Elimination, J. Inst. Math. Appl., 10, pp 118-124.

Curtis, A.R. and Reid, J.K. (1971) The Solution of Large Sparse Unsymmetric Systems of Linear Equations, J. Inst. Math. Appl., 8, pp 344-353.

Cuthill, E. and McKee, J. (1969) Reducing the Bandwidth of Sparse Symmetric Matrices, ACM Proc. 24th National Conf., New York.

Dahlquist, G. and Bjorck, A. (1974) Numerical Methods, Prentice-Hall.

Duff, I.S. (Ed.) (1981) Sparse Matrices and Their Uses, Academic.

Duff, I.S. and Reid, J.K. (1976) An Implementation of Tarjan's Algorithm for the Block Triangularisation of a Matrix, Harwell Report CSS 29.

Dufort, E.C. and Frankel, S.P. (1953) Stability Conditions in the Numerical Treatment of Parabolic Differential Equations, Math. Comp., 7, pp 135-152.

Evans, D.J. (Ed.) (1974) Software for Numerical Mathematics, Academic.

Faddeev, D.K. and Faddeeva, V.N. (1963) Computational Methods of Linear Algebra, Freeman, San Francisco.

Fletcher, C.A.J. (1976) The Application of the Finite Element Method to Two-Dimensional Inviscid Flow, WRE. Tech. Note 1606.

Forsythe, G.E. (1967) Today's Computational Methods of Linear Algebra, SIAM Review, 9, pp 489-515.

Forsythe, G. and Moler, C.B. (1967) Computer Solution of Linear Algebraic Systems, Prentice-Hall.

Fox, L. (1964) An Introduction to Numerical Linear Algebra, Clarendon Press, Oxford.

Fox, L. (1977) Finite Difference Methods for Elliptic Boundary-Value Problems in "The State of the Art in Numerical Analysis" (Ed. Jacobs, D.A.H.) Academic.

Fox, L. and Mayers, D.F. (1968) Computing Methods for Scientists and Engineers, Clarendon Press, Oxford.

George, J.A. (1972) Block Elimination of Finite Element Systems of Equations in "Sparse Matrices and Their Applications" (Eds. Rose, D.J. and Willoughby, R.A.) pp 41-52, Plenum, New York.

George, J.A. (1971) Computer Implementation of the Finite Element Method, Stanford Comp. Sc. Dept., Tech. Report STAN-CS-71-208, California.

George, J.A. (1977) Solution of Linear Systems of Equations: Direct Methods for Finite Element Problems in "Sparse Matrix Systems" (Ed. Barker, V.A.) Springer-Verlag.

Gibbs, N.E., Poole, W.G. and Stockmeyer, P.K. (1976) An Algorithm for Reducing the Bandwidth and Profile of a Sparse Matrix, SIAM J. Num. Anal., 13, pp 236-250.

Gregory, R.T. and Karney, L.K. (1969) A Collection of Matrices for Testing Computational Algorithms, Wiley.

Gustavson, F.G. (1972) Some Basic Techniques for Solving Sparse Systems of Linear Equations in "Sparse Matrices and Their Applications" (Eds. Rose, D.J. and Willoughby, R.A.), pp 41-52, Plenum, New York.

Hall, M. (1956) An Algorithm for Distinct Representatives, Amer. Math. Monthly, 63, pp 716-717.

Hamming, R.W. (1971) Introduction to Applied Numerical Analysis, McGraw-Hill.

Hamming, R.W. (1973) Numerical Methods for Scientists and Engineers, 2nd Ed., McGraw-Hill.

Hood, P. (1977a) Frontal Solution Program for Unsymmetric Matrices, Int. J. Num. Meth. Eng., 10, pp 379-399.

Hood, P. (1977b) Note on Frontal Solution Program for Unsymmetric Matrices, Int. J. Num. Meth. Eng., 11, p. 1202.

Householder, A.S. (1964) The Theory of Matrices in Numerical Analysis, Blaisdell, New York.

Irons, B.M. (1970) A Frontal Solution Program for Finite Element Analysis, Int. J. Num. Meth. Eng., 2, pp 5-32.

Irons, B.M. (1980) Private Communication.

Isaacson, E.A. and Keller, H.B. (1966) Analysis of Numerical Methods, Wiley.

Jacobs, D.A.H. (Ed.) (1977) The State of the Art in Numerical Analysis, Academic.

Jennings, A. (1966) A Compact Storage Scheme for the Solution of Simultaneous Equations, Comput. J., 9, pp 281-285.

Jennings, A. (1977) Matrix Computation for Engineers and Scientists, Wiley.

Jennings, A. and Tuff, A.D. (1971) A Direct Method for the Solution of Large Sparse Symmetric Simultaneous Equations in "Large Sparse Sets of Linear Equations (Ed. Reid, J.K.),Academic.

Lee, R.L. (1980) Private Communication.

Lowenthal, F. (1975) Linear Algebra with Differential Equations, Wiley.

Mann, K.J. (1981a) Finite Element Galerkin Program for Atmospheric Convection, CIT Publication. (To Appear).

Mann, K.J. (1981b) Finite Element Least Squares Program for Atmospheric Convection, CIT Publication. (To Appear).

Martin, H.C. and Carey, G.F. (1973) Introduction to Finite Element Analysis, McGraw-Hill.

Meis, T. and Marcowitz, U. (1981) Numerical Solution of Partial Differential Equations, Springer-Verlag.

Noble, B. (1969) Applied Linear Algebra, Prentice-Hall.

Pizer, S.M. (1975) Numerical Computing and Mathematical Analysis, Science Research Associates (SRA), Chicago.

Ralston, A. (1965) A First Course in Numerical Analysis, McGraw-Hill.

Reid, J.K. (Ed.) (1971) Large Sparse Sets of Linear Equations, Academic.

Reid, J.K. (1977a) Solution of Linear Systems of Equations: Direct Methods (General) in "Sparse Matrix Systems" (Ed. Barker, V.A.), Springer-Verlag.

Reid, J.K. (1977b) Sparse Matrices in "The State of the Art in Numerical Analysis" (Ed. Jacobs, D.A.H.), Academic.

Rose, D.J. and Willoughby, R.A. (Eds.) (1972) Sparse Matrices and Their Applications, Plenum.

Sargent, R.W.H. and Westerberg, A.W. (1964) "Speed up" in Chemical Engineering Design, Trans. Inst. Chem. Eng., 42, pp 190-197.

Tarjan, R.E. (1972) Depth First Search and Linear Graph Algorithms, SIAM J. Comput., 1, pp 146-160.

Tewarson, R.P. (1973) Sparse Matrices, Academic.

Varah, J.M. (1972) On the Solution of Block-Tridiagonal Systems Arising from Certain Finite-Difference Equations, Math. Comp., 26, pp 859-868.

Von Rosenberg, D.U. (1969) Methods for the Numerical Solution of Partial Differential Equations, Elsevier.

Westlake, J. (1968) A Handbook of Numerical Matrix Inversion and Solution of Linear Equations, Wiley.

Wilkinson, J.H. (1974a) Error Analysis and Related Topics. The Classical Error Analyses for the Solution of Linear Systems, Bull I.M.A., Vol. 10, no. 9/10.

Wilkinson, J.H. (1961) Error Analysis of Direct Methods of Matrix Inversion, J. Assoc. Comput. Mach., 8, pp 281-330.

Wilkinson, J.H. (1974b) Numerical Linear Algebra on Digital Computers, Bull I.M.A., Vol. 10, no. 9/10.

Wilkinson, J.H. (1963) Rounding Errors in Algebraic Processes, Prentice-Hall.

Wilkinson, J.H. (1965) The Algebraic Eigenvalue Problem, Oxford University Press.

Wilkinson, J.H. and Reinsch, C. (1971) Handbook for Automatic Computation Vol. II, Linear Algebra, Springer-Verlag.

Young, D.M. and Gregory, R.T. (1973) A Survey of Numerical Mathematics Vol. II., Chpt. 12, pp 779-834, Addison-Wesley.

11. Commercial Packages

IMSL Available through International Mathematical and Statistical Libraries, Inc., 6th Floor - NBC Building, 7500 Bellaire Boulevard, Houston, Texas 77036, USA.

LINPACK Available through IMSL Inc. (as above).

NAG Available through NAG Central Office, 7 Banbury Rd., Oxford OX2 6NN, England.

TWODEPEP Available through IMSL Inc. (as above).

Appendix A Listing of Fortran program for the solution of large linear systems with a tridiagonal coefficient matrix

```
C ******************************************************************
C
      SUBROUTINE TRIDIA(N,A,B,C,D,X,BETA,Y)
C
C     This subroutine solves the matrix equation:-
C             A * X = D         (1)
C     where  A is a tridiagonal matrix.
C
      DIMENSION A(N),B(N),C(N),D(N),X(N)
      DIMENSION BETA(N), Y(N)
C
C
C ------------------------------------------------------------------
C     GLOSSARY OF TERMS
C ------------------------------------------------------------------
C
C
C     A(N)    array of entries of 'diagonal' on the left of
C         the leading diagonal. (Note: A(1)=0.0 must be
C         read in with the data).
C     AMULT    is set equal to constant for each row.
C     B(N)    array of diagonal entries.
C     B(I)    component of B(N), it contains the entry
C         from the 'Ith' row.
C     BETA(N)    an intermediate array arising from the
C           LU decomposition of A.
C     C(N)    array of entries of 'diagonal' on the right of
C         the leading diagonal.(Note: C(N)=0.0 must be
C         read in with data).
C     D(N)    array of entries from the R.H.S of equation 1
C     X(N)  array of the solution to tridiagonal matrix.
C     X(I)   contains the solution to the 'Ith' equation
C     Y(N)    an intermediate array arising from the
C           LU decomposition af A.
C
C ------------------------------------------------------------------
C
C
C
      BETA(1) = B(1)
      Y(1) = D(1)
C
C     Forward substitution begins.
      DO 1 I=2,N
           AMULT = A(I)/BETA(I-1)
           BETA(I) = B(I) - AMULT*C(I-1)
           Y(I) = D(I) - AMULT*Y(I-1)
    1 CONTINUE
C
C     Backward substitution begins.
      X(N) = Y(N)/BETA(N)
      DO 2 J=2,N
           I=N-J+1
           X(I) = (Y(I) - C(I)*X(I+1))/BETA(I)
    2 CONTINUE
C
C     Outputing the data.
      WRITE(1,3)
    3 FORMAT(//,5X,'For the TRIDIAGONAL matrix',///,
     *5X,'A(I)',6X,'B(I)',6X,'C(I)',9X,'D(I)',8X,'THE SOLUTION',
     *2X,'X(I)',/,4X,25('-'),8X,5('-'),8X,18('-'))
      DO 5 K=1,N
           WRITE(1,4) A(K),B(K),C(K),D(K),X(K)
    4      FORMAT(3(3X,F7.3),6X,F7.3,17X,F8.4)
    5 CONTINUE
      RETURN
      END
```

Appendix B Listing of Fortran program for the solution of large linear systems with random sparseness

```
C
C
C     SUBROUTINE FOR SPARSE LINEAR SYSTEMS
C     ------------------------------------
C
C
C                         DESIGNED BY:  C. FLETCHER
C                              Weapons Research 1976
C
C                       IMPLIMENTED BY:  C. HALL  1981
C
C                         SUPERVISOR  : K. J. MANN
C                                Mathematics Dept.  C. I. T.
C
C                            COMPUTER:   PRIME (C. I. T. )
C
C     SOLVES SYSTEMS OF N LINEAR EQUATIONS STORING ONLY NON ZERO
C     ELEMENTS. USES SPARSE GUASS ELIMINATION
C
      SUBROUTINE SPAREL(AH, IH, N, LCMAX, JEND, IDIM, B, X,
     *IELIM, JPT, AHP, IHP, IHPDIM)
C
C ----------------------------------------------------------------
C
C     GLOSSARY OF TERMS
C
C ----------------------------------------------------------------
C     AH        :  Non-zero elements of Sparse matrix.
C     AHP       :  Temporary location of 'fill-in' terms.
C     B         :  Array containing right-hand side of equation.
C     I         :  Current work row.
C     IA        :  Is the position of the largest coefficient in
C                  the Ith row.
C     IDIM      :  Is the increased dimensions of AH and IH.
C     IELIM(I)  :  Contains the position of largest coefficient
C               :  in row I.
C     IH        :  Column position of corresponding terms in AH.
C     IHOLD     :  Temparary storage location of coefficient
C                  in row I for finding  IELIM.
C     IHP       :  Column position of corresponding terms in AHP
C     IHPDIM    :  The dimensions of arrays IHP and AHP.
C     J         :  No. of equation in second sweep. (DO 13).
C     JCBIG     :  Is the maximum value of JCP
C     JCM       :  Column counter for deleting the location of PIVOT
C                  in row J.
C     JCMH      :  Temporary stores JCM during row  J.
C     JCP       :  Column position of term deleted from row J.
C     JEND(i)   :  Contains the No. of non-zeros in row  i.
C     JPT       :  No. of non-zeros in each column.
C     JSA       :  Counter in the reverse sweep.
C     KEND      :  No. of non-zero's in row  I.
C     L         :  Column position in the Ith equation.  (DO 8).
C     LA        :  Column counter in row  J.
C     LCBIG     :  Is the maximum No. of terms in AH.
C     LCMAX     :  Is the location of the last entry in  AH.
C     LCT       :  Column counter for finding IHOLD.
C     LHOLD     :  Temporary location of LCT. (for row I).
C     M         :  Column position in the Jth equation. (DO 9)
C     MCT       :  LCT - KEND.
C     MEND      :  No. of non-zeros in row J.
C     MHOLD     :  Temporary storage of  MCT + 1.
C     N         :  No. of equations.
C     NCT       :  Column counter for eliminating X(IHOLD) from Jth rou
C
C     PIVOT     :  Is the multiplying factor applied to all rows
C                  about the pivotal row.
C     X         :  Array containing the solution.
C ----------------------------------------------------------------
C
      DIMENSION B(N), X(N), JEND(N), AH(IDIM), IH(IDIM)
      DIMENSION IELIM(N), JPT(N), AHP(IHPDIM), IHP(IHPDIM)
C
C
      NM1=N-1
      LCBIG=0
      JCBIG=0
      LCT=0
```

```
C
C     I is the equation from which  X(IHOLD) will be eliminated
      DO 26 I=1,NM1
           LCT=LCT+1
           KEND =JEND(I)
           PIVOT =AH(LCT)
           IHOLD=IH(LCT)
           DO 1 K=2,KEND
                LCT=LCT +1
                IF(ABS(AH(LCT)).LT.ABS(PIVOT)) GO TO 1
                PIVOT  =AH(LCT)
                IHOLD=IH(LCT)
    1      CONTINUE
C
C          Normalize the Ith equation with the largest coefficient.
           MCT =LCT-KEND
           IELIM(I)=IHOLD
           DO 2 K=1,KEND
                MCT=MCT+1
                AH(MCT)=AH(MCT)/PIVOT
    2      CONTINUE
           B(I)=B(I)/PIVOT
C
C          Eliminate  X(IHOLD) from all subsequent equations
           MCT=LCT
           IP1=I+1
           JCP=0
           JCM=0
           DO 13 J=IP1,N
                JCPH=JCP
                JCMH=JCM
                MEND=JEND(J)
                MHOLD=MCT+1
                NHOLD=MCT+MEND
                NCT=MCT
                IF(IH(MHOLD).GT.IHOLD)GO TO 10
                LHOLD=LCT-KEND
                DO 9 M =1,MEND
                     MCT=MCT+1
                     IF(IH(MCT).GT.IHOLD)GO TO 10
                     IF(IH(MCT).LT.IHOLD)GO TO 9
                     PIVOT=AH(MCT)
                     NCT=MHOLD-1
                     B(J)=B(J)-B(I)*PIVOT
                     DO 8 L=1,KEND
                          LA=LHOLD+L
    3                     NCT=NCT+1

                          IF(NCT.GT.NHOLD)GO TO 6
                          IF(IH(LA).LT.IH(NCT))GO TO 6
                          IF(IH(LA).EQ.IH(NCT))GO TO 4
                          JD=NCT-JCM
                          AH (JD)=AH(NCT)
                          IH(JD)=IH(NCT)
                          GO TO 3
    4                     IF(IH(LA).NE.IHOLD)GO TO 5
                          JCM=JCM+1
                          GO TO 8
C
C                         Drop XIHOLD) from the Jth equation.
    5                     AH(NCT)=AH(NCT)-AH(LA)*PIVOT
                          GO TO 7
C
C                         Include   X(IH(LA)) in the Jth equation
C                         and store temporarily in  AHP/IHP.
    6                     JCP =JCP +1
                          IHP(JCP) = IH(LA)
                          AHP(JCP) = - AH(LA) * PIVOT
                          NCT =NCT - 1
                          GO TO 8
    7                     JD = NCT -JCM
                          AH(JD) = AH(NCT)
                          IH(JD) = IH(NCT)
    8                CONTINUE
    9           CONTINUE
   10           MCT = MHOLD - 1 + JEND(J)
   11           NCT = NCT + 1
                IF(NCT .GT. NHOLD) GO TO 12
                JD = NCT - JCM
                AH(JD) = AH(NCT)
                IH(JD) = IH(NCT)
                GO TO 11
   12           JEND(J) = JEND(J) - JCM + JCMH
                JPT(J) = JCP - JCPH
```

```
13      CONTINUE
        LCMAX =LCMAX -JCM
        MCT = LCMAX + 1
        MHOLD = MCT
        LCMAX = LCMAX + JCP
        IF(LCMAX .LE. LCBIG) GO TO 14
        LCBIG = LCMAX
14      IF(JCP .LE. JCBIG) GO TO 15
        JCBIG =JCP
15      CONTINUE
C
C       Test that the Dimensions of AHP/IHP and AH/IH
C       have not been exceeded.
        IF (JCP .LE. IHPDIM) GO TO 17
        WRITE(1,16) JCP,I
16      FORMAT('  JCP TOO LARGE, = ',I8,' I = ',I5,)
        RETURN
17      IF(LCMAX .LE. IDIM) GO TO 19
        WRITE(1,18) LCMAX,I,J
18      FORMAT('  LCMAX TOO LARGE  = ',I8,' I = ',I5,' J =   I5)
        RETURN
C
C       Do a reverse pass thru  AH/IH to fill in new terms which
C       have been stored temporarily in  AHP/IHP

19      DO 25 JA =IP1,N
            J = IP1 + N - JA
            MEND = JEND(J)
            JCPH = JCP
            JSA = 0
            MCT = MHOLD
            IF(JCP .LE. 0) GO TO 24
            DO 22 M = 1,MEND
                MCT = MCT -1
                IF(JSA .GE. JPT(J)) GO TO 21
20              IF(IH(MCT) .GE. IHP(JCP)) GO TO 21
                JSA = JSA + 1
                JD = MCT +JCP
                AH(JD) = AHP(JCP)
                IH(JD) = IHP(JCP)
                JCP = JCP - 1
                IF(JCP .LE. 0) GO TO 24
                IF(JSA .LT. JPT(J)) GO TO 20
21              JD = MCT + JCP
                AH(JD) = AH(MCT)
                IH(JD) = IH(MCT)
22          CONTINUE
23          IF(JSA .GE. JPT(J)) GO TO 24
            JSA = JSA + 1
            JD = MCT + JCP - 1
            AH(JD) = AHP(JCP)
            IH(JD) = IHP(JCP)
            JCP = JCP - 1
            GO TO 23
24          MHOLD = MHOLD - JEND(J)
            JEND(J) = JEND(J) -JCP + JCPH
25      CONTINUE
26 CONTINUE
C
C   Backsubstitution to obtain the  X's.
   IELIM(N) = IH(LCMAX)
   B(N) = B(N)/AH(LCMAX)
   LCT = LCMAX + 1
   DO 28 I = 1,N
        NA = N + 1 -I
        KEND = JEND(NA)
        IA = IELIM(NA)
        X(IA) =B(NA)
        DO 27 K = 1,KEND
            LCT = LCT -1
            IF(IH(LCT) .EQ. IA) GO TO 27
            IB = IH(LCT)
            X(IA) = X(IA) - X(IB) * AH(LCT)
27      CONTINUE
28 CONTINUE
   WRITE(1,29) LCT,LCBIG,JCBIG
29 FORMAT('  LCT = ',I5,'  LCBIG = ',I5,' JCBIG = ',I5)
   RETURN
   END
```

NUMERICAL SOLUTIONS OF
PARTIAL DIFFERENTIAL EQUATIONS
J. Noye (Editor)

ITERATIVE METHODS FOR SOLVING LARGE SPARSE LINEAR SYSTEMS

Leonard H. Colgan

School of Mathematics and Computer Studies,
South Australian Institute of Technology,
South Australia, Australia.

1. INTRODUCTION

Because of the substantial increase in speed and efficiency of modern digital computers, numerical methods involving iterative processes have regained much popularity. For example, the discretization of partial differential equations, using either finite differences or finite elements, invariably leads to the problem of solving a large system of linear equations. In particular, elliptic partial differential equations in two dimensions, or more so in three dimensions, yield linear systems which, because of the very large number of equations with very few non-zero coefficients, suggest an iterative method rather than a direct method. The available literature on iterative techniques is vast, while the number of substantially different methods is equally vast. Hence a suitable selection has been made, taking into account the simplicity of description, ease of implementation on a computer, efficiency in the use of storage and time, availability of software, and most importantly, personal preference. Consequently, rather than detailing a succession of methods and their many variants, a small representative selection has been included, with a list of reference material for those interested enough to seek further information.

A brief summary of the contents of this article is as follows: The first section introduces a model problem for subsequent discussion, and this is followed by the necessary matrix preliminaries. Linear stationary iterative methods of the first degree are then described, along with their corresponding convergence criteria. The particular methods detailed are the basic Jacobi, Gauss-Seidel, Successive Overrelaxation (SOR), and Symmetric Successive Overrelaxation (SSOR) methods. Of these, it is fair to say that only Successive Overrelaxation has any practical claim for direct implementation. However, the Jacobi and SSOR methods are included because they can be accelerated substantially using Chebyshev or conjugate gradient acceleration procedures, which comprise the next two sections. Then follows a description of how the size of the system can possibly be reduced using a so-called "red-black" ordering on rectangular domains. Alternating-Direction Implicit (ADI) methods are represented by the Peaceman-Rachford method. Throughout these discussions, it should be understood that variants of these basic methods can yield further improvements. For example it is possible to use block-iterative methods (or line-iterative methods) in certain circumstances, and these can produce worthwhile time savings when compared to the fundamental point-iterative methods. These are described extensively in many references in the bibliography. Finally, a brief comparison of selected iterative methods is included.

A special acknowledgement must be made to Professor David M. Young of the Center for Numerical Analysis, at the University of Texas at Austin. In preparing a review article such as this, it was necessary to select notation, definitions, particular methods and examples from the list of reference material. Professor

Young's influence in this area is especially significant and much of the following material can be attributed directly to him.

2. THE MODEL PROBLEM

Consider the two-dimensional Poisson equation

$$\frac{\partial^2 u}{\partial x^2} + \frac{\partial^2 u}{\partial y^2} = -1, \text{ with } u(x,y) = 0$$

on the boundary of the unit square $0 \le x \le 1$, $0 \le y \le 1$.

Using a uniform grid of mesh-size h, we can approximate the given elliptic partial differential equation at a point (x,y) by the usual finite difference formula

$$\frac{u(x,y-h)+u(x-h,y)-4u(x,y)+u(x+h,y)+u(x,y+h)}{h^2} = -1.$$

The "model problem" corresponds to $h = 1/40$. However, for simplicity, suppose we were to use $h = \frac{1}{4}$. Then there would be 9 unknowns corresponding to the function values at the internal nodes. Let these be denoted by u_i, $i=1,..,9$, where the natural ordering is used.

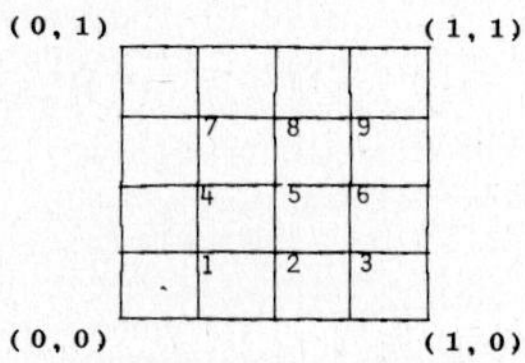

This leads to the matrix equation

$$\begin{pmatrix} 4 & -1 & 0 & -1 & 0 & 0 & 0 & 0 & 0 \\ -1 & 4 & -1 & 0 & -1 & 0 & 0 & 0 & 0 \\ 0 & -1 & 4 & 0 & 0 & -1 & 0 & 0 & 0 \\ -1 & 0 & 0 & 4 & -1 & 0 & -1 & 0 & 0 \\ 0 & -1 & 0 & -1 & 4 & -1 & 0 & -1 & 0 \\ 0 & 0 & -1 & 0 & -1 & 4 & 0 & 0 & -1 \\ 0 & 0 & 0 & -1 & 0 & 0 & 4 & -1 & 0 \\ 0 & 0 & 0 & 0 & -1 & 0 & -1 & 4 & -1 \\ 0 & 0 & 0 & 0 & 0 & -1 & 0 & -1 & 4 \end{pmatrix} \begin{pmatrix} u_1 \\ u_2 \\ u_3 \\ u_4 \\ u_5 \\ u_6 \\ u_7 \\ u_8 \\ u_9 \end{pmatrix} = h^2 \begin{pmatrix} 1 \\ 1 \\ 1 \\ 1 \\ 1 \\ 1 \\ 1 \\ 1 \\ 1 \end{pmatrix}$$

The coefficient matrix is symmetric and positive definite, sparse, and has a band-width of 7.

In the actual model problem with $h = 1/40$, we have a large sparse linear system $A\underset{\sim}{u} = \underset{\sim}{b}$, where the matrix A is of order 1521, and the vector $\underset{\sim}{u}$ contains the unknown values u_i, $i=1,..,1521$, numbered in the natural ordering. A has a band-width of 79, and there are at most five non-zero entries in each row of A. (A symmetric sparse storage scheme would be suitable.)

Apart from equations derived from points adjoining a boundary, we have to solve a system ($h = 1/40$) of the form

$$-u_{i-39} - u_{i-1} + 4u_i - u_{i+1} - u_{i+39} = h^2. \tag{1}$$

In general, numerical solutions of (elliptic) partial differential equations lead to linear systems with the coefficient matrix having a structure somewhat similar to the one generated by the model problem. However, finite element schemes often produce a matrix which is less sparse and with less pattern.

3. MATRIX PRELIMINARIES

Let B be a real N×N matrix, and $\underset{\sim}{x}$ an N-dimensional real vector.

(a) Then $\rho(B)$, the spectral radius of B, is the maximum of the moduli of the eigenvalues of B. Moreover, let m(B) and M(B) (or, if unambiguous, merely m and M) represent the smallest and largest eigenvalues of B, so that $m(B) \le \lambda \le M(B)$ whenever all the eigenvalues λ are real.

(b) For any matrix norm $\|\ \|$, $\rho(B) \le \|B\|$.

(c) We define the 2-norm of $\underset{\sim}{x}$ by

$$\|\underset{\sim}{x}\|_2 = (\underset{\sim}{x},\underset{\sim}{x})^{\frac{1}{2}} = \left(\sum_{i=1}^{N} x_i^2\right)^{\frac{1}{2}}.$$

The 2-norm of B is given by

$$\|B\|_2 = \{\rho(B^T B)\}^{\frac{1}{2}},$$

and this matrix norm is subordinate to the vector norm in that

$$\|B\|_2 = \max_{\underset{\sim}{x} \ne \underset{\sim}{0}} \frac{\|B\underset{\sim}{x}\|_2}{\|\underset{\sim}{x}\|_2}.$$

Clearly $\|B\underset{\sim}{x}\|_2 \le \|B\|_2 \|\underset{\sim}{x}\|_2$. Also, if B is symmetric, then $\|B\|_2 = \rho(B)$.

(d) For any nonsingular matrix W, we define the W-norm of the vector $\underset{\sim}{x}$ and of the matrix B by

$$\|\underset{\sim}{x}\|_W = \|W\underset{\sim}{x}\|_2$$

and

$$\|B\|_W = \|WBW^{-1}\|_2 .$$

Hence, if WBW^{-1} is symmetric, then $\|B\|_W = \rho(B)$.

(e) B is symmetric positive definite if and only if B is symmetric and $\underset{\sim}{x}^T B\underset{\sim}{x} > 0$ for all $\underset{\sim}{x} \ne 0$.

Then all the eigenvalues of B are positive, and there exists a unique symmetric positive definite matrix, denoted by $B^{\frac{1}{2}}$, whose square is equal to B. Also, B is symmetric positive definite if and only if there exists a nonsingular matrix X such that $B = X^T X$.

(f) B is irreducible (indecomposable) if it cannot be put into the form $\begin{pmatrix} F & 0 \\ G & H \end{pmatrix}$, where F and H are square, by simultaneous row and column permutations.

(g) A matrix B is said to have "Property A" if, by appropriate simultaneous row and column permutations, B can be written in the form

$$\begin{pmatrix} D_1 & H \\ K & D_2 \end{pmatrix},$$

where D_1 and D_2 are square diagonal matrices.

(h) $B = \{b_{ij}\}$ has <u>weak diagonal dominance</u> if

$$|b_{ii}| \geq \sum_{\substack{j=1 \\ j \neq i}}^{N} |b_{ij}| \quad (1 \leq i \leq N),$$

and for at least one i, the inequality is strict.

Then if B is a real symmetric matrix which is irreducible, $b_{ii} > 0$ for $i=1,..,N$, and B has weak diagonal dominance then B is symmetric positive definite.

The matrix A in the linear system $A\underset{\sim}{u} = \underset{\sim}{b}$ derived from the model problem is irreducible, has "Property A", has weak diagonal dominance and is symmetric positive definite.

4. BASIC ITERATIVE METHODS

Consider a large sparse linear system

$$A\underset{\sim}{u} = \underset{\sim}{b}, \tag{2}$$

where A is a given real, nonsingular N×N matrix, and $\underset{\sim}{b}$ is a given real N-component vector. Such systems that arise from discretizations of partial differential equations yield a matrix A that is usually symmetric positive definite, and so this assumption shall be made. In any case, such a condition is often necessary to guarantee convergence of the iterative methods considered.

<u>A linear stationary iterative method of the first degree</u> can be constructed (theoretically) in the following manner. Suppose there exists a nonsingular matrix Q which in some sense can be imagined as being an approximation to A, but, unlike A, is readily invertible. More particularly, it is possible to solve conveniently any linear system of the form $Q\underset{\sim}{x} = \underset{\sim}{c}$. For example, Q may be a diagonal, tri-diagonal, or lower triangular matrix, or the product of two or more such matrices. Q is called a <u>splitting matrix</u>.

The system (2) is clearly equivalent to

$$Q\underset{\sim}{u} = (Q-A)\underset{\sim}{u} + \underset{\sim}{b}$$

which leads to the iterative process

$$Q\underset{\sim}{u}^{(n+1)} = (Q-A)\underset{\sim}{u}^{(n)} + \underset{\sim}{b}.$$

Although Q^{-1} is not usually found explicitly, this can be written as

$$\underset{\sim}{u}^{(n+1)} = B\underset{\sim}{u}^{(n)} + \underset{\sim}{k}, \tag{3}$$

where $B = I - Q^{-1}A$ and $\underset{\sim}{k} = Q^{-1}\underset{\sim}{b}$.

Because the <u>iteration matrix</u> B is constant, this process is called stationary. (Information concerning nonstationary methods, such as Richardson's method, can be found in Forsythe and Wasow; or Young [2], [4].)

The basic method (3) is <u>convergent</u> if the sequence $\underset{\sim}{u}^{(0)}, \underset{\sim}{u}^{(1)}, \underset{\sim}{u}^{(2)}, \ldots$ converges to the true solution $\underset{\sim}{u}$ for all initial vectors $\underset{\sim}{u}^{(0)}$, and this limit is independent of the choice of $\underset{\sim}{u}^{(0)}$.

A necessary and sufficient condition for convergence is that $\rho(B)<1$.

The <u>error vector</u> after n iterations is

$$\underset{\sim}{\varepsilon}^{(n)} = \underset{\sim}{u}^{(n)} - \underset{\sim}{u}.$$

The <u>residual vector</u> is

$$\underset{\sim}{r}^{(n)} = \underset{\sim}{b} - A\underset{\sim}{u}^{(n)}, \quad \text{(from (2))}$$

and the <u>pseudo-residual vector</u> is

$$\underset{\sim}{\delta}^{(n)} = B\underset{\sim}{u}^{(n)} + \underset{\sim}{k} - \underset{\sim}{u}^{(n)}$$

$$= \underset{\sim}{u}^{(n+1)} - \underset{\sim}{u}^{(n)}, \quad \text{(from (3)), } \underline{\text{in this case}}.$$

Since $\underset{\sim}{u} = B\underset{\sim}{u} + \underset{\sim}{k}$, then from (3) we have

$$\underset{\sim}{\varepsilon}^{(n+1)} = B\,\underset{\sim}{\varepsilon}^{(n)} \text{ and hence}$$

$$\underset{\sim}{\varepsilon}^{(n)} = B^n\,\underset{\sim}{\varepsilon}^{(0)}.$$

Consequently, $\dfrac{\|\underset{\sim}{\varepsilon}^{(n)}\|_2}{\|\underset{\sim}{\varepsilon}^{(0)}\|_2} \le \|B^n\|_2$, for any $\underset{\sim}{\varepsilon}^{(0)} \neq \underset{\sim}{0}$.

The <u>average rate of convergence</u> for n iterations is defined by

$$R_n(B) = -\frac{1}{n}\ln \|B^n\|_2.$$

If $R_n(B_1) < R_n(B_2)$, then B_2 is iteratively faster for n iterations than B_1.

Then, to reduce an initial error by a prescribed factor, say 10^{-6}, it would require approximately

$$n \simeq \frac{\ln 10^6}{R_n(B)} \text{ iterations.}$$

However, the average rate of convergence, $R_n(B)$, is rarely available.

The <u>asymptotic rate of convergence</u>, denoted by $R(B)$, is defined by

$$R(B) = \lim_{n\to\infty} R_n(B),$$

and it can be shown that

$$\underline{R(B) = -\ln \rho(B)}. \tag{4}$$

(Since $\|B^n\|_2 \ge [\rho(B)]^n$ for all $n\ge1$, then $R(B) \ge R_n(B)$ for any positive n for which $\|B^n\|_2 < 1$.)

This asymptotic rate of convergence $R(B)$, for a convergent matrix B, is certainly the simplest practical way in modern usage of measuring the rapidity of convergence. Hence, a crude estimate of the number of iterations required to reduce the norm of the error by a factor 10^{-6} is given by

$$n \simeq \frac{\ln 10^6}{R(B)}. \tag{5}$$

However, this estimate can often yield misleadingly low values for n. [For example, see Varga.]

4.1 Jacobi Method

For the model problem, this is equivalent to rewriting (1) in the iterative form

$$u_i^{(n+1)} = \tfrac{1}{4}(u_{i-39}^{(n)} + u_{i-1}^{(n)} + u_{i+1}^{(n)} + u_{i+39}^{(n)} + h^2).$$

We can express the matrix A as the matrix sum

$$A = D + L + U,$$

where D is the diagonal matrix such that $d_{ij} = a_{ii}$, and L and U are respectively strictly lower and upper triangular N×N matrices, whose entries are those respectively below and above the main diagonal of A. We can then rewrite (2) as

$$D\underset{\sim}{u} = -(L+U)\underset{\sim}{u} + \underset{\sim}{b},$$

which in turn yields the Jacobi iteration method

$$\underset{\sim}{u}^{(n+1)} = -D^{-1}(L+U)\underset{\sim}{u}^{(n)} + D^{-1}\underset{\sim}{b}.$$

Hence our general notation corresponds to a splitting matrix $Q_J \equiv D$, an iteration matrix

$$B_J \equiv -D^{-1}(L+U) = I - D^{-1}A, \text{ and } \underset{\sim}{k}_J \equiv D^{-1}\underset{\sim}{b}.$$

It should be pointed out that it is possible to normalize the initial matrix problem so that the diagonal elements are unity (see Varga), and this avoids numerous division operations during the iterative process.

The Jacobi method will converge providing $\rho(B_J) < 1$. However, this is not guaranteed, even though the matrix A might be symmetric positive definite, because, although it can be shown that the eigenvalues of B_J are real and < 1, it is possible that $m(B_J) < -1$. (The Jacobi method will converge if both A and 2D-A are symmetric positive definite.)

A useful sufficient condition for convergence is that A is irreducible and has weak diagonal dominance with $d_{ii} > 0$ (a stronger condition than positive definiteness).

Another sufficient extra condition besides positive definiteness is for the matrix A to have "property A". In this case the eigenvalues of B_J are either zero or exist as real pairs $\pm\lambda_i$ for $|\lambda_i| < 1$. This would imply that $\rho(B_J) = M(B_J) < 1$.

Because of its importance in the iterative methods still to be discussed, we denote $\lambda_J = M(B_J)$ to be the largest positive eigenvalue of B_J.

The Jacobi iteration matrix for the model problem possesses all these properties. In fact, it can be shown that $\lambda_J = \cos\pi h = \cos\pi/40 \simeq .9969$.

Hence the approximate number of iterations required to reduce the error by a factor 10^{-6} is

$$n \simeq \frac{\ln 10^6}{-\ln\cos\pi/40} \qquad \text{i.e. } n \simeq 4475$$

[With $\underset{\sim}{\varepsilon}^{(0)} = \underset{\sim}{0}$, it actually takes 4479.]

We note that the convergence rate of the Jacobi method appears to be very slow. It can be shown for any generalized Dirichlet problem that the number of iterations required to achieve a specified level of convergence is proportional to $1/h^2$ for a small mesh-size h. The fact that the Jacobi method can be accelerated, using Chebyshev acceleration, to yield a proportionality of 1/h, is adequate reason for its inclusion here. The next two basic methods, Gauss-Seidel and Successive Overrelaxation, although individually faster than Jacobi, cannot be accelerated in the same way.

NOTE: Suppose our basic Jacobi iteration was of the form

$$u_i^{(n+1)} = a_1 u_{i-k}^{(n)} + a_2 u_{i-1}^{(n)} + a_3 u_{i+1}^{(n)} + a_4 u_{i+k}^{(n)} + c, \text{ for } a_i > 0, \ i=1,..,4.$$

This might correspond to a general elliptic partial differential equation where, say,

$$k = \frac{1}{h} - 1,$$

and u(x,y) is given on the boundary of the unit square. Then it can be shown that

$$\lambda_J = 2(\sqrt{a_1 a_4} + \sqrt{a_2 a_3}) \cos\pi h.$$

Our model problem corresponds to $a_i = \frac{1}{4}$, i=1,..,4.

4.2 Gauss-Seidel Method

For the model problem, this is equivalent to rewriting (1) in the iterative form

$$u_i^{(n+1)} = \tfrac{1}{4}(u_{i-39}^{(n+1)} + u_{i-1}^{(n+1)} + u_{i+1}^{(n)} + u_{i+39}^{(n)} + h^2);$$

i.e. the latest estimates for all components of $\underset{\sim}{u}$ are always used in subsequent computations. In matrix form,

$$\underset{\sim}{u}^{(n+1)} = D^{-1}(-L\underset{\sim}{u}^{(n+1)} - U\underset{\sim}{u}^{(n)} + \underset{\sim}{b})$$

$$(D+L)\underset{\sim}{u}^{(n+1)} = -U\underset{\sim}{u}^{(n)} + \underset{\sim}{b}$$

$$\underset{\sim}{u}^{(n+1)} = -(D+L)^{-1}U\underset{\sim}{u}^{(n)} + (D+L)^{-1}\underset{\sim}{b}.$$

Hence the splitting matrix is $Q_{GS} \equiv D+L$, the iteration matrix is $B_{GS} \equiv -(D+L)^{-1}U = I - (D+L)^{-1}A$, and $\underset{\sim}{k}_{GS} \equiv (D+L)^{-1}\underset{\sim}{b}$.

It can be shown that if A is symmetric positive definite, then the Gauss-Seidel iteration <u>will</u> converge. (The eigenvalues need not be real, and complex eigenvalues tend to produce an irregular convergence.)

However, if A is "consistently ordered" (see Kincaid and Young; or Varga, for a precise definition) and has "Property A", then the eigenvalues of B_{GS} are related to those of B_J in the following way: the eigenvalues of B_{GS} will be real, and for each nonzero pair $\pm\lambda_i$ of B_J, there corresponds eigenvalues of $0, \lambda_i^2$ for B_{GS}.

Hence, in this case,

$$\rho(B_{GS}) = [\rho(B_J)]^2 = \lambda_J^2,$$

and the Gauss-Seidel iteration converges about twice as fast as the Jacobi iteration. It would require about 2240 iterations to achieve the accuracy specified in (5); i.e. $R(B_{GS}) = 2\,R(B_J)$.

The number of iterations is again, of course, proportional to $\frac{1}{h^2}$.

The Gauss-Seidel method is included here mainly to introduce the next procedure, Successive Overrelaxation. Before doing so, a brief comment on a stopping criterion is warranted. Ideally, we might wish to guarantee that

$$\frac{\|\underset{\sim}{\varepsilon}^{(n)}\|_2}{\|\underset{\sim}{u}\|_2} = \frac{\|\underset{\sim}{u}^{(n)} - \underset{\sim}{u}\|_2}{\|\underset{\sim}{u}\|_2} < 10^{-6}$$

say, with an added safeguard condition that $n \le n_{max}$, for some prescribed n_{max}.

However, we cannot measure either $\|\underset{\sim}{\varepsilon}^{(n)}\|_2$ or $\|\underset{\sim}{u}\|_2$ precisely. We can find the pseudo-residual $\underset{\sim}{\delta}^{(n)}$ at each step, but this can only be used effectively if we have a good estimate for the spectral radius of the iteration matrix. It should be pointed out that a stopping test of the form

$$\frac{\|\underset{\sim}{u}^{(n)} - \underset{\sim}{u}^{(n-1)}\|_2}{\|\underset{\sim}{u}^{(n)}\|_2} < 10^{-6}$$

does <u>not</u> imply the accuracy specified above.

4.3 Successive Overrelaxation (SOR) Method

For the model problem, this is equivalent to rewriting (1) in the iterative form

$$u_i^{(n+1)} = \omega\{\tfrac{1}{4}(u_{i-39}^{(n+1)} + u_{i-1}^{(n+1)} + u_{i+1}^{(n)} + u_{i+39}^{(n)} + h^2)\} + (1-\omega)u_i^{(n)}.$$

The parameter ω is called a <u>relaxation parameter</u>. The new estimate $u_i^{(n+1)}$ can be regarded as a weighted average of a term obtained in a manner similar to the Gauss-Seidel concept and the current estimate $u_i^{(n)}$. This iteration is linear, stationary and completely consistent with (2), in the sense that if it converges, it gives the true solution $\underset{\sim}{u}$.

In matrix form,

$$\underset{\sim}{u}^{(n+1)} = \omega\{D^{-1}(-L\underset{\sim}{u}^{(n+1)} - U\underset{\sim}{u}^{(n)} + \underset{\sim}{b})\} + (1-\omega)\underset{\sim}{u}^{(n)}$$

$$(D+\omega L)\underset{\sim}{u}^{(n+1)} = [-\omega U + (1-\omega)D]\underset{\sim}{u}^{(n)} + \omega\underset{\sim}{b}$$

$$\underset{\sim}{u}^{(n+1)} = (D+\omega L)^{-1}[-\omega U + (1-\omega)D]\underset{\sim}{u}^{(n)} + (D+\omega L)^{-1}\omega\underset{\sim}{b}.$$

The splitting matrix is $Q_{SOR} \equiv \frac{1}{\omega}D + L$, the iteration matrix is

$$B_{SOR} \equiv (D+\omega L)^{-1}[-\omega U+(1-\omega)D] = I - (\tfrac{1}{\omega}D + L)^{-1}A,$$

and

$$\underset{\sim}{k}_{SOR} \equiv (D+\omega L)^{-1}\omega\underset{\sim}{b} = (\tfrac{1}{\omega}D + L)^{-1}\underset{\sim}{b}.$$

For $\omega = 1$, the method is merely Gauss-Seidel. It can be shown that if A is symmetric positive definite and irreducible, then $\rho(B_{SOR}) < 1$ if and only if $0 < \omega < 2$. (See Young [4].) i.e. if A is symmetric positive definite, this iterative method converges if and only if the relaxation parameter $\omega \in (0,2)$. Moreover, $\rho(B_{SOR}) \geq |\omega-1|$, with equality only if all eigenvalues of B_{SOR} are of modulus $|\omega-1|$. (See Kahan; or Varga.)

For $0 < \omega < 1$, the method is termed <u>underrelaxation</u>.

For $1 < \omega < 2$, the method is termed <u>overrelaxation</u>, and this is usually the case.

Suppose the matrix A is symmetric positive definite and consistently ordered. Then we know the eigenvalues of B_J are real with modulus less than one. Then the eigenvalues λ of B_J are related to the eigenvalues μ of B_{SOR} as follows:

The eigenvalues of B_J are $\pm\lambda_1, \pm\lambda_2, \ldots, \pm\lambda_{N1}$ ($\lambda_i > 0$) and N2 zero eigenvalues (2N1+N2=N).

Corresponding to $\pm\lambda$, $\lambda>0$, there are two eigenvalues μ of B_{SOR} satisfying the relationship

$$\mu + \omega - 1 = \omega\lambda\sqrt{\mu}, \tag{6}$$

or equivalently,

$$\mu = \left(\frac{\omega\lambda \pm \sqrt{\omega^2\lambda^2-4(\omega-1)}}{2}\right)^2.$$

For $\omega > 1$, this usually produces complex eigenvalues μ. The remaining N2 eigenvalues of B_{SOR} are all equal to $(1-\omega)$.

In particular, we know that $\omega = 1$ corresponds to the Gauss-Seidel iteration. Then equation (6) confirms the earlier statement concerning the rates of convergence of the Jacobi and Gauss-Seidel methods under the conditions being assumed; i.e. $\rho(B_{GS}) = [\rho(B_J)]^2 = \lambda_J^2$.

There is an optimal value of ω, denoted by ω_{opt}, that minimises the spectral radius of B_{SOR}, depending on λ_J;

i.e. ω_{opt} <u>minimises</u> $\max\limits_{-\lambda_J \leq \lambda_i \leq \lambda_J} \left|\frac{\omega\lambda_i \pm \sqrt{\omega^2\lambda_i^2-4(\omega-1)}}{2}\right|^2$.

It can be shown that

$$\omega_{opt} = \frac{2}{1+\sqrt{1-\lambda_J^2}}, \tag{7}$$

and for a general relaxation parameter ω,

$$\rho(B_{SOR}) = \begin{cases} \left(\frac{\omega\lambda_J + \sqrt{\omega^2\lambda_J^2-4(\omega-1)}}{2}\right)^2, & \text{if } 0 < \omega < \omega_{opt} \\ \omega - 1, & \text{if } \omega \geq \omega_{opt}. \end{cases}$$

For convenience of notation, when $\omega = \omega_{opt}$, let $\rho(B_{SOR}) = \bar{\mu}$. If we were to know ω_{opt}, which we can certainly find via (7) if we know λ_J, then the rate of convergence of the SOR method is about $2\sqrt{2}$ times the square root of the rate of convergence of the Jacobi method (and hence about twice the square root of the rate of convergence of the Gauss-Seidel method). In particular, the number of iterations required to obtain the error reduction previously prescribed is now proportional to 1/h.

For the model problem with $h = \frac{1}{40}$, $\lambda_J = \cos \pi h$.

Hence, from (7), $\omega_{opt} = \frac{2}{1+\sin \pi h}$ and the spectral radius of B_{SOR} with $\omega = \omega_{opt}$ is

$$\bar{\mu} = \omega_{opt} - 1 = \frac{1 - \sin\pi h}{1 + \sin\pi h} \simeq .8545,$$

and

$$n \simeq \frac{\ln 10^6}{-\ln(\omega_{opt} - 1)},$$

which implies 88 iterations.

For effective use of the SOR method, it is essential that we have an accurate estimate for λ_J, and thereby ω_{opt}. If we overestimate ω_{opt}, the decrease in the asymptotic rate of convergence is not too serious. However an underestimate of ω_{opt} produces a considerable reduction in the convergence rate. If the matrix A has "Property A", it is interesting that the value of ω_{opt} is independent of the order in which the unknowns are actually determined for most practical orderings. Estimates for λ_J can sometimes be obtained a priori, or else an adaptive procedure can be employed for the automatic determination of λ_J as the actual iterative procedure is being carried out. (See Young [4]; or Hageman.)

A possible adaptive procedure then can easily be incorporated into an SOR code using the fact that given either λ_J (the spectral radius of B_J) or $\bar{\mu}$ (the spectral radius of B_{SOR} with $\omega = \omega_{opt}$), all three quantities λ_J, $\bar{\mu}$ and ω_{opt} can be determined via relation (7) as well as

$$\lambda_J = \frac{\bar{\mu} + \omega_{opt} - 1}{\omega_{opt} \sqrt{\bar{u}}} \tag{8}$$

which is derivable from (6) and the formula for $\bar{\mu} = \rho(B_{SOR})$.

The following adaptive procedure is not rigorous in that it requires $\omega < \omega_{opt}$ for all values of ω used while we seek the optimum value. However, in practice, it has been quite successful.

(i) Choose a value ω so that $\omega < \omega_{opt}$ is certain; for example, we could start with $\omega = 1$.

(ii) Apply a few steps of SOR, and then use the ratio of consecutive pseudo-residuals

$$R^{(n)} = \frac{\|\underset{\sim}{\delta}^{(n)}\|_2}{\|\underset{\sim}{\delta}^{(n-1)}\|},$$

where $\underset{\sim}{\delta}^{(n)} = \underset{\sim}{u}^{(n+1)} - \underset{\sim}{u}^{(n)}$, as an estimate for $\bar{\mu}$.

(iii) Use this estimate for $\bar{\mu}$, and the current ω, to calculate a corresponding estimate for λ_J via (8).

(iv) Use this λ_J estimate to calculate an improved value for ω_{opt} via (7). Then return to (ii).

Provided we are sensible in deciding how often to apply this adaptive process, the value of ω_{opt} can be obtained fairly easily without adding much cost to the overall Successive Overrelaxation method. This process theoretically relies on the estimate ω approaching ω_{opt} from below. As it does so, the convergence rate of the actual SOR method improves noticeably.

A stopping test which is very nearly equivalent to

$$\frac{\|\underset{\sim}{\varepsilon}^{(n)}\|_2}{\|\underset{\sim}{u}\|_2} < 10^{-6}, \text{ say, is as follows:}$$

After the nth iteration, define

$$H = \begin{cases} R^{(n)}, \text{ if } \omega - 1 < R^{(n)} < 1, \\ \omega - 1, \text{ if } R^{(n)} < \omega - 1. \end{cases}$$

Then the SOR process terminates when

$$\frac{\|\underset{\sim}{\delta}^{(n)}\|_2}{(1-H)\|\underset{\sim}{u}^{(n+1)}\|_2} < 10^{-6}.$$

This test is found to be satisfied within a couple of iterations after the desired accuracy has actually been achieved.

4.4 Symmetric Successive Overrelaxation (SSOR) Method

Each iteration of the SSOR method can be regarded as two distinct operations. Firstly, it uses a forward SOR iteration in which the unknowns are computed successively in their natural ordering, and then a backward SOR iteration in which the unknowns are computed in the opposite order. The same value of the relaxation factor ω, $0 < \omega < 2$, is used in both parts of each SSOR iteration.

Let B_{BSOR} be the iteration matrix associated with the backward sweep. Then each SSOR iteration, depending upon a relaxation factor ω, can be written in a matrix form such as

$$\underset{\sim}{u}^{(n+½)} = B_{SOR}\, \underset{\sim}{u}^{(n)} + \underset{\sim}{k}_1$$

$$\underset{\sim}{u}^{(n)} = B_{BSOR}\, \underset{\sim}{u}^{(n+½)} + \underset{\sim}{k}_2,$$

where

$$B_{SOR} = (D+\omega L)^{-1}[-\omega U+(1-\omega)D]$$

and

$$B_{BSOR} = (D+\omega U)^{-1}[-\omega L+(1-\omega)D].$$

Hence, the iteration matrix for the Symmetric Successive Overrelaxation method is given by

$$B_{SSOR} \equiv (D+\omega U)^{-1}[-\omega L+(1-\omega)D](D+\omega L)^{-1}[-\omega U+(1-\omega)D].$$

It can be shown that the corresponding splitting matrix is

$$Q_{SSOR} \equiv \frac{1}{\omega(2-\omega)}(D+\omega L)D^{-1}(D+\omega U).$$

Interestingly, if A is a symmetric positive definite matrix (i.e. $L^T = U$), and $0 < \omega < 2$, then the splitting matrix Q is itself symmetric positive definite. This fact can be used to prove that, if $0 < \omega < 2$, all the eigenvalues λ of B_{SSOR} are real and lie in the interval $0 \le \lambda < 1$. (See Young [4].) The rate of convergence of this SSOR method is not particularly sensitive to the precise choice of the relaxation factor ω, and so the exact optimum value of ω is not necessary. A number of good values for ω have been expounded, depending upon the information available. In any case, it is known that the SSOR method converges satisfactorily provided

$$\rho(D^{-1}LD^{-1}U) \le \tfrac{1}{4}\,.$$

This particular condition is satisfied for the model problem, and for most examples of generalised Dirichlet problems provided the natural ordering of the mesh points is employed.

If we know that $\rho(D^{-1}LD^{-1}U) \le \frac{1}{4}$, and we also know λ_J, the largest eigenvalue of the corresponding Jacobi iteration matrix, then a good value for ω would be

$$\omega^* = \frac{2}{1+\sqrt{2(1-\lambda_J)}}. \qquad \text{(See Young [4].)}$$

Moreover, for this value ω^*, it can be shown that

$$\rho(B_{SSOR}) \le \frac{1-\sqrt{\frac{1-\lambda_J}{2}}}{1+\sqrt{\frac{1-\lambda_J}{2}}}.$$

In a more general situation, we might be able to find an upper bound for $\rho(D^{-1}LD^{-1}U)$, say $\bar{\beta}(\ge\frac{1}{4})$, and an <u>upper bound</u> for λ_J, say $\bar{\lambda}_J$. In this case, a good value for ω would be

$$\omega^* = \frac{2}{1+\sqrt{1-2\bar{\lambda}_J+4\bar{\beta}}}$$

with a corresponding bound for $\rho(B_{SSOR})$.

As a typical comparison between the effectiveness of the SOR method using ω_{opt}, and the SSOR method using ω^*, consider the model problem. Here $\lambda_J = \cos\pi h$, and we know $\bar{\beta} \le \frac{1}{4}$.

Hence

$$\omega^* = \frac{2}{1+\sqrt{2(1-\cos\pi h)}} = \frac{2}{1+2\sin\frac{\pi h}{2}},$$

which is very close to $\omega_{opt} = \dfrac{2}{1+\sin\pi h}$.

Also, $\rho(B_{SOR}) = \dfrac{1-\sin\pi h}{1+\sin\pi h}$, whereas

$$\rho(B_{SSOR}) \le \frac{1-\sin\frac{\pi h}{2}}{1+\sin\frac{\pi h}{2}}.$$

Comparing these, we can see that the SOR method converges about twice as quickly as the SSOR method. This condition is true for most linear systems derived from generalized Dirichlet problems for which $\bar{\beta} \le \frac{1}{4}$, and the number of iterations necessary is approximately proportional to 1/h, for small h. Besides needing twice as many iterations, the SSOR method requires twice as much work per iteration. However, these disadvantages can be overcome in two ways. Firstly, it is possible to extrapolate the SSOR method, which can square the spectral radius of the iteration matrix. But most importantly, since the eigenvalues λ of B_{SSOR} are real with $0 \le \lambda < 1$, we can accelerate the convergence of the SSOR method by employing either Chebyshev acceleration or conjugate gradient acceleration (see ahead).

A substantial improvement in the convergence rate can be achieved in this manner. This is not possible with the SOR method. Similarly, we can accelerate the Jacobi method, but not Gauss-Seidel.

5. CHEBYSHEV ACCELERATION (See Hageman and Young; Kincaid and Young; or Varga.)

Consider a basic stationary iterative method of the form

$$\underset{\sim}{u}^{(n+1)} = B\,\underset{\sim}{u}^{(n)} + \underset{\sim}{k},$$

where the eigenvalues of the iteration matrix B are real.

For example, we might use the Jacobi iteration with $B_J = I - D^{-1}A$, or the Symmetric Successive Overrelaxation iteration with

$$B_{SSOR} = (D+\omega U)^{-1}[-\omega L+(1-\omega)D](D+\omega L)^{-1}[-\omega U+(1-\omega)D].$$

Let m = m(B) and M = M(B) represent respectively the smallest and largest eigenvalues λ of B, so that

$$m \le \lambda \le M, \text{ for all } \lambda.$$

The Chebyshev acceleration technique does not assume that the matrix A of the original large sparse linear system is necessarily positive definite. Instead, it merely requires that the iteration matrix B is similar to a symmetric matrix with real eigenvalues less than one.

i.e. $M < 1$.

Even if the basic method diverges because $m < -1$, this acceleration process converges and is still very effective. In all cases, it produces a considerable improvement in the rate of convergence.

Fundamentally, we require the existence of a nonsingular symmetrization matrix W such that WBW^{-1} is symmetric with eigenvalues $\lambda < 1$.

Equivalently, we require W such that

$$W(I-B)W^{-1} = I - WBW^{-1}$$

is symmetric positive definite.

For example, suppose A is symmetric positive definite, and let $W = D^{\frac{1}{2}}$. (The basic Jacobi method will diverge if $m(B_J) < -1$.) Then

$$W(I-B_J)W^{-1} = D^{\frac{1}{2}}(D^{-1}A)D^{-\frac{1}{2}}$$
$$= X^TX$$

where $X = A^{\frac{1}{2}}D^{-\frac{1}{2}}$.

Similarly, for the SSOR method, let

$$W = \frac{1}{\sqrt{\omega(2-\omega)}} D^{-\frac{1}{2}}(D+\omega U), \quad 0 < \omega < 2.$$

Then $W(I-B_{SSOR})W^{-1}$ is symmetric positive definite. In fact, we could also choose $W = Q^{\frac{1}{2}}_{SSOR}$ or $W = A^{\frac{1}{2}}$ in this case.

Even though it is not essential, it is convenient to assume that A is symmetric positive definite in order to analyze the effectiveness of Chebyshev acceleration. It is a polynomial acceleration, and is termed a semi-iterative method with respect to the corresponding basic iterative method. An outline of the theory behind Chebyshev acceleration is as follows:

Suppose we have a basic iterative method that generates a sequence of vectors $\underset{\sim}{v}^{(i)}$; i.e. we have an initial vector $\underset{\sim}{v}^{(0)}$, and then

$$\underset{\sim}{v}^{(1)} = B\,\underset{\sim}{v}^{(0)} + \underset{\sim}{k} \text{ etc, such that}$$
$$\underset{\sim}{v}^{(n+1)} = B\,\underset{\sim}{v}^{(n)} + \underset{\sim}{k}.$$

At the nth iteration, we wish to choose constants α_{ni}, i=0,1,..,n and then define

$$\underset{\sim}{u}^{(0)} = \underset{\sim}{v}^{(0)}; \; \underset{\sim}{u}^{(n)} = \sum_{i=0}^{n} \alpha_{ni}\, \underset{\sim}{v}^{(i)},$$

in such a way that the sequence of vectors $\underset{\sim}{u}^{(n)}$ tends towards the true solution $\underset{\sim}{u}$ of $A\underset{\sim}{u} = \underset{\sim}{b}$ in some optimal fashion. If the initial vector $\underset{\sim}{v}^{(0)}$ was in fact $\underset{\sim}{u}$, then clearly $\underset{\sim}{v}^{(n)} = \underset{\sim}{u}$ for all $n \geq 0$, and so we require in this case that $\underset{\sim}{u}^{(n)} = \underset{\sim}{u}$ for all $n \geq 0$.

Hence we have the added condition that $\sum_{i=0}^{n} \alpha_{ni} = 1$, for each $n \geq 0$.

Since $\underset{\sim}{u} = \sum_{i=0}^{n} \alpha_{ni}\underset{\sim}{u}$, we have

$$\underset{\sim}{\varepsilon}^{(n)} = \underset{\sim}{u}^{(n)} - \underset{\sim}{u} = \sum_{i=0}^{n} \alpha_{ni} (\underset{\sim}{v}^{(i)} - \underset{\sim}{u})$$

$$= \sum_{i=0}^{n} \alpha_{ni} B^i \underset{\sim}{\varepsilon}^{(0)} \quad \text{since } \underset{\sim}{u}^{(0)} = \underset{\sim}{v}^{(0)}$$

$$= \left(\sum_{i=0}^{n} \alpha_{ni} B^i \right) \underset{\sim}{\varepsilon}^{(0)}$$

i.e. $\underset{\sim}{\varepsilon}^{(n)} = P_n(B)\, \underset{\sim}{\varepsilon}^{(0)}$, where the polynomial $P_n(x)$ is given by

$$P_n(x) = \sum_{i=0}^{n} \alpha_{ni} x^i, \text{ for } n \geq 0.$$

The only restriction we have imposed so far on the polynomials $P_n(x)$ is that $P_n(1) = 1$.

Let W by the symmetrization matrix corresponding to the matrix B. Since WBW^{-1} is symmetric, then $WP_n(B)W^{-1}$ is also symmetric, and hence $\|P_n(B)\|_W = \rho(P_n(B))$ (see Matrix Preliminary (d)).

Thus,

$$\|\underset{\sim}{\varepsilon}^{(n)}\|_W \leq \|P_n(B)\|_W \|\underset{\sim}{\varepsilon}^{(0)}\|_W$$

implies

$$\frac{\|\underset{\sim}{\varepsilon}^{(n)}\|_W}{\|\underset{\sim}{\varepsilon}^{(0)}\|_W} \leq \rho(P_n(B)).$$

At the nth iteration, our objective is therefore to choose the constants α_{ni} so as to minimise $\rho(P_n(B))$.

Let $\lambda_1 = m, \lambda_2, \ldots, \lambda_{N-1}, \lambda_N = M < 1$ be the eigenvalues of B. The eigenvalues of $P_n(B)$ are $P_n(\lambda_1), \ldots, P_n(\lambda_N)$. Hence, for each n, we wish to minimise

$$\max_{i=1,\ldots,N} |P_n(\lambda_i)|.$$

Instead of this problem, consider a similar minimisation problem which determines the virtual spectral radius $\bar{\rho}(P_n(B))$; we minimise

$$\max_{m \leq \lambda \leq M} |P_n(\lambda)|,$$

given that $P_n(1) = 1$.

The solution of this problem is classical, and can be expressed in terms of Chebyshev polynomials, $T_n(x)$. Because of a linear transformation, and the condition $P_n(1) = 1$, we find that

$$P_n(x) = \frac{T_n\left(\frac{2x-M-m}{M-m}\right)}{T_n\left(\frac{2-M-m}{M-m}\right)}, \quad n \geq 0.$$

Clearly, this expression could be used to generate the coefficients α_{ni}, but this would be cumbersome, and, besides, the necessary storage of all the vectors $\underset{\sim}{v}^{(0)}, \underset{\sim}{v}^{(1)}, \ldots, \underset{\sim}{v}^{(n)}$ would be unmanageable. This Chebyshev acceleration process can be reduced to a second degree iteration by making use of the second order recurrence relation for Chebyshev polynomials, viz.

$$T_0(x) = 1$$

$$T_1(x) = x$$

$$T_{n+1}(x) = 2x\, T_n(x) - T_{n-1}(x),\ n \geq 1.$$

Using this recurrence, and a certain amount of algebra, the <u>Chebyshev acceleration</u> process can be expressed simply as the second-degree semi-iterative method:

Let m, M < 1 be the smallest and largest eigenvalues respectively of the general iteration matrix B. Define constants

$$\gamma = \frac{2}{2-M-m}\ ;\ \sigma = \frac{M-m}{2-M-m}.$$

Then, given an arbitrary $\underset{\sim}{u}^{(0)}$, for $n \geq 0$ let

$$\underset{\sim}{u}^{(n+1)} = \omega_{n+1}\{\gamma(B\underset{\sim}{u}^{(n)} + \underset{\sim}{k}) + (1-\gamma)\underset{\sim}{u}^{(n)}\} + (1-\omega_{n+1})\underset{\sim}{u}^{(n-1)},$$

where the sequence of acceleration parameters ω_n is defined by

$$\omega_1 = 1;\ \omega_2 = \frac{1}{1-\frac{1}{2}\sigma^2};\ \omega_{n+1} = \frac{1}{1-\frac{1}{4}\sigma^2\omega_n} \text{ for } n \geq 2.$$

If m and M are correctly chosen, then Chebyshev acceleration converges considerably faster than the basic method. One possible disadvantage is that it requires storage of two iteration vectors at each stage. A first-degree form does exist, but the acceleration parameters then tend to become large, possibly making the iteration unstable due to rounding error.

It can be shown that the sequence of ω_n parameters tends towards a limit ω_∞, which is given by

$$\omega_\infty = \frac{2}{1+\sqrt{1-\sigma^2}}.$$

$$\text{Let } r = \omega_\infty - 1 = \frac{1-\sqrt{1-\sigma^2}}{1+\sqrt{1-\sigma^2}}.$$

From the derivation of the acceleration procedure, and knowing the maximum value of the Chebyshev polynomials over the relevant interval, it can then be shown that the virtual spectral radius is

$$\bar{\rho}(P_n(B)) = \frac{2r^{n/2}}{1+r^n}.$$

Of course, $\rho(P_n(B)) \leq \frac{2r^{n/2}}{1+r^n} \simeq 2r^{n/2}$ for large n. Consider the model problem, where $M = -m = \cos\pi h$, $h = 1/40$, when we apply Chebyshev acceleration to the basic Jacobi method. Then $\gamma = 1$, $\sigma = M$, and so

$$r = \frac{1-\sin\pi h}{1+\sin\pi h}.$$

To achieve the accuracy prescribed earlier, we require

$$n \simeq \frac{2\ln(2\times10^6)}{-\ln r} \simeq 185 \text{ iterations},$$

which can be confirmed by experiment.

The number of iterations is proportional to $1/h$ for small h, and there are approximately twice as many as for the Successive Overrelaxation method using ω_{opt}.

If we were to use Symmetric Successive Overrelaxation, and apply Chebyshev acceleration, a similar analysis shows that the number of iterations is now proportional to $1/\sqrt{h}$ for small h, and in the case of the model problem, only 26 iterations are required.

However, this needs to be balanced by taking into account the amount of work involved in each iteration, along with the total time and storage requirements.

It should be noted that the formulae for the constants γ and σ, as well as for the acceleration parameters ω_n, involve the smallest and largest eigenvalues m, M of the basic iteration matrix B.

Also, a stopping test that is frequently used in order to terminate the Chebyshev acceleration procedure at about the correct iteration is of the form

$$\frac{1}{1-M}\frac{\|\underset{\sim}{\delta}^{(n)}\|_2}{\|\underset{\sim}{u}^{(n)}\|_2} < 10^{-6}, \text{ or similar, where } \underset{\sim}{\delta}^{(n)} = B\underset{\sim}{u}^{(n)} + \underset{\sim}{k} - \underset{\sim}{u}^{(n)}.$$

Hence, if the values of m, M are not known, estimates will have to be found. Suppose these are $\bar{m}$, $\bar{M}$ respectively. It can be shown that Chebyshev acceleration is not really sensitive to errors involved in $\bar{m}$, especially if $\bar{m} \leq m$, but it is very sensitive to errors in $\bar{M}$. In particular, as $\bar{M}$ approaches M from below, the rate of convergence improves dramatically. Adaptive procedures have been developed which generate accurate estimates for $\bar{M}$ while the actual acceleration process is in operation. In the case of Jacobi with Chebyshev acceleration, it is often sufficient to let $\bar{m} = -\bar{M}$, whilst if SSOR is the basic method, $\bar{m}$ is usually left as zero throughout. Of course, in this latter case, a good value ω^* of the overrelaxation parameter also needs to be found adaptively. Details of these various adaptive procedures can be found in Hageman and Young; or Grimes, Kincaid, MacGregor and Young. For more information on Chebyshev acceleration, and its comparison with the Successive Overrelaxation iterative method, one can refer to Golub; or Golub and Varga.

6. CONJUGATE GRADIENT ACCELERATION

In 1952, Hestenes and Stiefel, "Methods of conjugate gradients for solving linear systems", in the J.Res.Nat.Bur.Standards, introduced the classic conjugate gradient concept. It is a particular case of earlier conjugate-direction methods, and can be derived by modifying the "steepest descent" optimisation technique. Moreover, for an Nth order system $A\underset{\sim}{x} = \underset{\sim}{b}$, it would converge theoretically in at most N iterations. However, after a certain amount of research flurry that immediately followed, little practical implementation of this conjugate gradient method was used until recently. Renewed interest has been sparked by broadening the scope of the original method of Hestenes and Stiefel. Rather than regarding the conjugate gradient method as an isolated one, it can be considered as possibly the simplest case of accelerating a basic iterative method using "conjugate gradient acceleration". More specifically, consider the basic stationary Richardson iterative technique, where the splitting matrix Q and the symmetrization matrix W (if A is symmetric positive definite) are both the identity matrix I, and the iteration matrix is I-A. Then the original conjugate gradient method can be regarded as conjugate gradient acceleration applied to this Richardson iteration. Moreover, a conjugate gradient acceleration can be applied under the same conditions as required for Chebyshev acceleration, and invariably it converges in fewer iterations. Also, the acceleration parameters do not involve eigenvalues, and hence can be generated automatically, although the largest eigenvalue M of the iteration matrix may eventually be required for a suitable stopping test similar to the ones outlined in earlier methods. The fact that the method theoretically terminates in N iterations is unimportant for large sparse problems, because any practical iterative technique necessarily must converge in considerably fewer steps than this.

For simplicity, we shall assume that A is symmetric positive definite. Actually, as is the situation with Chebyshev acceleration, it is sufficient to require that a symmetrization matrix W exists for the iteration matrix B. e.g. Richardson (W=I), Jacobi ($W=D^{\frac{1}{2}}$), SSOR ($W = \frac{1}{\sqrt{\omega(2-\omega)}} D^{-\frac{1}{2}}(D+\omega U)$), under suitable assumptions.

Let $\underset{\sim}{r}^{(n)} = \underset{\sim}{b} - A\,\underset{\sim}{u}^{(n)}$ be the residual vector after n iterations. Suppose $\underset{\sim}{u}^{(0)}$ is arbitrary. Then the original <u>conjugate gradient method</u> is of the form

$$\underset{\sim}{u}^{(n+1)} = \underset{\sim}{u}^{(n)} + \alpha_n \underset{\sim}{p}^{(n)},$$

where $\underset{\sim}{p}^{(n)}$ are suitably chosen <u>direction vectors</u>, and are generated iteratively by the relation

$$\underset{\sim}{p}^{(n)} = \underset{\sim}{r}^{(n)} + \beta_{n-1}\underset{\sim}{p}^{(n-1)}, \text{ given } \underset{\sim}{p}^{(0)} = \underset{\sim}{r}^{(0)}.$$

The parameters α_n, β_n are determined by

$$\alpha_n = \frac{(\underset{\sim}{r}^{(n)},\underset{\sim}{r}^{(n)})}{(\underset{\sim}{p}^{(n)},A\underset{\sim}{p}^{(n)})}, \text{ or equivalently } \frac{(\underset{\sim}{p}^{(n)},\underset{\sim}{r}^{(n)})}{(\underset{\sim}{p}^{(n)},A\underset{\sim}{p}^{(n)})},$$

and

$$\beta_n = \frac{(\underset{\sim}{r}^{(n+1)},\underset{\sim}{r}^{(n+1)})}{(\underset{\sim}{r}^{(n)},\underset{\sim}{r}^{(n)})}, \text{ or equivalently } \frac{-(\underset{\sim}{r}^{(n+1)},A\underset{\sim}{p}^{(n)})}{(\underset{\sim}{p}^{(n)},A\underset{\sim}{p}^{(n)})}.$$

Hestenes and Stiefel suggest that the first expressions for α_n and β_n are simpler to compute, but the second values for α_n and β_n tend to yield better results. In practice, the iterations would proceed thus:

given $\underset{\sim}{u}^{(n)}$, $\underset{\sim}{p}^{(n-1)}$, compute $\underset{\sim}{r}^{(n)}$, followed by, in order, β_{n-1}, $\underset{\sim}{p}^{(n)}$, α_n and hence $\underset{\sim}{u}^{(n+1)}$.

The validity of the classic conjugate gradient method lies in the fact that the residuals are orthogonal,

i.e. $(\underset{\sim}{r}^{(i)}, \underset{\sim}{r}^{(j)}) = 0$ for $i \neq j$; and the direction vectors are "$A^{\frac{1}{2}}$ - orthogonal", or "conjugate",

i.e. $(A^{\frac{1}{2}}\underset{\sim}{p}^{(i)}, A^{\frac{1}{2}}\underset{\sim}{p}^{(j)}) = 0$ for $i \neq j$, or equivalently, $(\underset{\sim}{p}^{(i)}, A\underset{\sim}{p}^{(j)}) = 0$ for $i \neq j$.

The property of orthogonal residuals has as an immediate consequence the theoretical termination property mentioned earlier.

Now, suppose we decide to eliminate the direction vectors $p^{(n)}$ from the formulae above. This yields a second-degree iterative process for $\underset{\sim}{u}^{(n+1)}$ involving the residual vectors;

$$\underset{\sim}{u}^{(n+1)} = \omega_{n+1}\{\gamma_{n+1}\underset{\sim}{r}^{(n)} + \underset{\sim}{u}^{(n)}\} + (1-\omega_{n+1})\underset{\sim}{u}^{(n-1)},$$

where

$$\gamma_{n+1} = \frac{(\underset{\sim}{r}^{(n)}, \underset{\sim}{r}^{(n)})}{(\underset{\sim}{r}^{(n)}, A\underset{\sim}{r}^{(n)})},$$

and the parameters ω_n are defined by the sequence

$$\omega_1 = 1$$

$$\omega_{n+1} = \frac{1}{1 - \dfrac{(\underset{\sim}{r}^{(n)}, \underset{\sim}{r}^{(n)})\gamma_{n+1}}{(\underset{\sim}{r}^{(n-1)}, \underset{\sim}{r}^{(n-1)})\gamma_n \omega_n}}, \quad \text{for } n \geq 1.$$

As can be seen, this has a form somewhat similar to the Chebyshev acceleration process. In fact, this can be regarded as a special case of the following more general conjugate gradient acceleration method.

Consider a general linear stationary iterative method of the first degree to solve $A\underset{\sim}{u} = \underset{\sim}{b}$, of the form

$$\underset{\sim}{u}^{(n+1)} = B\,\underset{\sim}{u}^{(n)} + \underset{\sim}{k},$$

corresponding to a splitting matrix Q, where $B = I - Q^{-1}A$. Let W be a non-singular symmetrization matrix such that $W(I-B)W^{-1}$ is symmetric positive definite. Let $\underset{\sim}{\delta}^{(n)} = B\,\underset{\sim}{u}^{(n)} + \underset{\sim}{k} - \underset{\sim}{u}^{(n)}$ be the pseudo-residual vector.

Then the conjugate gradient acceleration procedure as applied to this basic method is given by:

Given an arbitrary $\underset{\sim}{u}^{(0)}$, for $n \geq 0$ let

$$\begin{aligned}\underset{\sim}{u}^{(n+1)} &= \omega_{n+1}\{\gamma_{n+1}(B\underset{\sim}{u}^{(n)} + \underset{\sim}{k}) + (1-\gamma_{n+1})\underset{\sim}{u}^{(n)}\} + (1-\omega_{n+1})\underset{\sim}{u}^{(n-1)} \\ &= \omega_{n+1}\{\gamma_{n+1}\underset{\sim}{\delta}^{(n)} + \underset{\sim}{u}^{(n)}\} + (1-\omega_{n+1})\underset{\sim}{u}^{(n-1)}, \qquad (8)\end{aligned}$$

where the sequences of acceleration parameters γ_n and ω_n are defined by

$$\gamma_{n+1} = \cfrac{1}{1 - \cfrac{\underset{\sim}{\delta}^{(n)T} W^T W B \underset{\sim}{\delta}^{(n)}}{\underset{\sim}{\delta}^{(n)T} W^T W \underset{\sim}{\delta}^{(n)}}} \quad \text{and}$$

$$\omega_1 = 1$$

$$\omega_{n+1} = \cfrac{1}{1 - \cfrac{\underset{\sim}{\delta}^{(n)T} W^T W \underset{\sim}{\delta}^{(n)} \gamma_{n+1}}{\underset{\sim}{\delta}^{(n-1)T} W^T W \underset{\sim}{\delta}^{(n-1)} \gamma_n \omega_n}} .$$

It can be shown that this procedure minimises the $[W^T W(I-B)]^{\frac{1}{2}}$ norm of the error vector $\underset{\sim}{\varepsilon}^{(n)}$ when compared with any general polynomial acceleration procedure applied to the same basic iterative method. In particular, it would often require fewer iterations than Chebyshev acceleration. It is worthwhile to point out a few relevant properties of this conjugate gradient acceleration process:

(a) If we consider the basic Richardson iteration with $Q = I$, $B = I-A$, $\underset{\sim}{k} = \underset{\sim}{b}$, $\underset{\sim}{\delta}^{(n)} = \underset{\sim}{r}^{(n)}$, $W = I$ etc., it can be seen immediately that the conjugate gradient acceleration reduces to the second-degree form of the classic conjugate gradient method.

(b) Unlike Chebyshev acceleration, the symmetrization matrix W is needed in this process. More precisely, we require $W^T W$, and upon inspecting W for the Jacobi and SSOR methods, this is usually simple or can be eliminated by a suitable preconditioning (scaling) of the original linear system. It can be shown that the pseudo-residuals are W orthogonal

i.e. $(W\underset{\sim}{\delta}^{(i)}, W\underset{\sim}{\delta}^{(j)}) = 0$, for $i \neq j$.

(c) The acceleration parameters do not require knowledge of the eigenvalues m(B), M(B). However, since a stopping test is of the form

$$\frac{1}{1-M} \frac{\|\underset{\sim}{\delta}^{(n)}\|_2}{\|\underset{\sim}{u}^{(n)}\|_2} < 10^{-6} \text{ say,}$$

an estimate for M(B) must be found. Adaptive procedures to find M(B) have been developed. Again, details can be found in Hageman and Young; or Grimes, Kincaid, MacGregor and Young.

(d) Since $B\underset{\sim}{\delta}^{(n)}$ is needed in the evaluation of the parameter γ_{n+1}, it is beneficial if we can avoid having to determine $B\underset{\sim}{u}^{(n)}$ also.

This can be achieved by using the second form (8) for the definition of $\underset{\sim}{u}^{(n+1)}$, and generating the pseudo-residual vectors $\underset{\sim}{\delta}^{(n)}$ by means of the second degree recursion

$$\underset{\sim}{\delta}^{(n+1)} = \omega_{n+1}\{\gamma_{n+1} B\underset{\sim}{\delta}^{(n)} + (1 - \gamma_{n+1})\underset{\sim}{\delta}^{(n)}\} + (1 - \omega_{n+1})\underset{\sim}{\delta}^{(n-1)},$$

which can be derived quite simply.

(e) There appears to be more work necessary in the calculation of the acceleration parameters on each iteration of the conjugate gradient acceleration than for the corresponding Chebyshev acceleration. Also, more storage is required. Nevertheless, for a wide range of problems, conjugate gradient acceleration uses less computer time, even when the eigenvalues for Chebyshev acceleration are known in advance. However, counter-examples to this can be constructed.

(f) Suppose the matrix A is not necessarily symmetric, but $A^T + A$ is symmetric positive definite. We can then use a splitting matrix

$$Q = \tfrac{1}{2}(A^T + A)$$

and so obtain a basic iterative method

$$Q\underset{\sim}{u}^{(n+1)} = \tfrac{1}{2}(A^T - A)\underset{\sim}{u}^{(n)} + \underset{\sim}{b},$$

where Q is symmetric positive definite. For general problems, Q may not be easily invertible, and consequently it may require an internal iteration procedure within each basic iteration before $\underset{\sim}{u}^{(n+1)}$ can be computed. Furthermore, the eigenvalues of the iteration matrix $B = \tfrac{1}{2}Q^{-1}(A^T - A)$ are purely imaginary. However, a <u>generalized conjugate gradient acceleration procedure</u> has been developed, corresponding to this basic iterative process. [See Concus and Golub.]

Let $\underset{\sim}{\delta}^{(n)} = Q^{-1}[\tfrac{1}{2}(A^T - A)\underset{\sim}{u}^{(n)} + \underset{\sim}{b}] - \underset{\sim}{u}^{(n)}$ be the pseudo-residual. Then this method can be expressed in the form

$$\underset{\sim}{u}^{(n+1)} = \omega_{n+1}(\underset{\sim}{\delta}^{(n)} + \underset{\sim}{u}^{(n)}) + (1 - \omega_{n+1})\underset{\sim}{u}^{(n-1)}, \text{ for } n \geq 0,$$

where the acceleration parameters are given by

$$\omega_1 = 1$$

$$\omega_{n+1} = \cfrac{1}{1 + \cfrac{(\underset{\sim}{\delta}^{(n)}, Q\underset{\sim}{\delta}^{(n)})}{(\underset{\sim}{\delta}^{(n-1)}, Q\underset{\sim}{\delta}^{(n-1)})\omega_n}}, \text{ for } n \geq 1.$$

7. REDUCED SYSTEMS (RED-BLACK SYSTEMS)

Consider the simpler version of the model problem where we used $h = \tfrac{1}{4}$. Suppose we number the internal nodes in the "red-black" ordering, as follows:

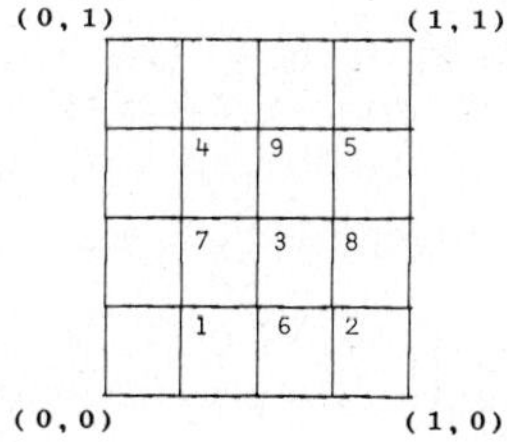

This will lead to the matrix equation

$$\begin{pmatrix} 4 & 0 & 0 & 0 & 0 & . & -1 & -1 & 0 & 0 \\ 0 & 4 & 0 & 0 & 0 & . & -1 & 0 & -1 & 0 \\ 0 & 0 & 4 & 0 & 0 & . & -1 & -1 & -1 & -1 \\ 0 & 0 & 0 & 4 & 0 & . & 0 & -1 & 0 & -1 \\ 0 & 0 & 0 & 0 & 4 & . & 0 & 0 & -1 & -1 \\ . & . & . & . & . & . & . & . & . & . \\ -1 & -1 & -1 & 0 & 0 & . & 4 & 0 & 0 & 0 \\ -1 & 0 & -1 & -1 & 0 & . & 0 & 4 & 0 & 0 \\ 0 & -1 & -1 & 0 & -1 & . & 0 & 0 & 4 & 0 \\ 0 & 0 & -1 & -1 & -1 & . & 0 & 0 & 0 & 4 \end{pmatrix} \begin{pmatrix} u_1 \\ u_2 \\ u_3 \\ u_4 \\ u_5 \\ .. \\ u_6 \\ u_7 \\ u_8 \\ u_9 \end{pmatrix} = h^2 \begin{pmatrix} 1 \\ 1 \\ 1 \\ 1 \\ 1 \\ .. \\ 1 \\ 1 \\ 1 \\ 1 \end{pmatrix}$$

In general, a system $A\underset{\sim}{u} = \underset{\sim}{b}$ is said to be a "red-black system" if it can be partitioned so that

$$A = \begin{pmatrix} D_1 & H \\ K & D_2 \end{pmatrix}, \quad \underset{\sim}{u} = \begin{pmatrix} \underset{\sim}{u}_1 \\ \underset{\sim}{u}_2 \end{pmatrix}, \quad \underset{\sim}{b} = \begin{pmatrix} \underset{\sim}{b}_1 \\ \underset{\sim}{b}_2 \end{pmatrix},$$

where D_1, D_2 are diagonal matrices of order N_1, N_2 respectively such that N_1, N_2 are the numbers of "red" and "black" points, and $N = N_1 + N_2$. If we assume A is symmetric positive definite, then $K^T = H$. Clearly, whenever finite difference discretizations are used over rectangular domains, we can arrange for the large sparse linear systems to be "red-black" systems.

Then $A\underset{\sim}{u} = \underset{\sim}{b}$ can be rewritten as

$$\begin{cases} D_1 \underset{\sim}{u}_1 + H \underset{\sim}{u}_2 = \underset{\sim}{b}_1 \\ K \underset{\sim}{u}_1 + D_2 \underset{\sim}{u}_2 = \underset{\sim}{b}_2, \end{cases}$$

which suggests an iteration of the form:

Given $\underset{\sim}{u}_2^{(0)}$, perform

$$\begin{cases} D_1 \underset{\sim}{u}_1^{(n)} = -H \underset{\sim}{u}_2^{(n)} + \underset{\sim}{b}_1 \\ D_2 \underset{\sim}{u}_2^{(n+1)} = -K \underset{\sim}{u}_1^{(n)} + \underset{\sim}{b}_2. \end{cases} \tag{9}$$

Since D_1^{-1}, D_2^{-1} are trivially found, the sequence of $\underset{\sim}{u}_2^{(n)}$ vectors can be computed along with $\underset{\sim}{u}_1^{(n)}$. Eliminating $\underset{\sim}{u}_1^{(n)}$ from the two equations, the above iterative process can theoretically be regarded as

$$\underset{\sim}{u}_2^{(n+1)} = D_2^{-1} K D_1^{-1} H \underset{\sim}{u}_2^{(n)} + (D_2^{-1} \underset{\sim}{b}_2 - D_2^{-1} K D_1^{-1} \underset{\sim}{b}_1).$$

The size of this system is N_2, which is about half the size of the original system. Using the general notation for basic iterative methods, this <u>Reduced System</u> can be expressed in the form

$$\underset{\sim}{u}_2^{(n+1)} = B_{RS} \underset{\sim}{u}_2^{(n)} + \underset{\sim}{k}_{RS},$$

where $B_{RS} \equiv D_2^{-1} K D_1^{-1} H$, $\underset{\sim}{k}_{RS} \equiv (D_2^{-1} \underset{\sim}{b}_2 - D_2^{-1} K D_1^{-1} \underset{\sim}{b}_1)$. In practice however, the form (9) would be used for the computations during each iteration, and it is clearly possible to store the $\underset{\sim}{u}_1^{(n)}$ vectors in the same computer locations as the $\underset{\sim}{u}_2^{(n)}$. Moreover, the amount of work involved in each step of the Reduced System iterative method is about the same as for the basic Jacobi method.

As $\underset{\sim}{u}_2^{(n)} \to \underset{\sim}{u}_2$, then $\underset{\sim}{u}_1^{(n)} \to \underset{\sim}{u}_1$.

Suppose $W = D_2^{\frac{1}{2}}$. Then

$$\begin{aligned} W B_{RS} W^{-1} &= D_2^{\frac{1}{2}}(D_2^{-1} K D_1^{-1} H) D_2^{-\frac{1}{2}} \\ &= D_2^{-\frac{1}{2}} K D_1^{-1} H D_2^{-\frac{1}{2}} \\ &= X^T X \quad \text{where } X = D_2^{-\frac{1}{2}} H D_2^{-\frac{1}{2}}, \text{ since } K^T = H. \end{aligned}$$

Since $W B_{RS} W^{-1} = X^T X$ is symmetric, then $||B_{RS}||_W = \rho(B_{RS})$, and the eigenvalues of B_{RS} are real and non-negative. Also, using this "red-black" ordering, the basic Jacobi iterative method can be written in the form

$$\begin{pmatrix} \underset{\sim}{u}_1^{(n+1)} \\ \underset{\sim}{u}_2^{(n+1)} \end{pmatrix} = \begin{pmatrix} 0 & -D_1^{-1} H \\ -D_2^{-1} K & 0 \end{pmatrix} \begin{pmatrix} \underset{\sim}{u}_1^{(n)} \\ \underset{\sim}{u}_2^{(n)} \end{pmatrix} + \begin{pmatrix} D_1^{-1} \underset{\sim}{b}_1 \\ D_2^{-1} \underset{\sim}{b}_2 \end{pmatrix},$$

and so

$$B_J \equiv \begin{pmatrix} 0 & -D_1^{-1} H \\ -D_2^{-1} K & 0 \end{pmatrix}.$$

Now,

$$B_J^2 = \begin{pmatrix} D_1^{-1} H D_2^{-1} K & 0 \\ 0 & D_2^{-1} K D_1^{-1} H \end{pmatrix} = \begin{pmatrix} B_{RS}^* & 0 \\ 0 & B_{RS} \end{pmatrix}, \text{ say.}$$

i.e. the $\underset{\sim}{u}_2$ values after two Jacobi iterations are the same as after one Reduced System iteration. It can be shown that $\rho(B_J^2) = \rho(B_{RS}^*) = \rho(B_{RS})$. When the matrix A is symmetric positive definite and has "Property A", we know that $\rho(B_J) < 1$. Hence, under these conditions, $\rho(B_{RS}) = [\rho(B_J)]^2 < 1$.

Because of these properties, we can clearly accelerate this Reduced System method using either Chebyshev acceleration or conjugate gradient acceleration. In all cases, the number of iterations is about one half as many as needed in the corresponding Jacobi process to achieve the same accuracy. The gains in time and storage requirements are of a similar order.

For example, the Chebyshev acceleration procedure applied to the reduced system could be carried out in the following way:

Given $\underset{\sim}{u}_2^{(0)}, \ldots, \underset{\sim}{u}_2^{(n)}$,

find $\underset{\sim}{u}_1^{(n)}$ using $D_1 \underset{\sim}{u}_1^{(n)} = -H \underset{\sim}{u}_2^{(n)} + \underset{\sim}{b}_1$, and the pseudo-residual $\underset{\sim}{\delta}_2^{(n)}$ by means of the expression

$$D_2 \underset{\sim}{\delta}_2^{(n)} = -K \underset{\sim}{u}_1^{(n)} + \underset{\sim}{b}_2 - \underset{\sim}{u}_2^{(n)}.$$

Then the Chebyshev acceleration is of the form

$$\underset{\sim}{u}_2^{(n+1)} = \omega_{n+1} \{\gamma \underset{\sim}{\delta}_2^{(n)} + \underset{\sim}{u}_2^{(n)}\} + (1-\omega_{n+1}) \underset{\sim}{u}_2^{(n-1)},$$

where constants $\gamma = \dfrac{2}{2-\lambda_J^2}$, $\sigma = \dfrac{\lambda_J^2}{2-\lambda_J^2}$, and the acceleration parameters ω_n are given recursively by

$$\omega_1 = 1; \ \omega_2 = \frac{2}{2-\lambda_J^2}; \ \omega_{n+1} = \frac{1}{1-\frac{1}{4}\sigma^2\omega_n}, \text{ for } n \geq 2.$$

This is, of course, equivalent to the general form, because $m = 0$ and $M = \lambda_J^2$. To obtain the prescribed accuracy for $\underset{\sim}{u}$, a suitable stopping test might be

$$\frac{\sqrt{2}}{1-\lambda_J^2}\,\frac{\|\underset{\sim}{\delta}_2^{(n)}\|_2}{\|\underset{\sim}{u}_2^{(n)}\|_2} < 10^{-6}, \text{ or similar.}$$

Again, if λ_J is not known, adaptive processes have been developed.

Other methods similar to a Reduced System with Chebyshev acceleration are also available, and are obtainable as modifications of the fundamental "red-black" ordering concept.

8. ALTERNATING-DIRECTION IMPLICIT (ADI) METHODS

PEACEMAN-RACHFORD METHOD

There are a number of different ADI methods. Only the Peaceman-Rachford method will be considered here (see Birkhoff, Varga and Young; or Varga).

Consider the simpler model problem corresponding to $h = \frac{1}{4}$, and the resulting linear system $A\underset{\sim}{u} = \underset{\sim}{b}$. Assume we are using the natural ordering.

Then the matrix A can be written as $A = A_1 + A_2$, where

$$A_1 = \begin{pmatrix} 2 & -1 & 0 & 0 & 0 & 0 & 0 & 0 & 0 \\ -1 & 2 & -1 & 0 & 0 & 0 & 0 & 0 & 0 \\ 0 & -1 & 2 & 0 & 0 & 0 & 0 & 0 & 0 \\ 0 & 0 & 0 & 2 & -1 & 0 & 0 & 0 & 0 \\ 0 & 0 & 0 & -1 & 2 & -1 & 0 & 0 & 0 \\ 0 & 0 & 0 & 0 & -1 & 2 & 0 & 0 & 0 \\ 0 & 0 & 0 & 0 & 0 & 0 & 2 & -1 & 0 \\ 0 & 0 & 0 & 0 & 0 & 0 & -1 & 2 & -1 \\ 0 & 0 & 0 & 0 & 0 & 0 & 0 & -1 & 2 \end{pmatrix}$$

and

$$A_2 = \begin{pmatrix} 2 & 0 & 0 & -1 & 0 & 0 & 0 & 0 & 0 \\ 0 & 2 & 0 & 0 & -1 & 0 & 0 & 0 & 0 \\ 0 & 0 & 2 & 0 & 0 & -1 & 0 & 0 & 0 \\ -1 & 0 & 0 & 2 & 0 & 0 & -1 & 0 & 0 \\ 0 & -1 & 0 & 0 & 2 & 0 & 0 & -1 & 0 \\ 0 & 0 & -1 & 0 & 0 & 2 & 0 & 0 & -1 \\ 0 & 0 & 0 & -1 & 0 & 0 & 2 & 0 & 0 \\ 0 & 0 & 0 & 0 & -1 & 0 & 0 & 2 & 0 \\ 0 & 0 & 0 & 0 & 0 & -1 & 0 & 0 & 2 \end{pmatrix}.$$

The elliptic partial differential equation from which this arose was

$$\frac{\partial^2 u}{\partial x^2} + \frac{\partial^2 u}{\partial y^2} = -1,$$

when $u(x,y) = 0$ on the boundary of the unit square $0 \leq x \leq 1$, $0 \leq y \leq 1$.

After using the finite difference discretization and multiplying by $(-h^2)$, we can see that A_1 is derived from the $\partial^2 u/\partial x^2$ term, and A_2 is derived from $\partial^2 u/\partial y^2$.

In a more general situation, suppose we discretize

$$a \frac{\partial^2 u}{\partial x^2} + b \frac{\partial^2 u}{\partial y^2} + c \frac{\partial u}{\partial x} + d \frac{\partial u}{\partial y} + e\,u = f,$$

on, say, a rectangular region.

Let the matrices H, V, Σ respectively contain the terms corresponding to

$$a \frac{\partial^2 u}{\partial x^2} + c \frac{\partial u}{\partial x}, \qquad b \frac{\partial^2 u}{\partial y^2} + d \frac{\partial u}{\partial y} \qquad \text{and} \qquad e\,u,$$

after multiplying by $(-h^2)$. Then $A = H + V + \Sigma$. Assume Σ is a non-negative diagonal matrix, and H, V are symmetric positive definite matrices with non-positive off-diagonal elements.

Let $A_1 = H + \frac{1}{2}\Sigma$, $A_2 = V + \frac{1}{2}\Sigma$, so that $A = A_1 + A_2$.

It is worth noting that most of the rigorous justification of the following method relies on H, V being symmetric positive definite, Σ being a multiple of the identity matrix, and $HV = VH$. This then implies that A_1 and A_2 are symmetric positive definite, and $A_1A_2 = A_2A_1$.

The equation $A\underset{\sim}{u} = \underset{\sim}{b}$ is clearly equivalent to each of the equations

$$(A_1 + \omega_{n+1} I)\underset{\sim}{u} = \underset{\sim}{b} - (A_2 - \omega_{n+1} I)\underset{\sim}{u}$$

and

$$(A_2 + \omega_{n+1} I)\underset{\sim}{u} = \underset{\sim}{b} - (A_1 - \omega_{n+1} I)\underset{\sim}{u},$$

for any ω_{n+1}.

This suggests the Peaceman-Rachford ADI method:

$$\begin{cases} (A_1 + \omega_{n+1} I)\underset{\sim}{u}^{(n+\frac{1}{2})} = \underset{\sim}{b} - (A_2 - \omega_{n+1} I)\underset{\sim}{u}^{(n)} \\ (A_2 + \omega_{n+1} I)\underset{\sim}{u}^{(n+1)} = \underset{\sim}{b} - (A_1 - \omega_{n+1} I)\underset{\sim}{u}^{(n+\frac{1}{2})} \end{cases}$$

The ω_n are acceleration parameters. Since $A_1 + \omega_{n+1} I$ is usually a tridiagonal matrix, and $A_2 + \omega_{n+1} I$ can be made tridiagonal after a suitable permutation, each of the implicit steps in the iterations can be carried out using the well-known Gaussian elimination algorithm for tridiagonal systems. In effect, the first equation is equivalent to horizontal sweeps, whereas the second equation should be regarded as vertical sweeps; hence the category, "Alternating-Direction Implicit Methods". The auxiliary vector $\underset{\sim}{u}^{(n+\frac{1}{2})}$ is not retained from one complete iteration to the next.

In practice, it is efficient to use acceleration parameters in the manner $\omega_1, \omega_2, .., \omega_k, \omega_1, \omega_2, .., \omega_k, \omega_1, \omega_2, .., \omega_k, \ldots$ for a suitable k. If $k = 1$, the process is stationary. The Peaceman-Rachford ADI method can theoretically be expressed in the form

$$\underset{\sim}{u}^{(n+1)} = B_{n+1}\underset{\sim}{u}^{(n)} + \underset{\sim}{k}_{n+1}, \quad n \geq 0,$$

where the iteration matrix B_{n+1} is given by

$$B_{n+1} = (A_2 + \omega_{n+1} I)^{-1}(A_1 - \omega_{n+1} I)(A_1 + \omega_{n+1} I)^{-1}(A_2 - \omega_{n+1} I).$$

Then $\underset{\sim}{\varepsilon}^{(n)} = \prod_{i=1}^{n} B_i \underset{\sim}{\varepsilon}^{(0)}$.

If $\rho(\prod_{i=1}^{n} B_i)$ is the spectral radius of this matrix product, then the average rate of convergence after n iterations is $R_n = -\frac{1}{n}\ln(\rho(\prod_{i=1}^{n} B_i))$.
Although optimal acceleration parameters ω_n to minimise this average rate of convergence are not available, good ones, that are simple to calculate, have been found to work quite satisfactorily.

Let α, β be found so that $0 < \alpha \le \mu \le \beta$ and $0 < \alpha \le \lambda \le \beta$ for all eigenvalues μ of A_1 and eigenvalues λ of A_2. Such bounds on the eigenvalues can easily be determined using the relevant theories.

Then two commonly used parameters are the following:

(1) Peaceman-Rachford Parameters
Find the smallest integer k such that $(\sqrt{2}-1)^{2k} \le \alpha/\beta$. Then, for i=1,2,..,k, let $\omega_i = \alpha(\beta/\alpha)^{(2i-1)/2k}$. These parameters are clearly not evenly spaced, but are geometrically spaced. (In the special case of the stationary Peaceman-Rachford method, k = 1, and so $\omega = \sqrt{\alpha\beta}$.)

It can then be shown that

$$\rho(\prod_{i=1}^{k} B_i) \le \left\{\frac{1-(\alpha/\beta)^{1/2k}}{1+(\alpha/\beta)^{1/2k}}\right\}^2 = \delta^2,$$

say, and so the average rate of convergence

$$R_k \simeq -\frac{2}{k}\ln\delta.$$

The condition $(\sqrt{2}-1)^{2k} \le \alpha/\beta$ is used to determine k because it is a consequence of the result that R_k is maximised when $\delta = \sqrt{2}-1$.

For example, consider the model problem. It can be shown that $\alpha = 4\sin^2\pi h/2$ and $\beta = 4\cos^2\pi h/2$. Hence

$$\frac{\alpha}{\beta} = \tan^2\frac{\pi h}{2},$$

and so we require $(\sqrt{2}-1)^k \le \tan\frac{\pi h}{2}$. For $h = \frac{1}{40}$, this yields k = 4. Then we obtain

$$\omega_1 \simeq 0.01385,\ \omega_2 \simeq 0.06986,\ \omega_3 \simeq 0.35245,\ \omega_4 \simeq 1.7781.$$

(2) Wachspress Parameters
Find the smallest integer $k \ge 2$ such that $(\sqrt{2}-1)^{2(k-1)} \le \frac{\alpha}{\beta}$. Then

$$\omega_i = \alpha\left(\frac{\beta}{\alpha}\right)^{(i-1)/(k-1)},\ \text{for } i=1,2,..,k.$$

For the model problem with h = 1/40, this yields k = 5. Then we obtain

$$\omega_1 = \alpha \simeq 0.006165,\ \omega_2 \simeq 0.03110,\ \omega_3 \simeq 0.1569,\ \omega_4 \simeq 0.7916,$$
$$\omega_5 = \beta \simeq 3.9938.$$

Numerical experiments suggest that the Wachspress parameters are superior to the Peaceman-Rachford parameters by a factor of about 2, provided that k is determined as described.

For both sets of parameters, the number of iterations required to obtain the specified accuracy varies according to ln(1/h) as the mesh-size h tends to zero. This is to be compared with 1/h for both

(i) the Successive Overrelaxation method, and
(ii) the Jacobi method with Chebyshev acceleration

and with $1/\sqrt{h}$ for the accelerated SSOR method. However, these other methods are perhaps applicable over a wider range of problems, and also the extra computations involved in the implicit solutions needed in each iteration of the Peaceman-Rachford method are appreciable. Hence, this ADI method is really only advantageous provided h is very small. In fact, experimentation suggests that, provided the ADI method is applicable, there exists a value of h, say h^*, such that for $h > h^*$ the SOR-type methods are preferable, but for $h < h^*$, the Peaceman-Rachford method uses less computer time.

9. COMPARISON OF ITERATIVE METHODS

Comments made about the convergence rates of the various iterative methods described in the preceding sections give a reasonable indication as to the suitability of each. Of course, the amount of time and machine storage space necessary to perform the iterations themselves must be taken into account. A number of references in the bibliography quote numerical results for a range of problems, and there is a reasonable degree of uniformity in their conclusions. For example, see Birkhoff, Varga and Young; or Westlake. Most of the iterative methods considered rely on the matrix A being symmetric positive definite, or else require (theoretically) the existence of a symmetrization matrix W. If A is not symmetric, newer methods that are basically extensions or variations of the classic methods have been formulated. Details can be found in Hageman and Young, or Manteuffel, or Alexsson and Gustafsson, or Widlund.

Of the methods outlined in the preceding sections, those that can seriously be considered for implementation to solve large sparse linear systems are

(1) Successive Overrelaxation method, which may need to be adaptive if ω_{opt} is unknown.

(2) Chebyshev acceleration of the basic Jacobi or SSOR methods. Again, adaptive techniques will be needed if the eigenvalues or relaxation parameters are not known beforehand.

(3) Conjugate gradient acceleration of the Jacobi or SSOR methods.

(4) The reduced system concept, using a "red/black" ordering, in conjunction with either Chebyshev or conjugate gradient acceleration.

(5) Peaceman-Rachford Alternating Direction method.

If the reduced system methods are applicable, numerical results indicate that they frequently use fewer iterations, less time and less storage than the other techniques. In particular, the reduced system combined with conjugate gradient acceleration has often proven to be optimal in each of these classifications. The SSOR method, with either Chebyshev or conjugate gradient acceleration also produces favourable results, and covers a wider class of problems. However, the actual programming implementation is more complicated. The Successive Overrelaxation method is preferred by many because of its simplicity in this regard, but usually considerably more iterations, and somewhat more execution time, are required. The Peaceman-Rachford method is really only superior when the region is rectangular (or similar), the mesh-size is small, and $k > 1$. It is advisable to select a value of k, the number of iteration parameters, which is too large than to use one which is too small. Savings have been found by actually choosing a value of k larger than the ones produced by the formulae described in that section.

BIBLIOGRAPHY

Axelsson, O., "Solution of linear systems of equations: iterative methods", in Lecture Notes in Math.: Sparse matrix techniques 572 (V.A. Barker, ed.) Springer-Verlag, N.Y., 1977.

Axelsson, O. and Gustafsson, I., "A modified upwind scheme for convective transport equation and the use of the conjugate gradient method for the solution of nonsymmetric systems of equations", Computer Sciences Dept. Report 77-12R, Chalmers Univ. Tech., Goteborg, Sweden, 1977.

Bartels, R. and Daniel, J.W., "A conjugate gradient approach to nonlinear elliptic boundary value problems in irregular regions", Lecture Notes in Math. 363, Springer-Verlag, N.Y., 1973.

Birkhoff, G., Varga, R.S. and Young, D., "Alternating direction implicit methods", Advances in Computers 3, 1962.

Buzbee, B.L., Golub, G.H. and Nielson, C.W., "The method of odd/even reduction and factorization with application to Poisson's equation", Siam J.Numer.Anal. 7, 1970.

Carré, B.A., "The determination of the optimal accelerating factor for successive overrelaxation", The Computer Journal 4, 1961.

Chandra, R., " Conjugate gradient methods for partial differential equations", doctoral thesis, Yale University, New Haven, Connecticut, 1977.

Concus, P. and Golub, G.H., "A generalized conjugate gradient method for non-symmetric systems of linear equations", Stan-CS-76-536, Computer Science Dept., Stanford University, Palo Alto, California, 1976.

Concus, P., Golub, G.H. and O'Leary, D.R. "A generalized conjugate gradient method for the numerical solution of elliptic partial differential equations", in Sparse Matrix Computation (J.R. Bunch and D.J. Rose, eds.), Academic Press, 1976.

Douglas, J. Jr., "On the numerical integration of $\frac{\partial^2 u}{\partial x^2} + \frac{\partial^2 u}{\partial y^2} = \frac{\partial u}{\partial t}$ by implicit methods", J.Soc.Ind.Appl.Math.3, 1955.

Douglas, J. Jr., "A survey of numerical methods for parabolic differential equations", Advances in Computers, Vol.2, (F.L.Alt, ed.), Academic Press, N.Y. 1961.

Douglas, J. Jr. and Rachford, H.H. Jr., "On the numerical solution of heat conduction problems in two or three space variables", Trans.Amer.Math.Soc. 82, 1956.

Ehrlich, L.W. "The Block symmetric successive overrelaxation method", doctoral thesis, Univ.Texas at Austin, Austin, Texas, 1963.

Engeli, M., Ginsburg, T., Rutishauser, H. and Stiefel, E., "Refined iterative methods for the computation of the solution and the eigenvalues of self-adjoint boundary value problems", Mitt.Inst.f.angew. Math.ETH Zurich, Number 8, 1959.

Evans, D.J., "The use of preconditioning in iterative methods for solving linear equations with symmetric positive definite matrices", J.Inst.Math.Appl.4, 1967.

Forsythe, G.E. and Wasow, W.R., "Finite Difference Methods for Partial Differential Equations", Wiley, N.Y. 1960.

Ginsburg, T. "The conjugate gradient method", Numer.Math.5, 1963.

Golub, G.H. "The use of Chebyshev matrix polynomials in the iterative solution of linear systems compared with the method of successive overrelaxation", doctoral thesis, Univ.Illinois, Urbana, Illinois, 1959.

Golub, G.H. and Varga, R.S. "Chebyshev semi-iterative methods, successive overrelaxation methods, and second-order Richardson iterative methods, PART I and PART II", Numer.Math.3, 1961.

Grimes, R.G., Kincaid, D.R., MacGregor, W.I. and Young, D.M. "ITPACK Report: Adaptive Iterative Algorithms using Symmetric Sparse Storage", CNA-139, Center for Numerical Analysis, Univ. of Texas at Austin, Austin, Texas, 1978.

Habetler, G.J. and Wachspress, E.L. "Symmetric Successive Overrelaxation in solving diffusion difference equations", Math.Comp.15, 1961.

Hageman, L.A. "The estimation of acceleration parameters for the Chebyshev polynomials and the successive overrelaxation methods", WAPD-TM-1038, Bettis Atomic Power Laboratory, Pittsburgh, Pennsylvania, 1972.

Hageman, L.A. and Young, D.M. "Applied Iterative Methods", Academic Press, N.Y., 1981.

Hageman, L.A., Luk, F. and Young, D.M. "The acceleration of iterative methods: preliminary report", CNA-129, Center for Numerical Analysis, Univ. of Texas at Austin, Austin, Texas, 1977.

Heller, J., "Simultaneous successive and alternating direction schemes", J.Soc. Indust.Appl.Math.8, 1960.

Henrici, P. "Discrete Variable methods in Ordinary Differential Equations", John Wiley and Sons, Inc., N.Y. 1962.

Hestenes, M.R. and Stiefel, E.L. "Methods of conjugate gradients for solving linear systems", NBS J.Res.49, 1952.

Kahan, W. "Gauss-Seidel methods of solving large systems of linear equations", doctoral thesis, Univ. of Toronto, Toronto, 1958.

Kincaid, D.R. "An analysis of a class of norms of iterative methods for systems of linear equations", doctoral thesis, Univ. of Texas at Austin, Austin, Texas, 1971.

Kincaid, D.R. and Young, D.M. "Survey of Iterative Methods", CNA-135, Center for Numerical Analysis, Univ. of Texas at Austin, Austin, Texas, 1978.

Manteuffel, T.A. "An iterative method for solving nonsymmetric linear systems with dynamic estimation of parameters", doctoral thesis, Univ. Illinois, Urbana, Illinois, 1975.

Peaceman, D.W. and Rachford, H.H. Jr. "The numerical solution of parabolic and elliptic differential equations" J.Soc.Indus.Appl. Math.3, 1955.

Price, H. and Varga, R.S. "Recent numerical experiments comparing successive overrelaxation iterative methods with implicit alternating direction methods", Report No. 91, Gulf Research and Development Co., Pittsburgh, Pennsylvania, 1962.

Schwarz, H.R., Rutishauser, H. and Stiefel, E. "Numerical Analysis of Symmetric Matrices", Prentice-Hall Inc. N.J., 1973.

Stone, H.L. "Iterative solutions of implicit approximations of multidimensional partial differential equations", Siam J.Num.Anal.5, 1968.

Tee, G.J. "Eigenvectors of the successive overrelaxation process and its combination with Chebyshev semi-iteration", Comput.J.6, 1963.

Varga, R.S. "Matrix Iterative Analysis", Prentice-Hall, N.J., 1962.

Wachspress, E.L. "Iterative solution of elliptic systems and applications to the Neutron diffusion equations of reactor physics", Prentice-Hall, N.J. 1966.

Westlake, J.R. "A handbook of numerical matrix inversions and solution of linear equations", John Wiley & Sons Inc., N.Y., 1968.

Widlund, O. "A Lanczos method for a class of nonsymmetric systems of linear equations", Siam J. Numer.Anal. No.4, 1978.

Wrigley, H.E. "On accelerating the Jacobi method for solving simultaneous equations by Chebyshev extrapolation when the eigenvalues of the iteration matrix are complex", Comput.J.6, 1963.

Young, D.M. [1], "Iterative methods for solving partial difference equations of elliptic type", doctoral thesis, Harvard Univ., Cambridge, Massachusetts, 1950.

Young, D.M. [2], "On Richardson's method for solving linear systems with positive definite matrices", J.Math.Phys.XXXII, 1954.

Young, D.M. [3], "Convergence properties of the symmetric and unsymmetric successive overrelaxation methods and related methods", Math.Comp.24, 1970.

Young, D.M. [4], "Iterative solution of large linear systems", Academic Press, N.Y. 1971.

Young, D.M. [5], "On the accelerated SSOR method for solving large linear systems", in Advances in Math. 23, 1977.

PART 2:

CONTRIBUTED PAPERS

NUMERICAL SOLUTIONS OF
PARTIAL DIFFERENTIAL EQUATIONS
J. Noye (Editor)

NUMERICAL SIMULATION OF TIDES IN THE GULF OF CARPENTARIA

Michele Rienecker

University of New South Wales
Kensington
New South Wales

John Noye

University of Adelaide
Adelaide
South Australia

The linearised long-wave equations have been used to predict tides within the Gulf of Carpentaria, using known tides along the open boundary in the Arafura Sea as input. By considering all dependent variables proportional to exp(-iωt), where ω is the circular frequency and t the time, time-dependence was removed from the system of governing equations.

The resulting system of elliptic partial differential equations was solved on a finite difference grid using a marching technique, which required the inversion of a small dense matrix rather than the inversion of a very large sparse one as required by usual direct methods of solution.

Comparison of computed values of amplitude and phase at the coast with recorded values for the dominant tidal components showed reasonable agreement. Computed co-tidal and co-amplitude charts for the same components for the whole of the gulf were similar to those obtained by Webb (1981) and Church and Forbes (1981) using different numerical models which covered slightly different regions.

1. INTRODUCTION

The Gulf of Carpentaria is a large, shallow sea with a surface area of about 193,000 km^2 and depths less than 70m. It is located in the North-Eastern part of Australia, and is bounded to the North by the coast of Papua-New Guinea. It is connected with the Indian Ocean through the Arafura Sea to the North-West, and with the Pacific Ocean through Torres Strait to the North-East.

A mathematical analysis of the response of this gulf to tidal forcing has been carried out by Williams (1972) and Buchwald and Williams (1975). They solved analytically the linearised, depth-integrated equations of motion and continuity, after removal of sinusoidal time dependence, on an L-shaped region of fixed depth. They treated Torres Strait as a land barrier since it is very shallow compared with the rest of the Gulf. A major omission in their model is the absence of a mechanism for dissipation of energy, as Miller (1966) has estimated that approximately 10% of the total lunar tidal flux from the deep oceans is dissipated in or near the Gulf of Carpentaria.

More recently, a numerical analysis of the tidal response of the gulf has been carried out by the present authors, by Webb (1981) and Church and Forbes (1981). These models used bottom friction as the mechanism for including energy dissipation. Webb solved the linearised, depth-averaged equation which described the surface elevation in the Gulf, including an area in the Arafura Sea out to the edge of the continental shelf, but closed off Torres Strait. After removing a sinusoidal time dependence, the resulting system of elliptic partial differential equations was solved using finite difference techniques on a rectangular grid. A special technique was incorporated to accommodate curved coastlines. Gaussian elimination, with back-substitution was used to solve the resulting system of linear algebraic equations, although there was a problem storing the large sparse matrix of coefficients involved.

In the work described in this paper, the linearised depth-averaged equations of motion and continuity were solved simultaneously over the gulf region, with an open boundary in the Arafura Sea to the West of Gove Peninsular extending in a northerly fashion to West Irian. Torres Strait is closed off - little tidal energy flows through it because it is choked with reefs. Although some flow through this Strait affects the phases in the Gulf, this effect is very small and unless the boundary input can be specified *accurately* it seems sensible to treat this very shallow region as a land boundary. After time-dependence was removed, a marching technique was used to solve these equations on a finite difference grid, so that the solution required only a small dense matrix to be inverted instead of the large sparse matrix of Webb's method.

By way of contrast, Church and Forbes (1981) solved the complete set of non-linear depth-averaged equations on a finite difference grid by time-stepping. Instead of specifying tidal amplitudes and phases at Torres Strait, they specified estimates of the tidal flow. Although this model has the advantage of being able to model residual circulation, it needs to be run for a long time to produce results for the tidal components. Estimates by Webb (1981) and a study by Rienecker (1980) show that non-linear effects are small for most of the Gulf of Carpentaria.

Comparison of computed values of amplitude and phase using the model described in this paper for the tidal components K1 and M2 near the coastal locations listed in the Australian National Tide Tables for 1981, showed reasonable agreement. The latter values were obtained by the analysis of recordings from tide-gauges near the coast, where the influence of local conditions may dominate. Computed co-tidal and co-amplitude charts for these same components over the whole of the gulf region were similar in nature to those obtained by Williams (1972), Webb (1981) and Church and Forbes (1981). For the diurnal tide, the main feature was an amphidromic point located just west of the centre of the Gulf, with the tides rotating clockwise around that point. For the semi-diurnal tide, a number of amphidromic points were found, extending North-West from Karumba. In this model, there are three such points; in Webb's model, there is only one point in the South-East of the Gulf.

2. THE TIDAL EQUATIONS

The general two-dimensional depth-averaged equations governing fluid motion have been derived by such authors as Nihoul (1975). Assuming certain restrictions such as the depth of water is large compared with the amplitude of the oscillation in the tide-height, these equations may be approximated by the linear forms:

$$\frac{\partial}{\partial x}(hU) + \frac{\partial}{\partial y}(hV) = -\frac{\partial Z}{\partial t} \tag{2.1a}$$

$$\frac{\partial U}{\partial t} + \frac{r}{h}U - fV = -g\frac{\partial Z}{\partial x} \tag{2.1b}$$

$$\frac{\partial V}{\partial t} + \frac{r}{h}V + fU = -g\frac{\partial Z}{\partial y} \tag{2.1c}$$

where $\underset{\sim}{q} = (U,V)$ is the depth-averaged horizontal velocity,

h is the depth at mean sea-level,

Z is the surface elevation above mean sea-level,

t is the time coordinate,

f is the Coriolis parameter, considered constant, given by the product $2\Omega\cos\theta$, where Ω is the angular velocity of the earth and θ is the co-latitude,

g is the acceleration due to gravity,

r is a bottom friction parameter, considered constant and

x,y are orthogonal Cartesian coordinates, positively increasing to the East and North respectively.

The assumptions implicit in the linearisation, with a discussion of their validity, may be found in Noye and Tronson (1978).

Equations (2.1) are used to model the tidal propagation in the Gulf of Carpentaria, where the motion is produced in response to an input $\mathrm{Re}\{\zeta_0(x,y)\exp(-i\omega t)\}$, of period $T = 2\pi/\omega$, across the open boundaries where $i = \sqrt{-1}$. The equations are solved subject to this input where the real parts of the solutions for Z, U and V will give the elevation and velocity fields at any instant in time.

Since the equations are linear, it follows that

$$Z(x,y,t) = \zeta(x,y)\exp(-i\omega t),$$

$$U(x,y,t) = u(x,y)\exp(-i\omega t),$$

$$V(x,y,t) = v(x,y)\exp(-i\omega t),$$

so that the explicit time dependence in (2.1) may be removed. The equations governing the spatial variation of the fluid motion are therefore

$$(-i\omega+r/h)u - fv = -g\ \partial\zeta/\partial x, \tag{2.2a}$$

$$(-i\omega+r/h)v + fu = -g\ \partial\zeta/\partial y, \tag{2.2b}$$

$$\partial(hu)/\partial x + \partial(hv)/\partial y = i\omega\zeta, \tag{2.2c}$$

subject to the boundary conditions that $\zeta_0(x,y)$ is defined on the open boundary, and on the closed boundary to be approximated by straight-line segments aligned S-N or E-W,

$$u(x,y) = 0 \text{ on S-N segments,}$$

$$v(x,y) = 0 \text{ on E-W segments.}$$

3. THE FINITE DIFFERENCE NUMERICAL MODEL

The numerical model uses finite difference approximations to Equations (2.2). To determine the appropriate finite difference forms, a two-dimensional rectangular grid aligned S-N and W-E is superimposed on a plan of the region of study, the boundaries thus being approximated by horizontal and vertical straight line segments. The grid points lie at the intersection of the lines drawn parallel to the x- and y-axes; the grid spacing in the x-direction is Δx and that in the y-direction is Δy.

The grid is composed of elements; within each element are a ζ-, a u- and a v-point arranged in a staggered fashion, as shown in Figure 1. This configuration has been used by many authors, including Platzman (1958) and Leendertse (1967), because of the simple form taken by the coastal boundary conditions. If ζ is evaluated at a grid-point with coordinates (x,y) then u is evaluated at $(x-\Delta x,y)$ and v at $(x,y-\Delta y)$; the depth, h, is specified at the position (x,y). Each element is identified by an ordered pair, (ℓ,j), with $1 \le \ell \le m$ and $1 \le j \le n$, ℓ increasing in the positive x-direction, j increasing in the positive y-direction. The corresponding values of ζ, u, v, h are denoted by $\zeta_{\ell,j}$, $u_{\ell,j}$, $v_{\ell,j}$ and $h_{\ell,j}$. Figure 1 shows the (ζ,j)th element and surrounding grid-points. Each element is further labelled according to which one of 12 classes it belongs. These classes identify the manner of allocation of values to ζ, u and v and are set out in Table 3.1

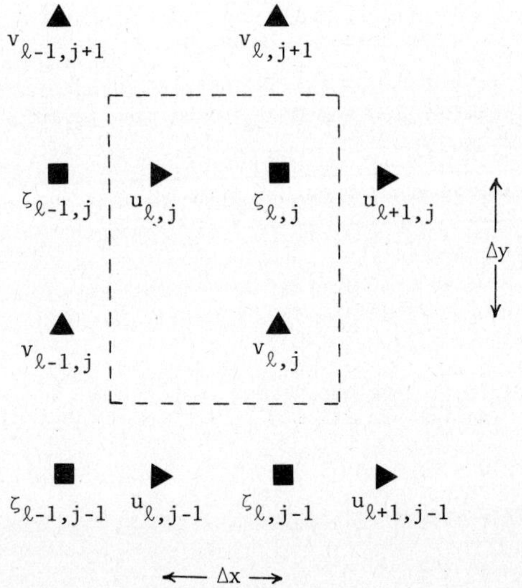

Figure 1 : The (ℓ,j)th element and surrounding grid-points.

From this table it can be seen that the coastal boundaries are always approximated in such a manner as to ensure that, if a land boundary is parallel to the y-direction, it passes through a u-point, and, if parallel to the x-direction, it passes through a v-point.

As far as is practicable, centred finite difference approximations to the derivatives in Equations (2.2) are used. Consider, for example, Equation (2.2a) which may be written at the point $(x-\Delta x,y)$,

TABLE 3.1 :

The classes of elements for the linear numerical model and their associated methods of allocation of values to ζ, u and v. /// indicates land, ——— indicates a solid boundary and — — — a Western open boundary, while ▶ depicts a u-point, ■ a ζ-point and ▲ a v-point.

Class	Element	Allocation
1.	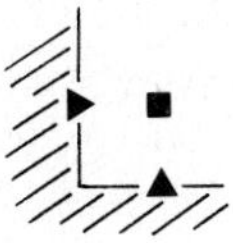	$u_{\ell,j} = 0$ $\zeta_{\ell,j}$ provisionally assigned $v_{\ell,j} = 0$.
2.	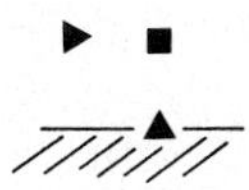	$u_{\ell,j}$, $\zeta_{\ell,j}$ as in Eq(3.3) $v_{\ell,j} = 0$.
3.	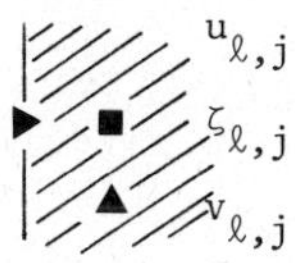	$u_{\ell,j}$ as in Eq(3.3a) $\zeta_{\ell,j}$, $v_{\ell,j}$ not calculated
4.	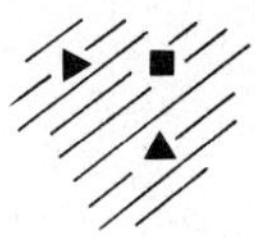	$u_{\ell,j}$, $\zeta_{\ell,j}$, $v_{\ell,j}$ not calculated
5.		$u_{\ell,j} = 0$ $\zeta_{\ell,j}$ provisionally assigned $v_{\ell,j}$ as in Eq(3.3c)
6.		$u_{\ell,j}$, $\zeta_{\ell,j}$, $v_{\ell,j}$ as in Eq(3.3)
7.	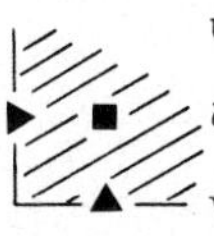	$u_{\ell,j}$ as in Eq(3.3a) $\zeta_{\ell,j}$ not calculated $v_{\ell,j} = 0$
8.	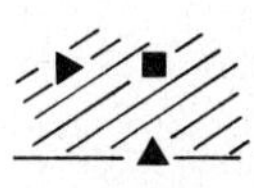	$u_{\ell,j}$, $\zeta_{\ell,j}$ not calculated $v_{\ell,j} = 0$.
9.		$u_{\ell,j}$ provisionally assigned $\zeta_{\ell,j}$ input data $v_{\ell,j}$ as in Eq(3.3c)
10.	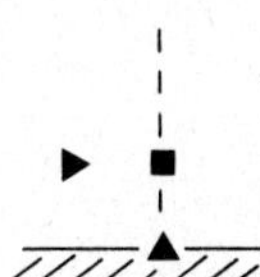	$u_{\ell,j}$ provisionally assigned $\zeta_{\ell,j}$ input data $v_{\ell,j} = 0$

$$\{-i\omega + rh^{-1}(x-\Delta x,y)\}u(x-\Delta x,y) - fv(x-\Delta x,y)$$
$$= - g \frac{\partial\zeta}{\partial x}(x-\Delta x,y). \qquad (3.1)$$

Now, using Taylor series expansions

$$\zeta(x,y) = \zeta(x-\Delta x,y) + \Delta x \frac{\partial\zeta}{\partial x}(x-\Delta x,y) + \frac{(\Delta x)^2}{2}\frac{\partial^2\zeta}{\partial x^2}(x-\Delta x,y) + O(\Delta x^3)$$

and

$$\zeta(x-2\Delta x,y) = \zeta(x-\Delta x,y) - \Delta x \frac{\partial\zeta}{\partial x}(x-\Delta x,y)$$
$$+ \frac{(\Delta x)^2}{2}\frac{\partial^2\zeta}{\partial x^2}(x-\Delta x,y) + O(\Delta x^3),$$

so that

$$\zeta(x,y) - \zeta(x-2\Delta x,y) = 2\Delta x \frac{\partial\zeta}{\partial x}(x-\Delta x,y) + O(\Delta x^3).$$

That is, if the (ℓ,j)th element is being considered,

$$\frac{\partial\zeta}{\partial x}(x-\Delta x,y) = \frac{1}{2\Delta x}(\zeta_{\ell,j} - \zeta_{\ell-1,j}) + O(\Delta x^2).$$

In order to avoid the inversion of a large $3mn \times 3mn$ complex matrix, a marching scheme is developed, which requires the approximation to $v(x-\Delta x,y)$ obtained by truncating the relation

$$v(x-\Delta x,y) = \tfrac{1}{2}(v_{\ell-1,j} + v_{\ell-1,j+1}) + O(\Delta x,\Delta y^2).$$

Hence, (3.1) may be written as

$$\{-i\omega + r\tilde{h}^{-1}_{\ell,j}\}u_{\ell,j} - \tfrac{1}{2}f(v_{\ell-1,j} + v_{\ell-1,j+1})$$
$$= - \frac{g}{2\Delta x}(\zeta_{\ell,j} - \zeta_{\ell-1,j}) + O(\Delta x,\Delta y^2), \qquad (3.2a)$$

where

$$\tilde{h}_{\ell,j} = \begin{cases} \tfrac{1}{2}(h_{\ell,j} + h_{\ell-1,j}), & \ell \neq 1 \\ h_{1,j}, & \ell = 1. \end{cases}$$

In the same manner, the finite difference approximations to Equations (2.2b) and (2.2c) are found to be

$$\{-i\omega + r\bar{h}^{-1}_{\ell,j}\}v_{\ell,j} + \tfrac{1}{2}f(u_{\ell,j} + u_{\ell,j-1})$$
$$= -\frac{g}{2\Delta y}(\zeta_{\ell,j} - \zeta_{\ell,j-1}) + O(\Delta y,\Delta x^2) \qquad (3.2b)$$

and

$$\frac{1}{2\Delta x}\{\tilde{h}_{\ell+1,j}u_{\ell+1,j} - \tilde{h}_{\ell,j}u_{\ell,j}\} + \frac{1}{2\Delta y}\{\bar{h}_{\ell,j+1}v_{\ell,j+1} - \bar{h}_{\ell,j}v_{\ell,j}\}$$
$$= i\omega\,\zeta_{\ell,j} + O(\Delta x^2,\Delta y^2), \qquad (3.2c)$$

where

$$\bar{h}_{\ell,j} = \begin{cases} \tfrac{1}{2}(h_{\ell,j} + h_{\ell,j-1}), & j \neq 1 \\ h_{\ell,j}, & j = 1. \end{cases}$$

These equations are rearranged to give, in explicit form, the equations for the evaluation of $u_{\ell,j}$, $v_{\ell,j}$ and $\zeta_{\ell,j}$ at all interior points in terms of values of u, v, ζ already computed, namely

$$u_{\ell,j} = \tilde{h}^{-1}_{\ell,j}\{\tilde{h}_{\ell-1,j}u_{\ell-1,j} - \frac{\Delta x}{\Delta y}(\bar{h}_{\ell-1,j+1}v_{\ell-1,j+1} - \bar{h}_{\ell-1,j}v_{\ell-1,j})$$
$$+ 2i\omega\Delta x\,\zeta_{\ell-1,j}\} \qquad (3.3a)$$

$$\zeta_{\ell,j} = \zeta_{\ell-1,j} - \frac{2\Delta x}{g}\{(-i\omega + r\tilde{h}^{-1}_{\ell,j})u_{\ell,j}$$
$$- \tfrac{1}{2}f(v_{\ell-1,j} + v_{\ell-1,j+1})\} \qquad (3.3b)$$

$$v_{\ell,j} = -\{-i\omega + r\bar{h}^{-1}_{\ell,j}\}^{-1}\{\tfrac{1}{2}f(u_{\ell,j} + u_{\ell,j-1})$$
$$+ \frac{g}{2\Delta y}(\zeta_{\ell,j} - \zeta_{\ell,j-1})\}. \qquad (3.3c)$$

The first equation is obtained by rearranging (3.2c) after replacing ℓ by ℓ-1, the second by rearranging (3.2a) and the third by rearrangement of (3.2b). These equations are explicit since, on calculation of any of the unknowns on the left-hand side, all the quantities on the right-hand side are known provided that the sweep through the grid follows increasing values of ℓ and, for each ℓ, increasing values of j.

The appropriate form of these equations for non-interior points are given in Table 3.1.

4. SOLVING THE FINITE DIFFERENCE EQUATIONS

The solution, using Equations (3.3) is found by the method called EVP by Roache (1972) or implicit-marching by Noye (1978). From Table 3.1 it can be seen that, for elements labelled 3 and 7, the value of $u_{\ell,j}$ is calculated according to Equation (3.3a); the desired result of such a calculation is, naturally, zero, since the associated grid-point lies on a land boundary. Hence, it is necessary that those values (called starting-values) designated as "provisionally assigned", namely $\zeta_{\ell,j}$ in elements labelled 1 and 5 and $u_{\ell,j}$ in elements labelled 9 and 10, should be such that a sweep through the grid produces the correct end-values for elements labelled 3 or 7 on calculation of (3.3). The correct starting values are determined by finding the end-values produced by specific provisional starting-values.

For a consistent scheme with a unique solution the number of starting values, say K, is the same as the number of end-values. The two sets of values are numbered in increasing order as they are encountered in the scheme. For any set of starting-values $\{s_\alpha, \alpha=1,..,K\}$, the sweep through the grid produces a corresponding set of end-values, $\{\varepsilon_\alpha, \alpha=1,..,K\}$. Since Equations (3.3) are linear, there is a simple linear relation between the s_α and the ε_α, namely

$$\underset{\sim}{\varepsilon} = \underline{\underline{A}}\underset{\sim}{s} + \underset{\sim}{\varepsilon}_0 \tag{4.1}$$

where

$$\underset{\sim}{s} = (s_1,..,s_K)^T,\ \underset{\sim}{\varepsilon} = (\varepsilon_1,..,\varepsilon_K)^T, \text{ and}$$

$\underset{\sim}{\varepsilon}_0$ is the end-vector produced by $\underset{\sim}{s} = \underset{\sim}{0}$.

$\underline{\underline{A}}$ is a K×K complex matrix whose columns are generated by starting-vectors of the form $\underset{\sim}{s}_\ell = (\delta_{k\ell})$ where $\delta_{k\ell}$ is the Kronecker delta. Thus, if $\underset{\sim}{\varepsilon}_\ell = \underline{\underline{A}}\underset{\sim}{s} + \underset{\sim}{\varepsilon}_0$, then $\underset{\sim}{a}_\ell = \underset{\sim}{\varepsilon}_\ell - \underset{\sim}{\varepsilon}_0$ where $\underset{\sim}{a}_\ell$ is the ℓth column of $\underline{\underline{A}}$.

Once these quantities have been determined, the correct starting-vector, $\underset{\sim}{s}^*$, is determined by substitution in (4.1) yielding

$$\underset{\sim}{s}^* = \underline{\underline{A}}^{-1}(\underset{\sim}{\varepsilon}^* - \underset{\sim}{\varepsilon}_0) \tag{4.2}$$

where ε^* is the desired end-vector. Hence, (K+1) sweeps through the grid produce the correct starting-values which are then used in a final run to determine the required values of ζ, u and v at interior grid points throughout the whole region.

5. STABILITY, CONSISTENCY AND CONVERGENCE

The concepts of stability, consistency and convergence are discussed in all books which deal with the numerical analysis of finite difference methods (for example, Noye, 1978, p.16).

Consistency is simply the requirement that, as Δx, $\Delta y \to 0$, the truncation errors (as evident in Equations (3.2)) must approach zero, so that the finite difference approximation approaches the partial differential equation in the limit.

Also, a finite difference approximation is stable if the difference between its theoretical solution and its actual numerical solution remains bounded. This difference arises because of round-off errors.

As Δx, $\Delta y \to 0$, we require that the discrete solution must approach the continuum solution, that is, the solution to the finite difference approximation must converge to the solution of the partial differential equation. This is generally hard to prove as the finite difference method is used solely because an exact solution to the partial differential equation is not known. Linear initial value problems may use Lax's Equivalence Theorem (see Noye, 1978, p.35) to relate consistency and stability to convergence. However, no analogous theorem exists for schemes such as the EVP method which are used to solve boundary value problems and cannot be classed as initial value problems. However, the finite difference method can be used to solve a similar, but more simple, problem for which an analytic solution is known, and a comparison between the two can be used as a guideline as to the likely convergence of the finite difference method when applied to more general problems.

5.1 Stability

To determine the usefulness of the finite difference model set up in Section 4, the error amplification properties of the scheme need to be analyzed. This examination can be carried out by means of a discrete perturbation analysis as in Noye (1978, p.85), which, although lacking the methodical formulation of the commonly used von Neumann stability analysis, has the advantage of providing a round-off error bound, rather than just the reassuring information that (for a time-stepping scheme) repeated progressions through the grid will not increase the error without bound.

In the analysis which follows, it is assumed that the depth h, is constant. The simplified finite difference equations are

$$u_{\ell,j} = u_{\ell-1,j} - \alpha\{v_{\ell-1,j+1} - v_{\ell-1,j}\} + \delta_3\zeta_{\ell-1,j} \tag{5.1a}$$

$$\zeta_{\ell,j} = \zeta_{\ell-1,j} - \delta_1\{\beta\, u_{\ell,j} - \tfrac{1}{2}f(v_{\ell-1,j} + v_{\ell-1,j+1})\} \tag{5.1b}$$

$$v_{\ell,j} = -\beta^{-1}\{\tfrac{1}{2}f(u_{\ell,j} + u_{\ell,j-1}) + \delta_2^{-1}(\zeta_{\ell,j} - \zeta_{\ell,j-1})\} \tag{5.1c}$$

where

$\alpha = \Delta x/\Delta y,$

$\beta = -i\omega + rh^{-1},$

$\delta_1 = 2\Delta x/g,$

$\delta_2 = 2\Delta y/g,$

$\delta_3 = 2i\omega\Delta x\, h^{-1}.$

Consider a perturbation of $\Delta\zeta_{\ell,j}$ in $\zeta_{\ell,j}$, then, from (5.1c) it can be seen that errors in $v_{\ell,j}$ and $v_{\ell,j+1}$ are produced, where

$$|\Delta v_{\ell,j}| = |\Delta v_{\ell,j+1}| = |\beta^{-1}\delta_2^{-1}\Delta\zeta_{\ell,j}|$$

From (5.1a), these in turn produce an error in $u_{\ell+1,j}$ which satisfies

$$|\Delta u_{\ell+1,j}| \le 2\alpha|\beta^{-1}\delta_2^{-1}\Delta\zeta_{\ell,j}| + |\delta_3\,\Delta\zeta_{\ell,j}|.$$

Finally, using these error bounds and (5.1b), the upper bound of the error in $\zeta_{\ell+1,j}$ is found to be

$$|\Delta\zeta_{\ell+1,j}| \le \{1 + 2\alpha\delta_1\delta_2^{-1} + \delta_1|\delta_3\beta| + |f\beta^{-1}|\delta_1\delta_2^{-1}\}|\Delta\zeta_{\ell,j}|$$

or

$$|\Delta \zeta_{\ell+1,j}| \dot{\leq} \{1 + |\theta|[1 + \phi^2]^{-\frac{1}{2}}\alpha + 2\alpha^2 + \frac{4\omega^2}{gh}(\Delta x)^2[1 + \phi^2]^{\frac{1}{2}}\}|\Delta \zeta_{\ell,j}| \qquad (5.2)$$

where

$$\theta = f\omega^{-1}$$

and

$$\phi = r\omega^{-1}h^{-1}.$$

The expression (5.2) indicates that the scheme is unstable from the point of view that an error introduced at any point is increased at each stage of the progression through the grid. However, the largest error occurs at the end-boundary in the x-direction (ℓ=m), and the error at any interior point is smaller than this end-error. Hence, by limiting m, the number of grid-steps in the x-direction, the round-off error at ℓ = m (and so for ℓ < m) can be kept less than some desirable value.

However, (5.2) also shows that the error is smaller for smaller values of Δx and α so that some compromise must be made between having a small value for Δx and a small number of grid-steps in the x-direction. Smaller values of α can be achieved with larger values of Δy, but this must increase the truncation error, as evident by (5.3), and, once again, some compromise must be made between having small round-off error propagation characteristics and having an acceptable truncation error.

If an acceptable value of Δx results in a grid which does not cover the region of interest, double precision can be used so that a useful number of significant figures can be retained at the end boundary of a larger grid; however, this increases the computer memory and time required for calculations. Nevertheless, it has been found that, for regions which are not too extensive, the amplification of round-off errors does not limit the use of Equations (3.3) in describing tidal propagation.

The results of the application of these equations to the Gulf of Carpentaria, Australia, are given in the next section. With model values Δx = 13km, α = 1/3, h ≥ 5m, $|\theta| \leq .5$ (using a latitude of 12½°S), the error amplification, given by

$$|\Delta \zeta_{\ell+1,j}| \leq 1.53|\Delta \zeta_{\ell,j}|,$$

is not too restrictive, and the maximum error in the end-values, using m = 25, was approximately 10^{-5}, which is acceptably small. For Δx=13km and α=2/3,

$$|\Delta \zeta_{\ell+1,j}| \leq 2.36|\Delta \zeta_{\ell,j}|$$

and the maximum error in the end-values was approximately 10^{-3}. In this case double precision was used and the accuracy was only just acceptable, so that no further refinement of the grid is really possible.

5.2 Consistency

The Taylor expansion approach used to obtain the finite-difference equations may be reversed for an analysis of the consistency of the scheme. Thus, for example, each variable in Equation (3.3a) may be expanded about the point (x,y), corresponding to the position where

$\zeta_{\ell-1,j}$ is calculated, to yield

$$\frac{\partial}{\partial x}(hu) + \frac{\partial}{\partial y}(hv) = i\omega\zeta - (\Delta x)^2\left[\frac{2}{3}u\frac{\partial^3 h}{\partial x^3} + \frac{h}{6}\frac{\partial^3 u}{\partial x^3} + \frac{\partial u}{\partial x}\frac{\partial^2 h}{\partial x^2} + \frac{1}{2}\frac{\partial h}{\partial x}\frac{\partial^2 u}{\partial x^2}\right]$$

$$- (\Delta y)^2\left[\frac{\partial v}{\partial y}\frac{\partial^2 h}{\partial y^2} + \frac{1}{2}\frac{\partial h}{\partial y}\frac{\partial^2 v}{\partial y^2} + \frac{2}{3}v\frac{\partial^3 h}{\partial y^3} + \frac{h}{6}\frac{\partial^3 v}{\partial y^3}\right] + O(\Delta x^3, \Delta y^3).$$

Hence, it can be seen that, as Δx, $\Delta y \to 0$, (3.3a) approaches Equation (2.2c). Similar results are obtained when the process is applied to the other two equations, indicating that the finite difference equations (3.3) are indeed consistent with the partial differential equations (2.2).

5.3 Convergence

The convergence of the system was tested using Equations (3.3) to find the solution for the exact situation as modelled analytically in Rienecker (1980) for a L-shaped region with an open boundary at the left end and different constant depths in the left-hand upper arm, the junction and the lower gulf section. The analytic results and the numerical results from grid-spacings $(\Delta x, \Delta y)$ and $(\Delta x, \Delta y/2)$ are presented in Tables 5.1 and 5.2. These show satisfactory convergence of numerical results from the model to the analytic results. The only area of disagreement is in the phase values near amphidromic points. Hence, accepting the fact that amphidromic points are singular regions in which any linear depth-integrated model is likely to be inaccurate (see Nihoul, 1977), the otherwise favourable agreement of the numerical model with the analytic model indicates that the solution provided by the linear numerical model will converge to the true solution for a more complicated boundary and bottom topography.

6. APPLICATION TO THE GULF OF CARPENTARIA

The model developed in the previous sections has been applied to the Gulf of Carpentaria and the adjacent waters. This Gulf is located in the North-Eastern part of Australia, between latitudes 10°S and 17°S and longitudes 135°E and 142°E. It has a roughly rectangular geometry and a relatively smooth bathymetry. It is a shallow area, the greatest depths being about 70m, the nearest deeper water occurring either east of Torres Strait or in the Timor Trench of the Arafura Sea. Figure 2 shows the overall geography of the area and the depth contours in metres (after Rochford, 1966).

The amplitudes and phases of the four main tidal components at various places are given in the Australian National Tide Tables, 1981. Those components studied here are

Solar Diurnal (K1) with period 23.9 hours,

Lunar Semi-Diurnal (M2) with period 12.4 hours.

An outline of the tidal features of the Gulf is given by Easton (1970), though his co-range lines are not very informative. He states that "the presence of a central amphidromic point is suggested by the diurnal and semi-diurnal components; further nodal points occur probably near Karumba and Groote Eylandt".

Cresswell's (1971) study supports the suggestion that "the diurnal wave travels clockwise around the perimeter of the Gulf, pivoting on some as-yet-unknown amphidromic point within the Gulf."

TABLE 5.1:

Comparison of the amplitudes (in m) for the M2 tide given by the analytic model in Rienecker (1980) and the linear numerical model described here. For each (x,y) value, the analytic result is given above the numerical result using a grid-spacing (Δx, Δy), followed by the numerical result using a grid-spacing of (Δx, Δy/2). The * indicates land.

.991	.970	.900	.735	.505	.275	.256	.487	.738	.949	1.096	1.165
.991	.970	.900	.739	.509	.280	.256	.482	.731	.941	1.086	1.153
.991	.970	.899	.737	.506	.275	.254	.484	.735	.945	1.091	1.159
.965	.943	.871	.703	.467	.211	.180	.437	.693	.904	1.051	1.123
.965	.944	.871	.706	.470	.215	.177	.432	.686	.895	1.041	1.111
.965	.944	.871	.704	.468	.211	.178	.436	.690	.900	1.046	1.117
.940	.916	.842	.676	.443	.180	.121	.388	.635	.837	.979	1.049
.940	.916	.842	.677	.444	.181	.118	.385	.630	.830	.969	1.038
.940	.916	.841	.676	.443	.179	.121	.388	.634	.834	.974	1.043
.916	.886	.806	.641	.419	.171	.087	.340	.567	.753	.884	.950
.916	.887	.807	.641	.418	.170	.089	.339	.565	.749	.877	.941
.916	.886	.806	.640	.417	.169	.091	.342	.568	.752	.881	.945
.899	.857	.754	.573	.375	.162	.070	.288	.489	.655	.771	.830
.899	.861	.763	.573	.370	.157	.080	.293	.491	.653	.767	.824
.899	.859	.758	.567	.370	.158	.077	.292	.491	.655	.769	.826
			.415	.301	.144	.043	.224	.395	.537	.638	.689
*	*	*	.379	.277	.127	.059	.234	.401	.540	.638	.687
			.404	.292	.137	.051	.229	.398	.538	.637	.686
			.320	.257	.147	.014	.139	.280	.399	.483	.525
*	*	*	.295	.234	.129	.007	.148	.285	.400	.482	.523
			.313	.249	.140	.019	.144	.282	.399	.482	.523
			.324	.278	.196	.092	.053	.158	.255	.325	.359
*	*	*	.308	.264	.184	.083	.054	.158	.252	.320	.353
			.319	.273	.191	.088	.058	.158	.253	.322	.355
			.376	.336	.267	.177	.082	.052	.123	.181	.209
*	*	*	.371	.331	.263	.175	.081	.045	.114	.171	.199
			.375	.335	.266	.176	.083	.054	.119	.176	.204
			.432	.395	.330	.247	.156	.069	.028	.071	.096
*	*	*	.436	.398	.335	.253	.163	.076	.014	.056	.081
			.435	.397	.333	.251	.160	.072	.028	.064	.089
			.467	.430	.367	.287	.199	.114	.041	.011	.034
*	*	*	.474	.438	.376	.298	.212	.128	.057	.012	.023
			.472	.435	.373	.293	.205	.120	.048	.011	.029

TABLE 5.2:

Comparison of the phases (in hrs) for the M2 tide given by the analytic model in Rienecker (1980) and the linear numerical model described here. For each (x,y) value, the analytic result is given above the numerical result using a grid-spacing (Δx, Δy), followed by the numerical result using a grid spacing of (Δx, $\Delta y/2$).

7.51	7.64	7.79	8.01	8.40	9.41	11.91	.66	1.04	1.23	1.33	1.38
7.51	7.64	7.77	7.99	8.38	9.37	11.83	.61	1.00	1.19	1.29	1.34
7.51	7.64	7.77	7.99	8.37	9.37	11.89	.64	1.02	1.21	1.31	1.36
7.52	7.62	7.72	7.87	8.14	8.93	12.39	1.00	1.27	1.39	1.46	1.49
7.52	7.62	7.72	7.86	8.12	8.90	12.33	.97	1.24	1.36	1.43	1.47
7.52	7.62	7.72	7.86	8.12	8.89	12.40	1.00	1.26	1.38	1.45	1.48
7.53	7.60	7.67	7.76	7.90	8.30	.70	1.35	1.50	1.57	1.62	1.65
7.53	7.60	7.66	7.75	7.89	8.27	.70	1.35	1.49	1.56	1.61	1.64
7.53	7.60	7.66	7.75	7.89	8.26	.77	1.37	1.50	1.57	1.62	1.65
7.54	7.58	7.63	7.68	7.73	7.78	1.65	1.68	1.72	1.75	1.79	1.82
7.54	7.58	7.62	7.66	7.70	7.71	1.78	1.71	1.74	1.76	1.79	1.82
7.54	7.58	7.62	7.67	7.71	7.72	1.77	1.71	1.74	1.77	1.80	1.83
7.55	7.56	7.58	7.63	7.66	7.53	2.49	1.95	1.91	1.92	1.95	1.98
7.55	7.56	7.56	7.60	7.61	7.41	2.63	2.00	1.95	1.95	1.97	2.01
7.55	7.56	7.57	7.61	7.63	7.45	2.58	1.99	1.94	1.95	1.97	2.00
			7.74	7.79	7.72	2.79	2.07	2.04	2.06	2.09	2.14
*	*	*	7.70	7.73	7.56	2.86	2.15	2.09	2.10	2.13	2.18
			7.72	7.77	7.64	2.86	2.12	2.08	2.09	2.12	2.17
			8.20	8.26	8.36	10.01	1.95	2.07	2.14	2.21	2.28
*	*	*	8.20	8.25	8.33	.43	2.06	2.14	2.20	2.26	2.33
			8.20	8.25	8.34	12.22	2.00	2.11	2.18	2.24	2.32
			8.72	8.76	8.89	9.38	.75	1.91	2.15	2.28	2.39
*	*	*	8.77	8.81	8.94	9.44	1.02	2.01	2.23	2.35	2.47
			8.75	8.78	8.91	9.43	.66	1.95	2.19	2.33	2.45
			9.01	9.02	9.09	9.26	9.89	.75	1.99	2.29	2.48
*	*	*	9.07	9.08	9.13	9.29	9.83	.78	2.12	2.42	2.61
			9.06	9.07	9.13	9.30	9.97	.48	2.04	2.37	2.57
			9.10	9.10	9.11	9.16	9.29	9.72	.66	2.22	2.56
*	*	*	9.16	9.15	9.15	9.18	9.25	9.47	12.28	2.55	2.88
			9.16	9.15	9.16	9.20	9.30	9.69	12.20	2.42	2.76
			9.10	9.07	9.05	9.03	9.03	9.05	9.17	2.20	2.74
*	*	*	9.16	9.13	9.10	9.06	9.01	8.94	8.74	6.93	4.20
			9.16	9.13	9.10	9.07	9.05	9.01	8.92	4.90	3.50

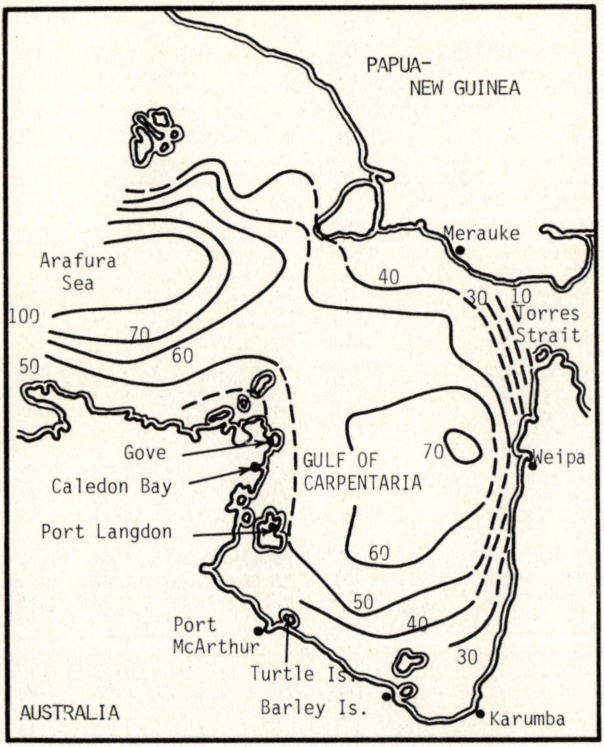

FIGURE 2 : The Gulf of Carpentaria and adjacent waters. The depth contours are shown in metres (after Rochford, 1966).

The boundary configuration of the numerical model can be seen in Figure 3. Δx is taken to be 13km and Δy to be 19.5km. The linear friction parameter was chosen to be .0015 for the M2 tide and .0003 for the K1 tide. Webb used the values .0010 and .0003 respectively.

The input along the open boundary for the numerical model was obtained from the analytic model described in Rienecker (1980), there being no data available across the open boundary. It could have been possible to determine input values from co-tidal and co-range charts as given by Easton (1970), but such data would be interpolations on diagrams which are themselves obtained by interpolation and extrapolation, and hence the input was not likely to be very accurate.

Figure 3 shows the response of the Gulf for the M2 tide as predicted by this linear numerical model. The amplitudes and phases for selected positions, as given in the Australian National Tide Tables for 1981 and the Admiralty Tide Tables 1978, Vol.III are shown on the Figure. The movement of the predicted tide around the perimeter of the Gulf agrees with observation in that the tide at Weipa lags behind that at Karumba, the change of phase from Bayley Island along the South-Western section of the Gulf is very slow and the tide in the Northern region around Caledon Bay lags behind that at Port McArthur in the Sout

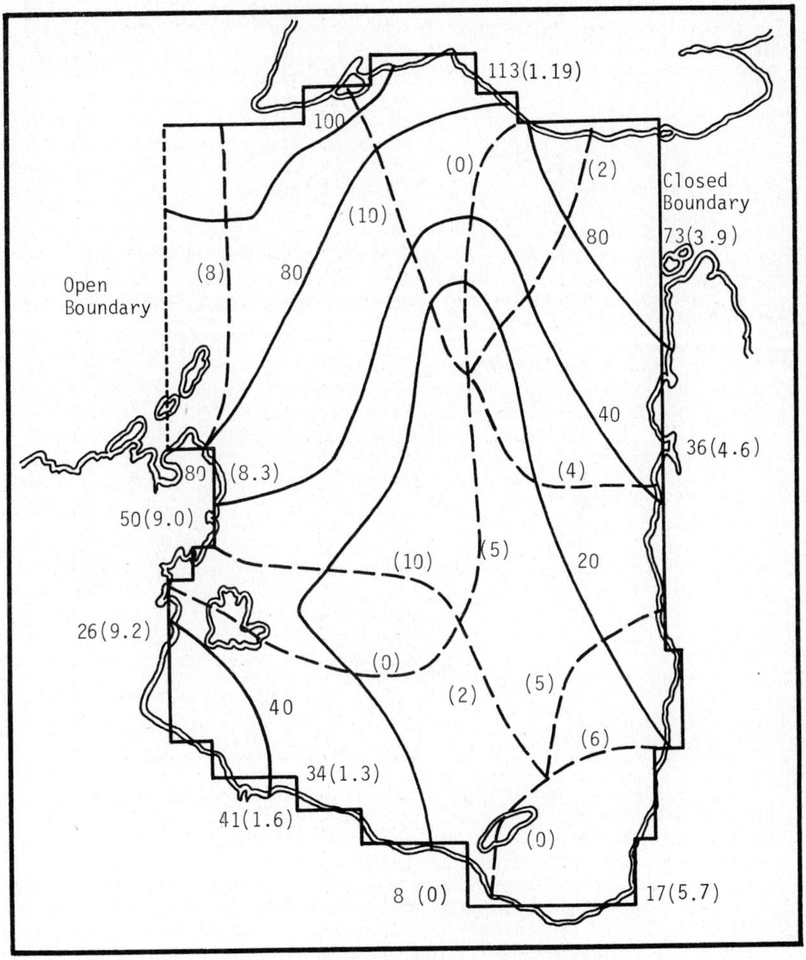

FIGURE 3 : The co-amplitude (cm) —— and co-phase (hr) --- lines for the M2 tide component predicted by the linear numerical model. Also shown are the recorded coastal amplitudes with the phases shown in brackets.

As with the other models of the Gulf, agreement with observation is worst along the Eastern boundary, possibly because of poor boundary conditions across Torres Strait and also possibly because the coastal features near, for example, Weipa are not represented in enough detail. The amplitude at Weipa is too large and the phase too small. At Karumba, the amplitude is correct but the phase is about 2 hours too large; the variation in phase is very rapid in this South-Eastern corner. The phases and amplitudes along the Southern and Western boundaries are in general good agreement with observation. The model predicts 3 amphidromic points as compared with 2 by Church and Forbes (at Groote Eylandt in the West and Mornington Island in the South-East) and 1 by Webb (at Mornington Island), though all models predict the central region to be of low amplitude and rapidly varying phase. The phase variation at the open boundary is consistent with the tide entering the Gulf region as a progressive wave and is similar to that found by Webb. The variation used by Church and Forbes is different, essentially being found by trial and error so that results were in accord with observation.

The response of the system to forcing by the K1 tide is shown in Figure 4. There is a single amphidromic point about which the rotation is clockwise, in keeping with observations. The position of this amphidrome is the same as found by Webb and slightly East of that predicted by Church and Forbes.

Comparison with data shows that the amplitude at Weipa is about 10cm too small but the phase is correct. The amplitude at Karumba is about 16cm too small. The phase here and along the Southern boundary is too small. The amplitudes along the Southern and Western boundaries are correct, except at Port Langdon where it is about 6cm too large.

Since no data is yet available away from the mainland or islands, there is no conclusive evidence as to the quantitative accuracy of any model in the interior of the Gulf. However, there is good agreement between this and other models, such as Williams (1972), Webb (1981) and Church and Forbes(1981) as far as the main features of the response of the Gulf to tidal forcing. The discrepancies with regard to specific observed tidal phenomena may possibly be due to lack of detail in coastal boundaries and in the bottom topography near the coast.

Another factor limiting the accuracy of the numerical models may still be the approximation to the real input along the open boundary. Further measurements in this area and also in Torres Strait would give better agreement between model results and recorded data within the Gulf. Additional data from the South-West of the Gulf is needed to check the positions and existence of amphidromic points for the M2 tide.

7. CONCLUSION

For the Gulf of Carpentaria, which has complicated tidal responses (some regions have dominant K1 tides, other dominant M2 and others mixed tides) highly accurate results will probably only be achieved with a fine grid, non-linear model with all major constituents interacting. However, this will be time-consuming and simpler models are needed to test the effects of variations in input, boundaries, grid-size and other parameters. The model presented here is ideal for providing such insights. Its usefulness lies in its simplicity, small storage and small running time requirements.

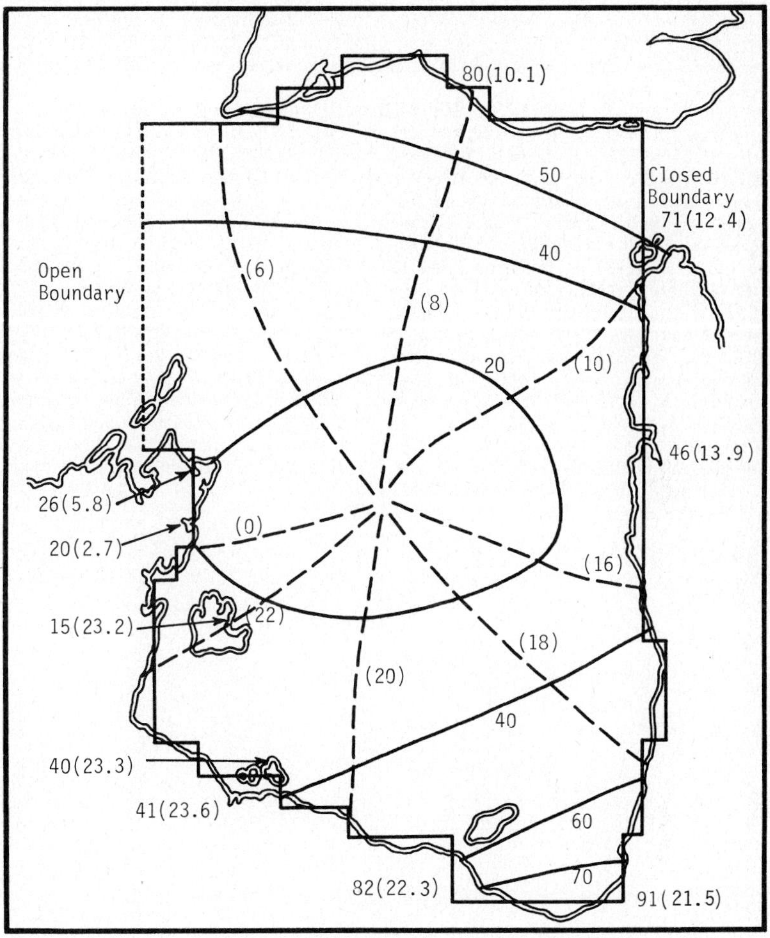

FIGURE 4 : The co-amplitude (cm) —— and co-phase (hr) --- lines forthe K1 tide component predicted by the linear numerical model. Also shown are the recorded coastal amplitudes with the phases shown in brackets and the boundaries of the numerical model.

8. REFERENCES

Buchwald, V.T. and Williams, N.V. (1975): *Rectangular Resonators on Infinite and Semi-Infinite Channels.* J.Fluid Mech. 67, 497-511.
Church, J.A. and Forbes, A. (1981): *Tidal Circulation in the Gulf of Carpentaria.* Aust.J.Mar.Fresh.Res. (in press).
Easton, A.K. (1970): *The Tides of the Continent of Australia.* Horace Lamb Centre for Oceanog.Res., Res.Paper No. 37.
Miller, G.R. (1966): *The Flux of Tidal Energy Out of the Deep Oceans.* J.Geophys.Res.71, 2485-2489.
Nihoul, J.C.J. (1975): *Hydrodynamic Models* in *Modelling of Marine Systems.* (Ed. J.C.J. Nihoul), Elsevier, Amsterdam, 292 pp.

Nihoul, J.C.J. (1977): *Three-dimensional model of Tides and Storm Surges in a Shallow Well-mixed Continental Sea.* Dyn.Atmos.Oceans 2, 29-47.
Noye, B.J. (1978): *An Introduction to Finite Difference Techniques* in *Numerical Simulation of Fluid Motion* (Ed. J. Noye), North-Holland Publ.Co., Amsterdam, 580pp.
Noye, B.J. and Tronson, K. (1978): *Finite Difference Techniques Applied to the Simulation of Tides and Currents in Gulfs* in *Numerical Simulation of Fluid Motion* (Ed. J. Noye), North-Holland Publ.Co., Amsterdam, 580pp.
Platzman, G.W. (1958): *A Numerical Computation of the Surge of 26th June, 1954 on Lake Michigan.* Geophysica 6, 407-438.
Rienecker, M.M. (1980): *Tidal Propagation in the Gulf of Carpentaria.* Ph.D. dissertation, University of Adelaide, 143pp.
Roache, P.J. (1972): *Computational Fluid Dynamics.* Hermosa Publishers, Alberquerque. 446pp.
Rochford, D.J. (1966): *Some Hydrological Features of the Eastern Arafura Sea and the Gulf of Carpentaria in August 1964.* Aust.J.Mar.Fresh.Res. 17, 31-60.
Webb, D.J. (1981): *A Numerical Model of the Tides in the Gulf of Carpentaria and the Arafura Sea.* Aust.J.Mar.Fresh.Res. 32, 31-44.
Williams, N.V. (1972): *The Application of Resonators and Other Methods to Problems in Oceanography.* Ph.D. dissertation, University of New South Wales.

The Admiralty Charts and the Australian National Tide Tables, published by the Hydrographer, R.A.N. were also used.

NUMERICAL SOLUTIONS OF
PARTIAL DIFFERENTIAL EQUATIONS
J. Noye (Editor)

A THREE DIMENSIONAL TIDAL MODEL FOR A SHALLOW GULF

John Noye

The University of Adelaide
South Australia

Robert May

Caulfield Institute of Technology
Melbourne
Australia

Michael Teubner

Northern Technical Services
Newport Beach, California
U.S.A.

Few numerical methods are available for calculating the three-dimensional description of sea circulation due to external tide forcing, and most of these apply only where the maximum variation of the sea-surface about mean sea level is small compared to the sea depth. By carrying out two transformations on the governing differential equations, one in the vertical and one in the horizontal, a new technique for calculating three-dimensional tidal currents and tide heights has been developed. This method applies to very shallow waters, where the tidal range is relatively large compared with the depth of the sea, as well as to deep waters.

1. INTRODUCTION

A detailed knowledge of the general circulation of estuaries and coastal seas is needed for many reasons, such as the tracking of oil slicks, industrial wastes and other pollutants which find their way into the sea. Until recently, prediction of the general circulation was achieved by solving the depth-integrated forms of the partial differential equations which govern the propagation of tides and storm surges. This method (see, for example, Heaps, 1969) was found to accurately predict sea-level variations but provided no information about the vertical structure of the flow.

Recently predictions of the horizontal velocity at any depth have been made by Heaps (1973), Leendertse et al (1973) and Caponi and Faller (1973). Heaps used finite difference techniques in the horizontal combined with spectral techniques in the vertical to solve linearised forms of the governing equations; Leendertse treated the sea as a series of layers, solving the non-linear depth-integrated equations by finite difference methods for each of these layers; and Caponi and Faller used a variation of the Marker and Cell technique to solve the complete three-dimensional tidal equations.

In all these three-dimensional models except that of Caponi and Faller (1973), it is necessary for the maximum variation to the sea surface about mean sea level to be small relative to the sea depth. Therefore they do not apply to the very shallow waters of coastal inlets or heads of gulfs, where the tidal range is large relative to the depth of water.

A new method of determining the three-dimensional tidal currents in very shallow homogeneous seas is described in this paper. The Reynolds' averaged form of the Navier-Stokes equations for a viscous fluid and the continuity equation for an incompressible fluid are transformed so that, at each instant, the irregular body of water in the sea is mapped onto a rectangular prism by means of finite difference techniques. A semi-implicit method is used in the vertical direction where the grid-spacing is small, in order to permit reasonable sized time steps to be used without the development of computational instability.

The method is particularly suitable for shallow gulfs, and has been applied to Spencer Gulf, South Australia. Computed tide elevations have been found to agree well with available data, while predicted current profiles have the same qualitative properties as tidal currents recorded in other shallow seas.

2. THE THREE-DIMENSIONAL TIDE EQUATIONS

The complete three-dimensional equations governing long wave motion have been developed by Nihoul (1975) and can be written as:

The equation of mass conservation for an incompressible fluid,

$$\frac{\partial u}{\partial x} + \frac{\partial v}{\partial y} + \frac{\partial w}{\partial z} = 0. \tag{2.1}$$

The equations of momentum conservation,

$$\frac{Du}{Dt} = -\frac{1}{\rho}\frac{\partial P}{\partial x} + \frac{1}{\rho}\frac{\partial \tau_{xx}}{\partial x} + \frac{1}{\rho}\frac{\partial \tau_{yx}}{\partial y} + \frac{1}{\rho}\frac{\partial \tau_{zx}}{\partial z} + fv, \tag{2.2}$$

$$\frac{Dv}{Dt} = -\frac{1}{\rho}\frac{\partial P}{\partial y} + \frac{1}{\rho}\frac{\partial \tau_{xy}}{\partial x} + \frac{1}{\rho}\frac{\partial \tau_{yy}}{\partial y} + \frac{1}{\rho}\frac{\partial \tau_{zy}}{\partial z} - fu, \tag{2.3}$$

$$\frac{Dw}{Dt} = -g - \frac{1}{\rho}\frac{\partial P}{\partial z} + \frac{1}{\rho}\frac{\partial \tau_{xz}}{\partial x} + \frac{1}{\rho}\frac{\partial \tau_{yz}}{\partial y} + \frac{1}{\rho}\frac{\partial \tau_{zz}}{\partial z} . \tag{2.4}$$

The symbols used in the above equations are:

- x,y,z are rectangular cartesian coordinates, with the x and y axes lying in the plane of the undisturbed sea surface and the z-axis directed vertically upwards,
- u,v,w are the corresponding velocity components,
- P is the pressure,
- ρ is the density (assumed constant),
- g is the acceleration due to gravity,
- f is the Coriolis parameter, which has the form $2\Omega \cos \theta$, where Ω is the angular velocity of the earth, and θ is the co-latitude,
- D/Dt is the operator of differentiation following the motion, namely $D/Dt \equiv \partial/\partial t + u\, \partial/\partial x + v\, \partial/\partial y + w\, \partial/\partial z$, and
- τ is the Reynolds stress, the subscripts indicating the direction of the stress and the orientation of the plane on which it acts: for example, τ_{xy} represents the Reynolds shear stress acting on the plane x = constant in the y-direction.

Writing the Reynolds' stress in Equations (2.2) and (2.3) in Boussinesq form, namely

$$\tau_{zx} = \rho N_z \, \partial u/\partial z, \; \tau_{zy} = \rho N_z \, \partial v/\partial z, \text{ etc.,} \tag{2.5}$$

where N_x, N_y, N_z are the appropriate eddy viscosity coefficients, yields

$$\frac{\partial u}{\partial t} + u\frac{\partial u}{\partial x} + v\frac{\partial u}{\partial y} + w\frac{\partial u}{\partial z} = -\frac{1}{\rho}\frac{\partial P}{\partial x} + \frac{\partial}{\partial x}\{N_x\frac{\partial u}{\partial x}\} + \frac{\partial}{\partial y}\{N_y\frac{\partial u}{\partial y}\} + \frac{\partial}{\partial z}\{N_z\frac{\partial u}{\partial z}\} + fv, \tag{2.6}$$

$$\frac{\partial v}{\partial t} + u\frac{\partial v}{\partial x} + v\frac{\partial v}{\partial y} + w\frac{\partial v}{\partial z} = -\frac{1}{\rho}\frac{\partial P}{\partial y} + \frac{\partial}{\partial x}\{N_x\frac{\partial v}{\partial x}\} + \frac{\partial}{\partial y}\{N_y\frac{\partial v}{\partial y}\} + \frac{\partial}{\partial z}\{N_z\frac{\partial v}{\partial z}\} - fu. \tag{2.7}$$

Also, Equation (2.4) can be simplified by assuming that the vertical acceleration Dw/Dt and the gradients of the Reynolds' stresses are small compared with g, so that it can be replaced by the hydrostatic approximation

$$\frac{\partial P}{\partial z} + \rho g = 0, \tag{2.8}$$

which yields, on integration over depth,

$$P = P_0 + \rho g(\zeta - z), \tag{2.9}$$

where P_0 is the atmospheric pressure and $\zeta(x,y,t)$ is the elevation of the sea surface above mean sea level. This equation is used to eliminate P from (2.6) and (2.7).

3. THE TRANSFORMATION

To enable Equations (2.1), (2.6-7) to be solved more easily using finite differences, they are first non-dimensionalised with regard to width and total water depth $h = d + \zeta$, where $d(x,y)$ is the distance of the sea floor below mean sea level.

Consider the transformation

$$\underset{\text{original system}}{(x,y,z,t)} \rightarrow \underset{\text{transformed system}}{(\chi,\lambda,\eta,T)}$$

in which the relationship between the systems is defined by

$$\chi = x, \; \lambda = (y-b)/B, \; \eta = (\zeta - z)/h, \; T = t, \tag{3.1a,b,c,d}$$

where b, B are defined in Figure 3.1, and depend only on x.

A circumflex is used to distinguish between the two coordinate systems - if ϕ is a function of position and time then

$$\phi(x,y,z,t) = \hat{\phi}(\chi,\lambda,\eta,T).$$

Associated with the new coordinate λ and η we define $\nu = D\lambda/DT$ and $\omega = D\eta/DT$. These are the components of a scaled form of velocity normal to the planes λ = constant and η = constant, respectively, and can be shown to be given by

$$\nu = -\frac{u}{B}\left(\frac{db}{dx} + \lambda\frac{dB}{dx}\right) + \frac{v}{B}, \tag{3.2}$$

$$\omega = \frac{1}{h}\left\{\frac{\partial\zeta}{\partial t} - \eta\frac{\partial h}{\partial t} + u\left(\frac{\partial\zeta}{\partial x} - \eta\frac{\partial h}{\partial x}\right) + v\left(\frac{\partial\zeta}{\partial y} - \eta\frac{\partial h}{\partial y}\right) - w\right\}. \tag{3.3}$$

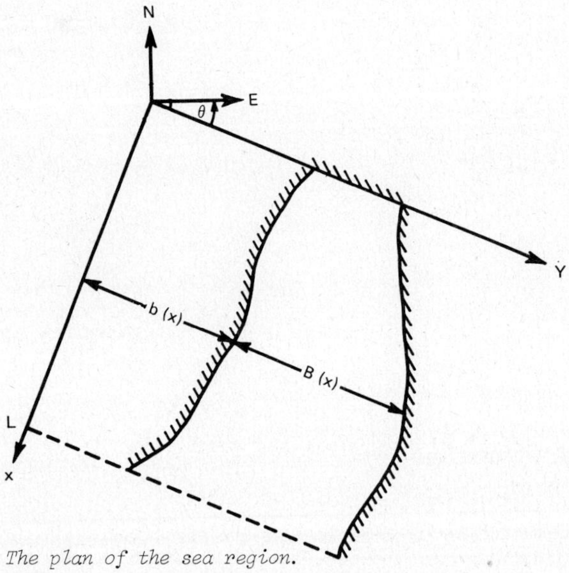

Figure 3.1 : The plan of the sea region.

4. THE TRANSFORMED EQUATIONS

4.1 Continuity Equation

Using the transformation defined by (3.1a-d), the continuity equation (2.1) becomes

$$\frac{\partial\hat{\zeta}}{\partial T} + \frac{1}{\hat{B}}\frac{\partial(\hat{B}\hat{h}\hat{u})}{\partial\chi} + \frac{\partial(\hat{h}\nu)}{\partial\lambda} + \hat{h}\frac{\partial\omega}{\partial\eta} = 0. \tag{4.1}$$

4.2 x-Momentum Equation

When the x-component of the momentum equation (Equation (2.6)) is transformed by means of (3.1), the terms involving the horizontal eddy viscosity become:

$$\frac{\partial}{\partial x}\{N_x\frac{\partial u}{\partial x}\} = \{\frac{\partial}{\partial\chi} - \frac{1}{\hat{B}}\frac{\partial\hat{\beta}}{\partial\chi}\frac{\partial}{\partial\lambda} + \frac{1}{\hat{h}}[\frac{\partial\hat{\gamma}}{\partial\chi} - \frac{1}{\hat{B}}\frac{\partial\hat{\gamma}}{\partial\lambda}\frac{\partial\hat{\beta}}{\partial\chi}]\frac{\partial}{\partial\eta}\}$$

$$\hat{N}_x\{\frac{\partial\hat{u}}{\partial\chi} - \frac{1}{\hat{B}}\frac{\partial\hat{u}}{\partial\lambda}\frac{\partial\hat{\beta}}{\partial\chi} + \frac{1}{\hat{h}}[\frac{\partial\hat{\gamma}}{\partial\chi} - \frac{1}{\hat{B}}\frac{\partial\hat{\gamma}}{\partial\lambda}\frac{\partial\hat{\beta}}{\partial\chi}]\frac{\partial\hat{u}}{\partial\eta}\}$$

$$= \frac{\partial}{\partial\chi}\{\hat{N}_x\frac{\partial\hat{u}}{\partial\chi}\} + \text{other terms}, \tag{4.2}$$

where

$$\hat{\beta} = \hat{b} + \lambda\hat{B}, \tag{4.3a}$$

$$\hat{\gamma} = \hat{\zeta} - \eta\hat{h}. \tag{4.3b}$$

The "other terms" involved in Equation (4.2) are generally small compared with $\partial/\partial\chi\ \{\hat{N}_x\ \partial\hat{u}/\partial\chi\}$, and are neglected (see Teubner, 1976).

Similarly

$$\frac{\partial}{\partial y}\{N_y\frac{\partial u}{\partial y}\} = \{\frac{1}{\hat{B}}\frac{\partial}{\partial\lambda} + \frac{1}{\hat{B}\hat{h}}\frac{\partial\hat{\gamma}}{\partial\lambda}\frac{\partial}{\partial\eta}\}\{\hat{N}_y\,[\frac{1}{\hat{B}}\frac{\partial\hat{u}}{\partial\lambda} + \frac{1}{\hat{B}\hat{h}}\frac{\partial\hat{\gamma}}{\partial\lambda}\frac{\partial\hat{u}}{\partial\eta}]\}$$

$$= \frac{1}{\hat{B}^2}\frac{\partial}{\partial\lambda}\{\hat{N}_y\frac{\partial\hat{u}}{\partial\lambda}\} + \text{other terms.} \tag{4.4}$$

Once again, the "other terms" are neglected. Thus the x-momentum equation becomes

$$\frac{\partial\hat{u}}{\partial T} + \hat{u}\frac{\partial\hat{u}}{\partial\chi} + \nu\frac{\partial\hat{u}}{\partial\lambda} + \omega\frac{\partial\hat{u}}{\partial\eta} = -g\frac{\partial\hat{\zeta}}{\partial\chi} + \frac{g}{\hat{B}}\frac{\partial\hat{\zeta}}{\partial\lambda}\frac{\partial\hat{\beta}}{\partial\chi} + \frac{\partial}{\partial\chi}\{\hat{N}_x\frac{\partial\hat{u}}{\partial\chi}\}$$

$$+ \frac{1}{\hat{B}^2}\frac{\partial}{\partial\lambda}\{\hat{N}_y\frac{\partial\hat{u}}{\partial\lambda}\} + \frac{1}{\hat{h}^2}\frac{\partial}{\partial\eta}\{\hat{N}_z\frac{\partial\hat{u}}{\partial\eta}\} + \hat{f}\hat{v}. \tag{4.5}$$

The pressure term $-\frac{1}{\rho}\frac{\partial P}{\partial x}$ in Equation (2.6) has been simplified by using Equation (2.9).

4.3 y-Momentum Equation

Once again neglecting the "other terms" associated with the transformations of the eddy viscosity terms $\partial/\partial x\ \{N_x\ \partial v/\partial x\}$ and $\partial/\partial y\ \{N_y\ \partial v/\partial y\}$, the y-momentum equation (2.7) becomes

$$\frac{\partial\hat{v}}{\partial T} + \hat{u}\frac{\partial\hat{v}}{\partial\chi} + \nu\frac{\partial\hat{v}}{\partial\lambda} + \omega\frac{\partial\hat{v}}{\partial\eta} = -\frac{g}{\hat{B}}\frac{\partial\hat{\zeta}}{\partial\lambda} + \frac{\partial}{\partial\chi}\{\hat{N}_x\frac{\partial\hat{v}}{\partial\chi}\}$$

$$+ \frac{1}{\hat{B}^2}\frac{\partial}{\partial\lambda}\{\hat{N}_y\frac{\partial\hat{v}}{\partial\lambda}\} + \frac{1}{\hat{h}^2}\frac{\partial}{\partial\eta}\{\hat{N}_z\frac{\partial\hat{v}}{\partial\eta}\} - \hat{f}\hat{u}. \tag{4.6}$$

The pressure term has again been simplified using Equation (2.9).

4.4 Simplification of the Continuity Equation

Equation (4.1) can be integrated over depth to give

$$\int_0^1 \frac{\partial\hat{\zeta}}{\partial T}\,d\eta + \frac{1}{\hat{B}}\int_0^1 \frac{\partial(\hat{B}\hat{h}\hat{u})}{\partial\chi}\,d\eta + \int_0^1 \frac{\partial(\hat{h}\nu)}{\partial\lambda}\,d\eta + \hat{h}\int_0^1 \frac{\partial\omega}{\partial\eta}\,d\eta = 0. \tag{4.7}$$

From the definition of ω (Equation (3.3)), it can be seen that $\omega = 0$ at $\eta = 0$ and $\eta = 1$, so Equation (4.7) becomes

$$\frac{\partial\hat{\zeta}}{\partial T} + \frac{1}{\hat{B}}\frac{\partial}{\partial\chi}\{\hat{B}\hat{h}\int_0^1 \hat{u}\,d\eta\} + \frac{\partial}{\partial\lambda}\{\hat{h}\int_0^1 \nu\,d\eta\} = 0. \tag{4.8}$$

Subtracting Equation (4.1) from this leads to

$$\frac{\partial\omega}{\partial\eta} = \frac{1}{\hat{h}\hat{B}}\frac{\partial}{\partial\chi}\{\hat{B}\hat{h}(\int_0^1 \hat{u}\,d\eta - \hat{u})\} + \frac{\partial}{\partial\lambda}\{\hat{h}(\int_0^1 \nu\,d\eta - \nu)\}. \tag{4.9}$$

Equations (4.8) and (4.9) are the two forms of the continuity equation used to solve for $\hat{\zeta}$ and ω respectively.

4.5 Velocity Relationship

Because the velocity $\hat{v}$ has been used in place of ν in (4.6) it is necessary to specify a relationship between them. It follows from Equation(3.2) that

$$\nu = \frac{1}{\hat{B}}(\hat{v} - \hat{u}\frac{\partial\hat{\beta}}{\partial\chi}). \qquad (4.10)$$

5. THE BOUNDARY CONDITIONS

The following boundary conditions are used in the solution of the three-dimensional tidal equations developed in the previous section:

On the sea floor a no-slip condition is used, giving

$$\hat{u} = \hat{v} = \omega = \nu = 0 \text{ at } \eta = 1. \qquad (5.1)$$

There can be no flow through any of the closed boundaries, so

$$\omega = 0 \text{ at } \eta = 0, \quad \nu = 0 \text{ at } \lambda = 0,1 \quad \text{and} \quad \hat{u} = 0 \text{ at } \chi = 0. \qquad (5.2)$$

Assuming no wind stress acts on the surface then

$$\frac{\partial\hat{u}}{\partial\eta} = \frac{\partial\hat{v}}{\partial\eta} = 0 \text{ at } \eta = 0. \qquad (5.3)$$

6. THE FINITE DIFFERENCE GRID AND APPROXIMATIONS

The finite difference grid used to solve the three-dimensional equations developed in Section 4 is an adaptation of the two-dimensional grid used by Heaps (1969) and is shown in Figure 6.1a. The elevation $\hat{\zeta}$ and the velocity components $\hat{u}$, $\hat{v}$, ν and ω are evaluated at separate points along alternate lines parallel to the coordinate axes, elevations at 0 points and velocities at X points. This diagram represents the "surface" or χ-λ plane, while Figure 6.1b shows a "section" parallel to the χ-η plane.

Using the grid described above, the three-dimensional equations derived in Section 4 have been approximated by finite difference equations. Forward differences have been used for all time derivatives, and centred differences have been used for all space derivatives. The circumflex has been omitted from this point, since the dependent variables are clearly functions of the new variables χ, λ, η and T.

(a) The depth-integrated equation of continuity (4.8):

$$\frac{1}{\Delta T}(\zeta^*_{i,j} - \zeta_{i,j}) + \frac{1}{2B_i\Delta\chi}\{\bar{B}_{i+1}(h_{i+1,j+1}U_{i+1,j+1} + h_{i+1,j}U_{i+1,j})$$

$$- \bar{B}_i(h_{i,j+1}U_{i,j+1} + h_{i,j}U_{i,j})\} + \frac{1}{2\Delta\lambda}\{(h_{i,j+1}V_{i,j+1} + h_{i+1,j+1}V_{i+1,j+1})$$

$$- (h_{i,j}V_{i,j} + h_{i+1,j}V_{i+1,j})\} = 0, \qquad (6.1)$$

where

$\phi_{i,j}$ is an approximation to $\phi(i\Delta\chi, j\Delta\lambda, T)$ for a general function $\phi(\chi,\lambda,T)$;

$\phi^*_{i,j}$ is an approximation to $\phi(i\Delta\chi, j\Delta\lambda, T+\Delta T)$;

$\Delta\chi, \Delta\lambda, \Delta T$ are the uniform grid spacings for the independent variables χ, λ, T respectively;

B_i is the width of the gulf across the i-th elevation line;

$\bar{B}_i$ $= \frac{1}{2}(B_i + B_{i-1})$, $i \neq 1$,

$= \frac{1}{2}(3B_1 - B_2)$, $i=1$;

$U_{i,j}, V_{i,j}$ are approximations to $\int_0^1 u \, d\eta$, $\int_0^1 v \, d\eta$ at the i,j velocity point, where Simpson's rule is used to perform the integration;

$h_{i,j}$ is the total depth of water at the velocity point (i,j).

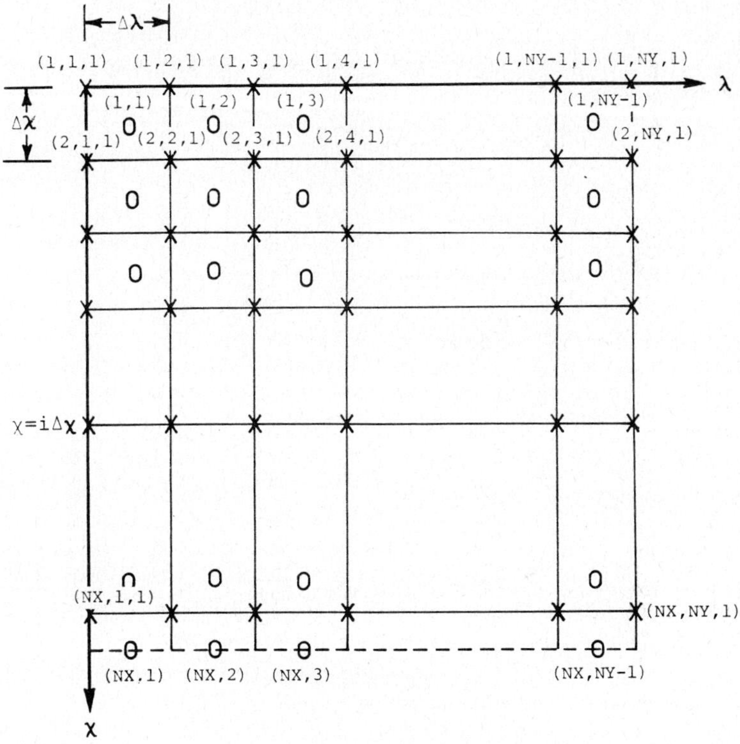

Figure 6.1a : Computation Grid in the plane η = 0

Equation (6.1) can be rearranged to give an explicit equation for the unknown $\zeta^*_{i,j}$ as follows:

$$\zeta^*_{i,j} = \zeta_{i,j} - \frac{\Delta T}{2\Delta\chi B_i}\{\bar{B}_{i+1}(h_{i+1,j+1}U_{i+1,j+1} + h_{i+1,j}U_{i+1,j}) - \bar{B}_i(h_{i,j+1}U_{i,j+1} + h_{i,j}U_{i,j})\} - \frac{\Delta T}{2\Delta\lambda}\{(h_{i+1,j+1}V_{i+1,j+1} + h_{i,j+1}V_{i,j+1}) - (h_{i+1,j}V_{i+1,j} + h_{i,j}V_{i,j})\} . \quad (6.2)$$

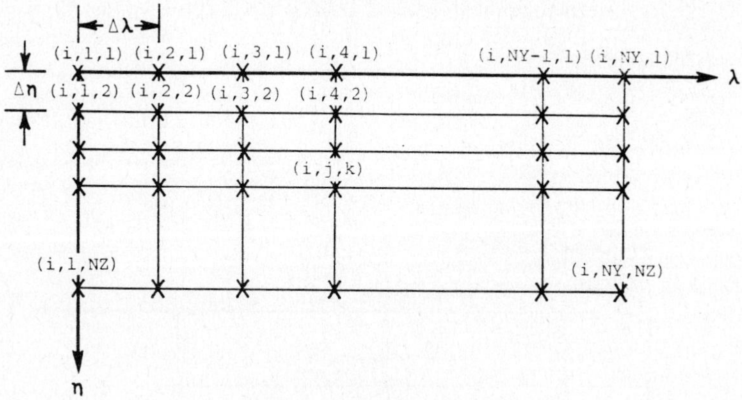

Figure 6.1b : Computation Grid in the plane $\chi = i\Delta\chi$

(b) The transformed x-momentum equation (4.5) becomes:

$$\frac{u^*_{i,j,k}-u_{i,j,k}}{\Delta T} + \frac{u_{i,j,k}}{2\Delta\chi}(u_{i+1,j,k}-u_{i-1,j,k})$$

$$+ \frac{v_{i,j,k}}{2\Delta\lambda}(u_{i,j+1,k}-u_{i,j-1,k}) + \frac{\omega_{i,j,k}}{2\Delta\eta}(u_{i,j,k+1}-u_{i,j,k-1})$$

$$= \frac{-g}{2\Delta\chi}(\zeta^*_{i,j}+\zeta^*_{i,j-1}-\zeta^*_{i-1,j}-\zeta^*_{i-1,j-1}) + \frac{gD_{i,j}}{2\bar{B}_i\,\Delta\lambda}(\zeta^*_{i,j}+\zeta^*_{i-1,j}-\zeta^*_{i,j-1}-\zeta^*_{i-1,j-1})$$

$$+ \frac{N_x}{(\Delta\chi)^2}(u_{i+1,j,k}-2u_{i,j,k}+u_{i-1,j,k}) + \frac{N_y}{(\bar{B}_i\,\Delta\lambda)^2}(u_{i,j+1,k}-2u_{i,j,k}+u_{i,j-1,k})$$

$$+ \frac{1}{(h^*_{i,j}\,\Delta\eta)^2}[M_{i,j,k}(u^*_{i,j,k+1}-u^*_{i,j,k})-M_{i,j,k-1}(u_{i,j,k}-u_{i,j,k-1})]$$

$$+ f\,v_{i,j,k}, \qquad (6.3)$$

where

$$M_{i,j,k} = N_z\,(i\Delta\chi, j\Delta\lambda, (k+\tfrac{1}{2})\Delta\eta, T),$$

$$D_{i,j} = \frac{b_i-b_{i-1}}{\Delta\chi} + (j-1)\frac{\Delta\lambda}{\Delta\chi}(B_i-B_{i-1}), \qquad (6.4)$$

b_i is the distance of the western boundary of the gulf from the x-axis at the i-th elevation line,

$u_{i,j,k}$ is an approximation to $u(i\Delta\chi, j\Delta\lambda, k\Delta\eta, T)$, and

$u^*_{i,j,k}$ is an approximation to $u(i\Delta\chi, j\Delta\lambda, k\Delta\eta, T+\Delta T)$.

The $\zeta^*_{i,j}$ are the values previously determined from Equation (6.2) and so are known. Note that the vertical eddy viscosity term has been approximated by a semi-implicit technique which permits the use of relatively large time steps without loss of stability of the finite difference system (see Roberts and Weiss, 1966).

Rearranging Equation (6.3) leads to

$$\begin{aligned} u^*_{i,j,k} = & \{\frac{1}{\Delta T} + \frac{M_{i,j,k}}{(h^*_{i,j}\Delta\eta)^2}\}^{-1}\{u_{i,j,k}[\frac{1}{\Delta T} - \frac{M_{i,j,k-1}}{(h^*_{i,j}\Delta\eta)^2}] \\ & - \frac{u_{i,j,k}}{2\Delta\chi}(u_{i+1,j,k} - u_{i-1,j,k}) - \frac{\nu_{i,j,k}}{2\Delta\lambda}(u_{i,j+1,k} - u_{i,j-1,k}) \\ & - \frac{\omega_{i,j,k}}{2\Delta\eta}(u_{i,j,k+1} - u_{i,j,k-1}) - \frac{g}{2\Delta\chi}(\zeta^*_{i,j} + \zeta^*_{i,j-1} - \zeta^*_{i-1,j} - \zeta^*_{i-1,j-1}) \\ & + \frac{gD_{i,j}}{2\bar{B}_i\Delta\lambda}(\zeta^*_{i,j} + \zeta^*_{i-1,j} - \zeta^*_{i,j-1} - \zeta^*_{i-1,j-1}) \\ & + \frac{N_x}{(\Delta\chi)^2}(u_{i+1,j,k} - 2u_{i,j,k} + u_{i-1,j,k}) + \frac{N_y}{(\bar{B}_i\Delta\lambda)^2}(u_{i,j+1,k} - 2u_{i,j,k} + u_{i,j-1,k}) \\ & + (h^*_{i,j}\Delta\eta)^{-2}(M_{i,j,k}u^*_{i,j,k+1} + M_{i,j,k-1}u_{i,j,k-1}) + f\,v_{i,j,k}\}. \end{aligned} \tag{6.5}$$

If the calculation is performed in the order of decreasing k, $u^*_{i,j,k+1}$ is known when $u^*_{i,j,k}$ is calculated. All other values on the right-hand side are known, so u can be found throughout the whole grid at time T+ΔT.

(c) The transformed y-momentum Equation (4.6): The terms in this equation have been approximated in a similar way to the transformed x-momentum equation, resulting in

$$\begin{aligned} v^*_{i,j,k} = & \{\frac{1}{\Delta T} + \frac{M_{i,j,k}}{(h^*_{i,j}\Delta\eta)^2}\}^{-1}\{v_{i,j,k}[\frac{1}{\Delta T} - \frac{M_{i,j,k-1}}{(h^*_{i,j}\Delta\eta)^2}] \\ & - \frac{u^*_{i,j,k}}{2\Delta\chi}(v_{i+1,j,k} - v_{i-1,j,k}) - \frac{\nu_{i,j,k}}{2\Delta\lambda}(v_{i,j+1,k} - v_{i,j-1,k}) \\ & - \frac{\omega_{i,j,k}}{2\Delta\eta}(v_{i,j,k+1} - v_{i,j,k-1}) - \frac{g}{2\bar{B}_i\Delta\lambda}(\zeta^*_{i,j} + \zeta^*_{i-1,j} - \zeta^*_{i,j-1} - \zeta^*_{i-1,j-1}) \\ & + \frac{N_x}{(\Delta\chi)^2}(v_{i+1,j,k} - 2v_{i,j,k} + v_{i-1,j,k}) + \frac{N_y}{(\bar{B}_i\Delta\lambda)^2}(v_{i,j+1,k} - 2v_{i,j,k} + v_{i,j-1,k}) \\ & + \frac{1}{(h^*_{i,j}\Delta\eta)^2}(M_{i,j,k}v^*_{i,j,k+1} + M_{i,j,k-1}v_{i,j,k-1}) - f\,u^*_{i,j,k}\}. \end{aligned} \tag{6.6}$$

(d) Determination of the ν-velocity Equation (4.10):

$$\nu^*_{i,j,k} = \frac{1}{\bar{B}_i}(v^*_{i,j,k} - u^*_{i,j,k} D_{i,j}), \tag{6.7}$$

where $D_{i,j}$ is defined in Equation (6.4).

(e) The second (general) continuity Equation (4.9) becomes:

$$\frac{\omega^*_{i,j,k+1} - \omega^*_{i,j,k}}{\Delta\eta} = \frac{1}{2h^*_{i,j}\bar{B}_i\Delta\chi}[h^*_{i+1,j}\bar{B}_{i+1}(U^*_{i+1,j} - u^*_{i+1,j,\bar{k}}) - h^*_{i-1,j}\bar{B}_{i-1}(U^*_{i-1,j} - u^*_{i-1,j,\bar{k}})] + \frac{1}{2h^*_{i,j}\Delta\lambda}[h^*_{i,j+1}(V^*_{i,j+1} - \nu^*_{i,j+1,\bar{k}}) - h^*_{i,j-1}(V^*_{i,j-1} - \nu^*_{i,j-1,\bar{k}})],$$

where

$$\phi_{i,j,\bar{k}} = (\phi_{i,j,k} + \phi_{i,j,k+1})/2. \tag{6.8}$$

Rearranging this equation gives

$$\omega^*_{i,j,k+1} = \omega^*_{i,j,k} + \frac{\Delta\eta}{2\Delta\chi h^*_{i,j}\bar{B}_i}[h^*_{i+1,j}\bar{B}_{i+1}(U^*_{i+1,j} - u^*_{i+1,j,\bar{k}}) - h^*_{i-1,j}\bar{B}_{i-1}(U^*_{i-1,j} - u^*_{i-1,j,\bar{k}})] + \frac{\Delta\eta}{2h^*_{i,j}\Delta\lambda}[h^*_{i,j+1}(V^*_{i,j+1} - \nu^*_{i,j+1,\bar{k}}) - h^*_{i,j-1}(V^*_{i,j-1} - \nu^*_{i,j-1,\bar{k}})]. \tag{6.9}$$

(f) The boundary conditions become:

On the bottom, $\eta = 1$,

$$u_{i,j,NZ} = v_{i,j,NZ} = \omega_{i,j,NZ} = \nu_{i,j,NZ} = 0. \tag{6.10}$$

On the surface, $\eta = 0$,

$$\omega_{i,j,1} = 0. \tag{6.11}$$

On the lateral boundaries, $\lambda = 0$ and 1,

$$\nu_{i,1,k} = \nu_{i,NY,k} = 0. \tag{6.12}$$

On the boundary, $\chi = 0$,

$$u_{1,j,k} = 0. \tag{6.13}$$

On the open boundary, $\zeta_{NX,j}$ is defined.

The derivatives $\partial\hat{\zeta}/\partial\chi$, $\partial\hat{\zeta}/\partial\lambda$ are required on the lateral boundaries $\lambda = 0$, $\lambda = 1$. For instance, in Equation (6.5) with $j = 1$, the pressure terms are approximated by

$$-\frac{g}{2\Delta\chi}(\zeta^*_{i,1}+\zeta^*_{i,0}-\zeta^*_{i-1,1}-\zeta^*_{i-1,0}) + \frac{gD_{i,1}}{2\bar{B}_i\Delta\lambda}(\zeta^*_{i,1}+\zeta^*_{i-1,1}-\zeta^*_{i,0}-\zeta^*_{i-1,0}), \tag{6.14}$$

where $\zeta^*_{i,0}$ and $\zeta^*_{i-1,0}$ are undefined. Of various finite difference approximations tried, the most acceptable was found to be linear extrapolation:

$$\zeta^*_{i,0} = 2\zeta^*_{i,1} - \zeta^*_{i,2}, \tag{6.15}$$

with a similar equation for $\zeta^*_{i-1,0}$.

A similar expression is used to approximate the pressure terms on the boundary $j = NY$ and $i = 1$.

Equations (6.5) and (6.6) also include centred difference approximations of second order spatial derivatives such as

$$\frac{\partial^2 u}{\partial\chi^2} \simeq (u_{2,j,k}-2u_{1,j,k}+u_{0,j,k})/(\Delta\chi)^2 \tag{6.16}$$

at the boundary $\chi = 0$. The second order accurate extrapolation procedure

$$u_{0,j,k} = 3u_{1,j,k} - 3u_{2,j,k} + u_{3,j,k} \tag{6.17}$$

is used, together with similar expressions for the other boundaries.

At the surface, $\eta = 0$, the spatial derivatives such as $\partial u/\partial z = 0$, are approximated to second order, giving expressions such as

$$u_{i,j,0} = u_{i,j,2}. \tag{6.18}$$

These are then used in Equations (6.5) and (6.6) when $k = 1$.

7. STABILITY OF THE NUMERICAL METHOD

The stability of the finite difference equations can be examined by considering the equations governing horizontal and vertical motions separately. They are:

The equations governing horizontal motions,

$$\hat{B}\frac{\partial\hat{\zeta}}{\partial T} + \frac{\partial(\hat{B}\hat{h}\hat{u})}{\partial\chi} + \frac{\partial(\hat{B}\hat{h}\hat{v})}{\partial\lambda} = 0, \tag{7.1a}$$

$$\frac{\partial\hat{u}}{\partial T} = -g\frac{\partial\hat{\zeta}}{\partial\chi} + \frac{g}{\hat{B}}\frac{\partial\hat{\zeta}}{\partial\lambda}\frac{\partial\hat{B}}{\partial\chi}, \tag{7.1b}$$

$$\frac{\partial\hat{v}}{\partial T} = -\frac{g}{\hat{B}}\frac{\partial\hat{\zeta}}{\partial\lambda}. \tag{7.1c}$$

The equations determining vertical motions,

$$\frac{\partial\hat{u}}{\partial T} + \omega\frac{\partial\hat{u}}{\partial\eta} = \frac{N_z}{\hat{h}^2}\frac{\partial^2\hat{u}}{\partial\eta^2}, \tag{7.2a}$$

$$\frac{\partial\hat{v}}{\partial T} + \omega\frac{\partial\hat{v}}{\partial\eta} = \frac{N_z}{\hat{h}^2}\frac{\partial^2\hat{v}}{\partial\eta^2}. \tag{7.2b}$$

The first system, represented by Equations (7.1a-c), is equivalent to the linearised two-dimensional tidal equations excluding terms which model friction effects and Coriolis forces. The second system incorporates the effect of the vertical velocity and eddy viscosity on the tidal accelerations, once again are examined separately because they are independent when linearised. The eddy viscosity terms associated with N_x and N_y have been excluded from the first system because Teubner (1976) found that their values had no effect on the stability.

7.1 The Horizontal Motions:
Equations (7.1a-c) become, if $\hat{h}$ takes the constant value H:

$$\hat{B}\frac{\partial\hat{\zeta}}{\partial T} + \frac{\partial\hat{Q}}{\partial\chi} + \frac{\partial\hat{P}}{\partial\lambda} = 0, \tag{7.3a}$$

$$\frac{\partial\hat{Q}}{\partial T} = -\hat{B}Hg\frac{\partial\hat{\zeta}}{\partial\chi} + gH\frac{\partial\hat{\beta}}{\partial\chi}\frac{\partial\hat{\zeta}}{\partial\lambda}, \tag{7.3b}$$

$$\frac{\partial\hat{P}}{\partial T} = gH\frac{\partial\hat{\beta}}{\partial\chi}\frac{\partial\hat{\zeta}}{\partial\chi} - \frac{gH}{B}\{1 + (\frac{\partial\hat{\beta}}{\partial\chi})^2\}\frac{\partial\hat{\zeta}}{\partial\lambda}, \tag{7.3c}$$

where $\hat{Q} = \hat{B}H\hat{u}$ and $\hat{P} = \hat{B}H\nu$.

The finite difference approximations to these equations are:

$$B_i(\zeta^*_{i,j} - \zeta_{i,j})/\Delta T + (Q_{i+1,j+1} + Q_{i+1,j} - Q_{i,j+1} - Q_{i,j})/2\Delta\chi$$
$$+ (P_{i+1,j+1} + P_{i,j+1} - P_{i+1,j} - P_{i,j})/2\Delta\lambda = 0, \tag{7.4}$$

$$(Q^*_{i,j} - Q_{i,j})/\Delta T = -B_i Hg(\zeta^*_{i,j} + \zeta^*_{i,j-1} - \zeta^*_{i-1,j} - \zeta^*_{i-1,j-1})/2\Delta\chi$$
$$+ gH(\frac{\partial\beta}{\partial\chi})_{i,j}(\zeta^*_{i,j} + \zeta^*_{i-1,j} - \zeta^*_{i,j-1} - \zeta^*_{i-1,j-1})/2\Delta\lambda, \tag{7.5}$$

$$(P^*_{i,j} - P_{i,j})/\Delta T = gH(\frac{\partial\beta}{\partial\chi})_{i,j}(\zeta^*_{i,j} + \zeta^*_{i,j-1} - \zeta^*_{i-1,j} - \zeta^*_{i-1,j-1})/2\Delta\chi$$
$$- \frac{bH}{B_i}\{1 + (\frac{\partial\beta}{\partial\chi})^2_{i,j}\}(\zeta^*_{i,j} + \zeta^*_{i-1,j} - \zeta^*_{i,j-1} - \zeta^*_{i-1,j-1})/2\Delta\lambda. \tag{7.6}$$

Let each dependent variable be replaced by

$$\zeta^n_{i,j} = E^n e^{i(i\theta_x + j\theta_y)}, \tag{7.7a}$$

$$Q^n_{i,j} = U^n e^{i(i\theta_x + j\theta_y)}, \tag{7.7b}$$

$$P^n_{i,j} = V^n e^{i(i\theta_x + j\theta_y)}, \tag{7.7c}$$

where E^n, U^n, V^n are the amplitudes of ζ, Q, P at the nth time step, θ_x θ_y are the x, y wave numbers of the Fourier components of an initial error distribution, and $i = \sqrt{-1}$.

Using the substitutions

$$\alpha = \frac{\Delta T}{2B\Delta\chi}, \quad \psi = \frac{\Delta T}{2B\Delta\lambda}, \quad \delta = \frac{BHg\Delta T}{2\Delta\chi},$$

where B is a typical value of B_i, and

$$\phi = \frac{gH\Delta T}{2\Delta\lambda}\nabla\beta, \quad \varepsilon = \frac{gH\Delta T}{2\Delta\chi}\nabla\beta, \quad \mu = \frac{gH\Delta T}{2B\Delta\lambda}\{1+(\nabla\beta)^2\},$$

where $\nabla\beta$ is a typical value of $(\partial\beta/\partial\chi)_{i,j}$, together with Equations (7.7a-c), Equations (7.4-7.6) can be written in matrix form as follows:

$$\begin{pmatrix} 1 & 0 & 0 \\ \delta(1+e^{-i\theta_y})(1-e^{-i\theta_x})-\phi(1+e^{-i\theta_x})(1-e^{-i\theta_y}) & 1 & 0 \\ -\varepsilon(1+e^{-i\theta_y})(1-e^{-i\theta_x})+\mu(1+e^{-i\theta_x})(1-e^{-i\theta_y}) & 0 & 1 \end{pmatrix} \begin{pmatrix} E^{n+1} \\ U^{n+1} \\ V^{n+1} \end{pmatrix}$$

$$= \begin{pmatrix} 1 & \alpha(1-e^{i\theta_x})(1+e^{i\theta_y}) & \psi(1-e^{i\theta_y})(1+e^{i\theta_x}) \\ 0 & 1 & 0 \\ 0 & 0 & 1 \end{pmatrix} \begin{pmatrix} E^n \\ U^n \\ V^n \end{pmatrix}. \tag{7.8}$$

This is of the form

$$A\underset{\sim}{X}^{n+1} = D\underset{\sim}{X}^n,$$

or

$$\underset{\sim}{X}^{n+1} = A^{-1}D\underset{\sim}{X}^n,$$

where $A^{-1}D$ is the amplification matrix for the vector

$$\underset{\sim}{X}^n = [E^n, U^n, V^n]^T. \tag{7.9}$$

The system represented by Equations (7.4-7.6) will be stable as long as the eigenvalues of the amplification matrix lie within the unit circle. The eigenvalues of this matrix are found to be

$$\lambda_1 = 1, \ \lambda_{2,3} = 1-2K \pm 2(K^2-K)^{1/2}, \tag{7.10}$$

where

$$K = \alpha\delta(1-\cos\theta_x)(1+\cos\theta_y) + \psi\mu(1+\cos\theta_x)(1-\cos\theta_y)$$
$$- 2\alpha\phi \sin\theta_x \sin\theta_y. \tag{7.11}$$

For $|\lambda_m| \leq 1$, m=1,2,3, Equation (7.10) requires that $0 \leq K \leq 1$. From analytic considerations and computational experimentation it was found that this condition is met if

$$(\alpha\phi)^2 \leq \alpha\psi\delta\mu, \tag{7.12}$$

and

$$0 \leq (\alpha\delta, \psi\mu) \leq \tfrac{1}{4}. \tag{7.13}$$

Condition (7.12) is always satisfied, while condition (7.13) requires

$$0 \leq \left(\frac{Hg(\Delta T)^2}{4(\Delta\chi)^2}, \frac{Hg(\Delta T)^2}{4(B\Delta\lambda)^2}\{1+(\nabla\beta)^2\}\right) \leq \tfrac{1}{4}. \tag{7.14}$$

The left inequality of condition (7.14) is always satisfied, so that this condition for stability becomes

$$\left(\frac{Hg(\Delta T)^2}{(\Delta\chi)^2}, \frac{Hg(\Delta T)^2}{B^2(\Delta\lambda)^2}\{1+(\nabla\beta)^2\}\right) \leq 1. \tag{7.15}$$

7.2 The Vertical Motions

The stability of Equations (7.2a,b) can be examined by considering only Equation (7.2a), as (7.2b) is of the same form. Equation (7.2a) becomes, with $\hat{h}$ taking the constant value H, and ω being a typical value of the scaled vertical velocity,

$$(u^*_{i,j} - u_{i,j})/\Delta T + \omega(u_{i,j+1} - u_{i,j-1})/2\Delta\eta$$

$$= \frac{N_z}{(H\Delta\eta)^2}(u_{i,j+1} - u_{i,j} - u^*_{i,j} + u^*_{i,j-1}). \tag{7.16}$$

That is

$$(1+r)u^*_{i,j} - ru^*_{i,j-1} = (r-s)u_{i,j+1} + (1-r)u_{i,j} + su_{i,j-1}, \tag{7.17}$$

where

$$r = \frac{N_z\,\Delta T}{(H\Delta\eta)^2} \quad\text{and}\quad s = \frac{\omega\Delta T}{2\Delta\eta}.$$

Applying the von Neumann technique of stability analysis yields the amplification factor for the error propagation associated with the finite difference equation (7.17),

$$G = \frac{1 + r(e^{i\beta}-1) - 2is\sin\beta}{1 + r(1-e^{-i\beta})},$$

where β is the phase angle associated with the chosen Fourier component of the error distribution. For stability, it is required that

$$|G| \leq 1 \text{ for all } \beta.$$

This condition is met if $s \leq r$, or

$$\frac{\omega H^2\Delta\eta}{2N_z} \leq 1. \tag{7.18}$$

8. THE VERTICAL EDDY VISCOSITY COEFFICIENT

Much data exists from field experiments for tidal flows in order to check numerically predicted vertical profiles of horizontal velocity. These results indicate the form of the vertical eddy viscosity coefficient which should be used in three-dimensional tidal modelling. For instance, functional forms for N_z have been obtained from tidal data by Bowden (1965), Nihoul (1975) and Wolf (1980).

A suitable vertically varying form of N_z has been used in tidal models by Cooper and Pearce (1977) and Nihoul (1978). It is proportional to the depth-mean horizontal current speed and has a parabolic vertical profile with a maximum value at the surface (e.g. Nihoul) or below the surface (e.g. Cooper and Pearce). The latter was chosen for this work, the form being

$$\hat{N}_z = k_1 + k_2 \hat{h} \hat{Q} (\tfrac{1}{2}+\eta)(1-\eta), \tag{8.1}$$

where

$$\hat{Q} = \{(\int_0^1 \hat{u}\, d\eta)^2 + (\int_0^1 \hat{v}\, d\nu)^2\}^{\frac{1}{2}},$$

and k_1, k_2 are constants.

9. APPLICATION TO SPENCER GULF

Spencer Gulf in South Australia is approximately 300 km long, 85 km wide at the mouth, and has an average depth of 30 m. Most of the northern half of the gulf is less than 10 m deep, while the southern half of the gulf is approximately 40 m deep. Salinity and temperature recordings show that its waters are well-mixed and can be considered homogeneous over depth. Both temperature and salinity are higher at the head of the gulf than at the mouth, but the resulting thermohaline current has been shown to be two orders of magnitude smaller than the tidally induced flow (Bullock, 1973).

Tide heights have been recorded for many years at the major ports of Whyalla, Port Lincoln, Port Augusta and Port Pirie, as well as at Wallaroo, Port Turton, Franklin Harbour and Port Victoria. Easton (1978) lists the tidal constants available for all these ports.

The mathematical model described has been applied to the propagation of the M2 tide along Spencer Gulf and the results have been described in detail in Noye, et al (1981).

Some typical cross-sections are shown in Figure 9.1, which also shows the layout of the grid-points at which velocities are computed (X) and the grid-points at which tide-heights are calculated (O). From co-tidal and co-range charts for the Southern Australian coast drawn by Easton (1970), the M2 tidal amplitude was taken to vary linearly from 15.0 to 17.0 cm across the mouth of the gulf with time lags which caried from 3.3 to 4.0 hr. The number of grid-points in the χ, λ, η directions of the rectangular prism were 15, 6 and 13 respectively, resulting in a total of 1170 velocity gridpoints being used.

The values of the constants chosen for the vertical eddy viscosity coefficient given by Equation (8.1) were $k_1 = 5 \times 10^{-4} m^2 sec^{-1}$, $k_2 = 10^{-2}$, while the constant values chosen for the horizontal eddy viscosity coefficients were $N_x = N_y = 5 m^2 sec^{-1}$.

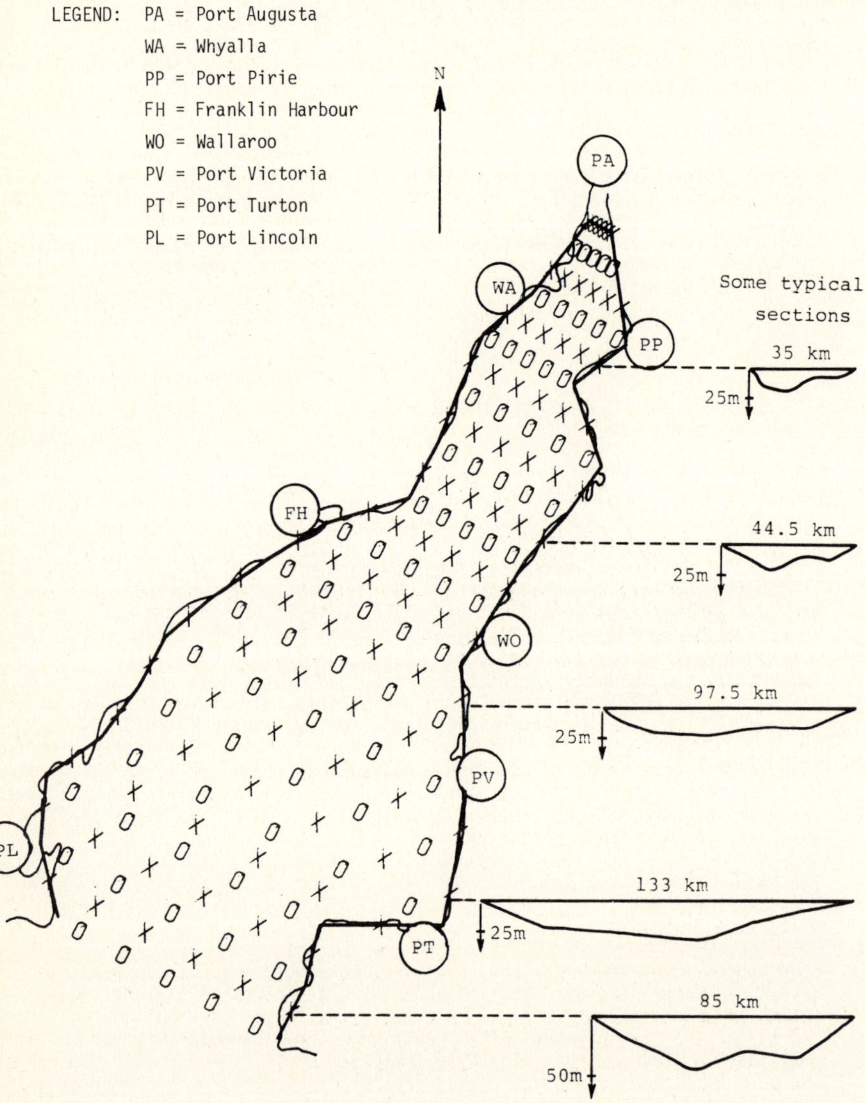

Figure 9.1 : Layout of grid-points, typical cross-sections and locations where tide-records are available, Spencer Gulf (after Noye, et al, 1981).

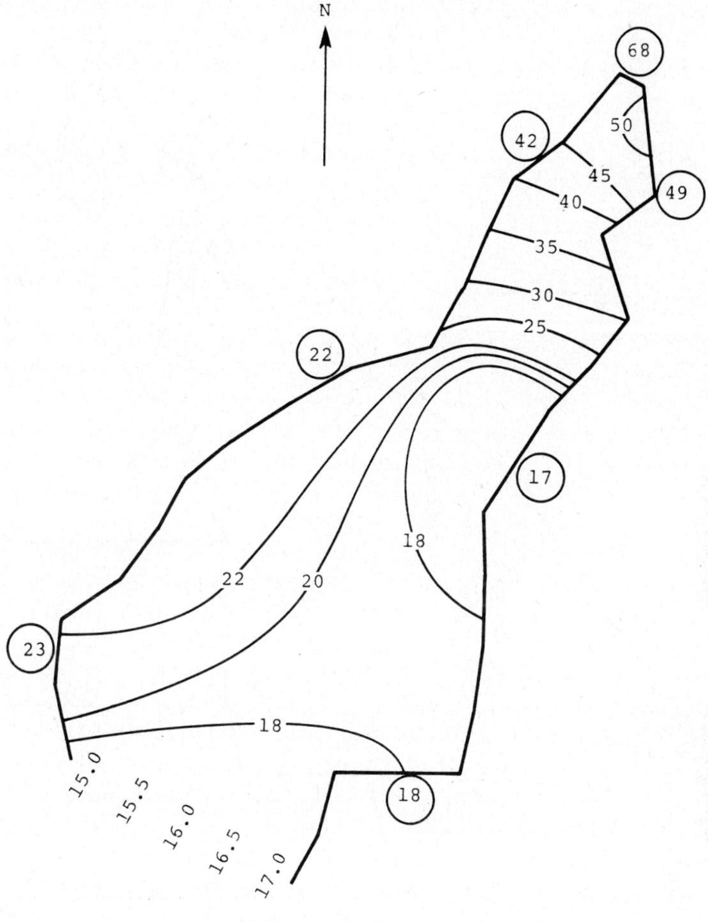

Figure 9.2 : Computed M2 co-amplitude lines in cm. M2 amplitudes obtained by analysis of coastal recordings are shown circled (after Noye, et al, 1981).

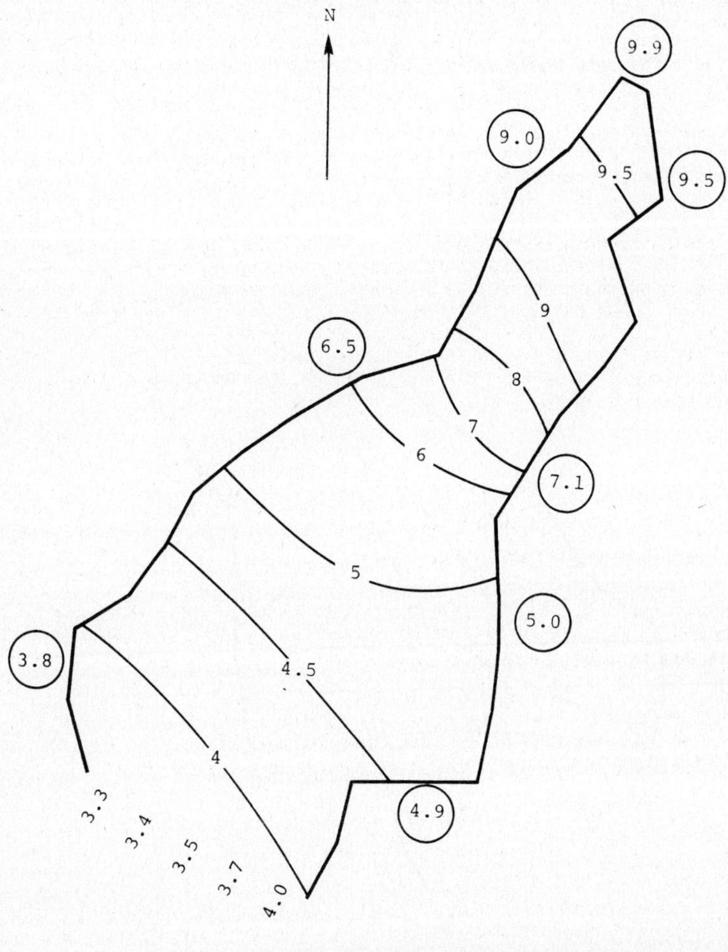

Figure 9.3 : Computed M2 co-tidal lines in hours. Values obtained by analysis of coastal recordings are shown circled (after Noye, et al, 1981).

Commencing with an initial condition of rest throughout the Gulf, the sea surface at the open boundary was set to oscillate with the circular frequency $1.405 \times 10^{-4} sec^{-1}$ of the M2 tide. After four tidal cycles had elapsed the starting transients were negligible and comparisons were made between tidal amplitudes and phases obtained from the numerical model and those found by analysis of tide records. The good agreement between recorded and computed amplitudes and phases at the locations listed previously is shown in Figures 9.2 and 9.3. These results are similar to those of the depth-integrated hydrodynamic numeric model of Noye and Tronson (1978), and fill the gaps in the co-range and co-tidal charts of the gulf drawn by Easton (1970).

There are no current records available for Spencer Gulf at the present time in order to check the computed horizontal velocity components. However, some interesting phenomena are evident in the computed results. For instance, a high shear in the bottom layers is evident at both full flood and ebb, while a reverse flow occurs near the bottom at slack water. As a consequence, the bottom stress leads the mean current throughout the tidal cycle. This agrees with experimental results obtained recently by an analysis of tidal current records obtained in the Irish Sea (see Wolf, 1980).

10. SUMMARY

A numerical method of computing the surface oscillations and three-dimensional current field for tidal propagation in shallow seas has been described. The method applies equally well to very shallow waters at the head of gulfs and to the main body of water in the deeper sections.

The method provides useful information about the vertical variation of the horizontal components of the induced currents. Results indicate that the bottom stress leads the horizontal components of the depth-mean velocity.

ACKNOWLEDGEMENT

The authors are grateful to the Australian Research Grants Committee for the financial assistance which enabled this work to be undertaken.

REFERENCES

Bowden, K.F., 1965. Horizontal mixing in the sea due to a shearing current. Journal of Fluid Mechanics, 21, 83-95.

Bullock, D.A., 1973. The general circulation of St. Vincent's Gulf and Spencer Gulf. In: B.J. Noye (Editor), Proceedings 2nd South Australian Regional Conference Physical Oceanography, 19-50.

Caponi, E.A. and Faller, A.J., 1973. A three-dimensional model for an estuary. Bulletin of the American Physical Society [2], 18: 1477.

Cooper, C.K. and Pearce, B.R., 1977. A three-dimensional numerical model to calculate currents in coastal waters utilizing a depth varying vertical eddy viscosity. Report No. 226, Ralph M. Parsons Laboratory, M.I.T., 147pp.

Easton, A.K., 1970. The tides of the continent of Australia, Ph.D. Thesis, The Flinders University of South Australia, 326pp.

Easton, A.K., 1978. A reappraisal of the tides on Spencer Gulf, Sth. Aust. Australian Journal of Marine and Freshwater Research, 29: 467-477.

Heaps, N.S., 1969. A two-dimensional numerical sea model. Philosophical Transactions of the Royal Society, Series A, 265: 93-137.

Heaps, N.S., 1973. Three-dimensional numerical model of the Irish Sea. Geophysical Journal of the Royal Astronomical Society 35: 99=120.

Leendertse, J.J., Alexander, R.C. and Liu, S-K, 1973. A three-dimensional model for estuaries and coastal seas, Vol.1: Principles of Computation. Rand Corp., R-1417-OWRR, Santa Monica, California.

Nihoul, J.C.J., 1975. Marine systems analysis. In: J.C.J. Nihoul (Editor), Modelling of Marine Systems, Elsevier Oceanography Series, 10: 3-40.

Nihoul, J.C.J., 1978. Three-dimensional modelling of marine circulation. In: C.A. Brebbia (Editor), Applied Numerical Modelling, Pentech Press, 73-81.

Noye, B.J., May, R.L. and Teubner, M.D., 1981. Three-dimensional numerical model of tides in Spencer Gulf. Ocean Management, 6, 137-147.

Noye, B.J. and Tronson, K.C.S., 1978. Finite difference techniques applied to the simulation of tides and currents in gulfs. In: B.J. Noye (Editor), Numerical Simulation of Fluid Motion, North-Holland Publishing Company, 285-356.

Roberts, K.V. and Weiss, N.O., 1966. Convective difference schemes. Mathematics of Computation, 20, 272-299.

Teubner, M.D., 1976. Tidal and termal propagation on the Port River Estuary. Ph.D. Thesis, The University of Adelaide, 167pp.

Wolf, J., 1980. Estimation of shearing stresses in a tidal current with application to the Irish Sea. In: J.C.J. Nihoul (Editor), Marine Turbulence, Elsevier Oceanography Series, 28: 319-344.

NUMERICAL SOLUTIONS OF
PARTIAL DIFFERENTIAL EQUATIONS
J. Noye (Editor)

NUMERICAL MODELLING OF LONG WAVES

ROBERT ARNOLD

Applied Mathematics Department
University of Adelaide

JOHN NOYE

Applied Mathematics Department
University of Adelaide

SUMMARY

A detailed report is given on the numerical modelling of the system of partial differential equations which govern long-wave motions in basins and seas. A diagnostic study is presented using a closed, wind affected basin. Finite difference techniques are used and the affects of using different grid spacings and various discretisations of the boundary conditions are described.

1. INTRODUCTION

In the past, many types of geophysical flows have been modelled using Reynold's averaged equations. For two-dimensional free surface flows in rivers, lakes or estuaries these equations may be written

$$\frac{Du}{Dt} = -g\frac{\partial\zeta}{\partial x} + \frac{1}{\rho}\frac{\partial\tau_{xz}}{\partial z}, \tag{1.1a}$$

$$\frac{\partial\zeta}{\partial t} + \frac{\partial}{\partial x}\int_{-h}^{\zeta} u\,dz = 0 \quad , \tag{1.1b}$$

where

$$\frac{D}{Dt} = \frac{\partial}{\partial t} + u\frac{\partial}{\partial x} + w\frac{\partial}{\partial z}.$$

Equation (1.1b) is a depth-averaged form of the continuity equation for an incompressible fluid,

$$\frac{\partial u}{\partial x} + \frac{\partial w}{\partial z} = 0. \tag{1.1c}$$

The variables used in the above equations are:

x, z	are cartesian coordinates, with x measured horizontally and z measured vertically upward from the mean free surface;
t	is the time;
$u(x,z,t)$	is the horizontal component of velocity;
$w(x,z,t)$	is the vertical component of velocity;
$\zeta(x,t)$	is the elevation of the free surface above the mean level;
ρ	is the density of the liquid, considered constant;
g	is the gravitational acceleration, considered constant;
$h(x,t)$	is the depth of liquid below the mean level;
$\tau_{xz}(x,z,t)$	is the shear stress acting in the x-direction along a plane parallel to the plane z = 0.

It should be noted that the velocities u and w in the above equations are not instantaneous values but are time-averaged values. The instantaneous velocity, $\underset{\sim}{q}$, has been replaced by the averaged velocity, $\bar{\underset{\sim}{q}}$, plus a fluctuating component, $\underset{\sim}{q}'$, that is,

$$\underset{\sim}{q} = \bar{\underset{\sim}{q}} + \underset{\sim}{q}'. \tag{1.2}$$

The velocities in Equation (1.1) are in fact the time-averaged values, but for convenience, the overbar has been omitted. The concept of this type of averaging is presented in Lamb (1932). Applying this averaging process to the Navier-Stokes equations gives the Reynold's equation (1.1).

A suitable equation describing the stress τ_{xz} in terms of u, v, w or ζ does not exist, so the system of equations given is not closed. By analogy with Newton's hypothesis for viscous laminar flow, the stress variable is usually written in

Boussinesq form as

$$\tau_{xz} = \rho N(x,z,t)\frac{\partial u}{\partial z} \tag{1.3}$$

where, by analogy with molecular viscosity, N is called the vertical eddy viscosity. Substitution of (1.3) into Equation (1.1a) gives

$$\frac{Du}{Dt} = -g\,\frac{\partial \zeta}{\partial x} + \frac{\partial}{\partial z}(N\frac{\partial u}{\partial z}). \tag{1.4}$$

Closure of this new set of equations is usually achieved by prescribing the vertical eddy viscosity (see Heaps (1972)). This is not very satisfactory since the eddy viscosity is a property of the flow, not of the fluid and should not be described 'a priori'.

Recently, closure has been achieved using an equation for the turbulent kinetic energy of a fluid motion. This equation is obtained by time-averaging the Navier-Stokes equations. For the two-dimensional system described here, this equation is found to be (see Launder & Spalding (1972), Launder and Spalding (1974), Jones & Launder (1972), Marchuk et al (1977))

$$\frac{DE}{Dt} = N(\frac{\partial u}{\partial z})^2 + \frac{\partial}{\partial z}(N\frac{\partial E}{\partial z}) - \varepsilon, \tag{1.5}$$

where E(x,z,t) is the turbulent kinetic energy and $\varepsilon(x,z,t)$ is the dissipation of turbulent kinetic energy. The vertical eddy viscosity is then given by

$$N = cE^2/\varepsilon, \tag{1.6}$$

where c = 0.08. Launder and Spalding (1972) present the following form for ε:

$$\varepsilon = c^{3/4}E^{3/2}/l, \tag{1.7}$$

where l is a mixing length scale. Following Vager and Kagan (1969), this length scale is determined by

$$l = \frac{-K\,E^{\frac{1}{2}}l^{-1}}{\frac{d}{dz}(E^{\frac{1}{2}}l^{-1})}, \tag{1.8}$$

where K is Von Karman's constant ($\simeq 0.41$).

The system of equations given by Equations (1.4)-(1.8) now forms a closed system. In this study these equations are applied to a long, closed basin of length L acted upon by a homogeneous surface wind stress $\tau_s = \tau_0\cos(\omega t)$, where the amplitude, τ_0, and circular frequency, ω, are both constants. The boundary conditions to this problem are:

the initial condition is a state of rest so

$$\zeta = u = w = 0 \text{ for } t = 0, \tag{1.9a}$$

there is no flow across the ends of the basin so

$$u = 0 \text{ at } x = 0,L, \tag{1.9b}$$

there is no flow across the surface of the liquid or the bottom of the basin so

$$w = \frac{\partial \zeta}{\partial t} + u\frac{\partial \zeta}{\partial x} \text{ at } z = \zeta \text{ and } w = 0 \text{ at } z = -h, \tag{1.9c}$$

there is a no slip condition on the bottom of the basin so

$$E = u = 0 \quad \text{at} \quad z = -h, \tag{1.9d}$$

there is no vertical flux of energy through the surface so

$$N\frac{\partial E}{\partial z} = 0 \quad \text{at} \quad z = \zeta, \tag{1.9e}$$

the surface stress τ_s is prescribed so

$$\rho N\frac{\partial u}{\partial z} = \tau_s \quad \text{at} \quad z = \zeta, \tag{1.9f}$$

on the bottom the mixing length is given by

$$l = K\, z_0 \quad \text{at} \quad z = -h, \tag{1.9g}$$

where z_0 is a roughness length.

A transformation to the vertical coordinate is now made to the governing equations to simplify the use of finite differences in the numerical solution. Since the surface and the bottom profile are variable, the transformation

$$\eta = \frac{z+h}{\zeta+h} \tag{1.10}$$

is used, so that $\eta = 0$ at the bottom and $\eta = 1$ at the surface. Using x,η,t as the new independent variables, equations (1.1b), (1.4) and (1.5) become

$$\frac{\partial\zeta}{\partial t} = \frac{\partial}{\partial x}[(h+\zeta)\int_0^1 u\, d\eta], \tag{1.11a}$$

$$\frac{Du}{Dt} = -g\frac{\partial\zeta}{\partial x} + \frac{1}{(\zeta+h)^2}\frac{\partial}{\partial\eta}(N\frac{\partial u}{\partial\eta}), \tag{1.11b}$$

$$\frac{DE}{Dt} = \frac{N}{(\zeta+h)^2}(\frac{\partial u}{\partial\eta})^2 + \frac{1}{(\zeta+h)^2}\frac{\partial}{\partial\eta}(N\frac{\partial E}{\partial\eta}) - \varepsilon. \tag{1.11c}$$

Note that, because of the transformation (1.10), the vertical component of velocity in the total derivative is $w' = \frac{\partial\eta}{\partial t} + u\frac{\partial\eta}{\partial x} + w\frac{\partial\eta}{\partial z}$. For convenience, the prime is dropped from this point.

The boundary conditions corresponding to (1.9) are:

$$\zeta = u = w = 0 \quad \text{at} \quad t = 0, \tag{1.12a}$$

$$u = 0 \quad \text{on} \quad x = 0, L, \tag{1.12b}$$

$$w = 0 \quad \text{on} \quad \eta = 0,1, \tag{1.12c}$$

$$E = u = 0 \quad \text{on} \quad \eta = 0, \tag{1.12d}$$

$$\frac{N}{(\zeta+h)}\frac{\partial E}{\partial\eta} = 0 \quad \text{on} \quad \eta = 1, \tag{1.12e}$$

$$\frac{N}{(\zeta+h)}\frac{\partial u}{\partial\eta} = \tau_s \quad \text{on} \quad \eta = 1, \tag{1.12f}$$

$$l = K\, z_0 \quad \text{on} \quad \eta = 0. \tag{1.12g}$$

Equation (1.8) is similarly transformed and the result integrated from $\eta = 0$ to $\eta = 1$ to yield

$$l = K[z_0(E/E_b)^{\frac{1}{2}} + (\zeta+h)E^{\frac{1}{2}}\int_0^{\eta} E^{-\frac{1}{2}}\, d\xi], \tag{1.13}$$

where E_b is the turbulent energy at the bottom $\eta = 0$.

Equations (1.1b) and (1.1c) are combined and the result transformed using (1.10) to give

$$(\zeta+h)\frac{\partial w}{\partial \eta} = \frac{\partial}{\partial x}[(\zeta+h)\int_0^1 u\, d\eta] - \frac{\partial}{\partial x}[(\zeta+h)u]. \tag{1.14}$$

On integration from $\eta = 0$ to $\eta = 1$ this gives

$$w = \frac{1}{(\zeta+h)}\{\eta\frac{\partial}{\partial x}[(\zeta+h)\int_0^1 u\, d\eta] - \frac{\partial}{\partial x}[(\zeta+h)\int_0^{\eta} u\, d\eta]\}. \tag{1.15}$$

Note that Equation (1.15) is a first order differential equation, but there are two boundary conditions, namely $w = 0$ at $\eta = 0,1$. The equation is therefor overdetermined, but the expression for w given by (1.15) satisfies both of the boundary conditions.

2. AN EXPLICIT NUMERICAL METHOD

The equations (1.11a,b,c), (1.13) and (1.15), subject to conditions (1.12), can now be solved numerically. Because the greatest changes in the horizontal velocity occur near the bottom and the surface, it is necessary that a fine grid spacing in the vertical direction is used near these. Since a coarse grid may be used midway between the bottom and the surface where the changes in velocity occur only gradually, the model was developed to accommodate variable grid spacings over depth. A constant grid spacing was used in the horizontal direction. The grid scheme used in the longitudinal section of the channel is shown in Figure 1.

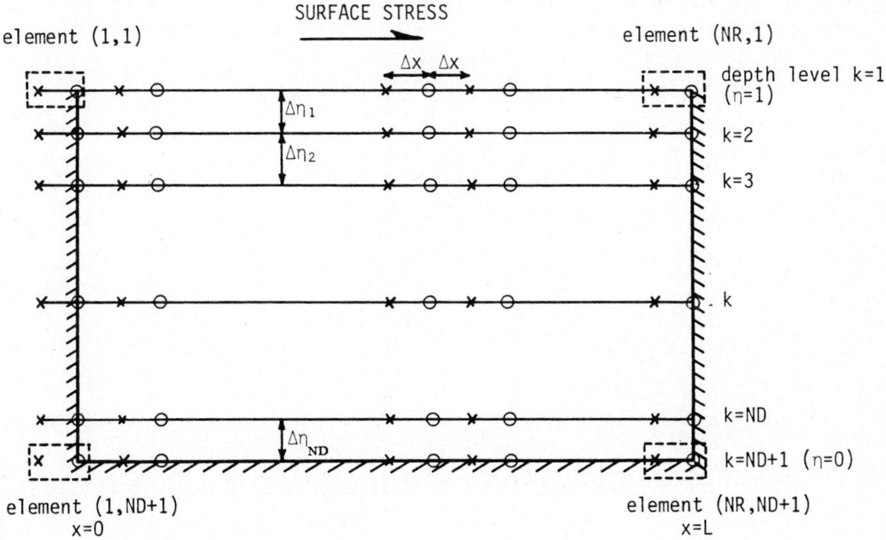

Figure 1 : Channel divided into NR horizontal elements with ND+1 depth levels.

An element of the difference scheme consists of two points labelled O and X respectively, situated a distance Δx apart. Corresponding points in each element are separated by the distance $2\Delta x$. At the O-points, the horizontal component of velocity, u, is calculated. All other variables are calculated at x-points. The notation $u|^n_{i,k}$ is used to represent the approximation for u(x,z,t) computed at the O-point in the ith element at the kth depth level and at time $n\Delta t$, where Δt is the constant time step used in the finite difference scheme.

Equations (1.11) were then discretised at the (i,k,n) grid point, in the forms:

$$\left.\frac{\partial\zeta}{\partial t}\right|^n_i = \frac{\partial}{\partial x}[(\zeta+h)\int_0^1 u\,d\eta]|^n_i, \tag{2.1a}$$

$$\left.\frac{\partial u}{\partial t}\right|^n_{i,k} + u\left.\frac{\partial u}{\partial x}\right|^n_{i,k} + w\left.\frac{\partial u}{\partial \eta}\right|^n_{i,k} = -g\left.\frac{\partial\zeta}{\partial x}\right|^{n+1}_i + \frac{1}{(\zeta+h)^2}\frac{\partial}{\partial\eta}(N\frac{\partial u}{\partial\eta})|^n_{i,k} + O\{\Delta t\}, \tag{2.1b}$$

$$\left.\frac{\partial E}{\partial t}\right|^n_{i,k} + u\left.\frac{\partial E}{\partial x}\right|^n_{i,k} + w\left.\frac{\partial E}{\partial \eta}\right|^n_{i,k} = \frac{N}{(\zeta+h)^2}(\frac{\partial u}{\partial\eta})^2|^n_{i,k} + \frac{N}{(\zeta+h)^2}\frac{\partial}{\partial\eta}(N\frac{\partial E}{\partial\eta})|^n_{i,k} - \varepsilon^n_{i,k}. \tag{2.1c}$$

Equation (2.1a) is used to calculate ζ at the new time level, n+1, with a knowledge of the values of ζ, u, w, E and ε calculated at the nth time level. All of these are used in Equations (2.1b) and (2.1c) to yield u and E at the (n+1)th time level. Equations (1.6), (1.7), (1.13) and (1.15) are then used to calculate N, l, ε and w at the new time level from the known values of u and E at that time.

We now investigate the manner in which each of the terms in the governing equations (2.1) can be discretised.

<u>The continuity equation (2.1a)</u>

$$\left.\frac{\partial\zeta}{\partial t}\right|^{n+1}_i = \frac{1}{\Delta t}\{\zeta^{n+1}_i - \zeta^n_i\} + O\{\Delta t\}, \tag{2.2a}$$

$$\int_0^1 u\,d\eta|^n_i = \bar{u}^n_i + O\{\Delta\eta_k\}, \tag{2.2b}$$

where

$$\bar{u}^n_i = \tfrac{1}{2}[\Delta\eta_1(u^n_{i,1} + u^n_{i,2}) + \ldots + \Delta\eta_{ND}(u^n_{i,ND} + u^n_{i,ND+1})],$$

and

$$\frac{\partial}{\partial x}[(h+\zeta)\int_0^1 u\,d\eta]|^n_i = \frac{1}{2\Delta x}[\bar{\bar{U}}^n_i - \bar{\bar{U}}^n_{i-1}] + O\{(\Delta x)^2\} \tag{2.2c}$$

where

$$\bar{\bar{U}}^n_i = \tfrac{1}{2}\bar{U}^n_i[h^{n+1}_i + \zeta^{n+1}_i + h^{n+1}_{i+1} + \zeta^{n+1}_{i+1}].$$

The appropriate boundary conditions for this equation are those in (1.12a), namely

$$\zeta^n_i = 0,\ \text{for } i=1(1)NR;\ t = 0. \tag{2.2d}$$

The integral in Equation (2.2b) has been estimated using the trapezoidal rule with uneven spacings and is only accurate to $O(\Delta\eta_k)$. This method of integration was used since it was the only way to calculate the $\int_0^\eta$ which appears in Equation (1.15).

<u>The u-equation of motion (2.1b)</u>

$$\left.\frac{\partial u}{\partial t}\right|^n_{i,k} = \frac{1}{\Delta t}\{u^{n+1}_{i,k} - u^n_{i,k}\} + O\{\Delta t\}, \tag{2.3a}$$

$$\left.u\frac{\partial u}{\partial x}\right|^n_{i,k} = \frac{u^n_{i,k}}{4\Delta x}\{u^n_{i+1,k} - u^n_{i-1,k}\} + O\{(\Delta x)^2\} \tag{2.3b}$$

$$\left.w\frac{\partial u}{\partial \eta}\right|^n_{i,k} = \tfrac{1}{2}\left(\frac{w^n_{i,k} + w^n_{i+1,k}}{r_k\Delta\eta_{k-1}\Delta\eta_k}\right)\cdot(\Delta\eta_{k-1}u^n_{i,k+1} - \Delta\eta_k(r_k-1)u^n_{i,k-1}$$

$$+ r_k(r_k-2)\Delta\eta_{k-1}u^n_{i,k}) + O\{\Delta\eta_k\Delta\eta_{k-1}\}, \tag{2.3c}$$

$$\left.g\frac{\partial \zeta}{\partial x}\right|^{n+1}_{i,k} = \frac{g}{2\Delta x}[\zeta^{n+1}_{i+1} - \zeta^{n+1}_{i} + O\{(\Delta x)^2\}, \tag{2.3d}$$

$$\left.\frac{1}{(\zeta+h)^2}\frac{\partial}{\partial\eta}\left(N\frac{\partial u}{\partial\eta}\right)\right|^n_{i,k} = \frac{4}{(\zeta^{n+1}_i+\zeta^n_{i+1}+h^n_{i+1}+h^n_i)^2}\Bigg\{\frac{u^n_{i,k-1}}{r_k(\Delta\eta_{k-1})^2}$$

$$[(r_k-1)(\tilde{N}^n_{i,k-1}+\tilde{N}^n_{i+1,k-1})-(r_k-2)(N^n_{i,k}+N^n_{i+1,k})]$$

$$+\frac{u^n_{i,k}}{r_k(\Delta\eta_{k-1})^2(\Delta\eta_k)^2}[-(\Delta\eta_{k-1})^2(\tilde{N}^n_{i,k}+\tilde{N}^n_{i+1,k})$$

$$-(r_k-1)(\Delta\eta_k)^2(\tilde{N}^n_{i,k-1}+\tilde{N}^n_{i+1,k-1}) + r_k(r_k-2)^2$$

$$(\Delta\eta_{k-1})^2(N^n_{i+1,k}+N^n_{i,k})] + \frac{u^n_{i,k+1}}{r_k(\Delta\eta_k)^2}$$

$$[(\tilde{N}^n_{i,k}+\tilde{N}^n_{i+1,k}) + (r_k-2)(N^n_{i,k}+N^n_{i+1,k})]\Bigg\} + O\{\Delta\eta_k\Delta\eta_{k-1}\}, \tag{2.3e}$$

where

$$r_k = \Delta\eta_k/\Delta\eta_{k-1} + 1 \quad \text{and} \quad \tilde{N}^n_{i,k} = \tfrac{1}{2}(N^n_{i,k+1} + N^n_{i,k}).$$

The approximation for $\partial u/\partial\eta|^n_{i,k}$ shown in Equation (2.3c) is a second order approximation to a derivative with respect to the depth for an unevenly spaced grid in the η direction. Often the formula

$$\left.\frac{\partial u}{\partial\eta}\right|^n_{i,k} \simeq \frac{u^n_{i,k+1} - u^n_{i,k-1}}{(\Delta\eta_k + \Delta\eta_{k-1})}, \tag{2.3f}$$

is used, but this is less accurate than (2.3c) since it involves a first order error.

The formula for $\partial/\partial\eta\ (N\ \partial u/\partial\eta)|^n_{i,k}$ is derived by first using the formula for $\partial/\partial\eta|^n_{i,k}$ given in Equation (2.3c) where the three depth levels are taken to be the kth, the level half-way between the kth and (k-1)th, denoted by (k-½), and the level denoted (k+½) which is half-way between the kth and (k+1)th levels. This gives the equation

$$\frac{\partial}{\partial\eta}\left(N\frac{\partial u}{\partial\eta}\right)\Big|^n_{i,k} = \frac{2\Delta\eta_{k-1}}{(\Delta\eta_{k-1}+\Delta\eta_k)\,\Delta\eta_k}\left(N\frac{\partial u}{\partial\eta}\right)\Big|^n_{i,k+\frac{1}{2}} - \frac{2\Delta\eta_k}{(\Delta\eta_{k-1}+\Delta\eta_k)\,\Delta\eta_{k-1}}\left(N\frac{\partial u}{\partial\eta}\right)\Big|^n_{i,k-\frac{1}{2}} + \frac{(\Delta\eta_k-\Delta\eta_{k-1})}{(\Delta\eta_{k-1}+\Delta\eta_k)}\left(N\frac{\partial u}{\partial\eta}\right)\Big|^n_{i,k} + O\{\Delta\eta_{k-1}\Delta\eta_k\}. \tag{2.3g}$$

The derivatives at the (k-½)th and (k+½)th levels are now approximated by central differences, namely

$$\begin{aligned} N\frac{\partial u}{\partial\eta}\Big|^n_{i,k+\frac{1}{2}} &= \frac{N^n_{i,k}}{\Delta\eta_k}\{u^n_{k+1} - u^n_k\} + O\{(\Delta\eta_k)^2\}, \\ N\frac{\partial u}{\partial\eta}\Big|^n_{i,k-\frac{1}{2}} &= \frac{N^n_{i,k-1}}{\Delta\eta_{k-1}}\{u^n_{i,k} - u^n_{i,k-1}\} + O\{(\Delta\eta_{k-1})^2\}. \end{aligned} \tag{2.3h}$$

The three-point formula given in Equation (2.3c) is again applied to the terms $(N\ \partial u/\partial\eta)^n_{i,k}$ which occur in Equation (2.3g) to give Equation (2.3e).

The boundary conditions for this equation are given by

$$u^n_{i,k} = 0 \text{ for } i=1(1)NR;\ t \geq 0;\ k = ND+1 \tag{2.3i}$$

and

$$u^n_{i,k} = 0 \text{ for } i=1,NR;\ t \geq 0;\ k=1(1)ND+1. \tag{2.3j}$$

The boundary condition for u at η = 0,1 is incorporated into the model by using a modified form of the u-equation at the k=1 level. All terms remain the same except for those which contain derivatives with respect to η, which are represented by a one-sided difference approximation. Hence, for the k=1 level the new terms are:

$$w\frac{\partial u}{\partial\eta}\Big|^n_{i,1} = \frac{(w^n_{i,1} + w^n_{i+1,1})\{u^n_{i,2} - u^n_{i,1}\}}{2\Delta\eta_1} + O\{\Delta\eta_1\}, \tag{2.3k}$$

and

$$\frac{1}{(\zeta+h)^2}\frac{\partial}{\partial\eta}\left(N\frac{\partial u}{\partial\eta}\right)\Big|^n_{i,1} = \frac{4(N^n_{i,1} + N^n_{i+1,1})\{u^n_{i,2} - u^n_{i,1}\}}{(h^{n+1}_i + h^{n+1}_{i+1} + \zeta^n_i + \zeta^n_{i+1})^2(\Delta\eta_1)^2} - \frac{4\tau_s}{(h^{n+1}_i + h^{n+1}_{i+1} + \zeta^n_i + \zeta^n_{i+1})\rho\Delta\eta_1} + O\{\Delta\eta_1\}, \tag{2.3l}$$

where the last term in Equation (2.3l) incorporates the boundary condition (1.12f). Equation (2.3l) is computed using the following intermediate step:

$$\frac{\partial}{\partial\eta}\left(N\frac{\partial u}{\partial\eta}\right)^n_{i,1} = \frac{2}{\Delta\eta_1}\left(N\frac{\partial u}{\partial\eta}\Big|^n_{i,1\frac{1}{2}} - N\frac{\partial u}{\partial\eta}\Big|^n_{i,1}\right) + O\{\Delta\eta_1\}. \tag{2.3m}$$

Using centred differences for the first term on the right hand side then gives the desired result.

<u>The turbulent kinetic energy equation (2.1c)</u>

$$\frac{\partial E}{\partial t}\Big|^{n+1}_{i,k} = \frac{1}{\Delta t}\{E^{n+1}_{i,k} - E^n_{i,k}\} + O\{\Delta t\}, \tag{2.4a}$$

$$u\frac{\partial E}{\partial x}\Big|^n_{i,k} = \frac{1}{8\Delta x}(u^n_{i,k} + u^n_{i-1,k})(E^n_{i+1,k} - E^n_{i-1,k}) + O\{(\Delta x)^2\}, \tag{2.4b}$$

$$w\frac{\partial E}{\partial\eta}\Big|^n_{i,k} = \frac{W^n_{i,k}}{r_k\Delta\eta_{k-1}\Delta\eta_k}\{\Delta\eta_{k-1}E^n_{i,k+1} - (r_k-1)\Delta\eta_k E^n_{i,k-1} + r_k(r_k-2)\Delta\eta_{k-1}E^n_{i,k}\}$$
$$+ O\{\Delta\eta_{k-1}\Delta\eta_k\}, \tag{2.4c}$$

$$\frac{N}{(\zeta+h)^2}\left(\frac{\partial u}{\partial\eta}\right)^2\Big|^n_{i,k} = \frac{N^n_{i,k}}{r_k\Delta\eta_{k-1}\Delta\eta_k(\zeta^{n+1}_i+h^{n+1}_i)^2}\{\Delta\eta_{k-1}\tilde{u}^n_{i,k+1} - (r_k-1)\Delta\eta_k\tilde{u}^n_{i,k-1}$$
$$+ r_k(r_k-2)\Delta\eta_{k-1}\tilde{u}^n_{i,k}\} + O\{\Delta\eta_{k-1}\Delta\eta_k\}, \tag{2.4d}$$

where $\tilde{u}^n_{i,k} = \frac{1}{2}(u^n_{i,k} + u^n_{i-1,k})$,

$$\frac{1}{(\zeta+h)^2}\frac{\partial}{\partial\eta}\left(N\frac{\partial E}{\partial\eta}\right)\Big|^{n+1}_{i,k} = \frac{2}{(\zeta^{n+1}_i+h^{n+1}_i)^2}\Big\{\frac{E^n_{i,k-1}}{r_k(\Delta\eta_{k-1})^2}[(r_k-1)\tilde{N}^n_{i,k-1} - (r_k-2)N^n_{i,k}]$$
$$+ \frac{E^n_{i,k}}{r_k(\Delta\eta_{k-1})^2(\Delta\eta_k)^2}[-(\Delta\eta_{k-1})^2\tilde{N}^n_{i,k} - (r_k-1)(\Delta\eta_k)^2\tilde{N}^n_{i,k-1}$$
$$+ r_k(r_k-2)^2(\Delta\eta_{k-1})^2N^n_{i,k}] + \frac{E^n_{i,k+1}}{r_k(\Delta\eta_k)^2}[\tilde{N}^n_{i,k} + (r_k-2)N^n_{i,k}]\Big\}$$
$$+ O\{\Delta\eta_k\Delta\eta_{k-1}\}, \tag{2.4e}$$

$$\varepsilon^n_{i,k} = c^{3/4}(E^n_{i,k})^{3/2}/l^n_{i,k}. \tag{2.4f}$$

The boundary condition (1.12d) implies

$$E^n_{i,k} = 0 \text{ for } i=1(1)\text{ NR};\ t \geq 0;\ k = \text{ND}+1. \tag{2.4g}$$

Boundary condition (1.12e) yields, in a similar fashion to the equations derived for u at k = 1, the following two derivatives:

$$w\frac{\partial E}{\partial\eta}\Big|^n_{i,1} = \frac{W^n_{i,1}}{\Delta\eta_1}[E^n_{i,2} - E^n_{i,1}] + O\{\Delta\eta_1\}, \tag{2.4h}$$

$$\frac{1}{(\zeta+h)^2}\frac{\partial}{\partial\eta}\left(w\frac{\partial E}{\partial\eta}\right)^{n+1}_{i,1} = \frac{2}{(\zeta_i^{n+1}+h_i^{n+1})^2}\frac{N^n_{i,1}}{(\Delta\eta_1)^2}\{E^n_{i,2}-E^n_{i,1}\} + O\{\Delta\eta_1\}. \qquad (2.4i)$$

The mixing length equation (1.8)

$$l^{n+1}_{i,k} = k[z_0(E^{n+1}_{i,k}/E^{n+1}_{i,ND+1})^{1/2} + (E^{n+1}_{i,k})^{1/2}(\zeta_i^{n+1}+h_i^{n+1})\bar{E}^{n+1}_{i,k}] \qquad (2.5a)$$

where

$$\bar{E}^{n+1}_{i,k} = [(E^{n+1}_{i,ND+1})^{-1/2} + (E^{n+1}_{i,ND})^{-1/2}]\Delta\eta_{ND} + \ldots + [(E^{n+1}_{i,k+1})^{-1/2}+(E^{n+1}_{i,k})^{-1/2}]\Delta\eta_k .$$

The boundary condition (1.12g) implies

$$l^{n+1}_{i,k} = k\, z_0 \text{ for } i=1(1)NR;\ t \geq 0;\ k = ND+1. \qquad (2.5b)$$

The depth-integrated continuity equation (1.15)

$$w^{n+1}_{i,k} = \frac{1}{(\zeta_i^{n+1}+h_i^{n+1})}\{(\Delta\eta_{ND}+\Delta\eta_{ND-1}+\ldots+\Delta\eta_k)\frac{(\bar{u}_i^{n+1}-\bar{u}_{i-1}^{n+1})}{2\Delta x} - \frac{(\bar{\bar{u}}^{n+1}_{i,k}-\bar{\bar{u}}^{n+1}_{i-1,k})}{2\Delta x}\} \qquad (2.6a)$$

where

$$\bar{\bar{u}}^{n+1}_{i,k} = \tfrac{1}{2}(h_i^{n+1}+\zeta_i^{n+1})[\Delta\eta_{ND}(u^{n+1}_{i,ND+1}+u^{n+1}_{i,ND}) + \ldots + \Delta\eta_k(u^{n+1}_{i,k+1}+u^{n+1}_{i,k})]. \qquad (2.6b)$$

Equation (2.6a) satisfies the required boundary conditions at $\eta = 0$ and $\eta = 1$.

The N-equation

$$N^{n+1}_{i,k} = c(E^{n+1}_{i,k})^2/\varepsilon^{n+1}_{i,k}. \qquad (2.7)$$

3. SOME VARIATIONS OF THE NUMERICAL METHOD

The equations given in Section 2 are explicit. They were programmed for a rectangular basin of length 40000m and constant depth 3.5m acted upon by a homogeneous oscillating wind stress with constant amplitude of $\tau_0 = 0.1$ dyne/m^2 and period of one day (compare the models of the Coorong, Walsh, 1973).

It was found that for even very small time steps, the method presented above is unstable. For values of $\Delta x = 250$m and $\Delta\eta_k = 0.02$ for k=1(1)51, stability was obtained only for an impractical value of $\Delta t < 30$ secs. An analysis of the data produced during the first few time steps indicated that, apart from $\partial\zeta/\partial t$, the dominant term in Equations (1.4) is the diffusive one, that is $\partial/\partial\eta(N\ \partial u/\partial\eta)$. In Equation (1.5), apart from $\partial E/\partial t$, the dominant term is $\{\partial/\partial\eta(N\ \partial E/\partial\eta) - \varepsilon\}$.

It was decided to incorporate these terms into the corresponding finite difference equations in an implicit manner; that is, they were differenced at the (n+1)th time level. Equation (2.1b) then becomes

$$\left.\frac{\partial u}{\partial t}\right|_{i,k}^{n} + u\left.\frac{\partial u}{\partial x}\right|_{i,k}^{n} + w\left.\frac{\partial u}{\partial \eta}\right|_{i,k}^{n} = -g\left.\frac{\partial \zeta}{\partial x}\right|_{i,k}^{n+1} + \frac{1}{(\zeta+h)|_i^{n+1}}\left(\frac{\partial}{\partial \eta}N^n\frac{\partial u^{n+1}}{\partial \eta}\right)\Big|_{i,k}$$

$$+ O\{\Delta t\}, \qquad (3.1)$$

where the last term is modelled by Equations (2.3e,l) as before, except each velocity, u, is calculated at the (n+1)th time level. Because the term $\varepsilon^n_{i,k}$ which occurs in Equation (2.1c) is of the same order of magnitude as the diffusion term, it was decided to also use an implicit method to calculate this term. In fact,

$$\varepsilon^n_{i,k} = \frac{c^{3/4}(E^n_{i,k})^{\frac{1}{2}}E^{n+1}_{i,k}}{l^n_{i,k}} \qquad (3.2)$$

was used in order to make the resulting finite difference scheme linear. Hence, Equation (2.1c) is modelled by

$$\left.\frac{\partial E}{\partial t}\right|_{i,k}^{n} + u\left.\frac{\partial u}{\partial x}\right|_{i,k}^{n} + w\left.\frac{\partial u}{\partial \eta}\right|_{i,k}^{n} = \frac{N}{(\zeta+h)}\left(\frac{\partial u}{\partial \eta}\right)^2\Big|_{i,k}^{n} + \frac{1}{(\zeta+h)|_i^n}\frac{\partial}{\partial \eta}\left(N\frac{\partial E}{\partial \eta}\right)\Big|_{i,k}^{n+1}$$

$$- \frac{c^{3/4}(E^n_{i,k})^{\frac{1}{2}}E^{n+1}_{i,k}}{l^n_{i,k}}. \qquad (3.3)$$

Both the u and E equations are now of the form

$$A^n u^{n+1}_{i,k-1} + B^n u^{n+1}_{i,k} + C^n u^{n+1}_{i,k+1} = D^n \qquad (3.4)$$

which may be solved by the very efficient Thomas Algorithm for tri-diagonal systems of linear equations (see Noye, 1981).

The performance of this model is far superior to the original explicit version. By experiment, it was found that the system was stable for Δx = 500m, Δt = 180 secs., using up to fifty depth levels.

The effect of changing the values of the three parameters Δx, Δt and $\Delta\eta_k$ was also examined. Firstly, ND the number of depth levels, was kept fixed at 8. A uniform depth spacing was used so that $\Delta\eta_k$ = 0.125 for k=1(1)ND. Four test runs were performed in which

(a) Δx = 1000m, Δt = 180 secs.,

(b) Δx = 500m, Δt = 180 secs.,

(c) Δx = 500m, Δt = 90 secs., and

(d) Δx = 250m, Δt = 30 secs.

The same values for the velocity components and the elevation were obtained at common gridpoints by each of (a)-(d) throughout the entire flow regime. This model seems very stable. The results obtained from RUN (a), using a rather coarse grid, are the same as those obtained using a very fine grid, RUN (d). The variation over depth of the horizontal velocity component, u, at x = L/4 after 1 wind cycle, that is, after 24 hours, is presented in Figure 2 by the line labelled [1].

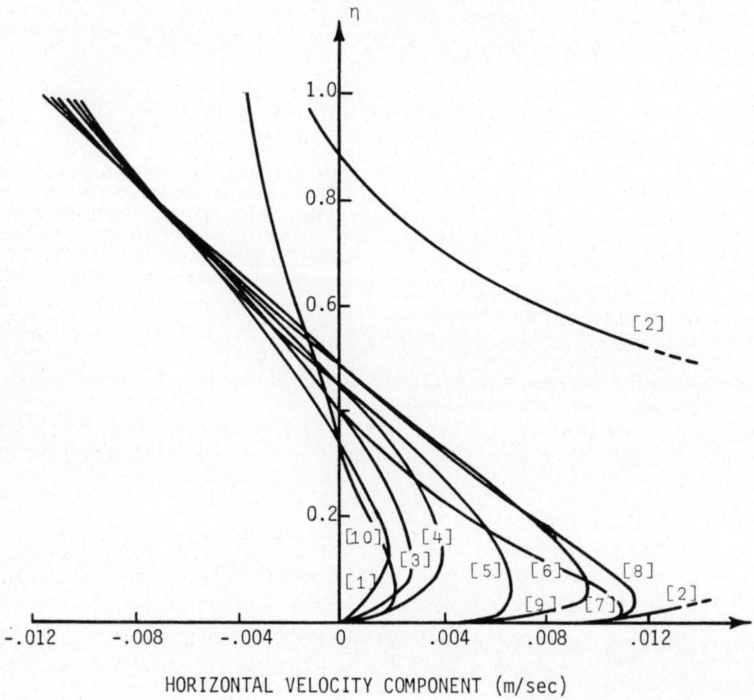

Figure 2 : Velocity profile obtained at $x = L/4$ after 1 wind cycle (i.e. after 24 hours) using various grid sizes.

[1] ND = 8, NR = 21 : Δx = 1000m, Δt = 180 secs., $\Delta\eta_k$ = 0.125, k=1,..,8

[2] ND = 8, NR = 41 : Δx = 500m, Δt = 180 secs., $\Delta\eta_1 = \Delta\eta_8 = 0.04$, $\Delta\eta_2 = \Delta\eta_7 = 0.08$, $\Delta\eta_3 = \Delta\eta_6 = 0.12$, $\Delta\eta_4 = \Delta\eta_5 = 0.26$

[3] ND = 10, NR = 41 : Δx = 500m, Δt = 150 secs., $\Delta\eta_k$ = 0.1, k=1,..,10

[4] ND = 16, NR = 41 : Δx = 500m, Δt = 150 secs., $\Delta\eta_k$ = 0.0625, k=1,..,16

[5] ND = 40, NR = 41 : Δx = 500m, Δt = 150 secs., $\Delta\eta_k$ = 0.025, k=1,..,40

[6] ND = 16, NR = 41 : Δx = 500m, Δt = 150 secs., $\Delta\eta_1=\Delta\eta_{16}=\Delta\eta_2=\Delta\eta_{15}=0.01$, $\Delta\eta_3=\Delta\eta_{14}=0.02$, $\Delta\eta_4=\Delta\eta_{13}=0.04$, $\Delta\eta_5=\Delta\eta_{12}=0.06$, $\Delta\eta_6=\Delta\eta_{11}=0.08$, $\Delta\eta_7=\Delta\eta_{10}=0.12$, $\Delta\eta_8=\Delta\eta_9=0.16$

[7] ND = 20, NR = 41 : Δx = 500m, Δt = 150 secs., $\Delta\eta_1=\Delta\eta_2=\Delta\eta_{20}=\Delta\eta_{19}=0.005$ $\Delta\eta_3=\Delta\eta_{18}=0.01$, $\Delta\eta_4=\Delta\eta_{17}=0.02$, $\Delta\eta_6=\Delta\eta_5=\Delta\eta_{16}=\Delta\eta_{15}=0.04$, $\Delta\eta_7=\Delta\eta_{14}=0.08$, $\Delta\eta_{10}=\Delta\eta_9=\Delta\eta_8=\Delta\eta_{13}=\Delta\eta_{12}=\Delta\eta_{11}=0.1$

[8] ND = 16, NR = 41 : Δx = 500m, Δt = 150 secs., $\Delta\eta_1=..=\Delta\eta_8=0.1$, $\Delta\eta_9=0.08$, $\Delta\eta_{10}=0.06$, $\Delta\eta_{11}=0.03$, $\Delta\eta_{12}=0.01$, $\Delta\eta_{13}=\Delta\eta_{14}=\Delta\eta_{15}=\Delta\eta_{16}=0.005$

[9] ND = 16, NR = 41 : Δx = 500m, Δt = 150 secs., $\Delta\eta_k$ as in [6] - using Simpson's Rule.

[10] ND = 16, NR = 41 : Δx = 500m, Δt = 150 secs., $\Delta\eta_k$ as in [6]. New boundary condition formulation, Equation (3.9).

Next, the effect of varying the number and size of the depth spacings was investigated. The lines labelled [2] to [8] in Figure 2 present the velocity profile at $x = L/4$ for various grid spacings over depth.

From this figure, it seems that the computed results are very sensitive with respect to the size of the grid spacing near the bottom. The smaller the spacing, the greater the velocity near the bottom. Also, as the grid spacing becomes smaller, so does the height above the bottom at which the maximum velocity occurs.

An extremely fine grid at the bottom (lines [6], [7] and [8] in Figure 2) produces velocity profiles quite logarithmic in nature at the bottom which is the shape the velocity profile is known to take in flow near solid boundaries.

Another important conclusion which can be made on the basis of these results is that a fine grid system is not needed at the surface of the channel. This fact enables one to save quite a deal of computer storage and still obtain equivalent results (compare lines [7] and [8] in Figure 2).

A very coarse vertical grid at mid-depth is not acceptable either, as is apparent from the result [2] obtained when using eight variable vertical grid spacings.

<u>Modelling of the individual terms</u>

The overall implicit manner in which Equations (2.1) were modelled seems to be a practical one, but there may be better ways of discretising individual terms. These will now be investigated.

Firstly, the way in which the vertical component of velocity is calculated may be improved. There are two reasons for using a scheme other than that proposed in Equation (2.5). Firstly, because the $\int_0^\eta$ term involves both odd and even numbers of variably spaced grid points, the trapezoidal rule was used to calculate the integrals. This is only a first-order accurate scheme, and the integrals will be approximated with an error of $O\{\Delta\eta_k\}$. If the differential equation (1.14) was chosen to calculate w, then only $\int_0^1 d\eta$ must be estimated, for which the more accurate Simpson's rule may be used. Secondly, it would save both space and time on the computer, if the integral of u from 0 to η did not have to be calculated.

Dividing both sides of Equation (1.14) by $(\zeta+h)$ gives

$$\frac{\partial w}{\partial \eta} = f(\eta) \tag{3.5a}$$

where

$$f(\eta) = \frac{1}{(\zeta+h)}\{\frac{\partial}{\partial x}[(\zeta+h)\int_0^1 u\, d\eta] - \frac{\partial}{\partial x}[(h+\zeta)u]\}. \tag{3.5b}$$

Equation (3.5) can be solved for a given x and t to give w at discrete points through a given column of water. Because Equation (2.5) is over-determined, with $w = 0$ prescribed at $\eta = 0$ and 1, the following scheme may be adopted: w was set to zero at the surface, and Equation (3.5) used to step down a given column of water. At the next time level, w was set to zero at the bottom and the stepping proceeded up the column.

The following scheme was used to step downwards:

$$w_{k+1} = w_k + \frac{\Delta\eta_k}{2}\{f(\eta_k) + f(\eta_{k+1})\}, \tag{3.6a}$$

which has a second order error. The integral over depth in $f(\eta_k)$ was computed using Simpson's rule for unequally spaced values, namely

$$\bar{u}_i^n = \int_0^1 u\, d\eta\Big|_{i,k}^n$$

$$= \frac{-1}{6}\sum_{k=1(2)}^{ND}\Big\{\frac{(\eta_{k+1}-\eta_{k-1})}{(\eta_k-\eta_{k-1})}(-2\eta_{k-1}+3\eta_k-\eta_{k+1})u_{i,k-1}^n+\frac{(\eta_{k+1}-\eta_{k-1})^3}{(\eta_k-\eta_{k-1})(\eta_{k+1}-\eta_k)}u_{i,k}^n$$

$$+\frac{(\eta_{k+1}-\eta_{k-1})}{(\eta_{k+1}-\eta_k)}(\eta_{k-1}-3\eta_k+2\eta_{k+1})u_{i,k+1}^n\Big\}, \qquad (3.6b)$$

which has a second-order error. In summary, an explicit expression for w when sweeping downwards is thus

$$w_{k+1}^n = w_k^n - \frac{\Delta\eta_k}{8(h_i^n+\zeta_i^n)\Delta x}\{(h_i^n+h_{i+1}^n+\zeta_i^n+\zeta_{i+1}^n)(u_{i,k}^n+u_{i,k+1}^n-2\bar{u}_i^n)$$

$$-(h_{i-1}^n+h_i^n+\zeta_{i-1}^n+\zeta_i^n)(u_{i-1,k}^n+u_{i-1,k-1}^n-2\bar{u}_{i-1}^n)\}, \qquad (3.7)$$

$k=1(1)ND-1$, commencing with $w_{i,1}^n = 0$.

At the next time level an upward sweep is used. The expression for w is

$$w_{k-1}^n = w_k^n - \frac{\Delta\eta_{k-1}}{2}\{f(\eta_k) + f(\eta_{k-1})\}, \qquad (3.8a)$$

which leads to

$$w_{k-1}^n = w_k^n + \frac{\Delta\eta_{k-1}}{8\Delta x(h_i^n+\zeta_i^n)}\{(h_i^n+h_{i+1}^n+\zeta_i^n+\zeta_{i+1}^n)(u_{i,k}^n+u_{i,k-1}^n-2\bar{u}_i^n)$$

$$-(h_{i-1}^n+h_i^n+\zeta_i^n+\zeta_{i-1}^n)(u_{i-1,k}^n+u_{i-1,k-1}^n-2\bar{u}_i^n)\}, \qquad (3.8b)$$

$k=ND+1(-1)2$, commencing with $w_{i,ND+1}^n = 0$.

With these forms for the w-equation it is found that the surface displacement, ζ, and the velocity profiles remain unchanged, since, for most times and at most places, the effect of the vertical velocity w appears to be negigible. It is still worthwhile to examine the actual values obtained for w. Although the method proposed above is 'more accurate' in that all approximations have second order errors, the answers obtained for w are less satisfactory than those obtained using the original method. For instance, the first method gives $w = 0$ at both $\eta = 0$ and $\eta = 1$ whereas with the second method, when sweeping down for example, a value of w at the bottom is obtained which is of the same order of magnitude as those values obtained in the inner regions of the flow.

The effect of using the more accurate Simpson's method for the integration (Equation (3.6)) in the model of the continuity equation was also examined. The line marked [9] in Figure 2 shows that exactly the same results were obtained by both methods. It seems that the variation of u is sufficiently smooth and the vertical grid spacing sufficiently small, particularly near the bottom, that the more simple trapezoidal rule is quite adequate.

The only other place where expressions with first order errors have been used occurs in the approximations used for modelling $\partial u/\partial\eta$ and $\partial E/\partial\eta$ at the surface (see, for example, Equations (2.3k), (2.3ℓ), (2.4h) and (2.4i)).

The following second order scheme was then tried:

$$\left.\frac{\partial u}{\partial \eta}\right|_{i,1}^{n} = \frac{-(r+1)}{\Delta\eta_1 r} u^n_{i,1} + \frac{r}{\Delta\eta_1(r-1)} u^n_{i,2} - \frac{1}{\Delta\eta_1 r(r-1)} u^n_{i,3}, \tag{3.9}$$

where

$$r = \frac{\Delta\eta_2}{\Delta\eta_1} + 1.$$

This yields the following expression for $\frac{\partial}{\partial\eta}(N\frac{\partial u}{\partial\eta})$ at the first depth level:

$$\begin{aligned}\frac{\partial}{\partial\eta}(N\frac{\partial u}{\partial\eta})^{n+1}_{i,1} &= u^{n+1}_{i,1}\{\frac{-4\tilde{N}^n_{i,1}}{(\Delta\eta_1)^2} - \frac{\Delta\eta_2 N^n_{i,2}}{\Delta\eta_1^2(\Delta\eta_1+\Delta\eta_2)}\} \\ &+ u^{n+1}_{i,2}\{\frac{4\tilde{N}^n_{i,1}}{(\Delta\eta_1)^2} - \frac{(\Delta\eta_2-\Delta\eta_1)N^n_{i,2}}{(\Delta\eta_1)^2\Delta\eta_2}\} \\ &+ u^{n+1}_{i,3}\{\frac{-N^n_{i,2}}{(\Delta\eta_1+\Delta\eta_2)\Delta\eta_2}\} - \frac{6\tau_s}{\Delta\eta_1\rho}(\zeta^{n+1}_i + h^{n+1}_i + \zeta^{n+1}_{i+1} + h^{n+1}_{i+1}).\end{aligned} \tag{3.10}$$

This formulation at level one destroys the tri-diagonal system. However, using the u-equation at depth level k = 1 which involves terms in $u^{n+1}_{i,1}$, $u^{n+1}_{i,2}$ and $u^{n+1}_{i,3}$, the term with $u^{n+1}_{i,3}$ may be eliminated from Equation (3.10), thus yielding an expression involving only terms in $u^{n+1}_{i,1}$ and $u^{n+1}_{i,2}$ in the first equation of the set of simultaneous algebraic equations. The Thomas algorithm can then be used on this modified set of equations. A similar treatment was also applied to the appropriate terms in the turbulent energy equation.

The velocity profiles obtained using this formulation are shown by [10] in Figure 2. This is a somewhat surprising result. It seems that this new formulation 'spread's the effect of the surface wind stress over the three top levels thus decreasing the value obtained for the surface velocity. Once again, it seems that the formally 'less accurate' first order method gives a more realistic result.

4. AN EXACT SOLUTION

The much simpler system of equations obtained assuming that N is a constant over depth is given by

$$\frac{\partial\zeta}{\partial t} = \frac{\partial}{\partial x}[H\int_0^1 u\, d\eta], \tag{4.1a}$$

$$\frac{\partial u}{\partial t} = -g\frac{\partial\zeta}{\partial x} + \frac{N}{(H)^2}\frac{\partial^2 u}{\partial\eta^2}, \tag{4.1b}$$

where the depth $h(x,\eta)$ is assumed to be a constant H and $\zeta \ll H$ is assumed to hold. The boundary conditions to this problem are the same as before, namely

$$u = 0 \text{ for } x = 0,L,$$

$$u = \zeta = 0 \text{ at } t = 0,$$

$$u = 0 \text{ on } \eta = 0,$$

$$\rho \frac{N}{H} \frac{\partial u}{\partial \eta} = \tau_s \text{ on } \eta = 1. \qquad (4.2)$$

An analytic solution to this problem with $\tau_s = \tau_0 e^{i\omega t}$ (real part intended) is given by Walsh (1974), namely

$$\zeta(x,t) = \frac{4b\tau_0}{L\rho g} \sum_{n=1}^{\infty} \left\{ \frac{\cos(K_{2n-1}x)}{(K_{2n-1})^2} [\sinh(bH) - R_{2n-1}\cosh(bH)] \right\} e^{i\omega t},$$

$$u(x,t) = \frac{4\tau_0}{L\rho Nb} \sum_{n=1}^{\infty} \frac{\sin(K_{2n-1}x)}{K_{2n-1}} \{[\sinh(bz) + \sinh(bH)] + R_{2n-1}[\cosh(bz) - \cosh(bH)]\} e^{i\omega t},$$

real parts intended, where $\varepsilon_p = g K_p/N$, $b^2 = i\omega/N$, $K_p = p\pi/L$ and

$$R_p = \frac{\sinh(bH)[Nb^5 + \varepsilon_p K_p bH] + \varepsilon_p K_p [1 - \cosh(bH)]}{\cosh(bH)[Nb^5 + \varepsilon_p K_p bH] - \varepsilon_p K_p \sin(bH)} \qquad (4.3)$$

The system of equations (4.1), (4.2) were modelled using the implicit scheme for calculating u with all the other individual terms being modelled as presented in Equations (2.2) and (2.3). When using a grid system defined by $\Delta x = 500$m, $\Delta t = 180$ secs., and $\Delta\eta_k = 0.125$ for k=1(1)8, results were obtained which were in error by less than 10^{-4}m/sec. for the velocity terms. When an even finer grid spacing was used, almost identical results were obtained by the numerical and analytic methods. For example, when comparing the horizontal velocity profile over depth at x = L/4 it was found that the grid spacing $\Delta\eta_k = 0.0625$, k=1(1)16, yielded results which were the same using both methods at all levels except four where the error was only 10^{-4}m/sec. When the variable spacing 16 depth level model was tested, all the values obtained were the same except for one which was in error by 10^{-4}m/sec.

5. CONCLUSION

As has been shown, a numerical scheme in which particular terms of the governing partial differential equations are discretised at the (n+1)th time level with others discretised at the nth time level, provides a very practical model. Indeed, for the set of equations (4.1), results obtained from the numerical scheme agreed exactly with the analytic values.

The results obtained by the numerical solution of the complete set of equations governing wind-induced motion in a closed basin (Equations (1.11), (1.12), (1.13) and (1.15)) suggest a number of conclusions.

Firstly, the results obtained depend critically on the grid spacing at certain boundaries. At the bottom, where a no-slip boundary condition is used, a fine grid spacing is required in the vertical direction. However, the model is less dependent on the grid spacing near the surface, where a derivative boundary condition on the horizontal velocity is used.

Secondly, it seems that wherever possible it is better to model an integral equation than a differential equation. For instance, by carrying out an integration, an explicit formulation for w is achieved (Equation (1.15)) which satisfies both boundary conditions for w. The results obtained using this expression were far more satisfactory than any scheme, no matter how accurate, proposed for modelling the overdetermined first-order differential equation (1.14).

Thirdly, the model employed second-order centred differences everywhere in space, except at the surface where one-sided differences with first order truncation errors were used. A scheme with second order errors for the vertical derivatives at the top depth level was proposed in Equation (3.9). This more complicated scheme introduced added algebra in order to preserve the tri-diagonal matrix, but was found to be less accurate than the simple approximation originally used.

Further modifications being tested include the use of central differencing in time wherever possible to reduce truncation errors and the extension of the model to three-dimensional problems.

6. REFERENCES

Heaps, N.S. (1972) "On the numerical solution of the three-dimensional hydrodynamical equations for tides and storm surges", Roy.Soc.des.Sciences de Liege, Ser.6(2).

Jones, W.P. and Launder, B.E. (1972) "Prediction of laminarization with a 2-equation model of turbulence", Int.J.Heat Mass Transfer, 15, 301-314.

Lamb, H. (1932) "Hydrodynamics", 6th Edition, Cambridge University Press.

Launders, B.E. and Spalding, D.B. (1972) "Mathematical Models of Turbulence", Academic Press.

Launder, B.E. and Spalding, D.B. (1974) "The numerical calculation of turbulent flows", Comp.Methods in Appl.Mech. and Eng., 3, pp.269-289.

Marchuk, G.I., Kochergin, V-P., Klimok, V.I. and Sukhorukov, V.A. (1977) "On the dynamics of the ocean surface mixed layer", J.Phys.Oceanog. 7, pp.865-875.

Noye, B.J. (1981) "Finite Difference methods for partial differential equations" in Numerical Solutions of Partial Differential Equations, ed. J. Noye, North-Holland, pp.1-135.

Vager, B.G. and Kagan, B.A. (1969) "The dynamics of the turbulent boundary layer in a tidal current", Izv.Akad.Nauk.SSSR, Fiz.Atmos.Okeana. 5(2), pp.88-93.

Walsh, P.J. (1974) "Long-wave wind effects on closed lakes, with special applications to the Murray Lakes, South Australia", Ph.D. Thesis, University of Adelaide.

NUMERICAL SOLUTIONS OF
PARTIAL DIFFERENTIAL EQUATIONS
J. Noye (Editor)

CLOUD-IN-CELL AND FINITE-DIFFERENCE MODELS FOR A STRATIFIED LAKE

John A. Church and Rory O.R.Y. Thompson

Division of Oceanography, CSIRO Marine Laboratories
P.O. Box 21, Cronulla, NSW 2230
Australia

Finite-difference techniques are used to solve the partial differential equations which govern the response of a stratified lake to wind stress. Both centred and upwind spatial representation of the advection terms are combined with Adams-Bashforth time integration. Some undesirable properties of these schemes are discussed and illustrated. A cloud-in-cell model that overcomes these deficiencies is discussed. Some early results of this model are presented.

1. INTRODUCTION

The response of a stratified lake to wind stress includes a wide range of time and length scales. One important phenomenon is the deepening of the well-mixed surface layer (the epilimnion) by entrainment of colder water from the region below the thermocline (the hypolimnion). This deepening is the subject of a recent paper by Spigel and Imberger (1980). They used a one-dimensional integral mixed layer model to determine the conditions when either stirring (due to turbulence generated near the surface) or turbulence associated with shear in the thermocline dominated. The response of the lake varied from one (with light wind) where entrainment proceeds very slowly and first mode internal oscillations are present to one (with strong wind) where stratification is destroyed before large scale baroclinic motions occur. Thompson and Imberger (1980) developed a two-dimensional finite-difference model in which mixing was controlled by the local Richardson number.

One of the partial differential equations that is necessary to describe the dynamics of a stratified lake is the advection equation

$$\frac{\partial T}{\partial t} + \frac{(uT)}{\partial x} + \frac{(vT)}{\partial y} = 0 . \qquad 1.1$$

When low order finite-difference techniques are used to solve this equation, numerical errors may arise. These errors can be particularly severe in the presence of sharp fronts, such as a sharp thermocline in a stratified lake. In the present example, space-centred representation of the advection terms, combined with Adams-Bashforth time integration results in negative numerical diffusion and the appearance of Gibbs phenomenon when there is flow perpendicular to the thermocline. When the space centred differences are replaced by upwind-differences, numerical diffusion results in a diffuse thermocline. There are numerous techniques which attempt to overcome these problems including higher order finite-difference techniques (Roache, 1976), spectral and pseudo-spectral techniques (Orszag and Israeli, 1974), Lagrangian methods, and the method which is applied here, the particle-in-cell (Harlow, 1963) or the cloud-in-cell (Birdsall and Fuss, 1969) method.

In the following, the problems associated with the low order finite-difference techniques for simulating the response of a stratified lake will be illustrated and the cloud-in-cell model described. Some initial results of the application of this model (including a short film[1]) will be presented. An essential part of all three models is the vertical mixing scheme suggested by Thompson (1980) and described briefly in Section 3.

2. THE PARTIAL DIFFERENTIAL EQUATIONS

We will consider a two-dimensional stratified lake of length L and depth H. An x-y co-ordinate system is chosen with the x-axis horizontal, the y-axis vertical and the origin at the bottom left-hand corner of the region. The non-linear equations for the conservation of momentum are (after applying the Boussinesq approximation)

$$\frac{\partial u}{\partial t} + \frac{\partial (uu)}{\partial x} + \frac{\partial (vu)}{\partial y} = - \frac{\partial p}{\partial x} \qquad 2.1$$

and

$$\frac{\partial p}{\partial z} = - g\frac{\rho}{\rho_0} \qquad 2.2$$

where p is the pressure divided by the mean density ρ_0. For the present, molecular diffusion and viscosity are neglected. For simplicity, we will consider only the effects of temperature on density. Thus the equation of state is

$$\rho = \rho_0(1+\alpha T) \qquad 2.3$$

where α is the coefficient of thermal expansion. For an incompressible fluid

$$\frac{\partial u}{\partial x} + \frac{\partial v}{\partial y} = 0. \qquad 2.4$$

The equation expressing the conservation of mass (after substituting equation 2.3 for the density) is

$$\frac{\partial T}{\partial t} + \frac{\partial (uT)}{\partial x} + \frac{\partial (vT)}{\partial y} = 0. \qquad 2.5$$

The initial conditions are those of zero velocity and zero thermocline displacement, i.e. a stratified lake before the imposition of any wind stress. The boundary conditions are wind stress τ at the surface, i.e.,

$$vu = \frac{\tau}{\rho_0} \quad \text{at} \quad y = H^- \qquad 2.6$$

and quadratic bottom friction, i.e.,

$$vu = - C_D\, u|u| \quad \text{at} \quad y = 0^+ \qquad 2.7$$

where C_D is the bottom drag coefficient. We will consider zero heat flux through the boundaries. We are interested only in the baroclinic motions so we can also specify

$$v = 0 \quad \text{at} \quad y = 0 \quad \text{and} \quad y = H \qquad 2.8$$

and

$$u = 0 \quad \text{at} \quad x = 0 \quad \text{and} \quad x = L. \qquad 2.9$$

Equations 2.1 to 2.5 may now be solved subject to the boundary conditions 2.6 to 2.9. We also find it convenient to use the stream function ψ where

$$u = - \frac{\partial \psi}{\partial y} \quad \text{and} \quad v = \frac{\partial \psi}{\partial x}. \qquad 2.10$$

Since there is no flow through the boundary we take $\psi = 0$ on the boundary.

3. TURBULENT MIXING

Parameterization of turbulent mixing in stably stratified fluids is a major problem in constructing models. Although there are recent advances, a complete understanding of this problem has not yet been attained. The most common approach is to parameterize the vertical viscosity and diffusivity by empirical formulae such as those suggested by Munk and Anderson (1948) or more recent closure schemes (e.g. Launder and Spalding, 1972). In a recent paper, Thompson (1980) suggested a simple mechanism which may allow modellers to reduce their dependence on arbitrary eddy viscosities and diffusivities. In this method, when the local Richardson number

$$R_i = \frac{g\,\frac{\partial\rho}{\partial y}}{\rho_0\left(\frac{\partial u}{\partial y}\right)^2} \qquad 3.1$$

falls below the critical value $R_{ic} = 0.25$, mixing proceeds rapidly whereas if the Richardson number is larger than 0.25 no mixing occurs. Numerically this can be modelled by turning on a large viscosity or more simply mixing the momentum and temperature completely when the Richardson number falls below critical. Mixing also occurs when the density gradient is unstable. Using this scheme, Thompson (1980) was able to simulate numerically some of the laboratory experiments of Thorpe (1973). This mechanism results in the patchy and sporadic appearance of turbulent regions with sharp edges as was observed by Woods (1969).

4. THE FINITE-DIFFERENCE MODELS

The finite-difference model is that of Thompson and Imberger (1980) and described only briefly here. After combining equations 2.3 and 2.2, the hydrostatic equation is integrated vertically. The finite-difference form used is

$$p^{i,j+1} = p^{i,j} - g\alpha T^{i,j}\,\Delta y \qquad 4.1$$

for j = 1 at the surface to J + 1 at the bottom. The time derivative in equation 2.1 is evaluated using the baroclinic pressure 4.1 and the advective terms. For the centred scheme,

$$\frac{\partial u^{i,j}}{\partial t} = \frac{F_M{}^{i-\frac{1}{2},j} - F_M{}^{i+\frac{1}{2},j}}{\Delta x} + \frac{G_M{}^{i,j+\frac{1}{2}} - G_M{}^{i,j-\frac{1}{2}}}{\Delta y} \quad \text{for } i = 2, I \quad \text{and } j = 1, J \qquad 4.2$$

where the horizontal momentum flux is

$$F_M{}^{i-\frac{1}{2},j} = \frac{1}{2}(p^{i,j} + p^{i,j+1}) + \tilde{u}^{i-\frac{1}{2},j}\,u^{i-\frac{1}{2},j} \qquad 4.3$$

with

$$\tilde{u}^{i-\frac{1}{2},j} = u^{i-\frac{1}{2},j} = \frac{1}{2}(u^{i,j} + u^{i-1,j}). \qquad 4.4$$

The vertical flux of momentum is

$$G_M{}^{i,j+\frac{1}{2}} = v^{i,j+\frac{1}{2}}\,\tilde{u}^{i,j+\frac{1}{2}} \qquad 4.5$$

where

$$v^{i,j+\frac{1}{2}} = \frac{1}{2}(v^{i-1,j+1} + v^{i,+1}) \qquad 4.6$$

and

$$\tilde{u}^{i,j+\frac{1}{2}} = u^{i,j+\frac{1}{2}} = \frac{1}{2}(u^{i,j} + u^{i,j+1}). \qquad 4.7$$

At the surface, the vertical momentum flux is the wind-stress

$$G_M^{i,\frac{1}{2}} = -\frac{\tau}{\rho} \tag{4.8}$$

and at the bottom is

$$G_M^{i,J+\frac{1}{2}} = -C_D u^{i,J}|u^{i,J}|. \tag{4.9}$$

The barotropic component of the horizontal velocity must be zero and therefore

$$\overline{u}^{i} = \frac{1}{J}\sum_{j=1}^{J} u^{i,j} \tag{4.10}$$

is subtracted from each velocity in the column. This ensures that the vertically integrated horizontal flux is zero and also that

$$v^{i,j+1} = v^{i,j} + \frac{\Delta y}{\Delta x}(u^{i+1,j} - u^{i,j}) \tag{4.11}$$

with $v^{i,1} = 0$ also implies $v^{i,J+1} = 0$. This approach was used by Bell and Thompson (1980) for ensuring constancy of the horizontal volume flux. The boundary condition at $i = \frac{1}{2}$ and $I + \frac{1}{2}$ is enforced by maintaining

$$u^{1,j} = -u^{2,j} \text{ and } u^{I+1,j} = -u^{I,j}.$$

The velocities u and v are now available for computing advection in equation 2.5. For the centred scheme

$$\frac{\partial T^{i,j}}{\partial t} = \frac{F_T^{i-\frac{1}{2},j} - F_T^{i+\frac{1}{2},j}}{\Delta x} + \frac{G_T^{i,j+\frac{1}{2}} - G_T^{i,j-\frac{1}{2}}}{\Delta y} \quad \text{for } i = 1 \text{ to } I \text{ and } j = 1 \text{ to } J \tag{4.12}$$

where the horizontal temperature flux is

$$F_T^{i+\frac{1}{2},j} = u^{i+\frac{1}{2},j}\,\frac{T^{i,j} + T^{i+1,j}}{2} \tag{4.13}$$

and the vertical temperature flux is

$$G_T^{i,j+\frac{1}{2}} = v^{i,j+\frac{1}{2}}\,\frac{T^{i,j} + T^{i,j+1}}{2}. \tag{4.14}$$

The time stepping of the equations is completed using the second-order Adams-Bashforth integration

$$f(t+\Delta t) = f(t) + \Delta t\left(\frac{3}{2}f_t(t) - \frac{1}{2}f_t(t-\Delta t)\right) \tag{4.15}$$

which Lilly (1965) found more accurate than the Lax-Wendroff scheme (Roache, 1976). Finally, the solution at the new time level is checked for instabilities due to temperature inversions or dynamic instabilities. Where instabilities occur, temperature and momentum are mixed vertically taking care to avoid any horizontal mixing.

The upwind model is the same as the centred model except for changing four lines of code defining the advective fluxes. In the horizontal momentum flux 4.3,

$$\tilde{u}^{i-\frac{1}{2},j} = \begin{cases} u^{i,j}, & \text{if } u^{i-\frac{1}{2},j} < 0 \\ u^{i-1,j}, & \text{otherwise} \end{cases} \tag{4.16}$$

with $u^{i-\frac{1}{2},j}$ from (4.4). In the vertical momentum flux (4.5),

$$\tilde{u}^{i,j+\frac{1}{2}} = \begin{Bmatrix} u^{i,j}, & \text{if } v^{i,j+\frac{1}{2}} > 0 \\ u^{i,j+1}, & \text{otherwise} \end{Bmatrix} \quad 4.17$$

instead of (4.7). The horizontal temperature flux (4.13) is replaced by

$$F_T^{i+\frac{1}{2},j} = u^{i+\frac{1}{2},j} \begin{Bmatrix} T^{i,j} & , \text{if } u^{i+\frac{1}{2},j} > 0 \\ T^{i+1,j}, & \text{otherwise} \end{Bmatrix}. \quad 4.18$$

The vertical temperature flux (4.14) is replaced by

$$G_T^{i,j+\frac{1}{2}} = v^{i,j+\frac{1}{2}} \begin{Bmatrix} T^{i,j} & , \text{if } v^{i,j+\frac{1}{2}} > 0 \\ T^{i,j+1}, & \text{otherwise} \end{Bmatrix}. \quad 4.19$$

That is, fluxes are computed by using the value from which the "wind" is blowing, rather than the (second-order) average value.

5. NUMERICAL PROBLEMS OF THE FINITE-DIFFERENCE TECHNIQUES

It is well recognized that there are problems in accurately representing the advection terms with finite-difference techniques (see for example Orszag, 1971). A simple example that illustrates this problem occurs when there is flow perpendicular to a sharp front such as a thermocline. For flow from the high to the low temperature region, the temperature of the first grid point past the front increases because of the advection of warm water. However, the temperature of the last grid point before the front also increases in the centred finite-difference scheme. This is because fluid of high temperature flows into the grid cell whereas the equal volume of fluid that must flow out does so at a much lower temperature. In this example, this causes static instabilities due to a region of higher temperature water at the base of the mixed layer. The surface layer is then mixed, redistributing the excess temperature throughout the surface layer and also redistributing the momentum. This is the same problem as Orszag (1971) illustrated for the advection of a passive tracer and Orszag's (1971) "wake of bad numbers" is similar to the inaccurate solution upstream of the thermocline in the present situation.

The problem discussed above can be overcome by changing from the centred space representation to the upwind representation. However, in this scheme, mass transferred from one grid cell to the next is immediately mixed throughout the cell. This mixing exhibits itself as numerical diffusion. For the upwind representation combined with the Adams-Bashforth time integration (referred to as the Crocco method (Roache, 1976, p. 74)), the numerical diffusion is

$$\alpha_e = \frac{1}{2} u^2 \Delta t. \quad 5.1$$

The ratio of the mass transport due to the advection term and due to the numerical diffusion is $uc/\left(\alpha \frac{\partial c}{\partial x}\right)$ where c is the concentration of tracer considered. When there are large variations such as across the thermocline) such that $\frac{\partial c}{\partial x} \sim \frac{c}{\Delta x}$, then this ratio is of order unity and the numerical diffusion is as large as the advection.

The above discussion illustrates some of the potential problems and in this example, these problems are shown to be important (Section 7). The essential problem is that each cell is forced to be homogeneous. Lagrangian grids overcome these problems but in many cases the grids become extremely distorted making the accuracy of the result questionable. These problems are overcome by the cloud-in-cell method described in the next section.

6. THE CLOUD-IN-CELL MODEL

The particle-in-cell method of Harlow (1963) or the cloud-in-cell method of Birdsall and Fuss (1969) uses the Lagrangian form rather than the Eulerian form of the equations. When equations 2.1 and 2.5 are rewritten in Lagrangian form there is considerable simplification due to the absence of the advective terms. The Lagrangian equations are

$$\frac{du}{dt} = -\frac{\partial p}{\partial x} \qquad 6.1$$

and

$$\frac{dT}{dt} = 0. \qquad 6.2$$

Cloud-in-cell methods are commonly used with a vorticity formulation of the equations (Leonard, 1980). Initially, the cloud-in-cell model used here was formulated in terms of vorticity but we could not obtain satisfactory solutions. These difficulties arose because it was not clear how vorticity was distributed when the mixing outlined in Section 3 occurred. To overcome this problem, the model was rewritten in terms of momentum.

In the cloud-in-cell method, the fluid consists of N finite sized clouds each carrying its own momentum and temperature. These particles are advected through the computational region. The size of the clouds is identical to the size of the Eulerian grid which is used to overlay the area. The information on the clouds is interpolated to the grid by the area overlap scheme described later. Time derivatives evaluated on the grid are interpolated back to the clouds. The clouds retain their identity as they move through the Eulerian grid thus eliminating problems of numerical diffusion but with no distortion problems associated with other Lagrangian methods. This simple technique also ensures conservation of mass and momentum. The cost is the necessity to interpolate from clouds to grid and from grid to clouds each time-step.

The first step in the model is to correct the grid velocities for the barotropic component as in the finite-difference models (equation 4.10). This correction is also expressed as a time derivative for interpolation to the clouds. The stream function ψ is then evaluated by vertically integrating equation 2.10 from the surface $j = 1$ to the bottom $j = J + 1$.

$$\psi^{i,j+1} = \psi^{i,j} + u^{i,j+\frac{1}{2}}\,\Delta y. \qquad 6.3$$

The hydrostatic pressure is then evaluated using the finite-difference form of equation 2.2; i.e. equation 4.1, and equation 6.1 evaluated using

$$\frac{du^{i,j}}{dt} = \frac{p^{i+1,j} + p^{i+1,j+1} - p^{i,j} - p^{i,j+1}}{2\Delta x} \qquad 6.4$$

for $i = 2, I-1$ and $j = 2, J$. At the surface and bottom, additional accelerations

$$\frac{\tau}{\rho\Delta y} \qquad \text{and} \qquad -\frac{C_D\, u^{i,j}\left|u^{i,j}\right|}{\Delta y}$$

respectively are included. The boundary conditions at $i = 1$ and I are enforced by leaving $u^{1,j}$ and $u^{I,j}$ equal to zero. Before the time derivatives are interpolated to the clouds, the grid values are checked for dynamic instabilities. Regions of dynamic instability are found by testing the grid Richardson number

$$R_i^{i,j} = \frac{0.5\ gd(T^{i-1,j} + T^{i,j} - T^{i-1,j+1} - T^{i,j+1})\Delta y}{(u^{i,j} - u^{i,j+1})^2} \qquad 6.5$$

against the critical value of 0.25. When $R_i^{i,j}$ falls below 0.25, the cloud list is searched to find all the clouds in the unstable region. The average momentum and temperature of the clouds are found and these averages assigned to each cloud. The grid temperature field is also searched for regions of static instability and a similar mixing procedure implemented. This method ensures that the mass and momentum summed over the clouds is conserved during the mixing. The clouds are moved to their new positions using velocities determined from the stream function and a two-step explicit method; i.e. for the x-coordinates of the clouds X_K

$$\left.\begin{aligned} X_k^{n+\frac{1}{2}} &= X_k^n + 0.5\ u(X_k^n, Y_k^n)\Delta t \\ X_k^{n+1} &= X_k^n + u(X_k^{n+\frac{1}{2}}, Y_k^{n+\frac{1}{2}})\Delta t \end{aligned}\right\} \qquad 6.6$$

and similarly for the y-coordinates where the superscript refers to the time level. The rates of change of the cloud momentums u_k and temperatures T_k are found using bi-linear interpolation of the time derivative fields from the grid values and new cloud values found by

$$\left.\begin{aligned} u_k^{n+1} &= u_k^n + \frac{du}{dt}(X_k^{n+\frac{1}{2}}, Y_k^{n+\frac{1}{2}})\Delta t \\ T_k^{n+1} &= T_k^n + \frac{dT}{dt}(X_k^{n+\frac{1}{2}}, Y_k^{n+\frac{1}{2}})\Delta t \end{aligned}\right\}. \qquad 6.7$$

Finally, new grid values for temperature and vorticity are determined by the area overlap scheme (Birdsall and Fuss, 1969) normalized by the sum of the areas; i.e.

$$u^{i,j} = \frac{\sum_{k=1}^{K} \alpha_{ij}^k\ u_k(X_k, Y_k)}{\sum_{k=1}^{K} \alpha_{ij}^k} \qquad 6.8$$

where α_{ij}^k is the area of the k'th cloud appearing in the cell centred at i,j divided by the area of the cell. Care must be taken in determining velocities used to advect the particles. The most common scheme (Leonard, 1980) is to obtain velocities by differencing the stream function (equation 2.10) and then bi-linearly interpolating these velocities to the particles. The divergence of the velocities so obtained is however, in general, non-zero and this leads to 'clumping' of particles. Zero divergence velocities are obtained using bi-cubic interpolation functions (Connor and Brebbia, 1976, Chapter 3) for ψ. These functions are evaluated using the values of ψ, ψ_x, ψ_y and ψ_{xy} at each grid point. The 'quilt' so obtained has ψ and its first derivatives continuous throughout the field. The velocities are now found from the first derivatives of ψ at the point of interest.

7. RESULTS AND DISCUSSION

The physical situation we are interested in modelling is that of a stratified lake. We consider an idealized two-dimensional (horizontal and vertical) lake of depth H 20 m and length L 10 km. At mid-depth there will be a thermocline several metres thick separating a well-mixed surface layer of 28.5°C and a bottom layer of 10.5°C. A horizontal grid length of 500 m and a vertical grid length of 1 m will be used in all of the models. For the present discussion, only two situations will be considered.

Firstly, one of strong wind (surface stress of 0.7 Nm^{-2}) where, according to the criteria of Spigel and Imberger (1980) and Thompson and Imberger (1981), the upwind end of the thermocline should surface, and rapid mixing occur, before the buoyancy restoring forces can cause first mode internal oscillations.

Secondly, a case of moderate wind (surface stress 0.07 Nm^{-2}) will be considered. For this wind strength, first mode internal oscillations should dominate with only limited mixing. The period of the oscillations for a two layer lake (where the layer depths are h_0 and h_1) is

$$T = \frac{2L}{\left(g' \frac{h_0 \, h_1}{h_0+h_1}\right)^{\frac{1}{2}}} \qquad \text{where} \quad g' = g \, \frac{\Delta\rho}{\rho} \, . \qquad 7.1$$

This is approximately 48 000 s in the present case. In both cases, the wind stress is in the direction of increasing x.

Some of the results of the finite-difference models will now be presented. These models exhibited some important differences from each other and from the study of Spigel and Imberger (1980). The results of the attempt to overcome these differences by the cloud-in-cell model will be presented next.

7a. *Finite-difference model results*

Some of the results of the centred finite-difference model were discussed by Thompson and Imberger (1980). The qualitative behaviour was as expected with first mode internal oscillations occurring for low wind stress. For high wind stress, the thermocline surfaces at the upwind end of the lake and rapid mixing occurs.

Despite the qualitative success of the model, there were some numerical problems. The most obvious of these problems occurs in the case of strong wind. In the temperature field after 20 000 s of model time, there are regions (the stippled areas in Figure 1a) where the temperature is outside the initial range of 10.5°C to 28.5°C. Clearly this phenomenon is physically impossible and the reasons for it were discussed in Section 5. In the temperature field of the upwind scheme (Figure 1b), the initial temperature range is not exceeded and the temperature gradients are everywhere less than in the centred scheme. Also, there has been more entrainment into the surface layer. Both of these phenomena are due to numerical diffusion (Section 5).

For moderate winds, internal oscillations of period approximately 50 000 s are present in both models. The temperature field for the centred and upwind models after one internal oscillation period (50 000 s) are shown in Figures 2a and b. It is very noticeable that for the upwind model the thermocline is far more diffuse than for the centred scheme. This difference is again due to the difference in the numerical diffusion of the two models. Since the stability of the lake to mixing is critically dependent on the vertical temperature gradient in the thermocline, this quantitative difference will be important in determining the mixing rate that occurs for a given wind stress. In both models, the interface is not sharp at the upwind end and there is a longitudinal surface temperature gradient of approximately 3°C. This is in contrast to the work of Spigel and Imberger (1980) who assume a sharp thermocline with no surface temperature gradients.

The stream function in the last half of the oscillation period is also considerably different in the two models. In the upwind model, negative values in the stream function are restricted to the lower half of the water column (Figure 3b) and thus there are positive velocities at the surface. In the centred scheme (Figure 3a), there is a region near the centre where the negative values of the stream function reach the surface. That is, there are negative velocities at the surface and since the wind stress is always positive, the term τu (the energy flux across the surface) will be negative.

7b. *Cloud-in-cell model results*

For the cloud-in-cell model, we used the same vertical and horizontal grid resolution as with the finite-difference models. Four clouds per cell (i.e. a total of 1 600 clouds for the 20 x 20 grid) were used. For this model, the average temperature of the particles is conserved exactly and at no stage does the temperature of any of the particles or grid points go outside the initial temperature range. The Lagrangian aspect of the model also eliminates the numerical diffusion associated with the advection terms in finite-difference models.

For the light wind case, the model does show first mode internal oscillations as predicted by Spigel and Imberger (1980). The thermocline is tilted by a flow in the direction of the wind in the surface layers and a return flow below the thermocline. In the last half of the cycle the thermocline returns to the horizontal as the flow below the thermocline reverses, the surface flow weakens and much of the return flow occurs along the upper surface of the thermocline. After approximately 50 000 s, the process repeats. Figures 4a and b show the temperature field after 25 000 s (approximately half an oscillation period and the time of maximum tilt of the thermocline) and after 50 000 s. At the upwind end, the thermocline is not as diffuse for the cloud-in-cell model as the finite-difference models and the surface horizontal temperature gradient is only 0.4°C, considerably less than in the finite-difference models. In both these aspects, the cloud-in-cell model is in closer agreement with the work of Spigel and Imberger (1980) than the finite-difference models. Runs with time steps of 100 and 200 s were compared for 50 000 s of model time. The only significant difference is that for the larger time step there is a greater shear near the surface, and this is only because the momentum input by the wind in one time step has not yet been mixed through the water column.

For the strong wind case, the upwind end of the thermocline surfaces after approximately 10 000s and entrainment proceeds rapidly. In contrast to the centred finite-difference model, the temperature (Figure 5) always lies within the initial temperature range. Slightly less mixing has occurred than in the upwind model. When the time step was increased from 40 to 200 s, slightly less mixing occurred.

8. CONCLUSIONS

We conclude that the cloud-in-cell model is in better agreement with the study of Spigel and Imberger (1980) than the finite-difference models. This is because the cloud-in-cell model eliminates problems of numerical dispersion and allows a more accurate simulation of sharp fronts for a given resolution than finite-difference models. The unrealistic situation of temperatures greater than or less than the initial temperature range that occurred in the centred finite-difference model does not occur in the cloud-in-cell model. However, the computing cost per time step is larger by a factor of 40 in the present application where the mixing criteria demand the use of short time steps and frequent searching of the cloud list.

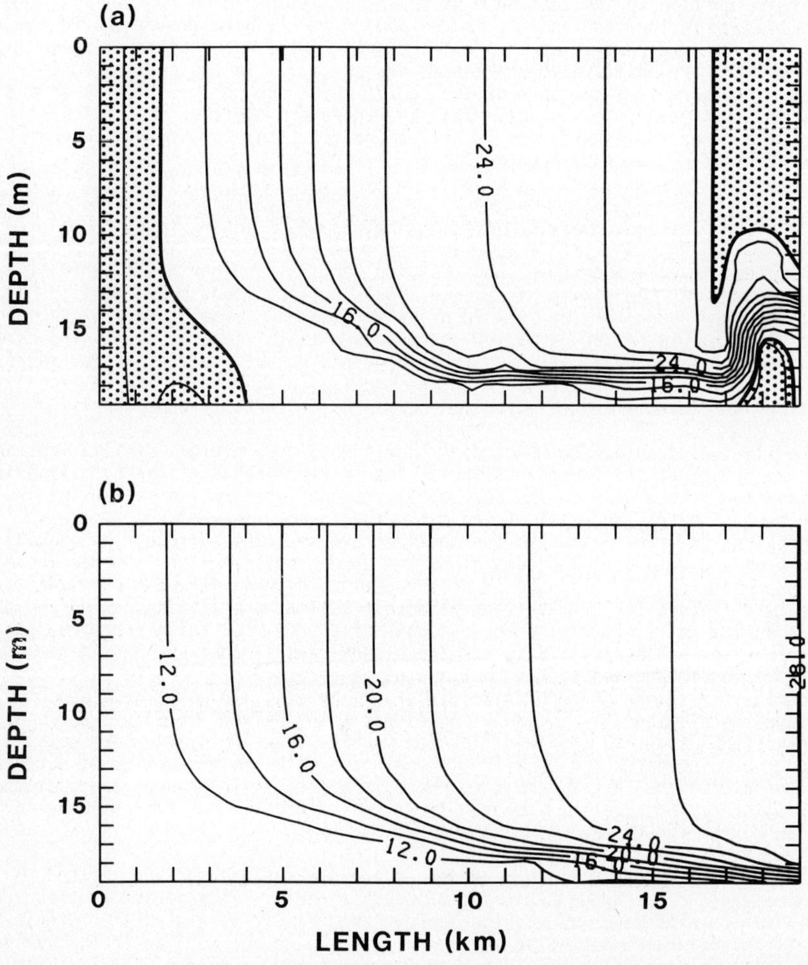

Figure 1. Temperature field for the strong wind case after 20 000 s.
(a) Centred finite-difference model. The stippled areas are regions where the temperature exceeds the initial range of 10.5°C to 28.5°C.
(b) Upwind finite-difference model. The depths and lengths are measured from one half a grid point inside the boundaries of the lake.

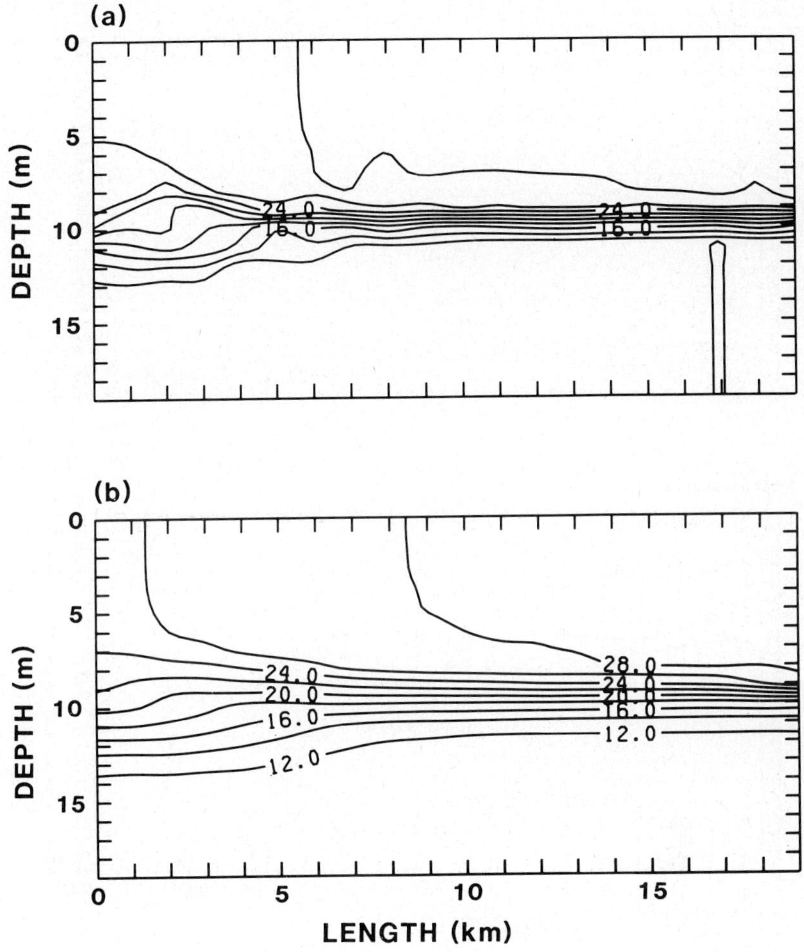

Figure 2. Temperature field for the light wind case after 50 000 s. (a) Centred finite-difference model. (b) Upwind finite-difference model. The depths and lengths are measured from one half a grid point inside the boundaries of the lake.

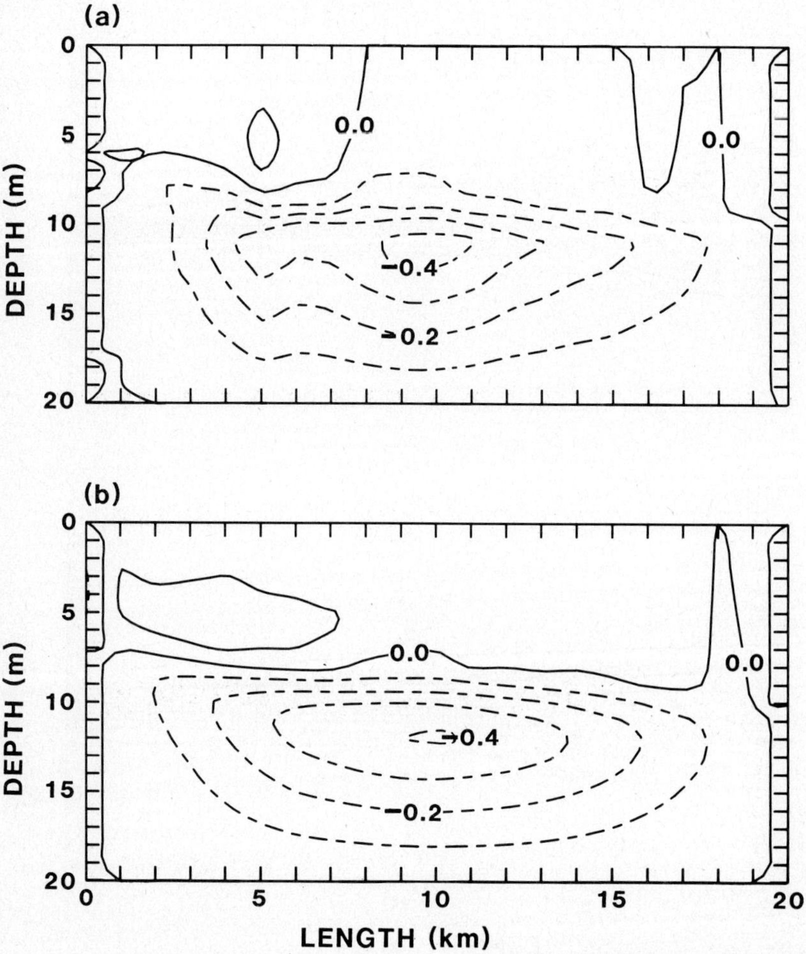

Figure 3. Streamfunction for the light wind case after 40 000 s. (a) Centred finite-difference model. (b) Upwind finite-difference model.

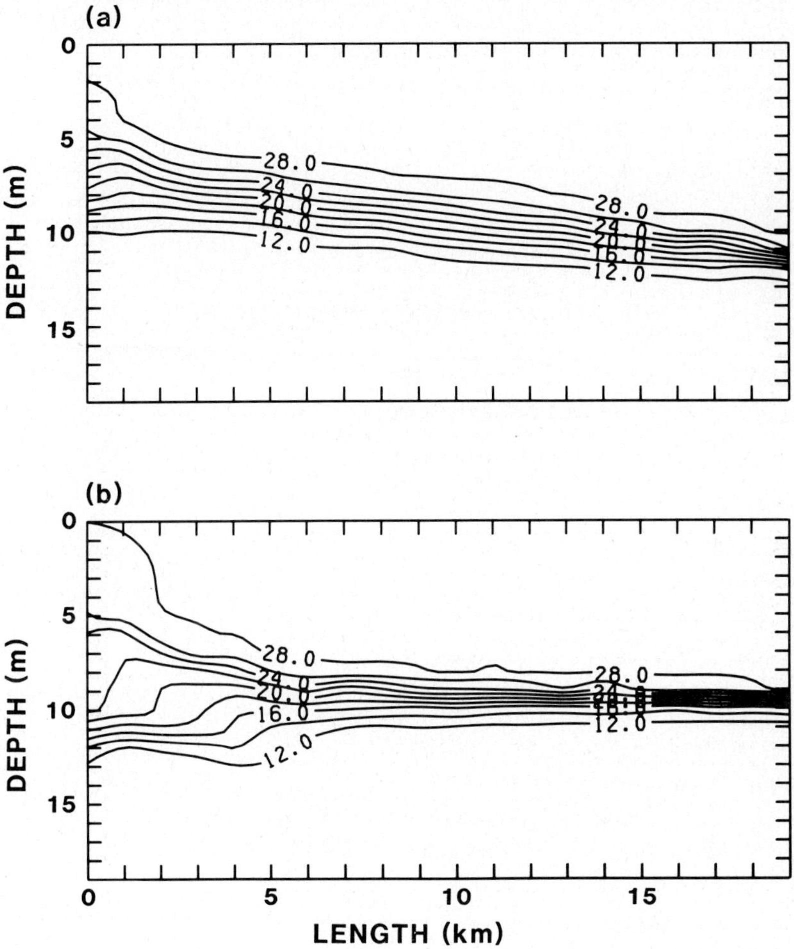

Figure 4. Temperature field for the light wind case for the cloud-in-cell model. (a) After 25 000 s. (b) After 50 000 s. The depths and lengths are measured from one half a grid point inside the boundaries of the lake.

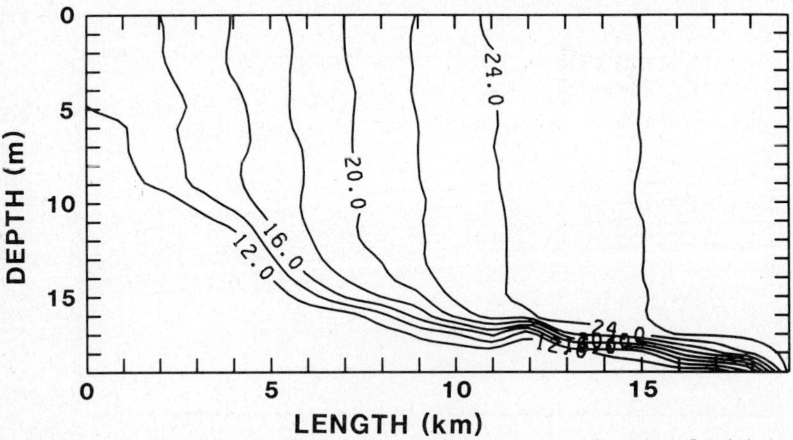

Figure 5. Temperature field for the strong wind case for the cloud-in-cell model after 20 000s. The depths and lengths are measured from one half a grid point inside the boundaries of the lake.

ACKNOWLEDGEMENTS

We thank Jörg Imberger for his hospitality and his initiation of this work. We thank V.A. Wadley for an inspirational conversation.

NOTE: [1]This film was presented at the Conference.

REFERENCES

Bell, Robert C., and Thompson, Rory O.R.Y. (1980), "Valley ventilation by cross winds", *Journal of Fluid Mechanics*, Vol. 96, pp. 757-767.

Birdsall, Charles K., and Fuss, Dieter (1969), "Clouds-in-clouds, clouds-in-cells physics for many-body plasma simulation", *Journal of Computational Physics*, Vol. 3, pp. 494-511.

Connor, J.J., and Brebbia, C.A. (1976), *Finite Element Techniques for Fluid Flow*, Newnes-Butterworths, London.

Harlow, Francis H. (1963), "The particle-in-cell method for numerical solution of problems in fluid dynamics", *Proceedings of Symposia on Applied Mathematics*, Vol. XV, *Experimental Arithmetic, High Speed Computing and Mathematics*, pp. 269-288.

Launder, B.E., and Spalding, D.B. (1972), *Mathematical Models of Turbulence*, Academic Press.

Leonard, A. (1980). "Vortex methods for flow simulation", *Journal of Computational Physics*, Vol. 37, pp. 289-335.

Lilly, Douglas K. (1965), "On the computational stability of numerical solutions of time-dependent non-linear geophysical fluid dynamics problems", *Monthly Weather Review*, Vol. 93, pp. 11-26.

Munk, W.H., and Anderson, E.R. (1948), "Notes on a theory of the thermocline", *Journal of Marine Research*, Vol. 7, pp. 276-295.

Orszag, Stephen A. (1971), "Numerical simulation of incompressible flows within simple boundaries: accuracy", *Journal of Fluid Mechanics*, Vol. 49, pp. 75-112.

Orszag, Stephen A., and Israeli, Moshe (1974), "Numerical simulation of viscous incompressible flows", *Annual Review of Fluid Mechanics*, Vol. 6, pp. 281-318.

Roache, Patrick J. (1976), *Computational Fluid Dynamics*, Hermosa Publishers, Albuquerque.

Spigel, Robert H., and Imberger, Jörg (1980), "The classification of mixed-layer dynamics in lakes of small to medium sizes", *Journal of Physical Oceanography*, Vol. 10, pp. 1104-1121.

Thompson, Rory O. Rafi Y. (1980), "Efficiency of conversion of kinetic energy to potential energy by a breaking internal gravity wave", *Journal of Geophysical Research*, Vol. 85, pp. 6631-6635.

Thompson, Rory O.R.Y., and Imberger, Jörg (1980), "Response of a numerical model of a stratified lake to wind stress", *Proceeding of the Second International Symposium on Stratified Flows*, Trondheim.

Thorpe, S.A. (1973), "Turbulence in stably stratified fluids: a review of laboratory experiments", *Boundary-Layer Meteorology*, Vol. 5, pp. 95-119.

Woods, J.D. (1969), "On Richardson number as a criterion for laminar turbulent laminar transition in the atmosphere and ocean", *Radio Sciences*, Vol. 4, pp. 1289-1298.

NUMERICAL SOLUTIONS OF PARTIAL DIFFERENTIAL EQUATIONS
J. Noye (Editor)

DISPERSION IN A MODEL OF A LAYERED AQUIFER-AQUICLUDE SYSTEM: A SIMPLE NUMERICAL TREATMENT

Malcolm R. Davidson

AAEC Research Establishment
Private Mail Bag, Sutherland 2233, NSW
AUSTRALIA

A mathematical model which predicts a dispersion of solute in underground water over time scales exceeding 10^4 years is described. In its simplest form, the model postulates a diffusion limited exchange of material between flowing groundwater and the stationary water in the confining aquiclude leading to a spread of solute in the mobile water. A simple numerical procedure for solving the governing equations allows the extension of the model to more general cases for which analytical solutions are unavailable.

INTRODUCTION

Groundwater contains a number of natural tracers such as dissolved ions, the stable isotopes of hydrogen and oxygen, radiogenic and atmospherically derived inert gases, natural and man-made tritium and carbon-14. Measurements of such tracers combine to answer a broad range of questions about the influences to which the groundwater has been exposed (Mazor, 1976). In particular, the concentrations of the stable isotopes of water and dissolved noble gases have been correlated with climatic trends (see e.g. Bath et al., 1979; Mazor, 1972); inferences can therefore be made about the climatic history at the recharge intake area over a period commensurate with the residence time of the groundwater by relating the observed tracer distribution to the response of a model aquifer system. Typically, aquifers with residence times of 10^4 - 10^5y are considered. The usual approach in such cases is simply to ignore dispersion processes completely and tacitly to assume a groundwater piston flow model.

There is already an extensive body of literature devoted to the description of transport processes in porous media. The dominant large scale mechanical dispersion in geologic systems is believed to be produced by gross changes in local flow directions caused by non-uniformities in hydraulic conductivities (Smith and Schwartz, 1980). However, although the predicted dispersion effects are orders of magnitude greater than those anticipated from laboratory studies, they can still be ignored when the length scale along the aquifer, over which concentration differences are to be investigated, is much greater than the dispersion distance $(D_L T)^{\frac{1}{2}}$. For instance, when the length scale is the distance travelled in time T by groundwater at velocity V (i.e. VT) and the dispersivity D_L/V is 3 m and V = 3 m/y, then VT = 100 $(D_L T)^{\frac{1}{2}}$ when T = 10^4y. Questions regarding the validity of choosing a constant value of D_L are considered by Smith and Schwartz (1980).

A different dispersion mechanism, which may be important on time scales where the usual mechanical dispersion (given by D_L) might be ignored, is described below. In its simplest form it postulates a transverse diffusion limited exchange of material between flowing groundwater in the aquifer (e.g. sandstone) and the stationary water in a confining aquiclude (e.g. clay, shale), leading to a longitudinal spread of solute in the mobile water. A characteristic time for the process is ℓ^2/D, where ℓ is the thickness of the aquiclude and D is the effective diffusion coefficient of the solute therein.

When D = 0.01 m^2/y for example, ℓ^2/D increases from 10^4y when ℓ = 10m to about 10^5y when ℓ = 30m. Davidson and Airey (to be published) use this model to interpret aspects of the distribution of climate sensitive natural tracers in groundwater in terms of paleoclimatic variations at recharge. The model has also been used by Skopp and Warrick (1974) and Turner (1958) to represent a porous medium with an internally distributed stationary phase (e.g. water in the boundary region of pore spaces) for which ℓ is very small. Other theoretical considerations involving stratified media have been made by Shamir and Harleman (1967), Fried and Combarnous (1971), and Al-Niami and Rushton (1979).

In this paper, a simple numerical procedure for solving the governing equations is presented. Calculations are made of the response to a step function input for which an analytical solution has given by Skopp and Warrick (1974). The method allows for the extension of the model to more general cases for which analytical solutions are unavailable. Following a comparison of these results with those for mechanical dispersion given by constant D_L, it is concluded that, although it may be appropriate to ignore D_L on time scales exceeding 10^4y, it may not be correct to assume a piston flow model in the presence of solute exchange between aquifer and aquiclude water, depending on the aquifer and aquiclude thicknesses.

PROBLEM DEFINITION

Consider an idealized homogeneous, isotropic, semi-infinite aquifer of height L with an adjacent confining aquiclude of thickness ℓ which is also homogeneous and isotropic. This layered system is represented schematically in Figure 1. Attention is focused on the saturated zone and it is assumed that no water reaches the aquiclude from above; all water is presumed to enter the system at the recharge zone (x = 0). Flow in the aquifer is taken to be one-dimensional with a constant pore velocity V (Darcy velocity/effective porosity) whereas water contained in the aquiclude is stationary.

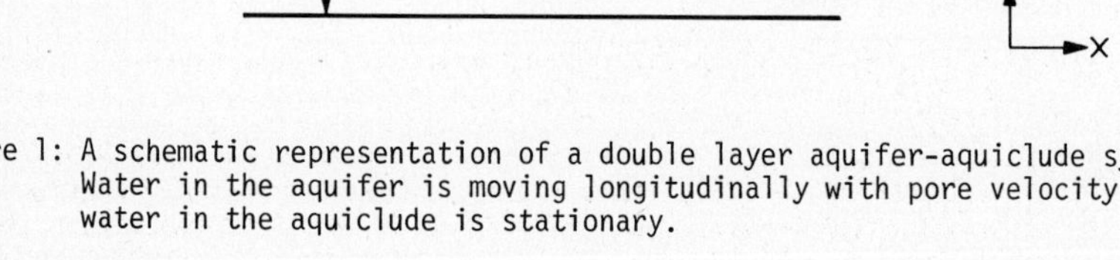

Figure 1: A schematic representation of a double layer aquifer-aquiclude system. Water in the aquifer is moving longitudinally with pore velocity V and water in the aquiclude is stationary.

This paper is concerned with the longitudinal convection and dispersion in the aquifer, from the recharge zone, of a non-reacting solute which can diffuse into and out of the adjacent aquiclude. The effects of radioactive decay of the solute and of adsorption onto the solid matrix of the system are ignored here but they can be easily included in the model. Solute concentrations are assumed to be low, so velocity components induced by mass transfer between layers can be ignored.

The convection-diffusion equations for the solute, in terms of rectangular coordinates (x,y) and time t, are

$$\frac{\partial C}{\partial t} + V\frac{\partial C}{\partial x} = D_T\frac{\partial^2 C}{\partial y^2} \tag{1}$$

and
$$\frac{\partial Q}{\partial t} = D\frac{\partial^2 Q}{\partial y^2}\ , \tag{2}$$

where longitudinal diffusion in the aquiclude and longitudinal mechanical dispersion in the aquifer have both been ignored (i.e. terms involving $\partial^2C/\partial x^2$ and $\partial^2Q/\partial x^2$ are not included). Henceforth, dispersion associated with the $\partial^2C/\partial x^2$ term will be called single layer dispersion while that governed by equations (1) and (2) will be called double layer dispersion.

Solute concentrations in aquifer and aquiclude water are given by C(x,y,t) and Q(x,y,t) respectively. D_T is the transverse dispersion coefficient in the aquifer and D is the molecular diffusion coefficient in the aquiclude. Boundary conditions describing continuity of solute concentration and flux at the interface y = L are

$$C(x,L,t) = Q(x,L,t) \tag{3a}$$

and
$$\mu_1 D_T \frac{\partial C}{\partial y}(x,L,t) = \mu_2 D\frac{\partial Q}{\partial y}(x,L,t) \tag{3b}$$

where μ_1 and μ_2 are the porosities of the aquifer and the aquiclude respectively. Zero flux conditions $\partial C/\partial y = 0$ and $\partial Q/\partial y = 0$ are chosen at y = 0 and $y = L + \ell$ respectively. Taking the average of equation (1) over the thickness L of the aquifer gives

$$\frac{\partial C_m}{\partial t} + V\frac{\partial C_m}{\partial x} = \frac{\mu_2}{\mu_1}\frac{D}{L}\left(\frac{\partial Q}{\partial y}\right)_{y=L}\ , \tag{4}$$

after applying the boundary conditions at y = 0 and y = L, and defining $C_m = \frac{1}{L}\int_0^L C\,dy$. The term on the right of equation (4) represents the flux of solute passing between aquifer and aquiclude.

Now diffusion in the aquiclude is much slower than transverse dispersion in the aquifer (i.e. $D \ll D_T$). Hence transverse gradients ($\partial C/\partial y$) in the aquifer will be much smaller than those in the aquiclude provided that the transverse concentration distribution in the aquifer is initially uniform or that sufficient time has elapsed for the $\partial C/\partial y$ to decay (on a time scale corresponding to diffusion in the aquiclude, transverse dispersion in the aquifer is instantaneous). Under these circumstances, C may be replaced by C_m in condition (3a) and the problem becomes identical to that described by Skopp and Warrick (1974), for which the Laplace transform solution has the form

$$\tilde{C}_m = \tilde{F}(s)\exp\left(-\frac{sx}{V} - \frac{x}{VL}\frac{\mu_2}{\mu_1}(sD)^{\frac{1}{2}}\tanh \ell(s/D)^{\frac{1}{2}}\right) \tag{5}$$

when $C_m(x,0) = Q(x,y,0) = 0$, $C_m(0,t) = F(t)$ and $\tilde{C}_m$ and $\tilde{F}$ are Laplace transforms of C_m and F respectively (e.g. $\tilde{C}_m = \int_0^\infty e^{-st} C_m dt$).

Skopp and Warrick (1974) inverted equation (5) for a step function input (F(t)=1) to give the solution

$$C_m = \frac{2}{\pi} \int_0^\infty \exp\left(-\frac{a\lambda}{2}\left(\frac{\sinh b\lambda - \sin b\lambda}{\cosh b\lambda + \cos b\lambda}\right)\right)$$

$$\cdot \left[\sin\left((t - \frac{x}{V}) \frac{\lambda^2}{2} - \frac{a\lambda}{2}\left(\frac{\sinh b\lambda + \sin b\lambda}{\cosh b\lambda + \cos b\lambda}\right)\right)\right.$$

$$\left. - \quad \sin\left(-\frac{a\lambda}{2}\left(\frac{\sinh b\lambda + \sin b\lambda}{\cosh b\lambda + \cos b\lambda}\right)\right)\right] \frac{d\lambda}{\lambda} \text{ for } t > \frac{x}{V} \tag{6}$$

and $\qquad C_m = 0 \text{ for } t \leq x/V$

where $\qquad a = (\mu_2 D^{\frac{1}{2}} x)/(\mu_1 VL) \quad , \quad b = \ell/D^{\frac{1}{2}}.$

An independent analysis shows that when $Dt/\ell^2 << 1$ (i.e. the elapsed time is much smaller than the characteristic transverse diffusion time for the aquiclude), this solution may be approximated by that corresponding to an aquiclude of infinite thickness; viz.

$$C_m = \text{erfc}(\tfrac{1}{2}\, a/(t - x/V)^{\frac{1}{2}}) \quad \text{for } t > \frac{x}{V} \tag{7}$$

$$\quad = 0 \qquad\qquad \text{for } t \leq \frac{x}{V} \quad .$$

In principle, it is now possible to determine the response to an arbitrary input function by means of a convolution integral. However, in practice it is more convenient to use a numerical method. The details of this technique are given below.

THE NUMERICAL METHOD

A convenient and rapid method of solving equations (2) and (4) for an arbitrary input function is the numerical technique described below. A similar method has been described by Bailey and Gogarty (1962) for the numerical determination of dispersion in straight tubes.

(i) Define nodal points (i = 1,...I) of constant spacing Δx along the length of the one dimensional model aquifer and, at each point, define nodes (j = 1,...J) of constant spacing Δy along the transverse (y) direction. The objective is to solve equations (2) and (4) by replacing them with difference equations at each nodal point and then marching through time t in discrete steps (n = 1,...N) of length Δt. Let $(C_m)_i^n$ be the value of C_m at position $x = i\Delta x$ and $Q_{i,j}^n$ be the value of Q at $x = i\Delta x$ and $y - L = j\Delta y$, all evaluated at time $t = n\Delta t$.

(ii) Let $\Delta x = V\Delta t$ and move the concentration distribution C_m forward a distance Δx by setting $(C_m)_i^* = (C_m)_{i-1}^n$. This is equivalent to solving

$$\frac{\partial C_m}{\partial t} + V\frac{\partial C_m}{\partial x} = 0 \quad . \tag{8}$$

(iii) Now solve equation (2) by replacing it with the following difference equation:

$$\frac{Q_{i,j}^{n+1} - Q_{i,j}^{n}}{\Delta t} = D\,\frac{Q_{i,j+1}^{n+1} - 2Q_{i,j}^{n+1} + Q_{i,j-1}^{n+1}}{(\Delta y)^2} \quad , \tag{9}$$

where $Q_{i,1}^{n+1} = (C_m)_i^*$ and $Q_{i,J}^{n+1} = Q_{i,J-1}^{n+1}$ are the corresponding difference equations for the boundary conditions. The solution of this implicit equation is a standard procedure (see e.g. Richtmyer and Morton, 1967).

(iv) Finally derive $(C_m)_i^{n+1}$ from the difference approximation to

$$\frac{\partial C_m}{\partial t} = \frac{\mu_2}{\mu_1}\frac{D}{L}\left(\frac{\partial Q}{\partial y}\right)_{y\,=\,L} \quad ,$$

viz.
$$\frac{(C_m)_i^{n+1} - (C_m)_i^*}{\Delta t} = \frac{\mu_2}{\mu_1}\frac{D}{L}\,\frac{Q_{i,2}^{n+1} - Q_{i,1}^{n+1}}{\Delta y} \quad . \tag{10}$$

The above procedure involves shifting the solute distribution $C(x,t)$ in the flowing water a distance $\Delta x = V\Delta t$ (so that no artificial numerical dispersion is introduced) followed by transverse diffusion in the stationary region, using fully implicit differencing and a subsequent readjustment of $C(x,t)$ according to the calculated flux across the interface. Results were checked against the analytical solutions (6) and (7) for a step input function; better than 1 per cent accuracy could always be achieved by choosing a small enough step size.

The numerical method described here enables the easy determination of the response of an aquifer-aquiclude system to an arbitrary input function from an arbitrary initial state. The technique may be used to extend the model to general multiple layer systems.

DISCUSSION

Figure 2 represents a comparison between double layer (solid curves) and single layer (dashed curves) dispersion in the case of a step input. The solid curves are determined numerically and the dashed curves are calculated from equation (10.6.22) in Bear (1972) for $V = 3$m/y and $D_L = 10\text{m}^2/\text{y}$, and plotted against dimensionless distance X and dimensionless time τ in Figure 2 assuming that $\ell = 10$m and $D = 0.01\text{m}^2/\text{y}$. Increasing ℓ has the effect of steepening the dashed curves.

First, it is noted that double layer dispersion becomes significant when τ is of order one (i.e. t of order ℓ^2/D) for $K\ (= (\mu_2\ell/\mu_1 L)) = 0.25$. This critical value of τ varies inversely with the exchange capacity coefficient K (as was expected). The next important observation is that it is indeed appropriate to ignore single layer dispersion when it is compared with the dispersion mechanism described here with K = 0.25 and $\tau > 1$, since the dashed curves become much steeper than the solid curves in Figure 2.

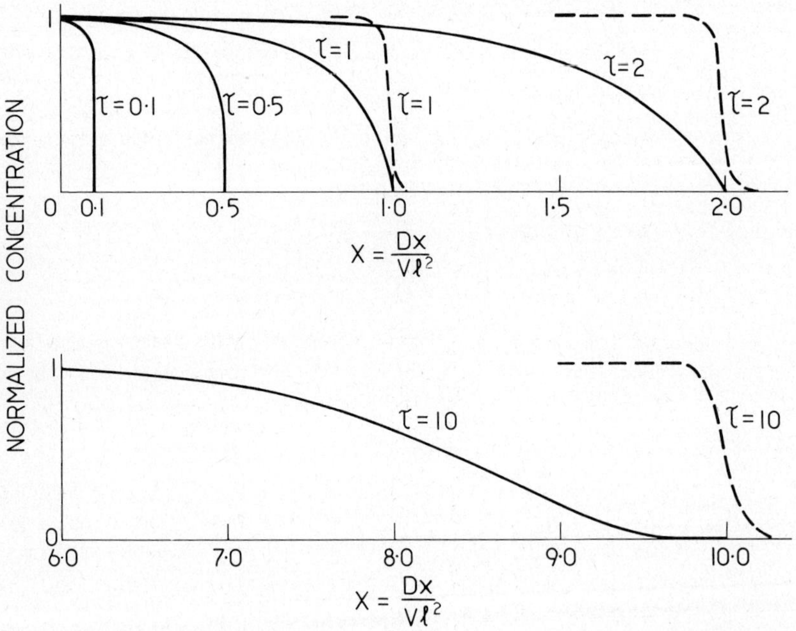

Figure 2: Response of the model aquifer to a step function input of solute concentration. The solid curves illustrate the dispersion due solely to a diffusion limited exchange of solute between flowing aquifer water and immobile aquiclude water when $K = (\mu_2 \ell/\mu_1 L) = 0.25$. The dashed curves represent only mechanical dispersion due to the non-uniformities in hydraulic conductivities for $V = 3$m/y, $D_L = 10\text{m}^2/\text{y}$, and are plotted against dimensionless variables $X = (Dx/V\ell^2)$ and $\tau = Dt/\ell^2$ assuming $\ell = 10$m, $D = 0.01$ m^2/y.

The choice of K = 0.25 in the model is based on the following parameter values: $\mu_2 = 0.05$, $\mu_1 = 0.20$ and $\ell/L = 1$ (i.e. identical thickness for both aquifer and aquiclude). The choice of μ_2 assumes an aquiclude such as shale. Alternatively, for a clay aquiclude we might choose $\mu_2 = 0.5$, $\mu_1 = 0.2$ and $\ell/L = 0.1$ to again obtain K = 0.25. For fixed L, double layer dispersion increases with increasing ℓ, but the effect decreases as L increases.

It is concluded that double layer dispersion can become significant over time scales which depend on the aquifer and aquiclude thicknesses and that it needs to be taken into account when interpreting tracer concentrations in terms of climatic parameters.

ACKNOWLEDGEMENT

The modelling ideas in this work arose out of some stimulating discussions with Peter Hartley of the AAEC Research Establishment. The author also wishes to thank Peter Airey (AAEC) for helpful discussions on model applications.

REFERENCES

[1] Al-Niami, A.N.S. and Rushton, K.R., Dispersion in stratified porous media: analytical solutions, Water Resources Research 15 (1979) 1044-1048.

[2] Bailey, H.R. and Gogarty, W.B., Numerical and experimental results on the dispersion of a solute in a fluid in laminar flow through a tube, Proc. Roy. Soc. A269 (1962) 352-369.

[3] Bath, A.H., Edmunds, W.M. and Andrews, J.N., Paleoclimatic trends deduced from the hydro-geochemistry of a triassic sandstone aquifer, U.K., in: Proc. Symposium on Isotope Hydrology Vol. II (IAEA, Vienna, 1979).

[4] Bear, J., Dynamics of fluids in porous media (Elsevier, N.Y., 1972).

[5] Davidson, M.R. and Airey, P.L., The effect of dispersion on the establishment of a paleoclimatic record from groundwater, to be published.

[6] Fried, J.J. and Combarnous, M.A., Dispersion in porous media, Advances in Hydrosciences 7 (1971) 169-282.

[7] Mazor, E., Paleotemperatures and other hydrological parameters deduced from noble gases dissolved in groundwaters; Jordan Rift Valley, Israel, Geochim. Cosmochim. Acta 36 (1972) 1321-1336.

[8] Mazor, E., Multitracing and multisampling in hydrological studies, in: IAEA Advisory Group on the Interpretation of Environmental Isotope and Hydrochemical Data in Groundwater Hydrology (Vienna, 1976).

[9] Richtmyer, R.D. and Morton, K.W., Difference methods for initial-value problems (2nd ed., Interscience, N.Y., 1967).

[10] Shamir, U.Y. and Harleman, D.R.F., Dispersion in layered porous media, J. Hydraul. Div. Am. Soc. Civil Eng. 93(HY5) (1967) 237-260.

[11] Skopp, J. and Warrick, A.W., A two-phase model for the miscible displacement of reactive solutes in soils, Soil Sci. Soc. Am. Proc. 38 (1974) 525-550.

[12] Smith, L. and Schwartz, F.W., Mass transport I. A stochastic analysis of macroscopic dispersion, Water Resources Research 16 (1980) 303-313.

[13] Turner, G.A., The flow-structure in packed beds - a theoretical investigation utilizing frequency response, Chem. Eng. Sci. 7 (1958) 156-165.

NUMERICAL SOLUTIONS OF
PARTIAL DIFFERENTIAL EQUATIONS
J. Noye (Editor)

A REVIEW OF THE METHODS OF SOLUTION OF HYPERBOLIC EQUATIONS FOR ELECTRON AND PLASMA MOTION IN UNIFORM AND NON-UNIFORM ELECTRIC FIELDS

R. Morrow

CSIRO Division of Applied Physics,
Sydney, Australia 2070

A review is undertaken of the methods of solution of the electron continuity equations which are hyperbolic. The methods examined are the Donor Cell, Runge-Kutta, Lax-Wendroff, the method of characteristics and the flux-corrected transport (FCT) algorithms of Boris and Book with the later modifications introduced by Zalesak. It is found that the "Phoenical LPE Shasta" FCT algorithm with the flux limiting algorithm of Zalesak gives the best results. An example is presented of the effect of space charge on the motion of ions and electrons in a plasma.

INTRODUCTION

The continuity equation for the drift of electrons in an electric field is

$$\frac{\partial N}{\partial t} = - \frac{\partial (NW)}{\partial x} \tag{1}$$

where t is the time variable, x is the space variable, N is the electron number density, and W the electron drift velocity which is a function of the electric field. The term on the right hand side of equation (1) is the drift term and is the most troublesome in solving this class of equation. Thus the purpose of this paper is to review all the methods of dealing with this term, neglecting source terms and diffusion terms which are not subject to the same problems.

The methods are presented in increasing order of accuracy and, generally, complexity. The Donor Cell method uses Euler integration in time and "upwind differencing" for the space variable, Chu (1978). Most texts stress the stability of this method without mentioning that it suffers from strong numerical diffusion with a diffusion coefficient dependent on the velocity, time step and mesh size, Roache (1972). Strang (1964) has suggested that a fourth-order Runge-Kutta method in the time variable coupled with fourth-order central differences in the space variable should be stable without diffusion. The practical use of high order methods for hyperbolic systems has recently been advocated by Turkel (1978). Thus the effect of using a fourth-order Runge-Kutta method in time with upwind differencing, simple central differencing and fourth-order central differencing of the space variable was tested.

The Lax-Wendroff method is often applied to shock waves, Roache (1972), and thus should work well in non-uniform fields. Both the one- and two-step methods were evaluated.

The hybrid method of characteristics due to Hartree, Ames (1977), which has been applied to a similar problem, Davies et al (1971), was also applied.

Finally the flux-corrected transport algorithms (FCT) of Boris and Book (1976 a and b) were applied. The particular FCT algorithm eventually adopted was the "Phoenical LPE Shasta" algorithm which is one of the most accurate of the range of algorithms they tested and is also convenient for use with source terms and boundary conditions. The original flux-correction algorithm developed by Boris and Book was applied as well as the new flux correction algorithm proposed by Zalesak (1979). As a practical example of the use of these methods for solving the motion of ions and electrons in non-uniform fields some results are presented which show the effect of space-charge electric fields on a drift-dominated plasma.

METHODS OF SOLUTION

a) Donor Cell - with Euler integration it is necessary to use upwind differencing for stability, as follows:

$$\begin{aligned} N_i^{j+1} &= N_i^j - \frac{\Delta t}{\Delta x}\left[U_i^j - U_{i-1}^j\right] \quad \text{for} \quad W_i^j \geqslant 0 \\ &= N_i^j - \frac{\Delta t}{\Delta x}\left[U_{i+1}^j - U_i^j\right] \quad \text{for} \quad W_i^j < 0 \end{aligned} \tag{2}$$

where $U_i^j = N_i^j W_i^j$, where j refers to the time step and i to the grid position.

b) The Runge-Kutta Method - a standard fourth-order Runge-Kutta integration was used for the time variable, Abramowitz and Stegun (1965). The space variable was treated in turn using upwind differences, simple central differences and fourth-order central differences, Wylie (1960), as suggested by Strang (1964). The equation used for fourth-order central differences is as follows:

$$\frac{\partial U_i^j}{\partial x} = \frac{1}{12\Delta x}\left(U_{i-2}^j - 8U_{i-1}^j + 8U_{i+1}^j - U_{i+2}^j\right), \tag{3}$$

c) The Lax-Wendroff Method - the standard two-step Lax-Wendroff method, Roache (1972), was applied as follows:

<u>Auxiliary Step</u>:

$$N_{i+\frac{1}{2}}^{j+\frac{1}{2}} = \tfrac{1}{2}\left(N_i^j + N_{i+1}^j\right) - \frac{\Delta t}{2\Delta x}\left(U_{i+1}^j - U_i^j\right) \tag{4}$$

<u>Main Step</u>:

$$N_i^{j+1} = N_i^j - \frac{\Delta t}{\Delta x}\left(U_{i+\frac{1}{2}}^{j+\frac{1}{2}} - U_{i-\frac{1}{2}}^{j+\frac{1}{2}}\right) \tag{5}$$

The original one-step Lax-Wendroff method was also applied, in this case it has the following form:

$$\begin{aligned} N_i^{j+1} = N_i^j &- \frac{1}{2}\frac{\Delta t}{\Delta x}\left[U_{i+1}^j - U_{i-1}^j\right] \\ &+ \frac{1}{4}\left(\frac{\Delta t}{\Delta x}\right)^2\left[\left(W_{i+1}^j + W_i^j\right)\left(U_{i+1}^j - U_i^j\right)\right. \\ &\left. - \left(W_i^j + W_{i-1}^j\right)\left(U_i^j - U_{i-1}^j\right)\right] \end{aligned} \tag{6}$$

d) The Hybrid Method of Characteristics - the method used is the hybrid method due to Hartree, Ames (1977), equation (1), neglecting the source term S, being written:

$$\frac{\partial N}{\partial t} + W\frac{\partial N}{\partial x} = - N\frac{\partial W}{\partial x}, \tag{7}$$

or

$$\frac{DN}{Dt} = - N\frac{\partial W}{\partial x}. \tag{7}$$

Equation (7) is then integrated along characteristic lines for only one time step to give a solution at regular grid points. The method, which is of second order, is outlined succinctly by Ames (1977) and some useful details are given by Davies and Evans (1973). Briefly the method involves an iterative determination of the best position of the characteristic line which terminates on a grid point at time $t + \Delta t$. This involves interpolation of density values between grid points at time t. The method is quite complex particularly when positive and negative ions are included, as each species has separate characteristic lines. All must be solved simultaneously because the equations are coupled.

e) Phoenical LPE Shasta - The Phoenical LPE Shasta algorithm is only one of the possible algorithms suggested by Boris and Book (1976 a). The complete details of this algorithm are outlined below:

1. Compute a transported and diffused intermediate solution, $\overset{*}{N}{}_i^{j+1}$, using an equation which in our notation is

$$\begin{aligned}\overset{*}{N}{}_i^{j+1} = N_i^j &- \tfrac{1}{2}\left[\varepsilon_{i+\frac{1}{2}}(N_{i+1}^j + N_i^j) - \varepsilon_{i-\frac{1}{2}}(N_i^j + N_{i-1}^j)\right] \\ &+ \left[\nu_{i+\frac{1}{2}}(N_{i+1}^j - N_i^j) - \nu_{i-\frac{1}{2}}(N_i^j - N_{i-1}^j)\right]\end{aligned} \tag{8}$$

where $\varepsilon_{i+\frac{1}{2}} = W_{i+\frac{1}{2}}.\ \Delta t/\Delta x$ and $W_{i+\frac{1}{2}}$ is the flow velocity half way between grid points i and $i+1$ calculated in this case by simple averaging. For the Phoenical LPE version of the Shasta algorithm

$$\nu_{i+\frac{1}{2}} = 1/6 + 1/3\ \varepsilon_{i+\frac{1}{2}}^2$$

2. Compute the raw antidiffusive fluxes which in our case for the Phoenical version are

$$\phi_{i+\frac{1}{2}} = \mu_{i+\frac{1}{2}}\left[\overset{*}{N}{}_{i+1}^{j+1} - \overset{*}{N}{}_i^{j+1} + (-N_{i+2}^j + 3N_{i+1}^j - 3N_i^j + N_{i-1}^j)/6\right] \tag{9}$$

where $\mu_{i+\frac{1}{2}} = (1 - \varepsilon_{i+\frac{1}{2}}^2)/6$

3. Compute corrected antidiffusive fluxes, $\bar{\phi}$, using

$$\bar{\phi}_{i+\frac{1}{2}} = S.\max\left\{0, \min\left[S.(\overset{*}{N}{}_{i+2}^{j+1} - \overset{*}{N}{}_{i+1}^{j+1}), \left|\phi_{i+\frac{1}{2}}\right|, S.(\overset{*}{N}{}_i^{j+1} - \overset{*}{N}{}_{i-1}^{j+1})\right]\right\} \tag{10}$$

where $|S| = 1$ and $S = \text{sign}\ (\overset{*}{N}{}_{i+1}^{j+1} - \overset{*}{N}{}_i^{j+1})$

4. Perform antidiffusion using

$$N_i^{j+1} = \overset{*}{N}{}_i^{j+1} - \bar{\phi}_{i+\frac{1}{2}} + \bar{\phi}_{i-\frac{1}{2}} \tag{11}$$

Thus N_i^{j+1} is the required solution at $t + \Delta t$, without any significant numerical diffusion or dispersion.

The new algorithm for flux limiting developed by Zalesak (1979) was also applied as follows:

The first flux constraint to be performed is

$$\bar{\phi}_{i+\frac{1}{2}} = 0 \text{ if } \phi_{i+\frac{1}{2}}(\overset{*}{N}{}^{j+1}_{i+1} - \overset{*}{N}{}^{j+1}_{i}) < 0$$
$$\text{and either } \phi_{i+\frac{1}{2}}(\overset{*}{N}{}^{j+1}_{i+2} - \overset{*}{N}{}^{j+1}_{i+1}) < 0 \qquad (12)$$
$$\text{or } \phi_{i+\frac{1}{2}}(\overset{*}{N}{}^{j+1}_{i} - \overset{*}{N}{}^{j+1}_{i-1}) < 0$$

which is said by Zalesak to have a minimal cosmetic effect; however, we find that troublesome peaks and troughs arise if the conditions of equation (12) are not applied.

The second stage of flux constraint is to calculate the following quantities:

P_i^+ = the sum of all antidiffusive fluxes into grid point i

$$= \max(0, \phi_{i-\frac{1}{2}}) - \min(0, \phi_{i+\frac{1}{2}}) \qquad (13)$$

P_i^- = the sum of all antidiffusive fluxes away from grid point i

$$= \max(0, \phi_{i+\frac{1}{2}}) - \min(0, \phi_{i-\frac{1}{2}}) \qquad (14)$$

$$Q_i^+ = N_i^{max} - \overset{*}{N}{}_i^{j+1} \qquad (15)$$

$$Q_i^- = \overset{*}{N}{}_i^{j+1} - N_i^{min} \qquad (16)$$

where N_i^{max} and N_i^{min} are defined by equations (21) and (22) below.

$$R_i^+ = \begin{Bmatrix} \min(1, Q_i^+ / P_i^+) & \text{if } P_i^+ > 0 \\ 0 & \text{if } P_i^+ = 0 \end{Bmatrix} \qquad (17)$$

$$R_i^- = \begin{Bmatrix} \min(1, Q_i^- / P_i^-) & \text{if } P_i^- > 0 \\ 0 & \text{if } P_i^- = 0 \end{Bmatrix} \qquad (18)$$

Correction factors $C_{i+\frac{1}{2}}$ are then calculated using

$$C_{i+\frac{1}{2}} = \begin{Bmatrix} \min(R_{i+1}^+, R_i^-) & \text{if } \phi_{i+\frac{1}{2}} \geqslant 0 \\ \min(R_i^+, R_{i+1}^-) & \text{if } \phi < 0 \end{Bmatrix} \qquad (19)$$

and the fluxes are corrected using

$$\bar{\phi}_{i+\frac{1}{2}} = C_{i+\frac{1}{2}} \cdot \phi_{i+\frac{1}{2}} \qquad \text{where } 0 \leqslant C_{i+\frac{1}{2}} \leqslant 1 \qquad (20)$$

The flux is then corrected using equation (11).

The choice of values of N_i^{max} and N_i^{min} is the key to flux limiting process and can be made in a variety of ways, Zalesak (1979). We follow Zalesak and use:

$$N_i^a = \max(N_i^j, \overset{*}{N}{}_i^{j+1})$$
$$N_i^{max} = \max(N_{i-1}^a, N_i^a, N_{i+1}^a, N_{i+\frac{1}{2}}^{peak}, N_{i-\frac{1}{2}}^{peak}) \quad (21)$$
$$N_i^b = \min(N_i^j, \overset{*}{N}{}_i^{j+1})$$
$$N_i^{min} = \min(N_{i-1}^b, N_i^b, N_{i+1}^b, N_{i+\frac{1}{2}}^{peak}, N_{i-\frac{1}{2}}^{peak}) \quad (22)$$

where $N_{j+\frac{1}{2}}^{peak}$ is the extrapolated peak or trough value found when the straight line, drawn through the values $\overset{*}{N}{}_{i-1}^{j+1}$ and $\overset{*}{N}{}_i^{j+1}$, intersects with the straight line through $\overset{*}{N}{}_{i+1}^{j+1}$ and $\overset{*}{N}{}_{i+2}^{j+2}$, between grid points i and i+1. (See Zalesak (1979) section V). It is the use of these peak values in flux limiting which allows sharp peaks to be preserved rather than clipped as found using equation (10), see results below.

INITIAL CONDITIONS

All the methods are applied using the same basic conditions employed by Davies et al (1971); these are outlined below. The electron densities are kept below the level where space charge effects need to be included, except for the results presented as a practical example below.

The electrons start near the cathode and move between two plane, parallel electrodes 3 cm apart. This space is represented by a grid of 101 points, hence $\Delta x = 0.03$ cm. The initial electron pulse is either triangular or square.

The uniform electric field is -5.58 kV/cm. The non-uniform electric field is taken, for convenience, to be linearly decreasing in amplitude from -11.16 kV/cm to -0.11 kV/cm, left to right in the figures.

The electron velocity is given by

$$W = 2.9 \times 10^5 \ E/p \ \text{cms}^{-1} \quad (23)$$

where p is pressure. The time step is $\Delta t = 5 \times 10^{-10}$s.

The Courant-Friedrichs-Lewy conditions, Potter (1977), requires that $\Delta t < \Delta x/W$ and this is maintained throughout the calculations for all the non-FCT methods. For the FCT methods the condition becomes $\Delta t < \Delta x/2W$.

In all methods except the FCT methods the electron density N is maintained positive or zero, i.e.

$$N \geqslant 0,$$

so that a calculated negative density is set equal to zero.

The Phoenical LPE Shasta method maintains positivity when the flux correction algorithms of Boris and Book (1976 a and b) is used and the above measure is not required. However, when the new flux correction algorithm of Zalesak is applied the calculated densities do dip slightly below zero, thus as a countermeasure, in the spirit of the FCT method, the transported and diffused solution is used when the density becomes negative, guaranteeing positivity.

RESULTS AND DISCUSSIONS

Solutions with a uniform electric field.

For the results of this section, the standard triangular electron pulse near the cathode was used.

a) The Donor Cell Method

As predicted, the results show strong numerical diffusion where the initial triangular electron pulse is unacceptably notified as it propagates, as shown in figure 1. The numerical diffusion can be shown to depend on the time step, the grid space size and the velocity.

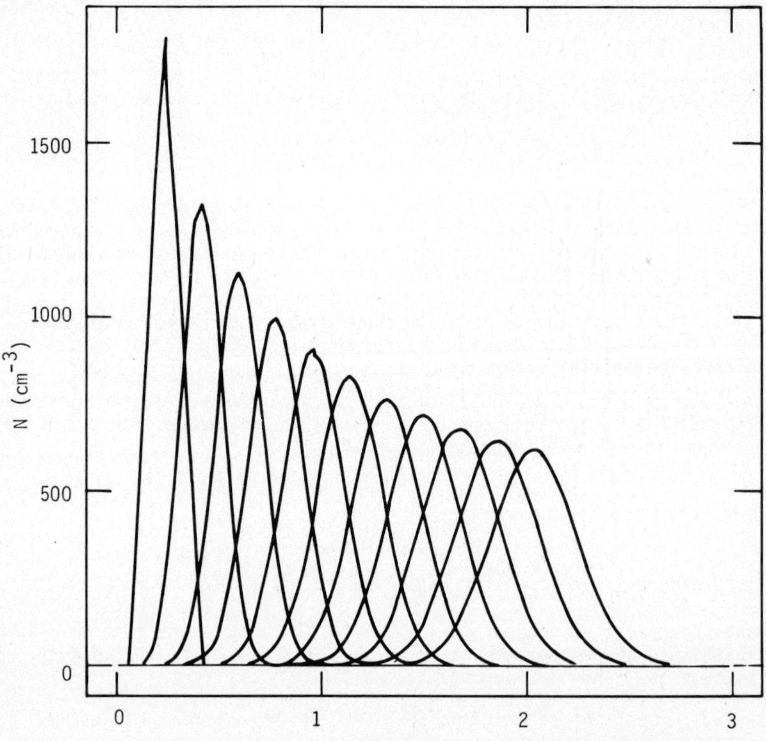

Figure 1
Calculations, using the Donor Cell method with triangular wave input at t = 0, where for 200 steps (100 ns) with curves drawn every 20 steps (10 ns), movement from left to right. Morrow (1981).

b) The Runge-Kutta method

The introduction of Runge-Kutta integration in time with upwind differencing had no effect on the numerical diffusion, and the results were comparable with those of figure 1.

When fourth-order central differences were used for the space variable, the results shown in figure 2 were obtained. These results show a small amount of dispersion, which removes the sharp corners, and very little dissipation. The position of the main peak coincides with the ideal position of the initial triangle displaced by a distance D = tW along the x axis. This ideal behaviour is marred by the appearance of the spurious smaller peaks behind the main pulse. As outlined by Roache (8), the spurious peaks are probably due to lack of dissipation in the equations. These small peaks would grow rapidly when ionization is included, leading to unacceptable results.

With simple central differences, the results are stable and very similar to figure 2, with the appearance of spurious peaks. However, there is some dissipation, and the peak heights slowly decay, as shown for the Lax-Wendroff method figure 3.

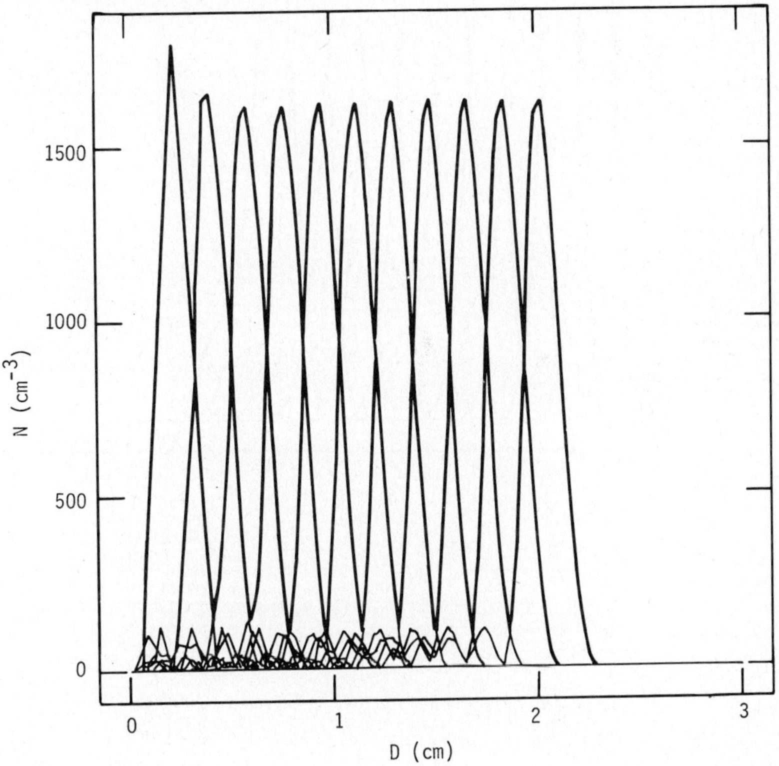

Figure 2
Calculations using the Runge-Kutta method with fourth-order central differences, conditions as for figure 1.
Morrow (1981).

c) The Lex-Wendroff method

The one- and two-step Lax-Wendroff methods gave almost indistinguishable results, both equally well represented by the curves of figure 3. In this case, there is a certain amount of dispersions and some dissipation. The dispersion is slightly asymmetric, distorting the pulse so that it leans slightly backwards with a longer leading edge, this delays the pulse slightly compared with the ideal position of the initial triangle.

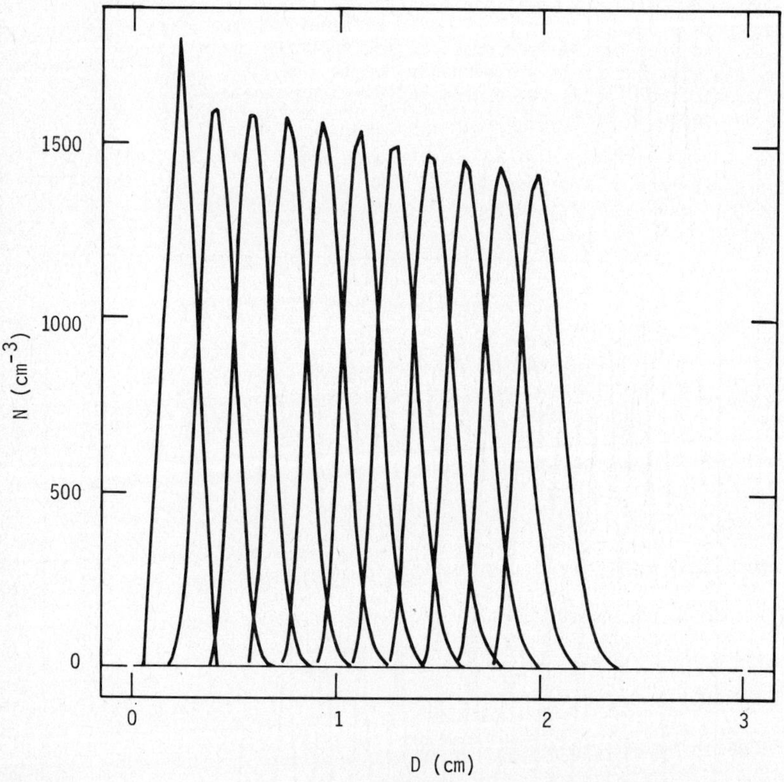

Figure 3
Calculations using either the one-step or two-step Lax-Wendroff methods, conditions as for figure 1. Morrow (1981).

d) The Hybrid method of characteristics

The results, shown in figure 4, are very similar to the Lax-Wendroff results, with slightly less dispersion and dissipation, the dispersion is symmetrical and the peak position is accurate.

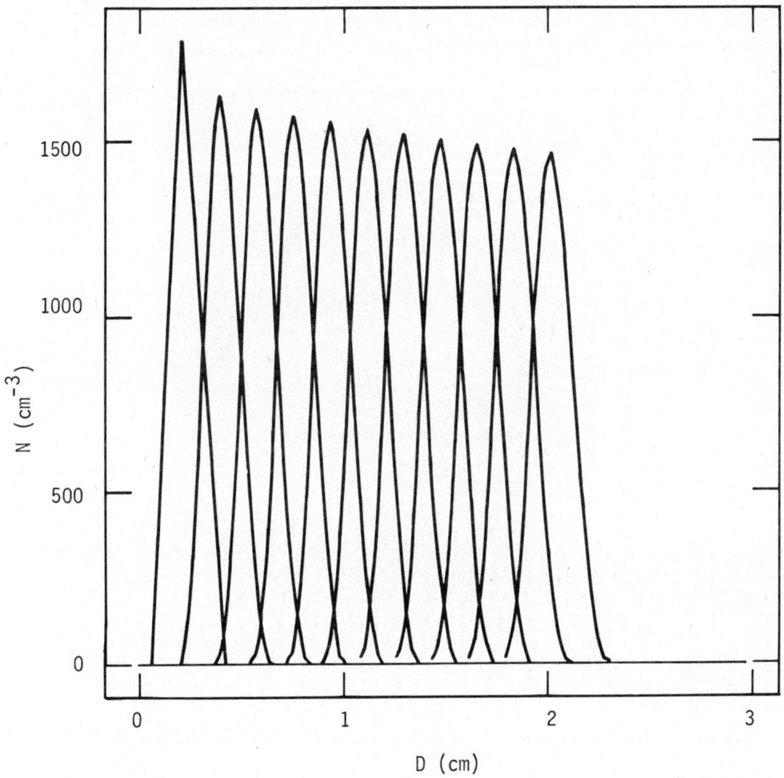

Figure 4
Calculations using the method of characteristics, with conditions as for figure 1. Morrow (1981).

The results for the four non-FCT methods discussed above can also be evaluated by comparing the final curves, at t = 100 ns, of figures 1, 2, 3 and 4 with the ideal curve for the initial triangle displaced a distance D = tW along the x axis, as shown in figure 5. The donor cell results show the devastating effect of the numerical diffusion. The Runge-Kutta method shows an excellent fit, marred by the appearance of a spurious peak centred at grid point 59. The Lax-Wendroff method give slightly asymmetric dispersion distorting the pulse so that it leans slightly backwards with a longer leading edge, this delays the pulse slightly. The method of characteristics gives symmetric dispersion with an accurate peak position.

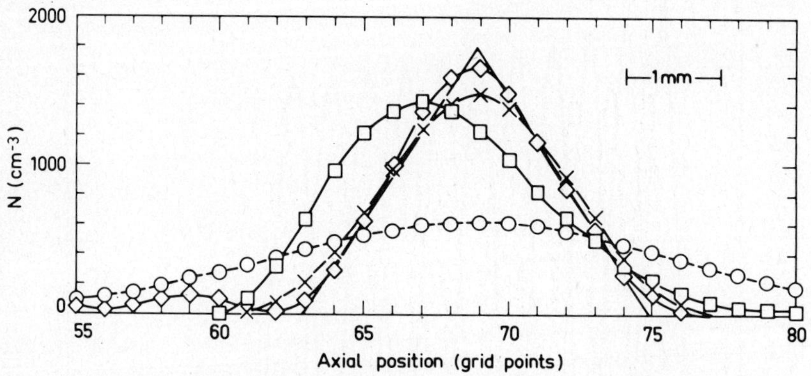

Figure 5

Simultaneous comparison of results from figures 1, 2, 3 and 4 after 200 steps (100 ns) with the initial triangular pulse of electrons displaced a distance D = tW along the x axis, where t = 100 ns.

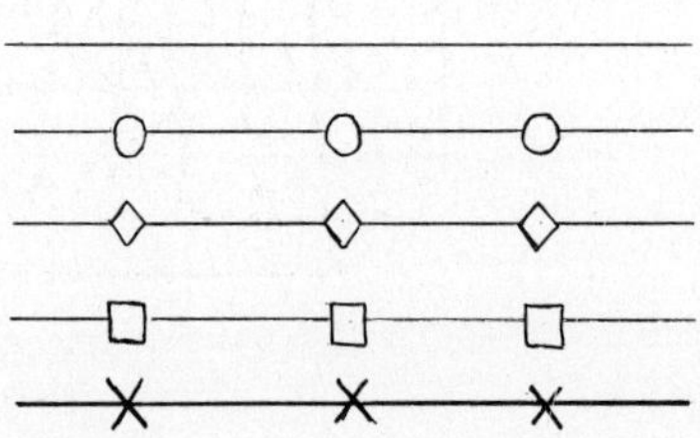

Morrow (1981).

e) Phoenical LPE Shasta

Using Zalesak's full flux correction algorithm, including flux limited by predicted maxima and minima between grid points and the flux constraint defined by equations (12), the results shown in figure 6 are obtained. The peak is as well preserved as that shown in figure 2 for the fourth order Runge-Kutta with fourth order central differences, without the associated spurious dispensive peaks. In fact when the curves of figure 6 are superimposed on those of figure 2 the almost exactly overlap.

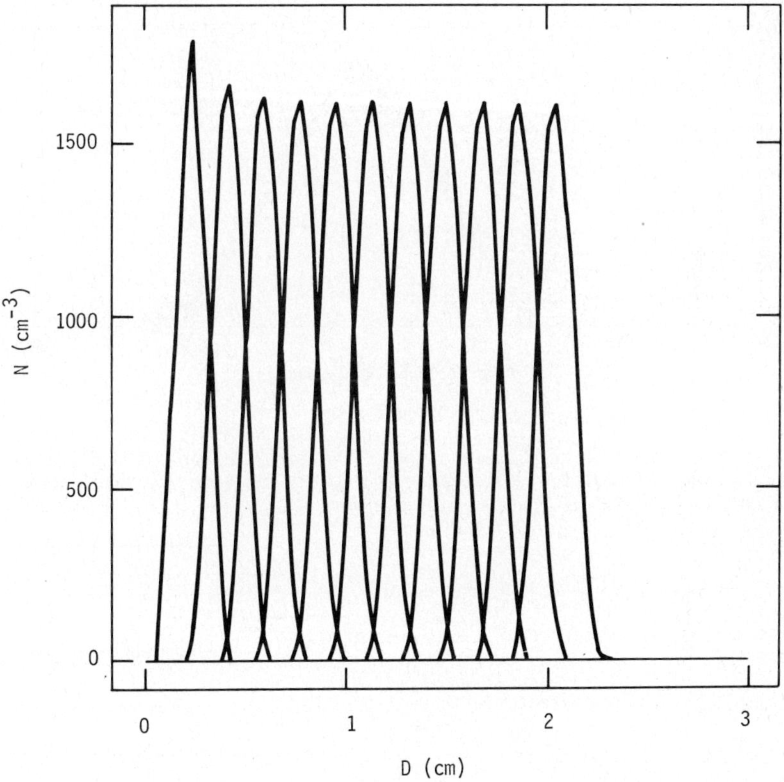

Figure 6
Calculations using the Phoenical LPE Shasta method with Zalesak's flux corrector, conditions as for figure 1. Morrow (1981).

The results of superimposing the final curve of figure 6, obtained using Zalesak's full flux corrector with peak preserver, on the initial triangle displaced a distance D = Wt along the x axis is shown in figure 7. Clearly the fit is excellent save for slight rounding of the peak. The result of using Zalesak's full flux corrector less the peak preserver, that is not using the predicted maxima between grid points as part of the flux limiter, is also shown in figure 7, the wave is degraded. The use of Boris and Book's original flux corrector, equation (10), is shown in figure 7 where the results show the characteristic "clipped" waveform, Zalesak (1979).

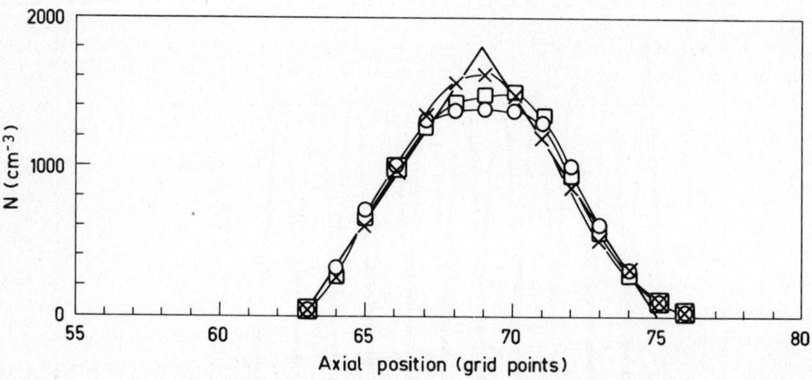

Figure 7

Simultaneous comparison of results from figure 6 after 200 steps (100 ns) with the initial triangular pulse of electrons displaced a distance D = tW along the x axis, where t = 100 ns.

(a) Initial pulse

(b) Boris and Book's original flux corrector

(c) Zalesak's full flux corrector less the peak preserver

(d) Zalesak's full flux corrector with peak preserver

Morrow (1981)

SOLUTIONS WITH A NON-UNIFORM ELECTRIC FIELD

The use of an electric field of decreasing amplitude spacially as outlined in the Initial Conditions section, resulted in all cases in the peak of the triangular wave increasing while the pulse width decreased. The same effect is illustrated more graphically using an initial square wave, as shown in figures 8, 9 and 10 for various methods.

For the two step Lax-Wendroff method shown in figure 8, dispersion causes a serious oscillation to develop on the leading edge; suppression of negative densities prevents similar oscillations appearing on the trailing edge. The result of numerically integrating the electron density with respect to the space variable using the trapezoidal rule, Abramowitz and Stegun (1965), for the initial and final distributions shows a net increase of 8% in the total electron density for figure 8.

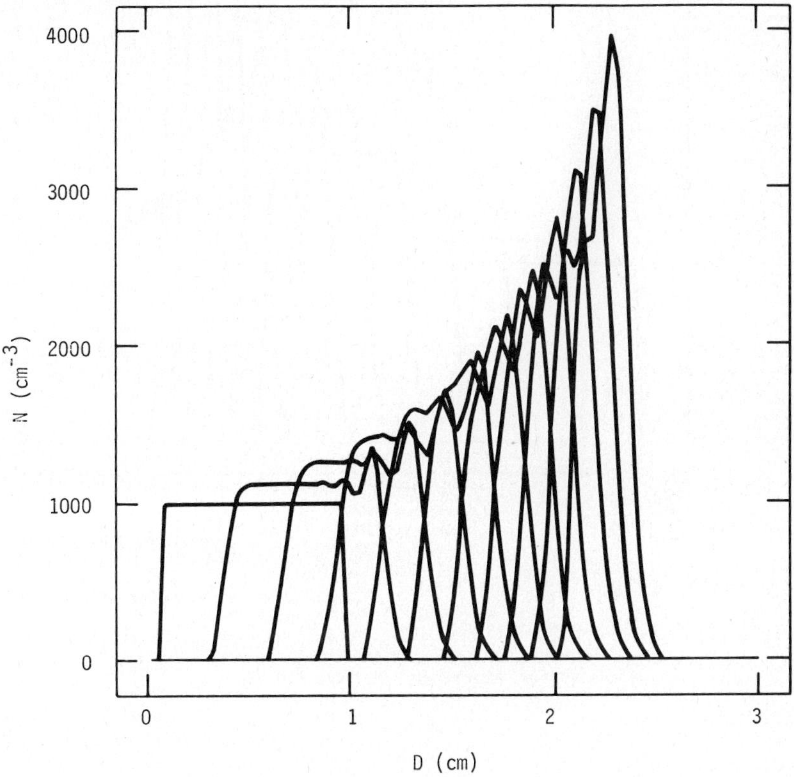

Figure 8
Calculations, using the Lax-Wendroff two-step method with a square-wave input at t = 0, were for 200 steps (100 ns) with curves drawn every 20 steps (10 ns). Movement is from left to right and the electric field decreases from left to right. Morrow (1981).

The method of characteristics gives better results with symmetrical distortion of the leading and trailing edges as shown in figure 9. Suppression of negative densities prevents negative distortions. The net density increased by 5% for figure 9.

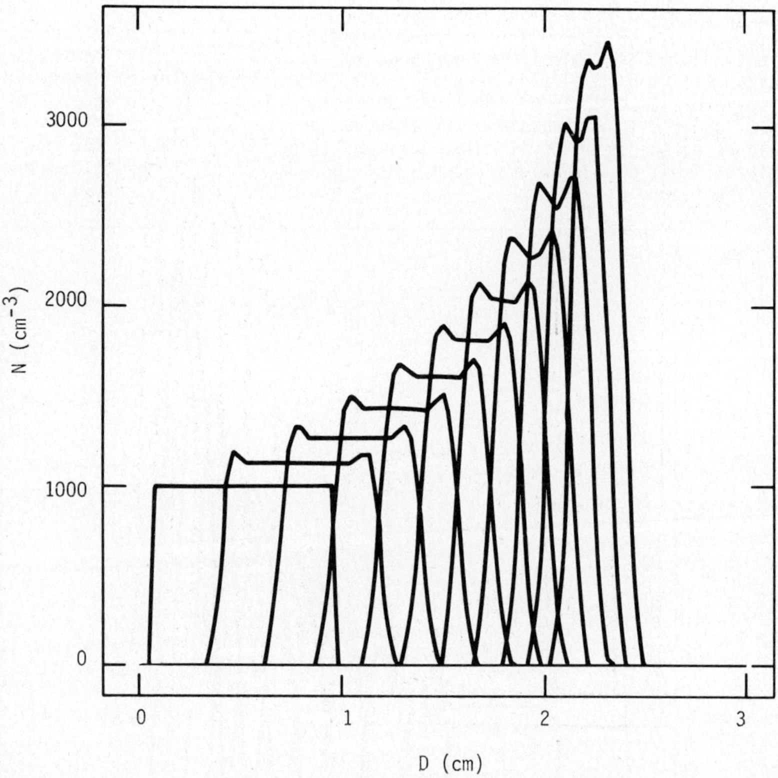

Figure 9
Calculations using the method of characteristics, with conditions as for figure 8. Morrow (1981)

Using the Phoenical LPE Shasta method the results are close to ideal as shown in figure 10. Integration of the electron density gives an increase of 1.4% in the first 30 ns after which the integrated density does not change by more than 1 part in 10^8 from 30 ns to 100 ns, a truly remarkable result since the distribution is transported and considerably modified in the interval from 30 ns to 100 ns.

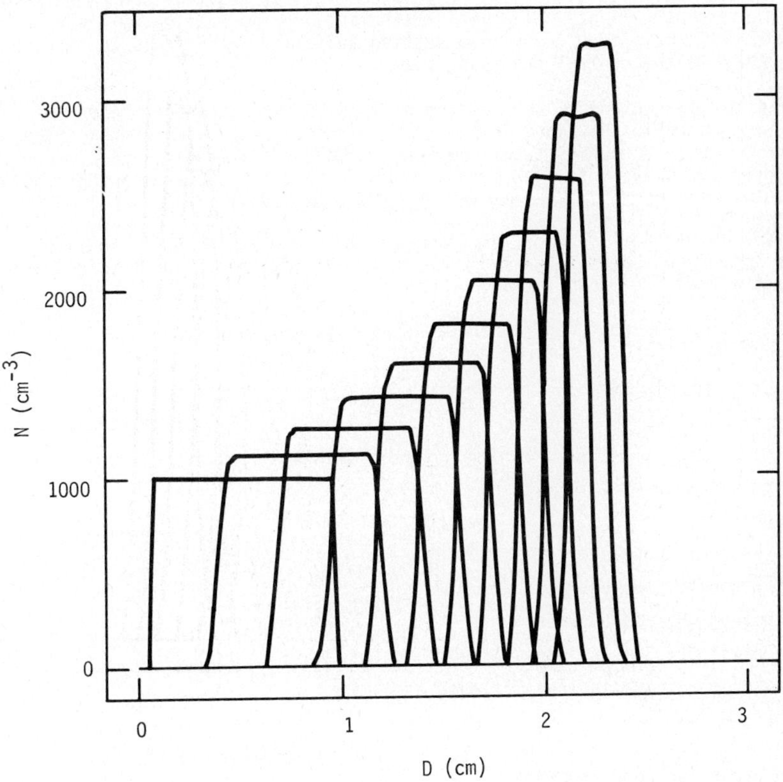

Figure 10
Calculations using the Phoenical LPE Shasta method for the conditions of figure 8. Morrow (1981).

COMPUTING TIMES

In Table 1 a comparison is made of the central processor execution times for calculations using the different algorithms to obtain the results shown in figures 1 to 7. The computer used was the CSIRO Cyber 76 in Canberra. Note that these times are only relative and do not refer to the simple application of the algorithm. In all cases the programs include positive and negative ion calculations as well as ionization, attachment, recombination and space charge effects. All have been removed by setting suitable coefficients to zero, but still contribute to the central processor time.

The Runge-Kutta method is the most complex method and therefore takes the longest time while the Donor Cell method is the simplest and fastest. The method of characteristics takes the second longest time for the simple problem of figure 5, however, for the more complex problems more iterations are required, while the time remains the same for the other methods.

The two Lax-Wendroff methods take similar times, while the Boris and Book algorithm, Phoenical LPE Shasta, takes only 10% more time, and the full Zalesak algorithms only 20% more time.

TABLE 1 Comparison of Central Processor (CP) execution times

Method	C.P. Time (s)
Euler with upwind differencing	7.3
Runge-Kutta	17.9
Lax-Wendroff, one-step	9.6
Lax-Wendroff, two-step	9.9
Method of Characteristics	*14.2
Phoenical LPE Shasta (Boris & Book)	10.9
Phoenical LPE Shasta (Full Zalesak treatment)	12.3

* Note that the method of characteristics is an iterative process and takes longer to converge for some problems.

A PRACTICAL EXAMPLE - SPACE CHARGE EFFECTS ON PLASMA MOTION.

THE PROBLEM

As a practical example of electron and ion motion in non-uniform fields some results are presented from the work of Morrow and Lowke (1981) which shows the effect of space-charge electric fields in a drift-dominated plasma. The continuity equations for ions and electrons are solved simultaneously using the two-step Lax-Wendroff method, including a small realistic amount of diffusion which tends to dampen the small oscillation inherent in the Lax-Wendroff method.

The equations to be solved are the continuity equations for electrons and ions in one-dimensional conservative form, neglecting source terms but including slight electron diffusion. The equations are:

$$\frac{\partial Ne}{\partial t} = -\frac{\partial (Ne\ We)}{\partial x} + \frac{\partial}{\partial x}\left(D\frac{\partial Ne}{\partial x}\right) \tag{24}$$

and

$$\frac{\partial Np}{\partial t} = -\frac{\partial (Np\ Wp)}{\partial x}$$

where Np and Ne are ion and electron densities, respectively, and Wp and We are ion and electron velocities respectively.

We is given by equation (23), Wp is given by:

$$\begin{aligned} Wp &= 2 \times 10^3 (1 - 4 \times 10^{-3} \times E/p) \text{ for } E/p \leqslant 80 \\ &= 1.25 \times 10^4 \times (E/p)^{\frac{1}{2}} - AP/E \text{ for } E/p > 80 \end{aligned} \tag{26}$$

(Davies et al 1971)

and the diffusion coefficient D by

$$D \approx 3.22 \times 10^{-3}\ \text{cm}^2\ \text{s}^{-1}$$

The effect of space charge, due to ion and electron concentrations, on the electric field is obtained by solving Poisson's equation:

$$\nabla E = -\frac{\zeta}{\varepsilon_0} \tag{27}$$

where ε_0 is the permitivity of free space and ζ is the charge density given by

$$\zeta = e(Np - Ne)$$

e is the electronic charge.

In order to correctly account for the space-charge field due to a finite charge distribution it is necessary to solve Poisson's equation in three dimensions, as discussed by Davies et al (1971). We use the method of Davies and Evans (1973) to account for these effects and solve for a cylinder of charge which gives the field at a point P due to a charge distribution $\zeta(x)$ by a simple geometric integration. We also include five image charge distributions beyond each electrode.

We allow for a field dependent diffusion coefficient using

$$\begin{aligned} \frac{\partial}{\partial x}\left(D\frac{\partial N}{\partial x}\right) = \frac{1}{2(\Delta x)^2} \big((D_{i+1} + D_i)(N^j_{i+1} - N^j_i) \\ - (D_i + D_{i-1})(N^j_i - N^j_{i-1}) \big) \end{aligned} \tag{28}$$

where N = Ne in this case.

INITIAL CONDITIONS

The initial applied electric field is uniform between parallel electrodes as outlined in the Initial Conditions section. The initial Gaussian electron and ion distributions are identical and overlap to give a neutral plasma with no field distortion. The distributions are cylindrical with a diameter of 0.5 cm over which the densities are constant while in the axial direction the density varies as

$$Ne = Np = No \exp \{-[(x - 1.5)/0.25]^2\} \text{ cm}^{-3} \tag{29}$$

RESULTS

The results of calculations are summarised in figure 11 for electron densities of 5×10^6 to 5×10^{11} cm^{-3}. In all cases the ion motion is negligible and the ion distribution is that given for t = 0 ns. As the electrons move charge neutrality no longer holds and in accordance with Poisson's equation the electric field is distorted.

In figure 11 (a) and (b) the density of peak value No of 5×10^6 cm^{-3} gives negligible field distortion and the Gaussian electron distribution is displaced to the right. The Lax-Wendroff method gives excellent results in this case.

In figure 11 (c) and (d) the peak density is 2.5×10^{10} cm^{-3} and field distortion is significant, modifying the electron motion in two ways. Near the centre of the ion distribution the electrons are slowed down due to the attraction of the ions, while further out the electrons repel each other and disperse.

When the peak density of ions and electrons is 5×10^{11} cm^{-3} slight initial changes in the electron distribution cause such large space-charge fields that the field is largely excluded from the centre of the plasma and the electrons stay largely locked to the ions, with only a slow leakage of electrons to the anode, as shown in figure 11 (e) and (f).

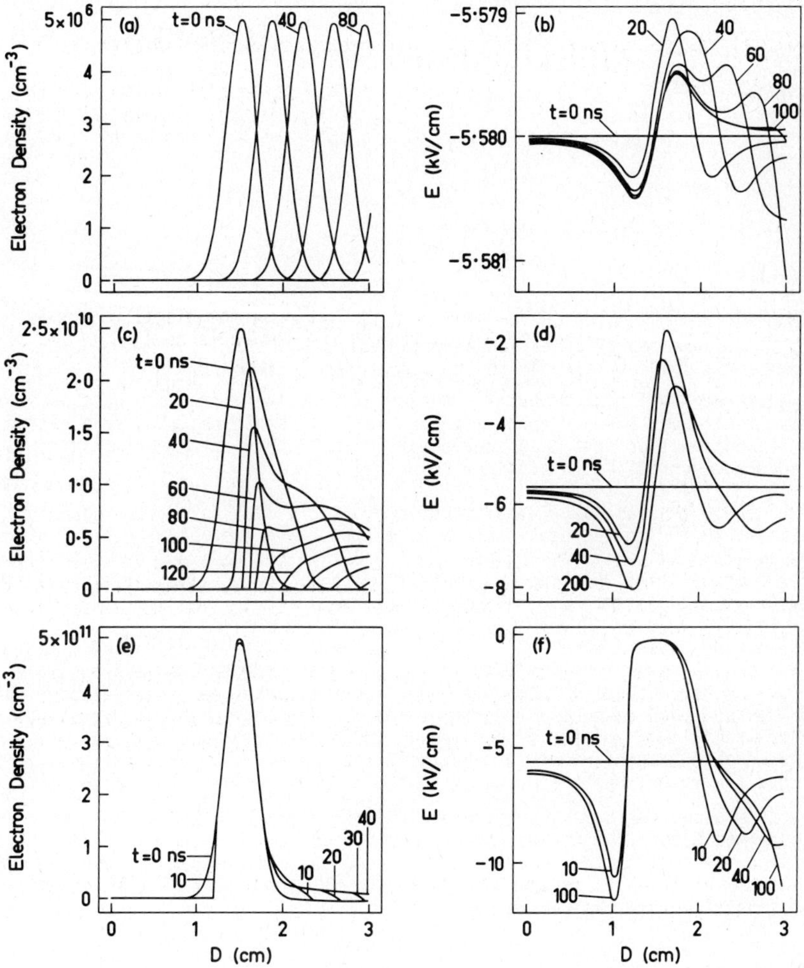

Figure 11

Electron densities and electric fields calculated for a plasma of various initial peak densities N_0, and applied field E_0 = -5.58 kV/cm. Successive electron distributions are at 20 ns intervals for figures (a) (b), (c) and (d) and 10 ns for figures (e) and (f). The ion distribution is represented by the distribution at t = 0.

Morrow and Lowke (1981).

(a) Electron motion, $N_0 = 5 \times 10^6$ cm^{-3}

(b) Electric field variation for case (a)

(c) Electron motion, $N_0 = 2.5 \times 10^{10}$ cm^{-3}

(d) Electric field variation for case (c)

(e) Electron motion, $N_0 = 5 \times 10^{11}$ cm^{-3}

(f) Electric field variation for case (e).

CONCLUSIONS

The Donor Cell method with upwind differencing is unacceptable due to numerical diffusion and the fourth-order Runge-Kutta method with fourth-order central differences is unacceptable due to the appearance of spurious peaks.

The two Lax-Wendroff methods and the method of characteristics all give satisfactory results in most cases. These methods exhibit distortions when propagating sharp discontinuities such as a square wave. The method of characteristics is the most accurate non-FCT method tested.

The FCT method "Phoenical LPE Shasta" using the full flux correction algorithm of Zalesak, including the peak-preserving method, gave the best results, which in all cases examined were close to ideal. The FCT method does not use significantly greater computer time than the Lax-Wendroff method, and uses less time than its closest rival, the method of characteristics.

The FCT method appears somewhat arbitrary, however it is soundly-based mathematically and simply recognises the fact that high-order methods are less accurate than lower-order methods in some instances, particularly near sharp discontinuities. The FCT method then recognises the error and replaces the higher-order solution with a suitably interpolated proportion of the higher order solution to which has been added a corresponding proportion the diffusive lower-order solution, sufficient to remove the error, which is usually in the form of dispersive ripples.

The results using the two-step Lax-Wendroff method to calculate space-charge effects were obtained before the FCT method was coded, and show how the Lax-Wendroff method can give quite satisfactory results. The use of the FCT method for the same problem has recently given very comparable results and allows calculations at significantly higher densities than the Lax-Wendroff method.

ACKNOWLEDGEMENTS

I would like to acknowledge much encouragement from, and many helpful discussions with Dr John J. Lowke of this laboratory. I must thank Dr Clive Fletcher for many helpful discussions. I would also like to thank Dr J. Atkinson and Dr J. Knight for their help.

REFERENCES

[1] Chu, C.K., Numerical Methods in Fluid Dynamics, Chia-Shun Yih, ed., Advances in Applied Mechanics, 286, (Academic Press, New York, 1978).

[2] Roache, P.J., Computational Fluid Dynamics (Hermosa, Albuquerque, 1972).

[3] Strang, G., Accurate Partial Difference Methods: II. Non-Linear Problems, Numerische Mathematik, 6 (1964), 37-46.

[4] Turkel, E., On the Practical Use of High Order Methods for Hyperbolic Systems, ICASE report No. 78-19, Nov. 16, 1978 (Institute for Computer Applications In Science and Engineering NASA Langley Research Centre, Hampton, Virginia, USA).

[5] Ames, W.F., Numerical Methods for Partial Differential Equations, (Academic Press, New York, 1977).

[6] a) Boris, J.P. and Book, D.L., Solution of Continuity Equations by the Method of Flux-Corrected Transport in: John Killeen, ed., Methods in Computational Physics, (Academic Press, New York 1976, 85-130.

[6] b) Boris, J.P. and Book, D.L., Flux-Corrected Transport III. Minimal-Error FCT Algorithms, J.Comput.Phys., 20, (1976) 397-431.

[7] Zalesak, S.T., Fully Multidimensional Flux-Corrected Transport Algorithms for Fluids, J.Comput.Phys., 31, (1979), 335-362.

[8] Abramowitz, M. and Stegun, I.A., Handbook of Mathematical Functions (Dover, New York, 1965).

[9] Wylie, C.R., Advances Engineering Mathematics (McGraw-Hill, New York, 1960).

[10] Davies, A.J. and Evans, C.J., Theory of Ionization Growth in Gases Under Pulsed and Static Fields, Report to CERN European Organization for Nuclear Research (Geneva, 1973).

[11] Davies, A.J., Davies, C.S. and Evans, C.J., Computer simulation of rapidly developing gaseous discharges, Proc. IEEE (1971) 816-823.

[12] Potter, D., Computational Physics (Wiley, Chichester/New York, 1977).

[13] Morrow, R., Numerical Solution of Hyperbolic Equations for Electron Drift in strong Non-Uniform Electric Fields, J.Comp.Phys-, in press.

[14] Morrow, R., and Lowke, J.J., Space-chage effects on drift dominated electron and plasma motion, J.Phys.D., in press.

NUMERICAL SOLUTIONS OF PARTIAL DIFFERENTIAL EQUATIONS
J. Noye (Editor)

COMPUTATIONS OF CHEMICAL REACTIONS IN TUBULAR FLOW

P. Lalousis

CSIRO Division of Applied Physics, Sydney, Australia 2070

In the chemical reaction A + B → C + D, involving the passage of reacting gases in a diluent through a cylindrical tube with excess reactant, [B] >> [A], the depletion of the reactant and the formation of the product can be described by two partial differential equations (PDEs) of the elliptic type. These PDEs are the continuity equations of the chemical species in a reacting flow system, including radial and axial diffusion, and are subject to boundary conditions such as the degree of absorption along the tube walls. In this paper, an iterative method of computing the PDEs is reported.

INTRODUCTION

In analysing first-order kinetics of chemical reactions in tubular flow reactors, previous researchers have used orthogonal series expansions , Walker (1961), and non-orthogonal series expansions, Duong and Steinberg (1980), in computing the continuity equation that describes the depletion of reactant A in the chemical reaction A + B → C + D with excess reactant [B] >> [A]. Table I of Duong and Steinberg (1980) gives a summary of work done on convective diffusion with chemical reactions; in most cases axial diffusion was not considered. Poirier and Carr (1971) have computed the continuity equation of the reactant using finite differences without axial diffusion; in the cases where axial diffusion was considered, the continuity equation was computed by approximate power series expansions.

The purpose of computing the continuity equation in tubular flow is to determine the correction to the chemical rate coefficient for an experimental environment in which reactant A is injected into a flowing carrier B, for which [B] >> [A]. The concentration of A along the tube will be simply

$$[A]_z = [A]_0 \exp\{-2kz/V\} \tag{1}$$

and the concentration of the product C will be (if there are no wall reactions)

$$[C]_z = [A]_0(1 - \exp\{-2kz/V\}).$$

It has been assumed that the axial distance is equal to the reaction time multiplied by the average flow velocity on the z-axis. But under conditions of tubular flow the carrier velocity has a parabolic profile, so reactants at the centre of the tube will travel faster than those near the tube wall, and concentrations of products near the tube wall will be greater than those near the tube axis. There is also the complication of wall reactions, but in spite of difficulties, the concentration of reactant A will still exhibit an exponential decay in the axial direction, Brown (1977), i.e.

$$[A]_z = [A]_0 \exp\{-2k^*z/V\} \quad (2)$$

where k^* is the apparent rate coefficient

By computing the continuity equations, the concentrations of reactants and products are obtained radially and axially. Further, fitting equation (2) to the computed concentrations of reactant yields k^*, the apparent rate coefficient, in terms of k, the true rate coefficient.

In this paper the PDEs, that describe the continuity equations of reactants and products, are computed using finite differences and the Gauss-Siedel iterative method with successive over-relaxation (SOR). More accurate values of k^* should be obtained with this procedure since the calculation is two dimensional including both radial and axial diffusion. This method is also more versatile in that it can handle different diffusion coefficients, different reactant and product absorption coefficients, k_w, and any initial profile of reactants. Computing both PDEs rather than just one, is required in the case of experimental environment for which only the concentrations of products (C or D) can be monitored.

MATHEMATICAL DESCRIPTION

For the steady state chemical reaction $A + B \rightarrow C + D$ in a tubular reactor system, with excess reactant $[B] \gg [A]$, and for which first-order kinetics apply, the depletion of reactant A is described by

$$V[1 - (\frac{r}{R})^2] \frac{\partial c}{\partial z} = D_1[\frac{1}{r}\frac{\partial}{\partial r}(r\frac{\partial c}{\partial r}) + \frac{\partial^2 c}{\partial z^2}] - kc \quad (3)$$

where $c = [A]$, the concentration of chemical species A, D_1 is the diffusion coefficient for A in the gas mixture, and boundary conditions

$$c = c_0 \text{ at } z = 0,\ c = 0 \text{ at } z = \infty$$

$$\partial c/\partial r = 0 \text{ at } r = 0, \text{ and}$$

$$D_1(\partial c/\partial r) = k_{w1}c \text{ at } r = R,$$

i.e. some loss of A by wall reaction. The formation of the product (C or D) is described by

$$V[1 - (\frac{r}{R})^2]\frac{\partial c}{\partial z} = D_2[\frac{1}{r}\frac{\partial}{\partial r}(r\frac{\partial c}{\partial r}) + \frac{\partial^2 c}{\partial z^2}] + kc_A \quad (4)$$

where $c = [C]$ is the concentration of chemical species C, $c_A = [A]$ is the concentration of chemical species A, D_2 is the diffusion coefficient for chemical species C in the gas mixture, and boundary conditions are

$$c = 0 \text{ at } z = 0,\ \partial c/\partial z = 0 \text{ at } z = \infty,\ \partial c/\partial r = 0 \text{ at } r = 0, \text{ and}$$

$$D_2(\partial c/\partial r) = -k_{w2}c \text{ at } r = R$$

i.e. loss due to wall reactions.

These PDEs are continuity equations of chemical species in a binary fluid mixture, (Bird, Steward and Lightfoot (1960).

NUMERICAL PROCEDURE

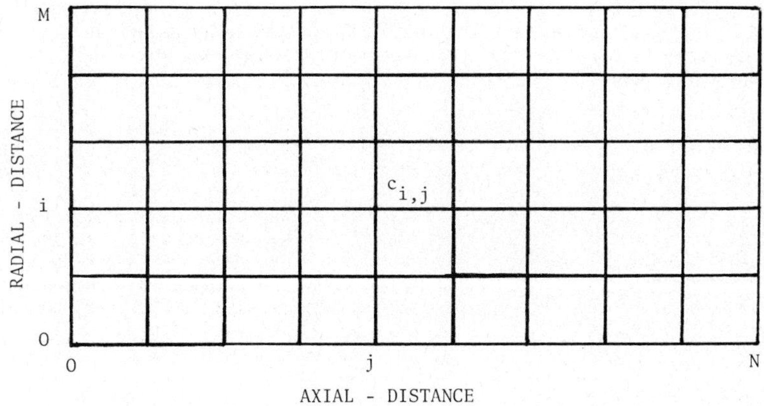

Figure 1
The finite-difference grid

The finite-difference grid used is shown in Figure 1. The ranges $0 \leqslant r \leqslant R$ and $0 \leqslant z \leqslant z_\infty$ have been divided into M and N intervals, respectively, of grid spacing δr and δz, respectively, thus $c_{i,j}$ is the concentration at point $(i\delta r, j\delta z)$.

The finite difference form of the radial diffusion term, is as follows, Crank (1975),

$$\frac{D}{r}\frac{\partial}{\partial r}\left(r\frac{\partial c}{\partial r}\right) = \frac{D}{2i(\delta r)^2}\left[(2i+1)\,c_{i+1,j} - 4ic_{i,j} + (2i-1)\,c_{i-1,j}\right]$$

for i = 1 to M; at r = 0 the boundary condition is $\partial c/\partial r = 0$ which gives

$$\frac{D}{r}\frac{\partial}{\partial r}\left(r\frac{\partial c}{\partial r}\right) = \frac{4D}{(\delta r)^2}\left[c_{1,j} - c_{0,j}\right]$$

for i = 0, the axial diffusion term in finite-difference form is

$$D\frac{\partial^2 c}{\partial z^2} = \frac{D}{(\delta z)^2}\left[c_{i,j+1} - 2c_{i,j} + c_{i,j-1}\right]$$

for j = 1 to N, where $D = D_1$ or D_2 and c = [A] or [C].

z_∞, at which c = 0 or $\partial c/\partial z = 0$, may be determined in two ways

(i) Assume z_∞ to be a fixed length, L, complete computations, then take $z_\infty = 2L$. Repeat the computations again, and if results do not change (within a given error tolerance) then $z_\infty = L$, otherwise, increase L and repeat the procedure until there is no change in the results.

(ii) Approximate z_∞ from equation (1) when concentration falls by a factor of, say, 100.

For PDE 1, (equation 3) the boundary condition $c = 0$ at z_∞ in finite-difference form $c_{i,N} = 0$, can be replaced by $\partial c/\partial z = 0$ at z_∞, i.e. $c_{i,N-1} = c_{i,N+1}$. Computations were completed with both these conditions, but there was no change in the results apart from the value of $c_{i,N}$ which can safely be disregarded in computing k*. The advective term is finite-differenced using a backward differencing scheme, (it was found that iterative methods for there problems always converged using such a scheme), as follows

$$\frac{\partial c}{\partial z} = \frac{c_{i,j} - c_{i,j-1}}{\delta z} .$$

The boundary condition at $r = R$ becomes

$$D\frac{\partial c}{\partial r} = D\,\frac{c_{M+1,j} - c_{M-1,j}}{2\delta r} = -\,k_w c_{M,j}$$

hence

$$c_{M+1,j} = \frac{-2\delta r k_w c_{M,j}}{D} + c_{M-i,j},$$

where $D = D_1$ or D_2, $k_w = k_{w1}$ or k_{w2}, $c = [A]$ or $[C]$.

Thus for equation (3) we have

$$c_{i,0} = c_0 \quad \text{for } i = 0 \text{ to } M,$$

$$c_{i,j} = \frac{1}{\varepsilon_i}[\alpha_i c_{i+1,j} + \beta_i c_{i-1,j} + \gamma_i c_{i,j+1} + \delta_i c_{i,j-1}]$$

for $i = 0$ to M, $j = 1$ to N, where

$$\alpha_i = \frac{D_1(2i+1)}{(\delta r)^2 2i} \quad \text{for } i = 1 \text{ to } M-1$$

$$= 4D_1/(\delta r)^2 \quad \text{for } i = 0$$

$$= 0.0 \text{ for } i = M ,$$

$$\beta_i = \frac{D_1(2i-1)}{(\delta r)^2 2i} \quad \text{for } i = 1 \text{ to } M-1$$

$$= 0.0 \text{ for } i = 0$$

$$= 2D_1/(\delta r)^2 \quad \text{for } i = M ,$$

$$\gamma_i = \frac{D_1}{(\delta z)^2} \quad \text{for } i = 0 \text{ to } M ,$$

$$\delta_i = \frac{D_1}{(\delta z)^2} + \frac{V}{\delta z}[1 - (\frac{r_i}{R})^2] \quad \text{for } i = 0 \text{ to } M ,$$

and

$$\varepsilon_i = \frac{2D_1}{(\delta r)^2} + \frac{2D_1}{(\delta z)^2} + k + \frac{V}{\delta z}\left[1 - \left(\frac{r_1}{R}\right)^2\right] \quad \text{for } i = 1 \text{ to } M\text{-}1$$

$$= \frac{4D_1}{(\delta r)^2} + \frac{2D_1}{(\delta z)^2} + k + \frac{V}{\delta z} \quad \text{for } i = 0$$

$$= \frac{2D_1}{(\delta r)^2} \quad \frac{2D_1}{(\delta z)^2} + k + \frac{k_{w1}(2M + 1)}{2M\delta r} \quad \text{for } i = M.$$

Similarly for equation (4) we have $c_{i,0} = 0$ for $i = 0$ to M,

$$c_{i,j} = \frac{1}{\varepsilon_i}\left[\alpha_i c_{i+1,j} + \beta_i c_{i-1,j} + \gamma_i c_{i,j+1} + \delta_i c_{i,j-1} + kc^A_{i,j}\right]$$

for $i = 0$ to M, $j = 1$ to N, where $c^A_{i,j}$ is the concentration of chemical species A at point $(i\delta r,\ j\delta z)$ obtained from equation 3,

$$\alpha_i = \frac{D_2(2i + 1)}{(\delta r)^2 2i} \quad \text{for } i = 1 \text{ to } M\text{-}1$$

$$= 4D_2/(\delta r)^2 \quad \text{for } i = 0$$

$$= 0.0 \text{ for } i = M,$$

$$\beta_i = \frac{D_2(2i-1)}{2i(\delta r)^2}$$

$$= 0.0 \text{ for } i = 0$$

$$= 2D_2/(\delta r)^2 \quad \text{for } i = M,$$

$$\gamma_i = D_2/(\delta z)^2 \quad \text{for } i = 0 \text{ to } M,$$

$$\delta_i = \frac{D_2}{(\delta z)^2} + \frac{V}{\delta z}\left[1 - \left(\frac{r_i}{R}\right)^2\right] \quad \text{for } i = 0 \text{ to } M,$$

and

$$\varepsilon_i = \frac{2D_2}{(\delta z)^2} + \frac{2D_2}{(\delta r)^2} + \frac{V}{\delta z}\left[1 - \left(\frac{r_i}{R}\right)^2\right] \quad \text{for } i = 1 \text{ to } M\text{-}1$$

$$= \frac{4D_2}{(\delta r)^2} + \frac{2D_2}{(\delta z)^2} + \frac{V}{\delta z} \quad \text{for } i = 0$$

$$= \frac{2D_2}{(\delta r)^2} + \frac{2D_2}{(\delta z)^2} + \frac{k_{w2}(2M+1)}{2M\delta r} \qquad \text{for } i = M.$$

Since equation (3) is independent of the products, it is computed first, followed by equation (4). Applying the Gauss-Seidel iterative method with SOR the $(n + 1)^{th}$ iterate of $c_{i,j}$ (for the reactant) is given by

$$c_{i,j}^{n+1} = \frac{\omega}{\varepsilon_i} [\alpha_i c_{i+1,j}^{n} + \beta_i c_{i-1,j}^{n+1} + \gamma_i c_{i,j+1}^{n} + \delta_i c_{i,j-1}^{n+1}] + (1 - \omega) c_{i,j}^{n}$$

where ω is the relaxation factor.

With the availability of FORTRAN 77 compilers, the programming implementation of this iterative procedure becomes a trivial task. A Distributed Array Processor, Webb (1980), is most suitable for this type of problem, in this paper all the computations were done on a Cyber 76 under FORTRAN 77, with level of optimization performed by the compiler equal to OPT=2.

Attempts were also made to compute the two PDEs using the finite elements program TWODEPEP, IMSL (1980), a powerful package for computing elliptic, parabolic and eigenvalue partial differential equation problem in two-dimensions. The two PDEs were coded into TWODEPEP's preprocessor format, but no convergence was obtained because the two PDEs have a non-symmetric Jacobian, and solving non-symmetric problems is still somewhat of an art.

RESULTS AND DISCUSSION

The numerical values of the various constants used range as follows:

D_1, D_2 = 10-300 $cm^2\ s^{-1}$
k = 10-300 s^{-1}
k_{w1}, k_{w2} = 0-10 s^{-1}
c_0 = 1, normalized, of chemical species A
V = 2000 cm s^{-1}
R = 1.088 cm

Taking $z_\infty \simeq 40 - 80$ cm was quite sufficient, since a further increase in z_∞ does not change the concentrations within 4 significant figures, for the above values of the various constants used. M and N, the number of radial and axial grid points, respectively, are always chosen such that $\delta r \approx \delta z$. Taking M = 5 and N = 184 for z_∞ = 40 cm was sufficient in that increasing M and N by a factor of 2 produced no change in the values of computed concentrations within 4 significant figures.

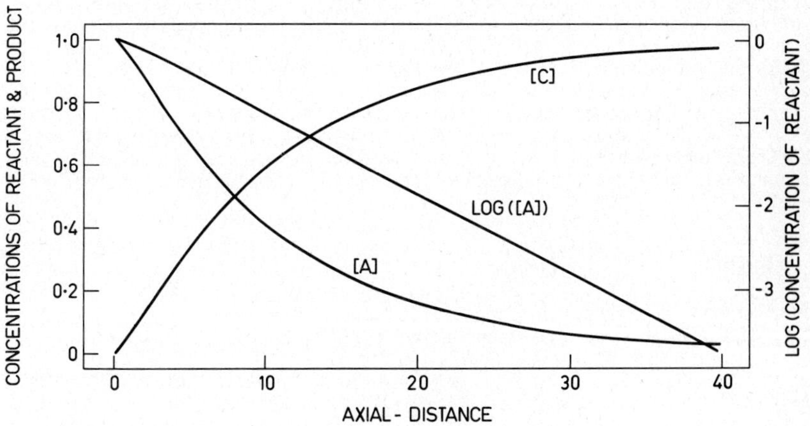

Figure 2
Plots of reactant and product concentrations, and log of the reactant concentration.

Figure 2 is a plot of the computed concentrations, normalized, of reactant and product against axial distance and also of the logarithm of concentration of reactant against axial distance, at $r = 0$, for the case $D_1 = D_2 = 100\ cm^2\ s^{-1}$, $k = 100\ s^{-1}$, $k_{w1} = k_{w2} = 0.0$, $z_\infty = 40$ cm. k^* for this particular case is 96.35 s^{-1}. If $k_{w1} = k_{w2} = 0$, i.e. no wall reaction, and $D_1 = D_2$ then the formation of the products should exhibit a growth of the form

$$c = c_0\ (1 - \exp\ (-2k^*z/V))$$

where k^* should be the same as in the decay of the reactant, fitting this type of equation to the computed concentrations of products for the same case as above we obtain $k^* = 96.35\ s^{-1}$, in good agreement with k^* from reactant.

CPU Time (in seconds)	Number of iterations	ω	M	N
1.43	200	1.4	5	184
3.35	450	1.0	5	184
14.51	600	1.4	10	368

Table 1
CPU time and number of iterations for various ω, M and N, in computing the PDEs.

Table 1, gives the Cyber 76 CPU (central processor unit) time in seconds, the number of iterations required such that any further iterations would not change the value of concentrations within 5 significant figures, for various values of ω, M and N for the case in Figure 2. The optimum value of the relaxation factor, ω, is determined empirically from various values of $\omega (0.1 \leqslant \omega < 2)$ since no formula exists for evaluating the optimum ω, Smith (1978). In this case optimum ω was about 1.4, for $\omega \geqslant 1.5$ the iteration process never converged. The other crucial fact is that the iteration process, with $\omega \leqslant 1.4$, always converges for any value of the various constants provided the advective term has been evaluated using backward differencing.

ACKNOWLEDGEMENTS

The author wishes to express his appreciation to I. Plumb and J.R. Steven for presenting the problem, and subsequent discussions throughout this work.

NOMENCLATURE

c	Concentration of reactant or product.
c_0	Initial concentration of reactant.
D_1, D_2	Molecular diffusion coefficients of reactant and product.
k	Chemical rate coefficient.
k^*	Apparent rate coefficient.
k_{w1}, k_{w2}	Chemical rate absorption coefficients by wall reactions of reactant and product.
M	Number of radial grid points.
N	Number of axial grid points.
r	Radial distance.
R	Radius of tube.
V	Velocity of flow on axis.
z	Axial distance.
z_∞	Axial distance at $z = \infty$
δr	Radial grid spacing.
δz	Axial grid spacing.
ω	Relaxation factor.

REFERENCES

[1] Walker, R.E., Chemical Reaction and Diffusion in a Catalytic Tubular Reactor, Phys. Fluids, 4, (1961), 1211

[2] VI-Duong Dang and Meyer Steinberg, Convective Diffusion with Homogeneous and Heterogeneous Reactions in Tube, J. Phys. Chem. 84, (1980), 214-219

[3] Poirier, R.V. and Carr R.W. Jrn., The Use of Tubular Flow Reactor for Kinetic Studies over Extended Ranges, J. Phys. Chem. 75, (1971), 1593

[4] Brown, R.L., Tubular Flow Reactors with First-order Kinetics, J. Res. Nat. Bur. Stand., 83, 1978, 1.

[5] Bird, R.B., Stewart, W.E. and Lightfoot, E.N., Transport Phenomena, Chapter 18, (Wiley, New York, 1960)

[6] Crank, H., The Mathematics of Diffusion, (Clarendon Press, Oxford, 1975)

[7] Webb, S.J., Solution of elliptic partial differential equations on the ICL Distributed Array Processor, ICL Technical Journal, November 1980

[8] TWODEPEP User's Manual, 2nd edition, IMSL 1980.

[9] Smith, G.D., Numerical Solutions of Partial Differential Equations: Finite Difference Methods (Claredon Press, Oxford, 1978)

NUMERICAL SOLUTIONS OF
PARTIAL DIFFERENTIAL EQUATIONS
J. Noye (editor)

NONLINEAR DIFFUSION WITH A BARRIER

R.D. Braddock
J.Y. Parlange
D.D. Lockington
P. Doilibi*

School of Australian Environmental Studies
Griffith University, Nathan, Queensland 4111,
Australia.
*School of Social and Industrial Administration
Griffith University.

The diffusion of soil-water from a source into a soil is the starting point of many problems in soil physics. Typically the diffusivities are highly nonlinear and vary rapidly with the concentration. Here, solutions are sought for the problem where a barrier interferes with the flux of soil-water. Finite difference techniques are used in conjunction with a predictor-corrector strategy to discretise the nonlinear diffusion equation. The presence of a precursor in the solution for the water concentration permits the solution to interact with the barrier. A solubility condition limits the application of the technique to a modest range of parameter values.

INTRODUCTION

The problem of describing the horizontal infiltration of water into a homogeneous soil is the starting point for the study of the movement of fluids and solutions through general soils. The horizontal motion of soil water is governed by the general diffusion equation

$$\frac{\partial \theta}{\partial t} = \frac{\partial}{\partial x} \left(D(\theta) \frac{\partial \theta}{\partial x}\right), \tag{1}$$

in one space variable x, where θ is the dimensionless soil water constant, time is denoted by t, and the diffusivity $D(\theta)$ is a function of θ (Philip (1969)). Diffusion processes arise naturally in the flow of water in porous media, of heat in conducting materials, the dispersal of chemical compounds in solution and in the transport of other quantities down their gradient. Thus (1) has wide applicability subject to the nature of the diffusivity D and to the nature of the boundary conditons relevant to the physical situation under consideration.

Of particular interest in this paper, are boundary conditions of the form

$$\begin{aligned} &\theta = 1, \quad x = 0, \quad t > 0, \\ &\frac{\partial \theta}{\partial x} = 0, \quad x = L, \quad t > 0, \\ &\theta = \theta_I, \quad 0 < x < L, \quad t = 0, \end{aligned} \tag{2}$$

where $\theta_I < 1$ is constant (see Section 2). These boundary conditions arise in the heating of a bar of length $2L$, at an initial temperature θ_I, by maintaining the ends at the normalised temperature $\theta = 1$. In section 2, these conditions will be used in the context of soil physics. In the context of heat conduction, the diffusivity D can be considered constant, i.e. $D(\theta) = D_o$, a constant for a wide range of practical temperatures and materials. In this case, eqn (1) is linear, solutions can be superimposed and a variety of analytic techniques are available (Widder (1975)). The above boundary conditions are readily handled and, using the separation of variables technique, the solution for $\theta_I = 0$, is

$$\theta = 1 - \sum_{m=o}^{\infty} \frac{4}{\pi(2m+1)} \exp(-\alpha_m^2 t) \sin\left(\frac{\pi x(2m+1)}{2L}\right), \tag{3}$$

where $\alpha_m = \pi D^{\frac{1}{2}} (2m+1)/2L$ (see Noye (1981)).

The case in which $D = D(\theta)$ is dependent on the concentration arises in soil physics, where the diffusion process is strongly dependent on the concentration (Philip (1969)). In this case, eqn (1) is strongly nonlinear, alternate methods of solution are required and numerical methods are particularly useful. The case in which the zero flux condition is moved to $x = L = \infty$, permits some analytic simplification of the problem as a similarity transform can be used to convert eqn (1) to an ordinary differential equation (Parlange et al., (1980)). Using this transformed equation, Braddock and Parlange (1981) were able to obtain very accurate numerical solutions for the soil-water profiles. The structure of these profiles has serious implications for the numerical solution of the finite problem, where the flux boundary condition is located at $x = L < \infty$.

The nature of the nonlinear diffusivities arising in soil physics will be discussed, together with the physical settings giving rise to boundary conditions of the form of eqn (2). The nature of the solutions of eqn (1) for $L = \infty$, will be discussed so as to highlight some of the difficulties in calculating numerical solutions of eqn (1) subject to the boundary conditions in eqn (2). Finite differences with a predictor-corrector strategy will be used in calculating the solutions.

THE BASIC MODEL

Consider the horizontal infiltration of water into a homogeneous soil at an initial soil-water content $\theta = \theta_I$, where θ_I is constant. This situation corresponds to ditch irrigation where the vertical flows can be ignored, and there is symmetry in the infiltration from the ditches which are located a distance 2L apart (see Figure 1). The same model is applicable to horizontal infiltration where the flow is halted by a vertical impervious barrier or wall (see Figure 1); the wall being a distance L from the ditch. In this case,

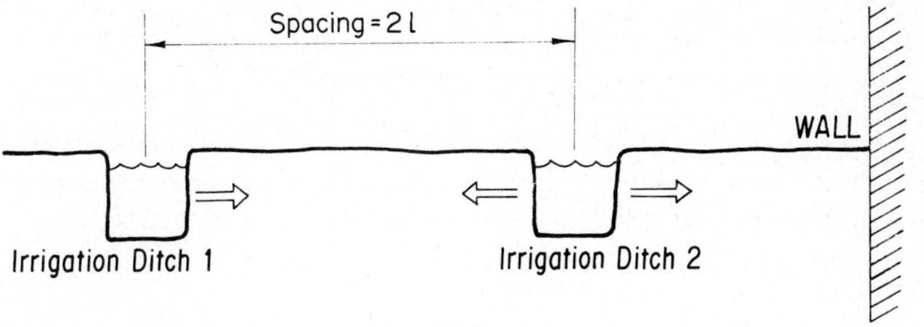

Figure 1
Ditch irrigation leading to horizontal infiltration into a homogeneous soil. The zero flux condition arises either from symmetry, or the presence of a barrier.

the impervious wall prohibits the flow of water and leads to the zero flux condition. For the two symmetrically placed ditches, the asymmetry of the opposing flows leads to the zero flux condition at $x = L$. Thus the boundary conditions in eqn (2) are applicable.

The flux $\underline{F}$ of soil-water flowing down its gradient is governed by Darcy's law (Philip (1969)), and

$$\underline{F} = - D \frac{\partial \theta}{\partial x} , \tag{4}$$

where D is the diffusivity. It is well known that $D = D(\theta)$ depends on the properties and structure of the transmitting medium, i.e., the pores in the soil matrix, and that $D(\theta)$ is strongly dependent on the water content. The diffusivity may also depend on position if the soil is not homogeneous, on the direction of flow if the proes have preferred shapes and orientations, and on whether the soil is being wetted or dried; the hysteresis effect (Parlange, (1980)). In general, these effects are minor compared with the dependence on θ, and they will be ignored. A consideration of the flow of soil water within the soil matrix using Darcy's law leads immediately to the diffusion equation, eqn (1) (Widder (1975)).

The nature of $D = D(\theta)$ can be determined by careful laboratory measurements and appropriate theoretical considerations (Bruce and Klute (1956), Passioura (1976)).

In practice, the experimental data is fitted by least squares to one of two functional forms;

$$D(\theta) \;\alpha\; \theta^n, \quad \text{the power law}$$

$$D(\theta) \;\alpha\; e^{n\theta}, \quad \text{the exponential law,}$$

where n is typically in the range from 4 to 10. One such fitted function is

$$D(\theta) = 9 \times 10^{-4} \exp(8.36\theta), \qquad (5)$$

(Hayhoe (1978)), and this is evaluated in Table 1. This table illustrates the basic difficulties in studying water movement in soils; the diffusivity varies by several orders of magnitude. In general, accurate experimental data is

TABLE 1

Normalised soil-water content θ	Diffusivity $D(\theta) = 9 \times 10^{-4} \exp(8.36\theta)$
0.0	0.0009
0.1	0.0021
0.2	0.0048
0.3	0.0111
0.4	0.0255
0.5	0.0588
0.6	0.1357
0.7	0.3122
0.8	0.7224
0.9	1.6668
1.0	3.8454

available for θ larger than approximately 0.4 and the behaviour of D for θ less than 0.4, is largely a matter for conjecture. Certainly the exponential and power law forms are fitted to data for θ moderate to large, and the behaviour for θ near zero is an artifact of the functional forms rather than the result of a theoretical derivation or experimental measurement.

SOLUTIONS FOR $L = \infty$

Consider eqn (1) subject to the boundary conditions in eqn (2), except that the flux condition is applied at $x = L \to \infty$. In this case, the similarity transform

$$\phi = x/\sqrt{t} \quad , \tag{6}$$

can be used to transform eqn (1) into an ordinary differential equation of the form (Philip (1969)),

$$\phi/2 + d\,(D(\theta)\ d\theta/d\phi)\,/\,d\theta = 0, \tag{7}$$

and the boundary conditions reduce to the form

$$\begin{aligned} \theta &= 1 \quad &&\text{at} \quad \phi = 0, \\ \theta &= \theta_I \quad &&\text{as} \quad \phi \to \infty. \end{aligned} \tag{8}$$

Equation (7) is a second order ordinary differential equation which, by appropriate algebraic manipulation, can be turned into an equation for $\phi = \phi(\theta)$, or an equation for $\theta = \theta(\phi)$. Note that the boundary conditions in eqn (8), imply that eqns (7) and (8) constitute a two end point boundary value problem. Generally this problem requires the use of shooting techniques for its solution. Using the capacity to select the independent variable, Braddock and Parlange (1981) were able to produce numerical estimates of the soil-water profiles $\theta = \theta(\phi)$. They claimed an accuracy of $O(10^{-7})$ for profiles determined using the diffusivity given in eqn (5); the profile thus obtained for $\theta_I = 0$ is shown in Figure 2. The main features of this profile are:

a) the steep wetting front for ϕ near 1, where the negative slope of the curve can be as large as $O(10^3)$.

b) the exponential tail for ϕ near 1.1. Braddock and Parlange (1981) showed that this tail was a property of the diffusivity $D(\theta{=}0) = 0.0009$ for the exponential form in Eqn (5).

The above properties remain in the solution profiles when other parameter values are used in the exponential diffusivity function $D(\theta) = Doe^{n\theta}$, although the constant Do is unnecessary since it is readily absorbed into a modified similarity variable by rescaling. Decreasing the value of n decreases the variability of $D(\theta)$ and the wetting front is not as steep. Conversely, increasing n increases the steepness of the wetting front as the variation in $D(\theta)$ increases. However, changing the value of θ_I, the initial water content, changes the residual diffusivity $D(\theta_I)$. As θ_I increases $D(\theta_I)$ also increases and the tail becomes more pronounced; this serves to decrease the sharpness of the wetting front. This tail arising from the constant residual diffusivity $D(\theta_I)$ is similar to the precursor of the main signal in eqn (3), obtained for the linear case D = constant.

For the power law diffusivity $D(\theta) \alpha \theta^n$ (any constant Do can be absorbed into a rescaled ϕ), and $\theta_I \neq 0$, the soil-water profiles are similar to those illustrated in Figure 2. Again, the structure of the tail depends on the residual

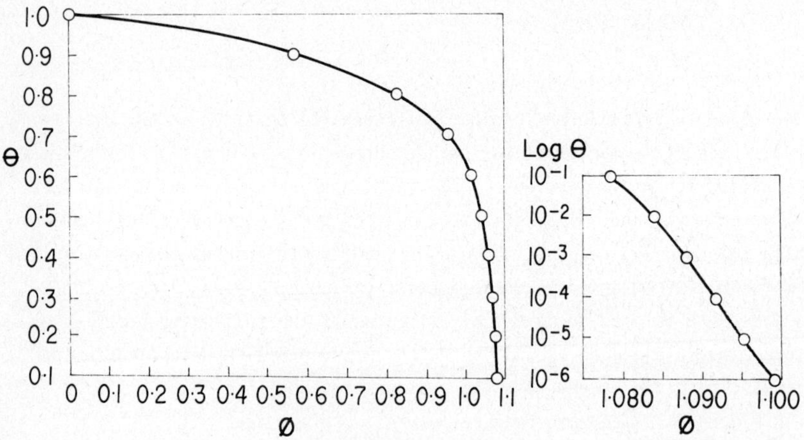

Figure 2
Profile of the soil water constant $\theta = \theta(\phi)$ for $D = 0.0009 \exp(8.36\theta)$, $\theta_I=0$ and $L = \infty$. An expanded log scale has been used to illustrate the tail.

diffusivity $D(\theta_I)$, although the wetting fronts exhibit somewhat steeper gradients. The case $\theta_I = 0$ is interesting in that $D(\theta_I = 0) = 0$ and thus for there to be a non-zero flux at the base of the front, then eqn (4) implies that the gradient $\left.\frac{\partial\theta}{\partial x}\right|_{\theta=0}$ must be infinite. In this case, there is no tail and the profile $\theta = \theta(\phi)$ becomes vertical at $\theta = 0$. This behaviour is borne out in the calculation of the profiles from eqn (7).

NUMERICAL METHODS FOR THE FINITE CASE

Now consider the solution of eqn (1) subject to the boundary conditions in eqn (2), with the further consideration that the variable x is suitably scaled so that the flux condition is imposed at L = 1. In this case, there is no similarity variable available for use, the problem is fully two dimensional and must be handled as a partial differential equation. For the exponential diffusivity $D = \exp(n\theta)$, a further rescaling of the form $\Theta = n\theta$ is possible and, on multiplying eqn (1) by n, there results

$$\frac{\partial \Theta}{\partial t} = \frac{\partial}{\partial x} \left(\exp(\Theta) \frac{\partial \Theta}{\partial x} \right), \tag{9}$$

subject to the boundary conditions

$$\begin{aligned} \Theta &= n \text{ at } x = 0, \quad t > 0, \\ \frac{\partial \Theta}{\partial x} &= 0 \text{ at } x = 1, \quad t > 0, \\ \Theta_I &= n\theta_I \text{ at } t = 0, \quad 0 < x < 1. \end{aligned} \tag{10}$$

Thus the diffusion equation is independent of the parameter n, but the boundary conditions now depend on n. In practice, n is in the range 4 to 10, and this stretching produces steeper wetting fronts for Θ. For the power law case, there is no equivalent transform to simplify the dependence on n.

Finite difference methods are readily applied to eqn (1), with $D = \theta^n$, and to eqn (9); each equation being linked to its appropriate boundary conditions. In the interests of generality, the methods will be described in application to eqn (1) for an unspecified functional form of $D(\theta)$; the reduction to the particular diffusivities above is obvious. The x - t plane is divided into a set of points using the points

$$\begin{aligned} x_i &= ih, \qquad i = 0, 1, \ldots\ldots, N + 1, \\ t_j &= jK, \qquad j = 0, 1, 2, \ldots\ldots, \end{aligned} \tag{11}$$

where h and k are the step spacings in x and t, respectively, $x_{N+1} = 1$, and thus $h = 1/(N + 1)$. The following notation will be used; i.e.,

$$\begin{aligned} \theta(x_i, t_j) &= \theta_{i,j}, \\ D(\theta_{i,j}) &= D_{i,j}. \end{aligned}$$

Equation (1) is written in the form

$$\frac{\partial \theta}{\partial t} = D(\theta) \frac{\partial^2 \theta}{\partial x^2} + \frac{dD}{d\theta} \left(\frac{\partial \theta}{\partial x}\right)^2 , \tag{12}$$

in preparation for discretisation. The finite differencing is carried out using a predictor-corrector strategy in that the time line j + ½ is used to step from the time line j to the time line j + 1. In the predictive step, the terms in eqn (12) are represented in the form

$$D(\theta)\,\frac{\partial^2\theta}{\partial x^2} = D_{ij}\left[\frac{1}{h^2}\left(\theta_{i+1,j+\frac{1}{2}} - 2\theta_{i,j+\frac{1}{2}} + \theta_{i-1,j+\frac{1}{2}}\right)\right]$$

$$\frac{dD}{d\theta}\cdot\left(\frac{\partial\theta}{\partial x}\right)^2 = \left(\frac{dD}{d\theta}\right)_{i,j}\left[\frac{1}{2h}\left(\theta_{i+1,\,j} - \theta_{i-1,\,j}\right)\right], \tag{13}$$

$$\frac{\partial\theta}{\partial t} = \left(\theta_{i,j+\frac{1}{2}} - \theta_{i,j}\right) / {}^{k}/_{2}$$

Here, the central difference formulation for $\frac{\partial^2\theta}{\partial x^2}$, is based on the $j + \frac{1}{2}$ time line, while $D(\theta)$ is evaluated on the j time line. This strategem provides a term which is linear on the $j + \frac{1}{2}$ time line. Further, the term $\frac{dD}{d\theta}\left(\frac{\partial\theta}{\partial x}\right)^2$ is evaluated on the j time line whilst a simple forward difference is used for $\frac{\partial\theta}{\partial t}$. The boundary conditions yield the relations

$$\theta_{o,j} = 1, \quad j = 0, 1, 2, \ldots,$$

$$\theta_{i,o} = \theta_I, \quad i = 0, 1, \ldots\ldots, N, N+1, \tag{14}$$

$$\frac{\theta_{N+2,j+\frac{1}{2}} - \theta_{N,j+\frac{1}{2}}}{2h} = 0, \quad j = 0, 1, 2, \ldots\ldots .$$

Combining eqns (13) and (14), yields the matrix form

$$\begin{bmatrix} -2\left(\frac{D_{1,j}}{h^2} + \frac{1}{k}\right), & \frac{D_{2,j}}{h^2}, & 0, 0, 0 & & & \\ \frac{D_{1,j}}{h^2}, & -2\left(\frac{D_{2,j}}{h^2} + \frac{1}{k}\right), & \frac{D_{3,j}}{h^2}, & 0, 0 & & \\ - - - - - - & & & & & \\ - - 0, 0, & \frac{D_{i-1,j}}{h^2}, & -2\left(\frac{D_{i,j}}{h^2} + \frac{1}{k}\right), & \frac{D_{i+1,j}}{h^2}, & 0, 0, & \\ - - - - - - & & & & & \\ - - - - - - - & 0, 0, & \frac{D_{N-1,j}}{h^2}, & -2\left(\frac{D_{N,j}}{h^2} + \frac{1}{k}\right), & \frac{D_{N+1,j}}{h^2} & \\ - - - - - - - - - & & 0, & \frac{2D_{N,j}}{h^2}, & -2\left(\frac{D_{N+1,j}}{h^2} + \frac{1}{k}\right) & \end{bmatrix} \begin{bmatrix} \theta_{1,j+\frac{1}{2}} \\ - \\ \theta_{2,j+\frac{1}{2}} \\ - \\ \theta_{i,j+\frac{1}{2}} \\ - \\ \theta_{N,j+\frac{1}{2}} \\ \theta_{N+1,j+\frac{1}{2}} \end{bmatrix} =$$

$$\begin{bmatrix} \frac{D_{1,j+\frac{1}{2}}}{h^2} - \frac{1}{k}, & \frac{D_{2,j+\frac{1}{2}}}{2h^2}, & 0, & - - - - - - - - \\ \frac{-D_{1,j+\frac{1}{2}}}{2h^2}, & \frac{D_{2,j+\frac{1}{2}}}{h^2} - \frac{1}{k}, & \frac{D_{3,j+\frac{1}{2}}}{2h^2}, & 0, - - - - - \\ - - - - & - - - - & - - - - & - - - - \\ -0, \frac{D_{i-1,j+\frac{1}{2}}}{2h^2}, & \frac{D_{i,j+\frac{1}{2}}}{h} - \frac{1}{k}, & -\frac{D_{i+1,j+\frac{1}{2}}}{2h^2}, & 0 - - - \\ - - - - & - - - - & - - - - & - - - - \\ - - -, 0, -\frac{D_{N-1,j+\frac{1}{2}}}{2h^2}, & \frac{D_{N,j+\frac{1}{2}}}{2h^2} - \frac{1}{k}, & \frac{D_{N+1,j+\frac{1}{2}}}{2h^2} & \\ - - - - - - - 0, & \frac{D_{N,j+\frac{1}{2}}}{h^2}, & \frac{D_{N+1,j+\frac{1}{2}}}{2h^2} - \frac{1}{k} & \end{bmatrix} \begin{bmatrix} \theta_{1,j} \\ \theta_{2,j} \\ - \\ \theta_{i,j} \\ - \\ \theta_{N,j} \\ \theta_{N+1,j} \end{bmatrix} +$$

$$\begin{bmatrix} \frac{D_{0,j+\frac{1}{2}}}{h^2} + \frac{1}{2h}\left(\frac{dD}{d\theta}\right)_{0,j+\frac{1}{2}} \\ 0 \\ - \\ 0 \\ - \\ 0 \\ -\frac{1}{2h}\left(\frac{dD}{d\theta}\right)_{N,j+\frac{1}{2}} \end{bmatrix}. \tag{18}$$

Equation (18) is a linear tridiagonal system for θ evaluated on the j+1 time line, where the right hand side can be evaluated in terms of the j and j+½ time lines. The solvability condition is identical to that in eqn (16) except that it is evaluated at the j+½ time line. The truncation error for the full predictor-corrector step is $O(h^2, k^{3/2})$ (Ames (1977)), and the equations are consistent. Whilst the fundamental equation (12) is nonlinear, the process of evaluating $D(\theta)$ and $\frac{\partial^2\theta}{\partial x^2}$ on different time lines has resulted in linear algebraic systems which are readily solved. The solvability condition, eqn (16) is a sufficient condition for solving the system.

RESULTS AND DISCUSSION

Given the initial distribution of θ, eqns (15) and (18) are readily deployed with a predictor-corrector strategy to produce solutions for $\theta(x,t)$ on successive time lines. In developing this numerical solution, a minor problem arises from the discontinuity in the initial data near $x = 0$ (see eqn (2)); this discontinuity occurs for small values of t only as the diffusion process quickly smooths it out. The discretisation along the x- axis implies that this initial distribution at $t = 0$, is modelled by a straight line approximation between the points ($x = 0$, $\theta = 1$) and ($x = h$, $\theta = \theta_I$) with $\theta = \theta_I$ for $h < x < 1$. Thus there are large numerical gradients of $\frac{\partial\theta}{\partial x}$ in the numerical model for $t = 0$. This defect can produce errors in the numerical estimate of the flux $\underline{F}$ (see eqn (4)) for small times t. When very small time steps k are used, this error in the initial flux does not produce any appreciable errors in the total mass passing $x = 0$. In the calculation, a relatively small k value was required to prevent an unwanted small oscillation in the solution $\theta(x,t)$. This oscillation arose for the larger values of the time step parameter k and this is due to the solvability properties of the matrix systems (eqns (15) and (18)). Note that eqn (16) is only a sufficient condition for solvability and that, since $\frac{\partial^2 D(\theta)}{\partial x^2}$ must be evaluated from the calculated solution $\theta(x,t)$ by numerical differentiation, eqn (16) is not a reliable criteria. Again, the matrix systems may be solvable in a theoretical sense, but the possible ill conditioning or nearness to singularity will introduce practical problems as manifested above. These small ripples on each time line are removed by selecting a suitably small time step k; this being done by experimenting with the calculations. The penalty of using a small time step is that the calculations become inordinately long and time consuming.

The solutions for $L = \infty$ can also be used to start the calculation of the solution near $t = 0$. For small times t, the barrier at $x = 1$, is playing little part in the diffusion process and it is effectively located at ∞. Thus the solution develops, at least initially, as if the barrier were not there and the solutions discussed in section 3 are accurate approximations. Whilst these solutions have precursors, they do not sense the barrier, at least in the initial stages. Eventually these precursors sense the barrier and they are the mechanism by which the soil-water profile adjusts to the barrier. The solutions for $L = \infty$ were used to check the accuracy and convergence of the finite difference scheme. Convergence, for small t, was very good and the calculations converged rapidly with decreasing k. The accuracy was also checked for the profile moving across towards the barrier. When the front had moved to $x \approx \frac{1}{2}$ for the finite difference scheme, then the profile was accurate to at least 3 significant figures, when compared with the same $L = \infty$ calculations.

Solutions for the exponential diffusivity, $D = \exp(n\theta)$ for $n = 1$, are shown in Figure 3, which show the water profiles for the dimensionless water content

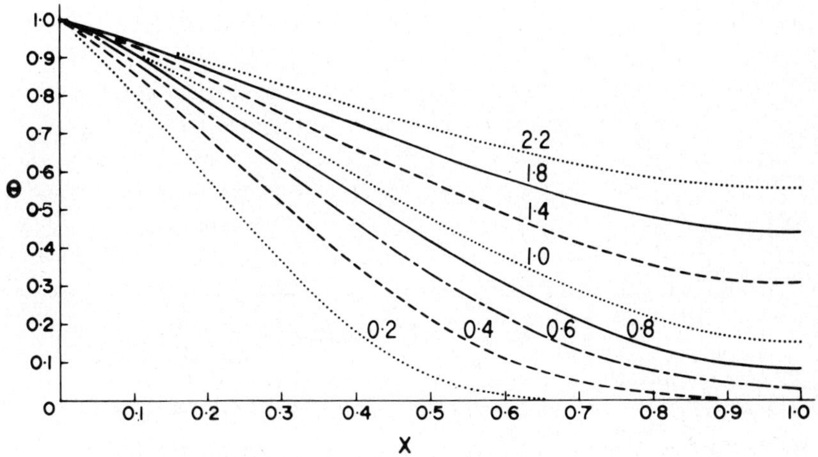

Figure 3
Profiles of the dimensionless soil-water content θ for an exponential diffusivity $D = e^{\theta}$, for $\theta_I=0$. The numbers against each curve indicate the nondimensional time in seconds.

$\theta(x,t)$, for $\theta_I = 0$, at various normalised times t (in seconds). The front in Figure 3 is not well developed, or very steep, this being due to the diffusivity having a relatively small range of values for $n = 1$; i.e. the nonlinearity is relatively weak. This feature is illustrated in Figure 4 which shows the soil-water concentration at the barrier, i.e., $\theta(x=1,t)$ as a function of time. This curve indicates that the precursor senses or arrives at the barrier at or near $t = 0.05$, and that it adjusts the finite difference solution to the presence of the barrier, before the main front arrives at or near $t = 0.1$. The relatively long time for the front to reach the barrier arises from the relatively small value of the diffusivity. Once the front contacts the barrier, the zero flux condition causes θ to increase and lessen the steepness of the front, i.e., the front weakens quickly.

The calculations are readily repeated for $n = 2$ and $n = 3$ but the cases for $n > 4$ require inordinate amounts of computer time and are too expensive. On the other hand, the cases for $\theta_I \neq 0$ are easier to handle; there are several reasons for this. First, the wetting front is steepest and has its greatest curvature at the junction with the precursor. For $\theta_I > 0$, the front does not

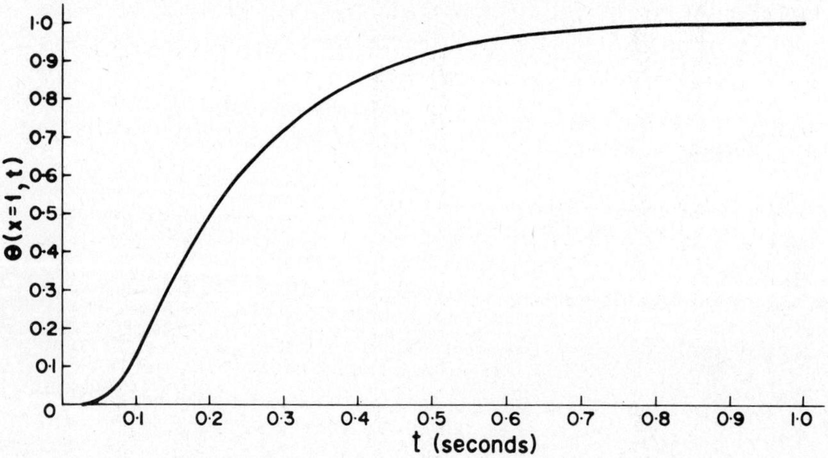

Figure 4
Soil-water concentration $\theta(x=1,t)$ at the barrier, as a function of time, for $D = e^{\theta}$.

attain the largest gradient or curvature and the solvability condition eqn (16) is not as demanding. Secondly, the residual diffusivity $D(\theta_I)$ is larger for $\theta_I > 0$, hence the precursor is more pronounced, senses the barrier sooner and permits a smoother transition. This larger residual diffusivity admits a smoother transition from the discontinuity at t = 0, x = 0 so that there is less error in determining the total flow in the first time steps.

With respect to the power law case, i.e., $D = \theta^n$, with $\theta = \theta_I$ for t = 0, then the residual diffusivity is $\theta_I{}^n$, $0 < \theta_I < 1$. This residual diffusivity is less than the corresponding residual diffusivity $e^{n\theta_I}$ obtained in the exponential case. For the power law diffusivity, the front is steeper, the precursor is less pronounced and the curvature at their junction, is steeper. Solving the power law case is far more demanding on the numerical procedures and solutions for small values of n are obtainable. For $\theta_I = 0$, the wetting front becomes vertical and there is no precursor. Here, the front moves out and has no precursor to sense the wall; eventually the front with $\frac{\partial\theta}{\partial x} = \infty$, touches the wall where $\frac{\partial\theta}{\partial x} = 0$. In this case, the formulation in eqn (12) is ill posed (Ames (1977)) and it is little wonder that the numerics have been stretched.

REFERENCES

Ames, W.F. (1977), *Nonlinear Partial Differential Equations in Engineering*, 2nd Edition, Barnes and Noble, Inc., New York.

Braddock, R.D., and Parlange, J.Y., (1980), "Some Precise Numerical Solutions of the Soil-Water Equation", *Soil Science Society of America, Journal*, Vol. 44, p.656.

Bruce, R.R., and Klute, A. (1956), "The Measurement of Soil-Water Diffusivity", *Soil Science Society of America, Proceedings*, Vol. 20, pp. 458-462.

Hayhoe, M.N. (1978), "Numerical Study of Quasi-analytic and Finite Difference Solutions of the Soil-Water Transfer Function", *Soil Science*, Vol. 125, pp. 68-74.

Noye, B.J. (1981), "Finite Difference Methods for Partial Differential Equations", in *Numerical Solutions of Partial Differential Equations*, ed. J. Noye, North-Holland, Amsterdam, pp. 1-135.

Parlange, J.Y. (1980). "Water Transport in Soils", *Annual Review of Fluid Mechanics*, Vol 12, p. 77.

Parlange, J.Y., Braddock, R.D. and Chu, B.S. (1980), First Integrals of the Diffusion Equation, An Extension of the Fujita Solutions, *Soil Science Society of America, Journal*, Vol 44, pp. 908-911.

Passioura, J.B. (1976), "Determining Soil-Water Diffusivities from one step Outflow Experiments", *Australian Journal of Soil Research*, Vol. 15, pp. 1-8.

Philip, J.R. (1969), "Theory of Infiltration", *Advances in Hydroscience*, Vol. 5, pp. 215-296.

Widder, H. (1975), *The Heat Equation*, Academic Press, New York.

Young, D. and Gregory, J.,(1972), *A Survey of Numerical Methods*, Addison Wesley, New York.

NUMERICAL SOLUTIONS OF
PARTIAL DIFFERENTIAL EQUATIONS
J. Noye (editor)

AN IMPLICIT FINITE DIFFERENCE SCHEME FOR THE DIFFUSION EQUATION SUBJECT TO THE SPECIFICATION OF MASS IN A PORTION OF THE DOMAIN

John R. Cannon[(1)]
Department of Pure & Applied Mathematics
Washington State University
Pullman, WA 99164, U.S.A.

John van der Hoek
Department of Pure Mathematics
The University of Adelaide
Adelaide, South Australia 5001.

An implicit finite difference scheme is derived together with analysis of convergence for the numerical solution for $u = u(x,t)$ of the diffusion equation

$$\frac{\partial u}{\partial t} = \frac{\partial^2 u}{\partial x^2}, \qquad 0 < x < 1, \quad 0 < t \leqslant T$$

subject to

$$u(x,0) = f(x), \quad 0 \leqslant x \leqslant 1$$

$$u(1,t) = g(t), \quad 0 < t \leqslant T$$

and the specification of mass

$$\int_0^b u(x,t)dx = M(t), \quad 0 \leqslant t \leqslant T$$

where $0 < b < 1$.

1. INTRODUCTION

We consider the problem of obtaining numerical approximations to the solution $u = u(x,t)$ of the problem

$$(1.1) \quad \frac{\partial u}{\partial t} = \frac{\partial^2 u}{\partial x^2}, \qquad 0 < x < 1, \quad 0 < t \leqslant T$$

$$(1.2) \quad u(x,0) = f(x), \qquad 0 \leqslant x \leqslant 1$$

$$(1.3) \quad u(1,t) = g(t), \qquad 0 < t \leqslant T$$

$$(1.4) \quad M(t) = \int_0^b u(x,t)dx, \quad 0 < t \leqslant T$$

and where f, g and M are known functions, and $0 < b < 1$ and T are positive constants. The boundary condition (1.4) is referred to as mass or energy specification and has been studied by CANNON [1], DECKERT and MAPLE [2], KAMYNIN [3], IONKIN [4], [5], CANNON and VAN DER HOEK [6], [7] and [8], where applications are indicated.

If $M(t)$ is differentiable, and u is a classical solution of (1.1)-(1.4) then (1.4) is equivalent to

$$(1.5) \quad M'(t) \equiv \frac{\partial M}{\partial t}(t) = \frac{\partial u}{\partial x}(b,t) - \frac{\partial u}{\partial x}(0,t), \quad 0 < t \leqslant T.$$

We also remark that by setting $v(x,t) = u(x,t) - 2\, b^{-1}(2-b)^{-1}(1-x)M(t)$, v satisfies, using (1.5)

---

(1) This research was supported in part by the National Science Foundation, Washington D.C., U.S.A.

$$(1.6)\qquad \frac{\partial v}{\partial t}(x,t) = \frac{\partial^2 v}{\partial x^2}(x,t) + F(x,t) \qquad 0 < x < 1, \quad 0 < t \leqslant T$$

$$(1.7)\qquad v(x,0) = h(x), \qquad 0 \leqslant x \leqslant 1$$

$$(1.8)\qquad v(1,t) = g(t), \qquad 0 < t \leqslant T$$

$$(1.9)\qquad \frac{\partial v}{\partial x}(b,t) = \frac{\partial v}{\partial x}(0,t), \qquad 0 < t \leqslant T$$

where $F(x,t) = -2\, b^{-1}(2-b)^{-1}M'(t)$, and $h(x) = f(x) - 2\, b^{-1}(2-b)^{-1}(1-x)M(0)$. The numerical method we describe below works equally well for the numerical solutions for equations of the form (1.6)-(1.9).

We shall assume that for sufficiently smooth data f, g and M that the solution to (1.1)-(1.4) belongs to the class $C^{4,2}([0,1]\times[0,T])$ that is, the solution u is continuous on $[0,1]\times[0,T]$ along with four derivatives with respect to x and two with respect to t. To this end we will assume f and M have bounded derivatives and g is bounded with $g(0) = f(1)$. The finite difference scheme is described in section 2 and error estimates are given in section 5. The error analysis uses methods similar to those of LEES [9] and the appropriate machinery for our problem is developed in sections 3 and 4.

2. AN IMPLICIT FINITE DIFFERENCE SCHEME

Using the representation formula for the solution of the heat equation (see WIDDER [10]), for any solution u of (1.1) and (1.2) we have

$$(2.1)\qquad \frac{\partial u}{\partial x}(b,t) = \int_0^1 \left\{\frac{\partial \theta}{\partial x}(b-\xi,t) - \frac{\partial \theta}{\partial x}(b+\xi,t)\right\} f(\xi)d\xi$$

$$- 2\int_0^t \frac{\partial^2\theta}{\partial x^2}(b,t-\tau)u(0,\tau)d\tau$$

$$+ 2\int_0^t \frac{\partial^2\theta}{\partial x^2}(b-1,t-\tau)u(1,\tau)d\tau$$

where

$$(2.2)\qquad \theta(x,t) = \sum_{m=-\infty}^{\infty} K(x+2m,t)$$

and

$$(2.3)\qquad K(x,t) = \begin{cases} \dfrac{1}{\sqrt{4\pi t}} \exp\left(-\dfrac{x^2}{4t}\right) & t > 0, \quad x \in \mathbb{R}. \\ 0 & t \leqslant 0, \quad x \in \mathbb{R}. \end{cases}$$

By an integration by parts, we see that

$$(2.4)\qquad \frac{\partial u}{\partial x}(b,t) = 2\,\theta(b,t)f(0)$$

$$- \{\theta(b-1,t)+\theta(b+1,t)\}\, f(1)$$

$$+ \int_0^1 \{\theta(b-\xi,t)+\theta(b+\xi,t)\}\, f'(\xi)d\xi$$

$$- 2\int_0^t \frac{\partial^2\theta}{\partial x^2}(b,t-\tau)u(0,\tau)d\tau$$

$$+ 2\int_0^t \frac{\partial^2\theta}{\partial x^2}(b-1,t-\tau)u(1,\tau)d\tau$$

Setting

$$D(t) = 2\theta(b,t)f(0) - \{\theta(b-1,t) + \theta(b+1,t)\}\ f(1)$$

(2.5) $$+ \int_0^1 \{\theta(b-\xi,t) + \theta(b+\xi,t)\}\ f'(\xi)d\xi$$

$$+ 2\int_0^1 \frac{\partial^2\theta}{\partial x^2}(b-1,t-\tau)g(\tau)d\tau - M'(t)$$

and

(2.6) $$N(t) = -2\frac{\partial^2\theta}{\partial x^2}(b,t) \qquad \text{for} \quad t > 0$$

equation (1.5) becomes

(2.7) $$\frac{\partial u}{\partial x}(0,t) = \int_0^t N(t-\tau)u(0,\tau)d\tau + D(t), \qquad 0 < t \leqslant T.$$

Subdivide the interval $\bar{I} = \{x|\ 0\leqslant x\leqslant 1\}$ into N equal subintervals of length h, and define

$$\bar{I}_h = \{x\in\bar{I}|\ x=ih,\ i=0,1,2,\dots,N\}.$$

Put $I_h' = \bar{I}_h - \{1\}$. Subdivide the interval $\bar{J} = \{t|\ 0\leqslant t\leqslant T\}$ into M equal subintervals of length k, and define

$$\bar{J}_k = \{t\in\bar{J}|\ t=jk,\ j=0,1,2,\dots,M\}.$$

The set $\bar{I}_h \times \bar{J}_k$ determines a mesh for $\bar{I}\times\bar{J}$. The parameter $\lambda = kh^{-2}$ is called the mesh parameter for $\bar{I}_h\times\bar{J}_k$.

We define first order difference quotients of a function $\phi = \phi(x,t)$:

$$\phi_x(x,t) = h^{-1}[\phi(x+h,t)-\phi(x,t)]$$
$$\phi_{\bar{x}}(x,t) = h^{-1}[\phi(x,t)-\phi(x-h,t)]$$
$$\phi_t(x,t) = k^{-1}[\phi(x,t+k)-\phi(x,t)]$$
$$\phi_{\bar{t}}(x,t) = k^{-1}[\phi(x,t)-\phi(x,t-k)].$$

We now define a function $v(x,t)$ on $(\bar{I}_h\cup\{-h\})\times\bar{J}_k$ by the following implicit finite difference scheme.

(2.8) $Lv(x,t) \equiv v_{\bar{t}}(x,t) - v_{x\bar{x}}(x,t) = 0$ for $(x,t)\in(\bar{I}_h-\{1\})\times(\bar{J}_k-\{0\})$.

(2.9) $v(1,t) = g(t)$ for all $t\in\bar{J}_k - \{0\}$.

(2.10) $v(x,0) = f(x)$ for all $x\in\bar{I}_h$

$v(-h,0) = v(0,0)$

(2.11) $$v_{\bar{x}}(0,t) = \sum_{t=k}^{t} kN(t-\tau)v(0,\tau) + D(t)$$

for all $t\in\bar{J}_k - \{0\}$.

Equations (2.8)-(2.11) determines $v(x,t)$ on $(\bar{I}_h\cup\{-h\})\times\bar{J}_k$ uniquely. $v(x,t)$ restricted to $\bar{I}_h\times\bar{J}_k$ is our approximation to the solution u of (1.1)-(1.4) evaluated on $\bar{I}_h\times\bar{J}_k$.

REMARKS

(i) We define $N(t) = 0$ for $t\leqslant 0$, so the summation in (2.11) could be taken from

$\tau = k$ to $\tau = t-k$.

(ii) $Lv(x,t) \equiv k^{-1}(v(x,t)-v(x,t-k)) - h^{-2}(v(x+h,t)-2v(x,t) + v(x-h,t))$ for $(x,t) \in (\bar{I}_h-\{1\}) \times (\bar{J}_k-\{0\})$.

3. PRELIMINARY RESULTS

Let $\mathcal{L}$ denote the linear space of all functions ψ defined on $\bar{I}_h$ satisfying $\psi(1) = 0$. On $\mathcal{L}$ we introduce two norms

$$\|\psi\|_0 = (h \sum_{x \in I_h'} |\psi(x)|^2)^{\frac{1}{2}}$$

$$\|\psi\|_1 = (\|\psi\|_0^2 + \|\psi_x\|_0^2)^{\frac{1}{2}}$$

and the associated inner product

$$(\varphi,\psi) = h \sum_{x \in I_h'} \phi(x)\psi(x)$$

for $\varphi,\psi \in \mathcal{L}$.

3.1 Lemma. Let $\varepsilon > 0$. For each $\psi \in \mathcal{L}$ and for each $x \in \bar{I}_h$ we have

(3.1) $$\psi(x)^2 \leqslant \varepsilon\|\psi_x\|_0^2 + \left(1 + \frac{16}{\varepsilon}\right)\|\psi\|_0^2.$$

Proof: Let $y \in I_h'$ and suppose in the first case that $y > x$. Then

$$\psi(y)^2 - \psi(x)^2 = \sum_{p=x}^{y-h} \{\psi(p+h)^2-\psi(p)^2\}.$$

But $\psi(p+h)^2 - \psi(p)^2 = h[\psi(p+h)+\psi(p)]\psi_x(p)$
and so by Cauchy's inequality

$$|\psi(p+h)^2-\psi(p)^2| \leqslant \varepsilon h|\psi_x(p)|^2 + \frac{8h}{\varepsilon}\{|\psi(p)|^2+|\psi(p+h)|^2\};$$

we then have

(3.2) $$\psi(x)^2 \leqslant \psi(y)^2 + \varepsilon\|\psi_x\|_0^2 + \frac{16}{\varepsilon}\|\psi\|_0^2.$$

We obtain the same inequality for $y \leqslant x$ in the same way. Multiplying (3.2) by h and summing y over I_h' and using $\sum_{y \in I_h'} h = 1$, we obtain (3.1).

3.2 Lemma. Let $\phi(t)$ and $\psi(t)$ be non-negative functions defined on $\bar{J}_k$, the latter being non decreasing. If

(3.3) $$\phi(t) \leqslant \psi(t) + Ck \sum_{\tau=0}^{t-k} \phi(\tau), \quad t \in \bar{J}_k - \{0\}$$

where $C \geqslant 0$, then

$$\phi(t) \leqslant \psi(t)\exp(Ct), \quad t \in \bar{J}_k.$$

Proof: This is Lemma 2 of [9].

Notation: In Lemma 3.3 we introduce a function $\phi(t)$ defined on $\bar{J}_k$ with values in the linear space $\mathcal{L}$. We shall let $\phi(x,t)$ denote $\phi(t)$ evaluated at $x \in \bar{I}_h$.

3.3 Lemma. Let $\phi(t)$ be a function defined on $\bar{J}_k$ with values in the linear space $\mathcal{L}$. If a, b are real numbers, then the following identity holds for all $t \in \bar{J}_k - \{0\}$.

$$(3.3)\qquad 2k \sum_{\tau=k}^{t} \left(a\hat{\phi}(\tau)+b\phi_{\bar{t}}(\tau), L\phi(\tau)\right)$$

$$= a\|\phi(t)\|_0^2 - a\|\phi(0)\|_0^2 + 2ak \sum_{\tau=k}^{t} \|\phi_x(\tau)\|_0^2$$

$$+ \left(b - \frac{ak}{2}\right) \|\phi_x(t)\|_0^2 - \left(b - \frac{ak}{2}\right) \|\phi_x(0)\|_0^2$$

$$+ 2bk \sum_{\tau=k}^{t} \|\phi_{\bar{t}}(\tau)\|_0^2 + k\left(b - \frac{ak}{2}\right) k \sum_{\tau=k}^{t} \|\phi_{x\bar{t}}(\tau)\|_0^2$$

$$+ 2ak \sum_{\tau=k}^{t} \hat{\phi}(0,\tau)\phi_{\bar{x}}(0,\tau) + 2bk \sum_{\tau=0}^{t} \phi_{\bar{t}}(0,\tau)\phi_{\bar{x}}(0,\tau)$$

where $\hat{\phi}(\tau) \equiv \frac{1}{2}(\phi(\tau)+\phi(\tau-k))$ for $\tau \in \bar{J}_k - \{0\}$, and where $\sum_{\tau=k}^{t} \psi(\tau)$ means the sum of the elements of the set $\{\psi(\tau) | \tau \in \bar{J}_k,\ k \leq \tau \leq t\}$.

Proof: The proof is a modification of that given in [9]. For completeness a proof is given in the appendix.

4. PRELIMINARY ESTIMATES

4.1 Lemma. Let $\phi(t)$ be a function defined on $\bar{J}_k$ with values in the linear space $\mathcal{L}$. Then

$$(4.1)\qquad 2k \sum_{\tau=k}^{t} \left(\phi(\tau), L\phi(\tau)\right)$$

$$= \|\phi(t)\|_0^2 - \|\phi(0)\|_0^2 + 2k \sum_{\tau=k}^{t} \|\phi_x(\tau)\|_0^2$$

$$+ k^2 \sum_{\tau=k}^{t} \|\phi_{\bar{t}}(\tau)\|_0^2 + 2k \sum_{\tau=k}^{t} \phi(0,\tau)\phi_{\bar{x}}(0,\tau).$$

Proof: Put $a = 1$ and $b = \frac{1}{2}k$ in Lemma 3.3 and observe that $\phi = \hat{\phi} + \frac{1}{2}k\phi_{\bar{t}}$.

4.2 Lemma. Let $\phi(t)$ be a function defined on $\bar{J}_k$ with values in the linear space $\mathcal{L}$.

Suppose in addition that $\phi(t)$ satisfies

$$(4.2)\qquad \phi_{\bar{x}}(0,t) = k \sum_{\tau=k}^{t} N(t-\tau)\phi(0,\tau) + F(t)$$

for $t \in \bar{J}_k - \{0\}$; then there exists a constant $C > 0$ so that

$$(4.3)\qquad \|\phi(t)\|_0^2 + k \sum_{\tau=k}^{t} \|\phi_x(\tau)\|_0^2 + k^2 \sum_{\tau=k}^{t} \|\phi_{\bar{t}}(\tau)\|_0^2$$

$$\leq \|\phi(0)\|_0^2 + Ck \sum_{\tau=k}^{t} \|\phi(\tau)\|_0^2 + 4k \sum_{\tau=k}^{t} \|L\phi(\tau)\|_0^2$$

$$+ 8k \sum_{\tau=k}^{t} F(\tau)^2.$$

Proof: By Lemma 4.1 we have

$$(4.4)\qquad \|\phi(t)\|_0^2 + 2k\sum_{\tau=k}^{t}\|\phi_x(\tau)\|_0^2 + k^2\sum_{\tau=k}^{t}\|\phi_{\bar{t}}(\tau)\|_0^2$$

$$= \|\phi(0)\|_0^2 - 2k\sum_{\tau=k}^{t}\big(\phi(\tau),L\phi(\tau)\big) - 2k\sum_{\tau=k}^{t}\phi(0,\tau)\phi_{\bar{x}}(0,\tau).$$

We make the following estimates:

$$(4.5)\qquad \Big|\sum_{\tau=k}^{t}\big(\phi(\tau),L\phi(\tau)\big)\Big| \leq 2\sum_{\tau=k}^{t}\|\phi(\tau)\|_0^2 + 2\sum_{\tau=k}^{t}\|L\phi(\tau)\|_0^2$$

$$(4.6)\qquad \Big|\sum_{\tau=k}^{t}\phi(0,\tau)\phi_{\bar{x}}(0,\tau)\Big| \leq 2\sum_{\tau=k}^{t}|\phi(0,\tau)|^2 + 2\sum_{\tau=k}^{t}|\phi_{\bar{x}}(0,\tau)|^2 .$$

By Lemma 3.1, for any $\varepsilon_1 > 0$

$$(4.7)\qquad |\phi(0,\tau)|^2 \leq \varepsilon_1\|\phi_x(\tau)\|_0^2 + \left(1+\frac{16}{\varepsilon_1}\right)\|\phi(\tau)\|_0^2 .$$

Let $|N(t)| \leq A$ for all $0 \leq t \leq T$; we have that

$$(4.8)\qquad |\phi_{\bar{x}}(0,t)|^2 \leq 2\left(k\sum_{\tau=k}^{t}N(t-\tau)\phi(0,\tau)\right)^2 + 2F(t)^2$$

$$\leq 2\sum_{\tau=k}^{t}k[N(t-\tau)]^2\sum_{\tau=k}^{t}k[\phi(0,\tau)]^2 + 2F(t)^2$$

$$\leq 2A^2T\sum_{\tau=k}^{t}k[\phi(0,\tau)]^2 + 2F(t)^2.$$

Substituting (4.7) into (4.8) yields

$$(4.9)\qquad |\phi_{\bar{x}}(0,t)|^2 \leq 2A^2T\varepsilon_1\sum_{\tau=k}^{t}k\|\phi_x(\tau)\|_0^2$$

$$+ 2A^2T\left(1+\frac{16}{\varepsilon_1}\right)\sum_{\tau=k}^{t}k\|\phi(\tau)\|_0^2 + 2F(t)^2$$

thus

$$(4.10)\qquad \sum_{\tau=k}^{t}k|\phi_{\bar{x}}(0,\tau)|^2 \leq 2A^2T^2\varepsilon_1\sum_{\tau=k}^{t}k\|\phi_x(\tau)\|_0^2$$

$$+ 2A^2T^2\left(1+\frac{16}{\varepsilon_1}\right)\sum_{\tau=k}^{t}k\|\phi(\tau)\|_0^2 + 2\sum_{\tau=k}^{t}kF(\tau)^2.$$

In (4.10) we have used the fact that

$$\sum_{\tau=k}^{t}k\sum_{S=k}^{\tau}k\|\phi_x(S)\|_0^2 \leq T\sum_{S=k}^{t}k\|\phi_x(S)\|_0^2$$

and a similar estimate

$$\sum_{\tau=k}^{t}k\sum_{S=k}^{\tau}k\|\phi(S)\|_0^2 \leq T\sum_{S=k}^{t}k\|\phi(S)\|_0^2.$$

Substituting (4.5)-(4.7), (4.10) into (4.4) yields

$$\|\phi(t)\|^2 + 2k\sum_{\tau=k}^{t}\|\phi_x(\tau)\|_0^2 + k^2\sum_{\tau=k}^{t}\|\phi_{\bar{t}}(\tau)\|_0^2$$

$$\leq \|\phi(0)\|_0^2 + \left(4+4\left(1+\frac{16}{\varepsilon_1}\right) + 8A^2T^2\left(1+\frac{16}{\varepsilon_1}\right)\right)\sum_{\tau=k}^{t}k\|\phi(\tau)\|_0^2$$

$$+ 4k \sum_{\tau=k}^{t} \| L\phi(\tau)\|_0^2 + \varepsilon_1(4+8A^2T^2) \sum_{\tau=k}^{t} k\|\phi_x(\tau)\|_0^2$$

$$+ 8k \sum_{\tau=k}^{t} F(\tau)^2 .$$

Choose $\varepsilon_1 = (4+8A^2T^2)^{-1}$, then

$$\|\phi(t)\|^2 + k \sum_{\tau=k}^{t} \|\phi_x(\tau)\|_0^2 + k^2 \sum_{\tau=k}^{t} \|\phi_{\bar{t}}(\tau)\|_0^2$$

$$\leqslant \|\phi(0)\|_0^2 + C(A,T) \sum_{\tau=k}^{t} k\|\phi(\tau)\|_0^2 + 4k \sum_{\tau=k}^{t} \|L\phi(\tau)\|_0^2$$

$$+ 8k \sum_{\tau=k}^{t} F(\tau)^2$$

where $C(A,T) = 4 + (4+8A^2T^2)\big(1+16(4+8A^2T^2)\big)$.

4.3 Lemma. For $k < (2C)^{-1}$ we have

$$\|\phi(t)\|_0^2 + 2k \sum_{\tau=k}^{t} \|\phi_x(\tau)\|_0^2 + 2k^2 \sum_{\tau=k}^{t} \|\phi_{\bar{t}}(\tau)\|_0^2 \tag{4.11}$$

$$\leqslant (2+2Cte^{Ct}) \|\phi(0)\|_0^2$$

$$+ 8Cte^{Ct} \sum_{\tau=k}^{t} \|L\phi(\tau)\|_0^2 + 16Cte^{Ct} \sum_{\tau=k}^{t} kF(\tau)^2$$

for all $t \in \bar{J}_k$, where C is the constant in equation (4.3).

Proof: For $k < (2C)^{-1}$ we get by (4.3) that

$$\|\phi(t)\|_0^2 + 2k \sum_{\tau=k}^{t} \|\phi_x(\tau)\|_0^2 + 2k^2 \sum_{\tau=k}^{t} \|\phi_{\bar{t}}(\tau)\|_0^2$$

$$\leqslant 2\|\phi(0)\|_0^2 + Ck \sum_{\tau=k}^{t-k} \|\phi(\tau)\|_0^2 + 8k \sum_{\tau=k}^{t} \|L\phi(\tau)\|_0^2$$

$$+ 16 \sum_{\tau=k}^{t} kF(t)^2 .$$

Dropping the second and third terms in the left hand side of this inequality and using Lemma 3.2, we get

$$\|\phi(t)\|_0^2 \leqslant \Big\{2\|\phi(0)\|_0^2 + 8k \sum_{\tau=k}^{t} \|L\phi(\tau)\|_0^2 + 16 \sum_{\tau=k}^{t} kF(\tau)^2\Big\} e^{Ct} \tag{4.12}$$

for all $t \in \bar{J}_k$; whence the result.

4.4 Lemma. Let $\phi(t)$ be a function defined on $\bar{J}_k$ with values in the linear space $\mathcal{L}$. Then

$$2k \sum_{\tau=k}^{t} \big(\phi_{\bar{t}}(\tau), L\phi(\tau)\big) \tag{4.13}$$

$$= \|\phi_x(t)\|_0^2 - \|\phi_x(0)\|_0^2 + 2k \sum_{\tau=k}^{t} \|\phi_{\bar{t}}(\tau)\|_0^2$$

$$+ k^2 \sum_{\tau=k}^{t} \|\phi_{x\bar{t}}(\tau)\|_0^2 + 2k \sum_{\tau=k}^{t} \phi_{\bar{t}}(0,\tau)\phi_{\bar{x}}(0,\tau)$$

for each $t \in \bar{J}_k - \{0\}$.

Proof: Put $a = 0$ and $b = 1$ in Lemma 3.3.

4.5 Lemma. Suppose $\phi(t)$ is a function defined on $\bar{J}_k$ with values in the linear space $\mathcal{L}$ and suppose that ϕ satisfies (4.2), then

$$(4.14)\quad k\sum_{\tau=k}^{t}\phi_{\bar{t}}(0,\tau)\phi_{\bar{x}}(0,\tau)$$
$$= \phi(0,t)k\sum_{\tau=k}^{t}N(t-\tau)\phi(0,\tau) - N(k)k\sum_{\tau=k}^{t}\phi(0,\tau+k)\phi(0,\tau)$$
$$- k^2\sum_{S=k}^{t-k}\sum_{\tau=S+k}^{t}N_{\bar{t}}(\tau-S)\phi(0,\tau-k)\phi(0,S)$$
$$+ k\sum_{\tau=k}^{t}\phi_{\bar{t}}(0,\tau)F(\tau),\quad \text{for } t\in\bar{J}_k-\{0\}.$$

Proof:

$$k\sum_{\tau=k}^{t}\phi_{\bar{t}}(0,\tau)\phi_{\bar{x}}(0,\tau)$$
$$= k\sum_{\tau=k}^{t}\phi_{\bar{t}}(0,\tau)\sum_{S=k}^{\tau}kN(\tau-S)\phi(0,S) + k\sum_{\tau=k}^{t}\phi_{\bar{t}}(0,\tau)F(\tau).$$

Now

$$k^2\sum_{S=k}^{t}\sum_{\tau=S}^{t}\phi_{\bar{t}}(0,\tau)N(\tau-S)\phi(0,S)$$
$$= k^2\sum_{S=k}^{t-k}\sum_{\tau=S+k}^{t}\phi_{\bar{t}}(0,\tau)N(\tau-S)\phi(0,S)$$
$$= k^2\sum_{S=k}^{t-k}\sum_{\tau=S+k}^{t}\{[\phi(0,\tau)N(\tau-S)]_{\bar{t}} - \phi(0,\tau-k)N_{\bar{t}}(\tau-S)\}\phi(0,S)$$

and

$$k^2\sum_{S=k}^{t-k}\sum_{\tau=S+k}^{t}[\phi(0,\tau)N(\tau-S)]_{\bar{t}}\phi(0,S)$$
$$= k\sum_{S=k}^{t-k}[\phi(0,t)N(t-S) - \phi(0,S+k)N(k)]\phi(0,S)$$
$$= \phi(0,t)k\sum_{S=k}^{t-k}N(t-S)\phi(0,S) - N(k)k\sum_{S=k}^{t-k}\phi(0,S+k)\phi(0,S)$$

whence the result.

4.6 Corollary. For $\phi(t)$ as in Lemma 4.5 we have the estimate

$$(4.15)\quad |k\sum_{\tau=k}^{t}\phi_{\bar{t}}(0,\tau)\phi_{\bar{x}}(0,\tau)|$$
$$\leqslant 2(\phi(0,t))^2 + C_1k\sum_{\tau=k}^{t}[\phi(0,\tau)]^2$$
$$+ k\sum_{\tau=k}^{t}|\phi_{\bar{t}}(0,\tau)||F(\tau)|$$

where $C_1 = 2TA^2 + 4A + 4TB$, A as before and

$B = \sup\{|N_{\bar{t}}(\tau)|;\ k\leqslant t\leqslant T\} < +\infty.$

Proof: (4.15) is a direct consequence of (4.14) noting that N and $N_{\bar{t}}$ are uniformly bounded on $\bar{J}_k - \{0\}$.

4.7 Lemma. Let $\phi(t)$ be the same as in Lemma 4.5 and suppose that F satisfies

(4.16) $\quad |F(t)| \leqslant Fk,$

where F is a positive constant, for all $t \in \bar{J}_k - \{0\}$. Let C be the constant in Lemma 4.3 and C_1 the constant in Corollary 4.6 then for

$$0 < k < \min\{\tfrac{1}{4}F^{-1}(1+64F)^{-1},\ (2C)^{-1},\ (8C_1)^{-1}\},$$

there exist constants C_2 and C_3 depending on C_1, C_2, F, T so that

$$\|\phi_x(t)\|_0^2 \leqslant C_2 e^{C_3 t}\{\|\phi(0)\|_0^2 + k\sum_{\tau=k}^{t}\|L\phi(\tau)\|_0^2 + k + \|\phi_x(0)\|_0^2\} \tag{4.17}$$

for all $t \in \bar{J}_k$.

Proof: By Lemma 4.4 we have

$$\begin{aligned}\|\phi_x(t)\|_0^2 &+ 2k\sum_{\tau=k}^{t}\|\phi_{\bar{t}}(\tau)\|_0^2 + k^2\sum_{\tau=k}^{t}\|\phi_{x\bar{t}}(\tau)\|_0^2 \\ &= \|\phi_x(0)\|_0^2 + 2k\sum_{\tau=k}^{t}(\phi_{\bar{t}}(\tau), L\phi(\tau)) \\ &\quad - 2k\sum_{\tau=k}^{t}\phi_{\bar{t}}(0,\tau)\phi_{\bar{x}}(0,\tau)\end{aligned} \tag{4.18}$$

for all $t \in \bar{J}_k - \{0\}$. Now

$$\begin{aligned}2k&\Big|\sum_{\tau=k}^{t}(\phi_{\bar{t}}(\tau), L\phi(\tau))\Big| \\ &\leqslant k\sum_{\tau=k}^{t}\|\phi_{\bar{t}}(\tau)\|_0^2 + 8k\sum_{\tau=k}^{t}\|L\phi(\tau)\|_0^2\end{aligned} \tag{4.19}$$

and by Corollary 4.6, there exists a constant $C' = 2C_1$ so that

$$\begin{aligned}&\Big|2k\sum_{\tau=k}^{t}\phi_{\bar{t}}(0,\tau)\phi_{\bar{x}}(0,\tau)\Big| \\ &\leqslant 2[\phi(0,t)]^2 + C'k\sum_{\tau=k}^{t}[\phi(0,\tau)]^2 + 2k\sum_{\tau=k}^{t}|\phi_{\bar{t}}(0,\tau)||F(\tau)| \\ &\leqslant 2[\phi(0,t)]^2 + C'k\sum_{\tau=k}^{t}\|\phi_x(\tau)\|_0^2 + 17C'k\sum_{\tau=k}^{t}\|\phi(\tau)\|_0^2 \\ &\qquad + 2k\sum_{\tau=k}^{t}|\phi_{\bar{t}}(0,\tau)||F(\tau)|\end{aligned} \tag{4.20}$$

where we have used Lemma 3.1 with $\varepsilon = 1$. Also

$$\begin{aligned}2k&\sum_{\tau=k}^{t}|\phi_{\bar{t}}(0,\tau)||F(\tau)| \\ &\leqslant 4Fk^2\sum_{\tau=k}^{t}[\phi_{\bar{t}}(0,\tau)]^2 + 4tFk\end{aligned}$$

where we have used

$$|\phi_{\bar{t}}(0,\tau)| \leqslant 2|\phi_{\bar{t}}(0,\tau)|^2 + 2.$$

Apply Lemma 3.1 to $\phi_{\bar{t}}$ with $\varepsilon = (4F)^{-1}$; then we have

$$[\phi_{\bar{t}}(0,\tau)]^2 \leq (4F)^{-1}\|\phi_{x\bar{t}}(\tau)\|_0^2 + (1+64F)\|\phi_{\bar{t}}(\tau)\|_0^2.$$

Hence

$$(4.21)\qquad 2k\sum_{\tau=k}^{t} |\phi_{\bar{t}}(0,\tau)||F(\tau)|$$

$$\leq k^2\sum_{\tau=k}^{t}\|\phi_{x\bar{t}}(\tau)\|_0^2 + 4kF(1+64F)k\sum_{\tau=k}^{t}\|\phi_{\bar{t}}(\tau)\|_0^2$$

$$+\,4tFk$$

$$\leq k^2\sum_{\tau=k}^{t}\|\phi_{x\bar{t}}(\tau)\|_0^2 + k\sum_{\tau=k}^{t}\|\phi_{\bar{t}}(\tau)\|_0^2 + 4tFk$$

as $4kF(1+64F) \leq 1$. Substitute (4.19)-(4.21) into (4.18) and obtain

$$\|\phi_x(t)\|_0^2 \leq C'k\sum_{\tau=k}^{t}\|\phi_x(\tau)\|_0^2 + 2[\phi(0,t)]^2$$

$$+\,17C'k\sum_{\tau=k}^{t}\|\phi(\tau)\|_0^2 + \|\phi_x(0)\|_0^2$$

$$+\,8k\sum_{\tau=k}^{t}\|L\phi(\tau)\|_0^2 + 4tFk.$$

Using the inequality (4.12) and

$$2[\phi(0,t)]^2 \leq \tfrac{1}{2}\|\phi_x(t)\|_0^2 + 130\|\phi(t)\|_0^2$$

which is Lemma 3.1 with $\varepsilon = \frac{1}{4}$, we have

$$\|\phi_x(t)\|_0^2 \leq 2C'k\sum_{\tau=k}^{t}\|\phi_x(\tau)\|_0^2 + 2\|\phi_x(0)\|_0^2$$

$$+\,\{(520+68C't)\|\phi(0)\|_0^2$$

$$+\,(2080k+272C't+16e^{-Ct})k\sum_{\tau=k}^{t}\|L\phi(\tau)\|_0^2$$

$$+\,Ftk(4160Fk+544C'tFk+4e^{-Ct})\}e^{Ct}.$$

Using $2C'k \leq \frac{1}{2}$ and Lemma 3.2 as in Lemma 4.3, we obtain the result.

5. ERROR ESTIMATES.

In this section we let v, defined on $(\bar{I}_h \cup \{-h\}) \times \bar{J}_k$ be the function satisfying (2.8) - (2.11) and let u, defined on $[0,1] \times [0,T]$, be a solution of (1.1)-(1.4) or equivalently (1.1)-(1.3) and (1.5). For $(x,t) \in \bar{I}_h \times \bar{J}_k$ define

$$(5.1)\qquad \phi(x,t) = u(x,t) - v(x,t).$$

For $x = -h$, $t \in \bar{J}_k - \{0\}$ define $u(x,t)$ by

$$(5.2)\qquad u(-h,t) = u(0,t) - h\,\frac{\partial u}{\partial x}(0,t).$$

This will allow us to calculate $L\phi$ at $(0,t)$ for each $t \in \bar{J}_k - \{0\}$. We shall assume throughout this section that $u \in C^{4,2}([0,1] \times [0,T])$.

5.1 Lemma. For ϕ given in (5.1) and (5.2) we have

$$(5.3)\qquad |L\phi(x,t)| \leq M(h^2 + k)$$

for $x \in \bar{I}_h - \{0,1\}$, $t \in \bar{J}_k - \{0\}$, where M

$$= \sup\left\{\left|\frac{\partial^4 u}{\partial x^4}(x,t)\right|, \left|\frac{\partial^2 u}{\partial t^2}(x,t)\right|;\ x \in [0,1],\ t \in [0,T]\right\} < \infty$$

(5.4) $\quad |L\phi(0,t)| \leqslant \tfrac{1}{2}\left|\frac{\partial^2 u}{\partial x^2}(0,t)\right| + Ch$

for $t \in \bar{J}_k - \{0\}$, where $C = \sup\left\{\left|\frac{\partial^3 u}{\partial x^3}(x,t)\right|;\ x \in [0,1],\ t \in [0,T]\right\}$

(5.5) $\quad \phi_{\bar{x}}(0,t) = k \sum_{\tau=k}^{t} N(t-\tau)\phi(0,\tau) + F(t)$

for $t \in \bar{J}_k - \{0\}$, where

$$F(t) = \int_0^t N(t-\tau)u(0,\tau)d\tau - \sum_{\tau=k}^{t} kN(t-\tau)u(0,\tau)$$

and

(5.6) $\quad |F(t)| \leqslant ATk$

where $A = \sup\{|N(t)|;\ 0 \leqslant t \leqslant T\}$ as before.

Proof: Equation (5.3) is a standard result - see for example AMES [11] (section 1.6).

By (5.2)

$$L\phi(0,t) = Lu(0,t)$$
$$= u_{\bar{t}}(0,t) = h^{-2}\left\{u(h,t) - u(0,t) - h\frac{\partial u}{\partial x}(0,t)\right\}$$

which yields (5.4). Equation (5.5) follows from (2.11) and (2.7). (5.6) follows immediately from (5.5).

5.2 Theorem. For each $u \in C^{4,2}([0,1] \times [0,T])$ a solution of equations (1.1)-(1.4), there exist constants $k_0 > 0$ and $C > 0$ depending on bounds for the derivatives of u on $[0,1] \times [0,T]$ up to order 4,2 and on T, bounds for N and its derivatives up to order at most 2 such that for every value of the mesh parameter λ but for $0 < k \leqslant k_0$

(5.7) $\quad \|(u-v)(t)\|_0^2 \leqslant C(h+kh+k)$

(5.8) $\quad \|(u-v)_x(t)\|_0^2 \leqslant C(h+kh+k)$

for all $t \in \bar{J}_k - \{0\}$, where v is given by (2.8)-(2.11).

Proof: Let ϕ be as in Lemma 5.1 then $\phi(x,0) = 0$ for $x \in \bar{I}_h$ and

$$|\phi_{\bar{x}}(x,0)| \leqslant h \sup\left\{\left|\frac{\partial^2 u}{\partial x^2}(x,0)\right|;\ x \in [0,1]\right\}.$$

(5.7) and (5.8) now follow from Lemma 5.1 and Lemmas 4.3 and 4.7.

5.3 Corollary. For u, v as in Theorem 5.2 and for k_0 and C defined there we have

(5.9) $\quad \sup_{x\in\bar{I}_h} \sup_{t\in\bar{J}_k} |u(x,t)-v(x,t)| \leqslant 3\sqrt{2}\, C^{\frac{1}{2}}(h+hk+k)^{\frac{1}{2}}$

whenever $0 < k \leqslant k_0$.

Proof: We use Theorem 5.2 together with Lemma 3.1 with $\varepsilon = 1$.

Remark. In (5.2) we could have extended u to $\{-h\} \times (\bar{J}_k - \{0\})$ by setting

$$u(-h,t) = u(0,t) - h\frac{\partial u}{\partial x}(0,t) + \frac{h^2}{2}\frac{\partial^2 u}{\partial x^2}(0,t)$$

for $t \in \bar{J}_k - \{0\}$. In this case (5.4) becomes

$$|L\phi(0,t)| \leq Ch$$

and as $F(t)$ becomes

$$\frac{h^2}{2}\frac{\partial^2 u}{\partial x^2}(0,t) + \int_0^t N(t-\tau)u(0,\tau)d\tau - \sum_{\tau=k}^{t} kN(t-\tau)u(0,\tau)$$

(5.6) becomes

$$|F(t)| \leq \left(AT + \frac{C}{\lambda}\right) k.$$

The estimates (5.7) and (5.8) are then replaced by

$$\|(u-v)(t)\|_0^2 \leq \bar{C}(h^2+k^2), \quad t \in \bar{J}_k - \{0\}$$

and

$$\|(u-v)_x(t)\|_0^2 \leq \bar{C}(h^2+k^2), \quad t \in \bar{J}_k - \{0\}$$

but here the constant $\bar{C}$ also depends on λ^{-1}.

6. APPENDIX

6.1 Proof of Lemma 3.3

Let $\phi(t)$ be a function defined on $\bar{J}_k$ taking values in the linear space L.

$$\begin{aligned}&(a\hat{\phi}(\tau) + b\phi_{\bar{t}}(\tau),\ L\phi(\tau))\\ &= (a\hat{\phi}(\tau) + b\phi_{\bar{t}}(\tau),\ \phi_{\bar{t}}(\tau) - \phi_{x\bar{x}}(\tau))\\ &= a(\hat{\phi}(\tau),\phi_{\bar{t}}(\tau)) - a(\hat{\phi}(\tau),\phi_{x\bar{x}}(\tau))\\ &+ b(\phi_{\bar{t}}(\tau),\phi_{\bar{t}}(\tau)) - b(\phi_{\bar{t}}(\tau),\phi_{x\bar{x}}(\tau)).\end{aligned}$$

Using the techniques of LEES [9] or directly we can prove

$$\hat{\phi}(x,\tau)\phi_{\bar{t}}(x,\tau) = \tfrac{1}{2}[\phi^2]_{\bar{t}}(x,\tau) \tag{6.1}$$

$$\hat{\phi}(x,\tau)\phi_{x\bar{x}}(x,\tau) = [\hat{\phi}\phi_{\bar{x}}]_x(x,\tau) - [\phi_x]^2(x,\tau) + \frac{k}{4}[\phi_x^2]_{\bar{t}}(x,\tau) + \frac{k^2}{4}[\phi_{x\bar{t}}]^2(x,\tau) \tag{6.2}$$

$$\phi_{\bar{t}}(\tau)\phi_{x\bar{x}}(x,\tau) = [\phi_{\bar{t}}\phi_{\bar{x}}]_x - \tfrac{1}{2}[\phi_x^2]_{\bar{t}} - \frac{k}{2}[\phi_{x\bar{t}}]^2. \tag{6.3}$$

As a consequence

$$(\hat{\phi}(\tau),\phi_{\bar{t}}(\tau)) = \tfrac{1}{2}(\phi(\tau),\phi(\tau))_{\bar{t}}, \tag{6.4}$$

$$\begin{aligned}(\hat{\phi}(\tau),\phi_{x\bar{x}}(\tau)) &= h\sum_{x\in I_h'}[\hat{\phi}\phi_{\bar{x}}]_x(x,\tau)\\ &\quad (\phi_x(\tau),\phi_x(\tau))\\ &+ \frac{k}{4}(\phi_x(\tau),\phi_x(\tau))_{\bar{t}} + \frac{k^2}{4}(\phi_{x\bar{t}}(\tau),\phi_{x\bar{t}}(\tau))\\ &= -\hat{\phi}(0,\tau)\phi_{\bar{x}}(0,\tau) - \|\phi_x(\tau)\|_0^2\\ &+ \frac{k}{4}\|\phi_x(\tau)\|_{0\bar{t}}^2 + \frac{k^2}{4}\|\phi_{x\bar{t}}(\tau)\|_0^2\end{aligned} \tag{6.5}$$

and

$$\begin{aligned}&(\phi_{\bar{t}}(\tau),\phi_{x\bar{x}}(\tau))\\ &= -\phi_{\bar{t}}(0,\tau)\phi_{\bar{x}}(0,\tau) - \tfrac{1}{2}\|\phi_x(\tau)\|_{0\bar{t}}^2 - \frac{k}{2}\|\phi_{x\bar{t}}(\tau)\|_0^2.\end{aligned} \tag{6.6}$$

Using (6.4)-(6.6) yields

(6.7) $\left(a\hat{\phi}(\tau)+b\phi_{\bar{t}}(\tau),\ L\phi(\tau)\right)$

$$= \frac{a}{2}\|\phi(\tau)\|^2_{0\bar{t}} + a\|\phi_x(\tau)\|^2_0 + \left(\frac{b}{2}-\frac{ak}{4}\right)\|\phi_x(\tau)\|^2_{0\bar{t}}$$

$$+ b\|\phi_{\bar{t}}(\tau)\|^2_0 + \left(\frac{bk}{2}-\frac{ak^2}{4}\right)\|\phi_{x\bar{t}}(\tau)\|^2_0$$

$$+ a\hat{\phi}(0,\tau)\phi_{\bar{x}}(0,\tau) + b\phi_{\bar{t}}(0,\tau)\phi_{\bar{x}}(0,\tau).$$

The Lemma now follows by multiply (6.7) by k and summing τ from $\tau = k$ to $\tau = t$ and using the identity

$$k\sum_{\tau=k}^{t}\psi_{\bar{t}}(\tau) = \psi(t) - \psi(0).$$

REFERENCES.

[1] CANNON, J.R. The Solution of the Heat Equation Subject to the Specification of Energy, Quart. Appl. Math. 21 (1963), 155-160.

[2] DECKERT, K.L. and MAPLE, C.G., Solution for Diffusion with Integral Type Boundary Conditions, Proc. Iowa Acad. Sci. 70 (1963), 354-361.

[3] KAMYNIN, L.I., A Boundary Value Problem in the Theory of Heat Conduction with a Non-classical Boundary Condition, Ž. Vyčisl. Mat. i Mat. Fiz. 4 (1964), 1006-1024. (English translation = USSR Computational and Mathematical Physics 4 (1964), 33-59.)

[4] IONKIN, N.I., Solution of a Boundary Value Problem in Heat Conduction with a Non-classical Boundary Condition, Differencial'nye Uravnenija 13 (1977), 294-304 (English translation = Differential Equations 13 (1977), 204-211).

[5] IONKIN, N.I., Stability of a Problem in Heat Transfer Theory with a Non-classical Boundary Condition, Differencial'nye Uravnenija 15 (1979), 1279-1283 (English translation = Differential Equations 15 (1980), 911-914).

[6] CANNON, J.R. and VAN DER HOEK, J., The Classical Solution of the One-Dimensional Two-Phase Stefan Problem with Energy Specification, to appear in Annali di Mat. Dura ed Appl.

[7] CANNON, J.R., and VAN DER HOEK, J., The One Phase Stefan Problem Subject to the Specification of Energy, to appear in J. Math. Anal. and Appl.

[8] CANNON, J.R., and VAN DER HOEK, J., The Existence of and a Continuous Dependence Result for the Solution of the Heat Equation Subject to the Specification of Energy, to appear in Supplemento Bolletino Unione Mathematica Italliana.

[9] LEES, M., A priori estimates for the solutions of difference approximations to parabolic partial differential equations, Duke J. Math. 27 (1960), 297-311.

[10] WIDDER, H., The Heat Equation (Academic Press, New York, 1975).

[11] AMES, W.F., Numerical Solutions for Partial Differential Equations (Academic Press, New York, 1977).

NUMERICAL SOLUTIONS OF
PARTIAL DIFFERENTIAL EQUATIONS
J. Noye (editor)

ERROR ESTIMATION USING A LEAST SQUARES FINITE ELEMENT SOLUTION FOR PARTIAL DIFFERENTIAL EQUATIONS

GRANT P. STEVEN

Department of Aeronautical Engineering
The University of Sydney
N.S.W., 2006
Australia

For the numerical solution to a partial differential equation to have some significance, it is important to be able to estimate the error in that solution. This work reports on a technique for 'a posteriori' error estimation on a Poisson equation. A least squares finite element solution technique is used to solve the equation. Results obtained so far indicate the possible existence of a single criterion which could indicate the piecewise percentage error in the prime variable as well as its derivatives.

INTRODUCTION

One of the ultimate goals of all engineers and physicists is to be able to make a model of a natural process in order that its behaviour can be predicted in advance. Such models take many forms. Of these forms, the one of interest in this work is a computer model. In the abstract a computer model can simply represent a mathematical formula but ultimately, to be of any use, it must represent a description of a physical process either in its mathematical form or as a direct simulation of the laws of nature involved in the study. For example, one could set up a finite difference or finite element mathematical model of a biharmonic equation and then set about solving this to represent plane stress elasticity problems. Alternatively a finite element model is readily set up which emulates the natural law of compatibility of deformation and sets about determining a solution based on the laws of statics, or in an approximate form, the Principle of Virtual Work.

The users of such computer modelling techniques are interested not only in the result, but the quality of the result, in the physical context of the problem under investigation. Often this means that not only is the accuracy and stability of the numerical technique under scrutiny, but also the accuracy of the physically significant part of the result is of interest. This statement may appear ambiguous and the following illustration will be presented. In solving an elliptic partial differential equation representing some force/flux physical situation (heat flow, ion flow, seepage flow), it may be possible to determine some indication of the errors involved in satisfying the governing differential equation. While these may be of interest, the engineer is more concerned with the physical part of the error, i.e. by how much is this numerical solution under or overestimating the flux across a particular part of the solution region. Another example from elastostatics could be that it may be possible to estimate by how much a numerical solution fails to satisfy the Principle of Minimum Potential which has been applied to obtain the solution, but this does not mean that an estimate of the error in a stress level at a particular location can be given.

Thus, to summarise the above points, it is a desirable goal to be able to give an 'a posteriori' error estimate of all significant physical aspects of any numerical modelling process. The ability to do this after a single solution is significant as in many situations bounds can be determined or convergence can be studied, but these involve several or at least two separate solutions, which for a major study, (stress analysis of an aircraft frame) could be a very extravagent process. Several researchers in finite element methods in recent years have turned their attention to this concept of 'a posteriori' error estimates. Babuska and Rheinboldt (1978a) and b)) develop mathematical formulations based on functional analysis and use adaptive finite element grids to aid their estimates. Kelly (1980) on the other hand uses interelement stress discontinuities to provide error estimates and reduced integration as a means of bounding solutions. Whilst most research workers in numerical modelling investigate accuracy using benchmark problems with known solutions c.f. Fried (1980), it is a much more difficult proposition to determine the accuracy in the general case where the exact solution is unknown.

In this present work the least squares finite element method is adopted to solve a simple partial differential equation with the intention of discovering whether, within the framework of the technique, there is a means of estimating errors. The Least Squares solution criterion has its basis in simple approximation theory and it was considered that some of its intrinsic features would make it well suited to the task. In following sections a brief history of the least squares method is presented, together with the necessary mathematical background to make use of it in a finite element context.

The ideas on error estimation are applied to an elliptic type partial differential equation. Such equations arise naturally by a combination of equations of motion or flow coupled with a conservation equation. With the least squares solution criterion the user is able to weight the solution towards either a more rigid satisfaction of the motion equations while relaxing the conservation equation or vice-versa. This means a better solution for fluxes could be obtained at the expense of conservation or vice versa. By using this weighting technique it is possible to examine convergence without doing several complete solutions on different grids. At the same time certain error estimators are calculated based on element residuals. For a benchmark problem these are examined to detect possible correlations with the known solution. The results of all the test runs conducted to date enable the author to make some provisional conclusions that it may be possible to give error estimates on both the prime variable and its derivatives on the basis of a single solution.

DESCRIPTION AND HISTORY OF THE LEAST SQUARES METHOD

To enable subsequent general discussion, a brief mathematical outline of the least squares method for a linear boundary value problem is presented:

$$Lu = f \text{ in domain } \Omega, \tag{1}$$

$$Bu = g \text{ on boundary } \Gamma. \tag{2}$$

In this L and B are differential operators, u is the solution sought and f and g are known functions. A series of approximate solutions are assumed u_n including n adjustable parameters used to minimise the error such that,

$$u(\underset{\sim}{x}) \simeq u_n(\underset{\sim}{a},\underset{\sim}{x}), \tag{3}$$

$\underset{\sim}{x}$ is the coordinate vector and $\underset{\sim}{a}$ the unknown parameter vector. Substituting the trial solution into the equations to be satisfied yields interior and boundary residuals,

$$R_\Omega = Lu_n - f \text{ in } \Omega, \tag{4}$$

$$R_\Gamma = Bu_n - g \text{ on } \Gamma. \tag{5}$$

To obtain the parameters $\underset{\sim}{a}$ a weighted sum of the residuals is integrated over Ω and Γ. This is a general description of the Method of Weighted Residuals (MWR) which is authoritatively treated in Findlayson's book (1972). Such MWR include

Galerkin's method, collocation, method of moments and the least squares method. For the least squares method, the weights used are the differentials of the residuals with respect to the parameter vector $\underset{\sim}{a}$.

$$\int_{\Omega} \frac{\partial R_{\Omega}}{\partial \underset{\sim}{a}} \left[R_{\Omega} \right] d\Omega \quad \int_{\Gamma} \frac{\partial R_{\Gamma}}{\partial \underset{\sim}{a}} \left[R_{\Gamma} \right] d\Gamma = 0. \tag{6}$$

Equation (6) is for a continuous problem and there are several possible variations in its form. There is also an equivalent formulation for discrete problems.

The least squares method (LSM) originated from Gauss (1794) (see T. Hall (1965)) but he did not publish his work on least squares estimation until 1809. In the meantime independent discoveries had been made by Legendre (1805) in France and Adrain (1808) in the United States. From this beginning and for all of the nineteenth century it was used in its discrete form for curve fitting and statistical data reduction. Boussinesq (1901) was the first to consider the application of the LSM to the solution of differential equations. A significant amount of work has been done since then on the solution of linear and non linear partial differential equations of all types. From a comprehensive survey on the use of the LSM the reader is referred to Eason (1976).

In a finite element context the use of the LSM as a solution criterion appears to have been first proposed by Brandes (1969) and later developed by Akin (1973), Akin and Sen Gupta (1973), Lynn (1974), Lynn and Arya (1973)(1974) and Milthorpe and Steven (1978).

As was stated above, the LSM is regarded as a category of the MWR. Least squares may also be interpreted as a variational principle, and in some cases is equivalent to other variational principles for the same problem. The interested reader is referred to Finlayson (1972) and Mikhlin (1964). However the most obvious, to this author at least, derivation of the method comes from approximation theory. Any attempt to fit a function f involves using an assumed model, say f_n, which is made as close to f as possible. A good measure of closeness is the L_2 norm or the integrated root mean residual defined as,

$$\| f - f_n \| = \left[\int_{\Omega} (f - f_n)\, d\Omega \right]^{\frac{1}{2}}.$$

This is minimised by the least squares method. Thus if $f_n = Lu_n$ it is clear that the least squares solution to partial differential equations is analogous to curve or surface fitting. Such an analogy provides a useful means of interpreting results of LS application to continuous problems.

LEAST SQUARES FOR FINITE ELEMENT SOLUTIONS.

A finite element least squares formulation follows similar steps to the general one of the last section, except that it is necessary to be most specific when describing the approximate solutions previously denoted as u_n in equation (3). In a finite element (FE) presentation, the field variable(s) $u(\underset{\sim}{x})$ is approximated locally within subdomains of the solution region Ω. Indeed, these subdomains are our finite elements and will be denoted by a superscript 'e'. The adjustable parameters within each element are the values of the variable at each node on the boundary and within the element enabling $u_n^e(\underset{\sim}{a},\underset{\sim}{x})$ to be represented as,

$$u_n^e(\underset{\sim}{a},\underset{\sim}{x}) = [N]\{u\}^n \tag{7}$$

where $\{u\}^n$ is a vector of the nodal variables or degrees of freedom (dof) and [N] is an interpolation matrix which describes the functional form of the approximation to $u^e(\underset{\sim}{x})$ within the element.

Since the residuals in equations (4) and (5) are scalar quantities, they can be regarded as the sum over all the elements in Ω and over all the surface elements on Γ.

$$\int_{\Omega} R_{\Omega} d\Omega = \sum_{\text{elements}} \int_{\Omega^e} R_{\Omega}^e d\Omega^e = \sum \int_{\Omega^e} (Lu_n^e - f) d\Omega^e \tag{8}$$

$$\int_{\Gamma} R_{\Gamma} d\Gamma = \sum_{\substack{\text{surface} \\ \text{elements}}} \int_{\Gamma^e} R_{\Gamma}^e d\Gamma^e = \sum \int_{\Gamma^e} (Bu_n^e - g) d\Gamma^e \tag{9}$$

Substituting for u_n^e from equation (7) and applying the least squares criterion where the squared residuals are differentiated with respect to the adjustable parameters, in this case $\{u\}^n$ leads to the expression,

$$\sum_{\text{elements}} \left\{ \left[\int_{\Omega^e} [H]^T [H] d\Omega^e \right] \{u\}^n - \left[\int_{\Omega^e} [H]^T \{f\} d\Omega^e \right] \right\}$$
$$+ \sum_{\substack{\text{surface} \\ \text{elements}}} \left\{ \left[\int_{\Gamma^e} [G]^T [G] d\Gamma^e \right] \{u\}^n - \left[\int_{\Gamma^e} [G]^T \{g\} d\Gamma^e \right] \right\} = 0. \tag{10}$$

where

$$[H] = [L][N],$$
$$[G] = [B][N].$$

This concludes the general formulation. For the present work it is more revealing to move one to the particular formulation for an elliptic partial differential equation, namely the Poisson equation. The reasons for this choice are that it is well documented and there are many solutions available. It is also simple to solve and offers a good means of conveying the meaning of the error estimators presented later.

In a finite element solution the admissible trial functions, the [N]'s in (7) must be continuous to one order less than the highest derivative in the solution criterion, see Zienkiewicz (1971). Thus for a Poisson equation this means the [N]'s must be C_1 continuous (i.e. continuous up to the first derivatives). Variational and Galerkin solution criteria achieve this by integration by parts and could thus only require C_0 continuity. The LSM has no such advantage and thus it initially appears to require the considerably more complex C_1 elements. This continuity restriction can be overcome by reducing the one second order partial differential equation to three first order ones by the introduction of auxiliary variables.

The Poisson equation to be solved is

$$\frac{\partial^2 \phi}{\partial x^2} + \frac{\partial^2 \phi}{\partial y^2} = f. \tag{11}$$

Introduce auxiliary functions α and β such that,

$$\alpha - \frac{\partial \phi}{\partial x} = 0,$$
$$\beta - \frac{\partial \phi}{\partial y} = 0,$$
$$\frac{\partial \alpha}{\partial x} + \frac{\partial \beta}{\partial y} = f,$$

$$\text{i.e.} \quad \begin{bmatrix} 1 & 0 & -\partial/\partial x \\ 0 & 1 & -\partial/\partial y \\ \partial/\partial x & \partial/\partial y & 0 \end{bmatrix} \begin{bmatrix} \alpha \\ \beta \\ \phi \end{bmatrix} = \begin{bmatrix} 0 \\ 0 \\ f \end{bmatrix},$$

i.e.

$$[L]\{u\} = \{f\}. \tag{12}$$

Boundary conditions can be expressed in three forms, not all of which may apply at any part of the boundary.

$$\frac{\partial\phi}{\partial n} = g_1, \quad \frac{\partial\phi}{\partial t} = g_2, \quad \phi = g_3. \tag{13}$$

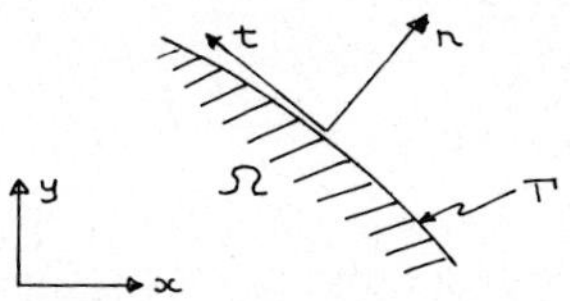

The boundary residual R_Γ will contain only these components of equation (13) which apply on any particular part of Γ. The auxiliary functions are now classed as variables in their own right, thus in general $\alpha \neq \frac{\partial\phi}{\partial x}$ and $\beta \neq \frac{\partial\phi}{\partial y}$ and the application of the least squares solution criterion endeavours to make the inequality an equality.

The next stage in the development of the solution is to decide on the type of finite element to be used. In this work a triangular element is selected for its natural simplicity. Within this element the base function ϕ is allowed to have a quadratic variation. To be consistent the auxiliary functions are taken to be linear, this equation (7) is expressed fully as,

$$\begin{aligned} \{u_n\}^e &= [N]\{u\}^n \\ \{u\}^n &= \{\alpha_1\beta_1\phi_1\alpha_2\beta_2\phi_2\alpha_3\beta_3\phi_3\phi_4\phi_5\phi_6\} \end{aligned} \tag{14}$$

$$[N] = \left[\begin{array}{ccc|ccc|ccc|ccc} L_1 & 0 & 0 & L_2 & 0 & 0 & L_3 & 0 & 0 & 0 & 0 & 0 \\ 0 & L_1 & 0 & 0 & L_2 & 0 & 0 & L_3 & 0 & 0 & 0 & 0 \\ 0 & 0 & L_1(2L_1-1) & 0 & 0 & L_2(2L_2-1) & 0 & 0 & L_3(2L_3-1) & 4L_1L_2 & 4L_2L_3 & 4L_3L_1 \end{array}\right] \tag{15}$$

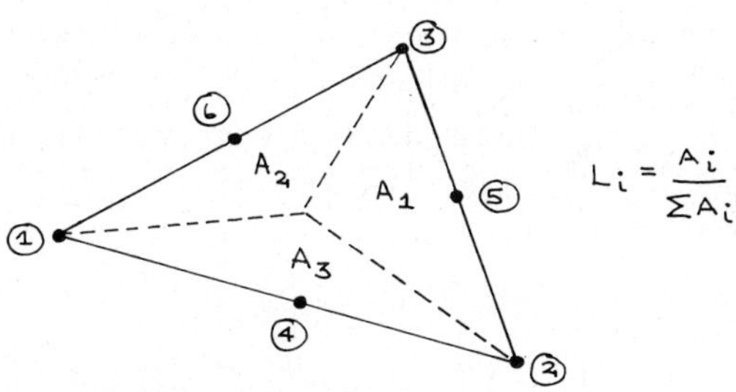

The L's are the natural coordinates within the element. For surface quantities the equivalent one dimensional forms are used.

In this work interest is focused on the desire to investigate the errors associated with the solution of the Poisson equation using a least squares solution criterion. It is well known that any numerical solution to a field problem gives increasingly poor results for increasing derivatives of the field variable. Unfortunately in many physical situations the derivatives are more physically significant than the base variable. For example, in a heat conduction situation the heat flux is generally more significant than the temperature or in fluid flow the velocities are of more interest to the engineer than the stream function or velocity potential. Thus it can be more important to obtain better values of these derivatives. In order to have some numerical influence in this area, the three constituant equations in (12) are individually weighted. The relative amounts of the weights can be used to indicate the significance placed on either the mathematical or physical meaning of a particular equation. The element residual is now expressed as;

$$R_\Omega{}^e = [W]^e\Big\{[L]\{u_n\}^e - \{f\}\Big\} \quad (16)$$

where

$$[W]^e = \begin{bmatrix} W_\alpha & 0 & 0 \\ 0 & W_\beta & 0 \\ 0 & 0 & W_\phi \end{bmatrix}$$

The diagonal matrix $[W]^e$ is the local set of weights indicating that, not only can different components of the governing p.d.e.'s be weighted but that this weighting can vary from one element to another.

Substituting equation (16) into (10) yields

$$\sum_{\text{elements}} \left\{\left[\int_{\Omega^e} [H]^T[H]d\Omega^e\right]\{u\}^n - \left[\int_{\Omega^e}[H]^T\{f\}d\Omega^e\right]\right\}$$
$$+ \sum_{\substack{\text{surface}\\ \text{elements}}} \left\{\left[\int_{\Gamma^e}[G]^T[G]d\Gamma^e\right]\{u\}^n - \left[\int_{\Gamma^e}[G]^T\{g\}d\Gamma^e\right]\right\} = 0 \quad (17)$$

where $[H] = [W]^e[L][N]$

$[G]$ = surface equivalent of $[H]$.

A more succinct form of (17) is;

$$\sum_{\text{elements}} \left[[K]^e\{u\}^n - \{F\}^e \right] + \sum_{\substack{\text{surface}\\ \text{elements}}} \left[[K_S]^e\{u\}^n - \{F_S\} \right] = 0 \quad (18)$$

$[K]^e$ = interior element "stiffness" matrix.

$[F]^e$ = interior element "force" vector.

$[K_S]^e$ = element boundary "stiffness" matrix.

$[F_S]^e$ = element boundary "force" vector.

Specific details of the form of these matrices and vectors are given in Appendix I.

CONVERGENCE AND ERROR BOUNDS

The most desirable feature of a convergent solution is that the difference between u and u_n in equation (3) decreases at every point in Ω and on Γ, as n increases. Uniform convergence implies that,

$$| u - u_n | < \varepsilon > 0.$$

Convergence in the mean implies that,

$$\| u - u_n \| < \varepsilon > 0.$$

Uniform convergence is a stronger form than convergence in the mean. In addition to convergence of u_n, it is also desirable that the residuals approach zero as n increases. In a finite element context with a fixed form of element interpolation, the parameter n is represented by either the total number of degrees of freedom or the number of elements for a given problem.

With the least squares method for continuous problems it has been shown that the residuals do converge, see Mikhlin's book. Thus residuals can be made as close to zero as required. This is not generally true of Rayleigh-Ritz or Galerkin methods, although these two do converge faster in the mean sense. Theoretical comparisons of various methods for convergence have been made but the results are inconclusive, see Nitsche (1969).

While it is comforting to know that convergence, for linear problems, is always present as n approaches infinity, it is more useful to be able to estimate the solution error, or failing that some bounds at a finite n, hopefully small. That is the purpose of this work and whilst it is not conclusive and refers only to specific problems, it nevertheless could provide a basis for further comparative studies towards a general formulation. For some problems it is possible to calculate a rigorous bound, Mikhlin (1964) has proved that for a Poisson equation the integrated mean square bound can be written as,

$$\int_{\Omega} \| u - u_n \| d\Omega \le \int_{\Omega} G R_{\Omega} d\Omega$$

where G is the Green's function for the problem. This is a very tight bounding condition since R_{Ω} has uniform convergence. However for most realistic problems the Green's function is unobtainable or only obtainable at an equivalent amount of computational effort, and so this approach to error estimation is discarded.

In this work several indicators of solution accuracy will be calculated and studied. Element residuals can be evaluated as,

$$\begin{aligned} R_{\alpha} &= \int_{\Omega^e} \left(\frac{\partial \phi}{\partial x} - \alpha \right) d\Omega^e, \\ R_{\beta} &= \int_{\Omega^e} \left(\frac{\partial \phi}{\partial y} - \beta \right) d\Omega^e, \\ R_{\phi} &= \int_{\Omega^e} \left(\frac{\partial \alpha}{\partial x} + \frac{\partial \beta}{\partial y} - f \right) d\Omega^e. \end{aligned} \tag{19}$$

From these an element mean square residual is calculated,

$$R_{ms} = (R_{\alpha}^2 + R_{\beta}^2 + R_{\phi}^2)^{\frac{1}{2}}. \tag{20}$$

Over all the elements the above quantities are summed to provide an overall indicator of solution quality.

While the relative changes in the above residuals for increasing numbers of elements or degrees of freedom will indicate convergence, they give no absolute indication of the error for a single solution. Thus the idea of an error measure is introduced. This is based on the work of Milthorpe (1981). Error measures on the three governing differential equations are defined as,

$$\begin{aligned} Em_{\alpha} &= R_{\alpha} \Big/ \int_{\Omega^e} \alpha d\Omega^e, \\ Em_{\beta} &= R_{\beta} \Big/ \int_{\Omega^e} \beta d\Omega^e, \\ Em_{\phi} &= R_{\phi} \Big/ \int_{\Omega^e} f d\Omega^e. \end{aligned} \tag{21}$$

with a mean square sum defined as,

$$Em_{ms} = (Em_{\alpha}^2 + Em_{\beta}^2 + Em_{\phi}^2)^{\frac{1}{2}}. \tag{22}$$

These measures can be summed over all the elements and used as an overall indicator.

BENCHMARK PROBLEM 1

The first problem to be considered is that of a square region of side 2 for which only the positive quadrant is analysed, see figure 1, which includes the boundary conditions.

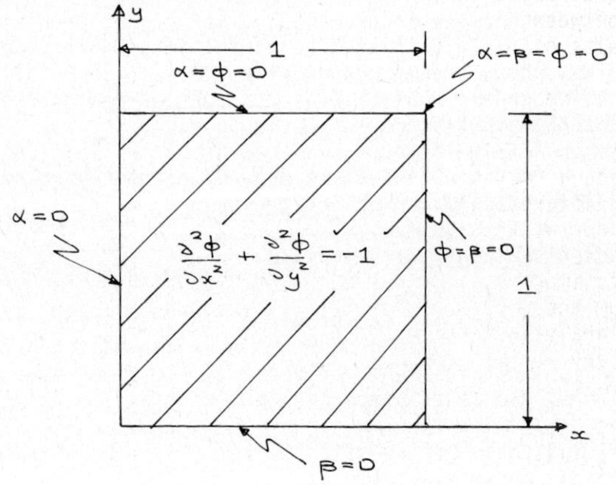

FIGURE 1: Solution region and boundary conditions for benchmark problem number 1.

The classical solution for this problem may be obtained by the Navier method and is;

$$\phi = \frac{16}{\pi^3} \sum_{n=1,3,5--} \frac{1}{n^3}(-1)^{\left(\frac{n-1}{2}\right)} \left[1 - \frac{\cosh\left(\frac{n\pi y}{2}\right)}{\cosh\left(\frac{n\pi}{2}\right)} \right] \cos\left(\frac{n\pi x}{2}\right)$$

$$\frac{\partial\phi}{\partial x} = -\frac{8}{\pi^2} \sum_{n=1,3,5--} \frac{1}{n^2}(-1)^{\left(\frac{n-1}{2}\right)} \left[1 - \frac{\cosh\left(\frac{n\pi y}{2}\right)}{\cosh\left(\frac{n\pi}{2}\right)} \right] \sin\left(\frac{n\pi x}{2}\right)$$

$$\frac{\partial\phi}{\partial y} = -\frac{8}{\pi^2} \sum_{n=1,3,5--} \frac{1}{n^2}(-1)^{\left(\frac{n-1}{2}\right)} \left[1 - \frac{\cosh\left(\frac{n\pi x}{2}\right)}{\cosh\left(\frac{n\pi}{2}\right)} \right] \sin\left(\frac{n\pi y}{2}\right)$$

$$\int_{-1}^{+1}\int_{-1}^{+1} \phi dxdy = \frac{1}{3}\left[1 - \frac{192}{\pi^5} \sum_{n=1,3,5} \frac{1}{n^5} \tanh\left(\frac{n\pi}{2}\right) \right].$$

It is interesting to note that the series expression for ϕ and its integral converge well. However the gradient terms have extremely poor convergence and give only 3½ figure accuracy after twenty five terms. Indeed while performing the calculations described herein with 64 bit computer floating point representation, it was often quicker and more accurate to obtain a finite element solution than it was to calculate the equivalent series solution values at each point.

Initially we investigate the effect of uniform equation weighting throughout the whole finite element mesh for three types of grid. The three grids are uniform with 8, 18 and 32 elements respectively. The weight on the conservation equation W_ϕ is set constant at one and the equations of motion weights W_α and W_β are made equal and varied between 0.1 and 10. Tables 1, 2 and 3 on the following pages summarise the results of these computer solutions. In these tables are presented the weights W_α and W_β, the summed mean square element residuals and error measures, the maximum value of ϕ at $(x,y) = (0,0)$, the L_2 norm of the difference between the series solution, ΣEm_ϕ and various other features.

These tables provide a considerable amount of information about the nature of the solutions for the different parameters. A feature to note for all three grids used, is the behaviour of the maximum values in relation to the weights. As the weights on the gradients go down, the function value improves and as the weights go up the gradients improve. These changes are illustrated graphically in Figures 2 and 3. From these we see that with diminishing weight the function approaches its "exact" analytical value for all grids and that the more elements there are the values at any weight are closer to the assymptote. A similar pattern exists for the gradients, Figure 3, but the approach to the assymptote is not nearly as well behaved. To obtain gradients it appears more satisfactory to ensure a good solution for the function then use the differentials of this rather than the auxiliary α and β values.

The poor behaviour of the auxiliary functions is also shown up in the values of ΣEm_α and ΣEm_β on tables 1, 2 and 3. Thus taking the above suggestion that gradients be obtained by differentiating the values of ϕ, we can examine the behaviour of Em_ϕ to see if its behaviour can provide an indication of the error. For the three grids with a range of weights the log of $\Sigma Em_\phi/NE$, where NE is the number of elements, is plotted against the log of the L_2 norm of the solution difference divided by the L_2 norm of the analytical solution. These plots are shown in Figure 4 with lines of constant percentage error in ϕ_{max} superimposed. This figure shows that provided the log $(\Sigma Em_\phi/NE)$ is less than -2 there will be an error in ϕ_{max} less than 0.1% and if log $(\Sigma Em_\phi/NE)$ is less than -1 the error in ϕ_{max} is less than 1%. Clearly these criteria are problem dependent.

$W_\alpha = W_\beta$, $W_\phi = 1$	ΣR_{MS}	ΣEM_{MS}	Φ_{MAX} (% ERROR)	$\lVert\phi-\phi_n\rVert/\lVert\phi\rVert$ (ΣEM_ϕ)	$\alpha_{MAX} = \beta_{MAX}$ (% ERROR)	$\lVert\alpha-\alpha_n\rVert/\lVert\alpha\rVert$ (ΣEM_α)	$\lVert\beta-\beta_n\rVert/\lVert\beta\rVert$ (ΣEM_β)	VOLUME (% ERROR)
0.1	0.124	834.6	.29492 (-0.082%)	0.15×10^{-5} (.00989)	-0.99753 (-47.71%)	0.234 (418.)	0.234 (418.)	.13991 (.475%)
0.2	0.121	209.8	.29473 (-0.0145%)	0.21×10^{-5} (.0386)	-0.99034 (-46.64%)	0.222 (105.8)	0.222 (105.8)	.13987 (.500%)
0.5	0.110	34.9	.29359 (0.372%)	0.59×10^{-5} (.206)	-0.94837 (-40.43%)	0.16 (18.12)	0.16 (18.12)	.13966 (.648%)
1.0	0.105	9.85	.29120 (1.18%)	0.59×10^{-4} (.545)	-0.86359 (-27.87%)	0.067 (5.28)	0.067 (5.28)	.13918 (.995%)
2.0	0.125	3.52	.28797 (2.27%)	0.24×10^{-3} (.923)	-0.76688 (-13.55%)	0.013 (1.84)	0.013 (1.84)	.13817 (1.708%)
5.0	0.157	2.21	.28038 (4.70%)	0.0017 (1.245)	-0.68858 (-1.96%)	0.0025 (0.98)	0.0025 (0.98)	.13368 (4.903%)
10.0	0.206	2.35	.26184 (11.14%)	0.014 (1.623)	-0.59854 (+11.63%)	0.0108 (0.76)	0.0108 (0.76)	.12113 (13.83%)

TABLE 1: Problem 1,8 elements. $\phi MAX = 0.294867$, $\alpha MAX = -0.6753$, Volume = 0.140577

$W_\alpha = W_\beta$ $W_\phi = 1$	ΣR_{MS}	ΣEM_{MS}	ϕ_{MAX} (% ERROR)	$\|\phi-\phi_n\|/\|\phi\|$ (ΣEM_ϕ)	$\alpha_{MAX} = \beta_{MAX}$ (% ERROR)	$\|\alpha-\alpha_n\|/\|\alpha\|$ (ΣEM_α)	$\|\beta-\beta_n\|/\|\beta_n\|$ (ΣEM_β)	VOLUME (% ERROR)
0.1	0.134	929.	.29467 (.0041%)	$.959*10^{-7}$ (.034)	-.99425 (-47.23%)	.209 (481.)	.209 (481.)	.14041 (.115%)
0.2	0.127	234.	.29448 (.069%)	$.221*10^{-6}$ (.131)	-.97823 (-44.85%)	.186 (120.4)	.186 (120.4)	.14038 (.135%)
0.5	0.102	39.4	.29355 (.386%)	$.536*10^{-5}$ (.594)	-.90090 (-33.40%)	.097 (21.6)	.097 (21.6)	.14025 (.232%)
1.0	0.088	11.2	.29231 (.804%)	$.243*10^{-4}$ (1.203)	-.79944 (-18.38%)	.024 (6.37)	.024 (6.37)	.14005 (.372%)
2	0.093	4.16	.29143 (1.103%)	$.492*10^{-4}$ (1.602)	-.73243 (-8.46%)	.0041 (2.17)	.0041 (2.17)	.13983 (.528%)
5.	0.099	2.71	.29016 (1.53%)	$.128*10^{-3}$ (1.78)	-.70202 (-3.95%)	.0015 (1.05)	.0015 (1.05)	.13905 (1.084%)
10.	0.108	2.69	.28622 (2.87%)	$.704*10^{-3}$ (1.94)	-.67569 (-0.056%)	.0006 (0.93)	.0006 (0.93)	.13651 (2.896%)

TABLE 2: Benchmark problem 1. 18 elements, 49 nodes solution. Series solution gives ϕ_{MAX} = 0.294867, $\alpha_{MAX} = \beta_{MAX}$ = -0.67531 and Volume = 0.140577.

$W_\alpha = W_\beta$ $W_\phi = 1$	ΣR_{MS}	ΣEM_{MS}	ϕ_{MAX} (% ERROR)	$\|\phi-\phi_n\|/\|\phi\|$ (ΣEM_ϕ)	$\alpha_{MAX} = \beta_{MAX}$ (% ERROR)	$\|\alpha-\alpha_n\|/\|\alpha\|$ (ΣEM_α)	$\|\beta-\beta_n\|/\|\beta\|$ (ΣEM_β)	VOLUME (% ERROR
0.1	.135	1012.	.29463 (0.019)	$.236*10^{-7}$ (.081)	-0.98982 (-45.6)	.195 (514.)	.195 (514.)	.14051 (0.047)
0.2	.123	258.	.29445 (.080)	$.209*10^{-6}$ (.294)	-0.96314 (-42.6)	.159 (135.)	.159 (135.)	.14049 (0.061)
0.5	.086	45.8	.29374 (.321)	$.345*10^{-5}$ (1.11)	-0.86069 (-27.4)	.0603 (25.6)	.0603 (25.6)	.14039 (.133)
1.0	.071	14.1	.29310 (.538)	$.100*10^{-4}$ (1.85)	-0.76772 (-13.6)	.0123 (7.93)	.0123 (7.93)	.14030 (0.197)
2.0	.072	5.43	.29278 (.647)	$.154*10^{-4}$ (2.25)	-0.71880 (-6.44)	.0022 (2.70)	.0022 (2.70)	.14022 (0.253)
5.0	.076	3.39	.29241 (.772)	$.266*10^{-4}$ (2.42)	-0.69664 (-3.15)	.00085 (1.017)	.00085 (1.017)	.14000 (0.410)
10.0	.079	3.24	.29127 (1.159)	$.887*10^{-4}$ (2.53)	-0.68374 (-1.25)	.00047 (0.98)	.00036 (0.98)	.13926 (0.937)

TABLE 3: Benchmark problem 1. 32 elements, 81 nodes solution. Series solution gives ϕ_{MAX} = 0.294867, $\alpha_{MAX} = \beta_{MAX}$ = -0.67531 and Volume = 0.140577.

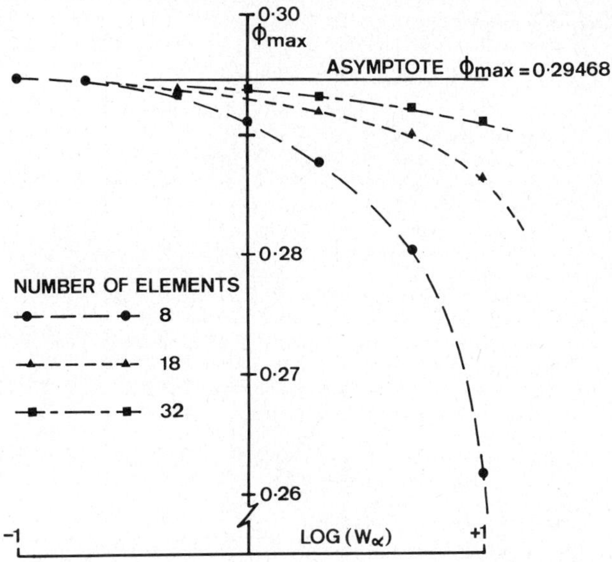

FIGURE 2: PLOTS OF ϕMAX VS LOG(W_α) FOR VARIOUS GRIDS SHOWING ASYMPTOTE FROM SERIES SOLUTION.

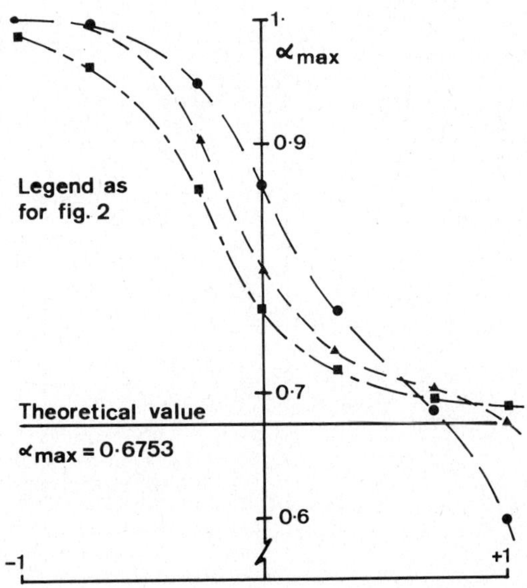

FIGURE 3: PLOTS OF αMAX VS LOG(W_α) FOR VARIOUS GRIDS SHOWING ASYMPTOTE.

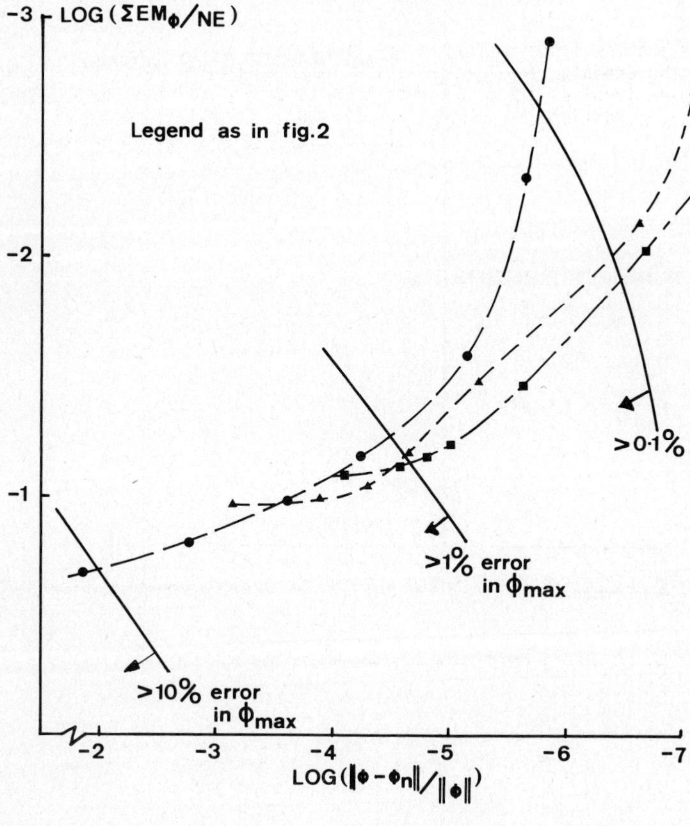

FIGURE 4: PLOTS OF $\mathrm{LOG}(\Sigma EM_{\phi}/NE)$ VS $\mathrm{LOG}(\| \phi \ \phi_n\| / \| \ \phi \|)$ FOR VARIOUS GRIDS. WEIGHT VALUES DESCEND FROM TOP RIGHT TO BOTTOM LEFT. APPROXIMATE POSITION OF ERROR BOUNDS SHOWN.

TEST PROBLEM 2

Our second test problem is more interesting and does not have an analytical solution, therefore no Green's function is available to calculate bounds to any solution. The problem involves one quarter of an I section which is shown, together with the boundary conditions in Figure 5 below.

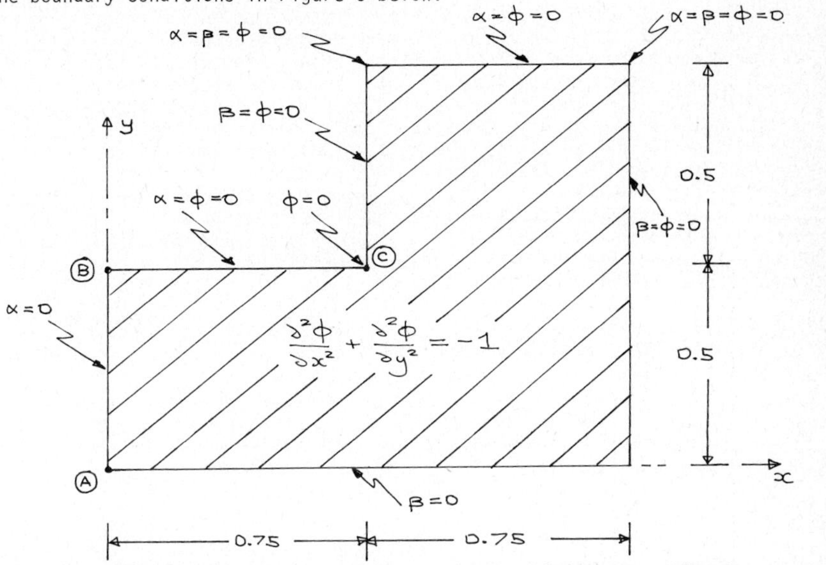

FIGURE 5: Solution region and boundary conditions for test problem 2.

Another feature of this problem is the re-entrant corner at Ⓒ in Figure 5, at this point there is an integrable singularity in the gradients. This can not be defined as a boundary condition and the α and β boundary values are not specified there. Since our continuum partial differential equation is solved in a discretised form for the finite element formulation, the model is essentially rounding off the corner with a radius related to the grid spacing at the corner. Thus with finer grids the results should demonstrate an increasing value in the gradients at this point.

For this problem four grids were modelled and each grid run with weights $W_\phi = 1$ and $W_\alpha = W_\beta = 0.2$, 1 and 5. Features of the results for all these runs are shown in Table 4. In this table $g_{\alpha\beta}$ is the value of $(\alpha^2 + \beta^2)^{\frac{1}{2}}$ at the points indicated, (Ⓐ Ⓑ or Ⓒ on Figure 5) and g_ϕ is the value of $\left[\left(\frac{\partial\phi}{\partial x}\right)^2 + \left(\frac{\partial\phi}{\partial y}\right)^2\right]^{\frac{1}{2}}$, other quantities have been previously defined.

For any grid the three values of $\phi_{Ⓐ}$, the maximum value recorded, are plotted against $\text{Log}W_\alpha$ in Figure 6. All the plots on this figure show there to be an asymptotic value for each grid. In order to estimate the asymptote we will assume the convergence to be exponential with $\text{Log}W_\alpha$ of the form,

$$\phi_n = \phi_{AS} - (\phi_0 - \phi_{AS})e^{-r},$$

where ϕ_{AS} is the asymptotic value, ϕ_0 is a starting value and ϕ_n any intermediate value. With the three values obtained in computer tests this expression can be solved to yield the value of ϕ_{AS} as,

$$\phi_{AS} = (\phi_2{}^2 - \phi_1\phi_3)/(2\phi_2 - \phi_1 - \phi_3),$$

where ϕ_1, ϕ_2 and ϕ_3 are the three values of $\phi_{Ⓐ}$ in ascending order. For each grid the asymptotic values are given in Table 4, together with the percentage errors in each of the three solution values relative to the asymptote. It is most significant to note that for the four grids the individual asymptotic values only differ by one digit in the fourth significant place. This leads to the conclusion that the correct value for the maximum is 0.1288 (7) and thus we are able to sketch in lines of constant error in Figure 6. To obtain this reliable asymptote it was necessary to do at least three solutions, albeit on the same mesh. It is still a desire to see if there is a characterisation of a single solution which gives some indication of the error. To this end Figure 7 has been plotted. This shows the Log $(\Sigma Em_{\phi}/NE)$,(NE = number of elements) plotted against $Log(W_{\alpha})$ for the four grids. Superimposed upon this are lines approximately indicating the errors in ϕ_{max} based upon their individual asymptotes. To a good approximation for all solutions with $Log(\Sigma Em_{\phi}/NE)$ less than -0.6 the relative error is less than 10%. The slopes of the curves all indicate that this accuracy criterion is achieved when the weight W is less than one. Indeed if this trend were continued it would appear that W_{α} should be as small as is numerically possible. By taking the gradients as $(\partial\phi/\partial x)$ and $(\partial\phi/\partial y)$ and not α and β it is also possible to state that the most accurate solution for ϕ will also give the most accurate solution for these partial derivatives.

Figures 8, 9 and 10 have been photographed directly off the computer video terminal and have been included to give the reader a visual impression of the solution to this problem.

GRID TYPE	$W_\alpha=W_\beta$, $W_\phi=1$	$\phi_{(A)}$	$g_{\alpha\beta}$ AT (B)	g_ϕ AT (B)	$g_{\alpha\beta}$ AT (C)	g_ϕ AT (C)	ΣR_α	ΣR_β	ΣR_ϕ	ΣEM_α	ΣEM_β	ΣEM_ϕ	ΣR_{MS}	ΣEM_{MS}	VOLUME	ϕ_{AS}	% ERROR
36 ELEMENT REGULAR	0.2	.12732	.4993	.5081	.3809	.5206	.1301	.1392	.1001	139.7	115.0	3.20	.284	237.0	.0760	.12897	1.279
	1.0	.12368	.4876	.4972	.2346	.5016	.0734	.0689	.1402	32.0	34.1	4.48	.211	57.9	.0739		4.101
	5.0	.11199	.4684	.4592	.3430	.4063	.0232	.0215	.278	21.9	5.60	8.90	.282	27.8	.0634		13.16
72 ELEMENT REGULAR	0.2	.12775	.4982	.5093	.371	.5893	.128	.128	.073	158.	1187.	4.52	.254	1309.	.0769	.12879	0.807
	1.0	.12523	.4829	.5013	.2573	.5716	.054	.054	.118	41.3	73.8	7.55	.170	103.8	.0753		2.764
	5.0	.11656	.4788	.5093	.3841	.5893	.0187	.0163	.206	41.2	8.60	13.2	.210	50.5	.0675		9.496
72 ELEMENT IRREGULAR	0.2	.12812	.5005	.5101	.1815	.677	.116	.116	.047	106.3	537.3	3.89	.214	600.	.0773	.12880	0.527
	1.0	.12617	.5036	.5029	.140	.654	.063	.058	.098	70.3	128.5	7.49	.156	181.8	.0757		2.041
	5.0	.11866	.4844	.4737	.384	.561	.015	.016	.202	63.8	104.6	16.05	.206	75.9	.0688		7.872
120 ELEMENT IRREGULAR	0.2	.12822	.5007	.5103	.1614	.7499	.1125	.1120	.038	466.7	660.6	5.24	.203	1053.	.0775	.12873	.396
	1.0	.12667	.5051	.5054	.1502	.7290	.0522	.0478	.089	108.0	92.8	11.3	.135	179.5	.0762		1.602
	5.0	.12043	.4898	.4856	.3982	.6414	.0141	.0134	.166	59.9	14.7	22.3	.170	76.9	.0706		6.447

TABLE 4: Summary of results for test problem 2.

CONCLUSIONS

The overall goal of this work was to discover if there existed a general means of establishing values for the error in the solution to a partial differential equation using a least squares finite element method. Not only the error in the solution process is required but specific errors in the value of the field variable

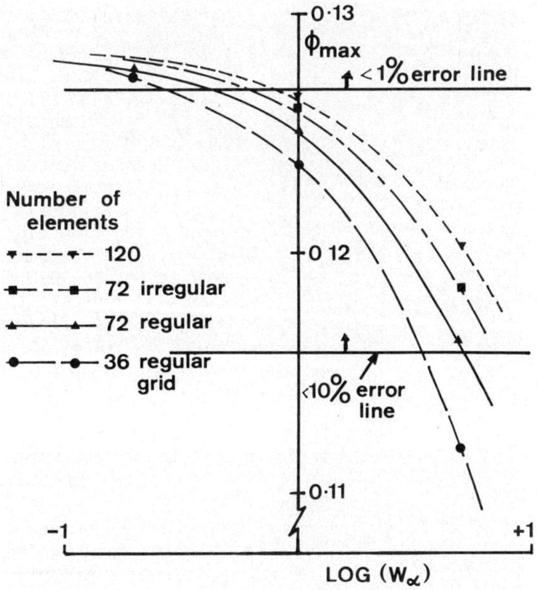

FIGURE 6: PLOTS OF ϕ_{MAX} VS LOG(W_α) FOR VARIOUS GRIDS SHOWING APPROXIMATE POSITIONS OF 10% AND 1% ERROR BOUNDS.

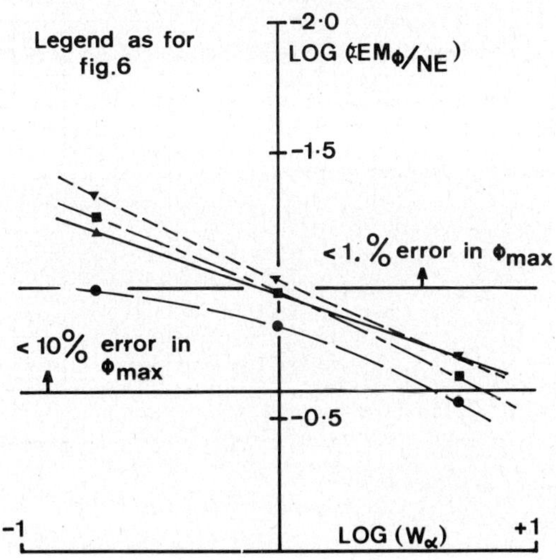

FIGURE 7: PLOTS OF LOG(ΣEM_ϕ/NE) AGAINST LOG(W_α) FOR VARIOUS GRIDS SHOWING APPROXIMATE POSITIONS OF 10% AND 1% ERROR BOUNDS.

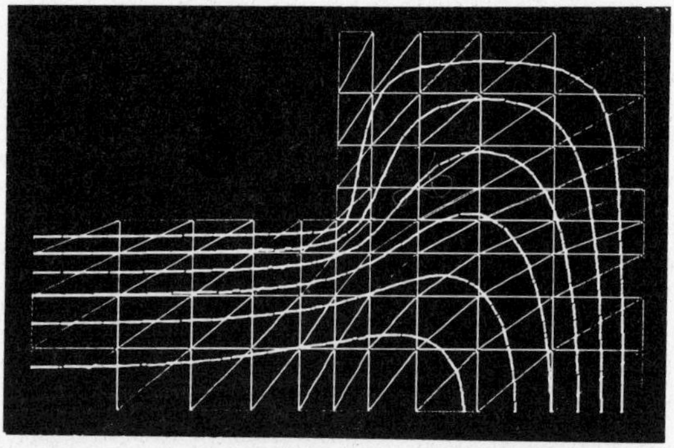

FIGURE 8: The 120 element mesh for Test Problem 2 showing solution contours with $W_\phi = 1$, $W_\alpha = W_\beta = 1$.

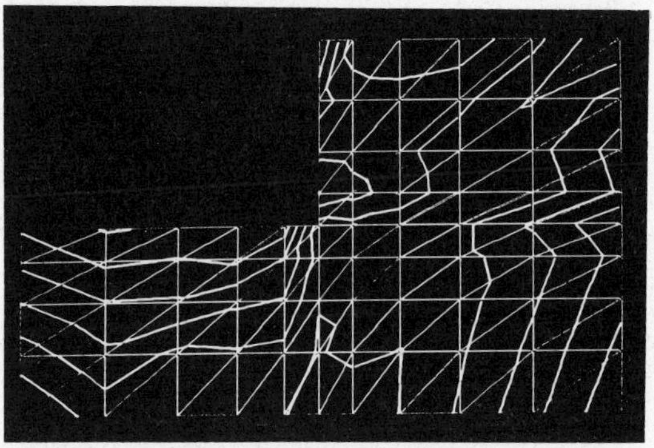

FIGURE 9: The 120 element mesh for Test Problem 2 showing contours of $g_{\alpha\beta} = (\alpha^2 + \beta^2)^{\frac{1}{2}}$.

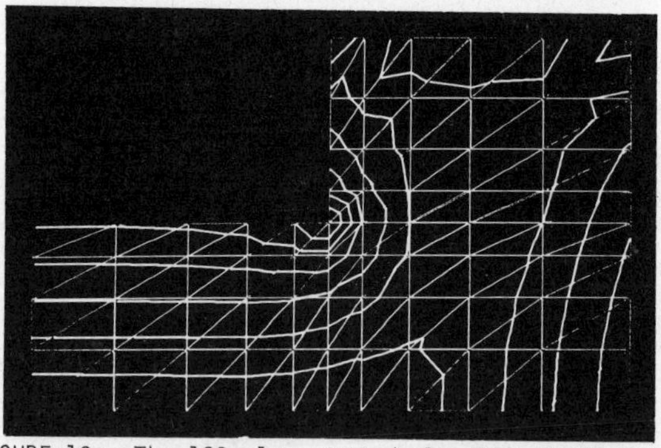

FIGURE 10: The 120 element mesh for Test Problem 2 showing contours of $g_\phi = \left[\left(\frac{\partial\phi}{\partial x}\right)^2 + \left(\frac{\partial\phi}{\partial y}\right)^2 \right]^{\frac{1}{2}}$.

and various integrals and differentials of it can all be of interest.

While this goal is a very ambitious one, certain conclusions can be drawn from the work reported herein and these must be followed up with investigations of other problems and other physical situations from the whole field of computer modelling. Two main conclusions can be drawn from the present work.

Firstly by judicious weighting, in a least squares formulation, towards the more significant part of the solution it is possible to gain a much better result. Judicious implies some knowledge of the physical process being studied or its mathematical formulation. For the Poisson equation that has been studied in this work the weighting should be applied to the conservation (continuity) equation and not to the equations of motion. For the same finite element mesh, the use of three solutions at different weights provides the means of using exponential extrapolation to obtain an asymptotic value, which for various grid solutions to the same problem, proved to be reliable. With a 'good' solution for the prime variable, its gradients are best represented by the partial differentials of the variable rather than take the auxiliary variables (α and β in this work).

Secondly, accepting that the best solution for the prime variable also gives the best solution for its gradients, studies were conducted to reveal whether there was a single 'a posteriori' indicator of the error in the prime variable solution. Both problems studied in this work demonstrated the feature that the quantity $\text{Log}(\Sigma \text{Em}_\phi/\text{NE})$ did provide a reliable indicator of the percentage error in the maximum value of the prime variable for the problems being solved. Since the overall solutions always appeared well behaved it seems logical to extrapolate this error bound to the error at all points. It was found that if $\text{Log}\ (\Sigma \text{Em}_\phi/\text{NE})$ was approximately less than -0.5 the error of ϕ_{max} was less than 10%, and if the Log were approximately less than -1 the error in ϕ_{max} was less than 1%. This result is very encouraging and it will be the subject of further research to investigate possible generalisations.

One final point to note is that all calculations for this work were conducted to 32 bit computer precision and whilst the author judges this sufficient for the size of problem studied, solutions at higher or lower precision may vary.

REFERENCES

[1] ADRAIN, R.: "Research Concerning the Probabilities of the Errors which Happen in Making Observations, etc.". The Analyst, Vol 1, No 4, pp 93-109 (1808).

[2] AKIN, J.E.: "A Least Squares Finite Element Solution of Non Linear Operators", in The Mathematics of Finite Elements and Applications, ed. Whiteman, Academic Press, New York, pp 153-162 (1979).

[3] AKIN, J.E. and SEN GUPTA, S.R.: "Finite Element Application of Least Squares Techniques", in Variational Methods in Engineering, Southampton.

[4] BABUSKA, I. and RHEINBOLDT, W.C.: "A - Posteriori Error Estimates for the Finite Element Method". Int. J. Num. Meth. Engng. 12, pp 1597 - 1615 (1978a).

[5] BABUSKA, I. and RHEINBOLDT, W.C.: "Error Estimates for Adaptive Finite Element Computations". SIAM J. Num. Anal. 15, (1978b).

[6] BOUSSINESQ, J.: "Théorie Analytique de la Chaleur", Vol 1, Gauthier-Villars, Paris, p 316, (1901).

[7] BRANDES, K.: "Diskrete Fehlerquadratmethode und Finite-Element Methode". Nuc. Engng. Res. 10, pp 503-504, (1969).

[8] EASON, E.D.: "A Review of Least-Squares Methods for Solving Partial Differential Equations". Int. J. Num. Meth. Engng. 10, pp 1021-1046, (1976).

[9] FINLAYSON, B.A.: The Method of Weighted Residuals and Variational Principles. Academic Press, New York, (1972).

[10] FRIED, I.: "On the Optimality of the Pointwise Accuracy of the Finite Element Solution". Int. J. Num. Meth. Engng 15, pp 451-476, (1980).

[11] GAUSS, K.F.: Theory of the Motion of the Heavenly Bodies Moving about the sun in Conic Sections. Translated by C.H. David, Little, Brown & Co., Boston 1857. Reprinted by Dover (1963).

[12] HALL, T.: Carl Friedrich Gauss. MIT Press, Cambridge, (1970).

[13] KELLY, D.W.: "Bounds of Discretization Error by Special Reduced Integration of the Lagrange Family of Finite Elements". Int. J. Num. Meth. Engng., 15, pp 1489-1506, (1980).

[14] LEGENDRE, A.M.: Nouvelles Methodes pour la Determination des Orbites des Cometes. Paris, Courcier, (1805).

[15] LYNN, P.P.: "Least Squares Finite Element Analysis of Laminar Boundary Layer Flows". Int. J. Num. Meth. Engng. 8, pp 865-876, (1974).

[16] LYNN, P.P. and ARYA, S.K.: "Use of Least Squares Criterion in the Finite Element Formulation". Int. J. Num. Meth. Engng. 6, pp 75-88, (1973).

[17] LYNN, P.P. and ARYA, S.K.: "Finite Elements Formulated by the Weighted Discrete Least Squares Method", Int. J. Num. Meth. Engng. 8, pp 71-90, (1974).

[18] MIKHLIN, S.G.: Variational Methods in Mathematical Physics. Macmillan, New York, (1964).

[19] MILTHORPE, J.F. and STEVEN, G.P.: "On a Least Squares Approach to the Integration of the Navier Stokes Equations", in Finite Elements in Fluids, Vol 3, ed. Gallagher et al. John Wiley, (1978).

[20] MILTHORPE, J.F.: "Solution of some Problems in Fluid Dynamics by Finite Element Methods". Doctorate thesis, University of Sydney, (1981).

[21] NITSCHE, J.: "Vergleich der Konvergenzgeschwindigheit des Ritzchen Verfahrens und der Fehlerquadratmethode", ZAMM, 49, pp 591-596, (1969).

[22] ZIENKIEWICZ, O.C.: "The Finite Element Method in Engineering Science". McGraw-Hill, (1971).

APPENDIX I

The element characteristic of $[K]^e\{u\}^n - \{F\}^e$ is shown in full matrix form on the next page.

Function Boundary Conditions

$$W_\phi(\phi - g_3) = 0 \quad \text{from equation (13)}$$

Surface interpolation of ϕ is

$$\phi = [N_S]\{\phi_S\}^n$$

$$\{\phi_S\}^n = \{\phi_1\ \phi_2\ \phi_3\}$$

$$[N_S] = [L_1(2L_1-1)\ L_2(2L_2-1)\ 4L_1L_2]$$

The L's are the natural coordinates along the external surface.

MATRIX IS SYMMETRIC

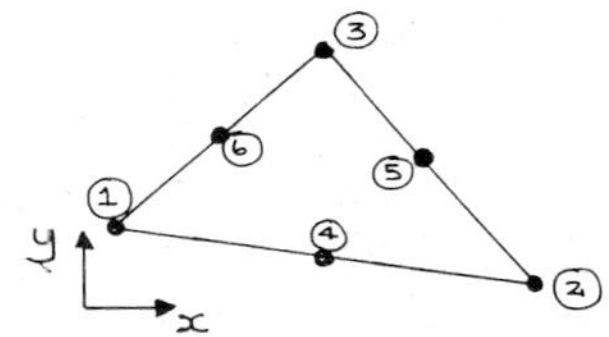

$$
\begin{bmatrix}
\frac{W_1}{6}+b_1^2W_3 & & & & & & & & & & & \\
W_3b_1c_1 & \frac{W_2}{6}+c_1^2W_3 & & & & & & & & & & \\
-W_1\frac{b_1}{3} & -W_2\frac{c_1}{3} & W_1b_1^2+W_2c_1^2 & & & & & & & & & \\
\frac{W_1}{12}+b_1b_2W_3 & W_3b_2c_1 & 0 & \frac{W_1}{6}+b_2^2W_3 & & & & & & & & \\
W_3b_1c_2 & \frac{W_2}{12}+c_1c_2W_3 & 0 & W_3b_2c_2 & \frac{W_2}{6}+c_2^2W_3 & & & & & & & \\
0 & 0 & -(W_1b_1b_2+W_2c_1c_2)/3 & -W_1\frac{b_2}{3} & -W_2\frac{c_2}{3} & W_1b_2^2+W_2c_2^2 & & & & & & \\
\frac{W_1}{12}+b_1b_3W_3 & W_3b_3c_1 & 0 & \frac{W_1}{12}+b_2b_3W_3 & W_3b_3c_2 & 0 & \frac{W_1}{6}+b_3^2W_3 & & & & & \\
W_3b_1c_3 & \frac{W_2}{12}+c_1c_3W_3 & 0 & W_3b_2c_3 & \frac{W_1}{12}+c_2c_3W_3 & 0 & W_3b_3c_3 & \frac{W_2}{6}+c_3^2W_3 & & & & \\
0 & 0 & -(W_1b_1b_3+W_2c_1c_3)/3 & 0 & 0 & -(W_1b_2b_3+W_2c_2c_3)/3 & -W_1\frac{b_3}{3} & -W_2\frac{c_3}{3} & W_1b_3^2+W_2c_3^2 & & & \\
-W_1\left(\frac{2}{3}b_2+\frac{1}{3}b_1\right) & -W_2\left(\frac{2}{3}c_2+\frac{1}{3}c_1\right) & 4(W_1b_1b_2+W_2c_1c_2)/3 & -W_1\left(\frac{1}{3}b_2+\frac{2}{3}b_1\right) & -W_2\left(\frac{1}{3}c_2+\frac{2}{3}c_1\right) & 4(W_1b_1b_2+W_2c_1c_2)/3 & -W_1\left(\frac{1}{3}b_2+\frac{1}{3}b_1\right) & -W_2\left(\frac{1}{3}c_2+\frac{1}{3}c_1\right) & 0 & \frac{8}{3}\left[W_1(b_1^2+b_2^2+b_1b_2)+W_2(c_1^2+c_2^2+c_1c_2)\right] & & \\
-W_1\left(\frac{1}{3}b_2+\frac{1}{3}b_3\right) & -W_2\left(\frac{1}{3}c_2+\frac{1}{3}c_3\right) & 0 & -W_1\left(\frac{2}{3}b_3+\frac{1}{3}b_2\right) & -W_2\left(\frac{2}{3}c_3+\frac{1}{3}c_2\right) & 4(W_1b_2b_3+W_2c_2c_3)/3 & -W_1\left(\frac{1}{3}b_3+\frac{2}{3}b_2\right) & -W_2\left(\frac{1}{3}c_3+\frac{2}{3}c_2\right) & 4(W_1b_2b_3+W_2c_2c_3)/3 & \frac{8}{6}\left[W_1(b_2b_3+b_2^2+2b_1b_3+b_1b_2)+W_2\binom{\text{SAME}}{\text{IN C's}}\right] & \frac{8}{3}\left[W_1(b_2^2+b_3^2+b_2b_3)+W_2(c_2^2+c_3^2+c_2c_3)\right] & \\
-W_1\left(\frac{1}{3}b_1+\frac{2}{3}b_3\right) & -W_2\left(\frac{1}{3}c_1+\frac{2}{3}c_3\right) & 4(W_1b_1b_3+W_2c_1c_3)/3 & -W_1\left(\frac{1}{3}b_1+\frac{1}{3}b_3\right) & -W_2\left(\frac{1}{3}c_1+\frac{1}{3}c_3\right) & 0 & -W_1\left(\frac{2}{3}b_1+\frac{1}{3}b_3\right) & -W_2\left(\frac{2}{3}c_1+\frac{1}{3}c_3\right) & 4(W_1b_1b_3+W_2c_1c_3)/3 & \frac{8}{6}\left[W_1(b_2b_1+b_1^2+2b_2b_3+b_1b_3)+W_2\binom{\text{SAME}}{\text{IN C's}}\right] & \frac{8}{6}\left[W_1(b_3b_1+b_3^2+2b_1b_2+b_2b_3)+W_2\binom{\text{SAME}}{\text{IN C's}}\right] & \frac{8}{3}\left[W_1(b_3^2+b_1^2+b_3b_1)+W_2(c_3^2+c_1^2+c_3c_1)\right]
\end{bmatrix}
\begin{bmatrix}
\alpha_1 \\ \beta_1 \\ \phi_1 \\ \alpha_2 \\ \beta_2 \\ \phi_2 \\ \alpha_3 \\ \beta_3 \\ \phi_3 \\ \phi_4 \\ \phi_5 \\ \phi_6
\end{bmatrix}
=
\begin{bmatrix}
b_1 f\Delta \\ c_1 f\Delta \\ 0 \\ b_2 f\Delta \\ c_2 f\Delta \\ 0 \\ b_3 f\Delta \\ c_3 f\Delta \\ 0 \\ 0 \\ 0 \\ 0
\end{bmatrix}
$$

IN THIS MATRIX

$W_1 = W_\alpha^2$

$W_2 = W_\beta^2$

$W_3 = W_\phi^2$

$b_1 = (y_2 - y_3)/2\Delta$

$b_2 = (y_3 - y_1)/2\Delta$

$b_3 = (y_1 - y_2)/2\Delta$

$c_1 = (x_3 - x_2)/2\Delta$

$c_2 = (x_1 - x_3)/2\Delta$

$c_3 = (x_2 - x_1)/2\Delta$

Δ = AREA OF TRIANGLE

NUMERICAL SOLUTIONS OF
PARTIAL DIFFERENTIAL EQUATIONS
J. Noye (Editor)

ANALYSIS OF ACOUSTIC RESONANCE IN A MOTOR VEHICLE PASSENGER COMPARTMENT USING THE BOUNDARY-INTEGRAL EQUATION METHOD

Howard J. Connell

Department of Mathematics
Royal Melbourne Institute of Technology
Melbourne
Victoria
AUSTRALIA

Interior noise is an important consideration in the design of small light-weight motor vehicles. Panel motion caused by road roughness may lead to excessive noise in the low frequency range (20 - 120 Hz). In this paper the boundary-integral equation method is applied to the problem of acoustic cavity resonance. Natural frequencies and corresponding acoustic modes are determined for a two-dimensional model of a motor vehicle passenger compartment. An assessment is made of the advantages and disadvantages of using this technique rather than the more usual finite element approach.

INTRODUCTION

The control of interior noise has become an important consideration in motor vehicle design with the use of light-weight unibody construction and other weightsaving measures (Wolf, Nefske and Howell (1976)). The major sources of interior noise have been identified as (1) the engine and transmission system, (2) road excitation and (3) aerodynamic excitation, though there are many other sources. Of particular interest here is road noise due to road roughness which induces vibration of roof and wall panels. The spectrum of road noise, which clearly depends on both the vibrational characteristics of body panels and the cavity acoustic properties of the passenger compartment, usually displays several large sharp peaks in the low frequency range ($\sim$ 100 Hz), while higher frequency peaks are less significant (Nefske and Howell (1976)).

Over the last decade, several researchers have used finite element techniques to analyze acoustic resonance in motor vehicle passenger compartments in an endeavour to reduce interior noise. Early work of Shuku and Ishihara (1973) was based on triangular elements within which the two dimensional pressure field $p(x,y)$ was represented by a cubic polynomial function of co-ordinates. Petyt, Lea and Koopmann (1976) modelled a light-weight van enclosure, using a three-dimensional twenty node isoparametric brick element with considerable accuracy in both predicted natural frequencies and mode shapes. Since motion of body panels may shift frequencies of acoustic modes and displace nodal lines, Wolf, Nefske and Howell (1976) modelled the whole elastic-acoustic system using the various NASTRAN plate and beam elements to model the flexible wall panels.

In recent years considerable research has been directed towards the estimation of the interior sound field in a motor vehicle passenger compartment due to vibrating panels. Jha and Cheilas (1978) have developed a procedure in which the body structure is treated as an assembly of point sources. The sound pressure field at a point inside the passenger compartment is then taken as the sum of the fields due to each of the point sources. Wolf, Nefske and Howell (1976) have modelled the coupled structural-acoustic system and used the forced vibration analysis offered by NASTRAN to predict successfully sound pressure levels in the passenger compartment of a motor vehicle.

In recent years boundary-integral equation methods have found application in elastodynamics (Cruse and Rizzo (1968)) and elastostatics (Rizzo (1967)). Such methods are claimed to offer important advantages over domain type methods such as finite elements and finite differences in that the resulting system of equations is much smaller and there is a considerable reduction in the data required. Further, it is suggested that the boundary-integral equation method gives added accuracy and is well suited to problems with infinite domains (Brebbia and Walker (1980)). De Mey (1976, 1977) and Hutchinson (1976) have shown that the boundary-integral equation method may be used to find the natural frequencies and corresponding acoustic modes of simple two-dimensional cavities.

In this paper we investigate the acoustic properties of a two-dimensional rigid-walled model of a motor vehicle passenger compartment using the boundary-integral equation method. This approach is compared to the more usual finite element method with respect to the computing resources required and suitability for interactive redesigning of the model.

PROBLEM FORMULATION

Let $P(x,y,t)$ be the acoustic pressure in the rigid-walled two-dimensional cavity which models the motor vehicle passenger compartment and occupies a region D in the xy plane with boundary curve C. In the usual simplified modelling of acoustic vibrations (Kinsler and Frey (1962)), the pressure $P(x,y,t)$ satisfies the wave equation

$$\frac{\partial^2 P}{\partial t^2} = c^2 \nabla^2 P \qquad (1)$$

within D while the normal derivative of pressure $\frac{\partial P}{\partial n}$ vanishes on the boundary C.

In the normal mode analysis we assume that the pressure varies sinusoidally with time. We take

$$P = p(x,y)\, e^{i\omega t} . \qquad (2)$$

The amplitude $p(x,y)$ of pressure variations then satisfies the Helmholtz equation

$$\nabla^2 p + k^2 p = 0 \qquad \text{in } D \qquad (3)$$

where $k = \frac{\omega}{c}$, subject to the boundary condition

$$\frac{\partial p}{\partial n} = 0 \qquad \text{on } C . \qquad (4)$$

The aim then is to find the discrete natural frequencies $\frac{ck}{2\pi}$ for which a solution of the field equation (3) satisfying the boundary condition (4) exists. These solutions describe the corresponding acoustic modes.

If $w(x,y;\, x_i,\, y_i)$ is a fundamental solution of the Helmholtz equation

$$\nabla^2 w + k^2 w = \delta(\underset{\sim}{r} - \underset{\sim}{r}(P_i)) \qquad (5)$$

where the point P_i has co-ordinates x_i and y_i, then using Green's second identity it may be easily shown

$$\lambda_i p(x_i,y_i) = \int_C \{p(x,y) \frac{\partial w}{\partial n}(x,y;\, x_i,y_i) - w(x,y;\, x_i,y_i) \frac{\partial p}{\partial n}(x,y)\}\, ds. \quad (6)$$

This parameter λ_i is 1 when P_i lies in the interior of D and $\frac{1}{2}$ when P_i is a point on the boundary curve where the tangent varies continuously. At a corner where the internal angle is ϕ_i then λ_i takes the value $\phi_i/2\pi$. If we now impose boundary condition (4) then equation (6) becomes

$$\lambda_i p(x_i,y_i) = \int_C p(x,y) \frac{\partial w}{\partial n}(x,y;\, x_i,y_i)\, ds. \quad (7)$$

DISCRETIZATION OF THE BOUNDARY

In applying the boundary-integral method we approximate the boundary curve C by straight line segments C_j $j = 1, 2, \ldots, N$ joining a selection of nodal points which subdivide C. Let $\bar{p}_{j-1}$ and $\bar{p}_j$ be approximate values of the pressure amplitude $p(x,y)$ at nodes $j-1$ and j. Along the line segment C_j joining nodes $j-1$ and j our trial solution $\bar{p}(x,y)$ is assumed to vary linearly according to

$$\bar{p} = \bar{p}_{j-1}(1-\xi_j) + \bar{p}_j \xi_j \quad (8)$$

where $\xi_j = \dfrac{s - s_{j-1}}{s_j - s_{j-1}}$. Here s_j denotes the distance of node j from node N along the straight line segments (a subscript value of 0 we associate with node N).

We now locate the singularity of the fundamental solution w at the boundary node P_i. If we substitute our trial solution into equation (7) we obtain

$$\sum_{j=1}^{N} \int_{C_j} \{\bar{p}_{j-1}(1-\xi_j) + \bar{p}_j \xi_j\} \frac{\partial w}{\partial n}(x,y;\, x_i,y_i)\, ds - \lambda_i \bar{p}_i = 0 \quad (9)$$

for $i = 1, 2, \ldots, N$.

Using a fundamental solution w given by

$$w(x,y;\ w_i,y_i) = \frac{1}{4} Y_0(k|\underset{\sim}{r} - \underset{\sim}{r}(P_i)|)$$

we then obtain a system of linear equations in the nodal values $\bar{p}_1, \bar{p}_2, \ldots, \bar{p}_N$ which we write

$$\sum_{j=1}^{N} G_{ij}(k)\,\bar{p}_j = 0 \qquad i = 1, 2, \ldots, N \qquad (10)$$

where the entry G_{ij} is given by

$$G_{ij} = \int_{C_{j+1}} \frac{1}{4}\frac{\partial}{\partial n}\{Y_0(k|\underset{\sim}{r} - \underset{\sim}{r}(P_i)|)\}(1 - \xi_{j+1})\,ds$$

$$+ \int_{C_j} \frac{1}{4}\frac{\partial}{\partial n}\{Y_0(k|\underset{\sim}{r} - \underset{\sim}{r}(P_i)|)\}\,\xi_j\,ds - \lambda_i\delta_{ij}\,. \qquad (11)$$

It is clear then that we must find the values of k for which zero is an eigenvalue of G. The corresponding eigenvectors then allow the construction of the appropriate acoustic modes using equation (6).

DEVELOPMENT OF THE NFREQ PROGRAM

A computer program named NFREQ has been written to implement the analysis of acoustic resonance of a two-dimensional rigid-walled cavity by the boundary-integral equation method in the manner described above.

As mentioned earlier, the boundary of the cavity is approximated by straight line segments joining successive nodes. Program NFREQ reads the number of boundary nodes and the co-ordinates of each node from a file of input data, and then computes the interior angle between consecutive line segments and related coefficient λ_i at node i for i = 1, 2, ..., N.

Gaussian quadrature is used to evaluate the integrals

$$\int_{C_j} (1-\xi_j)\frac{\partial w}{\partial n}(x,y;\ x_i,\ y_i)\,ds$$

$$\int_{C_j} \xi_j\frac{\partial w}{\partial n}(x,y;\ x_i,\ y_i)\,ds$$

for each element. The code has been organized in such a way that integration rules of higher order may be introduced without difficulty. Results obtained to date would seem to indicate that a four-point Gaussian rule is optimal. Execution times are considerably greater when higher order rules are used while accuracy is not substantially improved.

The generation of coefficient matrix $G(k)$ for a given value of k is carried out using an element-wise assembly technique. The entries of G are initially set to zero. The assembly of matrix G is achieved by adding in the contributions of each element in turn. For a typical element occupying segment C_j we add

$$\begin{cases} \int_{C_j} \xi_j \frac{\partial w}{\partial n}(x,y;\ x_i,y_i)\,ds & i \neq j-1,\ j,\ j+1 \\ 0 & i = j-1,\ j+1 \\ -\frac{\lambda_i}{2} & i = j \end{cases}$$

to G_{ij}, and add

$$\begin{cases} \int_{C_j} (1-\xi_j) \frac{\partial w}{\partial n}(x,y;\ x_i,\ y_i)\,ds & i \neq j-2,\ j-1,\ j \\ 0 & i = j-2,\ j \\ -\frac{\lambda_{j-1}}{2} & i = j-1 \end{cases}$$

to $G_{i,\ j-1}$.

For a given value of k, the eigenvalue of smallest magnitude and corresponding eigenvector of nodal values are found using inverse iteration (Dahlquist and Bjoerck (1974)). A secant style iterative technique is employed in the main program to find a value of k for which zero is an eigenvalue of $G(k)$. The efficiency of this approach is considerably enhanced by using the eigenvector from the current stage as the initial vector for inverse iteration at the next stage.

After finding a satisfactory solution for k, program NFREQ reads a data file containing the co-ordinates of interior points where the pressure $\bar{p}$ is to be computed. These values are determined in a subroutine and the estimates written out on a further file.

To facilitate the use of program NFREQ a program named GFILE has been written to allow the data file which describes the discretization of the boundary to be interpreted graphically on a Tektronix Graphics Display terminal so that the accuracy of the data defining the cavity may be checked.

Finally, the acoustic modes computed by NFREQ are interpreted graphically, using the University of Calgary CONTOUR package to display the contours of pressure amplitude.

RESULTS

Using the suite of programs described above, the first two longitudinal modes corresponding to frequencies of 97.9 and 167.5 Hz were computed for a two-dimensional model of the passenger compartment of the General Motors Holden Sunbird. The contours of pressure amplitude p corresponding to each of these modes are shown in Figures 1 and 2.

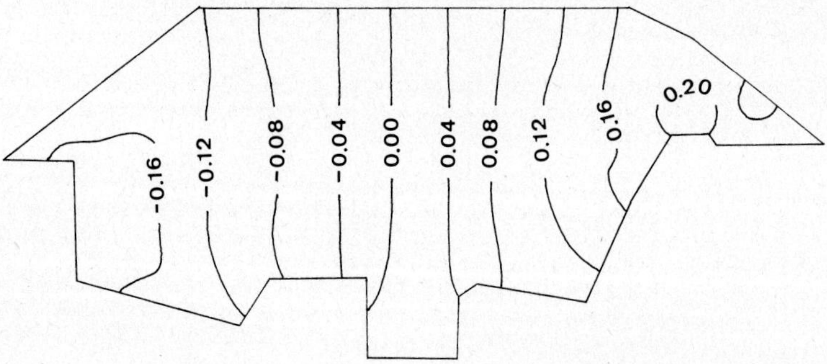

Fig. 1. First Sunbird acoustic mode (97.9 Hz).

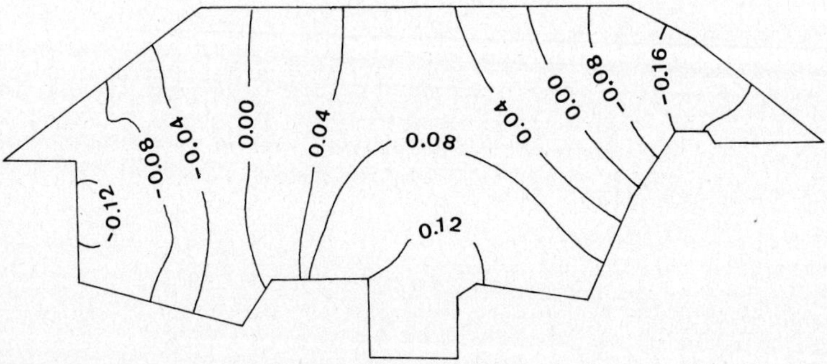

Fig. 2. Second Sunbird acoustic mode (167.5 Hz).

These results have been compared with those obtained using the IMSL TWODEPEP package which is available at RMIT. TWODEPEP employs a two-dimensional finite element approach based on the use of a six-node triangular element and offers a number of sophisticated features such as automatic mesh refinement. Results for the model of the Holden Sunbird indicate resonant frequencies of 93.9 and 156.9 Hz for the first two longitudinal modes.

A further comparison has been made with the experimental work of Sheers and Mitchell (1978) who found resonant frequencies of one Sunbird passenger compartment occurring at 125 and 135 Hz. The discrepencies here would seem to indicate the need to model the passenger compartment in a more realistic fashion (by considering effects such as absorption at the boundary, and the structural and cavity acoustic interaction).

DISCUSSION

Clearly the boundary-integral equation technique could be used to find higher frequency acoustic modes. However, it has been established that most of the sharp resonances occur at low frequencies around 100 Hz by experimental studies. The computation time required to find a natural frequency and corresponding acoustic mode is around 50 central processor seconds on the Cyber 170-720 at RMIT.

In comparing the performance of the NFREQ boundary-integral equation program with that of the TWODEPEP finite element package, the following advantages and disadvantages emerge :

Advantages of the boundary-integral equation approach

1. Reduced internal memory requirements

The boundary-integral equation approach requires only the boundary curve to be discretized (unlike the finite element method where the whole cavity is discretized). The solution is first found on the boundary at boundary nodes and then quite separately developed in the interior of the cavity. As a consequence, internal memory requirements are substantially reduced. It would seem that for a wide class of realistic problems out-of-core solution techniques would not be needed.

2. Suitability for design optimization

As mentioned above, the boundary element method requires only the discretization of the boundary of the cavity. As a consequence, design changes only require the modification of the file of boundary node data. Interactive redesigning is easily achieved. In contrast, when the finite element method is employed, it is necessary to tie changes to the boundary curve to the discretization of the cavity. Interactive redesigning would then seem feasible when a sophisticated mesh generator is available. It should be noted however, that since TWODEPEP requires only a very coarse initial triangulation (sufficient to define the region accurately) this advantage of the boundary-integral equation method is less significant here.

3. Ease in computing interior detail

Once the boundary-integral equation method has been used to determine the solution at boundary nodes, very little computational effort is required to determine the solution at any set of interior points. In particular, once the coarse features of an acoustic mode have been viewed, more detail in regions of interest can be generated very quickly without re-solving the problem. This feature is very useful in interactive redesigning.

Disadvantages of the boundary-integral equation approach

1. Increased execution times

The increased execution times in using the boundary-integral equation method (typically the determination of an acoustic mode using program NFREQ took three times as much central processor time as did the use of the TWODEPEP package), arise from the heavy use of the Bessel function Y_0 and the use of inverse iteration (for the eigenvalue of smallest magnitude) in conjunction with a secant style search.

2. Problem of exterior solutions

The boundary-integral equation method described in this paper locates solutions to both the interior and exterior problems. It is necessary to select the natural frequencies of the car passenger compartment from an intertwining set of values for which $G(k)$ is singular.

CONCLUSIONS

The boundary-integral equation method provides a useful alternative to the more frequently used finite element approach in analyzing acoustic cavity resonance of a car passenger compartment. The method has particular advantages when interactively redesigning for the optimization of some design feature.

The results obtained to date suggest a number of directions for further development. It seems clear that greater agreement with experiment would be achieved by modelling the structural acoustic interaction and sound absorption at the cavity walls. In addition, there is a need to develop boundary-integral equation procedures for three-dimensional cavities.

REFERENCES

[1] Brebbia, C.A. and Walker, S., Boundary Element Techniques in Engineering (Butterworth, London, 1980).

[2] Cruse, J.A. and Rizzo, F.J., A direct formulation and numerical solution of the general transient elastodynamic problem, 1, J. Math. and Appl. 22 (1968).

[3] Dahlquist, G. and Bjoerck, A., Numerical Methods (Prentice-Hall, New Jersey, 1974).

[4] De Mey, G., Calculation of eigenvalues of the Helmholtz equation by an integral equation, Int. J. Num. Meth. Engng 10 (1976) 59-66.

[5] De Mey, G., A simplified integral equation method for the calculation of the eigenvalues of the Helmholtz equation, Int. J. Num. Meth. Engng 11 (1977) 1340-1342.

[6] Hutchinson, J.R., Determination of membrance vibration characteristics by boundary-integral equation method, in : Brebbia, C.A. (ed.), Recent Advances in Boundary Element Methods, (Pentech, London, 1978).

[7] Jha, S.K. and Cheilas, N., Acoustic characteristics of car cavity and estimation of interior sound field produced by vibrating panel, ASME publication (December, 1976).

[8] Kinsler, L.E. and Frey, A.R., Fundamentals of Acoustics (2nd edn.) (Wiley, New York, 1962).

[9] Nefske, J. and Howell, L.J., Automobile interior noise reduction using finite element methods, SAE Paper 780365 (Detroit, 1976).

[10] Petyt, M. Lea, J. and Koopmann, G.H., A finite element method for determining the acoustic modes of irregular shaped cavities, Journ. Sound and Vibration 45(4) (1976) 495-502.

[11] Rizzo, F.J., An integral equation approach to boundary value problems of classical elastostatics, J. Appl. Math. 25(83) (1967).

[12] Sheers, K. and Mitchell, J.L., Passenger car noise reduction using a computer-based Fourier analyser, SAE Paper 78026 (Melbourne, 1978).

[13] Shuku, J. and Ishihara, K., The analysis of the acoustic field in irregularly shaped rooms by the finite element method, Journ. Sound and Vibration 29 (1973) 67-76.

[14] Wolf, J.A., Nefske, D.J. and Howell, L.J., Structural-acoustic finite element analysis of the automobile passenger compartment, SAE paper 760184 (Detroit,1976).

NUMERICAL SOLUTIONS OF
PARTIAL DIFFERENTIAL EQUATIONS
J. Noye (editor)

A DOUBLE CHEBYSHEV SERIES METHOD FOR ELLIPTIC PARTIAL DIFFERENTIAL EQUATIONS

T.S. Horner

Department of Mathematics
The University of Wollongong
Wollongong, N.S.W., Australia

It is shown how a Chebyshev series in two variables can be used to solve elliptic partial differential equations of Laplace type with Dirichlet boundary conditions. A simple recurrence relation is obtained for the coefficients in the series solution. The recurrence relation is of the form of the 5-point molecule of the finite difference method and a sparse matrix occurs in the solution of the associated algebraic equations. Efficient evaluation schemes for function values and derivatives are also presented.

INTRODUCTION

Chebyshev series have been used in the solution of differential equations by many authors, including, for ordinary differential equations, Lanczos (1957), Clenshaw (1957, 1962), Clenshaw and Elliott (1960), Fox and Parker (1968), Snell (1970), Boateng (1975), Morris and Horner (1977), Horner (1980), and for partial differential equations, Elliott (1961), Scraton (1965), Mason (1967a, 1967b, 1969), Knibb and Scraton (1971), Dew and Scraton (1973, 1975), Knibb (1975), Boateng (1975).

Recently, Haidvogel and Zang (1979), and Horner (1979) have solved elliptic partial differential equations using Chebyshev series in two variables. Haidvogel and Zang ultimately find the coefficients in the Chebyshev series by solving an algebraic problem with a full coefficient matrix, whereas the present method, based on Horner (1979), generates a sparse matrix, not unlike that obtained with the 5-point finite difference method.

Once the Chebyshev series is known, the values of the solution and its derivatives, can be found globally, and are not restricted to points of a grid as in the finite difference method. The formulae of Section 2 have been developed for these evaluations.

1. PRELIMINARY REMARKS ABOUT FUNCTIONS OF ONE VARIABLE IN TERMS OF CHEBYSHEV POLYNOMIALS

Let the function $f(x)$ $(\equiv f^{(0)}(x))$, and its derivatives $f'(x)$, $f''(x)$, ..., to the required order, be expressed as uniformly convergent series in the form

$$f^{(r)}(x) = \sum_{i=0}^{\infty}{}' a_i^{(r)} T_i(x) , \quad r = 0,1,\ldots , \tag{1.1}$$

where $a_0^{(r)}$, $a_1^{(r)}$, $a_2^{(r)}$, ... are constants, with the superscripts merely indicating the order of the derivative $f^{(r)}(x)$, and with the notation $\sum'$ indicating that $\frac{1}{2}a_0^{(r)}$, rather than $a_0^{(r)}$, is to be taken in the summation.

Then, using the fundamental theorem of calculus, in the form

$$\int f^{(r)}(x)dx = f^{(r-1)}(x) + \text{constant} \ ,$$

the coefficients in the series for successive derivatives can be related by

$$a_{i-1}^{(r)} - a_{i+1}^{(r)} = 2i\ a_i^{(r-1)} \ . \tag{1.2}$$

When ordinary differential equations are solved in Chebyshev series, the general equation (1.2) can be used with another equation derived from the particular equation being solved, in order to find the coefficients in the series solution. See Clenshaw (1957), Horner (1980). In Section 3 it is seen how these ideas can be extended to certain partial differential equations.

Following Clenshaw (1962), Smith (1965), Hunter (1970), Horner (1980), the following formulae can be found to evaluate the finite sum $\sum_{i=0}^{n}{}' a_i^{(r)} T_i(x)$ representing an approximation to $f^{(r)}(x)$.

$$\left.\begin{array}{l} \text{With } b_{k+1}^{-1} = \tfrac{1}{2}a_k \ , \quad k = 0,1,\ldots,n \ , \\ \text{let } b_{n+2-r}^{r} = b_{n+1-r}^{r} = 0 \\ \text{and calculate} \\ \qquad b_k^r = 2\ b_{k+1}^{r-1} + 2x\ b_{k+1}^r - b_{k+2}^r \ , \quad k = n-r,\ldots,1,0 \ . \\ \text{Then} \\ \qquad \dfrac{f^{(r)}(x)}{r!} = \tfrac{1}{2}\left(b_0^r - b_2^r\right) \ , \\ \text{for } r = 0,\ldots,n \ . \end{array}\right\} \tag{1.3}$$

Analogous results are now derived for functions of two variables.

2. FUNCTIONS OF TWO VARIABLES IN TERMS OF CHEBYSHEV POLYNOMIALS

Let the function $f(x,y)$ be expressed as a uniformly convergent double series of Chebyshev polynomials in the form

$$f(x,y) = \sum_{i=0}^{\infty}{}' \sum_{j=0}^{\infty}{}' a_{ij}\ T_i(x)T_j(y)$$

$$= \sum_{i,j=0}^{\infty}{}'' a_{ij}\ T_i(x)\ T_j(y) \ . \tag{2.1}$$

The notation $\sum_i{}'$ now signifies that $\tfrac{1}{2}a_{ij}$ must be taken whenever $i = 0$, and $\sum_j{}'$ means that $\tfrac{1}{2}a_{ij}$ must be taken whenever $j = 0$. When $i = j = 0$, the term $\tfrac{1}{4}a_{00}$ is taken in the summation. For the double summation over i and j , it is convenient to use the notation $\sum_{i,j}{}''$, having the combined meanings above.

The first few terms in the above expansion can be exhibited as

$$\tfrac{1}{4}a_{00}\, T_0(x)\, T_0(y) + \tfrac{1}{2}a_{01}\, T_0(x)\, T_1(y) + \tfrac{1}{2}a_{02}\, T_0(x)\, T_2(y) + \ldots$$
$$+ \tfrac{1}{2}a_{10}\, T_1(x)\, T_0(y) + a_{11}\, T_1(x)\, T_1(x) + a_{12}\, T_1(x)\, T_2(y) + \ldots$$
$$+ \tfrac{1}{2}a_{20}\, T_2(x)\, T_0(y) + a_{21}\, T_2(x)\, T_1(y) + a_{22}\, T_2(x)\, T_2(y) + \ldots$$
$$+ \ldots .$$

Writing

$$b_i = \sum_{j=0}^{\infty}{}' a_{ij}\, T_j(y) ,$$

$$c_j = \sum_{i=0}^{\infty}{}' a_{ij}\, T_i(x) ,$$

then (2.1) can be interpreted

$$f(x,y) = \sum_{i=0}^{\infty}{}' b_i T_i(x) , \tag{2.2}$$

or

$$f(x,y) = \sum_{j=0}^{\infty}{}' c_j T_j(y) . \tag{2.3}$$

THE NUMERICAL VALUE OF $\sum'' a_{ij}\, T_i(x)\, T_j(y)$, AND ITS DERIVATIVES

Basu (1973) gives the two-dimensional analogue of (1.3), $r = 0$. Using his notation, but slightly rearranging the formulae, results for function and derivative values are given below.

Thus, as in (2.2),

let $$S \equiv S_{mn}(x,y) = \sum_{i=0}^{m}{}' \sum_{j=0}^{n}{}' a_{ij}\, T_i(x)\, T_j(y) = \sum_{i=0}^{m}{}' b_i\, T_i(x) ,$$

where $$b_i = \sum_{j=0}^{n}{}' a_{ij}\, T_j(y) .$$

Let

$$g_{m+2} = g_{m+1} = 0 ,$$

and

$$\left.\begin{aligned} & d_{i,n+2} = d_{i,n+1} = 0 , \\ & d_{ij} = a_{ij} + 2y\, d_{i,j+1} - d_{i,j+2} , \quad j = n,\ldots,0 , \\ & b_i = \tfrac{1}{2}[d_{i,0} - d_{i,2}] , \\ & g_i = b_i + 2x\, g_{i+1} - g_{i+2} . \end{aligned}\right\} \quad i = m,\ldots,0 . \tag{2.4}$$

Then $$S \equiv S_{mn}(x,y) = \tfrac{1}{2}[g_0 - g_2] .$$

Alternatively, as in (2.3)

let $$S \equiv S_{mn} = \sum_{j=0}^{n}{}' c_j\, T_j(y) ,$$

where $$c_j = \sum_{i=0}^{m}{}' a_{ij}\, T_i(x) .$$

Let

$$g_{n+2} = g_{n+1} = 0 ,$$

and

$$\left.\begin{aligned} & d_{m+2,j} = d_{m+1,j} = 0 , \\ & d_{ij} = a_{ij} + 2x\, d_{i+1,j} - d_{i+2,j} , \quad i = m,\dots,0 , \\ & c_j = \tfrac{1}{2}[d_{0,j} - d_{2,j}] , \\ & g_j = c_j + 2y\, g_{j+1} - g_{j+2} . \end{aligned}\right\} \quad j = n,\dots,0 . \qquad (2.5)$$

Then $S \equiv S_{mn}(x,y) = \tfrac{1}{2}[g_0 - g_2]$.

In calculating S , either of (2.4) or (2.5) is satisfactory. In the formulae below, the results are extended in order to evaluate derivatives, with the work based on (2.4).

Write

$$g_i^{p,q} = \frac{1}{p!q!}\frac{\partial^{p+q} g_i}{\partial x^p \partial y^q} , \qquad b_i^{p,q} = \frac{1}{p!q!}\frac{\partial^{p+q} b_i}{\partial x^p \partial y^q} ,$$

$$d_{i,j}^{p,q} = \frac{1}{p!q!}\frac{\partial^{p+q} d_{i,j}}{\partial x^p \partial y^q} .$$

Then differentiation of (2.4) leads formally to the set of equations

$$\left.\begin{aligned} & g_{m+2}^{p,q} = g_{m+1}^{p,q} = 0 , \\ & d_{i,n+2}^{p,q} = d_{i,n+1}^{p,q} = 0 , \\ & d_{ij}^{p,q} = 2d_{i,j+1}^{p,q-1} + 2y\, d_{i,j+1}^{p,q} - d_{i,j+2}^{p,q} , \quad j = n,\dots,0 , \\ & b_i^{p,q} = \tfrac{1}{2}\left[d_{i,0}^{p,q} - d_{i,2}^{p,q}\right] , \\ & g_i^{p,q} = b_i^{p,q} + 2g_{i+1}^{p-1,q} + 2x\, g_{i+1}^{p,q} - g_{i+2}^{p,q} , \\ & S^{p,q} = \tfrac{1}{2}\left[g_0^{p,q} - g_2^{p,q}\right] , \end{aligned}\right\} \quad i = m,\dots,0 , \qquad (2.6)$$

where $S^{p,q} = \frac{1}{p!q!}\frac{\partial^{p+q}}{\partial x^p \partial y^p} f(x,y)$.

Moreover, $\{d_{ij}\}$ and $\{b_i\}$ are independent of x , and hence, <u>for $p > 0$</u> , the above become

$$\left.\begin{aligned} & g_{m+2}^{p,q} = g_{m+1}^{p,q} = 0 , \\ & g_i^{p,q} = 2g_{i+1}^{p-1,q} + 2x\, g_{i+1}^{p,q} - g_{i+2}^{p,q} , \quad i = m,\dots,0 , \\ & S^{p,q} = \tfrac{1}{2}\left[g_0^{p,q} - g_2^{p,q}\right] , \end{aligned}\right\} \qquad (2.7)$$

leading to all derivatives (including mixed derivatives) involving differentiation with respect to x . Obvious similarities exist between equations (1.3) and (2.7).

In these formulae, the various derivatives are found by letting $p,q = 0,1,\ldots,$ with the convention that

$$g_i^{0,0} = g_i \; ; \qquad d_i^{0,0} = d_i \; ;$$

$$b_i^{0,0} = b_i \; ; \qquad d_{i,j+1}^{0,-1} = \tfrac{1}{2}a_{ij} \; ;$$

$$d_{i,j+1}^{p,-1} = g_{i+1}^{-1,q} = 0 \;\; (p > 0) \; .$$

With $p \neq 0$, equation (2.7) leads to all derivatives except those purely with respect to y . These latter derivatives with $p = 0$, $q > 0$, can be obtained through the general scheme (2.6) above. However, if a number of them is wanted, it is more efficient to start again from (2.5) to obtain results analogous to (2.6), and hence analogous to (2.7). With $p, q = 0,1,2,$ the results (2.4), (2.6), (2.7) will be used to find function and derivative values, following the solution of second order partial differential equations.

3. METHOD OF SOLUTION OF PARTIAL DIFFERENTIAL EQUATION

HOMOGENEOUS EQUATION

Firstly, the equation

$$\frac{\partial^2 u}{\partial x^2} + \mu \frac{\partial^2 u}{\partial y^2} = 0 \; , \qquad \mu > 0 \; , \tag{3.1}$$

is considered on a region R subject to given boundary conditions on ∂R , the boundary of R . When $\mu = 1$, (3.1) is the two dimensional Laplace equation, but at first it is more convenient to consider the general form above.

Let the solution of (3.1) be expressed

$$u(x,y) = \sum_{i,j=0}^{\infty}{}'' \; a_{ij} \, T_i(x) \, T_j(y) \; , \tag{3.2}$$

using the notation already explained in section 2.

Series for the partial derivatives of $u(x,y)$, with respect to x , can be written

$$\frac{\partial u}{\partial x} = \sum_{i,j}{}'' \; a_{ij}^{(x)} \, T_i(x) \, T_j(y) \; ,$$

$$\frac{\partial^2 u}{\partial x^2} = \sum_{i,j}{}'' \; a_{ij}^{(xx)} \, T_i(x) \, T_j(y) \; ,$$

with similar series for $\dfrac{\partial u}{\partial y}$ and $\dfrac{\partial^2 u}{\partial y^2}$.

The sets of coefficients $\left\{a_{ij}^{(x)}\right\}$, $\left\{a_{ij}^{(xx)}\right\}$, $\left\{a_{ij}^{(y)}\right\}$, $\left\{a_{ij}^{(yy)}\right\}$ are constants and have the same significance in their respective series as the sets of coefficients $\left\{a_i^{(r)}\right\}$ in the series of section 1.

Substitution of the series for the partial derivatives, into equation (3.1), yields

$$\sum_{i,j}{}'' a_{ij}^{(xx)} T_i(x) T_j(y) + \mu \sum_{i,j}{}'' a_{ij}^{(yy)} T_i(x) T_j(y) = 0 ,$$

and hence

$$\left.\begin{array}{l} a_{ij}^{(xx)} + \mu a_{ij}^{(yy)} = 0 , \\ i,j = 0,1,2,\dots . \end{array}\right\} \tag{3.3}$$

on equating coefficients of like pairs of Chebyshev polynomials.
Then,

$$a_{i-1,j}^{(xx)} - a_{i+1,j}^{(xx)} + \mu \left\{a_{i-1,j}^{(yy)} - a_{i+1,j}^{(yy)}\right\} = 0 ,$$

and applying equation (1.2) to the series coefficients for the partial derivatives with respect to x ,

$$2\, i\, a_{ij}^{(x)} + \mu \left\{a_{i-1,j}^{(yy)} - a_{i+1,j}^{(yy)}\right\} = 0 .$$

Again, $$2\left\{a_{i-1,j}^{(x)} - a_{i+1,j}^{(x)}\right\} + \mu \left\{\frac{a_{i-2,j}^{(y\,y)} - a_{ij}^{(yy)}}{i-1} - \frac{a_{ij}^{(yy)} - a_{i+2,j}^{(yy)}}{i+1}\right\} = 0$$

and $$4i(i+1)(i-1)a_{ij} + \mu \left\{(i+1)a_{i-2,j}^{(yy)} - 2i\, a_{ij}^{(yy)} + (i-1)a_{i+2,j}^{(yy)}\right\} = 0 .$$

Further use of equation (1.2), now with regard to the derivatives in y , yields the final result

$$\begin{aligned} &i(i+1)(i-1)\left\{(j+1)a_{i,j-2} - 2j\, a_{ij} + (j-1)a_{i,j+2}\right\} \\ &+ \mu\, j(j+1)(j-1)\left\{(i+1)a_{i-2,j} - 2i\, a_{ij} + (i-1)a_{i+2,j}\right\} = 0 . \end{aligned} \tag{3.4}$$

There is an obvious resemblance to the equation

$$\begin{aligned} &u(x-h,y) - 2u(x,y) + u(x+h,y) \\ &\qquad + \mu \frac{h^2}{k^2} \{u(x,y-k) - 2u(x,y) + u(x,y+k)\} = 0 , \end{aligned}$$

obtained using central difference approximations for the partial derivatives. However, the important contrast is that the unknowns in the present method are the coefficients $\{a_{ij}\}$ in a series expansion, while in the finite difference method they are functions values $\{u(x,y)\}$ at the points $\{(x,y)\}$ on a grid in the region R . When the coefficients in the present series are known, the global solution can then be used to obtain function values, and derivatives, (using (2.4)) at any point of the region R . In many cases the Chebyshev coefficients $\{a_{ij}\}$ decrease rapidly in magnitude, and a smaller number of them is wanted in a Chebyshev polynomial approximation to the solution $u(x,y)$, than the number of grid points required for similar accuracy using finite difference methods.

If suitably large values of m,n are chosen for the polynomial $\sum_{i=0}^{m}{}' \sum_{j=0}^{n}{}' a_{ij}T_i(x)T_j(y)$ to give a satisfactory approximate solution to equation (3.1), then the problem becomes the algebraic one of finding the coefficients of the set $\{a_{ij}\}$.

Thus it is assumed that

$$a_{ij} = 0 \ , \quad i > m \ , \quad j > n \ . \tag{3.5}$$

Then letting i range from 2 to m and j from 2 to n, a total of $(m-1)(n-1)$ equations is obtained from (3.4).

Putting $i = 0,1$ and $j = 0,1$, merely leads to results consistent with the convention that

$$a_{\pm i,\pm j} = a_{ij} \ .$$

The remaining $(m+1)(n+1) - (m-1)(n-1) = (2m+2n)$, equations required for finding the coefficients $\{a_{ij}\}$, $i = 0,\ldots,m$, $j = 0,\ldots,n$ are obtained from boundary conditions. Dirichlet conditions will be considered in this paper.

It can be seen from equation (3.4) that calculation of the coefficients involves both the i- and j-subscripts being incremented in steps of 2. Thus in implementing (3.4), there are four independent sets of calculations, before boundary conditions are applied.

If $u(x,y)$ is purely odd or even, then a number of the calculations can be omitted.

THE DIRICHLET PROBLEM ON THE SQUARE

The solution of (3.1) corresponding to known values of $u(x,y)$ on the boundary of the square $-1 \leqslant x \leqslant 1$, $-1 \leqslant y \leqslant 1$, is considered.

Let $u(x,1) = \alpha(x) \equiv \tfrac{1}{2}\alpha_0 T_0(x) + \alpha_1 T_1(x) + \ldots + \alpha_m T_m(x)$

$$\left.\begin{aligned} &= \sum_{i=0}^{m}{}' \alpha_i T_i(x) \ , && -1 < x < 1 \ , \\ u(1,y) = \beta(y) &\equiv \sum_{j=0}^{n}{}' \beta_j T_j(y) \ , && -1 < y < 1 \ , \\ u(x,-1) = \gamma(x) &\equiv \sum_{i=0}^{m}{}' \gamma_i T_i(x) \ , && -1 < x < 1 \ , \\ u(-1,y) = \delta(y) &\equiv \sum_{j=0}^{n}{}' \delta_j T_j(y) \ , && -1 < y < 1 \ , \end{aligned}\right\} \tag{3.6}$$

with $\alpha(x)$, $\beta(y)$, $\gamma(x)$, $\delta(y)$ being known functions.

Now $u(x,1) = \sum_{i=0}^{m}{}' \left\{ \sum_{j=0}^{n}{}' a_{ij} \right\} T_i(x)$

and $u(x,-1) = \sum_{i=0}^{m}{}' \left\{ \sum_{j=0}^{n}{}' (-1)^j a_{ij} \right\} T_i(x)$

and equating coefficients of like Chebyshev polynomials gives

$$\left.\begin{aligned} \alpha_i &= \sum_{j=0}^{n}{}' a_{ij} = \tfrac{1}{2}a_{i0} + a_{i1} + \ldots + a_{in} \ , \\ \gamma_i &= \sum_{j=0}^{n}{}' (-1)^j a_{ij} = \tfrac{1}{2}a_{i0} - a_{i1} + \ldots + (-1)^n a_{in} \end{aligned}\right\} \ i = 0,\ldots,m. \tag{3.7}$$

Similarly

$$u(1,y) = \sum_{j=0}^{n}{}' \left\{ \sum_{i=0}^{m}{}' a_{ij} \right\} T_j(y) ,$$

$$u(-1,y) = \sum_{j=0}^{n}{}' \left\{ \sum_{i=0}^{m} (-1)^i a_{ij} \right\} T_j(y) ,$$

giving

$$\left.\begin{aligned} \beta_j &= \sum_{i=0}^{m}{}' a_{ij} = \tfrac{1}{2}a_{0j} + a_{1j} + \ldots + a_{mj} , \\ \delta_j &= \sum_{i=0}^{m}{}' (-1)^j a_{ij} = \tfrac{1}{2}a_{0j} - a_{1j} + \ldots + (-1)^m a_{mj} \end{aligned}\right\} j=0,\ldots,n, \quad (3.8)$$

These equations can be re-written in the alternative forms

$$\left.\begin{aligned} \tfrac{1}{2}(\alpha_i + \gamma_i) &= \tfrac{1}{2}a_{i0} + a_{i2} + \ldots \\ \tfrac{1}{2}(\alpha_i - \gamma_i) &= a_{i1} + a_{i3} + \ldots \end{aligned}\right\} i=0,\ldots,m , \quad (3.9)$$

$$\left.\begin{aligned} \tfrac{1}{2}(\beta_j + \delta_j) &= \tfrac{1}{2}a_{0j} + a_{2j} + \ldots \\ \tfrac{1}{2}(\beta_j - \delta_j) &= a_{1j} + a_{3j} + \ldots \end{aligned}\right\} j=0,\ldots,n . \quad (3.10)$$

In either case there are $2m + 2n + 4$ "boundary equations" to be satisfied by the coefficients. However only $2m + 2n$ equations are independent. For example, with $i = 0,1$ and $j = 0,1$ in (3.9) and (3.10), then

$$\left.\begin{aligned} \tfrac{1}{2}(\alpha_0 + \gamma_0) &= \tfrac{1}{2}a_{00} + a_{02} + \ldots \\ \tfrac{1}{2}(\alpha_0 - \gamma_0) &= a_{01} + a_{03} + \ldots \\ \tfrac{1}{2}(\alpha_1 + \gamma_1) &= \tfrac{1}{2}a_{10} + a_{12} + \ldots \\ \tfrac{1}{2}(\alpha_1 - \gamma_1) &= a_{11} + a_{13} + \ldots \end{aligned}\right\} \quad (3.11)$$

$$\left.\begin{aligned} \tfrac{1}{2}(\beta_0 + \delta_0) &= \tfrac{1}{2}a_{00} + a_{20} + \ldots \\ \tfrac{1}{2}(\beta_0 - \delta_0) &= a_{10} + a_{30} + \ldots \\ \tfrac{1}{2}(\beta_1 + \delta_1) &= \tfrac{1}{2}a_{01} + a_{21} + \ldots \\ \tfrac{1}{2}(\beta_1 - \delta_1) &= a_{11} + a_{31} + \ldots \end{aligned}\right\} \quad (3.12)$$

If the complete sets of equations (3.9) and (3.10) are regarded as a means of calculating a_{i0}, a_{i1}, when $i = 0,\ldots,m$ and a_{0j}, a_{1j} when $j=0,\ldots,n$, then (3.11) and (3.12) show how a_{00}, a_{01}, a_{10} and a_{11} can each be calculated in two different ways.

The values of a_{00}, a_{01}, a_{10}, a_{11} as calculated by (3.11) must be consistent with those calculated by (3.12). This is ensured if the boundary conditions are continuous at the vertices $(\pm 1, \pm 1)$, namely

$$\alpha(1) = \beta(1) , \quad \beta(-1) = \gamma(1) ,$$

$$\gamma(-1) = \delta(-1) , \quad \delta(1) = \alpha(-1) .$$

When this condition exists then the number of independent boundary equations is reduced by 4 to $2m + 2n$ and the equations (3.4), (3.7), (3.8) taken together (or (3.4), (3.9), (3.10)), are sufficient to calculate the $(m + 1)(n + 1)$ coefficients in the set $\{a_{ij}\}$.

It will be assumed for the present that the boundary conditions are continuous at the vertices. In section 5, discontinuities at the vertices are considered.

Writing equations (3.4), (3.9), (3.10) together leads to the set of $(m + 1)\times(n + 1)$ equations

$$\begin{aligned}
&\tfrac{1}{2}a_{i0} + a_{i2} + \dots = \tfrac{1}{2}(\alpha_i + \gamma_i) , \quad i=0,\dots,m , \\
&a_{i1} + a_{i3} + \dots = \tfrac{1}{2}(\alpha_i - \gamma_i) , \quad j=0,\dots,m , \\
&\left.\begin{aligned}
&\tfrac{1}{2}a_{0j} + a_{2j} + a_{4j} + \dots = \tfrac{1}{2}(\beta_j + \delta_j) \\
&a_{1j} + a_{3j} + a_{5j} + \dots = \tfrac{1}{2}(\beta_j - \delta_j) \\
&\mu f(j,i)a_{i-2,j} - g(i,j)a_{ij} + \mu f(-j,-i)a_{i+2,j} \\
&\quad + f(i,j)a_{i,j-2} + f(-i,-j)a_{i,j+2} = 0 \\
&\qquad i=2,\dots,m ,
\end{aligned}\right\} j=2,\dots,n , \qquad (3.13)
\end{aligned}$$

where $f(i,j) = i(i+1)(i-1)(j+1)$

and $g(i,j) = \{f(i,j) + f(-i,-j)\} + \mu\{f(j,i) + f(-j,-i)\}$.

Alternatively the equations can be re-ordered as

$$\begin{aligned}
&\tfrac{1}{2}a_{0j} + a_{2j} + a_{4j} + \dots = \tfrac{1}{2}(\beta_j + \delta_j) , \quad j=0,\dots,n , \\
&a_{1j} + a_{3j} + a_{5j} + \dots = \tfrac{1}{2}(\beta_j - \delta_j) , \quad j=0,\dots,n , \\
&\left.\begin{aligned}
&\tfrac{1}{2}a_{i0} + a_{i2} + \dots = \tfrac{1}{2}(\alpha_i + \gamma_i) \\
&a_{i1} + a_{i3} + \dots = \tfrac{1}{2}(\alpha_i - \gamma_i) \\
&f(i,j)a_{i,j-2} - g(i,j)a_{ij} + f(-i,-j)a_{i,j+2} \\
&\quad + \mu f(j,i)a_{i-2,j} + \mu f(-j,-i)a_{i+2,j} \\
&\qquad j=2,\dots,n .
\end{aligned}\right\} i=2,\dots,m , \qquad (3.14)
\end{aligned}$$

Table 1 shows the form of equations (3.13) when $m = 5$, $n = 3$.

$$\begin{bmatrix} \tfrac{1}{2}I & 0 & I & 0 \\ 0 & I & 0 & I \\ A_2 & 0 & B_2 & 0 \\ 0 & A_3 & 0 & B_3 \end{bmatrix} \begin{bmatrix} \underset{\sim}{a}_0 \\ \underset{\sim}{a}_1 \\ \underset{\sim}{a}_2 \\ \underset{\sim}{a}_3 \end{bmatrix} = \begin{bmatrix} \underset{\sim}{b}_0 \\ \underset{\sim}{b}_1 \\ \underset{\sim}{b}_2 \\ \underset{\sim}{b}_3 \end{bmatrix}$$

$A_j = \text{diag}\ \{0, 0, f(2,j), f(3,j), f(4,j), f(5,j)\}$, $j = 2,3$,

$$B_j = \begin{bmatrix} \tfrac{1}{2} & 0 & 1 & 0 & 1 & 0 \\ 0 & 1 & 0 & 1 & 0 & 1 \\ \mu f(j,2) & 0 & -g(2,j) & 0 & \mu f(-j,-2) & 0 \\ 0 & \mu f(j,3) & 0 & -g(3,j) & 0 & \mu f(-j,-3) \\ 0 & 0 & \mu f(j,4) & 0 & -g(4,j) & 0 \\ 0 & 0 & 0 & \mu f(j,5) & 0 & -g(5,j) \end{bmatrix} \quad j = 2,3 ,$$

$$\underset{\sim}{a}_j^T = [a_{0j}, a_{1j}, a_{2j}, a_{3j}, a_{4j}, a_{5j}] , \quad j = 0,1,2,3 ,$$

$$\underset{\sim}{b}_0^T = [\tfrac{1}{2}(\alpha_0 + \gamma_0), \tfrac{1}{2}(\alpha_1 + \gamma_1), \ldots, \tfrac{1}{2}(\alpha_5 + \gamma_5)] ,$$

$$\underset{\sim}{b}_1^T = [\tfrac{1}{2}(\alpha_0 - \gamma_0), \tfrac{1}{2}(\alpha_1 - \gamma_1), \ldots, \tfrac{1}{2}(\alpha_5 - \gamma_5)] ,$$

$$\underset{\sim}{b}_j^T = [\tfrac{1}{2}(\beta_j + \delta_j), \tfrac{1}{2}(\beta_j - \delta_j), 0, 0, 0, 0] , \quad j = 2,3$$

I is the identity matrix of order 6.

Table 1. Representation of equations (3.13), when $m = 5$, $n = 3$.

4. DISCUSSION OF TWO NUMERICAL RESULTS

(a) LAPLACE'S EQUATION (1)

Consider
$$\left.\begin{aligned} &\frac{\partial^2 u}{\partial x^2} + \frac{\partial^2 u}{\partial y^2} = 0 , \\ &u(x,\pm 1) = \cos(\tfrac{1}{2}\pi x) , \\ &u(\pm 1,y) = 0 . \end{aligned}\right\} \qquad (4.1)$$

The analytic solution is

$$u(x,y) = \cos(\tfrac{1}{2}\pi x)\cosh(\tfrac{1}{2}\pi y)/\cosh(\tfrac{1}{2}\pi)$$

being the real part of $\cos(\tfrac{1}{2}\pi z)/\cosh(\tfrac{1}{2}\pi)$, $z = x + iy$.

The problem is symmetric about both axes, so that in the solution (3.2), only even values of i, j occur.

With the notation of section 3,

$$\alpha(x) = \gamma(x) = \cos(\tfrac{1}{2}\pi x) ,$$
$$\beta(y) = \delta(y) = 0 .$$

The coefficients $\{\alpha_i\}$, $\{\gamma_i\}$, for the data (3.6) can be obtained using methods of Horner (1980), for finding the Chebyshev expansion of $\cos(\tfrac{1}{2}\pi x)$ on $(-1,1)$.

Using (3.4), (3.5), (3.9), (3.10), the approximate solution (3.2), with $m = n = 6$, is then

$$u(x,y) = \sum_{i,j=0}^{6}{}'' a_{ij} T_i(x)T_j(y) ,$$

and the calculated non-zero values of $\{a_{ij}\}$ are given in Table 2.

(2.4) can now be used to evaluate $u(x,y)$ at any point (x,y) of the square with vertices $(\pm 1, \pm 1)$. Comparison with the analytic solution at the selected points where $x = 0(0.2)1$, $y = 0(0.2)1$, shows a maximum difference of $\sim 10^{-5}$ between the approximate values and the analytic values.

i \ j	0	2	4	6
0	1.293257	.283627	.013479	.000268
2	-.684173	-.150045	-.007131	-.000142
4	.038348	.008410	.000400	.000008
6	-.000817	-.000179	-.000009	-.000000

Table 2. Non-zero coefficients in the approximate solution

$$u(x,y) = \sum_{i,j=0}^{6}{}'' a_{ij} T_i(x)T_j(y) . \quad \text{(Equation (4.1))}$$

(2.6) can be used to find the values of $\left|\frac{\partial^2 u}{\partial x^2} + \frac{\partial^2 u}{\partial y^2}\right|$ at the same points.

The maximum deviation from zero is $\sim 10^{-2}$ at (1,1), and the smallest deviation is $\sim 10^{-4}$ at (0,0). These values are calculated in order to provide some measure of the accuracy of the solutions, but the remarks in the previous paragraph indicate that the solution is very much better than this measure might indicate.

With m and n increased to 10, the non-zero coefficients in the approximate solution $u(x,y) = \sum_{i,j=0}^{10}{}'' a_{ij} T_i(x)T_j(y)$ can be found, and (2.4) gives values of $u(x,y)$ at the points previously used, in agreement to 7 decimal places with the values from the analytic solution. The maximum deviation of $\left|\frac{\partial^2 u}{\partial x^2} + \frac{\partial^2 u}{\partial y^2}\right|$ from zero is now less than 5×10^{-8}.

Thus in this first example, the methods of section 3 are shown to produce an extremely satisfactory numerical solution.

(b) LAPLACE'S EQUATION (2)

Consider
$$\left.\begin{aligned} &\frac{\partial^2 u}{\partial x^2} + \frac{\partial^2 u}{\partial y^2} = 0 , \\ &u(x,1) = 0 , \\ &u(x,-1) = \tfrac{1}{2}(1 - x^2) , \\ &u(\pm 1,y) = 0 . \end{aligned}\right\} \qquad (4.2)$$

This problem is symmetric about the y-axis, so that only even terms in x occur in the series solution. Again the analytic solution can be used to check

x	Values on y = 1	Values on y = -1
0	.000298	.002526
±0.2	.000089	.000741
±0.4	.000329	.002737
±0.6	.000010	.000053
±0.8	.000190	.000748
±1	0	1

y	Values on x = ±1
1	0
0.8	.002687
0.6	.002230
0.4	.000033
0.2	.002131
0	.001412
-0.2	.001301
-0.4	.003099
-0.6	.002273
-0.8	.002129
-1	1

Table 4. Values of $\left|\frac{\partial^2 u}{\partial x^2} + \frac{\partial^2 u}{\partial y^2}\right|$ on boundaries.

$m = n = 20$. (Equation (4.2))

The recurrence relation is obtained as a simple modification of (3.4), by adding to its left and right hand sides, terms which reflect the steps of its derivation, applied to the coefficients of u and g . Thus

$$\begin{aligned}
& i(i+1)(i-1)\left\{(j+1)a_{i,j-2} - 2j\, a_{ij} + (j-1)a_{i,j+2}\right\} \\
& + \mu\, j(j+1)(j-1)\left\{(i+1)a_{i-2,j} - 2i\, a_{ij} + (i-1)a_{i+2,j}\right\} \\
& + \frac{\lambda}{4}\Big[(i+1)\left\{(j+1)a_{i-2,j-2} - 2j\, a_{i-2,j} + (j-1)a_{i-2,j+2}\right\} \\
& \qquad - 2i\left\{(j+1)a_{i,j-2} - 2j\, a_{ij} + (j-1)a_{i,j+2}\right\} \\
& \qquad + (i-1)\left\{(j+1)a_{i+2,j-2} - 2j\, a_{i+2,j} + (j-1)a_{i+2,j+2}\right\}\Big] \\
= {} & \tfrac{1}{4}\Big[(i+1)\left\{(j+1)b_{i-2,j-2} - 2j\, b_{i-2,j} + (j-1)b_{i-2,j+2}\right\} \\
& \quad - 2i\left\{(j+1)b_{i,j-2} - 2j\, b_{ij} + (j-1)b_{i,j+2}\right\} \\
& \quad + (i-1)\left\{(j+1)b_{i+2,j-2} - 2j\, b_{i+2,j} + (j-1)b_{i+2,j+2}\right\}\Big] ,
\end{aligned} \tag{5.2}$$

where $g(x,y) = \sum'' b_{ij}T_i(x)T_j(y)$.

$$\text{Consider} \quad \left.\begin{aligned}
& \frac{\partial^2 u}{\partial x^2} + \frac{\partial^2 u}{\partial y^2} - 3u = 1 , \\
& u(x,\pm 1) = 0 , \\
& u(\pm 1,y) = 0 .
\end{aligned}\right\} \tag{5.3}$$

For the numerical solution

$$u(x,y) = \sum_{i,j=0}^{m,n}{}'' a_{ij} T_i(x)T_j(y)$$

the data involves

$$\left.\begin{array}{l} \alpha(x) = \beta(y) = \gamma(x) = \delta(y) = 0 , \\ \left.\begin{array}{l} \lambda = -3 \\ g(x,y) = 1 \end{array}\right\} . \end{array}\right\}$$

With $m = n = 20$, the coefficients in the solution (3.2) are less than 10^{-6} in magnitude for $i,j > 16$, and lead to the numerical values given in Table 5.

y \ x	.00	.20	.40	.60	.80	1.00
1.00	.00000	.00000	.00000	.00000	.00000	.00000
.80	-.07650	-.07486	-.06950	-.05876	-.03880	.00000
.60	-.12506	-.12203	-.11216	-.09286	-.05876	.00000
.40	-.15431	-.15027	-.13721	-.11216	-.06950	.00000
.20	-.16982	-.16517	-.15027	-.12203	-.07486	.00000
.00	-.17466	-.16982	-.15431	-.12506	-.07650	.00000

Table 5. Values of $u(x,y)$, $x,y = 0(0.2)1$. $m = n = 20$. (Equation (5.3))

(b) DISCONTINUITIES AT THE VERTICES

In some circumstances it is possible to cope with discontinuities at the vertices $(\pm1, \pm1)$, by observing that for any solution u of the homogeneous equation

$$\frac{\partial^2 u}{\partial x^2} + \mu \frac{\partial^2 u}{\partial y^2} = 0 ,$$

derivatives of u are also solutions, namely

$$\frac{\partial^2}{\partial x^2}\left\{\frac{\partial^{r+s}u}{\partial x^r \partial y^s}\right\} + \mu \frac{\partial^2}{\partial y^2}\left\{\frac{\partial^{r+s}u}{\partial x^r \partial y^s}\right\} = 0 . \tag{5.4}$$

In particular when $r = 0$, $s = 2$ or $r = 2$, $s = 0$, then the equation (5.4) can be used to modify boundary conditions involving u, to give conditions involving

$$\frac{\partial^2 u}{\partial x^2} \text{ or } \frac{\partial^2 u}{\partial y^2} .$$

These ideas have been used by Knibb and Scraton (1971) in the solution of parabolic partial differential equations in Chebyshev series.

Consider a problem

$$\left.\begin{array}{l} \dfrac{\partial^2 u}{\partial x^2} + \mu \dfrac{\partial^2 u}{\partial y^2} = 0 \\ \text{for which} \quad u(x,1) = 0 \ , \\ \qquad u(1,y) = 0 \ , \\ \qquad u(x,-1) = \gamma(x) \ , \\ \qquad u(-1,y) = 0 \ , \end{array}\right\} \tag{5.5}$$

and $\gamma(\pm 1) \neq 0$.

Then introduce the function $\bar{\gamma}(x)$, such that

$$\frac{d^2\bar{\gamma}}{dx^2} + \mu \gamma = 0 \ , \qquad \bar{\gamma}(\pm 1) = 0 \ . \tag{5.6}$$

$$\text{Then} \quad u = \frac{\partial^2 \bar{u}}{\partial y^2} = -\frac{1}{\mu}\frac{\partial^2 \bar{u}}{\partial x^2}$$

where $\bar{u}(x,y)$ is the solution of the problem

$$\left.\begin{array}{l} \dfrac{\partial^2 \bar{u}}{\partial x^2} + \mu \dfrac{\partial^2 \bar{u}}{\partial y^2} = 0 \ , \\ \bar{u}(x,1) = 0 \ , \\ \bar{u}(1,y) = 0 \ , \\ \bar{u}(x,-1) = \bar{\gamma}(x) \ , \\ \bar{u}(-1,y) = 0 \ , \end{array}\right\} \tag{5.7}$$

in which there are no discontinuities at the vertices. Of course $\dfrac{\partial^2 \bar{u}}{\partial y^2}\left(= -\dfrac{1}{\mu}\dfrac{\partial^2 \bar{u}}{\partial x^2}\right)$ will still be discontinuous at the vertices.

As an example take $\mu = 1$, $\gamma(x) = 1$, giving $\bar{\gamma}(x) = \frac{1}{2}(1-x^2)$ in (5.7). Then (5.7) is effectively the problem already solved as (4.2). For that problem the required second derivatives can be calculated using (2.6), (2.7). Thus the solution $u(x,y)$ of (5.5) is known because

$$u = \frac{\partial^2 \bar{u}}{\partial y^2} = -\frac{\partial^2 \bar{u}}{\partial x^2} \ ,$$

where $\bar{u}(x,y)$ is now the notation for the solution of (4.2).

As noted in previous discussions about the values of $\left|\dfrac{\partial^2 u}{\partial x^2} + \dfrac{\partial^2 u}{\partial y^2}\right|$ on the boundary of the region, it is found that $\dfrac{\partial^2 \bar{u}}{\partial y^2}$ and $-\dfrac{\partial^2 \bar{u}}{\partial x^2}$ differ near the boundary. An empirical rule is that the solution of (5.5) is given by $\dfrac{\partial^2 \bar{u}}{\partial y^2}$ "near" $x = \pm 1$ and by $-\dfrac{\partial^2 \bar{u}}{\partial x^2}$ "near $y = \pm 1$, and by either expression in the interior of the region. It should be emphasised that we have not obtained a new <u>series</u> for $u(x,y)$, the solution of (5.5), but have used the series solution of (4.2), together with the ability to calculate function <u>values</u> and derivative <u>values</u> using (2.6).

6. CONCLUSION

It has been seen that a simple recurrence relation (3.4) can be obtained for the coefficients in a Chebyshev series solution of elliptic partial differential equations of the form

$$\frac{\partial^2 u}{\partial x^2} + \mu \frac{\partial^2 u}{\partial y^2} + \lambda u = g(x,y) .$$

On the square region with vertices $(\pm 1, \pm 1)$, Dirichlet boundary conditions have been matched with the Chebyshev series to obtain a complete solution. Function and derivative values are then found at any point of the solution region, using (2.6).

The quoted numerical results were obtained by solving the algebraic equations for the series coefficients using a relaxation method. The test for convergence was that the <u>relative</u> change, from one iteration to the next, should be less than 10^{-6} in magnitude, for every coefficient a_{ij} in the solution. The iteration count for calculating a set of coefficients was usually greater than the count required to obtain the same number of grid-point values using finite differences. However the latter method has only $O(h^2)$ accuracy, so that much finer subdivision, and possibly extrapolation, is needed to obtain a solution of the original problem with accuracy and applicability (using (2.6)), comparable to that available from a small number of Chebyshev coefficients.

REFERENCES

[1] Basu, N.K., On double Chebyshev series approximation, SIAM J. Numer. Anal. 10 (1973) 496-505.

[2] Boateng, G.K., A practical Chebyshev collocation method for differential equations, Int. J. Comput. Math., section B 5 (1975) 59-79.

[3] Clenshaw, C.W., The numerical solution of ordinary differential equations in Chebyshev series, Proc. Camb. Phil. Soc 53 (1957) 134-149.

[4] Clenshaw, C.W., Chebyshev series for mathematical functions, Mathematical Tables, Volume 5, National Physical Laboratory.

[5] Clenshaw, C.W. & Elliott, D., A numerical treatment of the Orr-Sommerfeld equation in the case of a laminar jet, Q. Jl. Mech. Appl. Math. 8 (1960) 300-313.

[6] Dew, P.M. and Scraton, R.E., An improved method for the solution of the heat equation in Chebyshev series, J. Inst. Maths Applics 11 (1973) 231-240.

[7] Dew, P.M. and Scraton, R.E., Chebyshev methods for the numerical solution of parabolic partial differential equations in two and three space variables, J. Inst. Maths Applics 16 (1975) 121-131.

[8] Elliott, D., A method for the numerical integration of the one-dimensional heat equation using Chebyshev series, Proc. Camb. Phil. Soc 57 (1961) 823-832.

[9] Fox, L. and Parker, I.B., Chebyshev Polynomials in Numerical Analysis (Oxford University Press, London, 1968).

[10] Haidvogel, D.B. and Zang, T., The accurate solution of Poisson's equation by expansion in Chebyshev polynomials, J. Comput. Phys. 30 (1979) 167-180.

[11] Horner, T.S., Chebyshev polynomials in the solution of partial differential equations, Preprint 7/79. Dept. of Math., Uni. of Wollongong (1979).

[12] Horner, T.S., Recurrence relations for the coefficients in Chebyshev series solutions of ordinary differential equations, Maths. Comput. 35 (1980) 893-905.

[13] Hunter, D.B., Clenshaw's method for evaluating certain finite series, Computer J. 13 (1970) 378-381.

[14] Knibb, D., The numerical solution of parabolic partial differential equations using the method of Lanczos, J. Inst. Maths Applics 11 (1975) 181-190.

[15] Knibb, D. and Scraton, R.E., On the solution of parabolic partial differential equations in Chebyshev series, Computer J. 14 (1971) 428-432.

[16] Lanczos, C., Applied Analysis (Prentice-Hall, New York; Pitman, London, 1957).

[17] Mason, J.C., Chebyshev polynomial approximations for the L-membrane eigenvalue problem, SIAM J. Appl. Math. 15 (1967) 172-186.

[18] Mason, J.C., A Chebyshev method for the numerical solution of the one-dimensional heat equation in Proc. 22nd National Conference of ACM (Academic Press, London, 1967).

[19] Mason, J.C., Chebyshev methods for separable partial differential equations in Proc. IFIP Congress, Edinburgh (North Holland Publishing Co., Amsterdam, 1969).

[20] Morris, A.G. and Horner, T.S., Chebyshev polynomials in the numerical solution of differential equations, Maths Comput. 31 (1977) 881-891.

[21] Scraton, R.E., The solution of linear differential equations in Chebyshev series, Computer J. 8 (1965) 57-61.

[22] Smith, F.J., An algorithm for summing orthogonal polynomial series and their derivatives, with applications to curve-fitting and interpolation, Maths Comput. 19 (1965) 33-36.

[23] Snell, J., The solution in Chebyshev series of systems of linear differential equations with general boundary conditions, Computer J. 13 (1970) 103-106.

NUMERICAL SOLUTIONS OF PARTIAL DIFFERENTIAL EQUATIONS
J. Noye (editor)

PARABOLIC PARTIAL DIFFERENTIAL EQUATIONS WITH UNKNOWN COEFFICIENTS - ELECTRICAL CONDUCTIVITY MODELLING

R.S. Anderssen

Division of Mathematics and Statistics
Commonwealth Scientific and Industrial Research Organization
Canberra City, ACT 2601

In electromagnetic induction and diffusion, problems sometimes reduce to solving parabolic partial differential equations with unknown coefficients. The indeterminacy associated with the coefficients is matched by an overdeterminacy in the boundary conditions. Though theory gives, in terms of existence, uniqueness, smoothness and posedness, insight about what can and cannot be done, what is actually done when solving such problems computationally depends crucially on the type of information required and the form of the available data in which such information is contained. To illustrate, the problem of determining the electrical conductivity of the Earth from geomagnetic data is examined.

§1. INTRODUCTION

In many applications, what is required is not the solution of some forward problem such as the partial differential equation describing heat flow or diffusion, but the solution of some inverse problem such as the determination of an unknown coefficient of a partial differential equation with respect to overdetermined boundary conditions. Such problems arise in groundwater flow {cf. Hunt (1976)} and electrical conductivity modelling for the Earth {cf. Rikitake (1966)}.

Except formally at the level of existence, uniqueness, smoothness and posedness, such problems cannot be solved exactly.

To successfully solve them computationally, it is not only necessary to exploit their mathematical structure for the construction of algorithms, but also necessary to cope with the fact that the available data is observational. As a direct consequence, the type of information required and the form of the available data in which such information is contained predetermines to a major extent the available options for the construction of algorithms.

Though exact solutions are known, the conditions under which they hold are usually so specialized that they cannot be utilized for practical purposes.

In order to illustrate the dominant role played by the information required and the form of the available data in determining what is done computationally, we examine the problem of determining the electrical conductivity of the Earth from geomagnetic data. For a detailed discussion of the geophysical significance of electrical conductivity, the interested reader is referred to Hutton (1976).

The problem of electromagnetic induction in a radially symmetric sphere is formulated mathematically in §2 as a parabolic partial differential equation with an unknown coefficient (the electrical conductivity as a function of radius) and overdetermined boundary conditions involving the geomagnetic data. Though, mathematically, the formulation connecting the electrical conductivity with the geomagnetic data is non-linear, it has an underlying linear structure which can be exploited computationally. The consequences of this are examined in §3. The constraints which apply, depending on whether the required information is for mineral exploration purposes or global electrical conductivity modelling, are discussed in §4. Finally, in §5, we examine the rather radical approach which must be adopted for the global electrical conductivity modelling problem if one wishes to ensure that the modelling inserts no predetermined assumptions about the structure of electrical conductivity as a function of depth.

§2. ELECTROMAGENTIC INDUCTION IN A RADIALLY SYMMETRIC SPHERE

In the study of the origin, composition and evolution of the Earth, temperature and pressure along with chemical composition are the key parameters. They determine the depth at which phase changes should occur with respect to an assumed composition and thereby, in association with seismic data, constrain the various chemical configurations which could be used in modelling the Earth's interior.

Reasonably reliable estimates of pressure can be obtained from estimates of the density (cf. Bullen (1975)).

Except for heat flow measurements, which only yield information about the behaviour of temperature in the Earth's crust, the only direct constraint on it is electrical conductivity.

Electrical conductivity can be determined from observations of the Earth's transient magnetic field. The electric currents flowing in the upper atmosphere and ionosphere generate an external transient magnetic field. The changing magnetic lines of force of this external field induce electrical (eddy) currents within the Earth as it acts as an electrical conductor. This is simply a geophysical realization of Faraday's law that changing magnetic lines of force induce electrical currents in conductors. The induced internal currents in turn generate a corresponding internal magnetic field. The size and distribution of these currents, and hence the internal magnetic field, depend only on the external magnetic field and the electrical conductivity of the Earth. Thus, the interrelationship between the external and internally induced components of the Earth's transient magnetic field can be used to determine its electrical conductivity.

Before such interrelationships can be estimated, it is necessary to separate surface observations of the Earth's transient magnetic field $\underset{\sim}{H}$ into external and internally induced components. This is usually done on the assumption that the Earth can be modelled as a sphere of radius R with a spherically symmetric electrical conductivity. Under suitable assumptions, it can be shown that the transient field $\underset{\sim}{H}$ measured at the surface of the Earth can be expressed as the gradient of a scalar magnetic potential Ω which, in spherical polar coordinates, r, θ and φ, has the representation

$$\Omega(r,\theta,\varphi,t) = \sum_{n=1}^{\infty} \sum_{m=-n}^{n} F(r,t) S_n^m(\theta,\varphi) \ , \tag{2.1}$$

where $S_n^m(\theta,\varphi) = P_n^m(\cos\theta)\exp(im\varphi)$ is a surface harmonic and t denotes time. Under the assumption that no currents flow across the surface of the sphere, it follows from Gauss (1839) that $F(r,t)$ can be separated into components of external and internal origin; namely

$$F(r,t) = R\left\{e_n^m(t)\left(\frac{r}{R}\right)^n + i_n^m(t)\left(\frac{r}{R}\right)^{-n-1}\right\}$$

where $e_n^m(t)(r/R)^n$ and $i_n^m(t)(r/R)^{-n-1}$ denote the magnetic fields generated by electric currents flowing outside and inside the sphere, respectively.

It is a simple matter to show how the determination of the $e_n^m(t)$ and the corresponding $i_n^m(t)$ can be reduced to a problem in linear regression using observations of $\underset{\sim}{H}$. For example, a macroscopic examination of the Earth's transient magnetic field shows that, at least for magnetic disturbances with periods greater than 2 days and less than 1 year, the global magnetic variation can be modelled as a single zonal harmonic of order 1; namely,

$$P_1^0(\cos\theta) = \cos\theta \ ,$$

where θ defines colatitude. With respect to this harmonic, the H and Z components of the Earth's transient magnetic field $\underset{\sim}{H}$ at colatitude θ and time t can be written as

$$H = \left(\frac{1}{r}\frac{\partial\Omega}{\partial\theta}\right)_{r=R} = -\{e_1^0(t) + i_1^0(t)\}\sin\theta \ , \tag{2.2a}$$

and

$$Z = \left(\frac{\partial\Omega}{\partial r}\right)_{r=R} = \{e_1^0(t) - 2i_1^0(t)\}\cos\theta \ . \tag{2.2b}$$

Thus, with respect to a series of fixed times t, the solution of (2.2a) and (2.2b) for the corresponding $e_1^0(t)$ and $i_1^0(t)$ reduces for each time t to an elementary least squares problems with respect to the observations of H and Z recorded at various observatories about the Earth (cf. Anderssen (1969) and Anderssen and Seneta (1969)).

Maxwell's equations are the starting point for determining how the electrical conductivity $\kappa(r)$ determines the form of the $i_n^m(t)$ for given $e_n^m(t)$. Under the assumption that the Earth can be modelled as a sphere of radius R with a spherically symmetric electrical conductivity, Lahiri and Price (1939) have shown that no currents flow across its spherical surface (which verifies Gauss' (1839) assumption) and that $\kappa(\rho)$, $e_n^m(t)$ and $i_n^m(t)$ satisfies the following parabolic partial differential equation

$$\frac{\partial}{\partial\rho}\left(\rho^2 \frac{\partial f_n^m}{\partial\rho}\right) = n(n+1)f_n^m + 4\pi R^2\rho^2\kappa(\rho)\frac{\partial f_n^m}{\partial t} , \quad f_n^m = f_n^m(\rho,t) ,$$

$$0 \le \rho \le 1 , \quad t \ge 0 , \qquad (2.3)$$

with overdetermined surface boundary conditions

$$n(n+1)f_n^m(1,t) = ne_n^m(t) - (n+1)i_n^m(t) , \quad t \ge 0 , \qquad (2.4)$$

and

$$\left[\frac{\partial}{\partial\rho}(\rho f_n^m(\rho,t))\right]_{\rho=1} = e_n^m(t) + i_n^m(t) , \quad t \ge 0 , \qquad (2.5)$$

internal boundary condition

$$f_n^m(0,t) \equiv 0 , \quad t \ge 0 , \qquad (2.6)$$

and initial condition

$$f_n^m(\rho,0) \equiv 0 , \quad 0 \le \rho \le 1 , \qquad (2.7)$$

where $\rho = r/R$. The overdetermined surface boundary conditions (2.4) and (2.5) compensate for the unknown coefficient $\kappa(\rho)$ in the partial differential equation (2.3) and thereby yield a valid formulation for the determination of $\kappa(\rho)$. The internal boundary condition (2.6) reflects the fact that, for the Earth, the internally induced currents fail to penetrate to its centre. The initial condition (2.7) is based on the fact that any relationship between $e_n^m(t)$ and $i_n^m(t)$ must reflect the fact that $e_n^m(t)$ drives $i_n^m(t)$; i.e., is *causal* in that no effect can precede its cause.

Thus, with respect to given estimates of $e_n^m(t)$ and $i_n^m(t)$, the problem of modelling the electrical conductivity of the Earth is reduced to solving (2.3)-(2.7) for the unknown coefficient $\kappa(\rho)$.

§3. STRUCTURAL ASPECTS OF THE ELECTRICAL CONDUCTIVITY MODELLING PROBLEM

It is clear from (2.3)-(2.7) that the relationship between $\kappa(\rho)$ and the components $e_n^m(t)$ and $i_n^m(t)$ of the magnetic field is non-linear. Though various formulations for this relationship have yielded useful insight about its theoretical structure (cf. Bailey (1970) and Weidelt (1972)), they have not been so useful computationally. In fact, it is possible to avoid the non-linearity when solving for $\kappa(\rho)$ computationally. This hinges on the fact that, with respect to a given $\kappa(\rho)$, the components $e_n^m(t)$ and $i_n^m(t)$ are linearly related.

Thus, the non-linearity can be avoided in any computational procedure for which the forward problem reduces to solving (2.3)-(2.7) for $i_n^m(t)$ with respect to given values for $\kappa(\rho)$ and $e_n^m(t)$. In fact, from the point of view of the underlying physical process involved, this is the most natural way to formulate the forward problem, since $e_n^m(t)$ drives the electromagnetic induction process with respect to the Earth's $\kappa(\rho)$. In addition, it is also the most natural way to formulate the forward problem computationally since, in almost any method which one is likely to use, the essential mechanism will be one of calculating, for a given estimate of $\kappa(\rho)$, the form of $i_n^m(t)$ which corresponds to a given $e_n^m(t)$. A comparison of the calculated $i_n^m(t)$ with the observed is then used as the basis for correcting, updating or accepting the current estimate of $\kappa(\rho)$.

Two classes of methods which explicitly exploit the linear relationship between $e_n^m(t)$ and $i_n^m(t)$ are direct parametric modelling and transfer function modelling.

3.1 Direct Parametric Modelling

On physical grounds, the obvious approach for solving (2.3)-(2.7) is *direct parametric modelling* which can be formulated as follows :

Step 1. Choose a parametric model for $\kappa(\rho)$; viz.

$$\kappa(\rho) = \kappa(\rho;\alpha,\beta,\ldots,\gamma) .$$

Step 2. With respect to this model and a given form for $e_n^m(t)$ ($e(t)$, say), solve (2.7) for

$$f_n^m(\rho,t) = f(\rho,t;\alpha,\beta,\ldots,\gamma;e(t))$$

with respect to the following initial and internal boundary conditions

$$f_n^m(\rho,0) \equiv 0 , \quad 0 < \rho < 1 ,$$

$$f_n^m(0,t) \equiv 0 , \quad t \geq 0 ,$$

and the surface boundary condition which depends only on $e(t)$

$$n(n+1)f_n^m(1,t) + (n+1)\left[\frac{\partial}{\partial\rho}(\rho f_n^m(\rho,t))\right]_{\rho=1} = (2n+1)e(t)$$

Step 3. Use (2.4) to determine

$$i(t;\alpha,\beta,\ldots,\gamma;e(t)) = \frac{n}{n+1}e(t) - nf(1,t;\alpha,\beta,\ldots,\gamma;e(t)) .$$

Step 4. With respect to an appropriately chosen fitting criterion, such as least squares, adjust the parameters of $\kappa(\rho;\alpha,\beta,\ldots,\gamma)$ so that a closer agreement between $i(t;\alpha,\beta,\ldots,\gamma;e(t))$ and $i(t) = i_n^m(t)$ results.

The dependence of the interrelationship between $e(t)$ and $i(t;\alpha,\beta,\ldots,\gamma;e(t))$ on the choice of $\kappa(\rho;\alpha,\beta,\ldots,\gamma)$ can now be examined.

Up to the early sixties, when the possibility of using computational techniques were first exploited, all attempts to estimate the electrical conductivity of the Earth were based on this strategy.

Chapman and Price (1930) applied this approach under the assumption that $e(t)$ and $i(t)$ were periodic and that

$$\kappa(\rho) = \kappa(\rho;d,\bar{\kappa})$$

could be modelled as an insulator of unknown depth d surrounding a conductor of unknown constant conductivity $\bar{\kappa}$. In essence, it anticipates Fourier series methods, but was not greatly successful for three key reasons :

(i) The lack of access to the modern theory on the Fourier analysis of time series which is needed to fully exploit such an approach;

(ii) A lack of awareness of (as well as facilities to process) the huge amount of data required to yield reliable estimates of electrical conductivity at depths greater than a few hundred kilometres;

(iii) The use of a two-parameter model which does not fully reflect the underlying physics of the electromagnetic induction process.

We shall return to these points below.

Lahiri and Price (1939) also applied this approach to derive explicit solutions. In their case, they assumed that $e(t)$ and $i(t)$ were aperiodic and

$$\kappa(\rho) = \kappa(\rho;q,\bar{\kappa},m) = \begin{cases} 0 , & q \le \rho \le 1 , \\ \bar{\kappa}\rho^{-m} , & 0 \le \rho \le q . \end{cases}$$

It was clear from this work that the number of conductivity models for which explicit analytic solutions could be constructed would be quite limited, and hence that conductivity modelling for the Earth would have to be solved computationally. Many of the early methods developed for computers were based on direct parametric modelling (cf. Anderssen (1968)).

3.2 Transfer Function Modelling

Though a formal solution for the electrical conductivity modelling problem is not known in the time domain, it is possible to construct one in the Fourier domain (cf. Eckhardt (1968) and Bailey (1970)). However, as we shall discuss in subsequent sections, its implementation poses certain constraints on the form of the geomagnetic data that can be used. Thus, it represents an alternative to direct parametric modelling. The latter can be applied in situations where the Fourier domain approach is not applicable.

The essence of the idea behind the transfer function modelling approach dates back to Chapman and Price (1930), but, as we shall see below, requires access to quite deep mathematical results about transfer functions which were not known in 1930.

The surface boundary conditions (2.4) and (2.5) are special cases of the following more general boundary conditions

$$\frac{n(n+1)}{\rho} f_n^m(\rho,t) = n e_n^m(\rho,t) - (n+1) i_n^m(\rho,t) \ , \tag{3.1}$$

and

$$\frac{1}{\rho}\frac{\partial}{\partial\rho}(\rho f_n^m(\rho,t)) = e_n^m(\rho,t) + i_n^m(\rho,t) \ , \tag{3.2}$$

which hold on all spheres of radius ρR , $0 < \rho < 1$, concentric with the original sphere. Clearly, $e_n^m(t) = e_n^m(1,t)$ and $i_n^m(t) = i_n^m(1,t)$. They prove that an external field in a given surface harmonic excites only an internal field with the same surface harmonic structure. That is, for a given conductivity, $\kappa(\rho)$, $i_n^m(\rho,t)$, is a function of only $e_n^m(\rho,t)$, which proves the tacit assumption of this fact in the discussion of §2. In addition, the linearity of Maxwell's equations, and hence (2.3)-(2.7) for fixed $\kappa(\rho)$, implies that $i_n^m(\rho,t)$ is linearly related to $e_n^m(\rho,t)$; namely ,

$$i_n^m(\rho,t) = \int_{-\infty}^{\infty} k_n^m(\rho,t) e_n^m(\rho,t-\tau)\, d\tau \tag{3.3}$$

where $k_n^m(\rho,\tau)$, *the impulse response of the Earth* in the identified spherical harmonic mode, satisfies the *principal of causality*; namely,

$$k_n^m(\rho,\tau) = 0 \ , \quad \tau < 0 \ . \tag{3.4}$$

Introducing the Fourier transform

$$\hat{g}(\omega) = \int_{-\infty}^{\infty} g(t)\exp(i\omega t)\, dt \ ,$$

(3.3) becomes

$$\hat{i}_n^m(\rho,\omega) = \hat{k}_n^m(\rho,\omega)\hat{e}_n^m(\rho,\omega) \ , \tag{3.5}$$

where $\hat{k}_n^m(\rho,\omega)$ is referred to as *the transfer function* or *geomagnetic response of the Earth* in the identified spherical harmonic mode. Taking Fourier transforms of (3.1), (3.2) and (2.3) and using (3.5), Eckhardt (1963), (1968) showed that the non-linear relation, which defines the dependence of $\kappa(\rho)$ on the interrelationship between $e_n^m(\rho,t)$ and $i_n^m(\rho,t)$ (viz. $\hat{k}_n^m$) , takes the form

$$\frac{\partial \hat{k}_n^m}{\partial\rho} = -\frac{4\pi\kappa(\rho)\, i\omega R^2 \rho(n+1)}{n(2n+1)}\left[\hat{k}_n^m - \frac{n}{n+1}\right]^2 - \frac{2n+1}{\rho} k_n^m \ . \tag{3.6}$$

One of the indirect advantages of this result is that it can be used to study the effect of linearizing the relationship between $\kappa(\rho)$ and the (observed) magnetic components $e_n^m(t)$ and $i_n^m(t)$. A number of authors (cf. Parker (1971)) have proposed linearization as the basis for solving the electrical conductivity modelling problem; not because it was naturally suited to this problem, but because it had proved useful in the solution of other unrelated geophysical problems. As we have already indicated and shall see below, linearization does not represent an alternative to direct parametric minimization and transfer function modelling (cf. Anderssen (1979)), as it is unable to exploit in a natural way the underlying linear structure already discussed and exploited in some detail.

Introducing the transformations

$$\tau = \tau(n) = \frac{n}{n+1}\rho^{2n+1} ,$$

$$S = S(n) = 4\left[\frac{(n+1)R}{n(2n+1)}\right]^2 \rho^{-4n}\kappa(\rho) ,$$

and

$$\varphi_n^m = \varphi_n^m(\rho,\omega) = \frac{(n+1)\tau}{n}\left[\hat{k}_n^m - \frac{n}{n+1}\right] ,$$

(3.6) becomes

$$1 + \frac{\partial}{\partial\tau}(\varphi_n^m(\tau,\omega)) = - i\pi\omega S(\tau)\,[\varphi_n^m(\tau,\omega)]^2 . \qquad (3.7)$$

Exploiting properties of causal response functions, Bailey (1970) showed that (3.7) could be formally solved to yield

$$\kappa(\rho) = \frac{1}{4}\left[\frac{(2n+1)n}{(n+1)R}\right]^2 \rho^{4n}\left[2\int_0^\infty \mathrm{Re}\{(\varphi_n^m)^2\}d\omega\right]^{-1} , \qquad (3.8)$$

with

$$\frac{\partial}{\partial\tau}(\varphi_n^m) = -\left\{1+i\pi(\varphi_n^m)^2\left[2\int_0^\infty \mathrm{Re}\{(\varphi_n^m)^2\}d\omega\right]^{-1}\right\}, \qquad (3.9)$$

This indicates that one method of solution is to choose data for $e_n^m(t)$ and $i_n^m(t)$ of a form which will allow an estimate of $\hat{k}_n^m(1,\omega)$ to be derived which is suitable for implementing a method based on (3.8)-(3.9). We shall return to this point in §4.

§4. THE DATA AND POSSIBLE INFERENCES

The available data consists of various types of three-component observations of the transient magnetic field vector $\underset{\sim}{H}$ at numerous observatories around the Earth. In one form or another, these three-component observatories contain the required observations of H and Z of (2.2a) and (2.2b). Though there is therefore no major difficulty associated with obtaining observations of H and Z from published observatory data for all observatories active during some specific period of time, the actual format of the data (record length and data rate) to be analysed has to be decided on the basis of the inferences required and their interrelationship with the data.

There are two clearly distinct types of inference required :

a. information about local conductivity structure for mineral exploration purposes;

and

b. the global structure of the Earth's electrical conductivity.

In the former, the problem is to determine from given data the geometry of some local underground conductor. The major investment is therefore in the construction of an algorithm which will yield the desired resolution if suitable data are available. Since the depths of interest will be less than five kilometers, the optimal data will be those collected from an array of instruments simultaneously recording some strongly defined geomagnetic disturbance such as a magnetic storm. Because it is only necessary to record such disturbances over short periods of time, the resulting records (time series) will not be ideal candidates for Fourier analysis. Time domain methods based on direct parametric minimization are to be preferred, but this type of problem will not be pursued here.

In the latter, what is required is electrical conductivity as a function of depth as it represents a possibly independent constraint on temperature and chemical composition. Because it is known that electrical conductivity increases with depth, it is necessary to analyse long rather than short lengths of geomagnetic records (a point overlooked, for one reason or another, by previous authors). However, in order to work efficiently with and obtain the required information from such long records, it will be necessary to work in the frequency, not the time, domain, even though the observations are made in the latter.

In order to derive good frequency domain estimates, it is necessary to ensure that :

(i) the data is collected on an even time grid;

(ii) the data appear to be a realization of a stationary stochastic process (a condition which is satisfied by long records of geomagnetic data);

and

(iii) the sampling rate is sufficiently high to ensure that difficulties associated with spectral folding have been eliminated.

Fortunately, such data are available in a readily accessible form. In the remainder of this paper, we shall concentrate on this problem.

The types of inferences which could be sought from such data are : (a) an estimate of the depth of penetration of geomagnetic data as a function of period, in order to establish some quantitative estimate of the depth at which geomagnetic data of a given period can be expected to sample; (b) a parsimonous model of conductivity as a function of depth to see what structure it yields with respect to a minimum of assumptions; and (c) from the point of view of a comparison with information derived about Earth structure from seismological data, confirmation as to whether or not electrical conductivity changes discontinuously at some specific depth.

In order to obtain such information, it is necessary to implement some form of the transfer function modelling approach discussed above. However, the requirement that the resulting model be parsimonious and depend on a minimum of initial assumptions necessitates the use of a modelling strategy which decouples the information in the transfer function at one frequency from that in other frequencies. What is therefore required is a procedure for relating individual frequency estimates of the transfer function, derived from the H and Z data, directly to estimates of depth and conductivity. This can be done using a procedure proposed by Schmucker (1970).

§5. TRANSFER FUNCTION MODELLING USING SCHMUCKER'S TWO-PARAMETER MODELLING

As we have already seen, transfer function modelling is based on the assumption that the given output $y(t)$ is a result of having applied a linear filter to the given input $x(t)$. An operator $\underset{\sim}{L}$ is said to be a *linear filter*, if it satisfies

$$\underset{\sim}{L}\exp(i\omega t) = \ell(\omega)\exp(i\omega t)\ ,\quad \ell(\omega) = A(\omega)\exp(i\theta(\omega))\ , \tag{5.1}$$

where $A(\omega)$ and $\theta(\omega)$ are referred to as the *amplitude* and *phase* of the linear filter at frequency ω. For example, differentiation can be interpreted as a linear filter with $\ell(\omega) = i\omega$. The essence of transfer function modelling thereby reduces to determining the amplitude and phase of $\underset{\sim}{L}$ which is assumed to transform the given input $x(t)$ into the given output $y(t)$; namely,

$$\begin{aligned} y(t) &= \sum b_j \exp(i(\omega_j t+\psi_j)) \\ &= \underset{\sim}{L}\, x(t) = \underset{\sim}{L} \sum a_j \exp(i(\omega_j t+\varphi_j)) \\ &= \sum a_j A_j \exp(i(\omega_j t+\varphi_j+\theta_j)) \end{aligned} \tag{5.2}$$

which implies that

$$b_j = a_j A_j\ ,\quad \psi_j = \varphi_j + \theta_j\ ,\quad j = 1,2,\ldots, \tag{5.3}$$

where the notation $A_j = A(\omega_j)$ and $\theta_j = \theta(\omega_j)$ has been used, and the existence of the above Fourier series expansions for $y(t)$ and $x(t)$ is assumed. It therefore follows that the required amplitude and phase at frequency ω_j are given by

$$A_j = b_j/a_j\ ,\quad \theta_j = \omega_j - \varphi_j\ ,\quad j = 1,2,\ldots\ . \tag{5.4}$$

This transfer function model is often referred to as *the geomagnetic response of the Earth*.

However, such an analysis can only be justified if

(i) the tools exist for determining estimates of the amplitude and phase from given observational data; and

(ii) the resulting estimates of amplitude and phase have a direct geophysical interpretation.

For the former, one has the methods of time series analysis which are discussed in Anderssen, Gustafson and Winch (1979) and Anderssen and Gustafson (1981). If the observations of H are taken as the input and the observations of Z as the output, then the latter is achieved by using Schmucker's two-parameter modelling strategy which is based on determining from one frequency estimate of amplitude and phase the depth and conductivity at the mean depth of the internally induced current system associated with that frequency. In fact, on writing

$$A_j \exp(i\theta_j) = a(\omega_j) - ib(\omega_j) \tag{5.5}$$

the depth $z^*(\omega_j)$ and conductivity $\sigma^*(\omega_j)$ of Schmucker's *perfect substitute conductor* are given by (cf. Schmucker (1970), p.68)

$$z^*(\omega_j) = 3185.5 \times a(\omega_j)$$

$$\sigma^*(\omega_j) = T/\{4.45\times 10^4 \times b^2(\omega_j)\}\ , \quad T = 1/\omega_j \text{ (hours)} \tag{5.7}$$

Because of its simple nature, this method will yield conductivity-depth estimates which are decoupled from each other. It was found by Anderssen, Devane, Gustafson and Winch (1979) that if not enough smoothing had been applied when estimating A_j and θ_j, the resulting conductivity-depth profile will be multi-valued; and, further, that the degree of multi-valuedness is decreased as the level of smoothing is increased. It is therefore natural to use the single-valuedness of the conductivity-depth profile as the context-dependent criterion for determining the level of smoothing to be applied in the Fourier analysis of the data.

In using such a method to analyse the two years of hourly values for 1964 and 1965 for 17 observatories about the Earth, Anderssen, Devane, Gustafson and Winch (1979) obtained unambiguous confirmation that a major discontinuous change in electrical conductivity occurs at a depth of about 700 km. It would be difficult to obtain such confirmation from standard methods as it would be necessary to allow for the presence of the discontinuity in the actual modelling.

REFERENCES

[1] Anderssen, R.S. (1968) Note on conductivity models for the Earth, *J. Geophys. Res.* 73 (1968), 6535-6543.

[2] Anderssen, R.S. (1969) On the solution of certain overdetermined systems of linear equations which arise in geophysics, *J. Geophys. Res.* 74 (1969), 1045-1053.

[3] Anderssen, R.S. (1979) The one-dimensional electromagnetic induction problem: model refinement, *Geophys. J. R. astr. Soc.* 57 (1979), 67-78.

[4] Anderssen, R.S., Devane, J.R., Gustafson, S-Å. and Winch, D.E. (1979), The qualitative character of the global electrical conductivity of the Earth, *Phys. Earth and Planet. Int.* 20 (1979), 15-21.

[5] Anderssen, R.S., Gustafson, S-Å. and Winch, D.E. (1979), Estimating the phase of the response of the Earth to long-period geomagnetic fluctuations, *Earth Planet. Sci. Letters* 44 (1979), 1-6.

[6] Anderssen, R.S. and Gustafson, S-Å. (1981) Estimating the amplitude and phase of the geomagnetic response of the Earth (in preparation).

[7] Anderssen, R.S. and Seneta, E. (1969) A new analysis for the geomagnetic D_{st} field of the magnetic storm on June 18-19, 1936, *J. Geophys. Res.* 74 (1969), 2768-2773.

[8] Bailey, R.C. (1970) Inversion of the geomagnetic induction problem, *Proc. Roy. Soc. London* 315 (1970), 185-194.

[9] Bullen, K.E. (1975) *The Earth's Density*, Chapman and Hall, 1975.

[10] Chapman, S. and Price, A.T. (1930) The electric and magnetic state of the interior of the Earth, as inferred from terrestrial magnetic variations, *Phil. Trans. Roy. Soc. A* 229 (1930), 427-460.

[11] Eckhardt, D.H. (1963) Geomagnetic induction in a concentrically stratified earth, *J. Geophys. Res.* 68 (1963), 6273-6278.

[12] Eckhardt, D.H. (1968) Theory and interpretation of the electromagnetic impedance of the Earth, *J. Geophys. Res.* 73 (1968), 5317-5326.

[13] Gauss, K.F. (1839) Allgemeine Theorie des Erdmagnetismus. *Resultate aus dein Beobachtungen des magnetischen Vereins in Jahre* 1838, Leipzig. (Reprinted in *Werke* (1877) 5, 119-193).

[14] Hunt, B. (1976) An aquifer simulation study in northern Canterbury, *New Zealand J. Sci.* 19 (1976), 265-276.

[15] Hutton, V.R.S. (1976) The electrical conductivity of the Earth and planets, *Rep. Prog. Phys.* 39 (1976), 487-572.

[16] Lahiri, B.N. and Price, A.T. (1939) Electromagnetic induction in non-uniform conductors, and the determination of the conductivity of the Earth from terrestrial magnetic variations, *Phil. Trans. Roy. Soc. London* 237 (1939), 509-540.

[17] Parker, E.L. (1971) The inverse problem of the electrical conductivity in the mantle, *Geophys. J.R. astr. Soc.* 22 (1971), 121-138.

[18] Rikitake, T. (1966) *Electromagnetism and the Earth's Interior*, Elsevier Publishing Co., Amsterdam-London-New York, 1966, pp. 308.

[19] Schmucker, U. (1970) Anomalies of geomagnetic variations in the south-western United States, *Bull. Scripps. Inst. Oceanog.* 13, University of California Press, 1970, pp.165.

[20] Weidelt, P. (1972) The inverse problem of geomagnetic induction, *Z. für Geophysik* 38 (1972), 257-287.

NUMERICAL SOLUTIONS OF
PARTIAL DIFFERENTIAL EQUATIONS
J. Noye (editor)

SOLUTION OF NEUTRON DIFFUSION LINEAR EQUATIONS BY THE METHOD OF IMPLICIT NON-STATIONARY ITERATION

J.M. Barry and J.P. Pollard

Australian Atomic Energy Commission,
Research Establishment,
Private Mail Bag, Sutherland, 2232 NSW
AUSTRALIA

The use of iterative techniques to solve the large order ($\approx 10^5$ unknowns) system of linear equations arising in the solution of multigroup three-dimensional neutron diffusion equations is considered. The main emphasis is on the method of implicit non-stationary iteration (MINI), a scheme that strengthens the implicit nature of the process with different (changing) accelerating parameters holding for each unknown. MINI compares favourably with conjugate gradient and relaxation schemes for solving each symmetric energy group equation. MINI proves to be an excellent means of accelerating the non-symmetric overall coupled energy group equations.

1 INTRODUCTION

The detailed analysis of the steady state of a nuclear reactor is an important part of the assessment of suitability of reactor design and an important first step in the study of reactor safety. Mockup designs are possible, but they tend to be very expensive and time consuming in their construction. Rather, an indirect approach via a mathematical model of the reactor and its neutron interactions is preferred. The effect of extensive variation of basic design parameters (spacing of fuel elements, extent of moderator, etc.) can then be carried out simply by changing the basic data for the model. Because the model is not amenable to analytic solutions, except in rather special circumstances, a numerical approach is usually applied on a digital computer. Here the main concern is with a steady state multigroup neutron diffusion approximation to the more accurate Boltzmann transport equation (Stacey, 1969) for problems in 1, 2 and 3 dimensions with (x,y,z) Cartesian coordinates. A finite difference discretization of the diffusion operator is used which leads to a matrix eigenvalue problem with a sparse matrix. Particular attention is then given to the solution of associated linear equations of high order (perhaps as high as 10^5) in an iterative manner in an attempt to keep down the computation time.

2 THE DIFFUSION EQUATION

The steady state multigroup neutron diffusion equation appropriate to this study is

$$\left(-\nabla . D_g \nabla + \sigma_{remg}\right) \phi_g = \sum_{g'=1}^{G} \left(\sigma_{sg'\to g} + \frac{\nu}{k}\, \sigma_{fg'}\, \chi_g\right) \phi_{g'} \;, \; g = 1,2 \ldots, G, \qquad (2.1)$$

i.e. leakage + removals = scatters + fission,

where g denotes different neutron energy groups (G in all) for the energy range 10^7 eV to 10^{-3} eV, with 'lumped' or group properties being used to reduce the problem (neutron interaction properties can vary enormously with the energy of the initiating neutron);

ϕ_g denotes spatially dependent neutron flux (neutron density times velocity), for group g neutrons and is the required quantity;

D_g is the group g diffusion coefficient for the material of the reactor at a particular point in space (and like most material properties is constant for one material but varies in a discontinuous fashion across interfaces between materials constituting the reactor);

σ_{remg} is the macroscopic removal cross section;

$\sigma_{sg'\to g}$ is the scattering transfer cross section for group g' neutrons being transferred to group g but with

$$\sigma_{sg\to g} = 0 \quad ;$$

$\sigma_{fg'}$ is the fission cross section;

ν is the number of neutrons emitted per fission;

χ_g is the proportion of all neutrons that are emitted by fission group g (the fission spectrum); and

k is the largest magnitude eigenvalue of the problem corresponding to the eigenfunction ϕ_g (the so-called multiplication).

Associated with equation (2.1) are the following boundary conditions

(i) External boundaries

$$D_g \frac{\partial \phi_g}{\partial n} + c\phi_g = 0 \quad \text{for each group } g, \qquad (2.2)$$

where n = outward normal,

and $c = \begin{cases} 0.47 \text{ for an actual outer boundary (usually),} \\ 0 \text{ for a symmetric boundary.} \end{cases}$

(ii) Internal boundaries separating different materials L and R

$$\phi_g\Big|_L = \phi_g\Big|_R \qquad (2.3)$$

and

$$n.D_g\nabla\phi_g\Big|_L = n.D_g\nabla\phi_g\Big|_R \quad , \qquad (2.4)$$

i.e. continuity of flux and current across the boundaries.

In addition to 'lumping' of the material properties into neutron energy groups, a certain amount of 'smearing' of the properties in space is also used, otherwise each nut, bolt and washer would need to be explicitly represented (see, for example, Pollard, 1974). In a practical calculation, as few as (G=) 4 energy groups may suffice for as few as 10 different materials.

3 A ONE-DIMENSIONAL PROBLEM

To convey the essentials of the numerical techniques to be used to solve realistic reactor problems, we will introduce a simple one-dimensional model problem. This problem will expose the reader to some of the ideas employed in more realistic studies but the cumbersome notation normally required will be minimized.

The 1D problem is for a homogeneous (one material) 2-group model with three internal solution points as illustrated in figure 3.1.

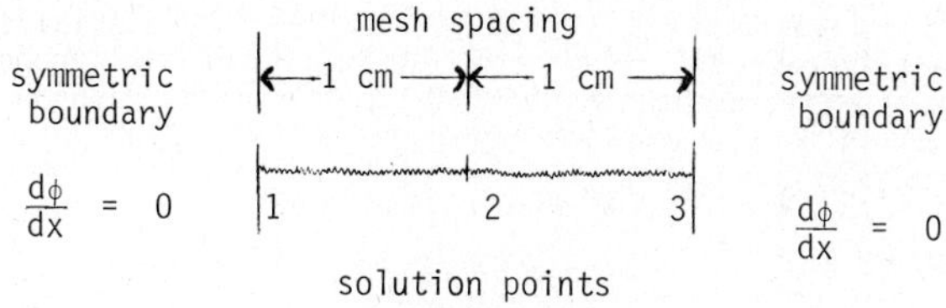

Figure 3.1
1D Layout

The material properties for this problem are given in table 3.1.

Table 3.1
Material properties

	g = 1	g = 2	units[1]
D	1	1	cm
σ_{rem}	1	2	cm^{-1}
$\sigma_{S1\to}$	0	1	cm^{-1}
$\sigma_{S2\to}$	3/4	0	cm^{-1}
$\nu\sigma_f$	0	5/4	cm^{-1}
χ	1	0	—

Clearly, for the model problem, the solution of equation (2.1) is

$$\begin{aligned} k &= 1, \\ \phi_1 &= \text{constant and} \\ \phi_2 &= \phi_1/2. \end{aligned}$$

It will be interesting to see how iterative methods approach this solution given non-constant trial solutions. As it turns out, iterative methods are unnecessary as the equations obtained from a finite difference approximation for this problem can be solved directly. However, in general, for 2 or 3D this is not so and therefore we continue with an iterative approach.

Using the notation

$$\phi_{i+3(g-1)} \quad = \quad \text{flux at point } i \text{ for group } g,$$

a finite difference approximation for the leakage operator gives, for example, about point 2 for group 1 flux:

$$-D_1 \frac{d^2\phi_1}{dx^2} \quad \simeq \quad -\phi_1 + 2\phi_2 - \phi_3$$

so in matrix notation the equation to be solved (with scattering terms of equation (2.1) moved to the LHS of the equation) is

$$\left[\begin{array}{ccc|ccc} 3/2 & -1 & 0 & -3/8 & 0 & 0 \\ -1 & 3 & -1 & 0 & -3/4 & 0 \\ 0 & -1 & 3/2 & 0 & 0 & -3/8 \\ \hline -1/2 & 0 & 0 & 2 & -1 & 0 \\ 0 & -1 & 0 & -1 & 4 & -1 \\ 0 & 0 & -1/2 & 0 & -1 & 2 \end{array}\right]\left[\begin{array}{c}\phi_1\\ \phi_2\\ \phi_3\\ \hline \phi_4\\ \phi_5\\ \phi_6\end{array}\right] = \frac{1}{k}\left[\begin{array}{ccc|ccc} 0 & 0 & 0 & 5/8 & 0 & 0 \\ 0 & 0 & 0 & 0 & 5/4 & 0 \\ 0 & 0 & 0 & 0 & 0 & 5/8 \\ \hline 0 & 0 & 0 & 0 & 0 & 0 \\ 0 & 0 & 0 & 0 & 0 & 0 \\ 0 & 0 & 0 & 0 & 0 & 0 \end{array}\right]\left[\begin{array}{c}\phi_1\\ \phi_2\\ \phi_3\\ \hline \phi_4\\ \phi_5\\ \phi_6\end{array}\right]$$

or
$$\begin{bmatrix} A_{11} & -A_{12} \\ -A_{21} & A_{22} \end{bmatrix}\begin{bmatrix} \Phi_1 \\ \Phi_2 \end{bmatrix} = \frac{1}{k}\begin{bmatrix} F_{11} & F_{12} \\ F_{21} & F_{22} \end{bmatrix}\begin{bmatrix} \Phi_1 \\ \Phi_2 \end{bmatrix} . \qquad (3.1)$$

Note that the complete matrix A has elements which satisfy

$$a_{ij} \leq 0,\ i \neq j,\ i = 1,2,\ldots,6;\quad j = 1,2,\ldots,6,$$

$$a_{ii} \geq -\sum_{j \neq i} a_{ji} \text{ (with inequality for at least one value of i)}$$

and is non-symmetric, although individual group partitions A_{11}, A_{12}, A_{21} and A_{22} are themselves symmetric. (For the more general 2 or 3D problem, the group partitions are irreducible, symmetric, positive definite and block tridiagonal.) As a consequence the solution fluxes are all positive.

Three iterative schemes are used to solve equation (3.1).

(i) An outer scheme which estimates the RHS using previous estimates of the flux and a Chebyshev extrapolation process (Hageman and Pfeifer, 1965) will not be considered further. For our model problem we will take the RHS as

$$S = (5/16, 5/8, 5/16, 0, 0, 0)^T \quad \text{(where T = transpose)} \qquad (3.2)$$

and concentrate on the other iterative schemes.

(ii) A group scheme which uses an estimate of Φ_2 in the calculation of Φ_1 - the so-called upscatter problem.

(iii) An inner scheme which solves for the flux along a line for a particular group - i.e. we estimate Φ_1 first then Φ_2.

To be definitive we will take a trial estimate of the flux as

$$(10^{-6}, 10^{6}, 10^{-6}, 10^{6}, 10^{-6}, 10^{6})^T \quad . \qquad (3.3)$$

3.1 Group Iteration

In its simplest form the group iteration consists of a Gauss-Seidel (GS) iteration

$$A_{11}\Phi_1^{(n)} = S_1 + A_{12}\Phi_2^{(n-1)} \qquad (3.4)$$

and
$$A_{22}\Phi_2^{(n)} = S_2 + A_{21}\Phi_1^{(n)} \qquad (3.5)$$

where n denotes the iteration count and S_1 and S_2 are partitions of the source given by equation (3.2). Here we will introduce the method of implicit non-stationary iteration (MINI - Barry and Pollard, 1977, 1979) appropriate to the group iteration. Instead of $\Phi_2^{(n-1)}$ on the RHS of equation (3.4), an extrapolated form is introduced which attempts to give a better estimate of the required quantity $\Phi_2^{(n)}$ than $\Phi_2^{(n-1)}$:

$$\Phi_2^{(n)} \simeq \Phi_2^{(n-1)} + \Gamma^{(n-1)}\left(\Phi_1^{(n)} - \Phi_1^{(n-1)}\right) \quad , \tag{3.6}$$

where $\Gamma^{(n-1)}$ is a diagonal matrix of positive terms $\underline{\gamma}_1^{(n-1)}$, $\underline{\gamma}_2^{(n-1)}$, and $\underline{\gamma}_3^{(n-1)}$; equation (3.4) then becomes

$$\left(A_{11} - A_{12}\,\Gamma^{(n-1)}\right)\Phi_1^{(n)} = S_1 + A_{12}\,\Phi_2^{(n-1)} - A_{12}\,\Gamma^{(n-1)}\,\Phi_1^{(n-1)} \quad . \tag{3.7}$$

Since A_{12} is also a diagonal matrix, the MINI difference to the basic GS scheme is small in terms of computational strategy but, as it turns out, big in terms of effectiveness of the solution process. However, we note the effective reduction that occurs to the diagonal terms of A_{11} which, it is anticipated , will increase the difficulty of the inner solution strategy. The increase turns out to be tolerable.

Barry and Pollard (ibid.) derive the rule for the calculation of the group extrapolating parameters $\underline{\gamma}_1^{(n-1)}$, etc. The appropriate form here is

$$\underline{\gamma}_i^{(n)} = \min\left\{|\delta_{i+3}^{(n)}|/|\delta_i^{(n)}|,\ |\delta_i^{(n)}|/|\delta_{i+3}^{(n)}|,\ \phi_{i+3}^{(n)}/\phi_i^{(n)}\right\} \quad , \tag{3.8}$$

$$i = 1,2,3,$$

where $\delta_i^{(n)} = \phi_i^{(n)} - \phi_i^{(n-1)}$, the iteration increment.

We note that from equation (3.8)

$$|\underline{\gamma}_i^{(n-1)}| \leqslant 1 \quad , i = 1,2,3, \tag{3.9}$$

which is a sufficient condition for convergence of the MINI scheme. For some problems as n increases, an asymptotically constant set of $\underline{\gamma}$s is obtained and this leads immediately to the solution.

On account of the non-symmetric nature of the overall matrix A, other methods discussed in the next section do not lend themselves to solving the upscatter problem. Traditionally, the GS method has been used, and is sometimes satisfactory, as the extent of upscatter for some problems is small. However, for fissile source problems or kinetic calculations (see Pollard 1974) an effective upscatter component can be quite large; an alternative approach such as MINI is then almost essential.

Table 3.2 shows the effectiveness of MINI compared with GS for the model problem in which the inner solution is solved directly. (In the next section iterative inner solution procedures are considered in relation to the procedures used for 2 and 3D.) The convergence criteria are taken to be

$$\max\left\{|\delta_1^{(n)}|/\phi_1^{(n-1)},\ |\delta_2^{(n)}|/\phi_2^{(n-1)}, \ldots,\ |\delta_6^{(n)}|/\phi_6^{(n-1)}\right\} \leqslant \varepsilon \quad , \tag{3.10}$$

$$\varepsilon = \text{error termination level, say } 10^{-4}.$$

We note that in general MINI uses only about a third more computation per iteration than GS, so a direct comparison of the number of iterations is reasonable.

Table 3.2

Number of iterations to converge the upscatter problem

GS	25
MINI	10

asymptotic $\underline{\gamma} \simeq (\frac{1}{2}, \frac{1}{2}, \frac{1}{2})^T$

3.2 Inner Iteration

The iterative solution of equations (3.4) and (3.5) may both be represented by the form

$$A\phi = S \quad , \tag{3.11}$$

where

$$A = \begin{bmatrix} a_1 & -1 & 0 \\ -1 & a_2 & -1 \\ 0 & -1 & a_3 \end{bmatrix} . \tag{3.12}$$

Again we could use a GS approach so that

$$\phi_i^{(n)} = \left(S_i + \phi_{i-1}^{(n)} + \phi_{i+1}^{(n-1)}\right)/a_i \quad , \quad i = 1,2,3 \tag{3.13}$$

where $\phi_0^{(n)}$ and $\phi_4^{(n)}$ are taken as zero. This equation could be written in matrix form as

$$A_L\phi^{(n)} = S + A_U\phi^{(n-1)} \quad , \tag{3.14}$$

with

$$A_L = \begin{bmatrix} a_1 & 0 & 0 \\ -1 & a_2 & 0 \\ 0 & -1 & a_3 \end{bmatrix} \quad \text{and } A_U = \begin{bmatrix} 0 & 1 & 0 \\ 0 & 0 & 1 \\ 0 & 0 & 0 \end{bmatrix} \tag{3.15}$$

and then

$$\delta^{(n)} = T\,\delta^{(n-1)} \quad , \tag{3.16}$$

where

$$\delta^{(n)} = \phi^{(n)} - \phi^{(n-1)} \quad \text{(the iteration increments)} \tag{3.17}$$

and

$$T = A_L^{-1} A_U = \begin{bmatrix} 0 & 1/a_1 & 0 \\ 0 & 1/a_1a_2 & 1/a_2 \\ 0 & 1/a_1a_2a_3 & 1/a_2a_3 \end{bmatrix} \tag{3.18}$$

is the so-called iteration matrix for the GS method. Convergence is determined by the largest magnitude eigenvalue of S, namely

$$\lambda_{GS} = \left(1/a_1 + 1/a_3\right)/a_2 \qquad (< 1) \quad . \tag{3.19}$$

The closer λ_{GS} is to one, the slower is the convergence (Varga, 1962).

For the group 2 partition the value is, simply, $\lambda_{GS2} = 1/4$ and for the group 1 partition when GS is used for the group iteration it is $\lambda_{GS1} \stackrel{*}{=} 4/9$. When MINI is used for the group iteration, however, this value increases to $\lambda_{GS1} = (16/21)^2$ (for the asymptotic situation) and convergence of an ordinary GS inner solution would be expected to slow down somewhat when the asymptotic situation is approached.

A very successful approach (Young 1971) for speeding up the GS approach is the method of successive overrelaxation (SOR). The procedure involves the calculation of

$$\phi_i^{(n)} = \omega\left(S_i + \phi_{i-1}^{(n)} + \phi_{i+1}^{(n-1)} - a_i\phi_i^{(n-1)}(\omega - 1)/\omega\right)/a_i \quad , \qquad (3.20)$$

$$i = 1,2,3,$$

where the optimum extrapolation parameter is given as

$$\omega = 2/\left(1 + (1 - \lambda_{GS})^{\frac{1}{2}}\right) \quad . \qquad (3.21)$$

Using the notation of the previous paragraph

$$\omega_2 = 1.07,$$

$$\omega_1 = 1.15,$$

and $\omega_1^{*} = 1.21,$

where the above estimate for λ_{GS1}^{*} has been used in the calculation of ω_1^{*} which should be more appropriate when MINI is used for the group iteration. (See Barry and Pollard (1979) for a more detailed discussion of this point.)

The MINI approach introduced in the previous section may also be used; it requires the calculation of

$$\phi_i^{(n)} = \left(S_i + \phi_{i-1}^{(n)} + \phi_{i+1}^{(n-1)} - \gamma_i^{(n-1)}\phi_i^{(n-1)}\right)/\left(a_i - \gamma_i^{(n-1)}\right) \quad , \qquad (3.22)$$

where the inner extrapolation parameters $\gamma_i^{(n-1)}$ are calculated as

$$\gamma_i^{(n)} = \min\left\{|\delta_{i+1}^{(n)}|/|\delta_i^{(n)}|,\ |\delta_i^{(n)}|/|\delta_{i+1}^{(n)}|,\ \phi_{i+1}^{(n)}/\phi_i^{(n)}\right\} \quad , \qquad (3.23)$$

$$i = 1,2$$

in much the same way as equation (3.8) gives the group parameters. (Of course the index, n, here refers to localized iteration for the one group and should not be confused with a similar counter for the group iteration.)

An asymptotic situation arises as iteration proceeds, resulting in

$$\gamma_1 = 1/(a_2 - 1/a_3)$$

and $\gamma_2 = 1/a_3$

which explains the prompt convergence since all eigenvalues of the (changing) iteration matrix vanish - see Barry and Pollard (1977) for more details. For the group 2 partition,

$$\gamma_1 = 2/7\ (= 0.29)$$
$$\gamma_2 = 1/2\ (= 0.50).$$

For the group 1 partition with GS group iteration,

$$\gamma_1 = 3/7\ (= 0.43)$$
$$\gamma_2 = 2/3\ (= 0.67)$$

and with MINI group iteration (and asymptotic γ's assumed)

$$\gamma_1 = 168/313 \ (= 0.54)$$
$$\gamma_2 = 16/21 \ (= 0.76).$$

For 2 and 3D problems the asymptotic situation is not obtained as readily, although MINI still performs well.

Table 3.3 shows the results of solving the model problem for various solution strategies. The inner convergence criterion is given by inequality (3.10).

Table 3.3

Number of group 1 plus group 2 inner iterations to converge inner problem

		inner method			
		GS	SOR with ω_1	SOR with ω_1^*	MINI
group method	GS	431	302	297	204
	MINI	219	161	139	102

The advantage of using MINI for both the inner iteration and the group solution can be seen from table 3.3. For more substantial 2 and 3D problems the advantage is not as pronounced but still exists, particularly for the group solution as will be discussed in the next sections.

4 HIGHER DIMENSIONAL PROBLEMS

The spatial portion of the few group one-dimensional model discussed in the previous section is most efficiently solved by direct means. When the number of spatial dimensions is higher, iterative schemes may become warranted. With the present trends in computer architecture, the cross-over point for relative advantage is a little obscured (Doherty 1980); however, for the class of problem dealt with in reactor physics, iterative schemes are more appropriate (if not essential). MINI may be applied to solve the many-dimensional problem in a block form, in a somewhat similar manner to the group iterative procedure of section 3.1. A description of MINI as it applies to the three spatial dimensional form of the neutron diffusion equation is given in an algorithmic form, which is somewhere between a computer language and mathematics. The authors believe that a full mathematical description as given (Barry and Pollard 1979) tends to ignore some fundamental computational aspects when the method is being evaluated on realistic large scale problems. In particular, aspects of how the data structures are organised between addressable memory and auxiliary storage are too easily overlooked. Such considerations should be the responsibility of the numerical analyst and not dismissed as being of relevance only to the computer scientist. The ultimate working computer algorithm may be very different in appearance from the mathematical formulation (this is particularly true of the incomplete Choleski conjugate gradient (ICCG) algorithm employed in the three-dimensional neutron diffusion program POW3D (Barry, et al. 1981). For the typical three-dimensional calculation for which POW3D is intended, only part of the problem may be held in real addressable memory at one point of time. In the implementation of MINI given here (as well as ICCG and SLOR), it is assumed that sufficient memory is available to treat each (x,y) plane as a separate block. The algorithm solves the three-dimensional problem

$$A \underset{\sim}{\phi} = \underset{\sim}{S}$$

and the following notation and structures are defined:

$\underset{\sim}{\phi}$ represents a storage vector (in auxiliary memory) containing the three-dimensional flux, ordered in terms of directly transferable subvectors

$$\underset{\sim}{\phi} = \left\{\underset{\sim}{\phi}_k; \quad k = 1,2,\ldots,N_z\right\}. \quad \text{Each } \underset{\sim}{\phi}_k \text{ itself is ordered as}$$

$$\underset{\sim}{\phi}_k = \left\{\underset{\sim}{\phi}_{jk}; \quad j = 1,2,\ldots,N_y\right\} \quad \text{and at the lowest level}$$

$$\underset{\sim}{\phi}_{jk} = \left\{\underset{\sim}{\phi}_{ijk}; \quad i = 1,2,\ldots,N_x\right\}, \quad \text{where } N_x,\ N_y,\ N_z \text{ represent}$$

the number of finite difference grid points in the respective spatial directions.

$\underset{\sim}{S}$ represents the right hand side, $\underset{\sim}{\delta}$ and $\underset{\sim}{\delta}^*$ represent in-plane and out of plane δ's respectively. All are oriented similarly to $\underset{\sim}{\phi}$.

$[\underset{\sim}{x}]$ represents a diagonal matrix formed with the vector $\underset{\sim}{x}$ as its diagonal.

$\leftarrow$ represents a fetch operation from auxiliary storage. The fetch operation applies both to block subvectors

e.g. $\underset{\sim}{x} \leftarrow \underset{\sim}{\phi}_i$

and to matrix blocks

e.g. $A \leftarrow A_{12}$,

where A_{12} represents the submatrix of A appropriate to the block structure used.

$\rightarrow$ represents storage.

diag (A) refers to an operation on the diagonal of A.

$\underset{\sim}{x}, \underset{\sim}{d}, \underset{\sim}{e}, \underset{\sim}{r}$ and $\underset{\sim}{y}$ are work vectors; A_1, A_2, and A_3 are block submatrices, all in primary memory. The vectors are all of length N_xN_y while the matrices are square of order N_xN_y. For the latter, however, only the non-zero components are actually stored.

The three-dimensional (spatial) form of MINI requires the use of two distinct γ's at each grid point. One is related to the forward line (the 'in plane' γ) and the other to the forward point (the 'out of plane' γ) as shown in figure 4.1. (A third γ, not shown, is necessary for the energy group MINI.) Previous iteration γ's (line, plane and group) are used except for the first iteration where $\gamma = 0$ (i.e. a GS iteration begins the process). Any improperly formed operations or data transfers in the description are assumed to be ignored.

*** 3-D MINI ALGORITHM ***

*** ASSUME ESTIMATES OF $\underset{\sim}{\phi}_k$ $(k = 1,2,\ldots,N_z)$ HAVE BEEN MADE AND STORED ***

<u>loop</u> n = 1, maximum number of iterations

$\underset{\sim}{b} \leftarrow \underset{\sim}{S}_1$

$\underset{\sim}{x} \leftarrow \underset{\sim}{\phi}_1$

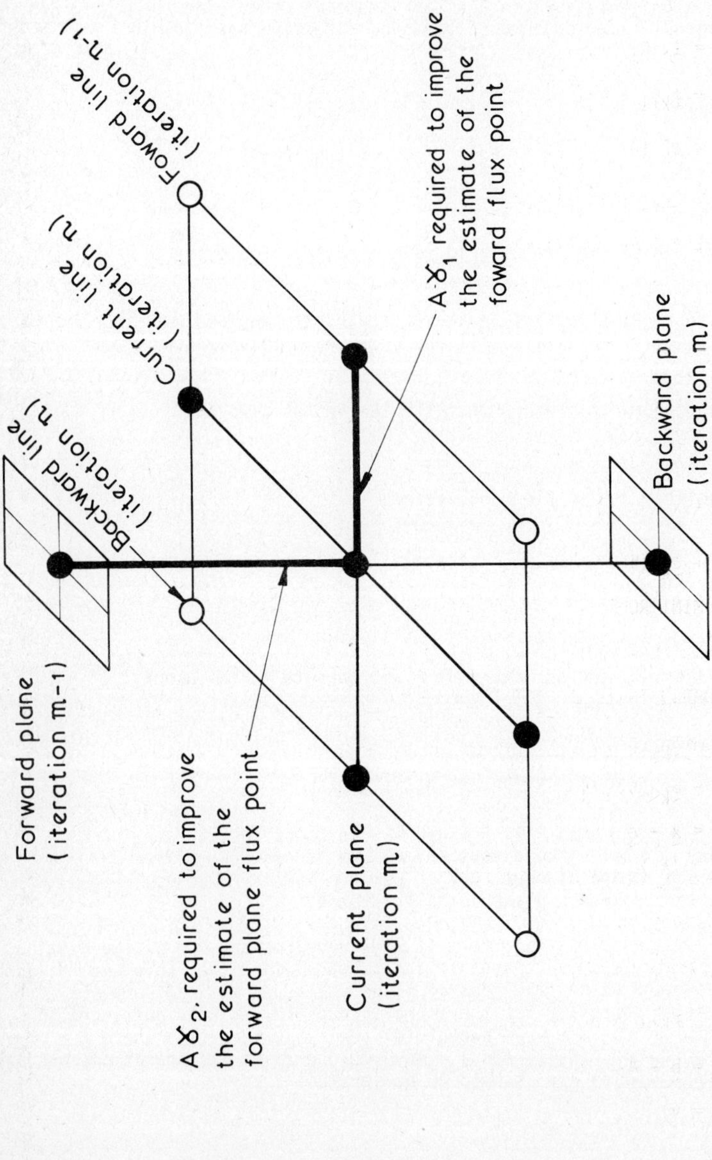

Figure 4.1

Y relationships for 3.D problem

```
A2 ← A11  ,  A3 ← A12
loop   k = 1, Nz
         y ← φ(k+1)
         d ← δ*(k) (out of plane δ's)
         e ← δ*(k+1) (out of plane δ's)
         for k ≠ Nz   loop ℓ = 1,NxNy
                          γ̄ℓ = min (eℓ/dℓ,dℓ/eℓ)
                          Γℓ = min (γ̄ℓ, |yℓ/xℓ|)
                      end loop ℓ
         b = b - A3 y + Γ^T A3 x
         diag (A2) = diag (A2) + diag ([Γ^T A3])
         e = x
         d ← δ(k) (in plane δ's)
*** INVOKE PLANE MINI ROUTINE TO SOLVE A x = b ***
         CALL PLANE MINI (A2, x, b, d)
*** SOLUTION RETURNED IN x, IN PLANE δ's IN d ***
         d → δ(k) (store in plane δ's)
         x → φ(k) (store solution for k^th z plane)
         d = x - e
         d → δ*(k) (store out of plane δ's)
         if (any |dℓ/eℓ| > ε) then set not-converged flag
         A1 = A3  ,  A2 ← A(k+1 k+1)  ,  A3 ← A(k+1 k+2)
         b ← S(k+1)
         b = b - A1 x
         x = y
    end loop k
if (not-converged flag unset) then finish
end loop n
finish with error condition
PROCEDURE PLANE MINI (A, x, b, d)
```

*** A, $\underset{\sim}{x}$, $\underset{\sim}{b}$, $\underset{\sim}{d}$ ARE BLOCK (LINE) ADDRESSABLE ***

<u>loop</u> n = 1, maximum iterations

<u>loop</u> j = 1,N_y

<u>for</u> j ≠ N_y <u>loop</u> i = 1,N_x

$$\bar{\gamma}_i = \min\left(|\{\underset{\sim}{d}_{j+1}\}_i/\{\underset{\sim}{d}_j\}_i|,\ |\{\underset{\sim}{d}_j\}_i/\{\underset{\sim}{d}_{j+1}\}_i|\right)$$

$$\Gamma_i = \min\left(\bar{\gamma}_i,\ |\{\underset{\sim}{x}_{j+1}\}_i/\{\underset{\sim}{x}_j\}_i|\right)$$

<u>end</u> <u>loop</u> i

$$\underset{\sim}{b}_j = \underset{\sim}{b}_j - A_{jj-1}\,\underset{\sim}{x}_{j-1} + \underset{\sim}{\Gamma}^T A_{jj+1}\,\underset{\sim}{x}_{j+1}$$

$$\text{diag}\,(A_{jj}) = \text{diag}\,(A_{jj}) + \text{diag}\,([\underset{\sim}{\Gamma}^T A_{jj+1}])$$

$$\underset{\sim}{d}_j = (A_{jj})^{-1}\,\underset{\sim}{b}_j \qquad \text{(solve tridiagonal system)}$$

<u>If</u> (any $|\{\underset{\sim}{d}_j\}_i| > \varepsilon$) <u>then</u> set not-converged flag

$$\underset{\sim}{x}_j = \underset{\sim}{x}_j - \underset{\sim}{d}_j$$

<u>end</u> <u>loop</u> j

<u>If</u> (not converged flag unset) <u>then</u> return

<u>end</u> <u>loop</u> n

<u>return</u> error condition

<u>end</u> procedure PLANE MINI

Neutron diffusion equations have traditionally been solved by the successive line overrelaxation (SLOR) technique, while more recently considerable interest has been rekindled in the conjugate gradient methods (Reid, 1971). MINI is consequently compared with SLOR and the incomplete Choleski decomposition variant of the conjugate gradient (ICCG) technique (Meijerink and van der Vorst, 1977). For the sake of brevity, the details of the SLOR and ICCG implementations are not given, but suffice it to say that they follow a data management structure based on (x,y) planes very similar to that detailed for MINI, even though ICCG is a global rather than a block scheme. For both SLOR and MINI, all floating point arithmetic operations are performed in double precision arithmetic on an IBM 3031 computer, while for ICCG single precision is employed for all but the partial LDL^T decomposition

Table 4.1

Number of floating point operations per iteration per point for 2-dimensional forms of the iterative schemes as implemented in POW3D

Type of Arithmetic	SLOR	MINI	ICCG
* /	10	12	20
+ -	6	7	17

of A. (Fortunately for the 7-point finite difference operator dealt with here, there is little computational overhead involved because it is necessary only to compute the diagonal components of D.) Single precision arithmetic was used because one advantage of the conjugate gradient method is that it has less sensitivity to rounding error.

Table 4.2

Number of I/O transfers and floating point operations per z iteration for the 3-dimensional forms of the iterative schemes as implemented in POW3D. N_x,N_y,N_z are the numbers of grid points in each direction, while I is the number of iterations necessary to converge the 2-dimensional subsystems.

Type of Operation	SLOR	MINI	ICCG
Matrix Block Transfers	N_z	N_z	$3\ N_z$
Vector Block Transfers	$3\ N_z$	$7\ N_z$	$20\ N_z$
* / Floating Point Arithmetic	$(3 + 10I)N_xN_y$	$(9 + 12I)N_xN_y$	$37N_xN_y$
+ - Floating Point Arithmetic	$(5 + 6I)N_xN_y$	$(5 + 7I)N_xN_y$	$32N_xN_y$

Table 4.3

Average time per iteration pass for 2-D Moata

Type of Iteration	Language	Average Time/Iteration on an IBM 3031 (sec)
SLOR	Assembler	0.69×10^{-3}
SLOR	FORTRAN	0.99×10^{-3}
MINI (double precision γ's)	Assembler	0.97×10^{-3}
MINI (double precision γ's)	FORTRAN	1.3×10^{-3}
MINI (single precision γ's)	FORTRAN	1.2×10^{-3}
ICCG	FORTRAN	0.15×10^{-3}

As a measure of the relative work and overhead involved with each scheme, the numbers of floating point arithmetic operations are tabulated in table 4.1 for one pass of the two-dimensional versions for each of the three schemes. The numbers of I/O transfer operations and floating point instructions are included for the three-dimensional versions in table 4.2. In table 4.3, the average time spent per iteration is recorded for several implementations of the three iterative methods. These are tabulated for a two-dimensional version of the Moata reactor, to be considered in section 5, in its full three-dimensional form. This model has the same finite difference form as the central (x,y) plane in the full version. Two results for the FORTRAN version of MINI are reported to show the difference in timing between the use of single and double precision arithmetic in calculating γ.

5 SOME THREE-DIMENSIONAL RESULTS

Results of a three-dimensional (spatial) k-eigenvalue calculation are presented for the thermal reactor Moata, a 100 kW Argonaut-type research reactor operated at Lucas Heights by the AAEC. This reactor is fuelled by highly enriched uranium clad with aluminium, cooled by light water and moderated by a graphite reflector. Details of the model solved are given in figure 5.1. Because of symmetry, only one eighth of the reactor need be considered. A 23 x 17 x 17 spatial finite difference scheme with edge mesh points is used, and eight energy groups are considered (seven of which involve upscatters) leading to a system of approximately 5 x 10^4 unknowns.

For this problem, the upscatter calculations are performed with either GS or MINI iterations. The calculations are also done with and without energy rebalance procedures. Energy rebalancing is a special form of coarse mesh rebalancing (Nakamura, 1977) in which the three spatial dimensions are collapsed to one point. Energy rebalancing is performed once on each outer (eigenvalue) iteration and is undertaken when the first energy group with upscatter into it is encountered.

The spatial portion of the problem is solved with the three iterative schemes. For the block techniques SLOR and MINI, the between plane inner iterations are converged to 0.1 times the accuracy of the outer Chebyshev extrapolation scheme and the in plane inner iterations to 0.07 of this accuracy. For the three-dimensional problem, SLOR is a two-level procedure (similar to that outlined for MINI). An optimal ω is required for the between plane (z) iterations as well as an ω for each (x,y) plane. Unlike the other two techniques, ICCG is global and no between plane iterations are required. The numbers of outer, z and x,y plane iterations and the central processing unit time on the IBM 3031 are recorded in table 5.1.

Examination of the results reveals that group rebalancing makes a considerable difference when the GS scheme is used for the energy group iterations, but has little effect when MINI is employed. The overhead involved in the group rebalance process is very small and the default procedure in POW3D is to have it enabled at all times. With MINI handling the energy groups, the improvement in performance is truly outstanding. For thermal reactors, upscatter may be significant and, in this particular model, five upscatter passes are performed.

Of the three spatial schemes, ICCG would at first seem disadvantaged by the results of this problem when the central processing unit (CPU) times are compared for the best combination of other parameters (MINI groups, energy balance on). Closer consideration is warranted however. To compare firstly ICCG and then the other two schemes, it is necessary to discount the times quoted for ICCG because the routine is written in FORTRAN, whereas the inner plane section for both SLOR and MINI has been coded in assembler language. This is best done by adjusting the MINI time upwards. The average time for individual iterations, table 4.3, leads to a correction factor of (0.013/0.0097) for a typical (x,y) plane of the Moata type. Of the 76.87 minutes spent on the MINI spatial iteration run, 27.63 minutes were used in actually solving the inner plane equations. This led to an adjusted result of 86.26 minutes which is very close to the 87.62 minutes required for ICCG. (The corrected figure for SLOR is 86.36.)

It would seem fair now to conclude that MINI, SLOR and ICCG are comparable in handling the spatial solution. Further consideration however is required. In the case of MINI and ICCG, because ICCG is global and MINI a block method, MINI

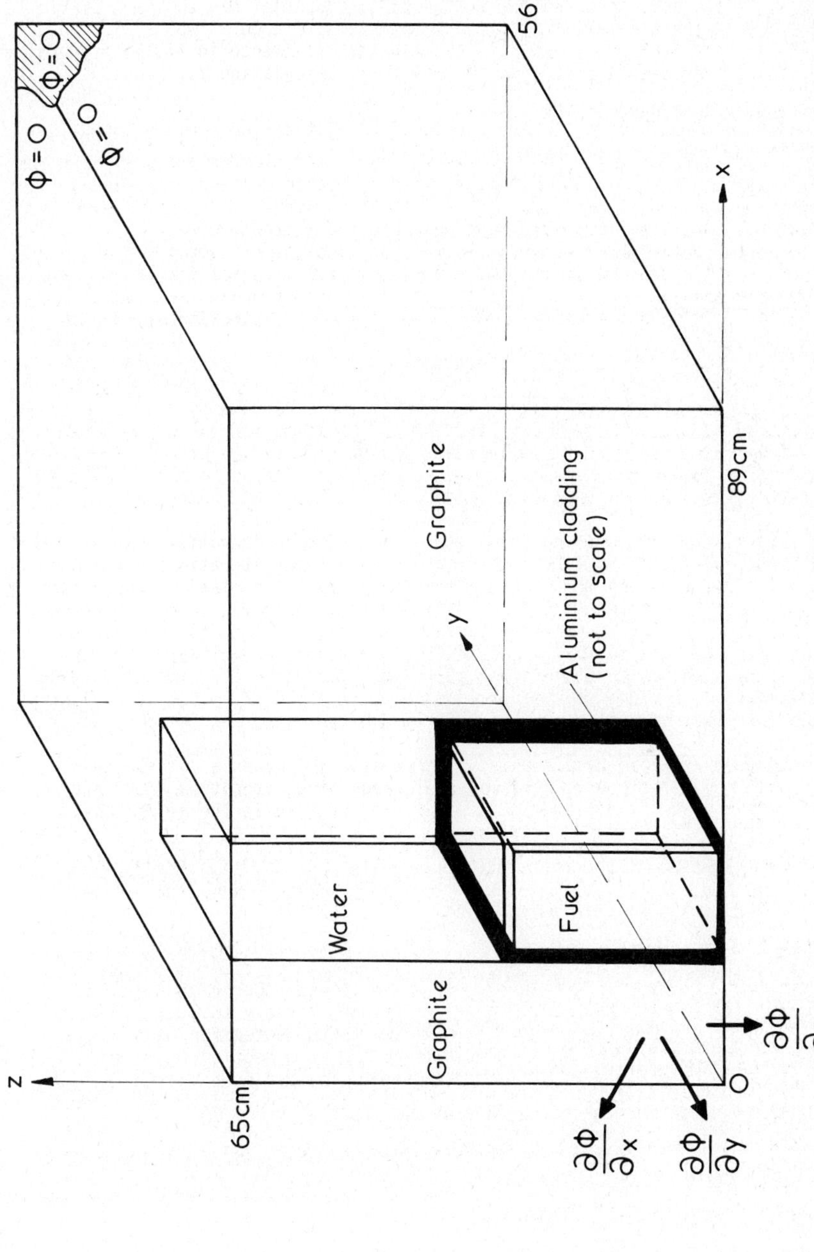

3-D Moata Reactor

Figure 5.1

appears to spend a greater proportion of its time working on individual planes than ICCG. For the best combination, MINI averages 2.05 iterations per (x,y) plane. One result of this appears to be that fewer z iterations are required for MINI than for ICCG (803 compared to 946). Because of the data structures, the I/O overhead required for ICCG is approximately three times as much as for MINI per z iteration (table 4.2). The cost of such transfers has a little effect on the CPU time, but more importantly it increases the elapsed time, which is by no means insignificant when transfers from auxiliary storage are involved. This time penalty is not reflected in the values quoted here.

Table 5.1

Results of the 3-dimensional 8-group Moata model
*Convergence never achieved in time available; extrapolation used to obtain these figures

Spatial Iterative Scheme	Energy Iterative Scheme	Group Balance	Outer Iterations	x,y Plane Iterations x 100	z Plane Iterations	Time (min)
SLOR	GS	OFF	28*	2200*	5500*	280*
SLOR	MINI	OFF	11	598	1608	93.40
SLOR	GS	ON	21	1647	4298	206.48
SLOR	MINI	ON	10	402	1079	73.73
MINI	GS	OFF	30*	1550*	3800*	280*
MINI	MINI	OFF	11	318	898	82.01
MINI	GS	ON	33	1077	2993	225.43
MINI	MINI	ON	10	281	803	76.87
ICCG	GS	OFF	28*		5410*	310*
ICCG	MINI	OFF	11		1364*	108.25
ICCG	GS	ON	21		3635	231.05
ICCG	MINI	ON	10		946	87.62

The assembler version of MINI (plane driven only) is 1.34 times as fast as the equivalent program with optimal FORTRAN. For a nuclear workhorse code such as POW3D, it is worthwhile making the effort to code the inner plane in this way, but to produce the equivalent code for ICCG would seem to be a significantly more difficult task and, at this stage, the authors are not convinced that the performance of three-dimensional ICCG warrants it.

The conjugate gradient portion of ICCG largely uses single precision arithmetic because the technique is believed to be less sensitive to round-off errors than either MINI or SLOR. The current version of MINI is wholly in double precision arithmetic, but it is possible that the γ determination could be performed adequately in single precision thus resulting in a small time saving (1/13, table 4.3). Although single precision arithmetic is sufficient for most reactor problems so far studied with ICCG, there are some in which it is impossible to obtain the accuracy required for the eigenvalue calculation in this mode.

When SLOR and MINI are compared for the best combination, a marginal superiority in timing is shown by SLOR. When the I/O transfers are totalled up, MINI appears to be slightly superior. In other calculations studied by the authors, MINI shows a more decisive advantage over SLOR on CPU time.

An additional problem is included here. For this problem, the mesh in the y direction for Moata is refined so that each internal grid point is doubled up on. The model now involves a mesh of 23 x 33 x 17 x 8 ($\approx 10^5$ unknowns). The reactor is solved for the optimal combination of parameters for the three iterative schemes and the results are reported in table 5.2. In this instance, SLOR is outclassed by the other iterative schemes, which again, on CPU times, appear to be somewhat comparable. Again MINI has significantly fewer z iteration passes and involves considerably less I/O overhead.

Table 5.2

Results of the 3-dimensional (refined y mesh) 8-group Moata model
*Convergence never achieved in time available; extrapolation used to obtain these figures

Spatial Iterative Scheme	Energy Iterative Scheme	Group Balance	Outer Iterations	x,y Plane Iterations x 100	z Plane Iterations	Time (min)
SLOR	MINI	ON	10*	3000*	8000*	680.0*
MINI	MINI	ON	10	304	833	154.4
ICCG	MINI	ON	10		1048	167.9

6 CONCLUSIONS

The three iterative methods, SLOR, ICCG and MINI compare favourably for solution of the spatial flux dependence in 'everyday' reactor calculations. However for refined problems with a large number of mesh points, MINI and ICCG are far superior to SLOR. Even so MINI has a decided advantage over ICCG as it makes less use of slow disk fetches. This means that MINI can be used to undertake more calculations in a given amount of time than ICCG. Further, the acceleration of solution of the group flux set of equations by MINI is outstanding compared with the only viable alternative - GS.

For the reasons stated, supported by the findings of the paper, MINI is the best choice for all round performance and should be the (default) solution method in a code such as POW3D.

7 FOOTNOTE

1. In reactor studies use of the old metric unit of centimetre has continued to this day.

8 REFERENCES

Barry, J.M., Harrington, B.V., & Pollard, J.P. (1981). AUS module POW3D. A general purpose multigroup neutron diffusion code. AAEC/E report, in preparation.

Barry, J.M., & Pollard, J.P. (1977). Method of implicit non-stationary iteration for solving neutron diffusion linear equations. Annals of Nuclear Energy, 4 (485-493).

Barry, J.M., & Pollard, J.P. (1979). Application of the method of implicit non-stationary iteration (MINI) to 3D neutron diffusion problems. Annals of Nuclear Energy, 6 (121-131).

Doherty, G. (1980). An outline of some numerical methods for the solution of diffusion equations. One day meeting, Diffusion and Moving Boundary Problems, University of Wollongong.

Hageman, L.A., & Pfeifer, C.J. (1965). The utilization of the neutron diffusion program PDQ-5. WAPD-TM-395.

Meijerink, J.A., & van der Vorst, H.A. (1977). An iterative solution method for linear systems of which the coefficient matrix is a symmetric M-matrix. Mathematics of Computation, 31 (148-162).

Nakamura, S. (1977). Computational methods in engineering and science with applications to fluid dynamics and nuclear systems. John Wiley, New York.

Pollard, J.P. (1974). AUS module POW - A general purpose 0, 1 and 2D multigroup neutron diffusion code including feedback free kinetics. AAEC/E269.

Pollard, J.P. (1975). Solution of neutron diffusion equations. In Anderssen, R.S., & Watts, R.O. (eds) Computational Methods in Mathematical Physics. University of Queensland Press.

Reid, J.K. (1971). On the method of conjugate gradients for the solution of large sparse systems of linear equations. In Reid, J.K. (ed.) Large Sparse Sets of Linear Equations. Academic Press, London and New York.

Stacey, W.M. Jr. (1969). Space-time nuclear reactor kinetics. Academic Press, New York.

Varga, R.S. (1962). Matrix iterative analysis. Prentice Hall, Englewood Cliffs, New Jersey.

Young, D.M. (1971). Iterative solution of large linear systems. Academic Press, New York.

NUMERICAL SOLUTIONS OF
PARTIAL DIFFERENTIAL EQUATIONS
J. Noye (editor)

ICCG APPLIED TO $Ax = \lambda Bx$

G. Doherty
University of Wollongong
P.O. Box 1144,
Wollongong,
Australia 2500

A variant of inverse iteration which uses the incomplete Cholesky conjugate gradient iterative method to solve the linear equations is presented. Enhancements of the basic method are discussed and results are given for a test problem.

1. INTRODUCTION

The generalised eigenvalue problem

$$Ax = \lambda Bx \tag{1}$$

arises in a number of applications where A is symmetric and sparse, and B is positive definite, symmetric and sparse. Using a Cholesky factorisation of B,

$$B = LL^T, \tag{2}$$

the generalised eigenvalue problem can be reduced to an ordinary eigenvalue problem,

$$Cy = (L^{-1} A L^{-T})y = y, \tag{3}$$

in the symmetric matrix C. A sparse B will usually give rise to a less sparse factor L, and the matrix C, if actually computed, would be full. If C is large, methods not requiring the explicit construction of C, such as variants of the power method [11] or the Lanczos algorithm [2] can be employed.

The method described here involves an incomplete Cholesky factorisation of B after the manner described by Meijerink and Van der Vorst [9] which restricts L to the same sparsity pattern as B. This requires storage for L equal to that required for A or B, which is often considerably less than that required to store the complete L, even under an optimal ordering. The method uses inverse iteration, optimisation on the subspace as in simultaneous iteration, incomplete Cholesky conjugate gradient solution of linear equations, and, occasionally, gradient reduction of the Rayleigh quotient.

2. INVERSE ITERATION

Inverse iteration is widely used to obtain eigenvectors corresponding to eigenvalues obtained to high precision by algorithms which produce the eigenvalues alone. A description of the process is given by Wilkinson [14]. As applied to the generalised eigenvalue problem it may be written

$$(A - \mu B)y_{n+1} = Bx_n, \tag{4}$$

$$x_{n+1} = y_{n+1}/\,\|y_{n+1}\|_B\ , \tag{5}$$

where $$\|y_{n+1}\|_B^2 = y_{n+1}^T\, B\, y_{n+1}\ . \tag{6}$$

The crucial computational step for the problems under discussion is contained in (4). In the usual applications of inverse iteration, μ is an (almost) exact eigenvalue of the problem and the solution of (4) would be carried out with pivoting since $A - \mu B$ is (almost) a singular matrix.

In our algorithm we propose to solve (4) by an iterative method. If μ is suitably chosen, $A - \mu B$ can be made positive definite (or negative definite) allowing us access to the lowest eigenvalues of the problem (or the highest). The incomplete Cholesky conjugate gradient algorithm applied to (4) requires $A - \mu B$ to be positive definite. Other variants of the conjugate gradient algorithm do not, but their performance on this problem has been disappointing.

If we restrict $(A - \mu B)$ to being positive definite, we cannot implement the usual iterative refinement of μ, viz.

$$\mu_n = (x_n^T\, A\, x_n)/(x_n^T\, B\, x_n)\ . \tag{7}$$

Instead, μ must approach λ_1, the smallest eigenvalue in the problem, from below. In our implementation, μ is fixed over a number of iterations so that the method resembles an inverse power method during such a phase.

Suppose that the eigenvalues and eigenvectors of the problem are λ_i and z_i respectively, with

$$A\, z_i = \lambda_i\, B\, z_i\ , \tag{8}$$

arranged to satisfy

$$\lambda_i \le \lambda_{i+1}\ . \tag{9}$$

It can easily be shown that if $\lambda_i \ne \lambda_j$,

$$z_j^T\, A\, z_i = z_j^T\, B\, z_i = 0\ , \tag{10}$$

and that it is possible to construct a B-conjugate basis for any subspace with $\lambda_i = \lambda_j$. The natural scalar product for this problem is therefore that involving B,

$$(x,y) = x^T B\, y\ . \tag{11}$$

Our iteration procedure is easily written in terms of the vectors z_i. If the starting guess is expanded

$$x_0 = \sum_i \alpha_i z_i\ , \tag{12}$$

then

$$y_1 = (A - \mu B)^{-1} B\, x_0\ ,$$

$$= \sum_i \alpha_i (\lambda_i - \mu)^{-1}\, z_i\ . \tag{13}$$

After n such steps,

$$x_n = \beta_n \sum_i \alpha_i (\lambda_i - \mu)^{-n} z_i \ , \qquad (14)$$

where β_n is the normalisation constant chosen so that

$$(x_n, x_n) = 1 \ . \qquad (15)$$

Equation 14 exhibits clearly the expected behaviour of the inverse iteration procedure. After some iteration the terms with λ_i nearest to μ dominate the expansion. The rate at which they dominate can be improved by moving μ closer to λ_1, and this is done as the calculation proceeds. In fact, we perform the inverse iteration step simultaneously on a number of independent vectors following the scheme introduced by Bauer [1].

3. SIMULTANEOUS ITERATION

The idea is to work with a set of m vectors $x_{o,j}$ rather than a single starting vector x_o. These vectors are arranged into an n×m matrix X for which the following relations are preserved at each step

$$X^T B X = I \ , \qquad (16)$$

and $$X^T(A - \mu B)X = \Gamma \ , \qquad (17)$$

where Γ is a diagonal matrix.

Relation (17) is achieved by performing a stabilised Gram Schmidt orthogonalisation, using the matrix B in the scalar product as already indicated (11). The m×m matrix

$$T = X^T(A - \mu B)X \qquad (18)$$

is then diagonalised using a FORTRAN version of Rutishauser's Jacobi routine [12]. Having obtained the orthogonal matrix U such that

$$U^T T U = \Gamma \ , \qquad (19)$$

X is replaced by XU so that both (16) and (17) are satisfied by the rotated vectors XU.

The matrix Γ contains the best estimates of $\lambda_i - \mu$ on the subspace generated by the current set of iterates X. These entries are sorted into ascending order, and the corresponding columns of X are shuffled if necessary. The vectors

$$Y = \Gamma^{-1} X \qquad (20)$$

are then used as the starting guesses for the next pass through the inverse iteration procedure.

The implementation of simultaneous iteration is very similar to Rutishauser's *ritzit* [11] scheme, except that here we perform the Jacobi rotations after every inverse iteration step because the rotation time is small compared with an inverse iteration step, whereas Rutishauser performed the rotation only occasionally since it requires computation comparable with multiplication by A.

4. GRADIENT REDUCTION OF THE RAYLEIGH QUOTIENT

Define $\eta(x) = (x^T A x)/(x^T B x)$. (21)

If A and B are symmetric

$$\nabla\eta = \frac{2}{x^T Bx} [Ax - \eta(x) B x] . \tag{22}$$

For the eigenvectors of our problem

$$A z_i = \lambda_i B z_i , \tag{23}$$

so,
$$\eta(z_i) = \lambda_i , \tag{24}$$

and
$$\nabla\eta(z_i) = 0 . \tag{25}$$

We can search for stationary points of η by starting at a point x_0, choosing a search direction d, and then searching down d until

$$\frac{d}{d\gamma} \eta(x_0 + \gamma d) = 0 \tag{26}$$

or

$$d^T \nabla\eta(x_0 + \gamma d) = 0 . \tag{27}$$

This method was fully described by Hestenes and Karush [3,4]. The method is suited to our type of problem since it only requires multiplication by the matrices A and B. It is better suited to locating the end points of the spectrum, which are genuine maximum and minimum points, than the interior eigenvalues, which are merely stationary points.

Equation (27) can be rewritten

$$\frac{d^T A(x_0 + \gamma d)}{d^T B(x_0 + \gamma d)} = \frac{(x_0 + \gamma d)^T A(x_0 + \gamma d)}{(x_0 + \gamma d)^T B(x_0 + \gamma d)} \tag{28}$$

yielding a quadratic equation in the step length γ. An obvious choice for the search direction d is the gradient of η at x_0 .

Writing

$$x_0 = \sum_i \alpha_i z_i , \tag{29}$$

d can be written (ignoring a multiplicative constant)

$$\begin{aligned} d &= \sum_i \alpha_i [A z_i - \eta(x_0)B z_i] , \\ &= \sum_i \alpha_i(\lambda_i - \eta)B z_i . \end{aligned} \tag{30}$$

It is possible to construct a block form of this iteration scheme without difficulty. We start with a set of vectors X satisfying

$$X^T A X = \Gamma , \tag{31}$$

$$X^T B X = I \ , \qquad (32)$$

constructed in the manner previously discussed. The value of $\eta(x)$ for each of the columns of X is simply the diagonal entry in Γ. Then for each column in turn, d is computed from

$$d = Ax - \eta(x)Bx \qquad (33)$$

giving, incidentally, a direction orthogonal to all the columns of X (from (31) and (32)). (28) is solved for direction in turn, and a new matrix X is constructed. This procedure can be very useful in our algorithms for resolving degenerate eigenvalue situations, but it was eclipsed by the Lanczos algorithm for general problems. It is not difficult to see why if we consider the simpler case $B = I$. Equation (30) then becomes

$$d = \sum_i \alpha_i(\lambda_i - \eta)z_i \ , \qquad (34)$$

and,

$$x + \gamma d = \sum_i \alpha_i[1 + \gamma(\lambda_i - \eta)]z_i \ . \qquad (35)$$

Suppose we are trying to compute z_1 and λ_1 and that η has converged quite close to λ_1. If the proportion of z_n is not to increase with respect to that of z_1

$$[1 + \gamma(\lambda_n - \eta)] \approx [1 + \gamma(\lambda_n - \lambda_1)] > -1$$

$$\text{i.e.} \quad \gamma \geq \frac{2}{(\lambda_n - \lambda_1)} \ . \qquad (36)$$

If γ satisfies this inequality, the z_2/z_1 ratio is $1 - 2(\lambda_2 - \lambda_1)/(\lambda_n - \lambda_1)$. This explains why the gradient algorithm works well on small problems where $\lambda_2 - \lambda_1 \approx \lambda_n - \lambda_1$, and poorly on large problems, where the far points λ_n of the spectrum limit the step size. Nevertheless, an occasional iteration of this procedure can be incorporated into a sequence of inverse iteration steps to some advantage, because inverse iteration reduces the components of the higher eigenvectors, while the gradient algorithm reduces components from the middle of the spectrum.

5. THE CONJUGATE GRADIENT ALGORITHM

In this section we replace the inverse iteration step

$$(A - \mu B)y = B\ x \qquad (37)$$

by the standard form in which the solution of linear equations is discussed, viz.

$$A\ x = b. \qquad (38)$$

Initially we will consider A to be sparse, symmetric and positive definite, though we would like to relax the positive definite restriction so that we would have more freedom in the choice of μ in (37). Define

$$R = \tfrac{1}{2}(b - Ax)^T A^{-1}(b - Ax). \qquad (39)$$

Since A and therefore A^{-1} is positive definite the solution of (38) minimises R.

This is obvious since

$$\nabla R = Ax - b \ . \tag{40}$$

A conjugate direction algorithm for the solution of (38) starts at a point x_o and proceeds to search along a set of directions d_i satisfying

$$d_i^T A d_j = 0 \quad \text{for} \quad i \neq j \tag{41}$$

so that R is minimised along each direction d_i in turn. Thus, to minimise R along a particular direction d we require

$$\frac{d}{d\gamma} R(x + \gamma d) = d^T \nabla R(x + \gamma d) = 0 \tag{42}$$

yielding

$$\gamma = d^T (b - Ax)/d^T A d. \tag{43}$$

If A is of order n, then after minimising along n directions starting from x_o, we have

$$\begin{aligned} x_n &= x_o + \sum_{i=1}^{n} \gamma_i d_i \\ &= x_o + \sum_i d_i{}^T(b - Ax_o)/d_i{}^T A d_i \\ &= A^{-1} b \ , \end{aligned} \tag{44}$$

since

$$A^{-1} = \sum_{i=1}^{n} (d_i{}^T A d_i)^{-1} d_i d_i{}^T \ . \tag{45}$$

Thus in exact arithmetic, the solution is obtained in n steps. A thorough discussion of the conjugate gradient algorithm, including an extension to a non-symmetric, indefinite A was given by Hestenes and Stiefel [5]. The method of generating conjugate directions d for the positive definite symmetric A was simply

$$\begin{aligned} d_{i+1} &= -\nabla R(x_i) + \beta d_i \\ &= (b - A x_i) + \beta d_i \ , \end{aligned} \tag{46}$$

with β chosen so that

$$d_i{}^T A d_{i+1} = 0 \ , \tag{47}$$

yielding

$$\beta = -d_i{}^T A(b - A x_i)/d_i{}^T A d_i \ . \tag{48}$$

Conjugate gradient methods have undergone a revival recently and better choices of conjugate directions have been developed for sparse matrices. Meijerink and Van der Vorst [9] developed the incomplete Cholesky variant for M-matrices which has been shown effective for other matrices arising in finite difference solutions of partial differential equations by Kershaw [6].

A is approximated by

$$A \approx L\ D\ L^T \tag{49}$$

where L has the same sparsity pattern as the lower triangle of A and D is a diagonal matrix. The factorisation is achieved by the usual Cholesky procedure

$$d_i = a_{ii} - \sum_{k=1}^{i-1} \ell_{ik}\ d_k\ \ell_{ik}\ , \tag{50}$$

$$\ell_{ii} = 1\ , \tag{51}$$

$$\ell_{ji} = \left[a_{ji} - \sum_{k=1}^{i-1} \ell_{ik}\ k_k\ \ell_{jk}\right] \Big/ d_i \tag{52}$$

by simply ignoring any elements for which $a_{ji} = 0$. When this procedure is applied to a symmetric M-matrix, such as that arising from a five point approximation to the ∇^2 operator, it is easy to show that the elements of D remain positive and that the numerical stability is similar to the exact LDL^T factorisation. Similar behaviour cannot be assured when the off diagonal elements are indefinite, and especially if A is not positive definite with negative diagonal entries.

Replace equation (46) by

$$d_{i+1} = -(LDL^T)^{-1}\ \nabla R(x_i) + \beta\ d_i \tag{53}$$

where β is chosen, as before, to preserve conjugacy. A simple motivation for this choice is that if we could choose

$$d_1 = -A^{-1}\ \nabla R(x_o) \tag{54}$$

the solution of the equations would be reached in one step, rather than after n steps. $(LDL^T)^{-1}$ is an easily computed approximation to A^{-1} which we hope would effect a comparable improvement in performance.

The empirical evidence is that the conjugate gradient scheme with this modification is computationally efficient, although the iteration scheme

$$x_{m+1} = x_m + (LDL^T)^{-1}(b - A\ x_m) \tag{55}$$

is not. The success of the algorithm is due to a clustering of the eigenvalues of $(LDL^T)^{-1}A$ which has been demonstrated empirically by Kershaw [6]. The clustering effect is most easily explained using matrix polynomials.

Let $$K = (LDL^T)^{-1}\ , \tag{56}$$

$$r_i = b - A\ x_i\ , \tag{57}$$

and $$h = A^{-1}b\ . \tag{58}$$

Then we may write

$$\begin{aligned} r_o &= b - A\ x_o \\ &= A(h - x_o) \end{aligned} \tag{59}$$

$$d_1 = K\, r_o$$
$$= KA(h-x_o) \quad (60)$$
$$r_1 = b - A\, x_1$$
$$= A(I - \gamma_1 KA)(h-x_o) \quad (61)$$
$$d_2 = KA[(1+\beta_1)I - \gamma_1 KA](h-x_o) \quad (62)$$

and so on. Hestenes and Stiefel [5] have shown that

$$x_m = x_o + \sum_{i=1}^{m} \gamma_i\, d_i \quad (63)$$

minimises R for all x of the form

$$x = x_o + \sum_{i=1}^{m} \eta_i\, d_i\ . \quad (64)$$

Examination of (64) and (59) to (62) shows that we could write

$$r(x) = b - A\, x$$
$$= A\, P_m(KA)(h - x_o)\ , \quad (65)$$

where $R_m(KA)$ is an arbitrary polynomial of degree m determined by the η_i. In particular, if the minimal polynomial of KA is of degree m, choosing $P_m(KA)$ to be this polynomial implies $r(x)$ is zero, and since x_m is minimising, that $r(x_m)$ is zero also. Thus if KA contains m distinct eigenvalues, and not n, the conjugate gradient algorithm converges in m steps, more or less independently of the size of the eigenvalues of KA.

We can discuss an entire class of methods of conjugate gradient type by considering

$$R = \tfrac{1}{2}(b - A\, x)^T H(b - A\, x)\ , \quad (66)$$

with

$$d_{i+1} = -K\, \nabla R + \beta\, d_i\ . \quad (67)$$

For example, $H = A^{-1}$, $K = I$ gives us the classic algorithm of Hestenes and Stiefel [5]; $H = I$, $K = I$ gives us their generalisation to the case where A is non symmetric and indefinite. $H = A^{-1}$, $K = (LDL^T)^{-1}$ gives us the incomplete Cholesky conjugate gradient algorithm. It is essential that H is positive definite so that the solution of the linear equations is the minimum point of R. It is desirable that K is positive definite since that ensures that each d_i is a descent direction, but this restriction on K does not appear essential. The various directions are chosen conjugate with respect to the matrix $A^T H A$, which if A is indefinite, is now positive semidefinite.

We have tried to extend the conjugate gradient algorithm to solve the case where A is symmetric but indefinite, using the following alternatives:

$H = I$, $K = I$ (the Hestenes and Stiefel procedure)

$H = I$, $K = (LDL^T)^{-1} A^{-1}$ (which need not be positive definite)

$$H = I \ , \ K = (LDL^T)^{-1}(LDL^T)^{-1} \quad \text{(which is positive definite)}$$

$$H = (LDL^T)^{-1}(LDL^T)^{-1} \ , \ K = I \quad \text{(an obvious alternative).}$$

In addition we have also tried the approximate LU factorisation and conjugate gradient procedure recommended by Kershaw [6], plus a variant where $U = L^T$ except for sign differences. The latter is possible since our A is symmetric, though indefinite.

The best performer of all the algorithms tried is the original Hestenes and Stiefel algorithm which involves no partial factorisation at all. Probably we should expect this. Since $(A-\mu B)$ is almost singular we would expect to have to pivot to perform a complete factorisation successfully. Indeed, all the algorithms work satisfactorily if μ is chosen $< \lambda_1$, though none are as good as the ICCG(0) algorithm, and the Hestenes and Stiefel algorithm is rather worse than the others. When μ is chosen in the interior of the spectrum, the incomplete factorisation is not helpful, and we are left with the original Hestenes and Stiefel algorithm, which performs disappointingly. Thus the isolation of internal eigenvalues near to a reference μ is much more time consuming that for eigenvalues at the ends of the spectrum.

6. IMPLEMENTATION

The computational sequence begins with a conservative estimate $\mu < \lambda_1$ and proceeds as follows:

(1) Unless restarting, generate a set of m random vectors X. (If better guesses for the eigenvectors are available the routine will use them.)

(2) Pass the matrix X through a stabilised Gram Schmidt orthogonalisation with respect to B, to obtain a set of vectors satisfying $X^TBX = I$.

(3) Compute the matrix $T = X^T(A - \mu B)X$, and diagonalise T using a FORTRAN version of Rutishauser's Jacobi routine to obtain $U^TTU = \Gamma$. The entries of Γ are the current estimates of $\lambda_i - \mu$. Test for convergence.

(4) Rotate the space X, shuffling the columns if required to obtain the eigenvalues in ascending order. After $X = XU$, then $X^TBX = I$ and $X^T(A-\mu B)X = \Gamma$.

(5) On most iterations, for each column of X in turn, solve $(A - \mu B)y = Bx$, using the incomplete Cholesky conjugate gradient iteration scheme. Replace x by y,

or

on the first iteration, and occasionally thereafter, for each column of X in turn, compute $d = Ax - \lambda_i Bx$, and then γ to satisfy $d^T\nabla(x + \gamma d) = 0$. Replace x by $x + \gamma d$.

(6) Return to step 2.

The present subroutine is not tuned for production work but is suitable for establishing the utility of the method for particular problems. The sparse matrices A, B and L are stored by column with an additional integer array to identify the row number, after the style described by Reid [10]. All real variables are stored in double precision and all computation is performed in double precision. Convergence is to 10 times the machine precision limit in the eigenvalue estimates $\lambda_i - \mu$, which will usually give λ_i to double precision accuracy.

Convergence of the inverse iteration steps is to .01 times the square root of the accuracy attained in the eigenvalues at the current step, with a maximum number of iterations which is steadily increased as the calculation progresses. Every fourth iteration, μ is replaced by $\mu + 0.9(\lambda_1 - \mu)$ until it is estimated that $(\lambda_2 - \mu) > 10^5(\lambda_1 - \mu)$, and that value of μ is retained until final convergence is obtained.

A test problem on which the method has been tried is the following

$$a_{ii} = 5,\ b_{ii} = 6\ ,$$

$$a_{i,i+1} = a_{i,i-1} = b_{i,i+1} = b_{i,i-1} = -1\ ,$$

$$a_{i,i+N} = a_{i,i-N} = b_{i,i+N} = b_{i,i-N} = -1\ ,$$

for $i = 1$ to N^2 for various values of N. The eigenvalues of this problem are

$$\lambda_{i,j} = \frac{5 - 2\ \cos\left(\frac{i\pi}{N+1} + \cos\ \frac{j\pi}{N+1}\right)}{6 - 2\ \cos\left(\frac{i\pi}{N+1} + \cos\ \frac{j\pi}{N+1}\right)}\quad i,j = 1,2,\ldots N\ .$$

Thus all the eigenvalues lie in the interval (0.5, 0.9) and the eigenvalues are present with various multiplicities. Since A and B commute, they can both be diagonalised by a single unitary transformation, but the numerical procedure does not take advantage of this fact.

This particular family of matrices has been studied extensively to produce optimal orderings for the complete Cholesky factorisation of B, but one significant advantage of this algorithm is that optimal ordering is not required. Removal of this requirement may simplify considerably the assembly of finite element matrices to use with this procedure. As always, there is a penalty for this flexibility, and it is simply the additional computation time devoted to the iterative solution of the linear equations.

The results for the test problem are given in the table below. The column headings are as follows:

N - the dimension of the matrix is N^2 .

M - the number of columns in the matrix X.

NC - the number of converged eigenvalues required.

NI - the number of (outer) inverse iteration steps.

NICCG - the total number of (inner) conjugate gradient steps.

TIME - UNIVAC 1106 CPU seconds (subject to usual noise problems).

For each subset, with different N , we would expect roughly twice the work of the previous subset since the matrix multiplication operation count is proportional to N^2. This expectation is met. The quantity NICCG/(M*NI) ranges between 4 and 6 as the matrix size increases, which is satisfactory since the order of the matrix has increased from 25 to 400 over the various subsets. The gradual increase in work required in the linear equation solution is offset by a bunching of the required eigenvalues, which actually operates to the advantage of this algorithm.

N	M	NC	NI	NICCG	TIME
5	3	2	34	479	15
5	4	2	31	581	18
5	5	2	27	552	20
5	6	2	26	709	23
5	6	3	28	796	25
5	6	4	43	1238	40
7	3	2	31	470	25
7	4	2	26	472	27
7	5	2	25	572	25
7	6	2	25	641	36
7	6	3	27	709	38
7	6	4	40	1048	53
10	3	2	26	322	33
10	4	2	24	384	38
10	5	2	24	502	53
10	6	2	24	621	67
10	6	3	24	621	76
10	6	4	34	805	110
14	3	2	24	355	80
14	4	2	24	471	110
14	5	2	22	544	120
14	6	2	24	723	154
14	6	3	24	723	155
14	6	4	34	997	216
20	3	2	24	374	163
20	4	2	24	554	213
20	5	2	24	725	277
20	6	2	22	787	290
20	6	3	24	862	308
20	6	4	29	1047	365

7. CONCLUSIONS

We expect an unwanted eigenvector z_j to be reduced in a computed eigenvector by a factor $((\lambda_1 - \mu)/(\lambda_j - \mu))^k$ after k iterations. Since μ is moved close to λ_1 as the iterations proceed, convergence to λ_1 is direct. However, the ICCG method does not permit us to choose $\mu > \lambda_1$, so convergence to higher eigenvectors is limited to $((\lambda_j - \lambda_1)/(\lambda_{m+1} - \lambda_1))^k$, where m is the number of vectors included in the working matrix X. This feature of the method is unsatisfactory but for the present, unavoidable. Obviously what we would like to do is replace μ by

$$\mu_j = (x_j^T A x_j)/(x_j^T B x_j)$$

when working with the approximation to eigenvector z_j. However this involves solving

$$(A - \mu B)y = B x_j$$

when $(A - \mu B)$ is symmetric but indefinite, and no satisfactory iterative procedure has been found for this step. The best appears to be the original proposal of Hestenes and Stiefel [10] which would be too slow for large matrices. Numerical experimentation continues, but an entirely different approach is probably required.

The distinguishing characteristics of this algorithm are

a) a roughly linear increase in computation with the order of the matrix, assuming a replicated sparsity pattern,

b) storage required for A, B, and a partial factorisation of $A - \mu B$, plus a number of working vectors of full length,

c) no optimal ordering required in the definition of A and B.

Clearly the method does not compete with reduction to standard form [8] for small problems. Larger problems can be tackled by a complete factorisation of B, with simultaneous iteration or a Lanczos method applied to $L^{-1}AL^{-T}$ which need not be formed explicitly. This method provides an alternative method for this class of problem. The method clearly is unsuitable for very large C.I. type problems [13].

It should be possible to identify problems for which the algorithm is suitable without much difficulty. A favourable ratio of $(\lambda_i - \lambda_1)/(\lambda_{m+1} - \lambda_1)$, rather than λ_{m+1}/λ_i for the power method, together with the ICCG working satisfactorily on the matrix $A - \mu B$, are the prime determining factors for success.

REFERENCES

1. Bauer, F.L: Das Verhafren der Treppeniteration und verwandte Verfahren zur Lösung algebraischer Eigen wert probleme. Z. angew. Math. Phys. 8, 214-235 (1957).

2. Golub, G.H. Underwood, R.R: The block Lanczos method for computing eigenvalues. Mathematical Software III (J.R. Rice ed.) pp.361-377. Academic Press, (1977).

3. Hestenes, M.R. Karush, W: A method of gradients for the calculation of the characteristic roots and vectors of a real symmetric matrix. J. Res. Nat. Bur. Stand. 47, 45-61 (1952).

4. Hestenes, M.R. Karush, W: Solutions of $Ax = \lambda Bx$. J. Res. Nat. Bur. Stand 47, 471-478 (1952).

5. Hestenes, M.R. Stiefel, E: Methods of conjugate gradients for solving linear systems. J. Res. Nat. Bur. Stand. 49, 409-436 (1952).

6. Kershaw, D.S: The incomplete Cholesky - Conjugate gradient method for the iterative solution of systems of linear equations. J. Comp. Physics 26, 43-65 (1978).

7. Lanczos, C: An iteration method for the solution of the eigenvalue problem of linear differential and integral operators. J. Res. Nat. Bur. Stand. 45, 255-282 (1950).

8. Martin, R.S. Wilkinson, J.H: Reduction of the symmetric eigenproblem $Ax = \mu Bx$ and related problems to standard form. Handbook for Automatic Computation, II, Linear Algebra. pp. 303-314, Springer, (1971).

9. Meijerink, J.A. Van der Vorst, H.A: An iterative solution method for linear systems of which the coefficient matrix is an M-matrix. Math. Comp. 31, 137, 148-162 (1977).

10. Reid, J.K: Two FORTRAN subroutines for direct solution of linear equations whose matrix is sparse, symmetric and positive definite. AERE (Harwell) R7119 (1972).

11. Rutishauser, H: Simultaneous iteration method for symmetric matrices. Handbook for Automatic Computation, II, Linear Algebra, pp. 284-302, Springer, (1971).

12. Rutishauser, H: The Jacobi method for real symmetric matrices. Handbook for Automatic Computation, II, Linear Algebra. pp. 202-211, Springer, (1971).

13. Shavitt, I. Bender, C.H. Pipano, A. Hosteny, R.P: The iterative calculation of several of the lowest or highest eigenvalues and corresponding eigenvectors of very large symmetric matrices. J. Comp. Physics 11, 90-108 (1973).

14. Wilkinson, J.H: The Algebraic Eigenvalue Problem. Oxford University Press, (1965).

NUMERICAL SOLUTIONS OF
PARTIAL DIFFERENTIAL EQUATIONS
J. Noye (editor)

RECENT DEVELOPMENTS IN HYBRID ELEMENTS

K.C. Watters and R. Jones

Aeronautical Research Laboratories
Defence Science and Technology Organisation
Australia

A method is described for efficiently obtaining accurate solutions to linear partial differential equations. General solutions from classical continuum mechanics are adapted to the discretisation procedure of the finite element method. The method is illustrated for problems of holes and cracks in plane elasticity. The region around the stress concentrator is enclosed by a special element to which are coupled standard elements forming the remainder of the structure. Extensions of the method to loaded holes and three dimensional cracks are outlined.

INTRODUCTION

Various hybrid elements to surround crack tips and holes for use in the stress analysis of structures are described in Refs. 1, 2, 3, and 4. Hybrid elements combine concepts of classical continuum mechanics with the discretisation of the finite element method. The idea is of general applicability to the approximate solution of partial differential equations (P.D.E.s) and is outlined below.

GENERAL THEORY

Consider a linear P.D.E. of the general form:

$$D\,[u] = 0 \tag{1}$$

where D is a differential operator

and u is a function which characterises the system.

There will in general be mixed boundary conditions (B.C.s) of the form:

$$G_1\,(u)\Big|_{\Gamma_1} = 0 \tag{2}$$

$$G_2\,(u)\Big|_{\Gamma_2} = F \tag{3}$$

If eqn. (1) has a general series solution of the form:

$$u = \sum_{n=1}^{N} a_n\, u_n \tag{4}$$

then, from this general solution, a particular solution can be selected which satisfies some of the B.C.s (say (2)). Then it is required to obtain a solution for the a_n's which approximately satisfies the remaining B.C.s, (3). If eqn. (1) is positive definite, then the B.C.s, (3), can be approximately satisfied by minimizing an energy functional (otherwise a Galerkin process must be used), viz:

$$I(u) = \int_V \langle D[u] , u \rangle \, dv. \tag{5}$$

As a result, in order that I obtain a minimum value the a_n's must satisfy:

$$\frac{\partial I}{\partial \underset{\sim}{\lambda}} = 0 \tag{6}$$

where $\underset{\sim}{\lambda}^T = \{a_1, \ldots, a_N\}$.

Since (1) is linear this yields:

$$\frac{\partial I}{\partial \underset{\sim}{\lambda}} = K \underset{\sim}{\lambda} \tag{7}$$

APPLICATION TO ELASTICITY

The solution procedure for eqns. (5) and (6) is illustrated below by some two dimensional (2-D) examples from elasticity theory. Referring to eqn. (1), for plane elastic problems,

$$D \equiv \Delta\Delta \qquad \text{and } u \equiv \phi$$

so that

$$\Delta\Delta\, \phi = 0 \tag{8}$$

where Δ is Laplace's operator

and ϕ is Airy's stress function.

Equation (8) has a general separated variable solution in 2-D polar co-ordinates given by:

$$\phi_n = \left[a_n r^n + b_n r^{-n} + c_n r^{2+n} + d_n r^{2-n} \right] \begin{Bmatrix} \sin n\theta \\ \cos n\theta \end{Bmatrix} \tag{9}$$

where n is unrestricted.

Special solutions apply for n = 0 and 1 and several other special solutions also exist (see Ref. 5). The first example for consideration is a crack tip. With the origin at the tip, and the crack surfaces stress free, the B.C.s are:

$$\sigma_{\theta\theta} = \tau_{r\theta} = 0 \text{ at } \theta = 0,\ 2\pi \text{ all } r$$

The particular solution from (9) which satisfies these B.C.s is:

$$\phi_n = r^{n+1}\left\{ a_n \left[\sin(n-1)\theta - \frac{n-1}{n+1}\sin(n+1)\theta \right] + b_n \left[\cos(n-1)\theta - \cos(n+1)\theta \right] \right\} \tag{10}$$

where $n = \tfrac{1}{2}, 1, 1\tfrac{1}{2}, 2, \ldots$
This solution was first developed by Williams (Ref. 6).

The second example is a circular hole, radius r_o. With the origin at the centre of the hole, and the hole surfaces stress free, the B.C.s are:

$$\sigma_{rr} = \tau_{r\theta} = 0 \quad \text{at } r = r_o \text{ all } \theta$$

Here the solution has the form:

$$\phi_n = \left\{ \left[-(n+1) r_o^{-2n} r^n + r^{-n} + n r_o^{-2n-2} r^{2+n} \right] a_n + \left[-n r_o^{-2n+2} r^n + (n-1) r_o^{-2n} r^{2+n} + r^{2-n} \right] b_n \right\} \left\{ \begin{matrix} \sin n\theta \\ \cos n\theta \end{matrix} \right\} \tag{11}$$

where $n = 0, 1, 2, 3, \ldots$
The requirement for single valuedness restricts n to integers.

When a hole or crack is part of some defined structure it remains to satisfy the external B.C.s of the structure. This is done using the finite element method. The stress concentrator or singularity is surrounded by a special element to which are coupled standard elements, forming the remainder of the structure. The general shape of such a special element is shown in Figure 1. The element boundary is a regular polygon. For that shape some simplification of the integral of eqn. (5) is possible.

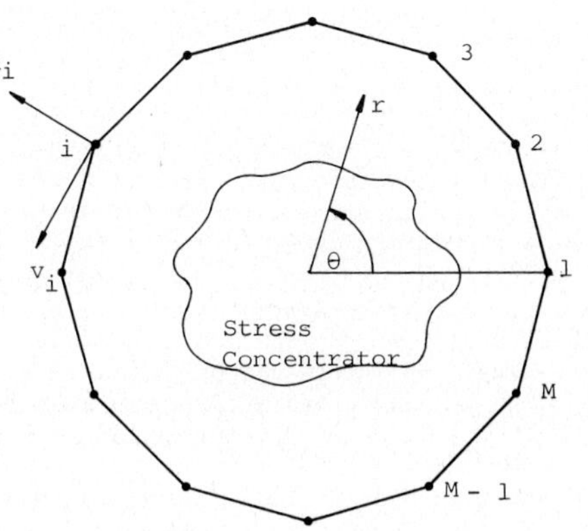

FIGURE 1: SPECIAL ELEMENT SHAPE

For an element with M nodes there are 2M degrees of freedom associated with u and v displacements at each node. From the stress function, ϕ_n, the displacements can be derived as:

$$\begin{Bmatrix} u \\ v \end{Bmatrix} = L \underset{\sim}{\lambda} \tag{12}$$

Appropriate rigid body motion parameters are included in $\underset{\sim}{\lambda}$ in addition to the stress function parameters.

The vector of nodal displacements, $\underset{\sim}{\delta}$, can be constructed by evaluating eqn. (12) at each of the nodal points; the results can be written in the form:

$$\underset{\sim}{\delta} = \begin{bmatrix} u_1 \\ v_1 \\ \cdot \\ \cdot \\ \cdot \\ u_M \\ v_M \end{bmatrix} = \begin{bmatrix} L_1 \\ \cdot \\ \cdot \\ \cdot \\ L_M \end{bmatrix} \underset{\sim}{\lambda} = M \underset{\sim}{\lambda} \quad \text{(say)} \tag{13}$$

from which,

$$\underset{\sim}{\lambda} = (M^T M)^{-1} M^T \underset{\sim}{\delta} \tag{14}$$

Equation (14) reduces to $\underset{\sim}{\lambda} = M^{-1} \underset{\sim}{\delta}$ if M is square. Otherwise a least squares fit between the parameters, $\underset{\sim}{\lambda}$, and the nodal degrees of freedom, $\underset{\sim}{\delta}$, is obtained.

The functional to be minimized in elasticity problems is the potential energy of the region. When considering the region of a finite element, the minimization procedure yields the stiffness matrix $\hat{K}$ which is related to the "primitive stiffness matrix" K as follows. From equation (7) we found that:

$$\frac{\partial I}{\partial \underset{\sim}{\lambda}} = K \underset{\sim}{\lambda}$$

so that making use of equation (14) we find:

$$\frac{\partial I}{\partial \underset{\sim}{\delta}} = ((M^T M)^{-1} M^T)^T K (M^T M)^{-1} M^T \underset{\sim}{\delta}$$

$$= \hat{K} \underset{\sim}{\delta} \quad \text{(say)} \tag{15}$$

The stiffness matrix, $\hat{K}$, for the special element can be assembled with the stiffnesses of the standard elements forming the remainder of the structure and a solution obtained using a standard finite element routine.

For the crack tip element the stress intensity factors K_I and K_{II} are given by parameters a_1 and b_1 respectively, which are readily recovered from the analysis. For the hole element the stress components at selectable points of interest are readily obtained by generating an appropriate stress recovery matrix.

Variants of the general element shape of Figure 1 are used when the problem for analysis contains symmetry. The element shape reduces to semicircular, quadrant, etc., as the degree of symmetry increases. Symmetry requirements influence the stress function and rigid body motion parameters appropriate to form $\underset{\sim}{\lambda}$ and the nodal degrees of freedom appropriate to form $\underset{\sim}{\delta}$.

EXAMPLE RESULTS

Several example analyses have been performed using the 2-D crack tip and hole elements. The results of these analyses have confirmed the expected accuracy of the method. For the problem of a transverse crack in the middle of a rectangular plate subject to uniform longitudinal tension, a series of analyses were performed to evaluate the influence of the circumscribing radius of the crack tip element, its number of nodes, and the cut-off point of the parameter series forming $\underset{\sim}{\lambda}$. The results of these analyses are reported in Ref. 1 and, for an optimal set of those parameters, the error in the calculated value for the stress intensity factor was 3.0%. This compares favourably with less accurate solutions by others which used more than twice the total number of nodes. Although the accuracy cited is for an optimal parameter choice, the accuracy is not highly sensitive to non-optimal parameter choices.

The corresponding problem of a circular hole in the middle of a rectangular plate subject to uniform longitudinal tension, was analysed in Ref. 3 using the hole element. The number of parameters forming $\underset{\sim}{\lambda}$ was constrained so that the matrix M was square, but the number of nodes and the aspect ratio of the circumscribing radius to the hole radius, could be varied. A coarse mesh, using 4 nodes for the hole element and an aspect ratio of 2.0, produced a value for the maximum stress concentration factor around the hole that was 3.1% in error with published accurate data. A more refined mesh using 10 nodes for the hole element, produced accuracies of the order of 1% for aspect ratios in the range 1.5 to 3.0.

FURTHER DEVELOPMENTS

A three dimensional (3-D) crack tip element has been developed (Ref. 2) in which the stress and displacement fields are given by the first term in their asymptotic expansions (Ref. 7). A state of plane strain exists near the crack front. The coefficients of the asymptotic expansions were arbitrarily assumed to vary linearly along the crack front. The stress and displacement fields satisfy the B.C.s of the crack but do not necessarily satisfy conditions of stress equilibrium and displacement compatibility. The crack front is taken as elliptically shaped and a parallelepiped shape is used for the element. The element spans the crack front such that a segment of the crack front passes approximately through the middle of it. Example analyses using the 3-D crack tip element have produced stress intensity factors within 4.6% of published accurate data.

A method has been developed to allow for applied loading on the surface of a hole (Ref. 4). The boundary conditions for the hole surface are then of the form:

$$\sigma_{rr} = \sigma_a(\theta),\ \tau_{r\theta} = \tau_a(\theta) \text{ at } r = r_o$$

The applied loading is represented as a finite Fourier series and the loaded hole B.C.s are satisfied by matching the Fourier coefficients with the stress function coefficients of eqn. (9). The applied loading does not change the stiffness matrix of the hole element but produces initial stresses and a set of equivalent nodal forces. Many example analyses have been performed (Refs. 3, 4) with applied loading on the hole, and errors in the maximum stress concentration around the hole were generally less than 5%.

One development which is currently being pursued is to prescribe a piecewise linear variation of displacement around the boundary of the special element. This will enforce displacement compatibility with adjoining standard elements. The boundary displacement distribution can be represented by a finite Fourier series whose coefficients can be determined analytically. The Fourier coefficients can be matched with the stress function coefficients yielding a relationship of the form:

$$\underset{\sim}{\lambda} = C\ \underset{\sim}{\delta} \tag{16}$$

Thus the matrix inversion of eqn. (14) has been eliminated. Apart from a saving in computing time, this is expected to improve the accuracy of the stiffness matrix evaluation. For non-uniform node distributions, the matrix product M^TM of eqn. (14) is ill-conditioned, and the numerical problems associated with its inversion can lead to an excessively stiff evaluation of the stiffness matrix. It is hoped that this development will allow non-uniform node distributions to be used successfully.

DISCUSSION

The hybrid element approach has been shown to produce accurate results in the vicinity of stress concentrators with more efficiency than fine meshes of standard elements. It is a boundary approach with all nodes lying along the outer boundary of the element. Indeed the volume integral of eqn. (5) can be transformed to a surface integral using Green's theorem. The region of the hybrid element can be made as large as is desired. If the regular shape of the element is dispensed with, the entire structure can be represented as one large hybrid element. The hybrid element approach then resembles the boundary points least squares method and the boundary integral method. Alternatively one large hybrid element of simple shape can be coupled with standard elements to model complex geometries.

REFERENCES

[1] Jones, R. and Callinan, R.J., The Use of Special Crack Tip Elements in Cracked Elastic Sheets, Int. J. Fracture 13 (1977) 51-64.

[2] Jones, R. and Callinan, R.J., Analysis and Repair of Flaws in Thick Structures, in: Francois, D. (ed.), Advances in Fracture Research, 1 (Pergamon Press, 1981).

[3] Watters, K.C., Developments in Hole Elements, M.Eng.Sc. Thesis, U. of Melbourne, (April 1980).

[4] Watters, K.C., Special Finite Elements for Loaded and Cracked Holes, ARL-STRUC-NOTE (to be published).

[5] Timoshenko, S., Theory of Elasticity (McGraw-Hill, 1934).

[6] Williams, M.L., On the Stress Distribution at the base of a Stationary Crack, J. of App. Mech. (March, 1957) 109.

[7] Sih, G.C. and Liebowitz, H., Mathematical Fundamentals-Fracture, 2 (Academic Press, N.Y., 1969).

CONFERENCE DELEGATES

1981 Conference on the Numerical Solutions of Partial Differential Equations, Queen's College, Melbourne University, Victoria, Australia.

R.S. ANDERSSEN, *Commonwealth Scientific and Industrial Research Organisation, Division of Mathematics and Statistics, P.O. Box 1965, Canberra City, Australian Capital Territory, 2601.*

K. ANKER, *Department of Mathematics, Caulfield Institute of Technology, 900 Dandenong Road, Caulfield East, Victoria, 3145.*

R.J. ARNOLD, *Department of Applied Mathematics, University of Adelaide, Adelaide, South Australia, 5000.*

K.R. AYYALARAJU, *Department of Mathematics, University of Melbourne, Parkville, Victoria, 3052.*

F.R. BARRINGTON, *Department of Mathematics, University of Melbourne, Parkville, Victoria, 3052.*

D.A. BARRY, *School of Australian Environmental Studies, Griffith University, Nathan, Queensland, 4111.*

J.M. BARRY, *Australian Atomic Energy Commission, Private Mail Bag, Sutherland, New South Wales, 2232.*

N.G. BARTON, *Commonwealth Scientific and Industrial Research Organisation, Division of Mathematics and Statistics, P.O. Box 218, Lindfield, New South Wales, 2070.*

R.C. BELL, *Commonwealth Scientific and Industrial Research Organisation, Division of Atmospheric Physics, P.O. Box 77, Mordialloc, Victoria, 3195.*

L. BENNETT, *Commonwealth Scientific and Industrial Research Organisation, Division of Mineral Physics, P.O. Box 136, North Ryde, New South Wales, 2113.*

P.J. BILLS, *32 Walkerville Terrace, Gilberton, South Australia, 5081.*

R.L. BISH, *Materials Research Laboratories, P.O. Box 50, Ascotvale, Victoria, 3032.*

D.R. BLACKMAN, *Department of Mechanical Engineering, Monash University, Clayton, Victoria, 3168.*

R.D. BRADDOCK, *School of Australian Environmental Studies, Griffith University, Nathan, Queensland, 4111.*

E. BRADFORD, *Applied Mathematics Division, Department of Scientific and Industrial Research, P.O. Box 1335, Wellington, New Zealand.*

K.C. BROWN, *Department of Mechanical Engineering, Melbourne University, Parkville, Victoria, 3052.*

P. CERONE, *Department of Mathematics, University of Wollongong, P.O. Box 1144, Wollongong, New South Wales, 2500.*

G.A. CHANDLER, *Commonwealth Scientific and Industrial Research Organisation, Division of Mathematics and Statistics, Private Bag No. 2, Glen Osmond, South Australia, 5064.*

S.S. CHOW, *Department of Pure Mathematics, Australian National University, P.O. Box 4, Canberra, Australian Capital Territory, 2600.*

J.A. CHURCH, *Commonwealth Scientific and Industrial Research Organisation, Division of Oceanography, P.O. Box 21, Cronulla, New South Wales, 2230.*

J.R. COLEBY, *Department of Applied Mathematics, University of Adelaide, Adelaide, South Australia, 5001.*

L.H. COLGAN, *Department of Mathematics, South Australian Institute of Technology, P.O. Box 1, Ingle Farm, South Australia, 5098.*

H.J. CONNELL, *Department of Mathematics, Royal Melbourne Institute of Technology, 124 La Trobe Street, Melbourne, Victoria, 3000.*

J. CROTTY, *Commonwealth Scientific and Industrial Research Organisation, Division of Geomechanics, P.O. Box 54, Mt. Waverley, Victoria, 3149.*

M.R. DAVIDSON, *Australian Atomic Energy Commission, Private Mail Bag, Sutherland, New South Wales, 2232.*

F.R. DE HOOG, *Commonwealth Scientific and Industrial Research Organisation, Division of Mathematics and Statistics, P.O. Box 1965, Canberra City, Australian Capital Territory, 2601.*

G. DOHERTY, *Department of Mathematics, University of Wollongong, P.O. Box 1144, Wollongong, New South Wales; 2500.*

J. DONALDSON, *Department of Mathematics, University of Tasmania, Hobart, Tasmania, 7000.*

L.A. DRAKE, *School of Earth Sciences, Macquarie University, North Ryde, New South Wales, 2113.*

A.K. EASTON, *Department of Mathematics, Swinburne College of Technology, John Street, Hawthorn, Victoria, 3122.*

D. ELLIOTT, *Department of Mathematics, University of Tasmania, Hobart, Tasmania, 7000.*

J.D. FENTON, *School of Mathematics, University of New South Wales, Kensington, New South Wales, 2033.*

G.F. FITZGERALD, *Department of Mathematics, Royal Melbourne Institute of Technology, Swanston Street, Melbourne, Victoria, 3001.*

C.A.J. FLETCHER, *Department of Mechanical Engineering, University of Sydney, Sydney, New South Wales, 2006.*

D.G. FORD, *Aeronautical Research Laboratories, P.O. Box 4331, G.P.O., Melbourne, Victoria, 3001.*

J.A. GEAR, *Department of Mathematics, University of Melbourne, Parkville, Victoria, 3052.*

R. GRIMSHAW, *Department of Mathematics, University of Melbourne, Parkville, Victoria, 3052.*

E. HAUSLER, *Department of Mathematics, Swinburne College of Technology, John Street, Hawthorn, Victoria, 3122.*

D.L. HILL, *Department of Applied Mathematics, University of Adelaide, Adelaide, South Australia, 5001.*

R.J. HILTON, *I.C.I. Australia Limited, Ballarat Road, Deer Park, Victoria, 3023.*

T.S. HORNER, *Department of Mathematics, University of Wollongong, P.O. Box 1144, Wollongong, New South Wales, 2500.*

S. HUNT, *7 Gordon Crescent, Blackburn, Victoria, 3130.*

D.R. JACKETT, *Commonwealth Scientific and Industrial Research Organisation, Division of Mathematics and Statistics, P.O. Box 1965, Canberra City, Australian Capital Territory, 2601.*

D.R. JENKINS, *C/- B.H.P. Limited, P.O. Box 188, Wallsend, New South Wales, 2287.*

R. JONES, *Aeronautical Research Laboratories, Box 4331, G.P.O., Melbourne, 3001.*

B.J. KACHOYAN, *23 Colvin Avenue, Carlton, New South Wales, 2218.*

P. KACHOYAN, *School of Mathematics, University of New South Wales, P.O. Box 1, Kensington, New South Wales, 2033.*

J. KAUTSKY, *School of Mathematics, Flinders University, Bedford Park, South Australia, 5042.*

D.J. KEWLEY, *Defence Research Centre, Salisbury, South Australia, 5108.*

J.H. KNIGHT, *Commonwealth Scientific and Industrial Research Organisation, Division of Mathematics and Statistics, P.O. Box 281, Lindfield, New South Wales, 2070.*

R. KOHOUTEK, *Department of Civil Engineering, Melbourne University, Parkville, Victoria, 3052.*

P. LALOUSIS, *Commonwealth Scientific and Industrial Research Organisation, Division of Applied Physics, Box 218, Lindfield, New South Wales, 2070.*

Y.C. LAM, *Department of Mechanical Engineering, University of Melbourne, Parkville, Victoria, 3052.*

K. McALLISTER, *School of Earth Sciences, Macquarie University, North Ryde, New South Wales, 2113.*

A.D. McCOWAN, *Department of Mechanical Engineering, Monash University, Clayton, Victoria, 3168.*

J.A. MACDONALD, *Materials Research Laboratories, P.O. Box 50, Ascot Vale, Victoria, 3032.*

K.J. MANN, *Department of Mathematics, Caulfield Institute of Technology, 900 Dandenong Road, Caulfield East, Victoria, 3145.*

R.L. MAY, *Department of Mathematics, Caulfield Institute of Technology, 900 Dandenong Road, Caulfield East, Victoria, 3145.*

J. MILTHORPE, *Department of Mechanical Engineering, University of Sydney, Sydney, New South Wales, 2006.*

W. MITCHELL, *Department of Applied Mathematics, University of Adelaide, Adelaide, South Australia, 5000.*

N. MOLLOY, *Department of Metallurgy, University of Newcastle, Newcastle, New South Wales, 2308.*

J.R. MOONEY, *Commonwealth Scientific and Industrial Research Organisation, Division of Computing Research, P.O. Box 1800, Canberra City, Australian Capital Territory, 2601.*

R. MORROW, *Commonwealth Scientific and Industrial Research Organisation, Division of Applied Physics, P.O. Box 218, Lindfield, New South Wales, 2070.*

B.J. NOYE, *Department of Applied Mathematics, University of Adelaide, Adelaide, South Australia, 5001.*

A. RAICHE, *Commonwealth Scientific and Industrial Research Organisation, Division of Mineral Physics, P.O. Box 136, North Ryde, New South Wales, 2113.*

M. RIENECKER, *School of Mathematics, University of New South Wales, P.O. Box 1, Kensington, New South Wales, 2033.*

P. ROBB, *Department of Mathematics, Swinburne College of Technology, John Street, Hawthorn, Victoria, 3122.*

T. RYALL, *Aeronautical Research Laboratories, P.O. Box 4331, G.P.O., Melbourne, Victoria, 3001.*

W.D. SCOTT, *Department of Environmental and Life Sciences, Murdoch University, Murdoch, Western Australia, 6150.*

J. SHEPHERD, *Department of Mathematics, Royal Melbourne Institute of Technology, Swanston Street, Melbourne, Victoria, 3000.*

H. SIPOLS, *Electricity Trust of South Australia, P.O. Box 6, Eastwood, South Australia, 5063.*

G.P. STEVEN, *Department of Aeronautical Engineering, University of Sydney, Sydney, New South Wales, 2006.*

A.N. STOKES, *Commonwealth Scientific and Industrial Research Organisation, Division of Mathematics and Statistics, 69 Yarrabank Road, South Melbourne, Victoria, 3205.*

I. TAGGART, *C/- B.H.P. Limited, P.O. Box 188, Wallsend, New South Wales, 2287.*

C. TARLOWSKI, *Commonwealth Scientific and Industrial Research Organisation, Division of Mineral Physics, P.O. Box 136, North Ryde, New South Wales, 2113.*

K. THALASSOUIDIS, *Department of Applied Mathematics, University of Adelaide, Adelaide, South Australia, 5001.*

TEH KOON KIAN, *Department of Mechanical Engineering, University of Melbourne, Parkville, Victoria, 3052.*

K.L. TEO, *School of Mathematics, University of New South Wales, P.O. Box 1, Kensington, New South Wales, 2033.*

J. TOMAS, *Department of Mechanical Engineering, Royal Melbourne Institute of Technology, Swanston Street, Melbourne, Victoria, 3000.*

G.D. TONG, *P.O. Box 149, Warracknabeal, Victoria, 3393.*

J. VAN DER HOEK, *Department of Pure Mathematics, University of Adelaide, Adelaide, South Australia, 5000.*

I. WALLIS, *Caldwell Connell Engineering, 60 Albert Road, South Melbourne, Victoria, 3205.*

L. WARDLE, *Commonwealth Scientific and Industrial Research Organisation, P.O. Box 54, Mt. Waverley, Victoria, 3149.*

W.W. WOOD, *Department of Mathematics, University of Melbourne, Parkville, Victoria, 3052.*

R.L. WOODWARD, *Materials Research Laboratories, P.O. Box 50, Ascotvale, Victoria, 3032.*

Z.S. WU, *Department of Applied Mathematics, University of New South Wales, P.O. Box 1, Kensington, New South Wales, 2033.*

A. YADAV, *Department of Civil Engineering, University of Melbourne, Parkville, Victoria, 3052.*

C. ZOPPOU, *Department of Mechanical Engineering, Melbourne University, Parkville, Victoria, 3052.*